3rd International Conference on Advanced Materials for Better Future (ICAMBF 2018)

IOP Conference Series: Materials Science and Engineering
Volume 578

Surakarta, Indonesia
15 - 16 October 2018

Editors:

Dimitris Achlioptas
Laszlo A. Vegh

ISBN: 978-1-5108-9812-7
ISSN: 1757-8981

Printed from e-media with permission by:

Curran Associates, Inc.
57 Morehouse Lane
Red Hook, NY 12571

Some format issues inherent in the e-media version may also appear in this print version.

Institute of Physics
Dirac House, Temple Back
Bristol BS1 6BE UK

Phone: 44 1 17 929 7481
Fax: 44 1 17 920 0979

techtracking@iop.org

Additional copies of this publication are available from:

Curran Associates, Inc.
57 Morehouse Lane
Red Hook, NY 12571 USA
Phone: 845-758-0400
Fax: 845-758-2633
Email: curran@proceedings.com
Web: www.proceedings.com

Table of contents

Volume 578

International Conference on Advanced Materials for Better Future 2018

15–16 October 2018, Solo Paragon Hotel, Indonesia

Accepted papers received: 26 June 2019
Published online: 4 October 2019

Preface

Papers

Materials Chemistry and Physics (MCP)

Drug and Biomaterials Technology (DBT)

Miscellaneous Topics (MT)

IOP Conf. Series: Materials Science and Engineering **578** (2019) 011001 doi:10.1088/1757-899X/578/1/011001

IOP Conf. Series: Materials Science and Engineering **578** (2019) 011001 doi:10.1088/1757-899X/578/1/011001

PREFACE

3rd International Conference on Advanced Material for Better Future 2018 (ICAMBF 2018)

The 3rd International Conference on Advanced Material for Better Future 2018 (ICAMBF 2018) was held on October 15-16, 2018 in Surakarta, Central Java, Indonesia. ICAMBF 2018 is an international scientific forum that was initiated to overcome material problems as well as factors affecting them that have emerged as the main issue and challenge attracting more attention from various scientists, government, and regulatory bodies worldwide. The ICAMBF 2018 was attended by leading scientists from countries near and abroad who work in the field of material.

This event based on theme Smart Materials for A Smart World that aims to provide a forum for scientists, researchers, and engineers within the nation and from around the world to present, discuss, share, and exchange their knowledge, experiences, and research finding on all aspects of material. Thus, this conference can give an excellent overview in all scope of fields including Functional Material, Material Chemistry and Physics, Advanced Nanomaterials, Polymer and Composites, Drug and Biomaterials Energy, Electrical, Optical and Magnetic Materials, Advanced Materials for Energy, Green Energy Materials, Batteries and Energy Storage Materials, Graphene and 2D Energy Materials, Separation Materials, Construction Materials, Metallurgy and Alloy Materials.

We are pleased to our outstanding invited speakers Prof. Ru-Jen Lin, Prof. Dr. Santiago Gómez-Ruiz, Prof. Dato' Dr. Burhanuddin Yeop Majlis, Prof. Santiago Gomez, Prof. Ari Handono Ramelan, Assoc. Prof. Dr. Alfi Khatib and other participated speakers for sharing the best knowledge in Advanced Material.

We also thanks to all researchers and student who participate in conference by submitting their papers to convey the current science discussed at ICAMBF 2018, all of the organizing committee wish you get awesome conference experience, and reviewers for their valuable revision of these papers.

Prof. Ir. Ari Handono Ramelan, M.Sc (Hons), Ph.D

Chairman of ICAMBF 2018

Dean of Faculty Mathematics & Natural Sciences

IOP Conf. Series: Materials Science and Engineering **578** (2019) 011002 doi:10.1088/1757-899X/578/1/011002

CONFERENCE COMMITTEE/ORGANIZER

Chairman	Prof. Ir. Ari Handono Ramelan, M.Sc., (Hons), Ph.D.	Sebelas Maret University, Indonesia
Co-Chairman	Dr. Sayekti Wahyuningsih, MSi	Sebelas Maret University, Indonesia
Secretary	Ahmad Marzuki, SSi, PhD	Sebelas Maret University, Indonesia
Scientific Committee	Prof. Ir. Ari Handono Ramelan, M.Sc., (Hons), Ph.D.	Sebelas Maret University, Indonesia
	Ahmad Marzuki, SSi, PhD	Sebelas Maret University, Indonesia
	Venty Suryanti. M.Phil. PhD	Sebelas Maret University, Indonesia
	Dr. Sayekti Wahyuningsih, MSi	Sebelas Maret University, Indonesia
	Dr.Desi Suci Handayani, M.Si.	Sebelas Maret University, Indonesia
	Prof. Dr. Santiago Gómez-Ruiz	Rey Juan Carlos University, Móstoles (Madrid), Spain
	Prof. Ko Tzu Hsiang	Lung Hwa University, Taiwan
	Prof. Dato Dr. Burhanuddin Yeop Majlis	Universiti Kebangsaan Malaysia (UKM), Malaysia
	Prof Roger Narayan	Joint Department of Biomedical Engineering, University of North Carolina/North Carolina State University, USA
	Dr. Agus Supriyanto, S.Si. M.Si	Sebelas Maret University, Indonesia
	Dr. Eng. Risa Suryana, S.Si. M.Si.	Sebelas Maret University, Indonesia
	Dr.rer.nat Witri Wahyu Lestari, M.Sc.	Sebelas Maret University, Indonesia
	Teguh Endah Saraswati, M.Sc, Ph.D	Sebelas Maret University, Indonesia
	Dr. Tetri Widiyani, M.Si.	Sebelas Maret University, Indonesia
	Dr. Artini Pangastuti, M.Si.	Sebelas Maret University, Indonesia

IOP Conf. Series: Materials Science and Engineering **578** (2019) 011003 doi:10.1088/1757-899X/578/1/011003

Peer review statement

All papers published in this volume of *IOP Conference Series: Materials Science and Engineering* have been peer reviewed through processes administered by the proceedings Editors. Reviews were conducted by expert referees to the professional and scientific standards expected of a proceedings journal published by IOP Publishing.

IOP Conf. Series: Materials Science and Engineering **578** (2019) 012001 doi:10.1088/1757-899X/578/1/012001

Color Stability of Anthocyanins Copigmentation from Red Rice (*Oryza sativa L.*) Bran by Spectrophotometry UV-Vis

E V Maylinda[1], A Rinadi[1], E A Putri[1], G Fadillah[2*] and S Wayuningsih[1]

[1] Chemistry Department, Faculty of Mathematics and Natural Sciences, Universitas Sebelas Maret, Surakarta, Indonesia

[2] Chemistry Department, Faculty of Mathematics and Natural Sciences, Universitas Islam Indonesia, Yogyakarta, Indonesia

*E-mail : ganjar.fadillah@uii.ac.id

Abstract. Red Rice (*Oryza sativa L.*) Bran is a waste that is often found in our environment. Pigment rice contains polar pigment anthocyanins (ACNs). Anthocyanins are a natural pigment that has low color stability and are easily degraded by environmental influences. In the present study, we have tried to increase the color stability of anthocyanins by copigmentation. Anthocyanins extraction was carried out using maceration techniques in 95% ethanol solvents. The extract of anthocyanins was characterized by spectrophotometer UV-Vis and Fourier transform infrared (FTIR). The results show that the spectrum UV-Vis and IR were obtained a characteristics peak at 470, 420, 340 nm and a hydroxyl group at 3400-3500 cm^{-1}, respectively. The various of Fe^{3+} and Al^{3+} as copigment and composition of anthocyanins to Fe^{3+} and Al^{3+} have been investigated. Anthocyanins stability was followed by measuring the loss of color by spectrophotometer UV-Vis, and the results show that Addition of Fe^{3+} and Al^{3+} could improve the anthocyanins stability.

1. Introduction

Rice bran (RB) is one of the by-products of rice milling and the main waste from the rice milling process. The rice milling waste is abundantly available in Indonesia which is known as one of the agricultural countries. However, the use of waste is still limited because it is considered to have low economic value. The utilization of the waste is currently only limited as animal feed. Recently researches regarding obtaining higher values from RB have been investigated. The previous studies have reported that the red rice bran has anthocyanins content as a red coloring pigment [1]. Anthocyanins are one of the many natural dyes in the red rice bran and are widely used as a natural dye. However, Munawaroh et al (2015) reported that the anthocyanins as a natural dye are easily degraded by several factors such as pH, temperature, presence oxygen, light, and enzyme [2].

The color stability of anthocyanins can be increased through complex formation by copigmentation reaction. Copigmentation is an interaction between the structure of anthocyanin and other molecules such as metals (Al^{3+}, Fe^{3+}, Sn^{2+}, Cu^{2+}) and other organic molecules such as flavonoids (flavones, flavanone, and flavonol). There is a co-fermentation between anthocyanin and other metal molecules that increase anthocyanin color and produce brighter colors and are protected from oxidation [3]. Fedenko et al (2017) have studied about copigmentation of anthocyanins. The results reported that the process could improve the color stability of anthocyanins through Me^{n+}-binding ability of anthocyanin structures [4]. This is because metals can stabilize the structure of anthocyanin pigments and can prevent the growth of microbes that can degrade the color. Some examples of suggested stabilization are intra-molecular promotional reactions of anthocyanins [5], ion formation between metal ions and anthocyanins [6] or decreased activity due to metal ions that inhibit flavylium ionic hydration [7]. In this study, we have tried to increase the color stability of anthocyanins through metal binding such as using $Al^{3+}$and Fe^{3+} as copigment. Copigmentation is known to be responsible for the color stability of anthocyanin and gives brighter and more stable colors than what would be expressed by an intact anthocyanin molecule through binding metal interaction.

2. Experimental

2.1 Materials

The materials are red rice bran, cyanidin 3-O-glucoside chloride as standard anthocyanin (Sigma Aldrich Singapore), ethanol (Merck), 1M NaOH, 1M HCl, 1% $FeCl_3.6H_2O$ (Sigma Aldrich Singapore), 0.01% $AlCl_3$ (Sigma Aldrich Singapore) and distilled water. The specific equipments that are used including pH meter Jenway, thermometers, shakers, Shimadzu analytical balance ATX224, Spectroquant® Prove 600 spectrophotometer UV-Vis, and spectrophotometer FTIR Shimadzu IRAffinity-1S.

2.2 Procedure

2.2.1 Anthocyanin Extraction from Red Rice Bran

Anthocyanin extraction from bran rice bran was carried out by maceration with ethanol solvent [8]. Amount of 20 g of red rice bran was dissolved with 80 mL of ethanol solvent with the addition of 20 mL HCl 1M then allowed to stand for 24 hours in a dark room. Then filtered with Whatman Filter paper no 42 so that anthocyanin extract was obtained and then analyzed by Spectroquant® Prove 600 spectrophotometer UV-Vis and spectrophotometer FTIR Shimadzu IRAffinity-1S.

2.2.2. *Complexation with Metals*

The anthocyanin extract was added with 0.01% $AlCl_3$ and 0.01% $FeCl_3.6H_2O$ with a concentration ratio of 1: 1, 1: 2, 1: 3, 1: 4, 1: 5 (w/w), respectively, and analyzed by UV-Vis and FTIR spectroscopy. The anthocyanin solution and metals $AlCl_3$ and $FeCl_3$ under optimum conditions were made in pH 1, 2, 3, 4, 5 then analyzed by Spectroquant® Prove 600 spectrophotometer UV-Vis and spectrophotometer FTIR Shimadzu IRAffinity-1S.

3. Results and Discussion

Anthocyanins are included in the flavonoid group of polyphenols. Generally, anthocyanins are unstable molecules as a natural dye. The color stability of anthocyanins depends on pH, temperature, presence oxygen and light. In this study, the extraction of natural pigment was carried out by the maceration process using 95% ethanol. Ethanol is a polar solvent so it can extract the anthocyanin from red rice bran. Anthocyanins have polar properties because they have hydroxyl groups, so the solvents used must be polar in accordance with the rules like dissolved like. The addition of acid to the solvent serves to denature the membrane of the bran so it dissolves the anthocyanin pigment and can be maximally extracted. The red rice bran and the results of extraction by maceration process are shown in Figure 1(b). The anthocyanin extract was characterized using a UV-VIS spectrophotometer at wavelengths of 350-700 nm. Spectra of anthocyanins characterization with UV-VIS as shown in Figure 1(a). Based on Figure 1(a), anthocyanins have a specific absorption at wavelengths 470, 420, and 340 nm. The presence of three absorption peaks obtained from sample extracts showed that the presence of anthocyanin compounds. The results show the presence of an aromatic group which is at a wavelength of 340 nm (benzoyl groups) and 400-500 nm (cynnamoil groups) [9].

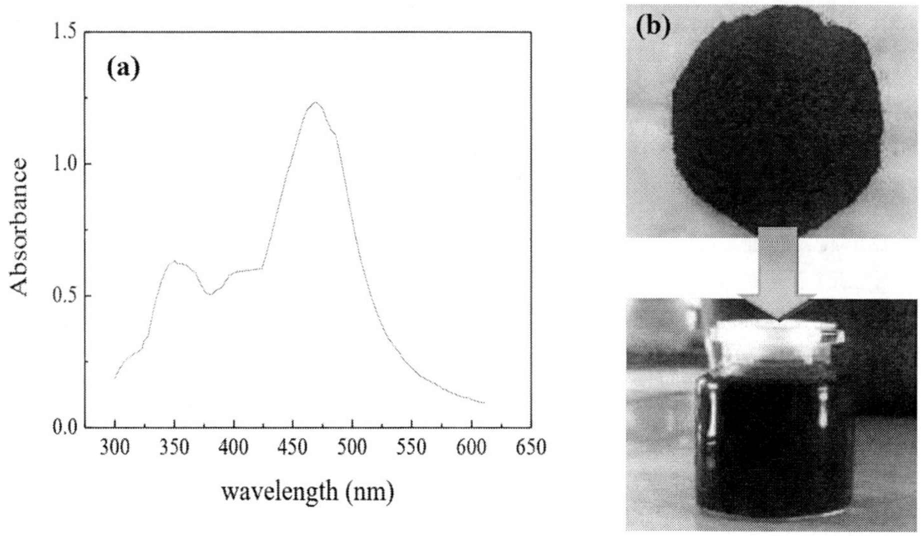

Figure 1. (a) UV-VIS spectra of characterization of anthocyanin, (b) red rice bran and extracted anthocyanins

Samples were also characterized by infrared spectroscopy as shown in Figure 2. From the IR data, it was confirmed that the existence of aromatic groups from UV data was strengthened with IR spectrum which shows the aromatic C=C and C-H bonds at wave numbers 1639 cm^{-1} and 642 cm^{-1}, respectively. In addition to aromatic groups from IR data, C-O bonds were also found which appeared at wave numbers 1046 cm^{-1}. The hydroxy groups were detected in the wave number 3422 cm^{-1} and the absorption of C-H stretching appeared at wave numbers 2928 - 2977 cm^{-1}. According to FTIR spectra and compared to standard anthocyanins (cyanidin 3-O-glucoside chloride), it is known that extracted anthocyanins have hydroxy groups and aromatic groups, so it can be concluded that the samples contain anthocyanin compounds.

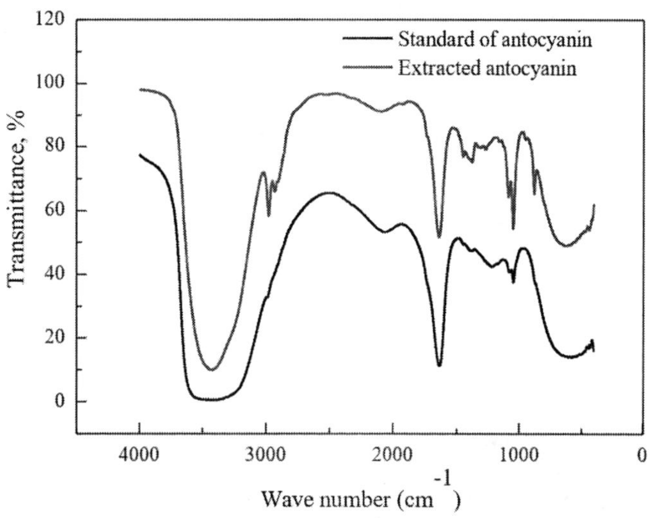

Figure 2. Infrared spectra of anthocyanin extract samples

As mentioned previously, copigmentation in this study was carried out by complexation reaction through metal binding using Al^{3+} and Fe^{3+}. Metal binding with anthocyanins structures can improve the stability of anthocyanin through the chetalization between hydroxyl groups and metals. Some suggested stabilization mechanisms are the promotion of intra-molecular reactions of anthocyanin, formation of ion pairs between metal ions and anthocyanin as illustrated in Figure 3 or decrease in water activity due to metal ions solvation which reduces the hydration of flavylium ions.

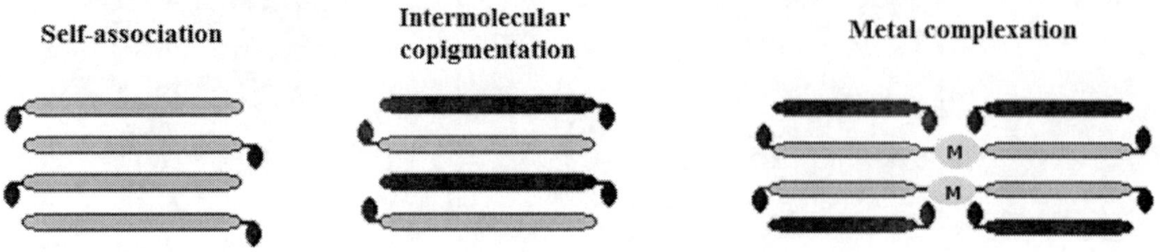

Figure 3. Mechanism of the formation of anthocyanin complexes with metals [4]

The effect of the ratio between anthocyanins to copigment (Al^{3+} and Fe^{3+}) was evaluated in this study. Anthocyanin and Al and Fe metal solutions were made by varying the volume anthocyanins to copigments (1:1, 1:2, 1:3, 1:4, and 1:5) respectively. The UV-Vis spectra is shown in Figure 4 and the result of anthocyanin copigmentation is shown in Table 1.

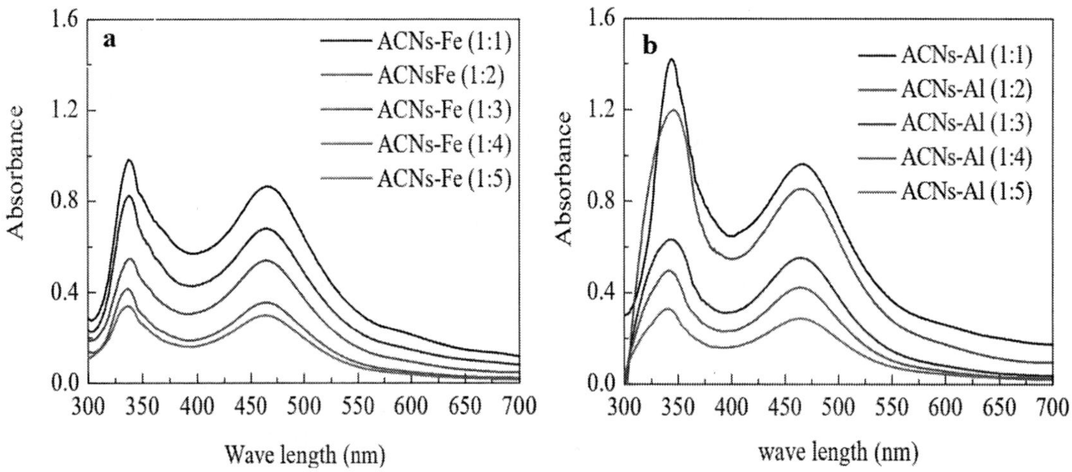

Figure 4. UV-Vis spectra with copigment (a) Fe metals and (b) Al metals

Table 1. Maximum metal wavelength

v/v	Fe		%	Al		%
	λ max	Abs	copigmentation	λ max	Abs	copigmentation
1:1	468	0.9657	74.53	466	0.8672	77.12
1:2	468	0.8672	59.08	465	0.5399	74.52
1:3	465	0.5627	67.47	464	0.67912	60.74
1:4	463	0.4275	37.84	464	0.3554	48.33
1:5	463	0.2902	25.92	463	0.2982	23.88

Based on Figure 4, the copigmentation is successfully formed both using metals Al^{3+} and Fe^{3+}. Some kinds of literature have been reported that the copigmentation can be observed as a bathochromic shift ($\Delta\lambda_{max}$) or a hyperchromic effect (ΔA). In this study, we found that the copigmentation process observed as a hyperchromic effect as shown in Figure 4 (UV-Vis spectra). The maximum %copigmentation of the anthocyanins is obtained as in Table 1. The maximum %copigmentation was obtained in the ratio of 1:1 both Fe and Al metals with a percentage of 4.53% and 77.12%, respectively. In this copigmentation process, there was a blue or hyperchromic shift. Hyperchromic effects occur due to increased color intensity after copigmentation [7]. High metal concentrations can increase a solvent solvation capacity which causes a hydration reaction. Therefore, the process can significantly reduce the color of the solution so the optimum of ration ACNs to metal was found in the ratio 1:1.

Then, copigmented anthocyanins were tested for color stability in various pH (1-5). Based on Figure 5, the results show that copigmented anthocyanins are still stable in acidic solution at low pH. The maximum results show the anthocyanins stable at pH 2.0. At low pH, the flavylium cationic anthocyanin was very dominant and the most stable. In acidic solution, the anthocyanins with metal copigment exist as four main structure, namely quinonoidalbase A, chalcone, AH^+ flavylium cation, and carbinol species. In addition, at low pH, the solubility of the metal will be higher so the metal can bind as well with the hydroxyl group of anthocyanin. Increasing the pH can decrease the color of anthocyanins because, at high pH, the dominant structure of anthocyanin will be in the form of carbinol [10,11]. The structure of carbinol has lost its conjugated bond and therefore does not adsorb visible light. The comparative studies of increasing color stability of anthocyanins with different copigment are shown in Table 2.

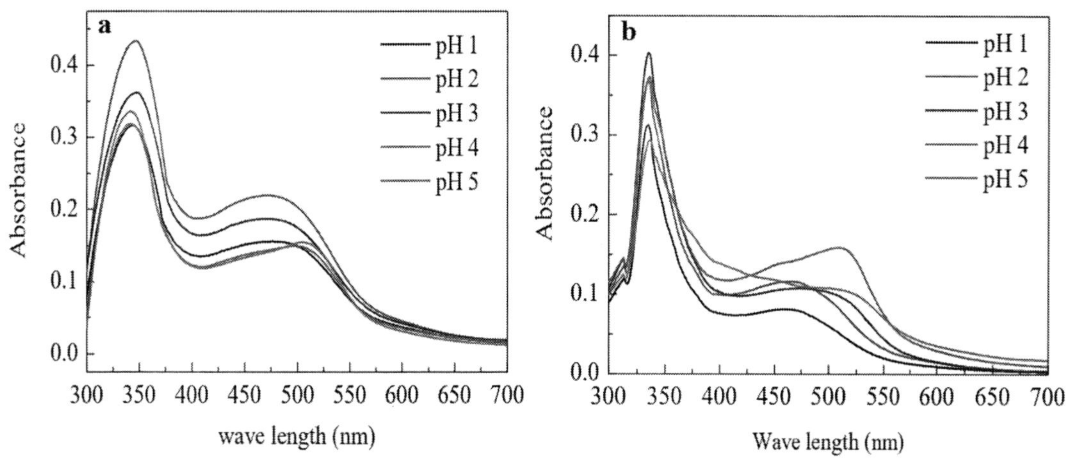

Figure 5. UV-Vis spectra with copigment (a) Fe metals and (b) Al metals

Table 2. %copigmentation of anthocyanins with different copigment

Copigments	Rasio anthocyanins to copigments	%copigmentation	References
Al^{3+}	1:1	77.12	Present study
Fe^{3+}	1:1	74.54	Present study
Ferrulic acid	-	59.4	[10]
Hydroxycinnamic acid	-	29.0	[11]

Refer to Table 2, Generally intermolecular copigmentation using organic molecules as copigment showed that %copigmentation is more lower than metals as copigments. Sigurdson et al (2016) investigated the chelation of Me^{n+} with anthocyanins at different pH (3-8). The results show that the metal as copigments has a good interaction through covalent bonding with the functional groups in anthocyanins [12].

4. Conclusion

In this study, we summarize that anthocyanins from red rice bran have been successfully extracted and increased the color stability through copigmentation methods. Anthocyanin compounds can be complexed with metals Al^{3+} and Fe^{3+} through metal and covalent binding between the metals and functional groups on anthocyanins. The optimum ratio anthocyanins to copigment were found at 1:1 with %copigmentation of 77.12% and 74.54% at pH 2.0. The current results may be of use in improving the color stability of anthocyanins from red rice bran. However, the more study is still needed to improve the color stability in the other several factors such as temperature, presence oxygen, and light.

References

[1] Sabir A, Rafi M and Darusman L K 2017 *Food Chem.* **221** 1717-1722

[2] Munawaroh H, Fadillah G, Saputri L.N.M.Z, Hanif Q A, Hidayat R, Wahyuningsih S 2016. *IOP Conf. Ser.: Mater. Sci. Eng.* **107** 012061

[3] Boulton R 2001 *A. J. Enol. nd Viticult.* **52** 67-8.

[4] Fedenko V S, Shemet S A and Landi M 2017 *J. Plant Physiol.* **212** 13-28

[5] Hoshino T, Matsumoto U, Goto T 1981 *Phytochemistry.* **20** 1971-1976

[6] Goto T, Hoshino T, Obha M 1976. *J. Agr. Biol. Chem.* **40** 1593-1596.

[7] Fahyuan H.D, Samsidar Farid F, Heriyanti Napitupulu S, Pakpahan S 2015 *JoP.* **1** 5-11.

[8] Rattanachitthawat S, Suwannalert P, Riengrojpitak S, Chaiyasut C, Pantuwatana S 2010 *J. Med Plants Res.* **4** 796-801.

[9] Syukri D, Darwis D and Santoni A 2013 *J. Chem. Pharm. Res.* **5** 1276-1282

[10] Fan L, Wang Y, Xie P, Zhang L, Li Y and Zhou J 2018 *Food Chem.* **275** 299-308

[11] Bimpilas A, Panagopoulou M, Tsimogiannis D and Oreopoulou V 2016 *Food Chem.* **197** 39-46

[12] Sigurdson G T, Robbins R J, Collins TM and Giusti M M 2016. *Food Chem.* **208** 26-34

IOP Conf. Series: Materials Science and Engineering **578** (2019) 012002 doi:10.1088/1757-899X/578/1/012002

Synthesis and Characterization of Some Bromochalcones Derivatives

E Susanti V H [1] and W A E Setyowati [1]
[1]Department of Chemistry FKIP Universitas Sebelas Maret, Surakarta, Indonesia

E-mail : elfisusantivh@staff.uns.ac.id

Abstract. Chalcones are considered to be precursors of flavonoids and have variety of biological activities. Three of bromochalcones derivatives were synthesized by a green chemistry approach using grinding technique and their structure also confirmed. Synthesis of bromochalcone, 4′-bromo-4-methoxychalcone, 4′-bromo-3,4-dimethoxychalcone, 4′-bromo-3,4,5-trimethoxychalcone were carried out using starting material of 4-bromoacetophenone through Claisen-Schmidt base catalyzed condensation. The structures of the synthesized compounds were confirmed by gas chromatography–mass spectrometry (GC-MS). The mass spectra of the compounds were showed corresponding molecular ion peak which was correlated with their molecular weight of that respected compound (4′-bromo-4-methoxychalcone = 317, 4′-bromo-3,4-dimethoxychalcone = 347, 4′-bromo-3,4,5-trimethoxychalcone = 377).

1. Introduction

Chalcone is a class of flavonoids, which are intermediates in flavonoid biosynthesis, so the role of these compounds is very important because it is an intermediate in the synthesis of other flavonoid derivatives [1]. Chalcone is found in various plant families but its numbers are very limited compared to other flavonoids because this compound is a minor category of flavonoids so that the percentage in small plants and the variation of its structure is relatively small [2]. Isolation of chalcone from plants is very difficult because of the enzyme chalcone synthetase (CSH) which will convert chalcone into flavones, so the isolation of chalcone from plants requires a relatively long time and is relatively expensive, so to get chalcone in various structures and a large amount can be done by synthesis [3].

Chalcone synthesis can be carried out through aldol condensation or more specifically the Claisen Schmidt condensation reaction, which is a condensation reaction of an aromatic aldehyde with an aromatic ketone in an acidic or alkaline atmosphere. This condensation is widely used in the reaction of the formation of carbon-carbon bonds because the reaction is simple, the raw material is easy to obtain, environmentally friendly and the reaction time is relatively not too long [3]. Chalcone compounds can be synthesized from acetophenone reacted with 3,4-dimethoxybenzaldehyde through Clasein-Schmidt condensation [3]. Claisen-Schmidt condensation reaction alkaline analysis will not form a mixture, because substituted benzaldehyde does not have H-α, so it does not form enolate. Benzophenone derivatives can form enolates and react with aldehydes faster than with themselves, because the aldehydes are more electrophilic than ketones [1]. The most widely used base catalyst is NaOH because of the greater yield [4]. Some base catalysts that have been used in the Claisen-Schmidt reaction are KOH (yield 88-94%) [5], $Ba(OH)_2$ (yield 88-98%) [6], NaOH (yield 90-96%) [7].

Schmidt Claisen can be carried out in two ways, conventional (with solvent) and grinding techniques (without solvents). Susanti (2014) stated that the use of solvents in condensation reactions in the synthesis of chalcone derivatives is hazardous to health and not environmentally friendly, so that a green chemistry approach is needed [8]. Green chemistry is a new method to reduce chemical hazards besides producing products in a more efficient and more efficient way. Grinding technique is a synthesis technique without using solvents, this technique is done by grinding all the reactants in a mortar so that the collisions between reactants occur and cause frictional energy from local heat which will accelerate the formation of chalcone [8]. In the previous study, hydroxy chalcone was synthesized from hydroxy acetophenone and benzaldehyde through Claisen-Schmidt condesion using conventional ethanol solvents. The results showed that the formation of hydroxy chalcone compounds required strong bases (50%), long reaction times (24 hours) and low yields (40-70%) [3,8], so that this study will be developed. the new design of synthesis of 3 bromo chalcone compounds from acetophenone substituted with bromo and methoxy substituted benzaldehyde (4-methoxybenzaldehyde, 3,4-dimethoxybenzaldehyde and 3,4,5-trimethoxybenzaldehyde) through a green chemistry approach of solvent-free synthesis using grinding techniques.

2. Experimental

2.1 Materials

4-bromoacetophenone, 4-methoxybenzaldehyde, 3,4-dimethoxybenzaldehyde, 3,4,5-trimethoxybenzaldehyde, sodium hydroxide, HCl, anhydrous Na_2SO_4, silica gel 60-F254, Iodine, ethanol, ether, EtOAc, $CDCl_3$, *n*-hexane, acetone. General chemicals were purchased from Merck or Aldrich, and were used without further purification. All solvents (Merck) were used without further purification.

2.2 Procedure

The bromo chalcone were synthesized as mentioned in the reference [8]. A mixture of 4-bromoacetophenones (10 mmol), benzaldehyde (10 mmol) and solid pallete of KOH (20 mmol) was ground with a mortar and pestle for 15 minutes at room temperature. The obtained solid mixture was diluted with cold water, acidified with cold HCl (10%), followed by drying over anhydrous Na_2SO_4. The bromochalcone synthesized were characterization used by GC-MS.

3. Results and Discussion

The composition of the reagents for the synthesis of bromochalcone refers to [3] The mole of each reagent used is 0.01 moles, while the base mole is 0.02 moles. In stoichiometry, it only takes 0.01 mol, but based on research to produce the product, it takes two basic moles of reagent moles. The use of KOH twice the mole of reagents can increase the impact so that the chance of reaction is greater. Therefore, the mole of KOH used as a catalyst for bromochalcone synthesis is twice the mole of reactants.

Based on the stoichiometric concept, the mass of reagents used were 4-bromoacetophenone 1.99 g, 4-methoxy benzaldehyde 1.36 g, 3.4-dimethoxybenzaldehyde 1.66 g, 3.4.5-trimethoxo benzaldehyde 1.96 g, and KOH 1.12 g.

Synthesis of 4'-bromo-4-methoxichalcone (Chalcone **1**) was started by mixing 4-bromoacetophenone, 4-methoxy benzaldehyde and KOH into pestle mortar and then grinding at room temperature for 15 minutes to obtain orange yellow solids. During these 15 minutes it is estimated that the reaction has taken place perfectly [3]. The reaction time in the grinding technique is shorter because there are many collisions at a faster speed between the reactant molecules and the amount of heat produced due to the friction of the reactant molecules during the grinding process.

4'-bromo-3,4-dimethoxichalcone (Chalcone **2**) was synthesized in the same way, using 4-bromoacetophenone, 3,4-dimethoxy benzaldehyde and KOH, while for synthesis 4'-bromo-3,4,5-trimethoxichalcone (Chalcone **3**) using 4-bromoacetophenone, 3,4,5-trimethoxy benzaldehyde and KOH. The synthesized Chalcone **1** in the form of pale white crystals is rather orange, Chalcone **2** in the form of yellow solids, while Chalcone **3** is in the form of a greenish yellow solid.

The mechanism of reaction of chalcone synthesis that occurs is as follows:

Stage I: Formation of carbanion from 4-bromoacetophenone

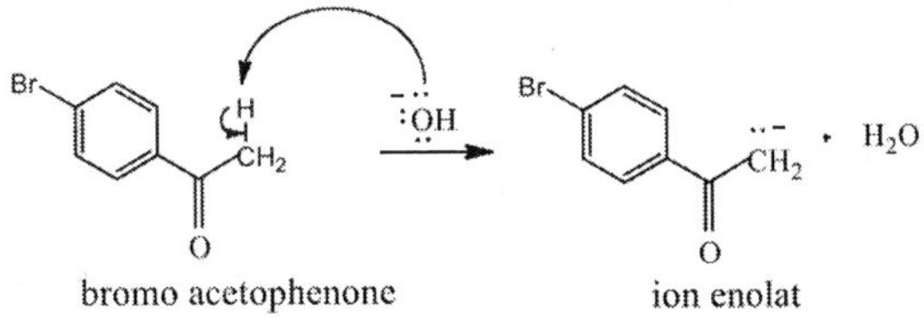

The first reaction stage is the formation of carbanion or enolate ions. 4-bromoacetophenone is an aromatic ketone which has hydrogen α. If treated with alkaline solutions such as KOH, the hydroxide ion from the base will attack hydrogen α from the ketone so that a carbanion is formed which can be stabilized by resonance and release the H_2O molecule.

Stage II: Nucleophilic addition reactions

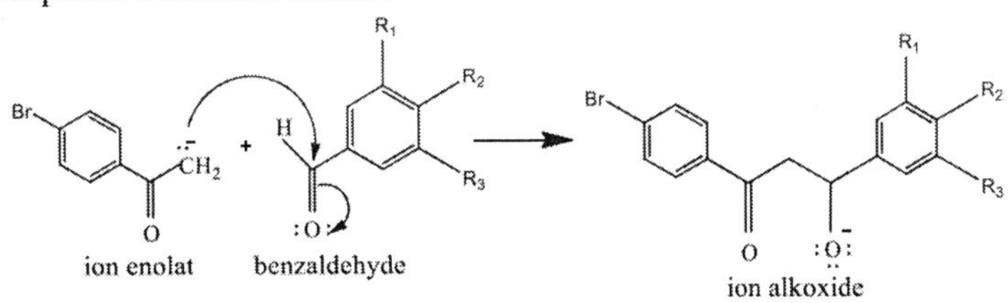

The second reaction stage is a nucleophilic addition reaction. Nucleophilic reactions occur when nucleophilic (an excess electron compound) attacks an atom of another compound and forms a new compound. At this stage the enolate or carbanion ion formed at stage 1 acts as a nucleophile that attacks the carbonyl group of benzaldehyde. An alkoxide ion is formed which has an excess of electron charge in the O atom.

Stage III: Formation of an aldol

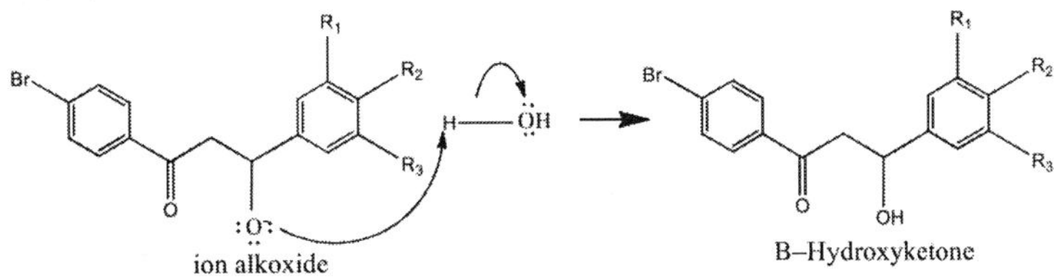

The third reaction stage is the formation of an aldol. Aldol is a compound formed from aldehydes and ketones where aldol takes protons from solvent molecules, H_2O. Alkoxide ions take hydrogen protons from H_2O molecules to form β-hydroxyketone (aldol). Then the hydroxide ion from H_2O binds to the potassium ion and returns to form a KOH base catalyst.

Stage IV: Dehydration reaction of aldol compounds

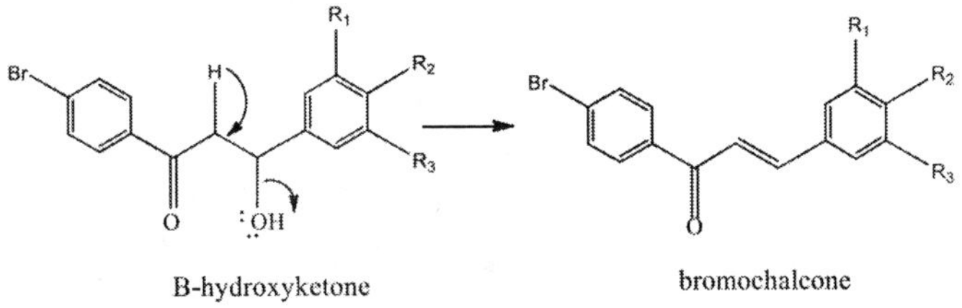

The fourth reaction stage is the dehydration reaction of the aldol compound. Dehydration reaction is a reaction of the release of water molecules. Carbonyl β-hydroxy like aldol is easily dehydrated, because the double bonds in the compound conjugate with the carbonyl group. According to [9], if dehydration produces a double bond that conjugates with an aromatic ring, dehydration takes place spontaneously. The β-hydroxyketone compound has hydrogen α to the carbonyl group which causes hydrogen α to be easily released in an alkaline atmosphere so that it can accelerate the dehydration of the aldol compound. The product is stable because there are double bonds conjugated with an aromatic ring. The dehydrated aldol compounds (loss of water) will form conjugated α, β unsaturated ketones or chalcone.

Overall the Claisen-Schmidt condensation reaction for bromo chalcone synthesis can be written as follows:

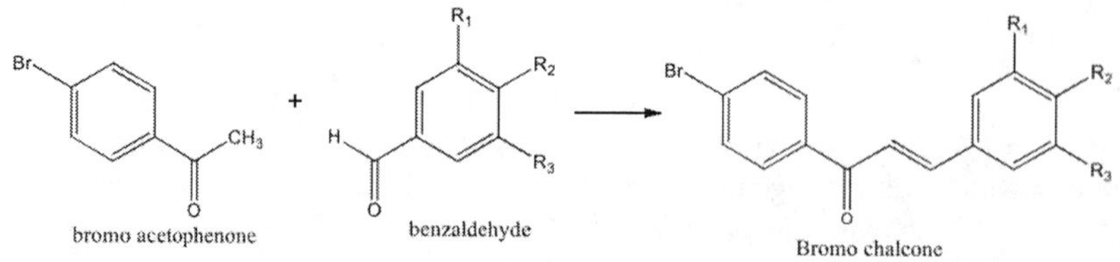

Characterization of synthesized compounds using GC-MS. Chalcone **1** gives a chromatogram with 3 peaks. The highest and dominant peak is the third peak with a tR of 37.7 minutes (Figure 1). The peak was identified as the peak of the chalcone 1, as evidenced by mass spectra showing the presence of molecular ions (M+) at m/z 316. This interpretation is supported also from the m/z analysis in theory using ChemDraw, m/z Chalcone **1** appears on m/z 316.01 (100.0%), 318.01 (97.7%), 317.01 (17.4%), 319.01 (17.1%), 320.01 (1.8%), 318.02 (1.4%).

Mass spectra analysis of Chalcone **1** shows fragmentation peaks, 348; 331; 317; 303; 288; 267; 223; 196; 191; 181; 152; 138; 119; 105; 89; 76; 63; 51; 39; and 28. Fragmentation of Chalcone **1** shows molecular ions (M+) releasing radical Br to form fragments with m/z 267. Fragment m/z 191 is the result of radical release of C_6H_4Br, while the m/z 164 fragment results from radical release of $C_7H_5O_2$. Molecular ion fragmentation releases C_6H_5O radicals giving fragments with m/z 191, while radical release $C_{10}H_{11}O_2$ produces fragments m/z 121.

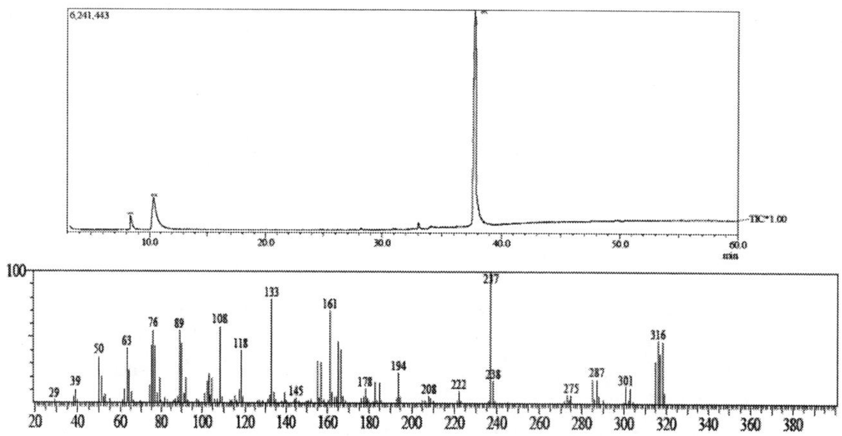

Figure 1. GC-MS Spectra of Chalcone 1

GC-MS Chalcone **2** analysis provided GC chromatogram as shown in Figure 2. The chromatogram showed 3 peaks, the third peak as the dominant peak. The third peak has a retention time of 40.71 minutes and an area of 46.88%. MS analysis shows that the third peak is Chalcone **2** with m/z 348. The m/z analysis with ChemDraw analysis gives data m/z 346.02 (100%), 348.02 (97.9%), 347.02 (18, 5%), 349.02 (18.2%), 348.03 (1.6%), 350.03 (1.6%). Based on the ChemDraw analysis of m/z 348 compounds formed by 97.9% greater than molecular weight of Chalcone **2**, that is at m / z 347, only obtained by 18.5%. The results of the GC-MS analysis concluded that the Chalcone **2** had been formed.

Molecular ion fragmentation Chalcone **2** (m/z 376) releases the OCH_3 radical and forms a cation with m/z 345. If the c radical is released, a cation will be formed with m/z 297. Then by releasing $C_{11}H_{13}O_3$

radical, a cation is formed with m/z 183, whereas cation with m/z 155 can be formed by releasing $C_{12}H_{13}O_4$ radical.

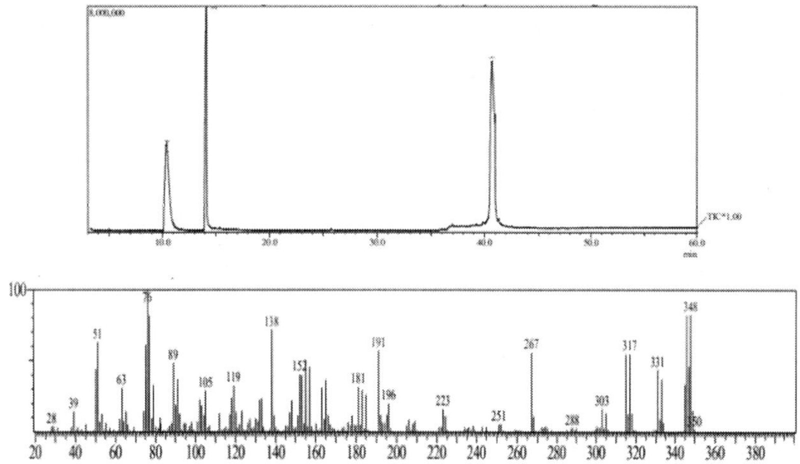

Figure 2. GC-MS Spectra of Chalcone 2

Chromatogram Chalcone **3** (Figure 3) shows 3 peaks, with peaks dominant at the third peak, area of 52.88%. The MS analysis shows that the Chalcone **3** has m/z 376. The price of m z is similar to the molecular weight of Chalcone **3** (377). The difference in MR and m/z can be checked using the ChemDraw application. ChemDraw application can help analyze the price of m/z from a compound, the price of m/z is more than one. This is because of the difference in abundance in each of the atoms that make up the molecule. Based on ChemDraw's analysis, m/z Chalcone **3** is 376.03 (100%); 378.03 (97.3%); 377.03 (19.5%); 379.03 (19.1%); 378,04 (2,7%); 380.04 (1.8%). Base on ChemDraw analysis, m/z 376.03 is the one with the highest relative abundance, this is in accordance with the m/z that appears on the MS spectra, 376.

Chalcone 3 gives fragmentation by releasing OH radicals to form fragments m/z 283. Fragments with m/z 269 are fragments resulting from the release of radical OCH_3, while fragments m/z 121 result from radical release of $C_{10}H_{11}O_2$. Fragmentation of molecular ions by releasing $C_7H_5O_2$ radicals produces m/z 164 fragments.

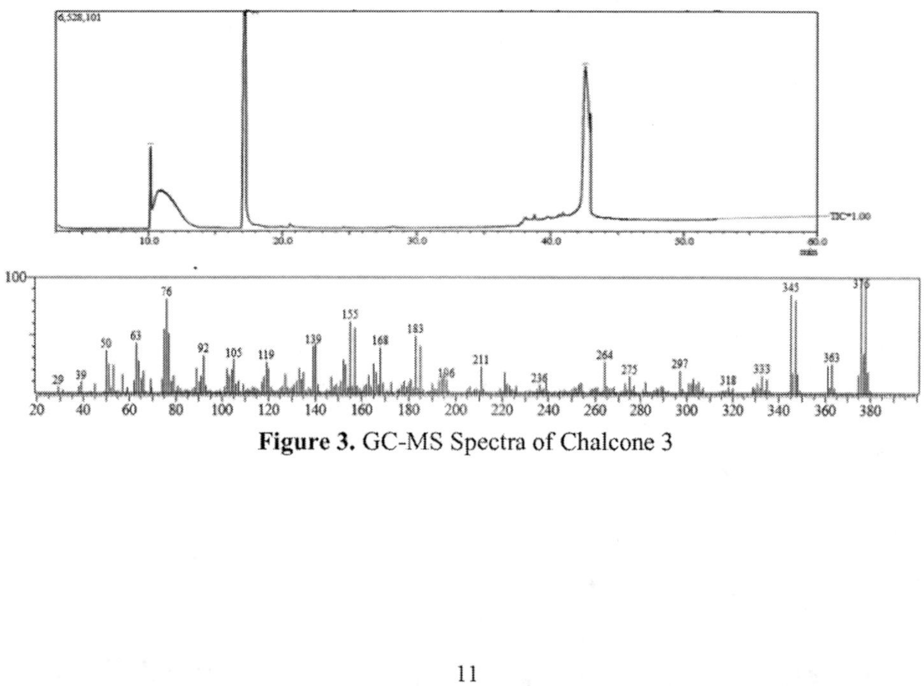

Figure 3. GC-MS Spectra of Chalcone 3

4. Conclusion

The bromochalcone derivatives (4'-bromo-4-methoxychalcone, 4'-bromo-3,4-dimethoxychalcone, 4'-bromo-3,4,5-trimethoxychalcone) can be synthesized from 4-bromoacetophenone and appropriate methoxy benzaldehyde through a green chemistry approach with grinding techniques.

Acknowledgment

The authors are thankful to DP2M DIKTI and Head of LPPM UNS, Dean of FKIP UNS, Head of Organic Laboratory UGM, and Chemistry Laboratory PMIPA FKIP UNS.

References

[1] Patil C B, Mahajan S K and Katti S A 2009 *J. Pharm. Sci. & Res.* **1**(3), 11-22
[2] Eryanti R, Zamri A, Jasril and Rahmita 2009 *Jurnal Natur Indonesia.* **12** (2), 173-177
[3] Susanti E V H, Matsjeh S, Wahyuningsih T D, Mustofa and Redjeki T 2012, *Indo. J. Chem.* **12**(2), 146-151
[4] Suzana, I M, Amalia K, Ekowati J, Rudyanto M, Poerwono H and Budiati T 2013 *Berkala Ilmiah* Kimia Farmasi **2**(1), 1-6
[5] Zangade S, Mokle S, Vibhute A and Vibhute Y 2011 *Chem. Sci. J.* **13**, 1-6
[6] Aguilera A, Alcantara A R, Marinas J M, and Sinisterr J V 1987 Can. J. Chem **65**, 1165-1171
[7] Mogilaiah K, Swami T K, Chandra A V, Srivani N and Vidya, K 2010 Indian J. Chem **49B**, 382-385
[8] Susanti E V H, Matsjeh S, Wahyuningsih, T D, Mustofa and Redjeki T 2014 *Indo. J. Chem* **4** (2) 174 – 178
[9] Budimarwanti C and Handayani S 2010 *Proceeding Seminar Nasional Kimia dan Pendidikan Kimia*, Yogyakarta: FMIPA Universitas Negeri Yogyakarta

Facile Synthesis of Carboxylated Activated Carbon Using Green Approach for Water Treatment

L Mahardiani[1*], S Saputro[1], F Baskoro[2,3], N M Zinki[1], and M Taufiq[1]
[1] Department of Chemistry Education, Faculty of Teacher Training and Education, Universitas Sebelas Maret, Jl. Ir. Sutami 36A, Kentingan, Surakarta, Indonesia
[2] Department of Applied Chemistry, National Chiao Tung University, Hsinchu 30010, Taiwan (ROC)
[3] Institute of Chemistry, Academia Sinica, Taipei 11529, Taiwan (ROC)

E-mail: lina80_ssa@yahoo.com

Abstract. A chemical treatment by using oxidizing agents is the most common methodology to modify or create functional groups on activated carbon. The modified activated carbon obtained from chemical oxidation needs further treatment and produced waste, therefore a new approach is needed. A facile green approach is reported. In this approach the activated carbon was oxidized using air without additional chemical treatment to increase the content of acidic surface functional groups, especially carboxyl groups. The formation of carboxyl groups was confirmed by Temperature Programmed Decomposition, Boehm titration, and Infra-red spectroscopy results. The Boehm titration result showed higher contents of carboxyl groups by three-fold. The carboxylated activated carbon synthesized by green approach showed excellent performance for water treatment.

1. Introduction

Activated carbon (AC) is a common term used to describe carbon-based materials produced from various carbonaceous rich materials, such as wood, coal, lignite and coconut shell [1]. Activated Carbon is inexpensive and have been applied for many purposes, and as an adsorbent is one of it. It has been proven that AC is an excellent adsorbent for organic and inorganic pollutant removal for both in aqueous or gaseous environment [2,3]. However, it is believed that it is important to enhance the effectiveness of AC as adsorbent for specific usage. Therefore, the modification of surface of AC is necessary to develop affinity for certain contaminant [4]. One of the methods to modify is by chemical activation since it able to create acidic functional groups onto the surface of activated carbon [5, 6]. Regardless of the advantage of chemical activation, the waste that produced after its activation can not be avoided. Recently the synthesize and modification of materials is based on the green chemistry principles. Green chemistry offers an alternative to the traditional environment protection agenda, such as preventing waste production and minimizing the use of reagents and solvents [7].

Air oxidation is one alternative to alter the surface [6,8]. This method is facile since AC was oxidized only by using the air to create the acidic functional groups such as carboxylic, phenolic, and lactone [9]. In addition, there was no waste produced and use no reagents or solvents during the oxidizing process. There are some experimental techniques that able to identify the formation of functional groups crafted onto the surface of AC. Infra-red spectroscopic methods also applicable to distinguished different functional groups based on the bands observed at different wavenumbers.

Generally, carbon will absorb almost all the radiation in the visible spectrum, since carbons are black. Therefore, it is not easy to get good spectra of functional groups on carbon material using FTIR. There are some techniques that able to conveniently identify the functional groups on the carbon materials.

The surface oxides on a carbon can have both acidic and basic properties. The Boehm titration are widely used to determine the number of the total acidic functional groups based on the volume of titrant consumed by the functional groups [10,11] and similar with basic functional groups. The temperature-programmed method is a recent analysis to determine the surface oxygen groups on the AC based on the CO and CO_2 released at different temperature [4,12]. It is assumed that each type of surface group decomposes to defined product, namely carbonyl and hydroxyl groups and ether-type oxygen will generate CO, while carboxyl groups will generate CO_2.

In this present work, we perform the study of the air oxidation on AC to modify its surface using green approach via air oxidation method and compare with AC treated with chemical activation, HNO_3. In addition, various techniques were conducted to confirm the formation of functional groups resulted from air oxidation. Further, we applied the modified AC for water treatment.

2. Experimental

2. 1 Chemicals and materials

Activated carbon (BA) was supplied by Ajinomoto Fine-Techno. Sodium bicarbonate (> 99.5%), sodium hydroxide (> 97%), nitric acid (>70%), potassium bromide (FTIR grade, crystal block), standard solution 0.05 mol/L HCl and 0.05 mol/L NaOH used for Boehm's titration, metil orange was purchased from Wako Pure Chemical Industries, Ltd.

2. 2 Methods

2. 2. 1 Synthesis of functional carbon sample

Treatment 1. Air oxidation: dried carbon (BA) about 4.0 g was spread into petri dish and heated under air flow in calcination apparatus. The calcination temperature was 425 °C with temperature rate 10 °C/min and kept the temperature for 3 h.

Treatment 2. HNO_3 activation: a 2.0 g of BA was placed into round bottom flask and mixed with 20 mL of HNO_3 for 3 h at 90 °C. The mixture was filtered and washed to neutral the pH using distilled water and further dried at 110 °C overnight and BA-HNO_3 was obtained.

2. 2. 2 Characterization

The surface functional groups of catalyst were detected by a Fourier transformed infrared (FTIR) spectrometer (Perkin Elmer Spectrum 100 S). The samples (0.1 wt %) were mixed with KBr and analyzed in the 4000 – 400 cm^{-1} wavenumber range, with resolution of 4 cm^{-1}, by averaging 64 scans.

Temperature Program Decomposition (TPD) spectra was obtained from BelCat II instrument (MicrotracBEL). This analysis was to determine the numbers of functional groups existed in the catalyst. For each run, 50 mg of catalyst was pretreated at 150 °C for 1 h before heated up to 1000 °C with ramp of 10 °C/min for 10 min under He flow. The concentration of CO and CO_2 were determined by the intensities of m/z at 28 and 44, respectively.

Boehm's titration was also conducted to determine the numbers of acidic functional groups were on the catalyst. For each analysis, 100 mg of catalyst was pretreated in 20 mL $NaHCO_3$ and NaOH solution for 24 h at room temperature. Filtrate obtained from filtration of pretreated sample were used for back titration. The quantity of acidity of various type were determined from the assumption that $NaHCO_3$ reacts with carboxylic groups and NaOH reacts with carboxylic, lactonic and phenolic groups.

2.2.3 Adsorption activity of modified AC

The adsorption activity of modified AC was conducted by contacting 0.1 g of AC with 20 mL $Pb(NO_3)_2$ 10 ppm and $Cu(NO_3)_2$ 10 ppm. Samples were collected at certain interval times then filtered before filtrate then was analyzed using Atomic Absorption Spectroscopy (AAS).

3. Results and Discussion

3.1 Temperature Program Decomposition

Thermal decomposition of activated carbon is useful to identify its surface chemical properties. Surface oxide of carbon are formed from various functional groups [11]. Carboxylic, lactone, phenol, carbonyl, anhydride, ether and quinone functional groups have been identified on carbon surface. The decomposition of those functional groups can release CO and CO_2 at different temperature by heating under inert conditions [4, 12]. Peaks of CO and CO_2 from TPD result can be distinctive accordingly with certain functional groups, namely CO peak was resulted from carboxylic anhydride, hydroxyl, ether or carbonyl, carbonyl, and pyrone-type structure decomposition, while CO_2 peak was resulted from carboxyl, carboxylic anhydride, peroxide, and lactone decompositions [13].

Untreated carbon (BA) showed a little functional group on its surface, which were 0.88 and 0.32 mmol g^{-1}, for CO and CO_2 respectively. The functional groups that existed on the surface of BA are hydroxyl, carbonyl and small amount of carbonyl. The functional groups existed on carbon were significantly increased after air oxidized and HNO_3 treatment (Table 1). The TPD spectra obtained showing a gradual increase both in CO and CO_2 released. However, the CO peak increased much more than the CO_2. This can be explained in terms of the mechanism for carbon gasification proposed by Zhuang et al. [14]. The reaction of free active site on the AC with oxygen into carbonyls, ethers and CO and further oxidized to CO_2 + CO or carbonyls, ethers are much more important compared with the reaction between free active site on the AC with oxygen into lactones and anhydrides. The amount of CO released was higher than that of CO_2 indicating that there are much more carbonyl, ether and phenol groups than lactones and anhydrides. The amount of CO was also obtained from the decomposition of carboxylic anhydrides into one CO and CO_2 molecule [15] in addition to the contribution of phenolic and quinones [16].

3.2 Boehm titration

The surface oxides on carbon can have both acidic and basic properties which coexist usually, but the concentration of basic site decreases with increasing of acidic property on the carbon surface. The acidic surface properties are caused by the presence of carboxyl groups (also for anhydride), lactone, and hydroxyl group of phenolic property [11]. The total acidity, which consists of lactone, hydroxyl and carboxylic functional groups, can be determined with titration of NaOH. In addition, carboxylic functional group alone can be determined with titration of $NaHCO_3$.

The number of acidic functional groups calculated by Boehm titration are shown in Table 1. The total acidic functional groups of BA was 0.35 mmol g^{-1} and 0.09 mmol g^{-1} of it was carboxylic functional group. Total functional groups were increased after the chemical treatment by using various oxidants or by using air (Table 1). The air oxidation result showed increased of total acidic functional groups, about seven-folds of total acidic functional groups and eleven-folds for carboxylic group. Functionalization by using HNO_3 was showed that total acidity of carbon greatly increased up to 3.38 mmol g^{-1}. This is because HNO_3 is really strong acid, therefore HNO_3 could made defects on surface of carbon that allow formation of phenolic, hydroxyl, carboxylic and anhydride in large amount. The differences on the value of surface oxygen functional groups on carbon between TPD and titration results may be due to the inaccessibility of some functional groups within the pores. In general, these results suggested that chemical activation were able to make carbon structure altered and they showed acidic character.

Table 1. Summary of chemical and physical properties of activated carbon as adsorbent

Sample	TPD		Boehm titration		N_2 isotherm	
	CO (mmol g^{-1})	CO_2 (mmol g^{-1})	Total acidic amounts (mmol g^{-1})	–COOH (mmol g^{-1})	S_{BET} (m^2 g^{-1})	Pore vol. (cm^3 g^{-1})
BA	0.88	0.32	0.35	0.09	1163	0.61
Air oxidized BA	5.29	1.16	2.64	1.00	1065	0.62
BA-HNO_3	3.42	2.34	3.38	2.03	990	0.51

3.3 Infra-red spectroscopy

In order to examine the types of functional groups formed in BA before and after functionalization with various oxidants, FT – IR spectroscopy was used (Figure 1). The FT – IR spectra for all samples show band around 1566 – 1582 cm^{-1} due to the presence of C–C stretching of aromatic C=C. In addition, there was peak around 1200 cm^{-1} assigned for physically adsorbed oxygen or aromatic–OH stretching. There was also strong peak observed at 1781 cm^{-1} which is attributed to the anhydride group resulting from C=O asymmetrical stretching mode and peak at 3500 cm^{-1} assigned for C–OH [17-19].

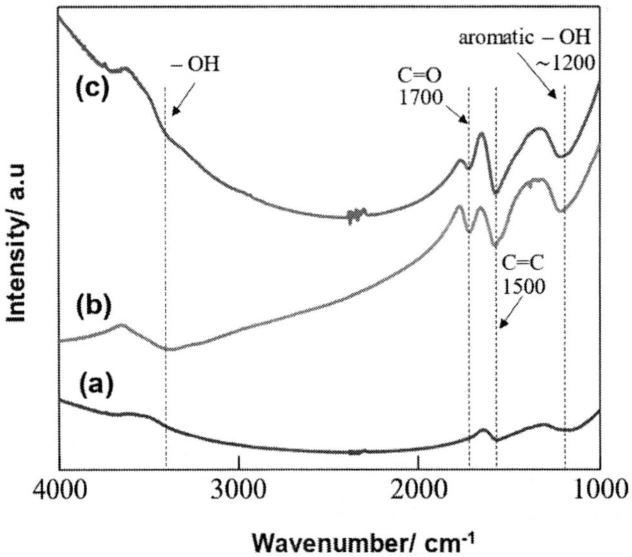

Figure 1. The infrared spectra of (a) BA, (b) BA – HNO_3, and (c) Air oxidized – BA

3.4 Adsorption activity of modified AC

As shown in figure 2, the adsorption of Pb(II) and Cu(II) using various AC adsorbents were evaluated. The modification of the AC surface influence the adsorption activity towards Pb(II) and Cu(II). The BA adsorbent showed lowest adsorption activity for Pb(II) and Cu(II), 20 and 35%, respectively. Meanwhile, the BA-HNO_3 showed the highest adsorption activity for Pb(II) solution (68%) but showed moderate adsorption activity towards Cu(II) about 40%. In other hand, the adsorption activity of air oxidized BA was not changed for Pb(II) and Cu(II), which are 60 and 68%, respectively. Thus, the AC with air oxidation or chemical activation shows higher activity [20, 21]. This is expected because the total of acidic amounts plays important role for this process. The more the acidic functional groups exist on the surface of AC, the more the Pb(II) and Cu(II) were adsorbed. However, the chemical activation method of AC becomes more selective towards certain species, but loses adsorption capacity to others [22]. This trend shows on the adsorption activity of BA-HNO_3 for Pb(II) and Cu(II). Meanwhile, air oxidized AC shows no difference of its adsorption activity for Pb(II) and Cu(II).

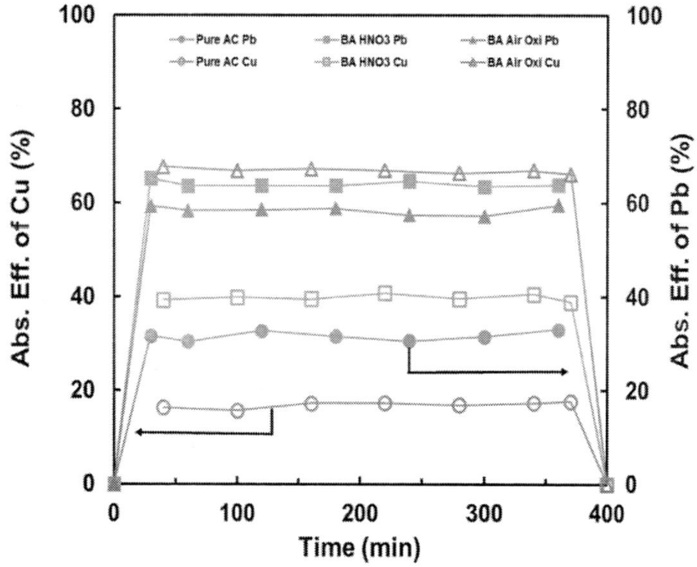

Figure 2. The adsorption activity of various AC adsorbent toward Pb(II) and Cu(II). Adsorption conditions: Pb(II) and Cu(II), 10 ppm, $m_{adsorbent}$, 0.1 g

4. Conclusion

The air oxidation is one of the methods to modify the surface of AC without producing waste and using the reagents or solvents. The modification results of air oxidation are comparable with that of chemical activation method. The modification results were confirmed by some techniques such as temperature programmed decomposition (TPD), Boehm titration, infra-red spectroscopy. TPD showed that evolution of CO and CO_2 of modified AC was increased significantly compared to untreated AC. TPD results was in accordance with Boehm titration where total acidic amounts of modified AC were increased almost ten-folds compared to untreated AC. Furthermore, the modified AC showed high adsorption activity toward Pb(II) and Cu(II).

Acknowledgment

The authors acknowledge funding from the Universitas Sebelas Maret, Surakarta – Indonesia via International Collaboration Grant.

References

[1] Acharya J, Sahu J N, Sahoo B K, Mohanty C R, and Meikap B C 2009 *Chem. Eng. J.* **150** 25
[2] Chingombe P, Saha B, and Wakeman R J 2005 *Carbon* **43** 3132
[3] O'Connell D W, Birkinshaw C, and O'Dwyer T F 2008 *Bioresour. Technol.* **99** 6709
[4] Figueiredo J L, and Pereira M F R 2010 *Catal. Today* **150** 2
[5] Moreno-Castilla C, Ferro-Garcia M A, Joly J P, Bautista-Toledo I, Carrasco-Marin F, and Rivera-Utrilla J 1995 *Langmuir* **11** 4386
[6] Otake Y and Jenkins R G 1993 *Carbon* **31** 109
[7] Sheldon R A, Arends I, and Hanefeld U 2007 *Green Chemistry and Catalysis* Wiley-VCH

IOP Conf. Series: Materials Science and Engineering **578** (2019) 012003 doi:10.1088/1757-899X/578/1/012003

[8] Silva A R, Freire C, de Castro B, Freitas M M A, and Figueiredo J L 2001 *Micro. Mesoporous Mater.* **46** 211
[9] Shrotri A, Kobayashi H, and Fukuoka A 2016 *ChemSusChem* **9** 1299
[10] Boehm H P 1994 *Carbon* **32** 759
[11] Boehm H P 2002 *Carbon* **40** 145
[12] Zhou J H 2007 *Carbon* **45** 785
[13] Figueiredo J L, Pereira M F R, Freitas M M A, and Orfao J J M 1999 *Carbon* **37** 1379
[14] Zhuang Q-L, Kyotany T, and Tomita A 1995 *Energy Fuels* **9** 630
[15] Calo J M, Cazorla-Amoros D, Linares-Solano A, Roman-Martinez M C, and Salinas-Martinez De Lecea C 1997 *Carbon* **35** 543
[16] Zielke U, Huttingerm K J, and Hoffman W P 1996 *Carbon* **34** 983
[17] Seo J-M and Baek J-B 2014 *Chem. Commun.* **50** 14651
[18] Severini F, Formaro L, Pegoraro M, and Posca L 2002 *Carbon* **40** 735
[19] Almeida R K S, Melo J C P, and Airoldi C 2013 *Micro. Mesoporous Mater.* **165** 168
[20] Collins J, Ngo T, Qu D, and Foster M 2013 *Carbon* **57** 174
[21] Guedidi H, Reinert L, Leveque J M, Soneda Y, Bellakhal N, and Duclaux L 2013 *Carbon* **54** 432
[22] Sweetman M J, May S, Mebberson N, Pendleton P, Vasilev K, Plush S E, and Hayball J D 2017 *Journal of Carbon Research* **3** 18

IOP Conf. Series: Materials Science and Engineering **578** (2019) 012004 doi:10.1088/1757-899X/578/1/012004

Synthesis and characterization of hematite (α-Fe_2O_3) from lathe waste using co-precipitation -calcination method

L M Khoiroh, M N Al-Chabib and A Prasetyo
Department of Chemistry, Faculty Science and Technology, Universitas Islam Negeri Maulana Malik Ibrahim Malang, Jl. Gajayana 50 Malang 65144, Indonesia

E-mail: lilikmfx@gmail.com

Abstract. Lathe waste contains high iron, therefore it has the potential to be a precursor in hematite (α-Fe_2O_3) synthesis. In this work, hematite was synthesized from lathe waste using the co-precipitation-calcination method with variation time of calcination: 1, 2, 3, 4 hours. The X-Ray diffraction data showed that the goethite (α-FeOOH) phase was obtained at precipitation stage then transformed to hematite (α-Fe_2O_3) phase after calcination and there is no impurities. The image of scanning electron microscopy showed that the distribution of particle size was not uniform. The color parameter showed that the highest chroma was achieved for 3 hours calcination.

1. Introduction

Iron craft industries produced lathe waste that contains 96 % of iron, which contributes to the environmental pollution significantly. The high iron content in lathe waste give opportunity to apply as precursor for iron compound synthesis. One of iron compound is hematite and this compound have advantages including non-toxic, having thermal stability [1], biodegradable [2], and durable [3]. In recent years, hematite becomes interesting in several applications as nanocatalyst, biomedical application, and especially in pigment application [3-6].

Many techniques have been developed to synthesized hematite, such as precipitation method [7], microwave-assisted calcination method [8], sol-gel method [9], hydrothermal technique [10], and thermal transformation technique [11]. Chemical precipitation method is a simple method, low cost, has high purity, short in preparation time, and high inhomogeneity [7, 11]. Synthesis of hematite by precipitation-calcination method results in high-crystallinity and nanoparticle sized product [7]. In synthesizing hematite, the compound change of structure, crystallite size, color, and distribution of morphology depend on the synthesis parameters, such as time of calcination [8]. Hematite have rhombohedral structure with crystallite size 40-41 nm, and lightness of 20.47 was calcined for 2 h [12], while for 3 h, hematite have crystallite size of 17 nm and lightness of 34.1 [11].

In this work, we investigated the effect of time calcination to structure, crystallite size, the color parameter, and morphology. We have used the co-precipitation-calcination method to synthesize hematite pigments. The sample products characterized using X-Ray Diffraction (XRD), color reader, and Scanning Electron Microscopy-Energy Dispersive Spectroscopy (SEM-EDS).

2. Experimental

2.1 Materials

All reagents used to synthesis were analytical grades and without further purification. Nitric acid (HNO_3), sodium bicarbonate ($NaHCO_3$), and sodium hydroxide (NaOH), were purchased from Merck, Germany. Lathe waste used for synthesizing was from iron craft industry of Malang, Indonesia.

2.2 Methods

Samples of iron lathe were dissolved into 200 mL of HNO_3 (Merck) and heated until formed a slurry. The slurry product was then used as a precursor for synthesizing hematite. The precursor was dissolved in 500 mL of aqua de-mineral then was added with 50 mL of sodium bicarbonate solution and sodium hydroxide until reach pH 6. The homogenous mixture was heated at 70°C for 1 hour under magnetic stirring until formed the yellowish precipitated. The obtained precipitated was aged, filtered, and washed with distilled water until a neutral pH was obtained, then was dried at room temperature. The red dry powder obtained with calcining at 750°C for 1, 2, 3, and 4 hours. The structures of red powder synthesized were observed by X-Ray Diffraction (XRD), the value of lightness, redness degree, and yellowness degree were studied by a color reader and the morphology, and the compositions of the element were observed by Scanning Electron Microscopy-Energy Dispersive Spectroscopy (SEM-EDS).

3. Result and Discussion

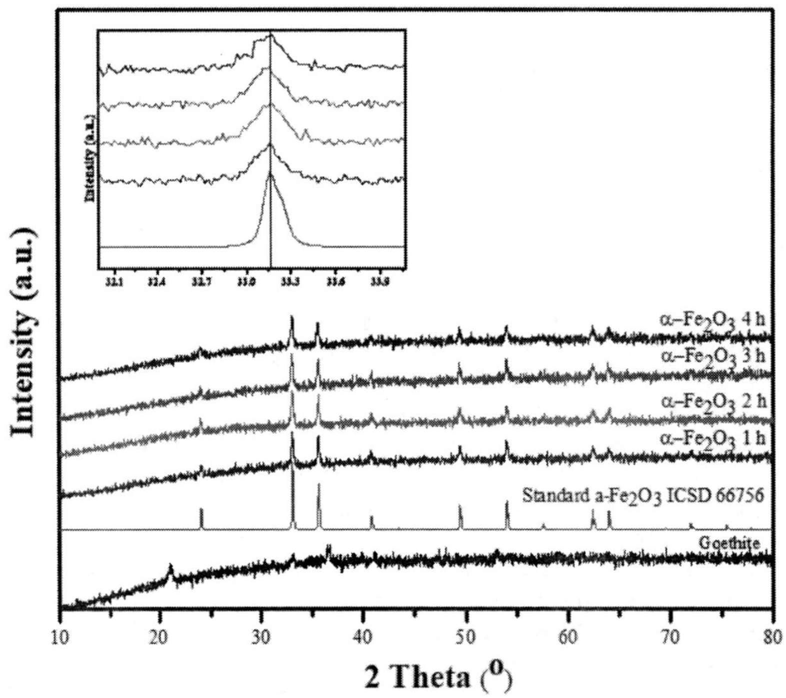

Figure 1. The X-Ray powder diffraction pattern of hematite obtained at calcination. Inset picture showed the peak position at around 33.1°

Figure 1 showed the XRD data of sample and showed that at precipitation stage at 70°C for 1 hour goethite phase was obtained and then transformed to hematite phase after calcination. XRD pattern in figure 1 shows that all sample after calcination has an agreement with pure of α-Fe_2O_3 (ICSD no

66756). Hematite successfully formed for 1-hour calcination and there was no changed phase when the calcination time was longer. All the reflection can be indexed in agreement with the expected rhombohedral structure of α-Fe_2O_3. The narrow peaks indicate that all of the hematite products are crystalline even though low crystallinity. The line broadening of the diffraction peak is affected crystallite size. Table 1 shows the average crystallites sizes are evaluated by Scherrer's equation at the highest intensity peak (104 planes). The average crystallite size associated with XRD data ranged between 14.84 and 24.34 nm, it means the size of hematite synthesized with longer calcination is increased due to a higher chance for crystal growth to take place in the calcination process. The highest crystallite size of α-Fe_2O_3 at 3 hours and decrease when calcined for 4 hours (20.83 nm).

Table 1. The crystallite size of hematite samples

The Compound	Crystallite Size (nm)
Hematite 1 h	14.83
Hematite 2 h	21.49
Hematite 3 h	24.34
Hematite 4 h	20.83

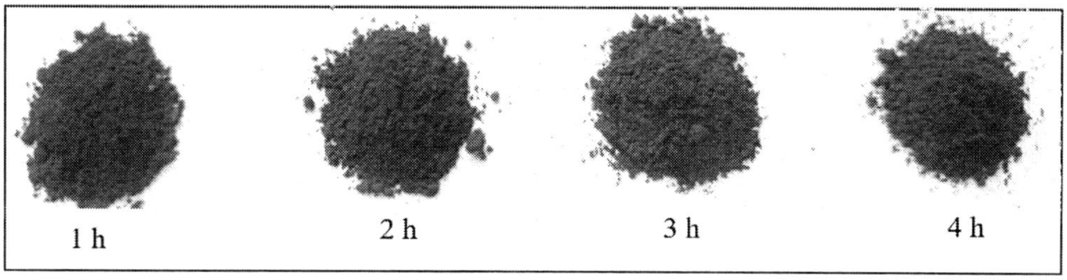

Figure 2. The visual aspect of hematite synthesized at variation of time calcination

Visual evaluation of calcined hematite at the variation of time showed dark red with no significantly different. The color parameter (L*) of hematite synthesized in figure 3 shows all samples high positive result for L* parameters. Hematite synthesized for 4 hours has the highest lightness. It was an increase in the intensity of lightness proportional to the time of calcination. Figure 4 shows that a* b* parameters of hematite synthesized accordance with standard [13]. From a*b* parameters, all hematite synthesized are more intense toward redness, except hematite which within 4 hours is showing more yellowness. Based on table 2 color parameter (C and H°), the highest chroma parameter reached for 3 hours and decrease for 4 hours, meanwhile, the decreasing was not significant. It indicates that hematite can synthesize from 1 to 4 hours calcination. The parameter C tends to have a continuous increase as in increasing crystallite size. Decreased C parameter is caused differences of crystallite size in order to cause a broader distribution of electron transition band affected a less vibrant appearance.

Table 2. Color parameter (C and H°) of hematite synthesized

Hematite (α- Fe_2O_3)	C	H°
Hematite 1 h	21.8	33.3
Hematite 2 h	21.8	32.5
Hematite 3 h	22.4	34.8
Hematite 4 h	21.4	35.5

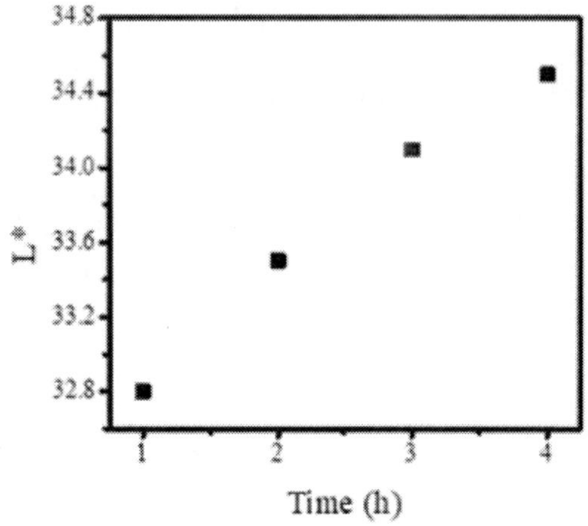

(a)

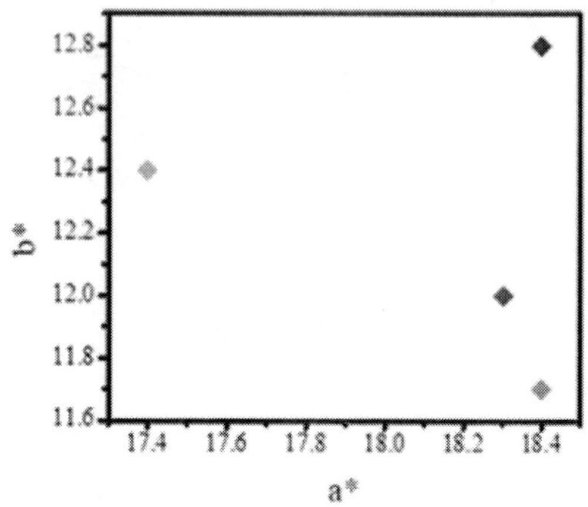

(b)

Figure 3. (a) Parameter of lightness at the variation of calcination time (b) Value of a*b* color space◆ 1 h◆ 2 h ◆3 h ◆4h

(a)

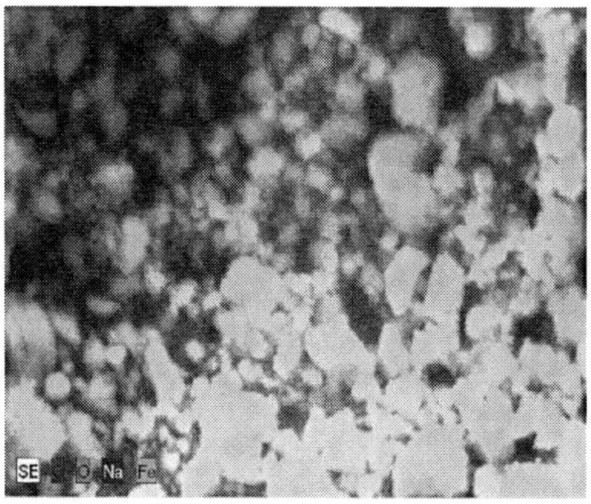

(b)

Figure 4. Synthesized hematite for 3 hours (a) SEM image of hematite (b) EDS image of hematite

The surface morphology of hematite obtained was studied using a Scanning Electron Microscope (SEM). Figure 5 showed the morphology of hematite calcined for 3 hours and can be seen that the particle shape of sample is not uniform and produces agglomerates. The results of energy dispersive spectroscopy analysis as tabulated at Table 3 and showed there are found impurities in sample such as carbon and sodium. The presence of sodium element is derived from precipitating agents ($NaHCO_3$) and carbon is derived from raw material (lathe waste).

Table 3. The composition of an element in hematite

Element	Composition (%)
Fe	50.84
O	36.58
C	11.61
Na	0.97

4. Conclusion

Hematite was synthesized from lathe waste using precipitation-calcination method have a rhombohedral structure. The highest crystallite size of hematite was obtained for 3 hours (24.34 nm). The parameter of a*b* of hematite synthesized is directly to redness due to the highest chroma for 3 hours. The morphology of hematite particle was not uniform and there is agglomeration.

Acknowledgment

The author gratefully thanks iron oxide team and acknowledge the financial support from Faculty of Science and Technology, Universitas Islam Negeri Maulana Malik Ibrahim Malang.

References

[1] Vujtek M, Zboril R, Kubinek R, and Mashlan M 2003 *Scie, Tech. and Educ. of Microsc.* **1** 1–8
[2] Ming L, Liling T, Tongyi L, Meeling C, Jia Z, Hui R T, and Shiqiang B 2014 *Phys. Chem.* **118** 10903-10910
[3] Prim S R, Folgueras MV, de Lima M A and Hotza D 2011 *J. Hazard. Mater.* **192** 1307 - 1313
[4] Nagaraj B, Kanchan M, and Yong R L 2017 *Adv. Nat. Sci. Nanosci. Nanotechnol.* **8** 025017
[5] Mou X, Wei X, Li Y and Shen W 2012 *Cryst. Eng. Comm.* **14** 5107-5120
[6] Mahnaz M, Mansor B A, Haron Md, Farideh N, Behzad N, Mohamd Z Ab R and Jamileh A 2013 *Molecules.* **18** 7533-7548
[7] Abdelmajid L, Brahim D, Abdellatif G and Salah A 2017 *Reslt. in Phys.* **7** 3007-3015.
[8] Tati N, Ferry I, Mikrajuddin A and Khairurrijal 2013 *Mater. Scien Forum.* **Vol 737** 197-203
[9] Olga O and Avairus K 2015 *Cer. Int.* **41** 4504 -4513
[10] Maiyong Z, Ying W, Dehai M, Xinghang Q, and Guowong D 2012 *J. Phys. Chem.* **116** 16276 – 16285
[11] Lilik MK, Diah M, Akhmad S, and Bambang I 2013 *Pure Appl. Chem. Res.* **2** (1) 27-34
[12] Septityana, Priyono, Rochman and Prastiyo 2013 *Youngs. Phis. Jour.* **1** 4
[13] R M Cornell and U Schwertmann 2003 *The Iron Oxides: Structure, Properties, Reactions, Occurrences, and Uses Second Edition* (Weinheim: WILEY- VCH)

IOP Conf. Series: Materials Science and Engineering **578** (2019) 012005 doi:10.1088/1757-899X/578/1/012005

Synthesis and characterization of chitosan-pectin as adsorbent of dyes

B Hastuti[1], S Hadi[2], S N Afifah[1], B Mulyani[1]
[1]Department of Chemistry Education, Faculty of Teacher Training and Education, Universitas Sebelas Maret
[2]Department of Pharmacy, Faculty of Mathematics and Natural Sciences, Universitas Sebelas Maret

E-mail: Budihastuti@staff.uns.ac.id

Abstract. Environmental pollution due to the disposal of colored substances can endanger both the environment and humans. For this reason, the presence of dyes pollutants needs to be reduced, where adsorption method has become one of the alternative. Research on synthesis of modified chitosan to pectin polyelectrolyte complex (PEC) Chi-Pec has been carried out in order to improve the characteristic of chitosan as adsorbent. The research was done by mixing chitosan solution in acetic acid with pectin solution in water for 3 hours stirring. The solution was then printed out into a polypropylene container and dried. The formed membrane was characterized using FTIR to identify functional groups, XRD and SEM to analyze their physical characteristic. The results showed that the Chi-Pec adsorbent had stability and resistance to the acidic environment. Characterization results using FTIR showed that the adsorbent has functional group of amine, carboxyl and hydrogen. The results of XRD show that the adsorbent was not crystalline. Furthermore, SEM data shows a solid structure of the adsorbent.

1. Introduction

Along with the establishment of Batik as a world heritage from Indonesia by UNESCO on October 2, 2009, nowadays the use of batik is not only on traditional events but also on various type of event [1]. This also has an impact on the growing development of the batik industry in Indonesia, especially in Solo. Most of them are home-based and small scale industries that do not have procedures for processing batik wastewater. It is have a bad impact in the form of water pollution due to processed dyes made in batik industry without processing [1]. Not only batik industry, other textile industries which in the process use dyes are also contributors to water pollution, especially rivers. In study of Septiono et. al., based on central java EPA, the biggest contributor to pollution in the Bengawan Solo River is textile industry, which is 41.5%. This number is indicated to come from activity industry, such as the garment industry, activities of dyeing process, spinning fabric and batik industry [2]. Therefore, the composition of dye in the aquatic environment must be reduced.

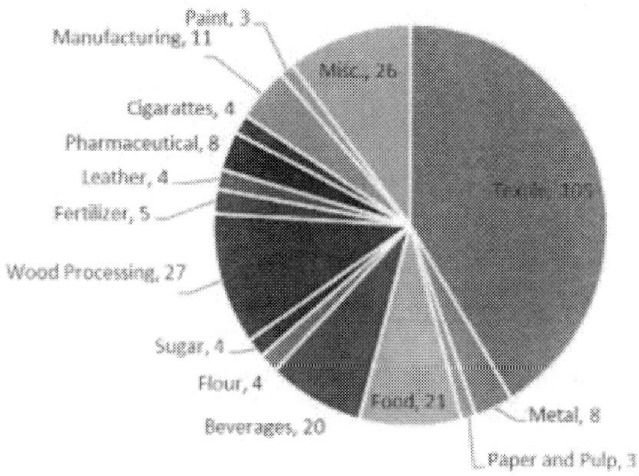

Figure 1. East Java EPA monitoring point in Begawan Solo and surrounding industry type. (Source : Center Java EPA, 2014)

Some dyes have properties that can interfered with the environment because some dyes are toxic to some organisms [3]. Methylene blue ($C_{16}H_{18}ClN_3S$) is one type of dye that widely used in the textile industry. If the body is contaminated with these dyes in the long time it can cause vomiting, nausea, anemia and hypertension [4]. One way to reduce dyes in aquatic environment is by absorption. One of these absorption can use a membrane derived from chitosan and pectin synthesis.

Chitosan is the second abundant natural polysaccharide after cellulose. Chitosan dissolves in acids and has an amine group which is polycation. Pectin is a natural polysaccharide that is water soluble and has a carboxyl methylate group [5]. The carboxyl group from pectin causes pectin to act as polyanion and react with chitosan which acts as polycation to form a PEC (Pec-Chi) membrane [6]. In research, Chen et al. (2010) succeeded in synthesizing PEC Pec-Chi membrane and after the analysis was obtained the results of PEC (Pec-Chi) tensile test where the value was greater than that of pure chitosan. The synthesis result of PEC Pec-Chi modification were also carried out to improve the mechanical properties of each polymer. Research on synthesis of modified chitosan to pectin polyelectrolyte complex (Pec-Chi) has been carried out in order to improve the characteristic of chitosan as adsorbent of dyestuff, especially in methylene blue.

2. Experimental

2.1. Material

Chitosan, Pectin, Acetic acid 5%, Aquadest or Water, Methylene blue ($C_{16}H_{18}ClN_3S$) powder.

2.2. Procedure

2.2.1. Preparation of adsorbent

Synthesis of The PEC (Pec-Chi) was made by coservation method that combine chitosan and pectin to form stable adsorbent membrane. Chitosan 0.2 gram was dispersed in 10 mL of acetic acid 5 % and then stirring until 3 hours. Besides that, Pectin 0.2 gram was dispersed in 10 mL water and then stirring until 3 hours. After that, Chitosan solution was added to pectin solution and then stirring until homogeneous. After all the solution is mixed, thenthe mixture was printed out into polypropylene container and dried in oven. After dry, the film can be peeled off from the container. The PEC(Pec-Chi) adsorbent membrane was ready.

3. Result and Discussions

3.1. FTIR Characterization

-COOH group is the dominant active group in pectin, while the dominant active group of chitosan is –NH_2 group. Fourier Transform Infrared (FTIR) Spectroscopy are used to see how interaction between components and the result of synthesis polyelectrolyte complex PEC (Pec-Chi) membrane. From figure 2, it can be seen on three spectra in the region 3000-3700 cm^{-1} can be observe a broad of hydrogen-bonded OH groups. In this area also can be observed NH group stretching vibrating which overlaps with OH vibration for chitosan spectra. In the area of 1500-1800 cm^{-1} can be observed the stretching vibration of the carbonyl group on the band at 1643 cm^{-1} for pectin (A). While in the chitosan (B) spectra, in the area of 1597 cm^{-1} show stretching vibration of amide group. In the area around 1600-1500 cm^{-1} in PEC (Pec-Chi)spectrum induces an amine group in wavelengh 1419 cm^{-1} that interacts with caboxyl group on 1597 cm^{-1}[7].

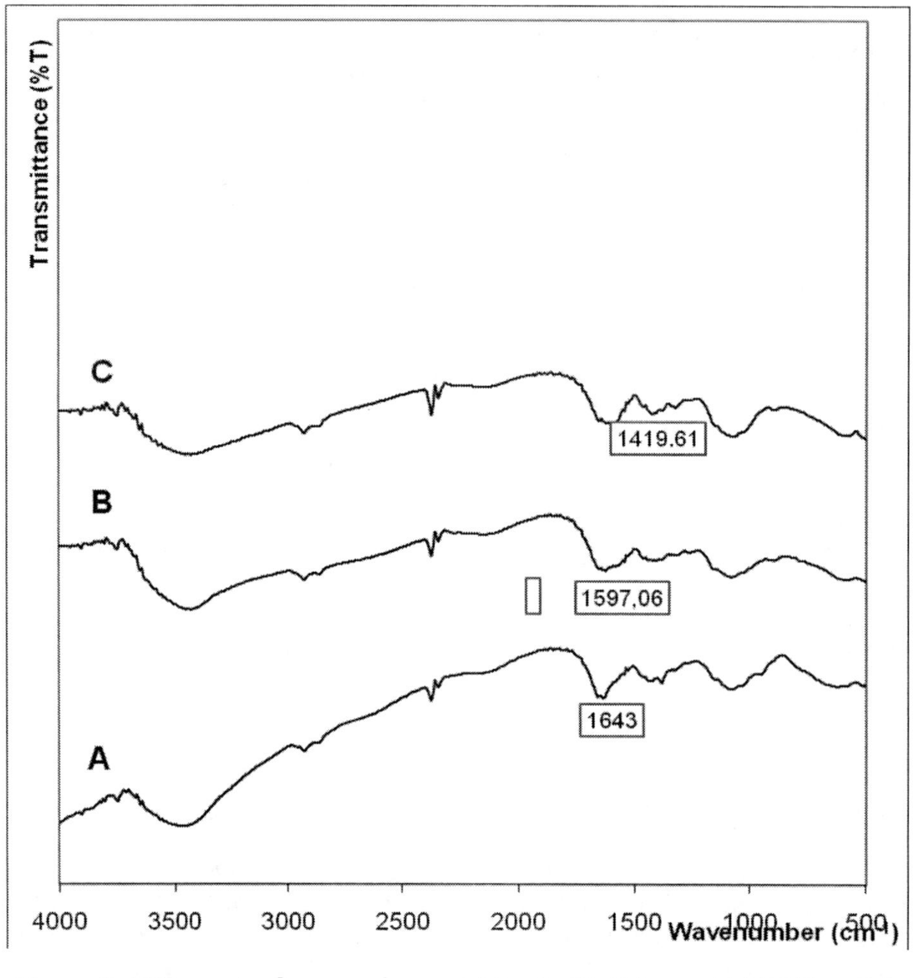

Figure 2. FTIR spectra from membrane: (a) Pectin (b) Chitosan (c) PEC(Pec-Chi)

3.2. Analysis of PEC Pec-Chi adsorbent by XRD

The phase form of the material from pure film (Pectin and Chitosan) and its modification, the membrane/film was tested using X-Ray Diffraction (XRD). XRD spectra result from pectin showed that the properties of pectin is amorphous, this was characterized by not sharp peaks found which appeared anomaly, at peak 2θ=20°. The characterized the phase of chitosan from the result of XRD test is semi-crystalline and have peak at 2θ=20° and 30°. In the PEC (Pec-Chi), between pectin and chitosan, the characteristics of chitosan is disappear and the halodisperse is indicated of amorphous. This can be explained by the strong interaction like hydrogen bond and ionic interaction between pectin and chitosan with has destroyed the close packing of chitosan molecule for those formation of regular crystalline [8]. The presence of pectin in PEC (Pec-Chi) decreases the crystalline level of chitosan which is seen by the comparison of peak of each spectra.

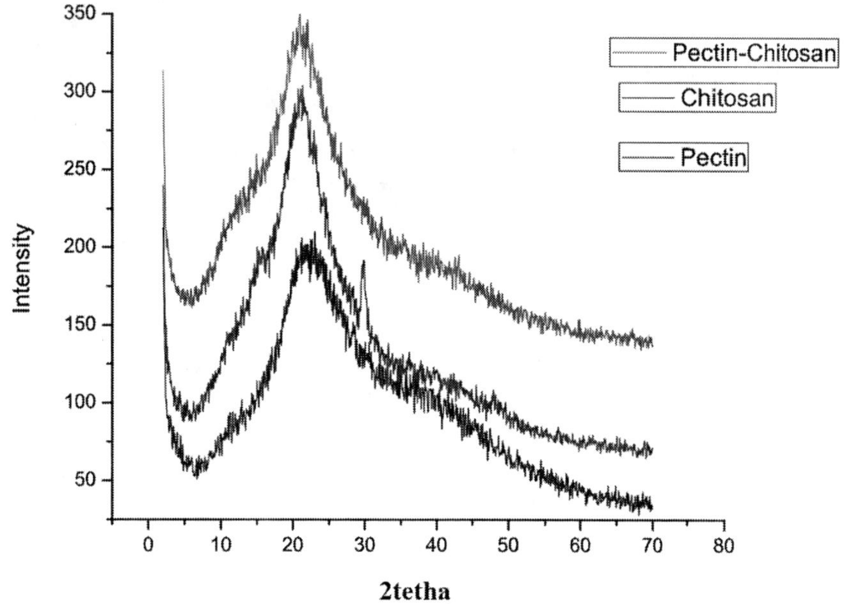

Figure 3. XRD test result from film: (a) Pectin (b) Chitosan (c) PEC (Pec-Chi) membrans

3.3. SEM Analysis

Morphology of the PEC (Pec-Chi) adsorbent film, it can be seen using SEM test, both form the surface and cross section. From the result of SEM for surface appearance with 500X magnification both chitosan and pectin have small and homogeneous pores (figure B and C). As form PEC (Pec-Chi) films, the pores that owned are smoother and smaller compared to chitosan or pectin films. PEC (Pec-Chi) film also has a high complexity when compared to its pure component films. But if the film is viewed from a cross section the pores are more rough while the result seen from the surface of the pores are smaller. So it implys that PEC (Pec-Chi) film was potential used as an adsorbent candidate for dyestuffs.

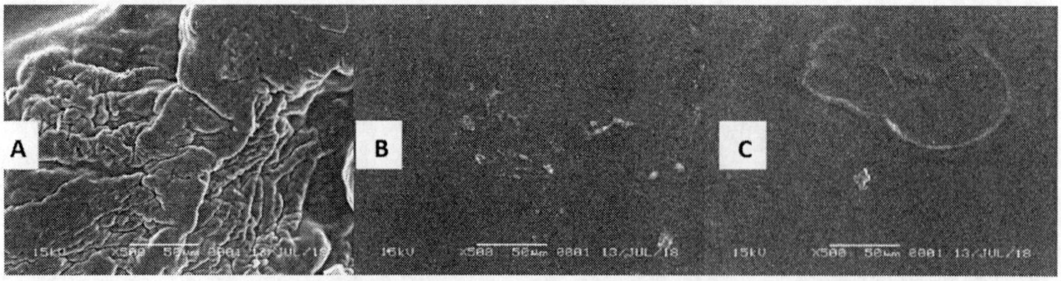

Figure 4. SEM test result of surface form film: (a) PEC (Pec-Chi) (b) Chitosan (c) Pectin

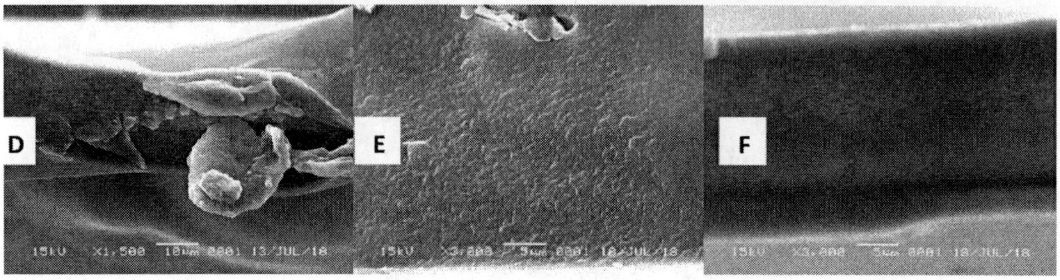

Figure 5. SEM test result of cross-section form film: (d) PEC (Pec-Chi) (e) Chitosan (f) Pectin

4. Conclusion

Pec-Chi polyelectrolyte complex (PEC)Pec-Chi were prepared by coaservation combine both polysaccharides Pectin and Chitosan. Based on FTIR spectroscopy, the adsorbent have active group comprise and interaction of carboxylate and amine groups. The result of characterization using XRD, show that the PEC adsorbent is amorphous. Based on SEM analysis, the film adsorbent has small pores with a high complexity.

References

[1] Kusumawati N, Wijiastuti A, Rahmadyanti E 2012 *J Environ. Sci. Eng.* **1** 672
[2] Septiono M A, Roosmini D, Salami H I R S, Ariesyadi D, Lufiandi 2016 *Industriaal activities and its effects to river water quality (case study citarum, bengawan solo and brantas).*
[3] Pathania D, Sharma S, Singh P 2017 *Arabian J. of Chem.* **10**, S1445–S1451
[4] Hameed B H 2009 *J Hazard. Mater* **161** 753-759
[5] Chen P, Kuo T Y, Kuo J R, Tseng Y P, Wang D M, Lai J Y, Hsieh H J 2010 *Carbohydr. Polym.* **82** 12361242
[6] Ayuni N P S, Yuningrat N W, Andriani K Y 2016 *Jurnal Sains dan Teknologi*. 706 – 717
[7] Marudova M, MacDougall A J, Ring S G 2004 *Carbohydate Research. 339*
[8] Ghaffari A, Navaee K, Oskoui M, Bayati K, Tehrani M R 2007. *European J. Pharm. and Biopharm* 175-186

IOP Conf. Series: Materials Science and Engineering **578** (2019) 012006 doi:10.1088/1757-899X/578/1/012006

Synthesis of Sodalite from Indonesian Kaolin with Conventional and Alkali Fusion Method

S N Khalifah[1,*], M Cahyawati[1], D K D Cahyani[1], A Arifah[1] and A Prasetyo[1]
[1]Department of Chemistry, Faculty Science and Technology, Universitas Islam Negeri Maulana Malik Ibrahim, Jalan Gajayana 50, 65144, Malang, Indonesia.

[*]E-mail: susikhalifah@gmail.com

Abstract. Sodalite has been synthesized from Indonesian kaolin. Kaolin was converted into more reactive metakaolin using conventional and alkali fusion methods. In the conventional method, kaolin is calcined at a temperature of 700 °C for 3 hours. Meanwhile, in the alkali fusion method, the mixture of kaolin and NaOH was calcined at a temperature of 600 °C for 2 hours. The effect of synthesis parameters such as aging time and crystallization temperature was investigated in conventional methods. Besides that, the crystallization time was investigated in the alkali fusion method. Transformation of kaolin into metakaolin and synthetic products was characterized using X-ray diffraction (XRD), X-ray Fluorescence (XRF), Fourier Transform Infrared (FTIR) and Scanning Electron Microscopy (SEM). The results showed that the kaolin structure contained a lot of quartz. In the conventional method, the kaolin structure cannot be converted into metakaolin. The synthesis product obtained from this method is a mixture of sodalite and quartz. Whereas, in the alkali fusion method, kaolin can be converted into metakaolin and sodium silicate. The synthesis product obtained is a pure sodalite.

1. Introduction

Sodalite is a type of zeolite which composed of β-cage that form micropore structure. β-cage (sodalite cage) consist of silica and alumina through association of simple 4 ring (S4R) and simple 6 ring (S6R) [1]. Sodalite has high stability compared with others zeolite that also have low Si/Al ratio such as NaA zeolite, NaX zeolite and NaY zeolite. High stability of sodalite was proposed by β-cage that establish sodalite framework with large density. Large density framework can decrease lattice energy. Therefore, it can increase stability of sodalite. Sodalite have much attention because it can be used in many fields such as ion exchange [2], optical material [3], hydrogen storage [1], nuclear waste isolation [3], infiltrating material for alumina and zirconia toughened alumina (ZTA) [4] and semiconductor [5].

Sodalite can be Synthesized using chemical precursor such as alumina, gamma alumina, aluminosilicate, silica gel, colloidal silica, sodium silicate and tetraethyl ortho silicate (TEOS) [5]. Synthesis using chemical precursor have been resulted pure product but also high cost synthesis. So, required low cost material that have large amount such as kaolin. One of kaolin that have abundant amount was Blitar kaolin. Blitar kaolin contains high alumina and silica. Therefore, it can be used as starting material for synthesis sodalite. Synthesis of sodalite from kaolin had been less reported, one of them produce pure sodalite [3] and with impurities nepheline [5].

IOP Conf. Series: Materials Science and Engineering **578** (2019) 012006 doi:10.1088/1757-899X/578/1/012006

Route for sodalite synthesis from kaolin begin with conversion kaolin to metastability phase such as metakaolin using conventional method or calcination at high temperature. Synthesis of metakaolin using conventional method was reported resulting metakolin with impurities Quartz [6]. Another method for synthesis metakolin is alkali fusion. Alkali fusion is a method which use NaOH to dissolve impurities such as Quartz and followed by calcination. Synthesis of metakaolin using alkali fusion method revealed pure metakaolin and few impurities.

This research will compare between conventional and alkali fusion methods to formed metakaolin, then followed by hydrothermal method for synthesis of sodalite. Variation of aging time and crystallization temperature was investigated in the conventional method. Furthermore, variation of crystallization time was investigated in the alkali fusion method.

2. Experimental

2.1. Material

Kaolin that supplied from Blitar (76.1% Si, 9.6% Al, 6.44% K, 1.84% Ca, 3.19% Ti, 2.27% Fe) East Java, Indonesia was ground to 200-300 mess as starting material, sodium hydroxide (Merck, >99%); Al_2O_3 (Merck, 99%).

2.2. Synthesis sodalite

Kaolin was converted to metakaolin using conventional method by calcination at a temperature 700 °C for 3 hours. The calcined product was ground and used as material for synthesis of zeolite. The alkali fusion for synthesis of metakaolin was done by crush 4.29 g kaolin with 8.58 g NaOH for 30 minutes. The mixture was fused at 600 °C for 1 hour. The fused product grounded in a mortar and used to synthesis of zeolite. The molar composition for synthesis sodalite of both method was: $10SiO_2$: Al_2O_3: 0.96 Na_2O: 180 H_2O.

The mixture then age at room temperature (~30 °C) for 10 days. Synthesis sodalite from Conventional product was investigated on crystallization temperature and aging time. Hydrothermal method for conventional product use crystallization temperature 90 °C, 100 °C, 110 °C and aging time 24 hours, 36 hours, and 46 hours. The alkali fusion product was investigated on Crystallization time at 6 hours, 12 hours and 24 hours. The results of synthesis filtered and washed with aquades to pH 9 or 10. Sodalite obtained then dried at 100 °C for 12 hours. The dried sample was kept on plastic bags for characterization using XRD, FTIR, and SEM.

2.3. Characterization

Chemical compositions of kaolin were determined using XRF (Philips Analytical's, PW2404, Almelo, The Netherlands). The phase identification was determined by XRD a Philips diffractometer with Cu-Kα radiation (Philips XRG 3100, Philips Analytical Inc) at 2 theta 5-60°. Infrared (IR) spectra were determined using Perkin Elmer Spectrum GX FTIR System spectrometer. Morphology of starting material and the synthesis product were observed by Hitachi S-570 Scanning Electron Microscopy (SEM) that operated at 15 kV.

3. Results and Discussion

3.1 Synthesis of metakaolin

The synthesis of metakaolin using conventional methods has been reported in previous studies where the results of XRD pattern indicated that the synthesis material still contain quartz as an impurities. The crystalline phase of material was dominated by quartz which reaches 90%. Quartz was stable at high temperatures hence calcination does not breakdown the quartz structure and no significant different on diffractogram pattern [7]. Therefore, Blitar kaolin includes low grade material. Furthermore, synthesis of metakaolin using alkali fusion method revealed sodium silicate that appeared in 2 theta 17°, 25°, 29.05°, 35.05°, 38°, and 49.05° (figure 1). Sodium silicate was formed by reaction of quartz on kaolin with NaOH at high temperatures.

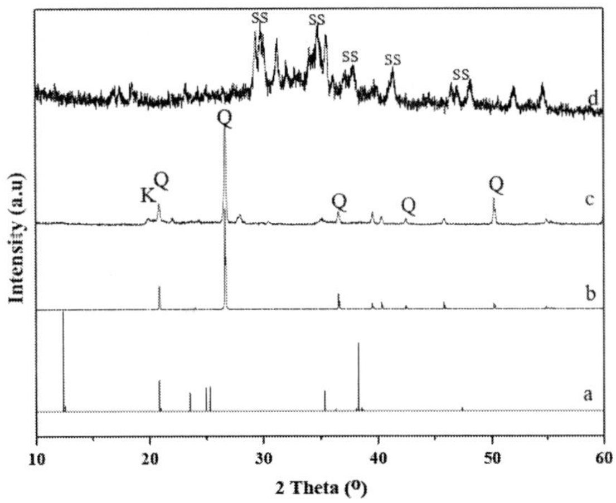

Figure 1. XRD pattern of (a) kaolin standard, (b) Quartz standard, (c) Blitar kaolin and (d) result of alkali fusion (Q = Quartz, K = kaolin, SS = sodium silicate).

The infared pattern of kaolin (figure 2) was shown vibration of Quartz as free Quartz at 785 cm^{-1}, while kaolin have vibration of Al-OH stretching at 3699 and 3450 cm^{-1}, H_2O bending at 1633 cm^{-1}, T-O-T asymmetric tetrahedral stretching at 1037 cm^{-1}, Al-OH octahedral stretching at 912 cm^{-1}, T-O-T (T= Si or Al) symmetric bending at 694 and 528 cm^{-1}, T-O-T asymmetric bending at 493 cm^{-1}.

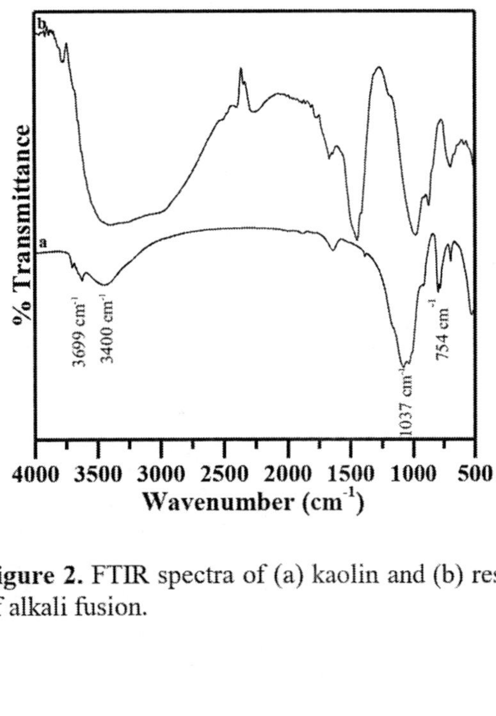

Figure 2. FTIR spectra of (a) kaolin and (b) result of alkali fusion.

Analysis using FTIR showed that metakaolin has similiar spectrum with kaolin (figure 2) but loses vibration of Al-OH octahedral at 3665 cm^{-1} as a result of convertion alumina octahedral to tetrahedral. Analysis using FTIR showed that alkali fusion process cause shifting of asymmetric T-O-T vibration at 1037 cm-1 to 984 cm^{-1} and loss vibration at 3699 cm^{-1}, 912 cm^{-1}, 754 cm^{-1} and the appearance vibration at 872 cm^{-1}. The shifting of vibration indicated transformation structure of octahedral alumina in kaolin to tetrahedral alumina in metakaolin. Alkali fusion process also cause demerge absorption from OH vibration at wavenumber 3400 cm^{-1} which is uptake of sodium silicate, the vibration enlarged because sodium silicate was hygroscopic, then vibration of asymmetric T-O-T strain at 984 cm^{-1}, T-O-T symmetric bending at 697 cm^{-1} and 520 cm^{-1}, asymmetric bending T-O-T at 461 cm^{-1} and asymmetric CO_3^{2-} strain of Na_2CO_3 at 1445 cm^{-1}. Asymmetric CO_3^{2-} vibration (Na_2CO_3) arises due to reaction of sodium silicate with carbon dioxide in the air. The FTIR spectra indicates alkali fusion process was not only produce sodium silicate but also metakaolin.

Morphology of kaolin (figure 3) was layered plate as a result of octahedral and tetrahedral sheet. Based on the research before, micrograph results from metakaolin that preparated from conventional method showed that the material was partially change to amorphous material. Moreover, based on the micrograph of alkali fusion product, layered kaolin was fully transformed to amorphous material.

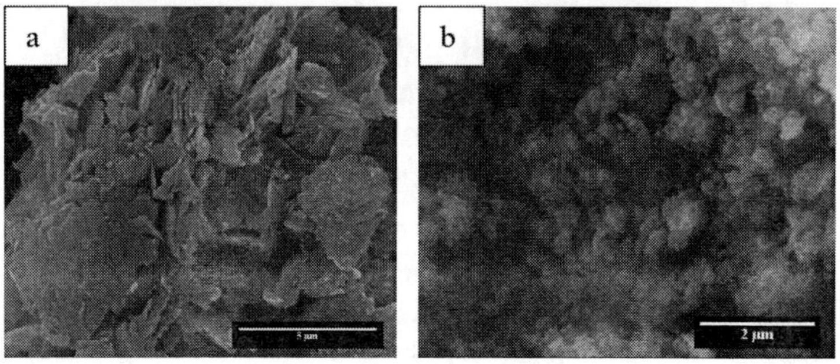

Figure 3. Morphology of (a) kaolin and (b) alkali fusion product.

3.2. Synthesis of sodalite

3.2.1. Synthesis of sodalite from kaolin prepared using conventional method

Metakaolin obtained from conventional methods was converted to sodalite and investigated in aging time. XRD pattern of product revealed sodalite denoted at 2 theta 14.5; 24.98; 32.5; 35.01; 43.9° and quartz as impurities (figure 4). Investigation on aging time does not show significant changes in crystallinity of product obtained.

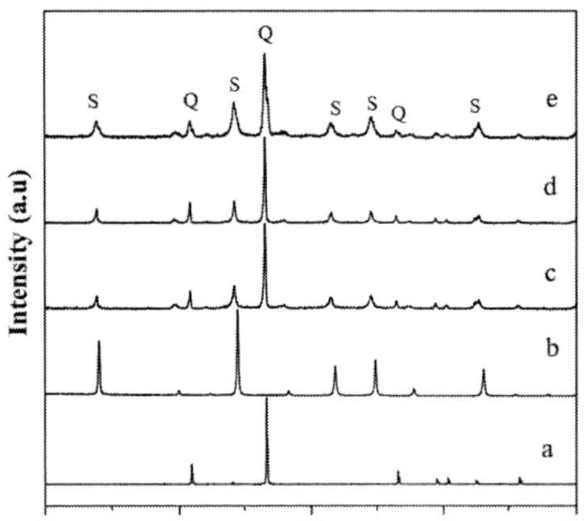

Figure 4. XRD pattern of (a) Quartz standard, (b) Sodalite standard, (c) aging for 24 hours, (d) 36 hours and (e) 48 hours (Q = Quartz, S = Sodalite) at crystallization temperature 100 °C.

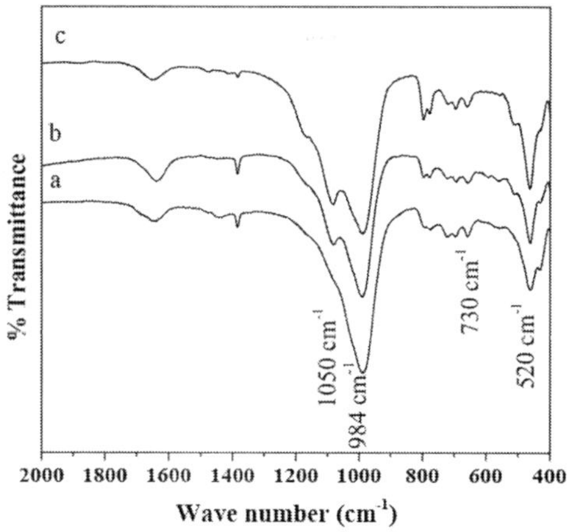

Figure 5. FTIR spectra of product with aging time (a) 24 hours, (b) 36 hours and (c) 48 hours at crystallization temperature 100 °C.

FTIR spectra of product (figure 5) was showed vibration of Si-O bending at 520 cm^{-1}, Si-O-Si symmetric stretching at 730 cm^{-1}, and Si-O-Si asymmetric stretching at 1050 cm^{-1} which refer to Quartz present. While sodalite appear at 984 cm^{-1} that refer to T-O-T asymmetric stretching. Increasing of aging time elevate vibration at 520 cm^{-1}, 730 cm^{-1}, and 1050 cm^{-1} due to increasing Quartz based on XRD analysis. Based on micrograph of sodalite (figure 6) have shown that increasing particel size proportionate with aging time which the particel size approximately 1.5 μm at aging time 48 hours.

Figure 6. Micrograph of product with aging time (a) 24 hours, (b) 36 hours and (c) 48 hours at crystallization temperature 100 °C.

The effect of crystallization temperature was also investigated on the conventional metakaolin method. The XRD pattern of resulting product was shown at figure 7. Based on the analysis using XRD the products which formed were sodalite and quartz. Crystallization temperature was not significantly influence the product obtained. Analysis using FTIR (figure 8) showed crystallization temperature decrease intensity of Quartz by decrease vibration at 520 cm^{-1}, 732 cm^{-1} and 1045 cm^{-1} denote vibration of Quartz even though XRD analysis was not showing different result. The micrograph of sodalite produced (figure 9) was shown decreasing of particel size by increasing crystallization temperature.

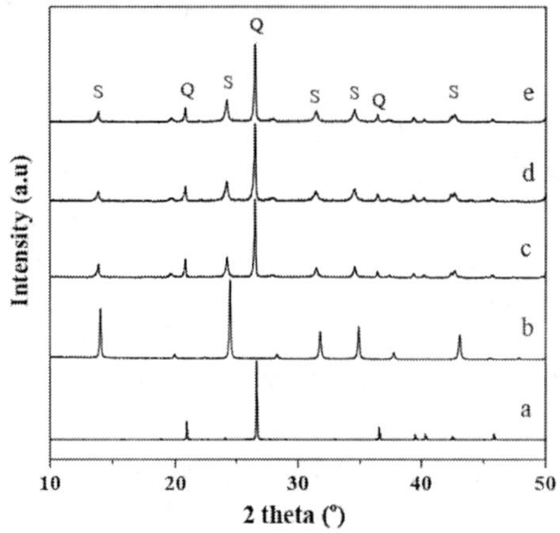

Figure 7. XRD pattern of (a) Quartz standard, (b) sodalite standard, crystallization temperature for (c) 90 °C, (d) 100 °C and (e) 110 °C at aging time 48 hours.

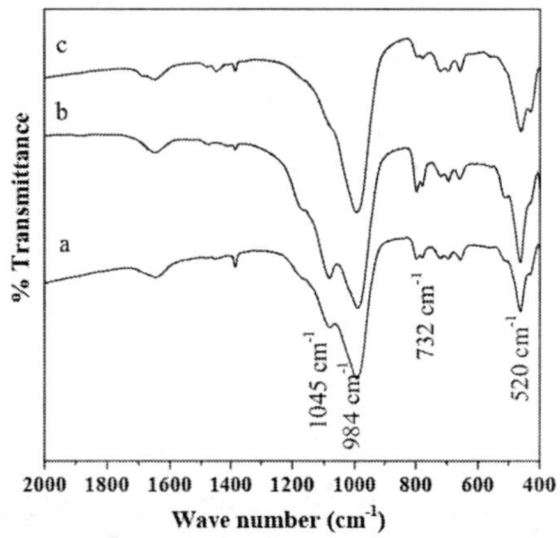

Figure 8. FTIR spectra of product with crystallization temperature for (a) 90 °C, (b) 100 °C and (c) 110 °C at aging time 48 hours.

Figure 9. Micrograph of product with crystallization temperature for (a) 90 °C, (b) 100 °C and (c) 110 °C at aging time 48 hours.

3.2.2 Synthesis of sodalite from kaolin prepared using alkali fusion method

The effect of crystallization time have been investigated on the synthesis sodalite from kaolin prepared using alkali fusion method. The diffractogram as XRD result was shown in figure 10. Based on the analysis using XRD can be known that product formed was pure sodalite due to reaction of quartz with NaOH on alkali fusion. Investigation on crystallization time showed that crystallization time affected width of peak consequence of smaller crystal of sodalite.

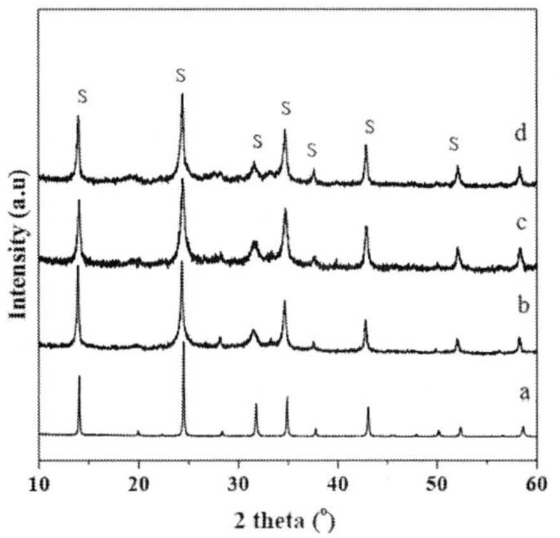

Figure 10. XRD pattern of sodalite standard (a), crystallization time for 6 hours (b), 12 hours (c) and 24 hours (d) (S = sodalite) at crystallization temperature 100 °C and aging time 48 hours.

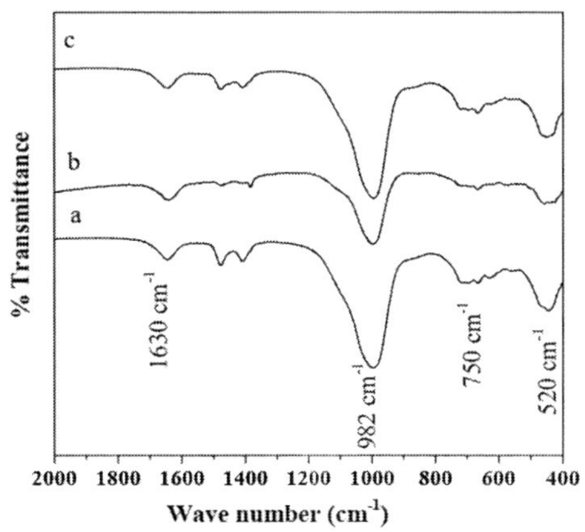

Figure 11. XRD pattern of sodalite with alkali fusion method at crystallization time (a) 6 hours, (b) 12 hours and (c) 24 hours in the crystallization temperature 100 °C and aging time 48 hours.

The FTIR spectra of sodalite (figure 11) showed that the product which formed was pure sodalite. Crystallization time also increases crystallinity of product referred by sharpening peak of O-H bending vibration at 1650-1600 cm^{-1} due to the presence of water absorbed physically. Other absorptions appear at wavenumbers 1250-950 cm^{-1} indicated the internal T-O-T asymmetric stretching, 720-650 cm^{-1} related to internal symmetric stretching of T-O-T, and 820-720 cm^{-1} referred to external symmetric stretching of T-O-T (Auerbach et al, 2003). Based on the micrograph results (figure 12) indicated that the product not only irregular particle, but also homogen. Crystallization time affect the size of particle compatible as XRD result, at 6 hours formed approximately 0.5 μm, and decline to 0,2 μm at 12 hours.

Figure 12. Micrograph of sodalite at crystallization time (a) 6 hours, (b) 12 hours and (c) 24 hours in the crystallization temperature 100 °C and aging time 48 hours.

4. Conclusion

This research reports that the transformation of Blitar kaolin using conventional methods has produced a product with quartz impurity. Whereas, when using the alkali fusion method, it produces metakaolin and sodium silicate. Sodalite synthesized from conventional methods has not been proven to be pure because there is still quartz impurities. Whereas, when using the alkali fusion method pure sodalite products have been obtained. Synthesis parameters such as the aging time, crystallization temperature, and crystallization time proven have effect on increasing the crystallinity and decreasing the particle size of the product obtained.

References

[1] Nanganoa L T, Mbadcam K J and Kang S 2016 *Inter. J. ChemTech Res.* **9** 725-7323
[2] Cruciani G, 2006 *J. Phys. Chem. Solids* **67** 1973–1994
[3] Mikuła A, Król M and Koleżyński A 2015 *Spectrochim. Acta A: Mol. Biomol Spectrosc.***144** 273-80
[4] Li J, Zeng X, Yang X, Wang C and Luo X 2015 *Mater. Lett.* **161** 157-9
[5] Naji G, Anita R A, Yahya R and Dabbagh A 2016 *Ceram. Int.* **42** 12253-61
[6] Chong S, Peterson J, Nam J, Riley B and Mccloy J 2017 *J. Am. Ceram. Soc.* **100** 2773
[7] Khalifah S N, Aini Z N, Hayati E K, Aini N and Prasetyo A 2017 *Mater. Sci. Eng.* **333**

Synthesis and characterization complex Cobalt(II) with sulfadiazine

M K Widowati[1] , S B Rahardjo[1*], S Hastuti[1], B F Ariyani[2]
[1]Chemistry Postgraduate Program, Universitas Sebelas Maret, Surakarta, 57126, Indonesia
[2]Department of Chemistry, Faculty of Mathematics and Natural Sciences, Universitas Sebelas Maret, Surakarta, 57126, Indonesia

E-mail: sentotbr@staff.uns.ac.id

Abstract. Complex of cobalt(II) with sulfadiazine has been synthesized in a ratio 6: 1 metal to ligand in methanol. The formation of a complex was characterized by a shift in the maximum wavelength (λmax) to the smaller wavelength (527 nm to 515 nm) which shows transitions $^4T_{1g}(F) \rightarrow {}^4A_{2g}$ dan $^4T_{1g}(F) \rightarrow {}^4T_{1g}(P)$. The percentage of cobalt in the complex measured by Atomic Absorption Spectrometer (AAS) showed the complex formula was $Co(Sulfadiazine)_3(NO_3)_2.(H_2O)_n$ (n = 2,3,4). The conductivity measurement indicated that the complex is 2 to 1 electrolyte. Thermal analysis with Differential Thermal Analyzer (DTA) indicates the presence of H_2O in the complex. The formula of the complex was estimated as $[Co(Sulfadiazine)_3(H_2O)_3](NO_3)_2.H_2O$. The complex was paramagnetic with μ_{eff} = 4.59 – 4.97 BM in high spin state. Infrared spectra show a shift in the absorption of primary NH_2 functional groups indicating that the functional group was coordinated to the central Co^{2+} atom monodentically. The geometry of the complex was probably octahedral.

1. Introduction

Inorganic chemistry research in the medical field has been widely developed through the synthesis of new compounds that are thought to have antibacterial activity [1]. Complex compounds consist of metal ions metals that have d orbitals that have not been fully filled by electrons with ligands that have donor atoms. Many new complex compounds that are successfully synthesized provide better results than their free ligands [2]. Complexes with cobalt metal have better antibacterial activity compared to complexes with copper and nickel metals [3]. Some cobalt complexes are also known to have higher antibacterial activity compared to other transition metals [4].

Sulfadiazine is a derivative of sulfonamides and polydentate ligand, so they can coordinate to metal ions [5]. The electron donor of sulfadiazine is O, N primary, N secondary, and N tertiary in the cyclic chain. The cobalt complex generally has octahedral and tetrahedral geometries, such as in the $[Co(H_2O)_4(sul)_2]$ (sul = *sulindac*) complex coordinating from two atoms from O from the ligand and four O atoms from H_2O molecules at Co^{2+} central ions forming geometry octahedral [6]. The Co complex can also have tetrahedral geometry as in the Co(II) with *(2-bromo-4-chloro-6 (cyclopropyliminomethyl)-phenol)* where two atoms O and two atoms N from ligand coordinates on Co^{2+} central ion [7]. Generally, Co (II) has a number of coordination of the complex with octahedral structures. Cobalt complexes may be formed with a variety of ligands, including sulfadiazine.

2. Experimental

2.1. Materials Chemicals

All chemicals used were purchased from E. Merck.

IOP Conf. Series: Materials Science and Engineering **578** (2019) 012007 doi:10.1088/1757-899X/578/1/012007

2.2. Physical measurements

Spectra UV-VIS was accomplished on Shimadzu UV-3601 spectrophotometer. The cobalt content was counted by Atomic Absorption Spectrometer (AAS) Shimadzu AA-6650. Infrared spectra were recorded with Prestige-21 Shimadzu spectrophotometers using KBr. Thermogravimetry (TG) and differential thermal analysis (DTA) Diamond Perkin Elmer analyzer used for analysis thermogravimetric and stability of the metal complex. Electrical molar conductivity (Λ*m) of 1 mM solution in DMF was account on Jenway CE 4071 conductivity meter at 25 °C. The magnetic moment was measured with Auto Sherwood Scientific 10169 Magnetic Susceptibility Balance.

2.3. Synthesis of Co(II) complex

Methanolic solution (10 mL) of $Co(NO_3)_2.6H_2O$ (3.492 g; 6 mmol) was added dropwise to methanolic solution (20 mL) of sulfadiazine (0.501 g; 1 mmol) with reflux for 24 hours. $Co(NO_3)_2.6H_2O$ solution initially purple and then after the addition of ligand sulfadiazine, the solution became brown. The solution was then concentrated 1/3 of the previous volume. The precipitate was filtered, then dried in a desiccator for 48 hours.

3. Results and Discussion

3.1. Electronic spectra

The formation of a complex is characterized by a shift in the maximum wavelength in electronic spectra. Spectra $Co(NO_3)_2.6H_2O$ and Co(II)-Sulfadiazine are shown in Figure 1. The maximum wavelength shift towards the wavelength smaller than 527.5 nm ($Co(NO_3)_2.6H_2O$) becomes 515.4 nm (Co(II) -sulfadiazine), indicates the formation of a complex between cobalt(II) and sulfadiazine which shows transitions $^4T_{1g}(F) \rightarrow {}^4A_{2g}$ dan $^4T_{1g}(F) \rightarrow {}^4T_{1g}(P)$. Shifting the wavelength uptake in a smaller direction indicates that the sulfadiazine ligand is stronger than H_2O so that it can replace the coordinated position of H_2O molecules in Co(II) ions[5].

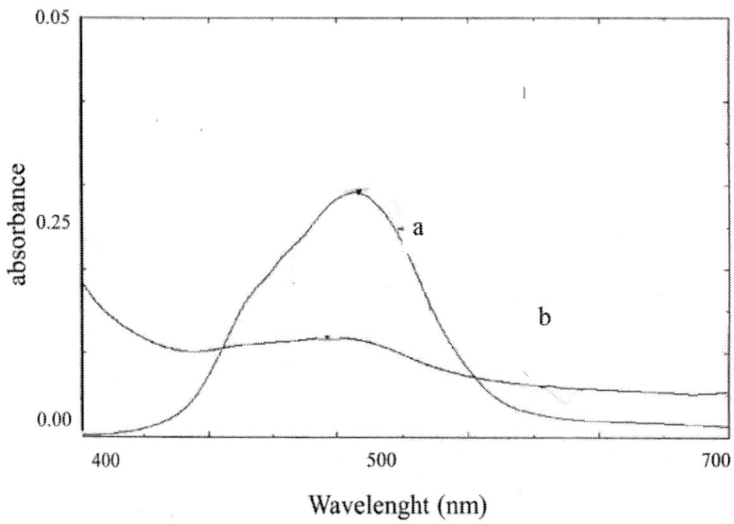

Figure 1. Electronic spectra of $Co(NO_3)_2.6H_2O$ (a) and Co(II)-sulfadiazine complex (b)

3.2. Analysis of the cobalt content with Atomic Absorption Spectroscopy (AAS)

Cobalt content in the complex counted with AAS bring out the value of 5.98 ± 0.19%. If the amount results are compared with theoretical cobalt content on various possible complex formulas as shown in Table 1, then the complex formula formed is estimated to be Co(Sulfadiazine)$_3$(NO$_3$)$_2$(H$_2$O)n (n = 2, 3, or 4). The difference in experimental results with the results of theoretical calculations of Co levels also occurs in the Co complex (N,N'-bis(2-hydroxybenzylidene)-1,1diaminobutane) [6] which has a difference of 0.02%.

Table 1. Amount of cobalt in the complex theoretically (%).

Formula	Molecular weight (g/mol)	% Cu
Co(Sulfadiazine)$_3$(NO$_3$)$_2$.(H$_2$O)$_2$	969.75	6.07 %
Co(Sulfadiazine)$_3$(NO$_3$)$_2$.(H$_2$O)$_3$	987.75	5.96 %
Co(Sulfadiazine)$_3$(NO$_3$)$_2$.(H$_2$O)$_4$	1005.75	5.85 %

3.3. Thermal analysis by TG/DTA

In Figure 2 (green curve) there is an endothermic peak at a temperature of 60.5 °C indicating the release of H_2O molecules in the complex Co(Sfdz)$_3$(NO$_3$)$_2$.(H$_2$O)$_n$ (n = 2,3,4) [7]. This is supported by data from the TGA (Thermogravimetric Analyzer) (blue curve) where there is a reduction in compound mass from 95.6% (temperature 62.6 °C) to 84.3% (temperature 177.8 °C). from the calculation mass reduction, it is estimated that the compound lost in this temperature range (from 62.6 °C to 177.8 °C) is four H_2O molecule. From the thermogravimetry (TG) and differential thermal analysis (DTA) can be expected that the empirical formula is **Co(Sulfadiazine)$_3$(NO$_3$)$_2$(H$_2$O)$_4$.**

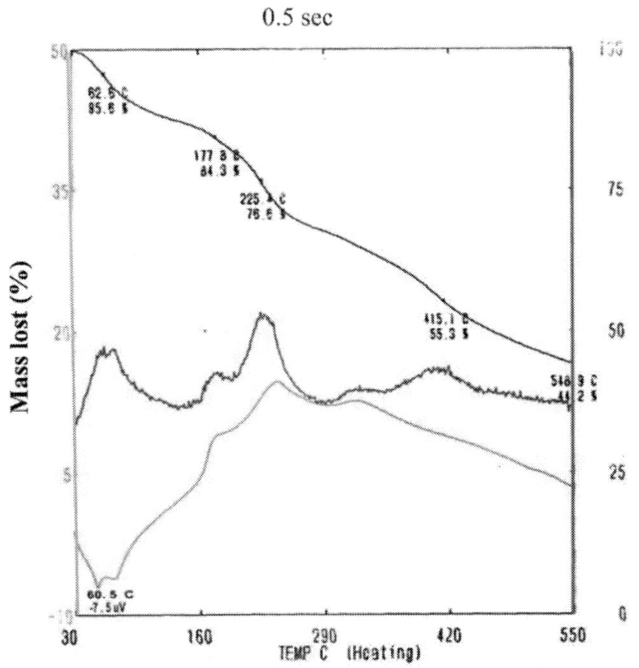

Figure 2. TG/DTA analysis of the complex

3.4. Analysis of molar conductivity

The results of measurements of the molar conductivity of solutions of standard compounds and complex solutions in methanol are represent in Table 2. The electrical conductivity of standard solutions were decide in DMF (10^{-3} M) at room temperature. The number of ions in the complex can be known by comparing the value of the molar conductive power of the complex solution with the standard molar conductivity. The value of electrical molar conductivity of the standard in Table 2 shows that the more ions in the solution the greater the value of the conductivity of the molars. The value of the molar conductivity of Co(II)-sulfadiazine complex approached the value of electrical conductivity of the solution of $Co(NO_3)_2.6H_2O$ and $Mn(SO_4)_2.6H_2O$ with a ratio cation : anion = 1: 1. This indicate that the ion NO^- acts as an anion because not coordinated to the central metal Co(II) [8-9]. Thus, the formula of the complex can be confirmed as $[Co(Sfdz)_3(H_2O)_3](NO_3)_2.H_2O$.

Table 2. Molar conductivity of standard and Co(II)-sulfadiazine in DMF

Solution	Λ_m (S cm^2 mol^{-1})	Total ions
DMF	1.15	-
$NaNO_3$	83.15	1:1
KNO_3	89.45	1:1
$Co(NO_3)_2.6H_2O$	128.50	2:1
$Mn(SO_4)_2.6H_2O$	156.05	2:1
$AlCl_3.6H_2O$	188.95	3:1
$Al(NO_3)_3.6H_2O$	222.25	3:1
$Co(Sfdz)_3(NO_3)_2.(H_2O)_4$	149.15	2:1

3.5. Infrared (IR) analysis
The wave number shift of sulfadiazine-free ligand functional groups and Co(II)-sulfadiazine complexes is shown in Table 3.

Table 3. Infrared absorption band of sulfadiazine and Co(II)-sulfadiazine

Functional groups	Wave number (cm^{-1})	
	Sulfadiazine	Co(II)-sulfadiazine
$-NH_2$ primary	3423.96	3423.15
	3355.91	3354.96
$-SO_2$ asymetric	1325.72	1326.32
$-SO_2$ simetryc	1156.63	1157.54
C=N	1651.97	1652.28
C=C	1492.88	1495.73

From Table 3 shows the shift of N primary from sulfadiazine-free ligand wave number to Co(II)-sulfadiazine complex towards a smaller direction, from 3355.91 cm^{-1} (sulfadiazine ligand) to 3354.96 cm^{-1} (complex Co(II)-sulfadiazine). The absorption spectra of the primary functional groups NH_2 sulfadiazine and Co(II)-sulfadiazine complexes are shown in Figure 3. The shift in wave numbers showed that the $-NH_2$ functional groups of sulfadiazine is coordinated on the Co(II) central atom.

3.6. Magnetic properties (μ_{eff})
The results of measuring the μeff of the Co-sulfadiazine complex were 4.61 BM. This value indicates that there are three unpaired electrons so that the complex is paramagnetic [10]. The octahedral complex of Co(II) represent a magnetic moment value is 4.3 to 5.2 BM [11-12], the value of μeff cobalt complex was present in octahedral geometry. Thus the complex structure $[Co(Sfdz)_3(H_2O)_3](NO_3)_2.H_2O$ is estimated as shown in Figure 3.

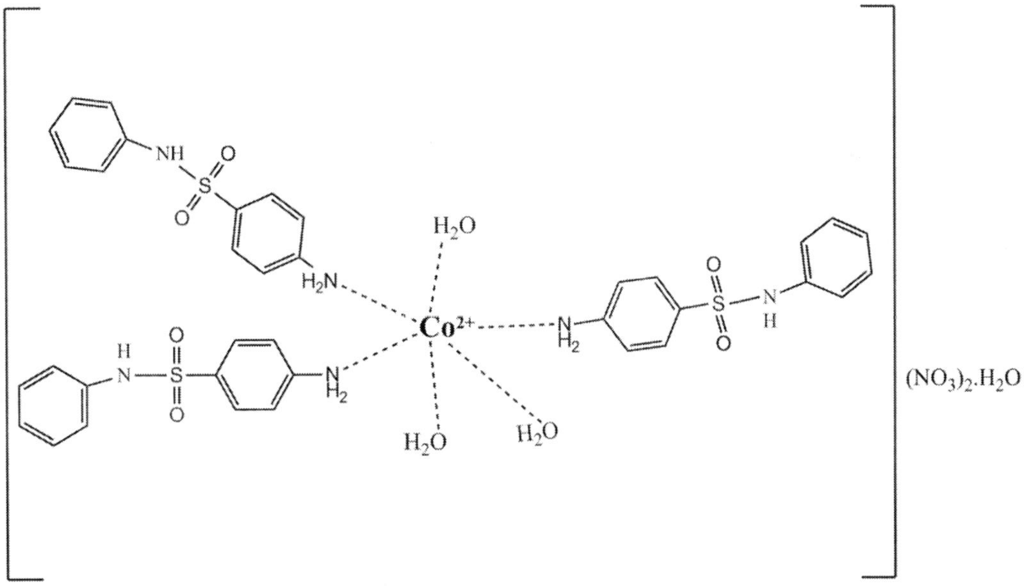

Figure 3. Suggested structure of $[Co(Sfdz)_3(H_2O)_3](NO_3)_2.(H_2O)$

4. Conclusion

Complex $[Co(Sfdz)_3(H_2O)_3](NO_3)_2.H_2O$ can be synthesized by addition sulfadiazine in methanol and Co(II) in methanol in a ratio 6: 1 of metal to ligand and stirring at 60°C during 24 hours. The results of the measurement of Co content in the Co(II)-sulfadizine complex and TG-DTA analysis obtained the complex formula $Co(Sfdz)_3(NO_3)_2(H_2O)_4$. The measurement of complex electrical conductivity results in a ratio of 2: 1 cation and anion. This means that NO^- is not coordinated on Co^{2+} and is located as an anion in the complex, so that the complex formula formed $[Co(Sfdz)_3(H_2O)_3](NO_3)_2.H_2O$. Infrared spectra shows a shift in wave number indicating that the $-NH_2$ group of sulfadiazine is coordinated on the Co(II) central atom. The measurement of complex magnetic moments yields an effect of 4.61 BM indicating an octahedral geometric complex.

References

[1] Bouchoucha A, Zaater S, Bouacida S, Merazig H, Djabbar S 2018 *J. Mol. Struct.* **1161** 345-355
[2] Ammar R, Alturiqi A, Nasser A, Alaghaz M, Zayed M 2018 *J. Mol. Struct.* **1168** 250-263
[3] El-Wakiel N, El-keiy M, and Gaber M 2015 *Spectrochim. Acta Part A Mol. Biomol. Spectrosc.* **146** 117-123
[4] Bandyopadhyay D, Layek M, Fleck M, Saha R and Rizzoli C 2017 *Inorg. Chim. Acta* **461** 174-182
[5] Rocha M, Piro O E, Echeverría G A, Pastoriza A C, Sgariglia M A, Soberón J R, Gil D M 2018 *J. Mol. Struct.*
[6] Materazzi S, Risoluti, R, Napoli, A. 2015. *Therm. Chim. Acta* **606** 90-94
[7] Shalash M A, Ali H I 2017 *Chem. Cent. J.* **11** 40
[8] Han Y J, Wang L, Li Q B, and Xue L W, 2017 *Acta Chim. Slov.* **64** (1) 179-185
[9] Suresh M S and V Prakash 2010 *Int. J. Phys. Sci.* **5** (14) 2203-2211
[10] Nithya P, Simpson J, Govindarajan S 2017 *Inorg. Chim. Acta*
[11] Kumar S, Kour G, Schreckenbach G, Andotra S, Hundal G, Sharma V, Jaglan S, Pandey S K 2017 *J. Mol. Struct.* **1141** 23-30
[12] Gull P, Hashmi A A 2014 *Eur. Acad. Res.* **2** 5064-5082

IOP Conf. Series: Materials Science and Engineering **578** (2019) 012008 doi:10.1088/1757-899X/578/1/012008

Synthesis and Characterization Complex Nickel(II) with Diphenylamine

Y A Rasyda[1], *S B Rahardjo[2], and F Nurdiyah[2]

[1]Chemistry Postgraduate Program, Universitas Sebelas Maret, Surakarta, 57126, Indonesia

[2]Department of Chemistry, Faculty of Mathematics and Natural Sciences, Universitas Sebelas Maret, Surakarta, 57126, Indonesia

*E-mail: sentotbr@staff.uns.ac.id

Abstract. $Ni(NO_3)_2.6H_2O$ and diphenylamine formed a complex compound in 1:6 mole ratio of metal to ligand in methanol. The forming of the complex was indicated by shifting of maximum wavelength in UV-Vis spectra. Analysis of the nickel percentage in the complex by Atomic Absorption Spectroscopy (AAS) showed an empirical formula of the complex was $Ni(diphenylamine)_4(NO_3)_2.6H_2O$. The conductivity measurement of the complex showed the charge ratio of cation and anion = 1:1. It meant that NO_3^- was coordinated to the Ni^{2+} center ion and likewise as a counter ion. The formula of complex was approximated $[Ni(diphenylamine)_4(NO_3)(H_2O)](NO_3).5H_2O$. Based on infrared spectra, it releaved diphenylamine as monodentate ligand which bind to nickel(II) through the secondary NH functional groups. Electronic spectra showed a transition peak at 25252 cm^{-1} and 13802 cm^{-1} which indicating octahedral geometry. The complex was paramagnetic with μ_{eff}= 2.99 BM.

1. Introduction

Coordination chemistry is basic chemistry of metal ions which receive electron pairs from ligands around it. Coordination compounds provide a wide range of colors and coordination numbers of metal ions exhibited their structure [1]. Nickel is one of transition metals and abundance in nature as nickel(II) because more stable than nickel(0), nickel(I), nickel(III), and nickel(IV) [2-4]. The geometry formed by the nickel (II) complex varies, including square planar [5,6], tetrahedral [6], trigonal bipyramidal [7], and octahedral [8]. Meanwhile, the most common of the nickel (II) complex geometry are square planar and octahedral. Each nickel(II) complex with certain ligands has different chemical and physical properties, thus making their investigation very interesting and challenging.

Diphenylamine is a derivative of aniline which has an important role in the pharmaceutical field. It has an electron lone pair in the –NH group as a donor site so that it can act as a ligand (Figure 1).

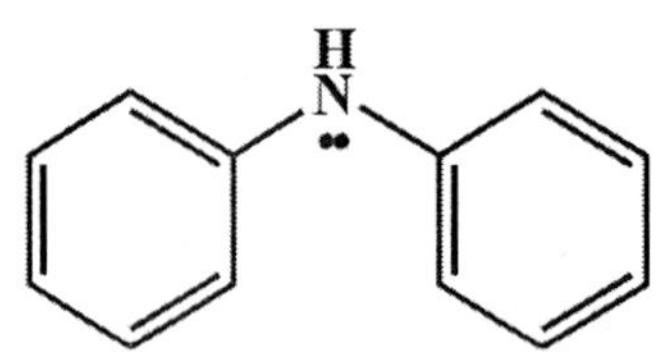

Figure 1. Diphenylamine structure

Some ligand compounds that have NH groups bond with their metal ions and form complexes as [Ni(CSH)(H_2O)] [6], [Ni(pbth)$_2$] [9], and Ni(3,3-thiodiproponic acid bis(4-amino-5-etylimino-2,3-dimethyl-1-phenyl-3-pyrazoline)(NO_3)]NO_3 [10]. At present, we are studying the synthesis, formation and properties as a result of characterization of nickel complexes with diphenylamine.

2. Experimental

2.1. Materials

All of the chemicals used were of analytical grade and used without further purification. They were purchased from E. Merck.

2.2. Physical measurements

The spectra UV-Vis of metal salt and metal complex was recordered in methanol solvent using UV-Vis Double Beam Shimadzu 1601 spectrophotometer. The nickel content was determined by Atomic Absorption Spectroscopy (AAS) Shimadzu AA-6650. Infrared spectra of ligand and complex were recordered as KBr pellets on Prestige-21 Shimadzu spectrophotometers. The magnetic moment was measured by Magnetic Susceptibility Balance (MSB) Auto Sherwood Scientific 10169. Molar conductivity was measured on conductivity meter Jenway CE-4071.

2.3. Synthesis of Ni(II) complex

The nickel(II) complex was prepared by dissolved diphenylamine (2.03 g, 12 mmol) in methanol (10 mL) followed by adding dropwise 10 mL methanolic solution of nickel nitrate hexahydrate (0.58 g, 2 mmol) with continuous stirring for ±3 h. Then, the resulting solution was evaporated until the volume was ± 10 mL. After ±48 h the green precipitated was filtered off, recrystallized, washed with methanol and dried out under vacuum (1.63 g, 84.33%).

3. Result and Discussion

3.1. Formation of the complex

Figure 2 shows the UV-Vis spectra of the nickel(II) complex and $NiNO_3.6H_2O$ in methanol. The maximum wavelengths shifted from 728 nm and 397 nm ($Ni(NO_3)_2.6H_2O$) toward smaller wavelengths 724 nm and 396 nm in the Ni(II)-diphenylamine complex. This shifting indicated the formation of the complex and also revealed that diphenylamine was a stronger ligand than H_2O [11].

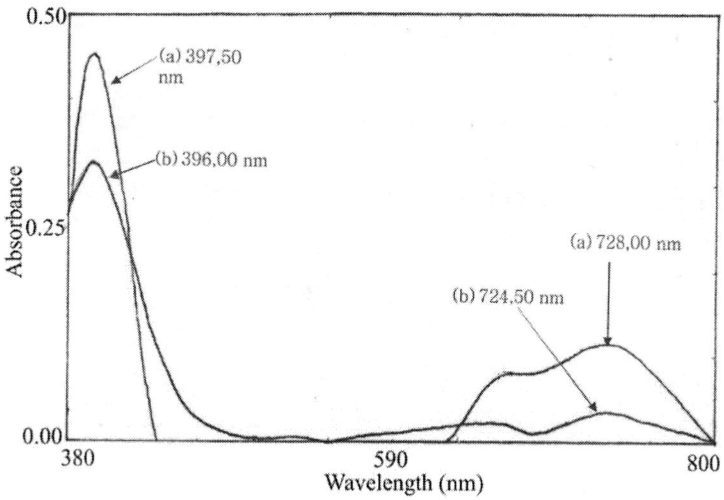

Figure 2. Electronic spectra (a) $Ni(NO_3)_2.6H_2O$ (b) Ni(II)-diphenylamine complex in methanol

3.2. Analysis of the nickel content with Atomic Absorption Spectroscopy

The content of nickel in the Ni(II)-diphenylamine complex measured by AAS presented a value of 5,92 ± 0,05%. Then, the possibility of complex formula can be obtained by comparing the nickel content of this measurement to the nickel content theoretically as shown in the Table 1. Thus, the formula of the nickel(II) complex could be approximated as Ni(diphenylamine)$_4$(NO_3)$_2$.6H_2O.

Table 1. Theoretical nickel content in the Ni(II)-diphenylamine complex

Empirical Formula	**Mr**	**% Ni**
Ni(dpa)$_4$ (NO_3)$_2$.4H_2O	931.63	6.30
Ni(dpa)$_4$ (NO_3)$_2$.5H_2O	949.63	6.18
Ni(dpa)$_4$ (NO_3)$_2$.6H_2O	**967.63**	**6.07**

3.3. Analysis of electrical conductivity

The measurements results of Ni(II)-diphenylamine complex and standard electrical conductivity in methanol (10^{-3} M) are shown in the Table 2. The Ni(II) complex electrical conductivity was 14.46 ± 0.16 S.cm^2.mol^{-1}. The cation and anion charge of Ni(II) complex could be known by comparing the electrical conductivity of the standard and Ni(II)-diphenylamine complex solution. So, it showed that the Ni(II)-diphenylamine complex charge ratio of cation and anion was 1:1 [12]. This meant that Ni(II)-diphenylamine complex was electrolyte where only one nitrate ion (NO_3^-) was located as counter ion and other nitrate ion coordinated on Ni(II) central ion.

Table 2. Electrical conductivity of standard and Ni(II)-diphenylamine complex in methanol

Solution	**Λ_M (S.cm^2.mol^{-1})**	**Cation : Anion Charge**
Metanol	0	-
$NiSO_4$.6H_2O	7.42	1 : 1
$CuSO_4$.5H_2O	8.60	1 : 1
$CuCl_2$.6H_2O	47.16	2 : 1
$NiCl_2$.6H_2O	53.16	2 : 1
$AlCl_3$.6H_2O	90.96	3 : 1
Ni(II)-difenilamin	**14.46 ± 0.16**	**1 : 1**

3.4. Infrared analysis

The infrared spectra of the complex revealed the characteristic bands of ligand functional groups. In comparison the infrared spectra of Ni(II) complex and diphenylamine, there was a shift towards the lower side in the vibration of the -NH group from 3406 and 3383 cm^{-1} (diphenylamine) to 3403 and 3379 cm^{-1} (Ni(II)-diphenylamine complex). This indicated the nitrogen atom (N) of the secondary amine group coordinated on the Ni(II) center ion. In the infrared spectra of complex, the band arised at 1385 cm^{-1} was the v(-NO_3) vibration which indicated the presence of the nitrate ion as counter ion as in the complex [Ni(3,3'-thiodipropionicacidbis(4-amino-5-ethylimino-2,3-dimethyl-1-phenyl-3-pyrazoline)NO_3]NO_3 [10]. Moreover, in the infrared spectra were also appeared absorption band of the nitrate group at 1260 cm^{-1} which showed that the nitrate ion coordinated on Ni^{2+} center ion [13].

3.5. Magnetic properties

The effective magnetic moment value (μ_{eff}) of the Ni(II)-diphenylamine complex was 2.99 ± 0.05 BM. This value indicated that there were two unpaired electron so that the complex was a paramagnetic complex. The values of μ_{eff} 2.9 to 3.3 BM in the nickel complex indicated an octahedral complex [14]. Hence, the Ni(II)-diphenylamine complex was suspected to have an octahedral geometry.

3.6. Electronic spectra

The electronic spectra of the complex were recordered in a methanol. The electronic spectra of the Ni(II) complex exhibited two bands of d-d transition at λ_{max} 13.802 and 25.252 cm^{-1}. These correspond to the $^3A_{2g}(F) \rightarrow {}^3T_{1g}(F)(v_2)$ and $^3A_{2g}(F) \rightarrow {}^3T_{1g}(P)(v_2)$ transitions, serially. These transition revealed that the nickel complex possessed an octahedral geometry [7], supported by their magnetic susceptibility values (2,9-3,3 BM). Therefore the probability of a complex formula was $[Ni(diphenylamine)_4(NO_3)(H_2O)](NO_3)_2.5H_2O$. The structure of the complex is suggested as shown in Figure 3, based on all above results.

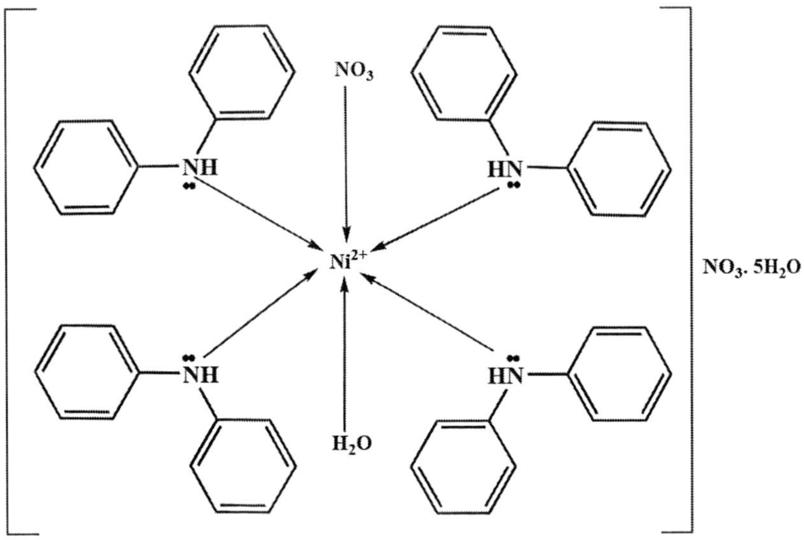

Figure 3. Suggested structure $[Ni(diphenylamine)_4(NO_3)(H_2O)](NO_3)_2.5H_2O$

4. Conclusion

The Ni(II)-diphenylamine complex could be synthesized from stirring $Ni(NO_3)_2.6H_2O$ and diphenylamine in methanol for 3 hour with 1:6 mole ratio of metal to ligand. The complex formula was $[Ni(diphenylamine)_4(NO_3)(H_2O)](NO_3)_2.5H_2O$. The functional group of ligand coordinated on the Ni(II) center ion was probably nitrogen atom from secondary NH group and forming an octahedral geometry. The complex was paramagnetic had two peaks at maximum wavelength of UV-Vis absorption at 724 nm and 396 nm.

Acknowledgement

The authors gratefully express acknowledgements to Ministry of Research Technology and Higher Education (RISTEKDIKTI) of the Republic of Indonesia for financial support.

References

[1] El-Sonbati AZ, Diab MA and Morgan SM 2016 *J. Mol. Liq.* **225** 195.
[2] Lee JD 1994 *Concise Inorganic Chemistry Fourth Edition* (London: Chapman and Hall)
[3] Hazari N, Melvin PR and Beromi MM 2017 *Nature Reviews* **1** 0025 1.
[4] Miessler GL, Fischer PJ and Tarr DA 2014 *Inorganic Chemistry Fifth Edition* Pearson (San Francisco: Education Inc.)
[5] Gűveli S, Bal-Demirci T, Űlkűseven B and Őzdemir N 2016 *Polyhedron* **110** 188.
[6] Alhazmi G AA and Elmetwally N 2017 *Arab. J. Chem.* **10** 1003.
[7] Craig GA, Sarkar A, Woodall CH, Hay MA, Marriot KE, Kamenev KV, Moggach SA, Brechin EK, Parsons S, Rajaraman G and Murrie M 2018 *Chem. sci.* **9** 6 1551.
[8] Wang X, Gao CQ, Gao ZY, Wu BL and Niu YY 2018 *J. Mol. Struct.* **1157** 395.

[9] Bharty MK, Kushawaha SK, Chaudhari UK, Dani RK, Maiti B and Butcher RJ 2017 *Transition Met. Chem.* **42** 3 243.
[10] Chandra S, Jain D, Sharma AK and Sharma P 2009 *Molecules* **14** 174.
[11] Rahardjo SB, Aditya HP 2017 *IOP Conf. Ser.: Mater. Sci. Eng.* **17** 012048.
[12] Rahardjo SB, Saraswati TE, Masykur A, Finantrena NNF, Syaima H 2018 *IOP Conf. Ser.: Mater. Sci. Eng.* **349** 01205.
[13] El-Saied FA, Salem TA, Shakdofa MME, Al-Hakimi AN, Radwan A S 2018 *Beni-Suef Univ J. Basic. Appl. Sci.* **7** 4 420.
[14] Omaka NO, Ekennia AC, Njoku NN, Onwudiwe DC 2017 *App. Organomet. Chem.* **15** 25.

IOP Conf. Series: Materials Science and Engineering **578** (2019) 012009 doi:10.1088/1757-899X/578/1/012009

Structure and Morphology of The (Ni, Co) Mo/Indonesian Natural Zeolite

G M H A Ichsan[1], **K D Nugrahaningtyas**[2,*], **D M Widjonarko**[2], and **F Rahmawati**[2]

[1]Master of Chemistry Program, Graduate School Sebelas Maret University
Jl. Ir. A. Sutami 36A, Kentingan, Surakarta 57126, Indonesia
[2]Department of Chemistry, Sebelas maret University
Jl. Ir. A. Sutami 36A, Kentingan, Surakarta 57126, Indonesia

*E-mail: khoirina@mipa.uns.ac.id

Abstract. The NiMo/Indonesian natural zeolite (NiMo/ZA) and CoMo/ZA catalysts are produced through the impregnation methods. These catalysts were characterized by using X-Ray Fluorescence (XRF), Fourier Transform Infra-Red (FTIR), X-Ray diffraction (XRD), and Scanning Electron Microscope (SEM). The XRF data show that the CoMo/ZA catalyst has a large amount of metal content more than NiMo/ZA. The FTIR spectra of catalyst show a shift of asymmetric stretching of TO4 group to a larger wavenumbers than the FTIR spectra of ZA. Meanwhile, the metal phase in the CoMo/ZA catalyst is CoO and MoO_3, and the metal phase in the NiMo/ZA catalyst is NiO and MoO_3.

1. Introduction

Indonesian natural zeolite and clay resources amounted more than 200 million tons [1]. While, natural zeolites that obtained from Java Island, Indonesia, such as West Java (Bayah and Cipatuja), Central Java (Wonosari, Yogyakarta), and east Java (Pacitan) are known to contain modernite [2,3]. It can use as a supporting metal in the preparation of heterogeneous catalysts [1–3].

Preparation processes such as modification and activation of the catalyst support may enhance its catalytic capacity. Many researchers have been modified of natural zeolites with several metals namely Ni, Ni-Mo, Co, Co-Mo [4–9], Pd, Cu, and Cr [3], respectively. Furthermore, Nickel, cobalt, and molybdenum are transition metals that frequently used as a metal catalyst, because they have high activity and selectivity.

The preparation of natural zeolite catalyst can be carried out by the impregnation method [3,4,9–11]. Nugrahaingtyas et al. [10] had modified natural zeolites with NiMo bimetallic catalysts using the impregnation method. The result showed an increase in acidity. Meanwhile, Sriningsih and coworkers [4] had compared the activity of the Mo/activated-natural zeolite (Mo/Z), NiMo/Z, Co/Z, and CoMo/Z catalysts. They obtained that the highest catalytic activity is CoMo/Z catalyst.

Modification of natural zeolite as a support for the catalyst, without initial treatment, has not been done. So, the purpose of this research made a bimetal catalyst with natural zeolite as a catalyst support, without any pre-treatment of natural zeolite. The preparation of the supported catalyst was carried out using the co-impregnation method. Another goal of this work is determined the character of NiMo/ZA and CoMo/ZA.

2. Experimental

2.1. Raw Material

The natural zeolite, which used in experiments, was supplied from Wonosari, Yogyakarta, Indonesia. As for other chemicals such as $Ni(NO_3)_2.6H_2O$, $Co(NO_3)_2.6H_2O$, $(NH_4)_6Mo_7O_{24}.4H_2O$, and ammonia 25% were purchased from E Merck, while the hydrogen and nitrogen gases were purchased from Samator Gas.

2.2. Catalyst Preparation

The NiMo/ZA and CoMo/ZA catalysts were prepared via the impregnation method. Ten grams of ZA was refluxed with a solution that contained the metal salt of $(NH_4)_6Mo_7O_{24}.4H_2O$ (0.98 gram) and Ni $(NO_3)_2.6H_2O$ (0.78 gram), 15 mL of distilled water, and 6 mL of ammonia. The reflux process was carried out at 60 °C for 2 hours. Then, the sample was evaporated at 80 °C with a pressure of 400 mBar, calcination by using nitrogen gas at a temperature of 550 °C for 3 hours at a flow rate of 15 mL/min, reduction by using hydrogen gas at a temperature of 400 °C for 2 hours at a flow rate of 15 mL/min. This catalyst is named NiMo/ZA. The same procedure was repeated in the preparation of the CoMo/ZA catalyst.

2.3. Catalyst Characterization

X-Ray Fluorescence (XRF) were recorded using BRUKER S2 Ranger for analysis metal composition. Functional groups were observed by using FTIR Shimadzu IR Prestige-21. All infrared spectra were collected data resolution of 4 cm^{-1} in the 4000-400 cm^{-1} range. X-Ray diffraction (XRD) patterns were recorded using Rigaku Miniflex600 for analysis crystallinity of catalysts. The morphology of catalysts was investigated by Scanning Electron Microscope (SEM) using JSM-6510LA. All morphology were collected data magnification in 20.000.

3. Results and Discussion

3.1. X-Ray Fluorescence (XRF)

Table 1 shows that the impregnation process has been successfully carried out. This statement is according to the appearance of metal that embedded in natural zeolites. However, it was found that the metal amount in the natural zeolite differed with the sum of metal added. This possibly because of the uneven metal dispersion, such as accumulation and agglomeration of metals in zeolites.

Meanwhile, there is a difference between Si/Al mole ratios in each catalyst. The Si/Al mole ratio decreased after metal loading, wherein the Si/Al mole ratio of the ZA material is 5.7, while the Si/Al mole ratio of NiMo/ZA and CoMo/ZA are 4 and 3.3, respectively. This confirmed that the amount of Al atoms decreased after the addition of NiMo and CoMo. These probably caused by an Al replacement on the zeolite framework.

Table 1. Chemical composition of natural zeolite, NiMo/ZA and CoMo/ZA

Metal element	Concentration (% w)		
	ZA	NiMo/ZA	CoMo/ZA
O	47.56	42.84	41.71
Si	31.48	22.69	22.33
Al	5.34	4.49	4.32
Co	-	-	3.06
Ni	-	3.41	-
Mo	-	4.26	5.92
Si/Al mole ratio	**5.7**	**4**	**3.3**

3.2. Fourier Transform Infra-Red (FTIR)

The spectra of natural zeolites before and after metal loading shows different patterns (Fig.1). The bands typical for natural zeolite structures are present at wave numbers of 794.67 cm^{-1} due to the vibration of the Si-O bending of the zeolite framework. The Si-O/Al-O vibration occurs at the wave number of 462.96 cm^{-1} for natural zeolite, 463.90 cm^{-1} for NiMo/ZA catalyst and 461.97 cm^{-1} for CoMo/ZA catalyst, respectively.

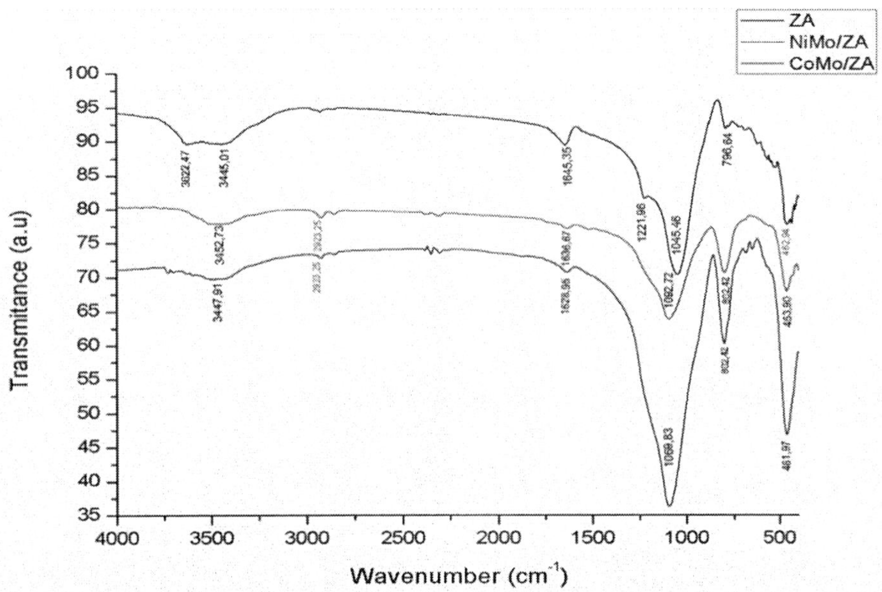

Figure 1. Spectra FTIR of (a) ZA (b) NiMo/ZA (c) CoMo/ZA

The band of 4000-3000 cm^{-1} shows the vibration of O-H hydroxyl stretching on the zeolite structure[12,13]. In addition, the OH bending of water molecules in the natural zeolite, NiMo/ZA, and CoMo/ZA catalyst is shown at the band of 1645.36 cm^{-1}, 1636.67 cm^{-1}, and 1628.95 cm^{-1}, respectively[14]. The shift occurs at the peak of 1100-1000 cm^{-1} due to asymmetric stretching vibrations of group TO_4 to larger numbers[15]. In the infrared spectra of the natural zeolite, the TO_4 asymmetric stretch vibration is shown in the 1045.46 cm^{-1}, whereas the NiMo/ZA catalyst has a shift of wave numbers at 1092.72 cm^{-1} and the shift at 1089.83 cm^{-1} occurs in CoMo/ZA catalysts.

3.3. X-Ray diffraction (XRD)

Figure 2 shows that the natural zeolite from Wonosari, Yogyakarta mainly consisted of modernite crystals [3]. The crystallinity of natural zeolites is poor, because there are still contained impurities in the natural zeolites, and the presence of tetrahedral framework of zeolite is still irregular. The diffractogram intensity of NiMo/ZA and CoMo/ZA catalysts are lower than natural zeolite (ZA). It indicated that the metals loading causes reduced crystallinity of NiMo/ZA and CoMo/ZA. Modification of natural zeolites with transition metals causes reduced crystallinity as indicated by a decrease in the intensity of some characteristic peaks[16].

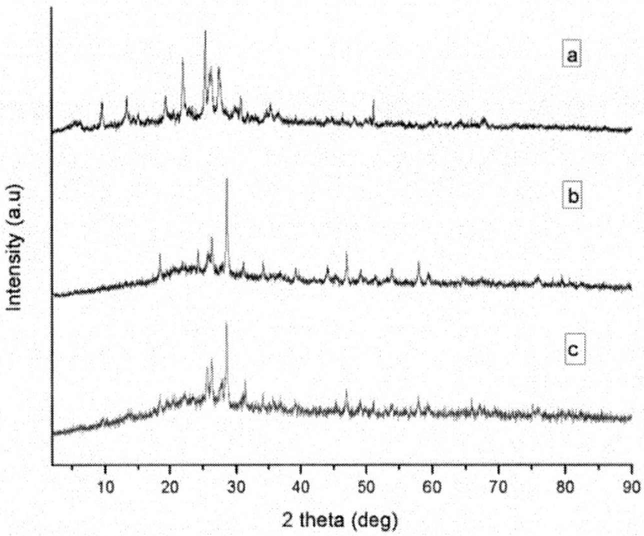

Figure 2. Diffractogram of (a) ZA (b)NiMo/ZA (c) CoMo/ZA

Furthermore, the appearance of new peaks in the diffractogram of NiMo/ZA and CoMo/ZA indicates that the metal impregnation process succeeded. The results of further analysis using Le Bail methods show that metal phases scattered in the natural zeolite are NiO, CoO and MoO3, respectively.

3.4. Scanning Electron Microscope (SEM)

Natural zeolite before modified with metals has an irregular morphological shape (Fig. 3). The changes of natural zeolite morphology due to metal oxide exposure to the natural zeolite. In figure 3b, showed that NiO metal oxides are scattered on the surface of natural zeolite which is identified by white spherical. Meanwhile, in figure 3c showed that CoO metal oxides with rhombic shape and MoO3 with hexagonal shape scattered on the surface of the natural zeolite.

Various surface morphologies indicated that the metals dispersion are not the same on each catalyst. Figure 3b and 3c showed that the surface characteristics of the natural zeolite completely disappeared. This is indicated that metals embedded and scattered on the surface of the natural zeolite.

Figure 3. Morphology of (a) ZA (b) NiMo/ZA dan (c) CoMo/ZA.

4. Conclusion

NiMo and CoMo metals can be added on the natural zeolite by the co-impregnation method and followed by calcination and reduction. The small amounts of transition metals loading into natural zeolites did not affect the zeolite crystallinity but significantly affected the morphology of the catalyst. Ni, Co, and Mo metals are as metal oxides NiO, CoO, and MoO3.

Acknowledgement

The authors are grateful to Ministry of Research, Technology and Higher Education of the Republic of Indonesia for support of this research under Penelitian Dasar Unggulan Perguruan Tinggi (PDUPT) Research Grant (Contract No.: 474/UN27.21/PP/2018).

References

[1] Arryanto Y, Amini S, Lu MGQ 2002 *J. Zeolit Indones. (Journal Indones. Zeolites)* **1**, 1–4.
[2] Kusuma RI, Hadinoto JP, Ayucitra A, Soetaredjo FE, Ismadji S 2013 *Appl. Clay Sci.* **74**, 121–126.
[3] Trisunaryanti W, Triyono, Rizki N C, Saptoadi H, Alimuddin Z, Syamsiro M, Yoshikawa K 2013 *IOSR Journal of Applied Chemistry* (IOSR-JAC) **3** (4), 29-34.
[4] Sriningsih W, Saerodji G M, Trisunaryant W, Triyono, Armunanto R, Falah I I 2014 *Procedia Environmental Sciences* **20** 215 – 224.
[5] Kubicka D, Kaluza L 2010 *Appl. Catal.* A **372**, 199-208.
[6] Horáček J, Kubička D 2017 *Fuel* **198**, 49–57.

[7] Li D, H Xu, and G D Guthrie 2000 *Journal Hydrotreatments of Catalysis* **189**(2) 281-296.
[8] Zhang X, Long J, Kong W, Zhang Q, Chen L, Wang T, Ma L & Li Y 2014 *Energy Fuels*. **8** 2562-2570.
[9] Sesario R, Dwi K, Rahmawati F, Heraldy E, Rachmadani. 2017 *AIP Conf. Proc.* **1868** 020002.
[10] Nugrahaningtyas KD, Widjonarko DM, Trisunaryanti W, Triyono 2009 *Proc. Int. Conf. CBEE 2009* 171–174.
[11] Nugrahaningtyas KD, Hidayat Y, Patiha, Prihastuti N, Yelvi B, Kalimah RUN 2017 *IOP Conf. Ser. Mater. Sci. Eng.* **176** 012024.
[12] Niwa M, Suzuki K, Morishita N, Sastre G, Okumura K, Katada N 2013 *Catal. Sci. Technol.* **3** 1919
[13] Niwa M, Suzuki K, Katada N 2006 *16th Saudi Arab. Jt. Symp.* 1–10.
[14] Korkuna O, Leboda R, Skubiszewska-Zięba J, Vrublevs'ka T, Gun'ko VM, Ryczkowski 2006 *Microporous Mesoporous Mater.* **87** 243–254.
[15] Mozgawa W 2000 *J. Mol. Struct.* 299–304.
[16] Kristiani A, Sudiyarmanto, Aulia F, Hidayati NL, Abimanyu H 2017 *MATEC Web Conf.* **101**, 01001.

IOP Conf. Series: Materials Science and Engineering **578** (2019) 012010 doi:10.1088/1757-899X/578/1/012010

Surface modification of activated carbon using an atmospheric pressure dielectric barrier discharge (DBD) plasma jet

R W Lubis[1], T E Saraswati[1*], U H Setiawan[1], K Kusumandari[2]
[1]Department of Chemistry, Faculty of Mathematics and Natural Sciences, Sebelas Maret University, Jl. Ir. Sutami 36A Kentingan, Surakarta 57126 Indonesia
[2] Department of Physics, Faculty of Mathematics and Natural Sciences, Sebelas Maret University, Jl. Ir. Sutami 36A Kentingan, Surakarta 57126 Indonesia

*Email: teguh@mipa.uns.ac.id

Abstract. We successfully modified an activated carbon layer using an atmospheric pressure plasma jet. The plasma was generated in dielectric barrier discharge (DBD) configuration using Cu wire and Cu tape separated by dielectric glass with an inner diameter of ~1 mm. An alternating current of 1.5 kV was supplied to the electrodes to generate the plasma using a mixture of Ar gas and ammonia vapor at varied flow speeds of 1 to 7 L/min for one-minute treatment. The changes to the wettability of the carbon surface were estimated from the contact angle of a water drop. After plasma treatment, a significant improvement in the surface hydrophilicity was observed. The contact angle decreased to ~6° from its initial contact angle of 74°. This hydrophilic property was likely due to the successful attachment of an amino group supplied by the NH_3 plasma. An amino-functional group was covalently bound to the carbon, which happened after etching by Ar plasma during the plasma jet processing. In addition, Fourier transform infrared (FTIR) profiles were taken to provide definitive proof of the suggested amino surface modification. Overall, the plasma treatment described in the present study may become a tool in surface treatment modification applications conducted in atmospheric conditions for carbon-based materials.

1. Introduction

Activated carbon is a carbon allotrope widely used in adsorption—for example, in wastewater processing for removing methylene blue [1], in the adsorption of heavy metal ions [2] or in the adsorption of CO_2 [3]. Activated carbon can be produced from carbonaceous material such as coal, fly ash, tree bark and nut shells [4], and it has a well-known large surface area, porous structures and a wide spectrum of surface functional groups [5]. The surface of activated carbon is known to have hydrophobic characteristics; however, in its application, hydrophilic surface characteristics are required because they provide more effective activity. This change in surface properties can be achieved using plasma treatment [6].

Atmospheric pressure plasma jets have received attention due to their ability to produce reactive species at ambient temperatures [6]. Plasma jets generate an ionized gas flow, forming plasma discharge shaped like a flare in the glass tube. The flare length can extend up to several centimetres, and it can be adjusted by increasing the gas flow, electrical current or pressure gradient [6]. However, a further increase in gas flow will reduce the length of the plasma jet [7]. Plasma jets are efficient tools for surface material modification [6] because the gas flow from the nozzle guides the reactive species straight to the surface of the material. Plasma jets are employed in various fields, such as surface treatment, cancer treatment, biomedical treatment, material processing and pollutant control [8, 9]. Moreover, plasma jets can be used to modify the surface characteristics of cotton fibres [10], metals (Al, Cu, and stainless steel) [11, 12] and polymers (PMMA, PE, PP, PET) [6, 13] to be more hydrophilic.

In the present study, we investigated thin-layer activated carbon surface modification using an atmospheric pressure dielectric barrier discharge plasma jet utilizing 1, 3, 6 and 7 L/minute flows of argon and ammonia gas. The results achieved were compared to activated carbon without plasma treatment. In order to provide definitive proof and understand the chemical reaction of the phenomena achieved, Fourier-transform infrared (FTIR) and ultraviolet visible (UV-Vis) spectroscopy studies were used to determine the surface characteristics before and after plasma treatment.

2. Experimental setup

2.1. Materials

Activated carbon from a local commercial supplier was deposited in a thin layer with double tape stuck on mica. After the activated carbon was flattened, an air spray was provided to clean and further flatten the thin layer. Ar gas (UHP, PT Samator), ammonia 25% (analytical grade, Merck), copper tape, copper wire and distilled water were the other materials used.

2.2. Plasma Jet

The atmospheric pressure dielectric barrier discharge plasma jet used in this work consisted of two electrodes and a 30 cm glass tube that had an inner diameter of 1.0 mm and an outer diameter of 5.0 mm. The tube was sealed with Teflon tape to prevent gas leakage. The Ar gas and ammonia vapor streams had flow rates of 1, 3, 6 and 7 L/minute. Copper wire 0.5 mm in thickness was used as an electrode inside the glass tube; copper tape was wrapped around the outside of the glass to serve as another electrode. High voltages were supplied to the electrodes using HVAC (Neon Pro) at 1.5 kV with a frequency of 50-60 Hz. The HVAC was connected to the regulator supply (Matsumoto Slide Regulator) to adjust the desired voltage. The configuration was installed vertically; the configuration scheme can be seen in Figure 1.

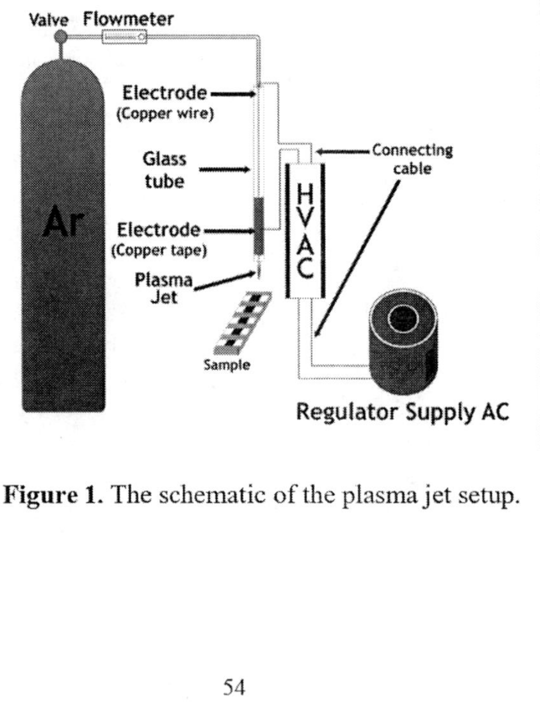

Figure 1. The schematic of the plasma jet setup.

2.3. Plasma treatment

The samples had a 1.0 x 1.0 cm square area and were placed 10 mm from the nozzle. The plasma treatment was carried out for 1 minute by manually directing the plasma jet at 5 points on the sample, firing at the central area for 20 seconds and the points to the top, left, bottom, and right of the centre of the sample for 10 seconds each, as shown in Figure 2. The morphological and chemical modifications of the activated carbon that were induced by the plasma jet process are discussed in the next section.

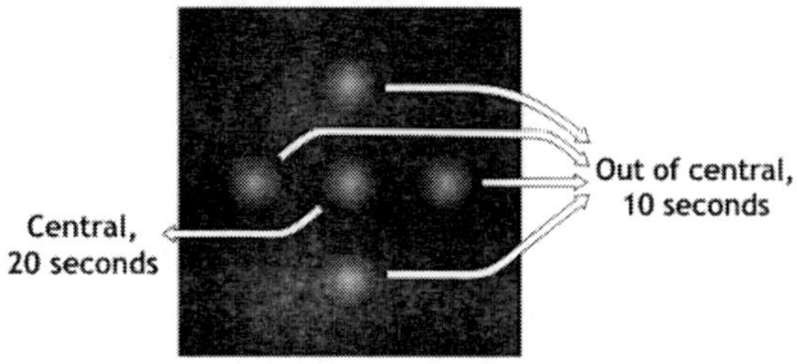

Figure 2. Area of the plasma treatment.

2.4. Treated surface analysis

The effect of plasma on the surface of activated carbon was studied by measuring the contact angle between the surface of the sample and a water droplet dropped right into the middle of the plasma treatment area. The surface characteristics before and after plasma treatment were analysed by Fourier-transform infrared spectroscopy (FTIR; IRPrestige-21 Shimadzu).

3. Results & Discussion

3.1 Contact angle

The changing of the surface characteristics of activated carbon after plasma treatment was indicated by the changing of the contact angle. Figures 3(a) to (e) show the observed contact angle of the activated carbon before and after treatment in various gas flows. The water contact angle (WCA) between activated carbon without plasma treatment and with plasma treatment showed significant differences—the untreated activated carbon had a WCA of 74°, while activated carbon with plasma treatment in the Ar/NH_3 gas flow 1, 3, 6 and 7 L/min had WCAs of 10°, 6°, 11° and 12°, respectively. The sample that experienced 3 L/min treatment was more hydrophilic than the others, showing that a flow of 3 L/min produces the best hydrophilic surface. Further increasing the gas flow rate only leads to a decrease in plasma length and the weakening of active species transportation along the tube, while the radial movement of the active species is enhanced [7]. Gas flow rates greater than 3 L/min also reduced the effectiveness of plasma formation because the gas remained in the tube for a shorter period, thus the plasma-activated species were rarely produced.

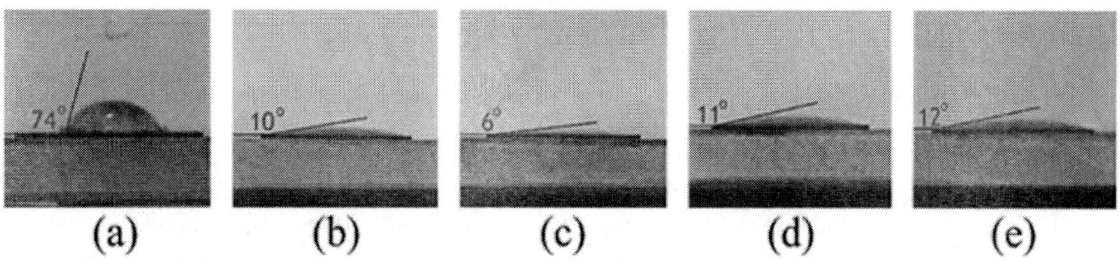

Figure 3. Comparison of the water contact angles of activated carbon before plasma treatment (a) and after plasma treatment in various gas flows (b-e; 1, 3, 6, 7 L/min, respectively).

The WCA of the activated carbon surface decreased after plasma treatment, indicating that the surface changed from hydrophobic to hydrophilic. This phenomenon likely occurred because plasma made with Ar/NH_3 gas produces reactive species such as -NH_2, which can attach to carbon atoms after carbon etching by reactive Ar plasma. Moreover, during the plasma treatment, the carbon etching may also have been followed by the introduction of oxygen-related polar groups [6, 10, 13, 11, 14, 15] that bonded with the carbon atoms. These oxygen-containing groups might have come from the atmospheric gas. Both amine- and oxygen-containing groups induced the formation of the hydrophilic surface of the sample.

3.2 FTIR Spectroscopy

In order to understand the chemical change on the surface of the activated carbon, an FTIR spectrometer was used to investigate the functional groups attached. In the spectra shown in Figure 4, the plasma-treated samples show several definitive peaks located at around 1400 cm^{-1}, 1600 cm^{-1} and 3240 cm^{-1}. Theses peaks correspond to the absorption of C—N amine, N—H bending and N—H stretching [16], respectively. The untreated activated carbon also shows a peak at 1400 cm^{-1}, which may indicate the peak of C=C stretching. C=C stretching generally has an absorption band around 1400-1600 cm^{-1}. The FTIR peak observed in the treated sample at 1400 cm^{-1} might overlap with C—N, as the peak at 1400 cm^{-1} in the treated sample spectra is stronger than that of the untreated sample.

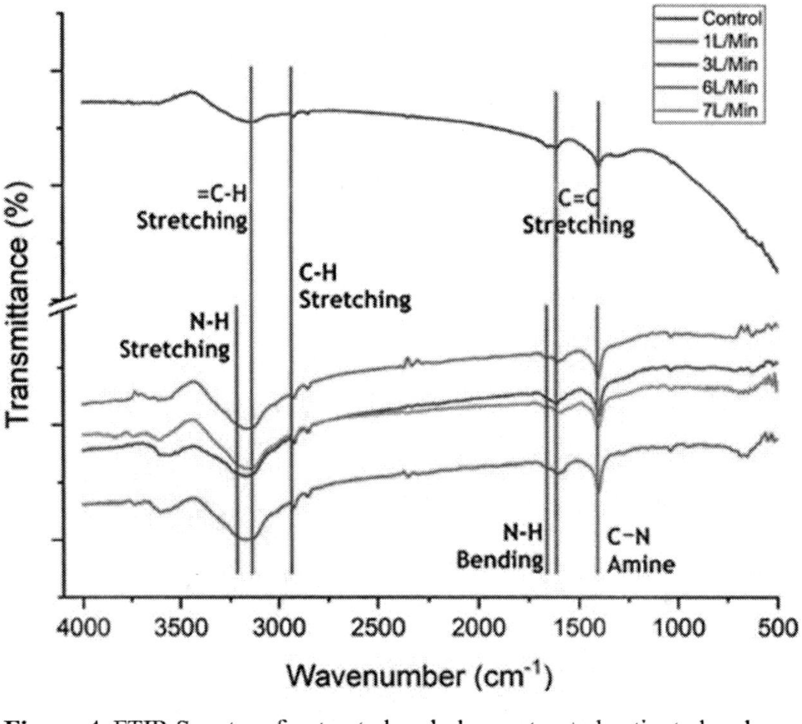

Figure 4. FTIR Spectra of untreated and plasma-treated activated carbon.

Sakudo et al. [17] explained that some of the treated surfaces were etched by plasma treatment. After the etching, the ionized particles from the plasma could attach to the sample surface, i.e., carbon atoms in the case of in this study. Even though the FTIR spectra profiles shown in Figure 4 are likely the same, however, the intensity of the FTIR absorption peaks in the treated sample are significantly higher than in the untreated one, indicating that the amounts of the represented functional groups increased after plasma treatment.

4. Conclusions

Atmospheric pressure plasma jet treatment was carried out to modify the surface of activated carbon using Ar/NH_3 plasma with gas flows of 1, 3, 6 and 7 L/min. The hydrophobic nature of activated carbon changed to produce a hydrophilic surface after plasma treatment, as indicated by a significant decrease in the contact angle from 74° to less than 15°. The changing of the surface characteristics was also confirmed using FTIR spectra, which showed the presence of C—N amines, N-H bending and N-H stretching, indicating the plausible attachment of hydrophilic amine groups covalently bonded to the carbon atoms.

Acknowledgements

This work was partly supported by a Grant-in-Aid of Research from the Ministry of Research, Technology and Higher Education, under project No. 543/UN27.21/PP/2018.

References

[1] Hameed B H, Din A T M and Ahmad A L 2007 *J. Hazard. Mater.* **141** 819-25
[2] Kobya M, Demirbas E, Senturk E and Ince M 2005 *Bioresour. Technol.* **96** 1518-21
[3] Shafeeyan M S, Daud W M A W, Houshmand A and Shamiri A 2010 *J. Anal. Appl. Pyrolysis* **89** 143-51
[4] Li L, Liu S and Liu J 2011 *J. Hazard. Mater.* **192** 683-90
[5] Song X, Liu H, Cheng L and Qu Y 2010 *Desalination* **255** 78-83
[6] Kostov K G, Nishime T M C, Castro A H R, Toth A and Hein L R O 2014 *Appl. Surf. Sci.* **314** 367-75
[7] Wang Y, Yin Z, Wang H and Zhao Z 2018 *Optik* **164** 165-70
[8] Laroussi M, Kong, M., Morfill, G., & Stolz, W. 2012 *Plasma Medicine: Applications of Low-Temperature Gas Plasmas in Medicine and Biology* (Cambridge: Cambridge University Press)
[9] Laroussi M 2008 *IEEE Transactions on Plasma Science* **36** 1612-4
[10] Zhang C, Zhao M, Wang L and Yu M 2017 *Vacuum* **137** 42-8
[11] Kim M C, Song D K, Shin H S, Baeg S H, Kim G S, Boo J H, Han J G and Yang S H 2003 *Surf. Coat. Technol.* **171** 312-6
[12] Toshifuji J, Katsumata T, Takikawa H, Sakakibara T and Shimizu I 2003 *Surf. Coat. Technol.* **171** 302-6
[13] Acsente T, Ionita M D, Teodorescu M, Marascu V and Dinescu G 2016 *Thin Solid Films* **614** 25-30
[14] Noeske M, Degenhardt J, Strudthoff S and Lommatzsch U 2004 *Int. J. Adhes. Adhes.* **24** 171-7
[15] Cheng C, Liye Z and Zhan R-J 2006 *Surf. Coat. Technol.* **200** 6659-65
[16] Cromwell N H, Miller F A, Johnson A R, Frank R L and Wallace D J 1949 *J. Am. Chem. Soc.* **71** 3337-42
[17] Sakudo N, Mizutani D, Ohmura Y, Endo H, Yoneda R, Ikenaga N and Takikawa H 2003 *Nucl. Instrum. Methods Phys. Res. B* **206** 687-90

IOP Conf. Series: Materials Science and Engineering **578** (2019) 012011 doi:10.1088/1757-899X/578/1/012011

Synthesis of nanosize manganese oxides from its ore using solvothermal method

F D Y P Abi[1], R Kusumaningrum[2], W B Widayatno[3], A S Wismogroho[3], A Noviyanto[2,4], N T Rochman[3] and M I Amal[5*]

[1] Departement of Chemistry, Faculty of Mathematics and Sciences, Universitas Katolik Widya Mandira, Kupang, East of Nusa Tenggara 71152 Indonesia

[2] Nano Center Indonesia, PUSPIPTEK South Tangerang, Banten 15314 Indonesia

[3] Research Center for Physics, Indonesian Institute of Sciences, PUSPIPTEK Tangerang Selatan, Banten 15314, Indonesia

[4] Department of Mechanical Engineering, Mercu Buana University, Jl. Meruya Selatan, Kebun Jeruk, West Jakarta 11650, Indonesia

[5] Research Center for Metallurgy and Materials, Indonesian Institute of Sciences, PUSPIPTEK Tangerang Selatan, Banten 15314, Indonesia

*E-mail: muha137@lipi.go.id

Abstract. Nano-sized manganese oxide was successfully synthesized from local manganese ore using solvothermal method. The manganese ore were obtained from Jereweh District, Sumbawa, West of Nusa Tenggara, Indonesia. The aims of this research are to study the effects of temperature and time to the morphology, size and phase of manganese oxide obtained by solvothermal method. In the first step, we performed the extraction of manganese precursor from manganese ore. The method to extract the manganese precursor was hydrometallurgy. The second step, manganese precursor that obtained in the first step was used to synthesis manganese oxide using solvothermal method. NaOH was used as precipitating agent. The phase, morphologies and particle size of manganese oxides were examined by x-ray diffractometer and scanning electron microscopy.

1. Introduction

Nano-sized particles have features in their properties and applications. And for several decades this research has been related to the synthesis and application of interesting nanoparticles to be studied in depth. Nanoparticles are 1-100 nm particles that have unique properties and are different from other sized particles. Some properties attached to a particle will increase many times when the particles are nanosize. This is because nano-sized particles are in a range of sizes that separate and are between atomic/compounds sizes and bulk materials [1].

Manganese has many types of oxidation state which affected the type of manganese oxide formation. The stable manganese oxide forms are MnO, Mn_3O_4, Mn_2O_3 and MnO_2. The each of manganese oxide type has its function in any field. As widely known, MnO_2 is applied in battery [2, 3]. Some studies reported that MnO_2 also has good antimicrobial activity [4] and enhanced the image contrast of Magnetic Resonance Image (MRI) instrument [5]. Hausmannite which have molecular formula as Mn_3O_4 reported can use in water treatment process [6]. The large specific surface area of MnO 1-D nanorod indicating that its can be used as catalyst, energy storage and biomedical images

[7]. Nanoparticle size of manganese oxide improves their surface area, change their properties to be great and different from that bulky material [7].

Nano-sized manganese oxide can be synthesized by several methods such as co-precipitation, sonochemical, solgel and hydrothermal. The hydrothermal/solvothermal is simple method, inexpensive, and promising to synthesis nanosize particles. Li et al reported the hydrothermal method can controlled the MnO_2 morphology transformation during different reaction time [8]. The type and morphology of manganese oxide nano-sized also be affected by the starting material that use in the reaction system. Manganese ore, especially from Sumbawa West of Nusa Tenggara is a huge potential resource that has a low economic value and not treated optimally. This study focuses on the synthesis of nano-sized manganese oxide by solvothermal which the manganese precursor is obtained from manganese ore directly.

2. Experimental

2.1 Materials

Nano-sized manganese oxide was synthesized immediately from manganese sulphate precursor as leaching process of manganese ore. Manganese ore that we used for this study were collected from Jereweh District, Sumbawa, West of Nusa Tenggara. As well, our previous study, the manganese ore were in pyrolusite (MnO_2) phase with the manganese content was 78.80% and iron was 17.77% as a major impurity [9]. Reagent that used in this study was sulphuric acid 95-97%, hydrogen peroxide 30%, ammonium hydroxide 25 % and sodium hydroxide, which were obtained from Merck. The other materials that we used in this study were Whatmann 42 D 110 mm filter paper, ethanol 96% and aqua demineralization.

2.2 Preparation of Manganese Precursor

Manganese rock was crushed into small chunk using hammer and was dried in the oven at 80 °C for 2 hours. The small chunk of manganese ore was milled using disk mill for 30 seconds and repeated three times until the powder was obtained. To get the homogeneous particle size, the powder of manganese ore was sieved to size of -200 mesh. Sieved powder of manganese ore was ready for use in the leaching process. The leaching process was adopted from Nayl et al studies which used sulphuric acid and hydrogen peroxide [10]. In the beginning, fifty grams of sulphuric acid 4 M was heated until temperature reach out 70°C in beaker glass then manganese ore and hydrogen peroxide 0.8 M was added slowly. The heating and stirring were continued for 90 minutes. After leaching process, the residues and liquor was separated using Buchner vacuum. The leach liquor was added ammonium hydroxide 25% drop by drop till pH solution at 7 and brown cake formed. This brown cake precipitated was iron impurities. Filtration was carried out to obtained manganese sulphate precursor which then used to synthesis nano-sized manganese oxide.

2.3 Manganese oxide nano-sized synthesis

Nano-sized manganese oxide was synthesized by solvothermal method. 45 mL of manganese sulphate precursor was mixed with sodium hydroxide 4M and the ratio was 1:1 (v/v). This mixture solution was stirred at room temperature for 15 minutes and transferred into autoclave teflon covered up with stainless steel. The hydrothermal was carried out at 150°C for 24 hours. The precipitated formed was washed using ethanol and aqua demineralization until pH 7.

2.4 Characterization

The phase that formed after solvothermal process was observed by X-ray diffraction (Smartlab, Rigaku, Japan). Microstructure of manganese oxide product were analysed by scanning electron microscopy (SEM, JIB-461OF, Jeol, Japan) with voltage and probe current was 10 kV and 3 mA, respectively. Average particle size of manganese oxide was calculated statistically by analysing SEM images

IOP Conf. Series: Materials Science and Engineering **578** (2019) 012011 doi:10.1088/1757-899X/578/1/012011

3. Results and Discussion

Jacob et al reported the thermodynamic data for Mn_3O_4, Mn_2O_3 and MnO_2 transformation by oxidation reaction approach as summarized in Table 1 [11]. This result showed that Mn_3O_4 are favourable instead of Mn_2O_3 and MnO_2 since the Gibbs free energy is the lowest among of those manganese oxides. This phenomenon indicating the oxidation reaction of Mn^{2+} has been occurring even it take very slow at room temperature. However, the manganese oxide obtained in the previous report [11] used chemical precursor. In this study, we used manganese sulphate from manganese ore without any further treatment.

Table 1. Thermodynamic Data of Manganese Oxide Oxidation [11]

	Mn_3O_4			Mn_2O_3			MnO_2		
T (°C)	ΔH_f^0 kJ mol^{-1}	S_T^0 JK^{-1} mol^{-1}	ΔG_f^0 kJ mol^{-1}	ΔH_f^0 kJ mol^{-1}	S_T^0 JK^{-1} mol^{-1}	ΔG_f^0 kJ mol^{-1}	ΔH_f^0 kJ mol^{-1}	S_T^0 JK^{-1} mol^{-1}	ΔG_f^0 kJ mol^{-1}
27	-1386.176	166.48	-1283.78	-961.536	113.70	-884.475	-521.449	52.75	-446.405
127	-1385.372	209.41	-1249.75	-961.512	114.38	-883.998	-521.450	53.09	-466.064
227	-1384.332	245.01	-1215.97	-960.818	144.93	-858.275	-521.220	70.11	-447.620
327	-1383.264	275.48	-1182.39	-960.126	169.79	-832.719	-520.615	84.82	-429.286
427	-1382.210	302.25	-1149.01	-959.359	191.17	-807.309	-519.895	97.56	-411.088

The solvothermal is one promising method to synthesis nano-sized material because simple and can take place at low temperature. Besides that, this method offers fine and single-phase compound. The oxidation reaction of Mn^{2+} already occurred spontaneously at sodium hydroxide addition at room temperature which is indicated by observation of this precursor colour changes from clear to brownish. This colour becomes solid during heating process. The oxidation reaction can be written in below (Hu et al, 2008):

$$4Mn^{2+} + 2H_2O + O_2 \rightarrow Mn^{3+} + 4OH^- \quad (1)$$

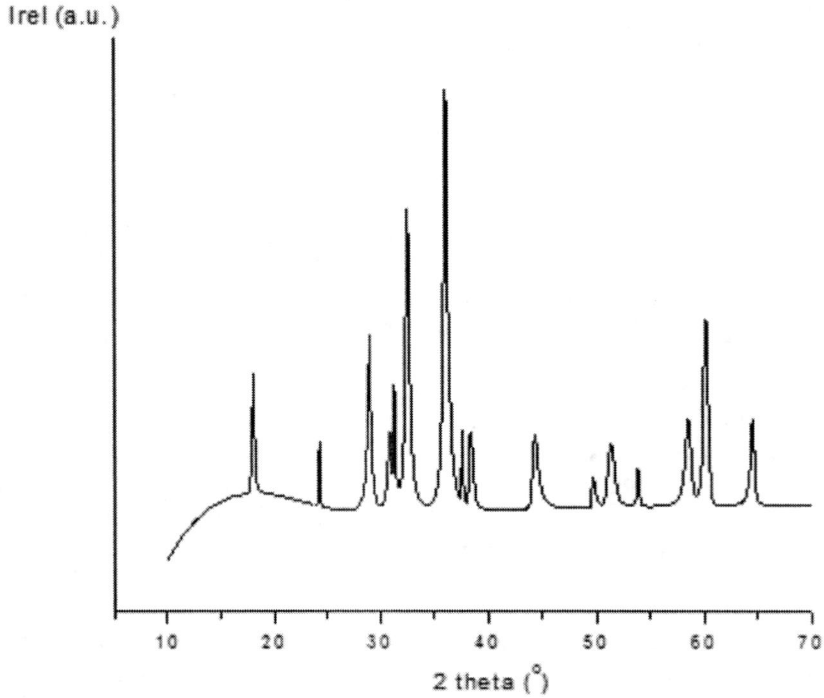

Figure 1. Diffractogram of synthesized particles using XRD.

Confirmation the phase of the product synthesis is showed in Figure 2. It was confirmed a single phase of Mn_3O_4 was observed without any trace of secondary phase. The main peak showed at 2 theta 36.05°; 32.54°; 29.03°; 60.02°; 29.03°; 18,06°; 64.47°; 38.34°; 44.31°; 51.29° and 30.98°. The reason is probably caused by limited oxygen supply in solvothermal system, encourage the existence of Mn^{2+} and Mn^{3+} combined to form Mn_3O_4. The single phase of Mn_3O_4 also formed by hydrothermal method in Hu et al studies [12]. This Mn_3O_4 seem has good crystallinity even it performed at low temperature. This indicates the pressure of reactor has a main role to determine its crystallinity. The morphology confirmed from image SEM showed that the Mn_3O_4 has a nano-sphere shape.

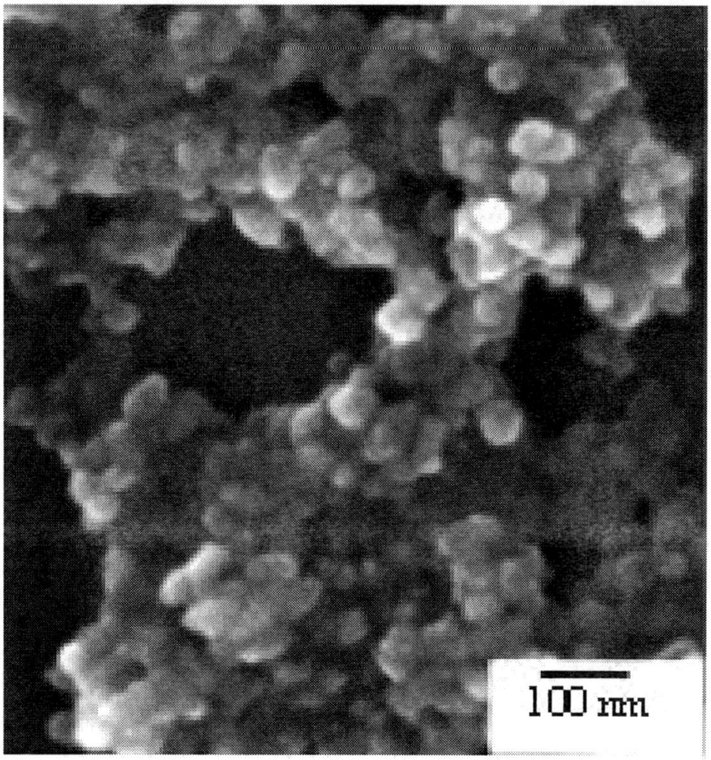

Figure 2. Scanning electron microscope image of Mn_3O_4.

The particle size of manganese oxide was 48 ± 12 nm. It showed that particle of manganese oxide experience agglomeration as a nature of nanoparticle. The nano-sized of Mn_3O_4 has good application as electrode material for supercapacitors [13]. The other studies reported that nanoparticle Mn_3O_4 with carbon black given higher specific capacitance [14]. The proposed of overall reaction in this systems follow reaction Equation 2.

$$6MnSO_4\,(aq) + 12NaOH\,(aq) + O_2\,(g) \rightarrow 2Mn_3O_4\,(s) + 6Na_2SO_4\,(aq) + 6H_2O\,(l) \qquad (2)$$

4. Conclusion

Mn_3O_4 nanoparticle was successfully synthesized from its ore by simple solvothermal method. The synthesized nanoparticle appears as a single phase of nano-sphere Mn_3O_4 with high crystallinity which had particles size was 48 ± 12 nm. This study offers the opportunity to process manganese ore from with potential of added value.

Acknowledgement
The funding for this research was provided by the National Intensive Research System Innovation Program 2018 (INSINAS), Ministry of Research and Technology of Indonesia based on contract number 06/INS-1/PPK/E4/2018.

References
[1] Schmid G 2004 *Nanoparticles from theory to application* (Weinheim:WILEY-VCH)
[2] Cheng F, Zhao J, Song W, Li C, Ma H, Chen J and Shen P, 2006, *J. Inorg. Chem.*, **45**, 2038-2044
[3] Zhang L L, Wang Z L, and Xu D, 2012, *Chin Sci Bull*, **57**: 4210–4214
[4] Rutz A., 2009, *Synthesis and Properties of Manganese Oxide Nanoparticles for Environmental Applications*, NNIN REU Research Accomplishments.
[5] Pan D, Schmieder A H, Wickline S A and Lanza G M, 2011, Tetrahedron, **67**, 8431-8444
[6] Chen H and He J, 2008, *J. Phys. Chem. C*, **112**, 17540–175450
[7] Zheng M, Zhang H, Gong X, Xu R, Xiao Y, Dong H, Liu X and Liu Y, 2013, *Nanoscale Res. Lett.*, **8**, 166
[8] Li Y, Zhou X, Zhou H, Shen Z and Chen T, 2008, Front. Chem. China, **3**, 128–132
[9] Kusumaningrum R, Rahmani S A, Widayatno W B, Wismogroho A S, Nugroho D W, Maulana S, Rochman N T, and Amal M I, 2016, *AIP Conf. Proc.*, **1964**, 020042-1–020042-6
[10] Nayl A A, Ismail L M, and Aly H F, 2011, *Int. J. Miner. Process.*, **100**, 116-123
[11] Jacob K T, Kumar A, Rajitha G, and Waseda Y, 2011, *J. High Temp. Mater. Proc.*, **30**, 459–472
[12] Hu C C, Wu Y T, and Chang K H, 2008, *J. Chem. Mater.*, **20**, 2890–2894
[13] Raj B G S, Asiri A M, Wu J J, Anandan S, 2015, *J. Alloy Compd.*, **636**; 234-240
[14] Ahmed K A M, Zeng Q , Wu K, Huang K, 2010, *J. Sol Stat Chem*, **183**, 744–751

IOP Conf. Series: Materials Science and Engineering **578** (2019) 012012 doi:10.1088/1757-899X/578/1/012012

Local Wood's Bio-Oil Upgrading Using Non-sulfided (Co, Mo)/USY Catalyst

K D Nugrahaningtyas[1,*] and E Prasetyorini[1]
[1]Department of Chemistry, Faculty of Mathematics and Natural Sciences, Sebelas Maret University, Jl. Ir. A. Sutami 36A, Kentingan, Surakarta 57126, Indonesia

*E-mail: khoirina@mipa.uns.ac.id

Abstract. In this present study, the non-sulfided (Co, Mo)/USY has used as a catalyst on the hydrotreating and hydrocracking (HC) reactions for upgrading the quality of bio-oil. The bio-oils were obtained from sengon, coconut, meranti, and bangkirai woody biomass by pyrolysis method at a temperature of 300 °C and atmospheric pressure. Meanwhile, the hydrotreating and HC reactions were done with a flow reactor at 300 °C, under H_2 gas flow (20 mL/min). The composition of bio-oil and upgrading bio-oil were analyzed by Gas Chromatography-Mass Spectrometry (GC-MS). The resulted show that the main component of bio-oil is levoglucosan, while the main product of upgrading bio-oil is methyl format. Bio-oil with the highest yield obtained from meranti on the HC reaction that used Co-Mo/USY catalyst.

1. Introduction

The supply of fossil fuels in the world requires alternative energy to replace fossil fuels. Biomass is one of the renewable fuels that have the potential to replace fossil fuels. One example of biomass is woody biomass from agriculture or forests. Woody biomass can convert into fuel by pyrolysis process which will produce bio-oil [1-4]. Based on the cellulose, hemicellulose and lignin composition, woody biomass divided into hardwood and softwood [5]. Therefore, the bio-oil composition, properties and characteristics that obtained from hardwood will be different with the bio-oil from softwood.

As an alternative fuel, bio-oils have the same combustion characteristics with light fraction oil [5]. Moreover, bio-oil is environmentally friendly because the nitrogen and sulfur components in biomass are less than fossil fuels [5-8]. Meanwhile, bio-oil has properties such as heating value [6], viscosity, corrosive, oxygen, water and ash content [9] poor than fossil fuels [6,9]. These characters cause bio-oil cannot use directly as fuel, so they need to be upgraded.

Bio-oil consists of many compounds such as ketone, phenols, aldehyde, ester, alcohol, acid, and hydrocarbon. The multi-chemical compounds in the bio-oil making its upgrading reaction is complicated. Oxygen removal in the bio-oil constituent components can be through several reactions. The reactions that occur can be decarbonization, decarboxylation [8], cracking, hydrocracking (HC) [10-12], hydrodeoxygenation (HDO), hydrogenation (HYD) and polymerization reactions [8,13]. Bimetallic catalyst has higher activity and lower rate deactivation that compare with the monometallic catalyst on bio-oil conversion [14]. HDO reaction of bio-oil that use bimetallic catalyst Fe-Co/SiO_2 has higher activity compared than monometallic catalyst Fe/SiO_2 and Co/SiO_2 respectively [8].

The sulfided CoMo and NiMo that supported by alumina (Al_2O_3) is a commercial catalyst for hydrotreating reactions [3]. The sulfided CoMo/Al_2O_3 has higher activity than the sulfided NiMo/ Al_2O_3, on HDO of anisole and guaiacol [15]. Ariyani and coworkers [16] have been used free-sulfide

CoMo/USY catalyst in HDO and HC of anisole at flow system reactor. Ariyani's research [16] result showed that the total yield of anisole HC and the oxygen-free products are 32.6% and 8.6%, respectively. Another researcher succeeded to convert anisole into oxygen-free compound (1,2,3,4,5-Penta methylbenzene) on Anisole HDO reaction in a batch reactor at 300 °C [17]. Nugrahaningtyas et al. [18,20], successfully converted 4-methoxy anisole into 1,4-dihydroxybenzene, used the same catalyst and reactor. Therefore, this research aimed to HDO and HC reaction of woody biomass-derived bio oil.

2. Experimental

The free-sulfide CoMo/USY has produced on the our previous study [19]. Woody biomass of sengon (*Paraserianthes falcataria*), coconut (*Cocos nucifera*), meranti (*Shorea leprosula*), and bangkirai (*Shorea laevis* Ridl), that obtained from local woody biomass Indonesia, was cut into 1-3 cm^3 in volume, then washed with distilled water, and oven at 110 °C for 4 hours. After that, the woody biomass pyrolyzed at a temperature of 300 °C. The bio-oil that obtained from pyrolysis processes then fractionally distilled to remove the water components. Upgrading bio-oil is carried out on a flow system reactor at a temperature of 300 °C and under H_2 gas flow rate of 20 mL/min, with CoMo/USY as free sulfide catalyst. Bio-oil and upgrading bio-oil were characterization by using GC-MS.

3. Results and Discussion

3.1. Sengon, coconut, meranti and bangkirai bio-oil component

The pyrolysis of sengon, coconut, meranti and bangkirai produces three types of products in the form of charcoal (biochar), bio-oil, and gas. The characteristics of bio-oil are dark brown, very thick and smell like smoke. The percentages of bio-oil that is produced from pyrolysis of sengon, coconut, meranti and bangkirai are 36.98%, 35.76%, 38.76% and 32.97%, respectively. The main component that is contained in bio-oil is levoglucosan. Levoglucosan is the main product of the pyrolisis of cellulose[20]. Bio-oil of each woody biomass has a different percentage of levoglucosan because of each woody biomass has a different cellulose content [5,21]. The higher levels of levoglucosan in bio-oil show the higher cellulose content in the woody biomass. The order of woody biomass that produces high amounts of levoglucosan is meranti, bangkirai, sengon, and coconut, respectively.

The percentage of levoglucosan component is also influenced by the temperature that used on the pyrolysis process [2]. The pyrolysis temperature at ± 300°C is optimum thermal decomposition of cellulose and hemicellulose components, but that is not optimum temperature for the lignin decomposition. That is according to Mohan et al. [5], which said that the decomposes of cellulose occurs at 240-350°C, hemicellulose decomposes at 200-260°C, while the lignin decomposition occurs at 280-500 °C.

Overall, the components of bio-oil derived from woody biomass are consisting of ethers, esters, carboxylic acids, aldehydes, ketones (cyclic and aromatic), alcohols (aliphatic and cyclic), furan, phenol, glucose, and impurity compounds such as S and N. Those compounds are the reaction product of cracking, dehydration, decarboxylation, ring opening, and degradation of cellulose, hemicellulose, and lignin [22]. Ketone, carboxylic acids, hydrocarbons, aromatics, and phenolic compounds are hemicellulose degradation products. The other products such as simple phenols which is substituted methyl groups, ketones (especially cyclic), aldehydes, and alcohol, are obtained from cellulose degradation. The lignin decomposed products are a complex phenol compound (a heavy molecular weight and substituted of methoxy group) and acid [22].

Table 1. Distribution of bio-oil compounds from pyrolysis of sengon, coconut, meranti and bangkirai woody biomass.

Compound	Percentage (%)			
	Sengon	Coconut	Meranti	Bangkirai
Ester	11.03	13.52	9.55	2.55
Ether	7.39	1.15	3.82	2.24
Carboxylic Acid	5.57	10.66	1.88	2.75
Ketone	14.85	7.85	8.78	18.21
Aldehyde	7.27	2.72	3.18	6.42
Alcohol	6.15	1.19	5.11	6.26
Phenolic	11.85	27.62	19.69	18.85
Hydrocarbons	-	-	1.05	1.22
Levoglucosan	30.71	17.57	46.94	35.17
Another Component Contain N and S	5.2	17.71	0.9	6.36

3.2. Product of Upgrade Bio-oil

3.2.1. Product Distribution of Upgraded Bio-oil

Table 2 shows the the liquid products composition of upgraded sengon bio-oil. Liquid product from upgraded sengon bio-oil is dominated by ester compounds (methyl format), and other products such as carboxylic acid, ketones, and phenols. There are not found levoglucosan, aldehyde, and impurities compounds such as N. Compounds of 2-hydroxy-3-methyl-2-cyclopentene-1-one and 2-methoxy phenol are only present in thermal reaction products. That's maybe, in the catalytic reaction, both compounds disappear and undergo further reaction into other compounds. According to Bui et al. [23], 2-hydroxy-3-methyl-2-cyclopentene-1-one compound can be converted into 2-cyclopentene-1-one through a cracked pathway of the methyl groups and HDO by broken of CO bonds, while 2-methyl-2-cyclopentenone is formed through the HDO pathway by CO bonds broken and transfer methyl groups. Meanwhile, 2-methoxy phenol can be converted into phenol by demethoxylation (DMO) pathway of the Csp2-OMe bond or through a demethylation pathway (DME) of the ArO-Me bond.

The composition of upgraded coconut bio-oil is presented in Table 3, where the product contains esters, carboxylic acids, and phenols. The main content bio-oil from coconut woody biomass after pyrolysis is levoglucosan (Table 1), while the liquid from upgraded coconut bio-oil is an ester (methyl format) and a high amount of phenol. The high phenol compound in this product was due to the amount of phenolic derivatives compounds that found in the coconut bio-oil before the upgrading reaction relatively higher than another woody biomass bio-oils. Whereas levoglucosan, alcohol, ketone, aldehyde, and heteroatom (S and N) compounds didn't found in liquid products of upgraded coconut bio-oil.

IOP Conf. Series: Materials Science and Engineering **578** (2019) 012012 doi:10.1088/1757-899X/578/1/012012

Table 2. Distribution of liquid product components from upgraded sengon bio-oil (% area).

Compound	Catalyst			
	Thermal	Co-Mo/ USY	CoMo/ USY	Mo-Co/ USY
Ester				
Methyl formate	75.01	75.01	86.41	89.75
Carboxylic Acid				
Propanoic acid	9.78	8.05	9.34	6.76
Ketone				
2-hydroxy-3-methyl-2-Cyclopenten-1-one	2.97	-	-	-
2-Cyclopenten-1-one	-	1.54	-	-
2-Methyl-2-cyclopentenone	-	2.9	-	-
Phenolic				
2-methoxy Phenol	2.98	-	-	-
Phenol	2.99	-	5.75	-

Table 3. Distribution of liquid product components from upgraded coconut bio-oil (% area).

Compound	Catalyst			
	Thermal	Co-Mo/USY	CoMo/ USY	Mo-Co/USY
Ester				
Methyl formate	71.15	81.19	82.3	77.53
Carboxylic Acid				
Propanoic acid	4.7	-	3.58	4.23
Phenyl acetate	3.61	-	-	-
Phenolic				
Phenol	18.69	16.96	12.27	16.39

Compound distribution of upgraded meranti bio-oil is presented in Table 4. The main compound product is an ester (methyl format), and other products are carboxylic acids, ketones, and phenols. Levoglucosan, ketones, alcohols, aldehydes, and impurities compounds that contained N elements disappeared in upgrading catalytic of meranti bio-oil. Meanwhile, ketones are only a product from upgrading bio-oil by thermal.

Table 4. Distribution of liquid product components from upgraded meranti bio-oil (% area).

Compound	Catalyst			
	Thermal	Co-Mo/ USY	CoMo/USY	Mo-Co/ USY
Ester				
Methyl formate	79.9	83.78	91.1	81.38
Carboxylic Acid				
Propionic acid	7.56	7.99	7.55	9.66
Ethylene diacetate	2.51	-	-	-
Ketone				
2-Methyl-2-cyclopentenone	6.13	-	-	-
Phenolic				
Phenol	2.55	6.87	-	7.61

Compounds distribution of thermal and catalytic upgraded bio-oil from bangkirai woody biomass presented in Table 5. Similar to the previous discussion, it is found that the main compound from upgraded bangkirai bio-oil is esters. Then, other products are carboxylic acids, ketones, furan, phenols, and bicyclic compounds. The thermal products of upgraded bangkirai bio-oil are ketones and furan compounds. Meanwhile, in the bangkirai bio-oil product which improved by using Co-Mo/USY as a catalyst, there is a high percentage of Juniper camphor and trans-caryophyllene which are both bicyclic compounds. The formation of this bicyclic compound shows a C-C coupling reaction [24]. The C-C coupling reaction products will increase at low hydrogen pressure due to low residence time [25]. The percentage of phenol compounds in the liquid product which is upgraded bio-oil from bangkirai woody biomass is high. That is caused by the content of phenolic compounds is relatively high at the feed of bangkirai bio-oil which is 27.4%. The presence of trans-Caryophyllene which is an oxygen-free bicyclic compound indicates the occurrence of a hydrodeoxygenation reaction.

Table 5. Distribution of liquid product components from upgraded bangkirai bio-oil (% area).

Compound	Catalyst			
	Thermal	Co-Mo/ USY	CoMo/ USY	Mo-Co/ USY
Ester				
Methyl formate	49.25	49.25	71.56	74.1
Carboxylic Acid				
Propanoic acid	9.83	-	8.31	11.14
3,7-Dimethyl-2,6-octadienoic acid	-	18.74	-	-
Ketone				
1-Hydroxy-2-Butanone	1.65	-	-	-
2,3-Pentanedione	2.57	-	-	-
2-hydroxy-3-methyl-2-Cyclopenten-1-one	2.75	-	-	-
Furan				
2-Acetylfuran	1.43	-	-	-
5 Methyl Furfural	3.81	-	-	-
2-Furancarboxaldehyde	2.63	3.34	-	-
Phenolic				
2-methoxy-Phenol	4.56	-	-	-
Phenol	16.44	6.45	18.72	13.35
Bicyclic				
trans-Caryophyllene	-	4.17	-	-
Juniper camphor	-	25.05	-	-

Esters which are the dominant products in liquid products of HDO bio-oil can come from the esterification reaction of carboxylic acids with alcohol compound in bio-oil [8,26]. In addition to esterification reaction, methyl formate can also be a hydrocracking product obtained through the C-C bond breaking of ester compounds in bio-oil, because USY has high acidity and has a large amount of Brønsted acid site [27] so that it leads to hydrocracking reactions. Esters which are the dominant products in upgraded bio-oil come from the esterification reaction of carboxylic acids with alcohol compound in bio-oil [8,26]. By the way, methyl formate can be obtained from hydrocracking pathway, through the C-C bond breaking of ester compounds in bio-oil, because USY has high acidity and has a large amount of Brønsted acid site. Pu and cowokers [27] say that the high acidity property leads to a hydrocracking reaction. Methyl formate is a levulinic acid derived compound that can use as an additive in fuels such as gasoline [11]. Meanwhile, levulinic acid is a product from the conversion of furan compounds [28] or/and a dehydration product of sugar and cellulose that used an acid catalyst [11]. The high amount of the product of low molecular weight carboxylic acid, because the carboxylic group is less reactive and requires an HDO reaction temperature higher than the reaction temperature in this study, which is more than 300 °C[11]. This shows that the reaction leads to a hydrocracking reaction, rather than a deoxygenation reaction. The nature of hydrocracking comes from a combination of the hydrogenation properties of Co sites and the cracking properties of the USY sites, while the deoxygenation reaction comes from Mo sites. The combination of hydrogenation and cracking properties are stronger than deoxygenation, and the content of oxygenated compounds is imbalanced with the number of Mo sites, cause the catalyst to be ineffective for deoxygenation reactions.

Alcohol, aldehydes, ketones, and ethers are more susceptible to HDO reactions than esters and carboxylic acids. Therefore, these compounds did not found in liquid products from bio-oil upgrading reactions. Hydrocarbon compounds are also not found in liquid products, because the hydrocarbon products formed are very light compounds, even as gases, that cannot be detected. According to Corma et al. [10], the product compounds can be C1-C4 hydrocarbon compounds, which are the product of cracking reactions using USY as catalysts. The disappearance of levoglucosan compound, which is the main compounds of bio-oil, indicated that HC or HDO reactions occur. Levoglucosan which had five oxygen components converts to methyl ester which has two elements of oxygen. Meanwhile, the impurities compounds which contain elements of nitrogen and sulfur had removed. It is indicated that the three catalysts have HDN and HDO activity.

The reaction mechanism in HDO bio-oil is very complicated. This is caused bio-oil contains highly complex organic compounds. So, it is possible that the bio-oil constituent compounds will compete with each other to absorb on the surface of the catalyst, and even react to each other. One example of an unstable compound in bio-oil is an aldehyde, which can react with other bio-oil components during storage [11]. Meanwhile, there are several bio-oil components that can inhibit the reaction on the surface of the catalyst, such as compounds that is containing sulfur or nitrogen.

3.2.2. Yield Total of Upgrading Bio-oil from Sengon, Coconut, Meranti, dan Bangkirai woody biomass

Figure 1 shows Mo-Co/USY, Co-Mo/USY and CoMo/USY catalyst activities in bio-oil upgrading reactions. The use of catalysts in the improvement of meranti and bangkirai bio-oil caused a significant increase in activity, but the increase in bio-oil coconut and sengon did not occur. This is possible caused of the 4-hydroxy benzenesulfonic acid compounds content in bio-oil coconut and sengon is poisoning or disabling the active site of the catalyst. The interaction between sulfur and the surface of the catalyst causes a decrease in the activity and selectivity of the catalyst. Sulfur will cover the active site of the catalyst and reduce the number of reactants that the active site can absorb and also weaken the bonds between metal and reactants [29].

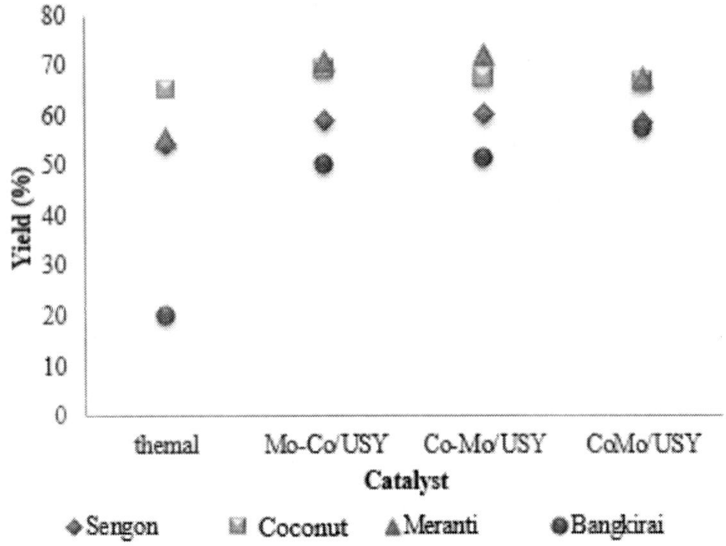

Figure 1. Correlation of catalyst types to the yield of liquid products from upgrading bio-oil of sengon, coconut, meranti and bangkirai woody biomass.

The activity of Co-Mo/USY catalyst was higher than CoMo/USY catalyst, in the upgrading reaction of bio-oil from sengon, coconut, and meranti woody biomass. This is possible because Co-Mo/USY catalysts contain more Mo metal than Co-Mo/USY catalysts. A large amount of Mo causes higher HDO activity, that is, by breaking C-O bonds. Meanwhile, the three types of bio-oil such as bio-oil from sengon, coconut, and meranti woody biomass, contain esters that are more amount than bangkirai, while esters are relatively more difficult to involvement in HDO reactions. Hence the increased reaction of sengon and meranti bio-oil requires a catalyst containing more Mo metal or catalyst with a higher Mo: Co ratio. However, Co-Mo/USY catalyst has lower activity than CoMo/USY in the reaction of upgrading bio-oil from bangkirai woody biomass.

The Mo-Co/USY, Co-Mo/USY and CoMo/USY catalysts have HDO activity for upgrading quality bio-oil with meranti> coconut> sengon> bangkirai sequences. The difference yield for each bio-oil is due to the difference of types and number of bio-oil components. HDO yield of bio-oil meranti is relatively higher than other bio-oils because in bio-oil meranti there is no sulfur-containing compound that can deactivate the catalyst. Besides the sulfur, compounds which contain nitrogen can also deactivate the catalyst [30]. The presence of compounds which contain nitrogen in meranti bio-oil is relatively less compared with other bio-oils, it means the compound components that can interfere with the reaction of HDO from bio-oil meranti decreased. In the reaction of HDO, bio-oil bangkirai has a relatively lower yield than other bio-oils. The low number of co-reactant compounds in bio-oil bangkirai affects the low yield of HDO from bio-oil bangkirai because co-reactant compounds will act as H donors for HDO reactions [31].

3.2.3. Selectivity of Co-Mo/USY, CoMo/USY, and Mo-Co/USY Catalyst toward Methyl Format

The reaction of upgrading bio-oil of sengon, coconut, meranti, and bangkirai woody biomass, using Mo-Co/USY, Co-Mo/USY and CoMo/USY catalysts, has the main product, methyl. However, in the upgrading reaction, Co-Mo/USY catalyst has lower selectivity than Mo-Co/USY and CoMo/USY catalyst caused by more varied types of particles or in Co-Mo/USY catalyst. The different of particles types provide the different catalytic active sites that allow the different types of reactions to take place. Therefore, the catalyst becomes less selective for one particular reaction type. In general, the

CoMo/USY catalyst has high selectivity of the methyl format formation, because its particles types are more uniform than the others.

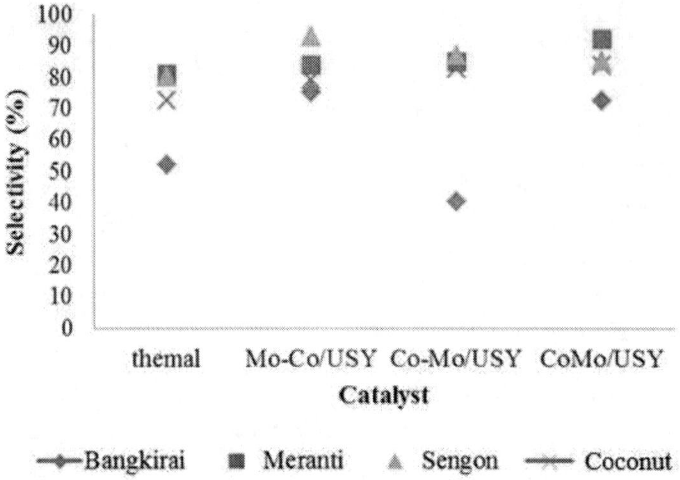

Figure 2. Correlation of catalyst type and the selectivity of the Methyl format on upgrading bio-oil from sengon, coconut, meranti, and bangkirai woody biomass.

The highest selectivity of methyl format shows that the catalytic upgrading of bio-oil tend to hydrocrack reactions, namely the cracking of C-C bonds. The hydrocracking reaction, which is more dominant in upgrading bio-oil, shows that the role of USY and Co metal are more dominant in the catalyst. While, the Mo metal as a deoxygenation site becomes less functional, so it is ineffective to break the CO bonds on the oxygenate compound from bio-oil. In accordance with Huber et al.[11], the reaction that occurs in upgrading bio-oil with zeolite catalyst is a cracking reaction. Besides that, the very high acidity of Mo-Co/USY (8.304 mmol/g), Co-Mo/USY (10.849 mmol/g) and CoMo / USY (11.206 mmol/g) catalysts[19] are also supporting factors for the dominance of bio-oil hydrocracking reactions. The high acidity of the catalyst shows the high number of Brønsted acid sites and Lewis acid sites on the catalyst. USY is a supporting material with a large number of Brønsted acid sites [27,32], where the Brønsted acid site can act as a proton donor [33]. Therefore, the large amount of hydrogen that can be donated by this catalyst will direct the reaction more dominantly against hydrocracking.

4. Conclusion

The upgrading of woody biomass-derived (bio-oil) has been carried out through HC levoglucosan reaction into methyl formate and the reaction of HDO from oxygenated phenol derivatives into phenol. Each bio-oil has different components so that it can cause each catalyst to have different optimal activity and selectivity in each bio-oil. The catalyst with optimal activity in the reaction of upgrading bio-oil from meranti woody biomass was Co-Mo/USY, which was equal to 72.39%. Meanwhile, the highest HC selectivity in the formation of methyl formate was Mo-Co/USY catalyst, which was 93.00%.

Acknowledgement

Thanks to the Directorate General of Higher Education, Indonesia for funding this project through the PENELITIAN DASAR UNGGULAN PERGURUAN TINGGI (PDUPT) GRANT under the contract number of 474 /UN27.21/PP/2018.

References

[1] Mazlan M A F, Uemura Y, Osman N B and Yusup S 2015 *Energy Convers. Manag.* **98** 208-14
[2] Jouhara H, Ahmad D, Boogaert I V D, Katsou E, Simons S and Spencer N 2018 *Therm. Sci. Eng. Prog.* **5** 117-43
[3] Balat M, Balat M, Kirtay E and Balat H 2009 *Energy Convers. Manag.* **50** 3147-57
[4] Shamsul N S, Kamarudin S K and Rahman N A 2017 *Renew. Sustain. Energy Rev.* **80** 538-49
[5] Mohan D, Pittman C U and Steele P H 2006 *Energy and Fuels* **20** 848-89
[6] Tan Y L, Abdullah A Z and Hameed B H 2017 *Bioresour. Technol.* **243** 85-92
[7] Bulushev D A and Ross J R H 2011 *Catal. Today* **171** 1-13
[8] Cheng S, Wei L, Julson J and Rabnawas M 2017 *Energy Convers Manag.* **150** 331-42
[9] Xiu S and Shahbazi A 2012 *Renew. Sustain. Energy Rev.* **16** 4406-14
[10] Corma A, Huber G W, Sauvanaud L and Connor P O 2007 *J. Catal.* **247** 307-27
[11] Huber G W, Iborra S and Corma A 2006 *Chem. Rev.* **106** 4044-98
[12] Bakhtyari A, Makarem M A and Rahimpour M R 2017 4 *Bioenergy Systems for the Future* (Elsevier Ltd.) pp 87-148
[13] Mortensen P M, Grunwaldt J, Jensen P A, Knudsen K G and Jensen A D 2011 *Appl. Catal. A: Gen.* **407** 1-19
[14] Alonso D M, Wettstein S G and Dumesic JA 2012 *Chem. Soc. Rev.* **41** 8075-98
[15] Putri I F and Nugrahaningtyas K D 2016 *J. Kim. dan Pendidik. Kim.* **1** 164-73
[16] Ariyani D, Nugrahaningtyas K D and Heraldy E 2018 *IOP Conf. Ser.: Mater. Sci. Eng.* 012058
[17] Nugrahaningtyas K D, Putri I F, Heraldy E and Hidayat Y 2018 *IOP Conf. Ser.: Mater. Sci. Eng*
[18] Nugrahaningtyas K D, Putri I F and Heraldy E 2017 *ALCHEMY J. Penelit. Kim.* **13** 66-74
[19] Nugrahaningtyas K D, Heraldy E, Hidayat Y and Rachmadani 2016 *Int. Conf. 6th BaSIC* 1-7
[20] Nolte M W, Saraeian A and Shanks B H 2017 *Green Chem.* **19** 3654-64
[21] Keijsers E R P, Yılmaz G and Dam J V 2013 *Carbohydr Polym.* **93** 9-21
[22] Stefanidis S, Kalogiannis K and Iliopoulou E 2013 *J. Anal. Appl. Pyrolysis* **105** 143-50
[23] Bui V N, Laurenti D, Afanasiev P and Geantet C 2011 *Appl. Catal. B: Environ.* **101** 239-45
[24] Gamliel D P, Baillie B P, Augustine E, Hall J, Bollas G M and Valla J A 2017 *Microporous Mesoporous Mater.* **261** 18-28
[25] Anand M, Sibi M G, Verma D and Sinha A K 2014 *J. Chem. Sci.* **126** 473-80.
[26] Tang Z, Lu Q, Zhang Y, Zhu X and Guo Q 2009 *Ind. Eng. Chem. Res.* **48** 6923-9
[27] Pu X, Liu N and Shi L 2015 Acid *Microporous Mesoporous Mater.* **201** 17-23
[28] Myllyoja J, Piilola R and Oyj N 2017 *United States Pat. Appl. Publ.* **1** 1-8
[29] Bambal A S, Guggilla V S, Kugler E L, Gardner T H and Dadyburjor D B 2014 *Ind. Eng. Chem. Res.* **53** 5846-57
[30] Argyle M D and Bartholomew C H 2015 *Catalyst* **5** 145-269
[31] Yu S Y, Waku T and Iglesia E 2003 *Appl. Catal. A: Gen.* **242** 111-21
[32] Huynh T M, Armbruster U, Kreyenschulte C R, Nguyen L H, Phan B M Q, Nguyen D A and Martin A 2016 *Catalysts* **6** 176
[33] Gunawardena D A and Fernando S D 2013 *Biomass Now-Cultivation and Utilization* (InTech Publisher) chapter 11 273-98

IOP Conf. Series: Materials Science and Engineering **578** (2019) 012013 doi:10.1088/1757-899X/578/1/012013

Identification of chemical structure of anthocyanin and other active substances of red color melinjo peels by FTIR and LC-MC analysis

E Tarmizi[1], A Z E Putri[1], M A Nusandari[1], Y A Husnil[1], R Saragih[2], L H Lalasari[3]
[1]Department of Chemical Engineering, Institut Teknologi Indonesia, Jl Raya Puspiptek Serpong-Tangsel, Indonesia
[2]Department of Food Technology, Institut Teknologi Indonesia, Jl. Raya Puspiptek Serpong-Tangsel, Indonesia
[3]Research Center for Metallurgy and Materials, LIPI, Kawasan Puspiptek GD. 470 Tangerang Selatan 15314, Indonesia

E-mail: ermitarmizi@gmail.com

Abstract. Research on red color melinjo peels containing anthocyanin active substance has been studied since 2011, starting from anthocyanin extraction stage (maceration) by using both polar and non-polar solvents, a variation of citric acid addition and stability of anthocyanin extract storage and application of anthocyanin for food and beverage. However, research on the identification of anthocyanin chemical structures and other active substances of red color melinjo peels is also necessary. The purpose of this study was to identify the structure of anthocyanin and other active substances from red color melinjo peels with FTIR and LC-MS analysis. The results obtained: from FTIR analysis shown wavelength, among others: 3425 cm^{-1}, 1725 cm^{-1}, 1632 cm^{-1} and 1219 cm^{-1}, it shows that inside the red color melinjo peels contained OH function group, C=O, C=C, and CO. From the LC-MS analysis shown at a minute of 1.62 observed BM = 302.236, which is the molecular weight of the Quercetin compound. The mass spectrum with the base peak height of m/z = 165, at min 3.83 has 100% abundance, whereas the 50% abundance at peak M+m/z = 242 has BM = 271 with the anthocyanin molecule of red color melinjo peels $C_{15}O_5H_{11}$. The anthocyanin type of red color melinjo peels is a cyanine compound.

1. Introduction

Anthocyanin is a natural coloring which belongs to the class of flavonoids which are widely used as dyes, especially drinks. These anthocyanin pigments can replace commercial synthetic dyes that are known to be toxic and carcinogenic [1]. In addition to acting as food and beverage coloring agents, anthocyanins are also believed in biological systems with the ability to bind free radicals and the ability to inhibit chemical reactions at the initiation stage which causes carcinogenesis [2]. According to Jackman and Smith [3], the source of anthocyanins is found in some plants, among others; strawberry, cherry, plum, cabbage, grapes, and red peppers. The anthocyanin concentration of the plant was calculated from the total anthocyanin content after being separated from the solvent. Mateus and Freitas [4] explained that there were 6 types of anthocyanins in foodstuffs such as pelargonidin, cyanidin, peonidin, delphinidin, petunidin, and malvidin. The type and chemical structure of anthocyanin are influenced by the raw materials used in obtaining the anthocyanin extract. Preliminary research has been carried out by Ermiziar et al [5] where the results of red color melinjo extract with ethanol solvent containing anthocyanin which has antioxidant activity (IC_{50}) of 28.43 mg. The function of these antioxidants is as anti-free radicals. Therefore, it is necessary to know other active substances besides anthocyanins from red color melinjo which can also function as antioxidants. Ermiziar et al [6] explained that red color melinjo waste containing anthocyanin can be used as a coloring agent for food and beverages. The red melinjo skin waste is obtained from the production of chips which is usually only used for vegetables and chips but the rest is wasted.

The concentration of anthocyanins contained in the skin melinjo red is influenced by the purity, chemical structure, and type of anthocyanin. Where in obtaining anthocyanin extract from melinjo red skin can be done by maceration process using polar solvents [6]. While the stages in determining the active substances contained in anthocyanins were carried out by characterization using FTIR and LC-MS analysis. The FTIR (Fourier Transform Infrared) analysis function is to determine the functional group based on the number of melinjo red skin anthocyanin waves. While the analysis of LC-MS (Liquid Chromatography - Mass Spectrometry) is an analytical technique used by combining the ability to separate sample components and detect sample components using mass spectrometry. LC-MS data can be used to provide information about molecular weight, chemical structure, identity and quantity of certain sample components [7]. In this study, the identification of chemical structures and other active substances from the red-skinned melinjo anthocyanin was carried out using FTIR and LC-MS from the anthocyanin extract samples obtained in previous studies. An example of anthocyanin produced in previous studies [5] is shown in Figure 1.

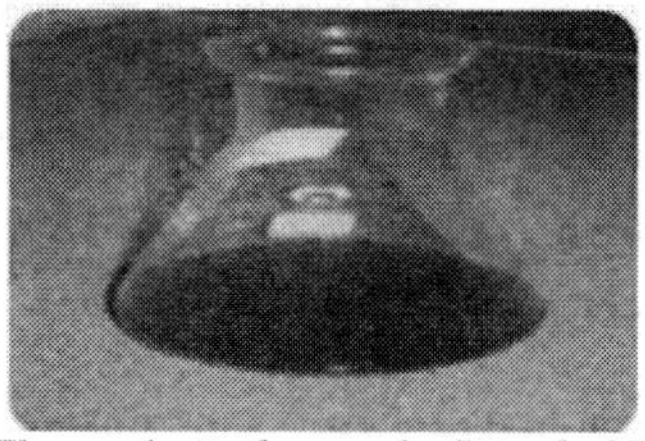

Figure 1. Anthocyanin Sample [5]

2. Experimental

The material used in the study was red color melinjo peels, 96% alcohol (Food Grade) and citric acid with the analytical grade. While the method used in anthocyanin isolation is maceration process as follows: 200 grams of red melinjo peels was cut into small pieces and put into 1000 mL of 96% alcohol solvent with a ratio (w/v) of red color melinjo peels / alcohol solvent of 1:5 for maceration limes for 3 days. In the maceration process citric acid was added with a concentration of 3%. The next step was filtering to separate red melinjo peels and filtrates containing mixtures of anthocyanin extract and solvent. The last step was to separate the solvent and anthocyanin extract using a rotary evaporator. The anthocyanin extract produced was then characterized using FTIR and LC-MS analysis.

3. Results and Discussion

3.1 Characterization of anthocyanin samples by FTIR analysis

The FTIR analysis is used to identify functional groups and compounds contained in anthocyanin samples produced from red color melinjo peels as shown in Figure 2.

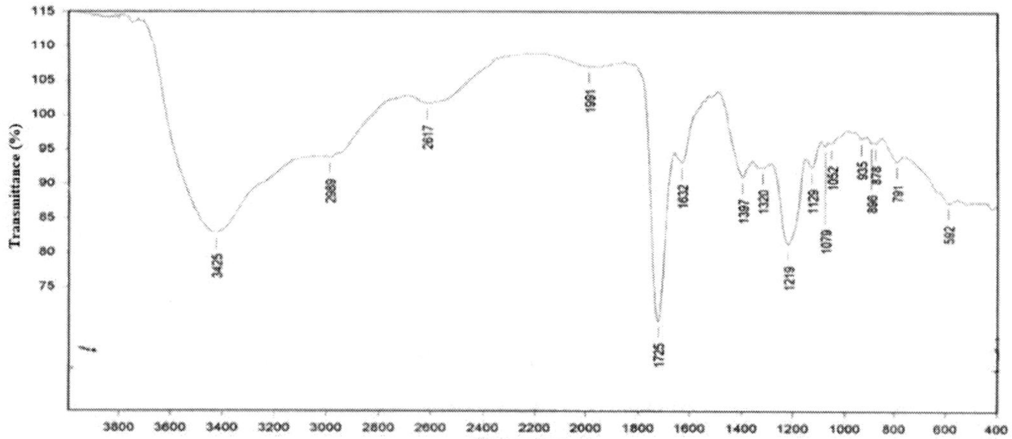

Figure 2. The FTIR spectrum of anthocyanin samples from red color melinjo peels

Figure 2 showed the FTIR spectrum results of red color melinjo peels which had a wave absorption band of 3425 cm^{-1}. According to Ghassempour (2008) [10], the vibration characteristics of hydroxyl groups were seen in the wave number area of 3700 cm^{-1}- 3100 cm^{-1}. So that the absorption shown at a wavelength of 3425 cm^{-1} is characteristic for absorption for OH groups. The spectrum in the area of 1725 cm^{-1} was stretching vibration C=O, absorption at 1632 cm^{-1} is the vibration of C=C, and the last absorption at 1219 cm^{-1} is the C-O stretch vibration. The results of FTIR analysis looked that the wave number range was the same as the wave number range of quercetin compounds. The obtained function groups can be seen in Table 1.

Table 1. The functional groups of FTIR analysis from anthocyanin samples

No.	Wave number		band	Function group
	Spectrum	Literature		
1	3425 cm^{-1}	3700-3100 cm^{-1}	widen	O-H
2	1725 cm^{-1}	1705-1725 cm^{-1}	sharp	C=O (ketone)
3	1632 cm^{-1}	1500-1600 cm^{-1}	medium	C=C (Aromatic)
4	1219 cm^{-1}	1080-1300 cm^{-1}	sharp	C-O

3.2 Characterization of anthocyanin samples by LC-MS analysis

Liquid Chromatography Mass - Spectrometry (LC-MS) is an analytical technique that combines the physical separation capability of liquid chromatography with the specificity of mass spectrometry detection. Liquid chromatography separates the sample components and then these ions are detected by a mass spectrometer. LC-MS data can be used to provide information about the molecular weight, chemical structure, identity and quantity of certain sample components [7]. The result of LC-MS analysis for anthocyanin samples from red color melinjo peels could be looked in Figure 3.

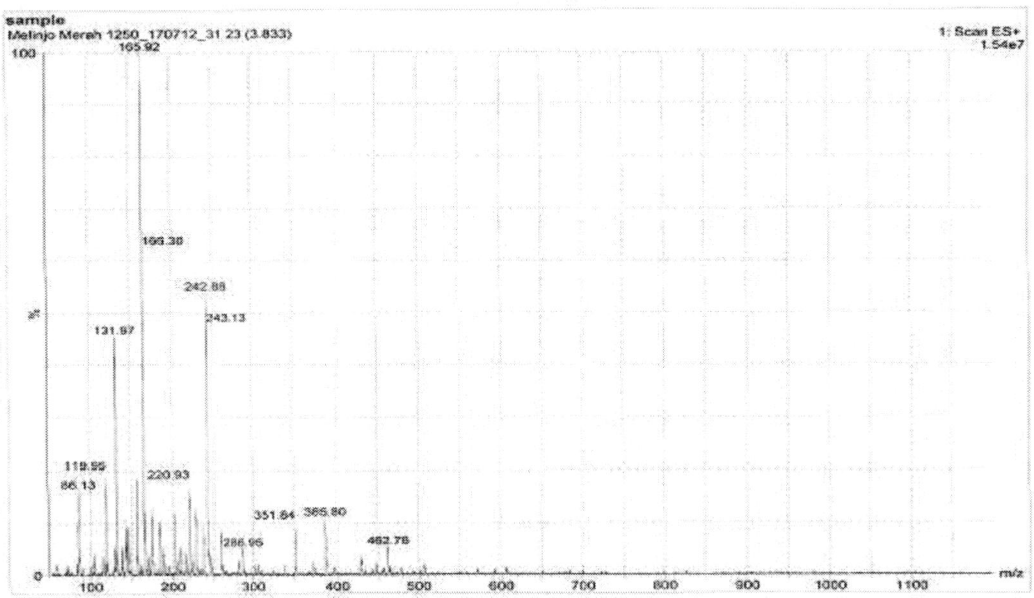

Figure 3. The LC-MS spectrum of anthocyanin spectrum from red color melinjo peels

Figure 3 showed the results of compound identification using the LC-MS method. The chromatogram results showed that anthocyanin extract from red color melinjo peels at 1.62 min produced a molecular weight of 302.236 m/z. This molecular weight was Quercetin compound. The following is a picture of the Quercetin structure

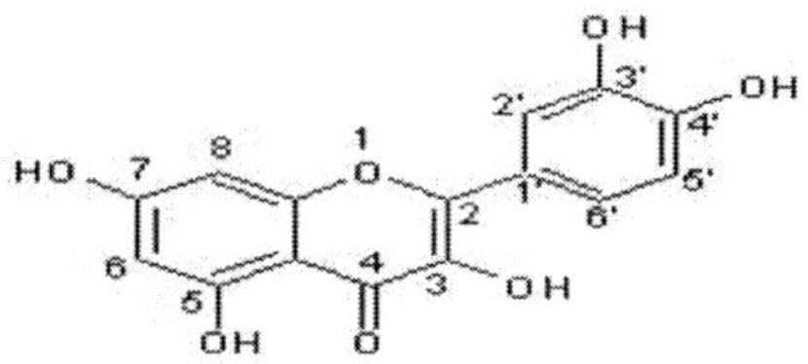

Figure 4. The chemical structure of Quercetin[8]

From The LC-MS spectrum in Figure 3 looked that the base peak height of m/z = 165, at 3.83 min has 100% abundance, whereas the 50% abundance at peak M+m/z = 242 has BM = 271 with the anthocyanin molecule of red color melinjo peels was $C_{15}O_5H_{11}$. This anthocyanin type of red color melinjo peels was a cyanine compound. The following is a picture of the cyanine structure

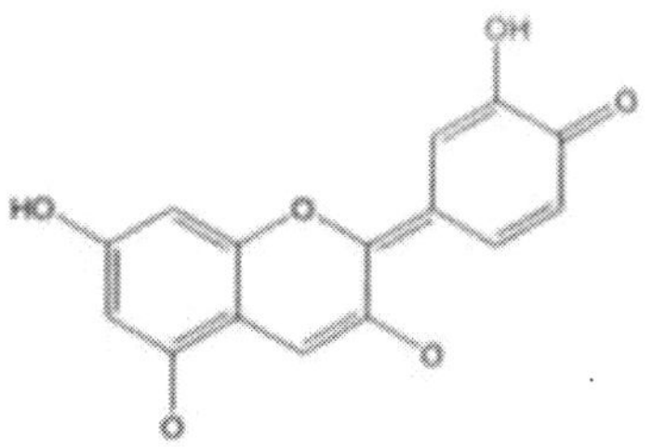

Figure 5. The chemical structure of cyanine[9]

4. Conclusion

The conclusion from the results of the characterization of anthocyanin samples using FTIR and LC-MS analysis was as follows: The FTIR analysis looked wavelength, among others: 3425 cm^{-1}, 1725 cm^{-1}, 1632 cm^{-1}, and 1219 cm^{-1}. It showed that anthocyanin from red color melinjo peels contained OH function group, C=O, C=C, and CO. From the LC-MS analysis shown at a minute of 1.62 has observed BM = 302.236, which is the molecular weight of the Quercetin compound. The mass spectrum with the base peak height of m/z = 165, at 3.83 min has 100% abundance, whereas the 50% abundance at peak M+m/z = 242 has BM = 271 with the anthocyanin molecule of red color melinjo peels was $C_{15}O_5H_{11}$. The anthocyanin type of red color melinjo peels was a cyanine compound.

Acknowledgments

This research was carried out at organic chemistry laboratorium at Department of Chemical Engineering, Institut Teknologi Indonesia. The authors are grateful for the main financial supports from RISTEK_DIKTI budget 2018.

References

[1] Francis F J 1999 *Colorants* (St Paul, MN: Eagen Press) pp 144.
[2] Smith M, Marley K, Seigler D, Singletary K and Meline B 2000 *J. Food Sci.* **65** 352-356.
[3] Jackman R L, Smith J L 1996 *Natural food colorant* (London: Blackie Academic and Professional).
[4] Mateus N and Freitas V 2009 *Anthocyanins as Food Colorants* (New York: Springer Science Business Media LLC).
[5] Tarmizi E, Saragih R, Lalasari L H 2017 *Applied Mechanics and Materials* **873** 83-88.
[6] Tarmizi E, Lalasari L H, Saragih R 2015 *AIP Conf. Proc.* **1699** 060018
[7] Technologies A 2001 *Basic of LC/MS Agilent* (USA)
[8] Alrawaiq N S and Abdullah A 2014 A *Int J Pharmtech Res.* **6**(3) 933-941.
[9] Einbond L S, Reynertson K A, Luo X, Basile M J, Kennelly E J 2004 *Food Chem.* **84**(1) 23-28.
[10] Ghassempour A, Heydari R, Talebpour Z, Fakhari A R, Rassouli A, Davies N, Aboul-Enein 2008 *J Liq Chromatogr Relat Technol.* **31**(17) 2686–2703.

The influence of scale-up treatment from red color melinjo peels mass on the produced anthocyanin characteristics

E Tarmizi[1], A Z E Putri[1], M A Nusandari[1], Y A Husnil[1], R Saragih[2], L H Lalasari[3]
[1]Department of Chemical Engineering, Institut Teknologi Indonesia, Jl Raya Puspiptek Serpong-Tangsel, Indonesia
[2]Department of Food Technology, Institut Teknologi Indonesia, Jl. Raya Puspiptek Serpong-Tangsel, Indonesia
[3]Research Center for Metallurgy and Materials, LIPI, Kawasan Puspiptek GD. 470 Tangerang Selatan 15314, Indonesia

E-mail: ermitarmizi@gmail.com

Abstract. Research on red dye foods containing anthocyanin derived from red color melinjo peels has been done since 2011 in a laboratory scale. However, to increase anthocyanin production to a larger scale (domestic industry) requires further research. The purpose of this study was to determine the effect of scale-up treatment from the amount of red color melinjo peels on the acquisition of anthocyanin extract. The study was conducted by the maceration method for 3 days, using 96% ethanol solvent with the addition of 3% citric acid. Comparison (m / v) of red color melinjo peels/ethanol was 1: 5 with red color melinjo peels mass of 100 g, 200 g, 300 g, 400 g, respectively. The results showed that the anthocyanin concentration from a red color melinjo peels mass of 100 g, 200 g, 300 g, 400 g was 2.1709 mg /g, 2.3046 mg/g, 2.3044 mg/g, 2.0039 mg/g with a density of 1.1596 g/ ml, 1.1604 g/ml, 1.1604 g/ml and 1.1596 g/ml, respectively.

1. Introduction

The increase in the use of food dyes which are prohibited, especially in market snacks makes consumers feel worried about aspects of food safety. Therefore, it is necessary to use alternative dyes in food. Where dye is one of the food addictive substances. To replace synthetic dyes that are no longer permitted, natural dyes should be used. The example of commercial synthetic dyes is sunset yellow FCF Rhodamine B, tartrazine and others. It is known that synthetic coloring materials need to be certified by the authorities before they can be used. This is a regulation from the Indonesian Ministry of Health to safeguard the use of commercial synthetic dyes. Synthetic dyes can cause several diseases when consumed exceeding the threshold value. Therefore, several studies have begun to actively introduce the use of color ingredients from nature such as anthocyanin pigment [1-5]. It was found in red color melinjo peels [6], blackberry and raspberry fruits [7]. Research has been done by Tarmizi,E et al [6] since 2011 that red color melinjo peels was as a natural food color with flavonoid content of 2.739 mg/g with antioxidant activity (IC_{50} value) of 28.430 mg and carotenoids of 241.220 ppm. The produced anthocyanin pigment is about 600.1 mg/100 g red color melinjo peels that the color is stable over two weeks shelf life.

Anthocyanin is classified as pigments called flavonoids. Flavonoid compounds including polar compounds that can be extracted also use polar solvents. Some polar solvents include ethanol, water, and ethyl acetate. In 2017, we still continued the research on isolation of anthocyanin from red color melinjo peels. It showed that the optimum maceration time for anthocyanin extraction was carried out for 3 days and the addition of 3% citric acid. The total anthocyanin concentration of red color melinjo peels at optimum conditions was 0.704 mg/gram extract. According to Wrostald [8], the maceration process of dragon fruit skin containing anthocyanin using 95% ethanol solvent and maceration times for 3 days. Vargas et al [9] explained the color of anthocyanin extract from dragon fruit produced is dark. While according to Wrostald et al [10], the determination of anthocyanin density is needed to express the length of the carbon chain in anthocyanin. Based on these data, this research will scale up the mass of red color melinjo peels to the acquisition of anthocyanin extract.

2. Experimental

The material used in this study is red color melinjo peels. The solvent used is 96% ethanol (food grade) and citric acid with the analytical grade. This research was carried out by maceration method that the steps was as follows: red color melinjo peels was cut into small pieces and put into 1000 mL of 96 % alcohol solvent with adding 3 % citric acid. This study used variations of red color melinjo peels mass of 100 g, 200 g, 300 g, and 400 g with maceration time for 3 days. The maceration products were filtered in separating wastes of melinjo peels and anthocyanin filtrates. Furthermore, the filtrate is evaporated using a rotary evaporator so that a red dye containing anthocyanin is obtained. To find out anthocyanin levels in melinjo skin, a spectrum absorbance was obtained from UV-VIS photometric spectrometry test with a wavelength of 400-500 nm. From absorbance data, the concentration of red dyes containing anthocyanin can be obtained. The following was a flow chart of research shown in Figure 1.

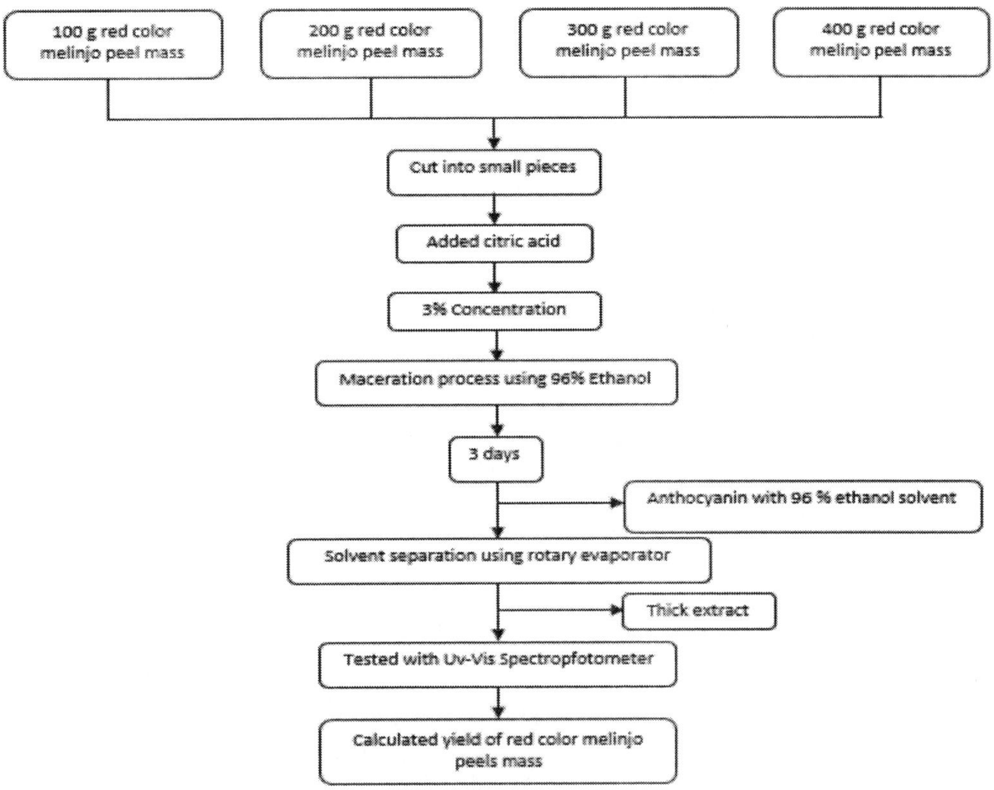

Figure 1. The flow chart of anthocyanin isolation from red color melinjo peels

3. Results and Discussion

The results of the analysis using the UV-VIS Photometric Spectrometry with a wavelength of 400-500 nm for anthocyanin samples at variations of red color melinjo peels mass of 100 g, 200 g, 300 g, and 400 g was shown as follows:

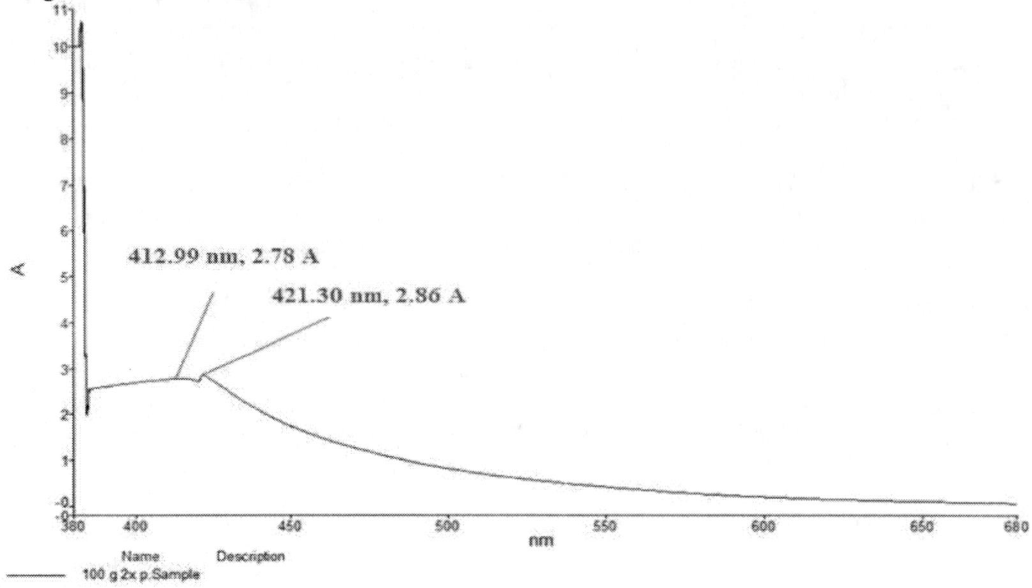

Figure 2. The UV-VIS spectrum of anthocyanin samples with a mass of 100 g

From Figure 2 could be seen that the wavelength range of 470-500 nm had sample absorbance values of 1.3 - 1. The highest absorbance value was 1.3 at a wavelength of 470 nm so that this value was used as data in the calculation of anthocyanin concentration. The calculation results showed that the concentration of anthocyanin was 2.1709 mg anthocyanin /1 gram extract.

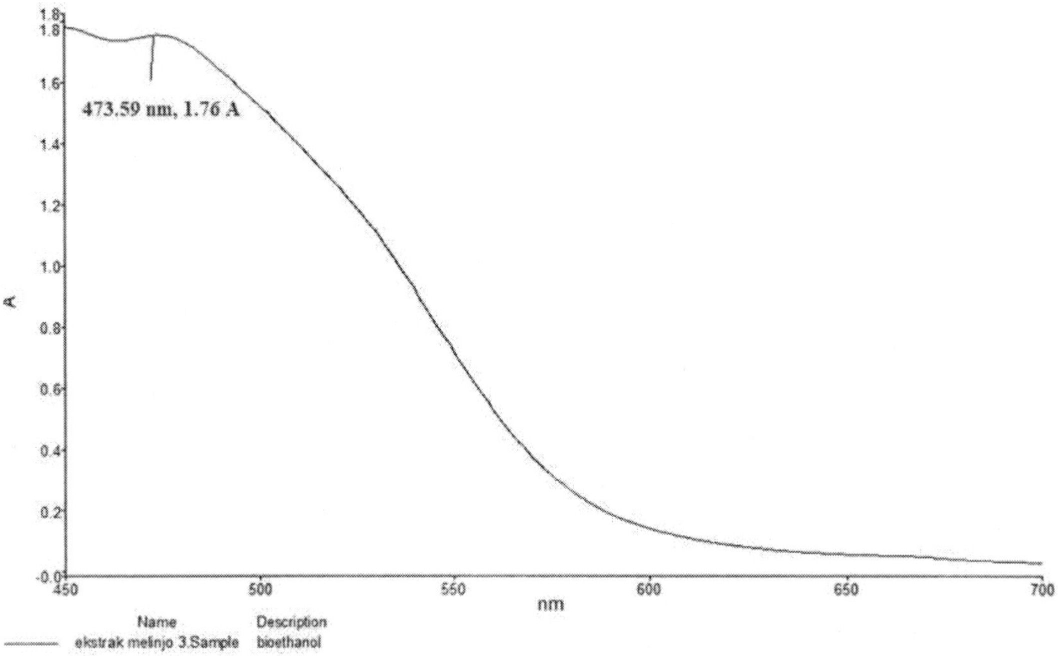

Figure 3. The UV-VIS spectrum of anthocyanin sample with a mass of 200 g

Figure 3 looked that the wavelength range of 470-500 nm had the sample absorbance value of 1.39 - 0.97. The highest absorbance value was 1.79 at a wavelength of 470 nm so that this value was used as data in the calculation of anthocyanin concentration. The results of the calculation of anthocyanin concentration were 2.3046 mg anthocyanin /1 g extract.

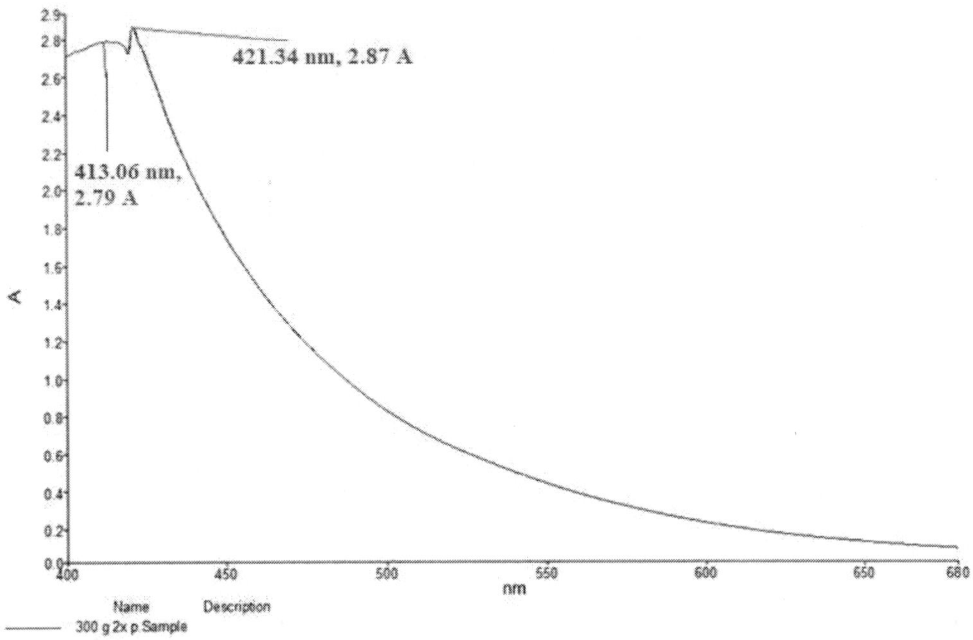

Figure 4. The UV-VIS spectrum of anthocyanin samples with a mass of 300 g

From Figure 4 showed that in the wavelength range of 470-500 nm, the sample absorbance value is 1.38-0.8. The highest absorbance value is 1.38 at a wavelength of 470 nm so that the value was used as data in the calculation of anthocyanin concentration. The results of the calculation of anthocyanin concentration were 2.3044 mg anthocyanin / 1 g extract.

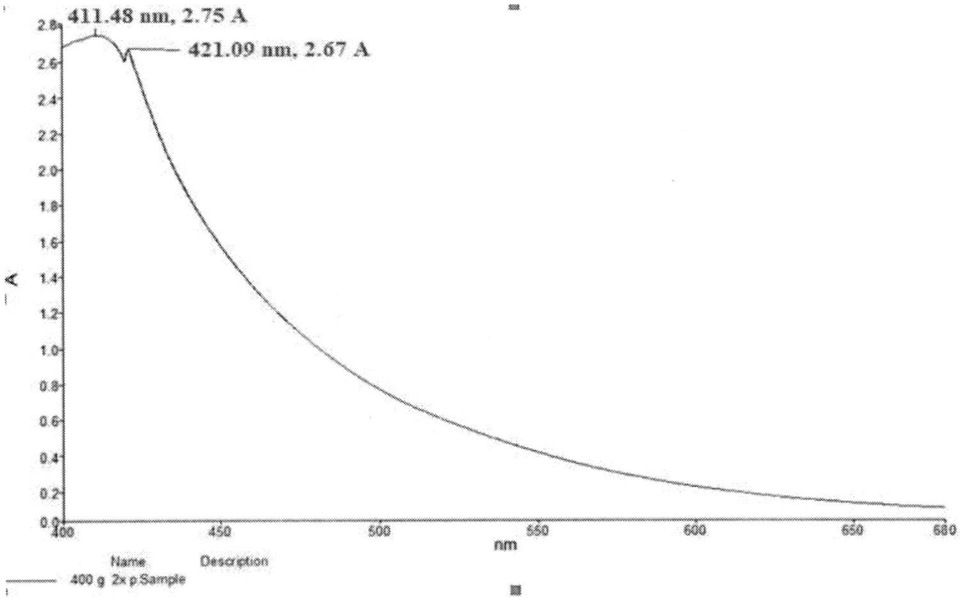

Figure 5. The UV-VIS spectrum of anthocyanin samples with a mass of 400 g

From Figure 5 can be seen that the wavelength range of 470-500 nm had the sample absorbance value of 1.2 - 0.8. The highest absorbance value was 1.2 at a wavelength of 470 nm so that this value was used as data in the calculation of anthocyanin concentration. The results of the calculation of anthocyanin concentration were 2.0039 mg anthocyanin/1 gram extract. The anthocyanin extraction data of red color melinjo peels with mass variations could be seen in Table 1.

Table 1. The concentration of anthocyanin samples in variations of red color melinjo peels mass

Melinjo peels Mass (g)	Concentration (mg/g)	Density (g/mL)
100	2,1709	1,1596
200	2,3046	1,1604
300	2,3044	1,1604
400	2,0039	1,1596

From Table 1 above decribed that the increase in red melinjo peel mass used in research from 100 g to 200 g caused an increase in the anthocyanin concentration extracted from 2.1709 mg/g to 2.3046 mg/g. Another condition was shown when the use of red melinjo peel mass was increased to 300 g, there was no significant change in the anthocyanin concentration value. The concentration of anthocyanin tended to decrease when red melinjo peel mass was increased to 400 g. Jackman et al [2] explained the decrease of anthocyanin concentration was possible happened by the presence of oxygen and various enzymes so that anthocyanin pigment degradated in the media. In this research, the presence of oxygen may be occurred when the red color melinjo peels were macerated by using alcohol in atmospheric condition to form anthocyanin extract, then this extract was separated from alcohol solvent by using rotary evaporator. Actually, the difference of anthocyanin concentration at variations of melinjo peels mass was not significant as shown in Table 1. This condition was hoped in the research for the scale-up of anthocyanin production from red color melinjo peels. The anthocyanin extracts isolated from red color melinjo peels has almost concentration as same as the previous research by Bowen Fobes [7] that he obtained the concentration of anthocyanin was approximatelly 1.46–21.99 mg/g blackberry and raspberry fruits. From these research data concluded that the mass of 200g was the best mass used for isolation of anthocyanin from red color melinjo peels.

(a) (b) (c) (d)

Figure 6. The visual observation of anthocyanin extract in mass variations (a) 100 g (b) 200 g (c) 300 g (d) 400 g

While according to Wrostald et al [10], the determination of density needed to express the length of the carbon chain in anthocyanin. From Table 1 above could be looked that there was no significant difference between the mass of 100 g, 200 g, 300 g, and 400 g. This shows that at a mass of 100 g, 200 g, 300g, and 400 g contained the same type of anthocyanin and there is no change in anthocyanin structure. The following is a visual observation of the color of anthocyanin extracts in variations of the mass of red color melinjo peels as shown in Figure 6.

Figure 6 was the visual observation result of the color of anthocyanin samples produced by using 96% ethanol solvent and an addition of 3% citric acid in the 3 days maceration process in mass variations of 100 g, 200 g, 300 g, and 400 g. It looked that there was no specific color change when the mass of red color melinjo peel increased to 400 g. It indicated that there is not a significant chemical reaction between anthocyanin and other constituents to form colorless compounds or new phases with brown color [2]. The color of anthocyanin samples were still dark red. This dark color indicates the presence of extracted anthocyanin pigments that was stable in the media.

4. Conclusion.

The conclusion could be drawn from this study was the maximum anthocyanin concentration of 2.3046 mg/g and the best density of 1.1604 obtained by 200 g of red color melinjo peel mass. Actually, the difference in the value of anthocyanin concentration and density at mass variations between 100 g, 200 g, 300 g, and 400 g was not too significant. Therefore, this research could continue to a large scale in producing natural dyes that have antioxidant properties.

Acknowledgments

This research was carried out at organic chemistry laboratorium at Department of Chemical Engineering, Institut Teknologi Indonesia. The authors are grateful for the main financial supports from RISTEK_DIKTI budget 2018.

References

[1] Pericles Markakis *1982 Book of Anthocyanins As Food Colors* (US: Academic Press).
[2] Jackman R, Tung M A, Yada R Y and Speers R A 1987 *J Food Biochem* **11** 201-247.
[3] Wrolstad R E 2006 *J Food Sci.* **69**(5) C419 - C425.
[4] Francis F J and Markakis P C 2009 *Crit Rev Food Sci Nutr* **28**(4) 274-314.
[5] Proctor, J. T. A. , Creasy, L. L. 1969 *Journal of Phytochemistry* **8** 2108
[6] Tarmizi E, Lalasari L H, and Saragih R 2015 *AIP Conf. Proc.* **1699** 060018
[7] Bowen-Forbes C S, Zhang Y and Nair M G 2010 *J. Food Compos. Anal.* **23** 554-560.
[8] Wrolstad R E 2000 "*Anthocyanin" In Natural Food Colorants (*New York: Marcel Dekker).
[9] Vargas M, Cortez J A, Duch E S, Lizama A P, and Mendez C H 2013 *Food Nutr Sci* **4**(12) 1221-1228.
[10] Wrostald R E, Durst R W, and Lee J 2005 *Trends Food Sci Technol.* **16**, 423-428.

IOP Conf. Series: Materials Science and Engineering **578** (2019) 012015 doi:10.1088/1757-899X/578/1/012015

Influence of Cd, Mn, and Co addition on the structural and optical properties of TiO_2 crystal

S Wahyuningsih[1], C Purnawan[2] and D N Aisyah[1]
[1]Inorganic Material Research Group, Faculty of Mathematics and Natural Sciences, Sebelas Maret University, Jalan Ir. Sutami No.36A Kentingan Surakarta
[2]Analytical and Instrumental Chemistry Research Group, Faculty of Mathematics and Natural Sciences, Sebelas Maret University, Jalan Ir. Sutami No.36A Kentingan Surakarta

E-mail: sayekti@mipa.uns.ac.id

Abstract. Synthesis of TiO_2-M (M = Cd, Mn, Co) has been done by the sol gel method. Solution of titanium (IV) isopropoxide was hydrolyzed using a solution of glacial acetic acid at a temperature range of 10-15 °C to obtain a sol, subsequently heated to form a white gel. Ion solution Cd (II), Mn (II) and Co (II) each added to the gel TiO_2 with Ti:M mole ratio (M= Cd, Mn, Co) = 3:1 (mol/mol). After calcination process at a temperature of 400 °C, we have obtained TiO_2 and TiO_2-M (Cd, Mn, Co). Characterization of the samples with XRD showed TiO_2-M (M = Cd, Mn, Co) produces crystalline of anatase TiO_2 phase. SEM-EDX results showed differences in the amount of metal contained within TiO_2-M (M= Cd, Mn, Co) that is equal to 8.62% (w/w) Cd; 3.26% (w/w) Mn; 0.721% (w/w) Co. The addition of transition metals ion as impurities as Cd (II), Mn (II), or Co (II) have inhibited the growth of TiO_2 crystals. The insertion of metal ions in the TiO_2 were also affected the optical properties shown by the energy band gap (Eg) shift in order of Eg TiO_2 (3.35 eV)>TiO_2-Cd (3.10 eV)>TiO_2-Co (3.00 eV)>TiO_2-Mn (2.5 eV). The eletronic structures of TiO_2 doped 3d transition metals showed that an electron occupied level energy occurs. Electron from dopand are somewhat delocalized, thus significantly contributing to the formaton of the valece band with O p and Ti 3d electrons. Based on a comparation with the absorption data, we show that t_2g state of the dopand plays a significant role in the photoresponse of TiO_2 under visible irradiaton.

1. Introduction

Photocatalytic degradation for waste treatment has been widely practiced. Commonly degradation methods was used as photocatalyst, electrocatalyst and photoelectrocatalyst [1,2,3,4]. Wastewater treatment with filtration, adsorption and biological degradation has also been carried out, but the physical method usually still leaves waste elsewhere [5]. Adsorption process using zeolites and chitosan still produce waste of adsorbent-filled adsorbent [6] so that the effort to convert waste into harmless environmental compounds less optimal.

The photocatalytic process by the TiO_2 photocatalyst was performed by exposure a photon source that capable to excite their electrons for the initiation of oxidation and reduction pollutant processes. TiO_2 with a high gap value (Eg) (~3.2-3.4 eV) requires photon energy in the ultra violet region (λ<380 nm). Efforts to increase the uptake of photon energy above 380 nm, it can be performed on the TiO_2 semiconductor through insertion of transition metal ions into TiO_2 structure. The insertion of transition metal ions to the TiO_2 crystal can be carried out by means of hydrogenation, metal doping [7,8,9], and nonmetallic doping [10,11].

The insertion of Cd^{2+} metal ions into the TiO_2 structure can decrease the band gap energy [12] and increase photocatalytic activity in visible light [13]. The decreasing of TiO_2 band gap by the addition of Co have increased photocatalytic activity on visible light was also reported [14,15] and Mn metals [8,9,16,17]. The effect of the insertion of Fe^{3+} metal ions on TiO_2 was capable influence the electronic level caused by the reduction potential of Fe^{3+} lying below the reduction potential of the excited state TiO_2 (TiO_2 (e^-)). In this study, we carried out the insertion of metal ions Cd^{2+}, Co^{2+} and Mn^{2+} through a sol gel process. The growth of TiO_2 crystals have inhibited by metal ions dopand. The character of the structure and optical properties were studied to discent the effect of metal insertion of Cd^{2+}, Co^{2+} and Mn^{2+} on the crystal structure and the ability of TiO_2-doped semiconductors to absorb a visible light.

2. Experimental

2.1. Materials

The materials used in this study were titanium tetraisopropoxide (TTIP) (E Merck), $CdSO_4.8H_2O$, $MnSO_4.1H_2O$, $CoSO_4.7H_2O$ (E merck), glacial acetic acid, aquades, and 98% ethanol (E Merck). Calcination of TiO_2 and TiO_2-M (M = Cd, Co, Mn) was performed by Furnace Thermo. Meanwhile, the analysis and characterization of materials were done by XRD (Bruker D8 Advantage), SEM-EDX (SEM 250 Quanta), UV Vis Spectrophotometer Pelkin Elmer Lambda 25, and FTIR (Shimadzu Prestige 21).

2.2. Synthesis of TiO_2-M (Cd, Mn, Co)

The titanium (IV) isopropoxide solution was hydrolyzed using a glacial acetic acid solution (temperature of 14 °C) at a ratio of 1: 10 (v/v) and then stirred continuously using a magnetic stirrer until a white sol gel was obtained, then stirred and heated at 90°C until obtained white gel [18]. After the gel obtained, the metal ions of Cd^{II}, Co^{II} and Mn^{II} in the aquades solvent were added to the TiO_2 sol gel until a stoichiometric ratio of Ti:M (M= Cd, Mn, Co)= 3:1 (mol/mol). The mixtures were subsequently cooled to room temperature for a perfect gelation process. The resulting gel were heated at 150 °C for ± 24 hours. The obtained TiO_2 xerogels were calcined at 400 °C for 2 hours. The resulting TiO_2-M (M= Cd, Mn, Co) crystalline powder were then analyzed by XRD, UV Vis, FTIR spectrophotometer and SEM-EDX.

3. Results and Discussion

Synthesis of TiO_2 with the insertion of metal (Cd, Mn, Co) was carried out using the sol-gel method. The insertion of metals was expected to replace several positions of Ti atoms in TiO_2. The calcination process was carried out at 400 °C for 2 hours to obtain TiO_2 crystals with an anatase phase. X-ray diffractograms of TiO_2 and TiO_2-M are shown in Figure 1.

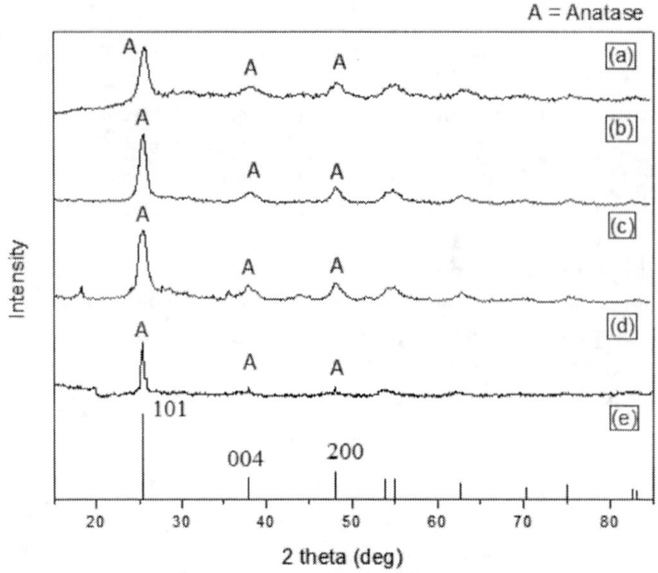

Figure 1. X-Ray Diffractogram of (a) TiO_2-Co (3:1) (b) TiO_2-Mn (3:1) (c) TiO_2-Cd (3:1) (d) TiO_2 (e) JCPDS TiO_2 No 78-2486

TiO_2-M synthesized using the sol-gel method shows the anatase phase crystals at 2θ= 25.49°; 2θ = 37.79° and 2θ = 48.04°. In Figure 1, A is a symbol of anatase phase formed in TiO_2-M and TiO_2. Based on the XRD spectra of TiO_2-M (3:1) (Figure 1), it can be seen that the main peaks are A (101), A (004), A (200) at 2θ= 25.49°; 2θ= 37.79 ° and 2θ= 48.04 ° which is the characterization of TiO_2. The XRD data is compatible with JCPDS No. 78-2486 after being processed with Jade MDI Program.
The insertion of metal ions Cd, Mn and Co shows the characteristics of TiO_2 Anatase. The diffractogram of TiO_2-Cd (3:1), TiO_2-Mn (3:1) and TiO_2-Co (3:1) only shows the peak of anatase at 2θ= 25.607° (d_{101}), 38.010° (d_{004}) and 48.190° (d_{200}). The insertion of Cd, Mn and Co in TiO_2 semiconductors inhibits the growth of TiO_2 crystals and causes a decrease in TiO_2-M crystal size (M = Cd, Co, Mn).

Table 1. Crystal sizes of TiO_2 and TiO_2-M (M = Cd, Co, Mn)

Material	Crystal Size (nm)
TiO_2	15.622
TiO_2-Cd	6.907
TiO_2-Co	7.223
TiO_2-Mn	7.404

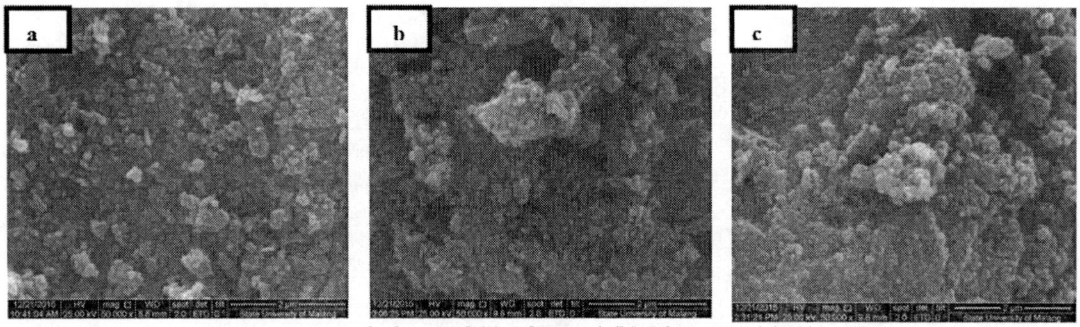

Figure 2. Morphology of (a) TiO_2-Cd (b) TiO_2-Mn (c) TiO_2-Co

SEM-EDX results showed that the percentage of M (M= Cd, Mn, Co) in TiO_2 is equal to 8.62%, 3.26% and 7.21%, respectively. The insertion of M (M= (Cd, Co, Mn) in TiO_2 crystallites affects the electronic properties showed by changes in band gap energy (Figure 3). The band gap energy from the calculation result is shown in Table 2.

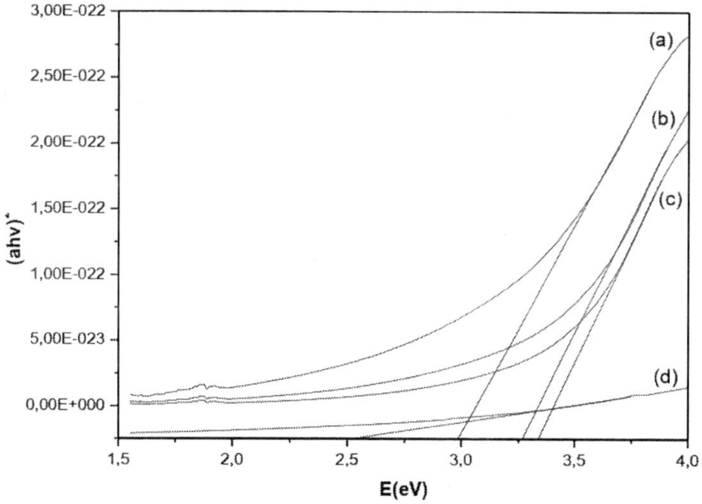

Figure 3. Tauc Plot Graph of (a) TiO_2-Co (b) TiO_2-Cd (c) TiO_2 (d) TiO_2-Mn

Table 2. Band gap energy and wavelengths of TiO_2-M (3:1)

Material	Band Gap (eV)	λ (nm)
TiO_2	3.35	370.15
TiO_2-Cd	3.15	393.65
TiO_2-Co	3.00	413.33
TiO_2-Mn	2.5	496.00

The insertion of metal to TiO_2 causes a decrease in band gap energy as shown in Table 2 with the band gap energy of TiO_2>TiO_2-Cd>TiO_2-Co>TiO_2-Mn. The wavelength is calculated using the equation λ_g= 1240/Eg. Visually, the insertion of the metal affects the band gap energy (Eg) result as in Figure 4.

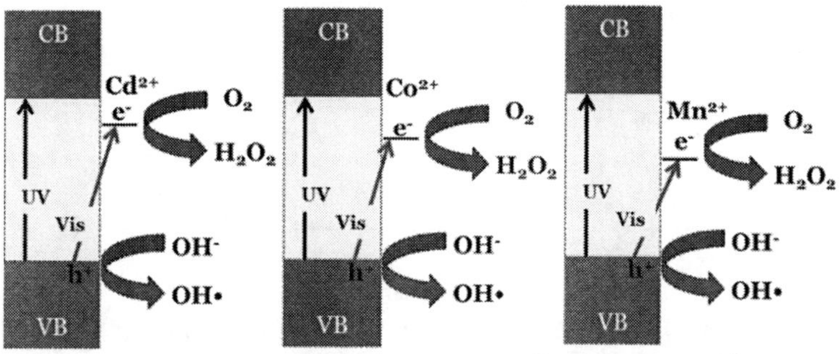

Figure 4. Scheme of insertion metal ions Cd^{2+}, Co^{2+}, and Mn^{2+} in TiO_2.

Infra red spectra of TiO_2 showed characteristic absorptions at 3405.47 cm^{-1}, 1627.99 cm^{-1}, and strong wide absorption band at 576.74-421.46 cm^{-1}. Absorption band at 3405.47 cm^{-1} showed the stretching vibration from H_2O in TiO_2 [19]. While the absorption band at 1625-1650 cm^{-1} showed the bending vibration of trapped water molecules in TiO_2 structure, and absorption band at 609.5-420.5 cm^{-1} represent O-Ti-O vibration [18]. The presence of Cd is showed at 1115-1057 cm^{-1} and 540-425 cm^{-1} which can be associated with the formation of Ti-O-Cd bonds. Two absorption bands at 1100 cm^{-1} and 980 cm^{-1} are the characteristic stretching vibration of the Ti-O-Cd in accordance with Ge *et al.* [20]. Both absorption bands show different intensities in the mole ratio of TiO_2 : Cd, where the absorption intensity looks sharp with the more addition of Cd. This observation is also occur at absorption band in the range 619.18–618.21 cm^{-1} as bending vibration of Cd.

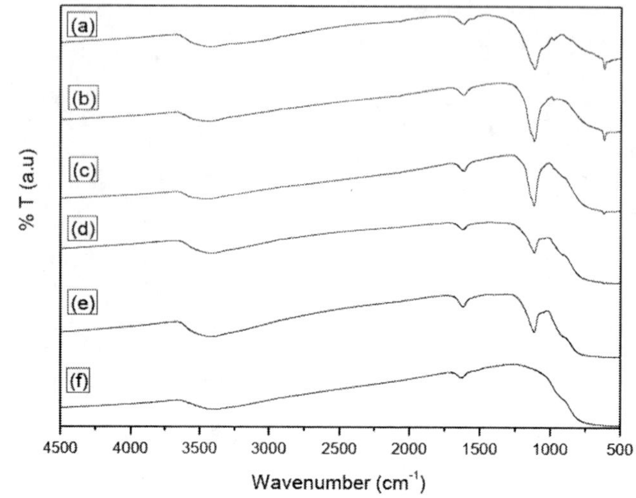

Figure 5. IR spectra of TiO_2-Cd, synthesized with mole ratio of TiO_2:Cd (a) 1:3 (b) 1:2 (c) 1:1 (d) 2:1 (e) 3:1, and (f) TiO_2

Similar characters are also showed in the absorption of TiO_2-Co (Figure 6). Weak wide absorption at 2840 cm^{-1} getting stronger with the increase of Co bonded, indicating absorption from OH on Co-O (bounded to H_2O). Absorption peak at 1000-1250 cm^{-1} is characteristic of Co, as evidenced by the increase absorption with the increasing of mole of Co [21].

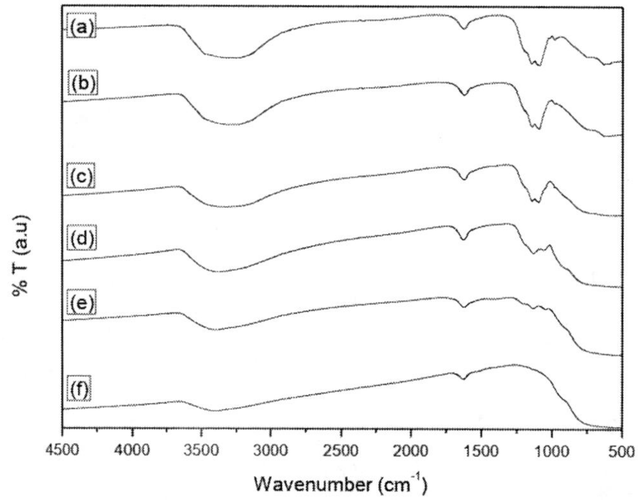

Figure 6. IR spectra of TiO_2-Co, synthesized with mole ratio of TiO_2:Co (a) 1:3 (b) 1:2 (c) 1:1 (d) 2:1 (e) 3:1, and (f) TiO_2

TiO_2-Mn spectra (Figure 7) show the characteristic absorption at 660 cm^{-1} which is a vibration of β-MnO_2. In addition, absorption band at 620 cm^{-1} and 530 cm^{-1} is indicated as vibrations of O-Ti-O and Mn-O-Ti. Absorption band at 1140–997 cm^{-1} is estimated to be characteristic of Mn-O because the intensity of the peak looks sharp within the addition of more Mn.

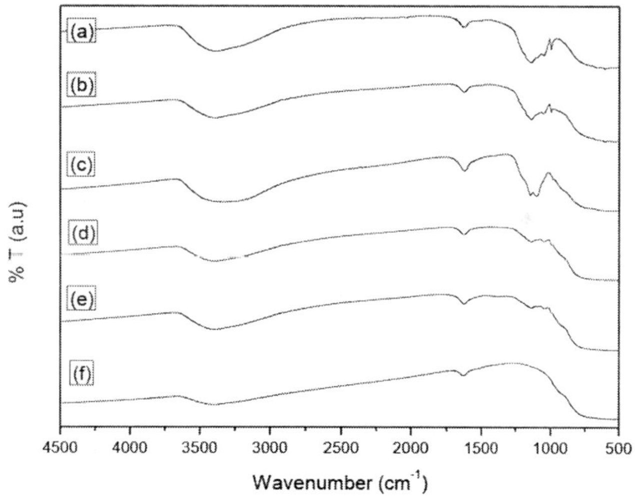

Figure 7. IR spectra of TiO_2-Mn, synthesized with mole ratio of TiO_2:Mn (a) 1:3 (b) 1:2 (c) 1:1 (d) 2:1 (e) 3:1, and (f) TiO_2

If we compare the FTIR spectra of the TiO_2-M (M = Cd, Co, Mn) (Figure 7), then the difference is seen in the M-O-Ti stretch vibration area at 1100-990 cm^{-1}. Absorption band in the area is characteristic of the insertion different dopants, and distinguish dopants added to TiO_2 crystal.

4. Conclusion

Synthesis of TiO_2 with the insertion of M (M= Cd, Co, Mn have been done using the sol-gel method. The insertion of Cd, Co, Mn decreases the crystallinity of TiO_2 as indicated by the decreasing in crystal size about ± 54%. The insertion of M (M=Cd, Co, Mn) in TiO_2 lattice were also affect the electronic properties of TiO_2 as a semiconductor with the band gap energy value (Eg) was decrease as Eg TiO_2> Eg TiO_2-Cd> Eg TiO_2-Co> Eg TiO_2-Mn.

References

[1] Rahmawati F, Wahyuningsih S and Irianti D 2014 *Bull. Chem. React. Eng. Catal.* **9** 45-52
[2] Wahyuningsih S, Purnawan C, Saraswati T E, Pramono E, Ramelan A H, Wisnugroho A and Pramono S 2014 *Journal of Enviromental Protection.* **5** 1630-1640
[3] Wijaya K, Sugiharto E, Fatimah I, Sudiono S, and Kurniaysih D 2006 *Teknoin.* **11** 199-209.
[4] Hamadanian M, A Reisi-Vanani and A Majedi 2010 *Journal of The Iranian Chemical Society.* **7**: S52-S58
[5] Mahardani, Nila Sari and Ferdyan Hijrah Kusuma 2010 *Journal Environment Chem.* **10** 6-17.
[6] Hwang A, Ming S 2009 *Journal of Hazardous Materials.* 1111-1121
[7] Chen X B, Y B Lou S Dayal, X F Qiu, R Krolicki, C Burda, C F Zhao and J Becker 2005. *Journal of Nanoscience and Nanotechnology.* **5** 1408–1420
[8] Binas V D, K Sambani, T Maggos, A Katsanaki, G Kiriakidis 2011 *Applied Catalysis B: Enviromental.* 113-114: 79-86
[9] Deng Q R, X H Xia, M L Guo and G Shao 2011 *Material Letters.* **65** 2051-2054
[10] Chen, Xiaobo and Burda, Ciemens 2008 *J. Am. Chem. Soc.* **130** 5018-5019
[11] Gole J L and Stout J D 2004 *Journal Physical Chemistry.* **108** 1230-1240
[12] Li Xin, Ting Xia, Changhui Xu, James Murowchick and Xiaobo Chen 2014 *Catalysis Today* **225** 64-73

[13] Shi J, Xiaoxia Y, Cui H, Xu Z, Fu M, Chen S and Wang L 2012 *Journal of Molucular Catalysis a: Chemical.* **356** 53-60
[14] Miao Y, Zhongbiao Z, Liang J, Yangmei S, Zhiying Y, Deliang D, Kai Z and Jiaqiang W 2014. *Powder Technology*. **266** 365-371
[15] Yang X, Cao C, Hohn K, Erickson L, Maghrang R, Hamal D and Klabunde K 2007 *Journal Catalys*. **252** 296-302
[16] Papadimitriou V S, Vassileios G S, Manolis N R, Panos P, Kyriaki S, Valentin T and George K. 2011 *Thin Solid Film.* **520** 1195-1201
[17] Wang L, Xi Zang, Peng Zang, Zetan Cao and Junhua Hu 2015 *Journal of Saudi Chemical Society* **19** 595-601
[18] Wahyuningsih S, C Purnawan, T E Saraswati, P A Kartikasari and N Praistia 2014 *Chemical Papers*. **68** 1248-1256
[19] Kuvarega A T, R W M Krause, B B Mamba 2011 *Journal of Physical Chemistry C.* **115** 22110-22120
[20] Ge L, F Zuo, J Liu, Q Ma, C Wang, D Sun, L Bartels, P Feng 2012 *Journal of Physical Chemistry C.* **116** 13708-13714
[21] Ganesh Ibram, A K Gupta, P P Kumar, P S Chandra Sekhar, K Radha, G Padmanabham and G. Sundararajan 2012 *Material Chemistry and Physics*. **135** 220-234

IOP Conf. Series: Materials Science and Engineering **578** (2019) 012016 doi:10.1088/1757-899X/578/1/012016

Textile Wastewater Treatment by Using Plasma Corona Discharge in a Continuous Flow System

M Amrial[1], K Kusumandari[1,*], T E Saraswati[2,*] and Y H Suselo[3]

[1] Department of Physics, Faculty of Mathematics and Natural Sciences, Universitas Sebelas Maret, Jl. Ir. Sutami 36A Kentingan Surakarta Indonesia 57126
[2] Department of Chemistry, Faculty of Mathematics and Natural Sciences, Universitas Sebelas Maret, Jl. Ir. Sutami 36A Kentingan Surakarta Indonesia 57126
[3] Department of Human Physiology and Biomedical Department, Faculty of Medicine, Universitas Sebelas Maret, Jl. Ir. Sutami 36A Kentingan Surakarta Indonesia 57126

*E-mail: kusumandari@staff.uns.ac.id ; teguh@mipa.uns.ac.id

Abstract. This study aimed to figure out plasma corona discharge for textile wastewater degradation with continuous flow system. Plasma reactor using AC voltage of 11.5 kV and two stainless steel electrode which set up parallelly 4.0±0.5 mm above wastewater. Continuous flow system means the wastewater flowing continuously with flow rate of 60 ml/min. Characteristics of absorbance, pH, temperature, Total Dissolved Solid (TDS), Total Suspended Solid (TSS) and conductivity were measured before and after plasma treatment. Absorbance, pH and TSS value of wastewater were decrease after plasma treatment,. While the TDS, conductivity and the temperature were relatively increase. This result indicate better quality of wastewater.

1. Introduction

Textile industry is one of contributors to high concentrated liquid pollutants. The industry produces waste which consists of complex molecules types of organic compounds which make waste colour, toxic and resistant to conventional treatment [1]. Conventional textile waste processing can be done by physics, chemistry and biology methods. This processing method is effective enough to degrade colours from waste, but there are some disadvantages of these methods. These methods require a lot of costs, chemicals and create secondary wastes like lots of mud [2].

Recently, plasma applications based on oxidation methods for waste degradation begin to be increased because of their flexibility and high oxidative capacity [3]. Wastewater treatment using plasma technology is cheaper and friendlier for environment than biological method because it does not produce secondary waste [3]. Plasma can oxidize waste in water, especially organic compounds. The reaction between ions and electrons in the plasma phase in waste produces UV and shock wave rays. The reaction produces very high energy and break down the water molecule (H_2O) to active species such as OH, O, H, and H_2O_2. The active species which are strong oxidants that can oxidize various kinds of compounds and can also kill bacteria from liquid waste [4].

In the previous study optimization of plasma discharge reactors was carried out with a continuous flow system for methylene blue degradation [5]. Furthermore, this research was carried out by processing textile waste by using plasma corona discharge in a continuous flow system.

2. Experimental

A sample was textile batik printing wastewater with amount of 1 litre. A pair of stainless steel was used as electrodes that each has a diameter of 2 mm with 2.000±0.005 cm apart. It places paralelly 4±0.5 mm above the sample. An AC voltage of 11.5 kV was used to generate plasma. Continuous flow system mean the wastewater was flowing during experiment. The flow rate of sample was 60 ml/min. The optimization of the reactor was evaluated from degradation of watewater with variation of treatment time with plasma. It was seen from absorbance, pH and temperature, total dissolved solid (TDS), total suspended soid (TSS) and conductivity before and after treatment using the plasma. The absorbance measured with UV-Vis spectrophotometer, while pH and sample temperature were measured using multi-parameter water quality meter. While TDS, TSS and conductivity were measured using TDS and conductivity meter with standard method of prepration.

3. Results and Discussion

Concentration of pollutan in the wastewaster is proportional with absorbance. It means the absorbance will high if it contains more pollutan and vice versa. Therefore, measurement the absorbance is one of easy method to point out the concentration of pollutan in the wastewater. Figure 1 shows the comparison between wastewater absorbance without and with plasma treatment for various time. Before plasma treatrment (0 minute), highest peak of absorbance is at wavelength of 450 nm. So, the next discussion will be focused at this wavelength. For 60 and 100 minutes of treatment, the absorbance sligthly decresase. And then, treatment for 180 minutes result in the more decrease of absorbance.

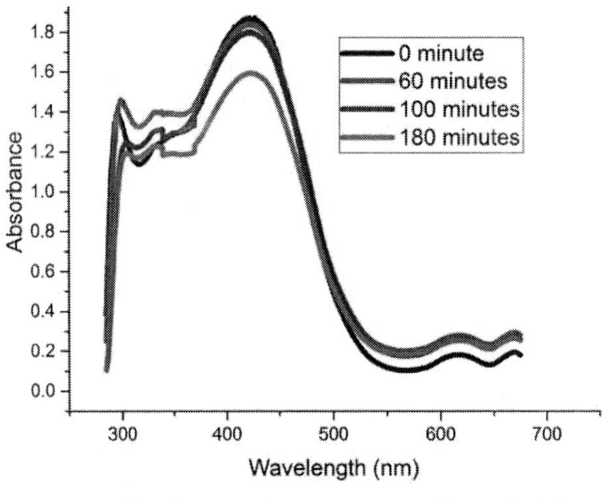

Figure 1. Absorbance of wastewater without (0 minute) and with (60,100, 180 minutes) plasma treatment.

For longer of treatment time, the more electrons are produced. Then electron ionizes the air and/or reacts with the wastewater. This simultan reaction resulting species active such as radicals and ions which will degrade the wastewater. Degradation of wastewater means the concentration of pollutans are decrease or in this paper shown by decrease of absorbance. So, longer time of treatment will produce more species active and more degradation will occure. Then, Table 1 shows percentage degradation of wastewater for various treatment times. However, value of degradation in this continuous flow system was smaller compare to the batch system for almost same time. It reasonable because in the continuous flow system, retention time of wastewater with plasma is shorter and it depends on the flow rate [5].

Table 1. Percentage degradation of wastewater with various treatment times.

Time of *treatment* (minute)	Percentage (%)
0	0
60	1.66
100	3.90
180	14.87

Then, physical characteristic of wastewater such as pH, temperature, Total Dissolved Solid (TDS), Total Suspended Solid (TSS) and conductivity were measured before and after plasma treatment for 2,3 and 4 hours, as shown in Figure 2.

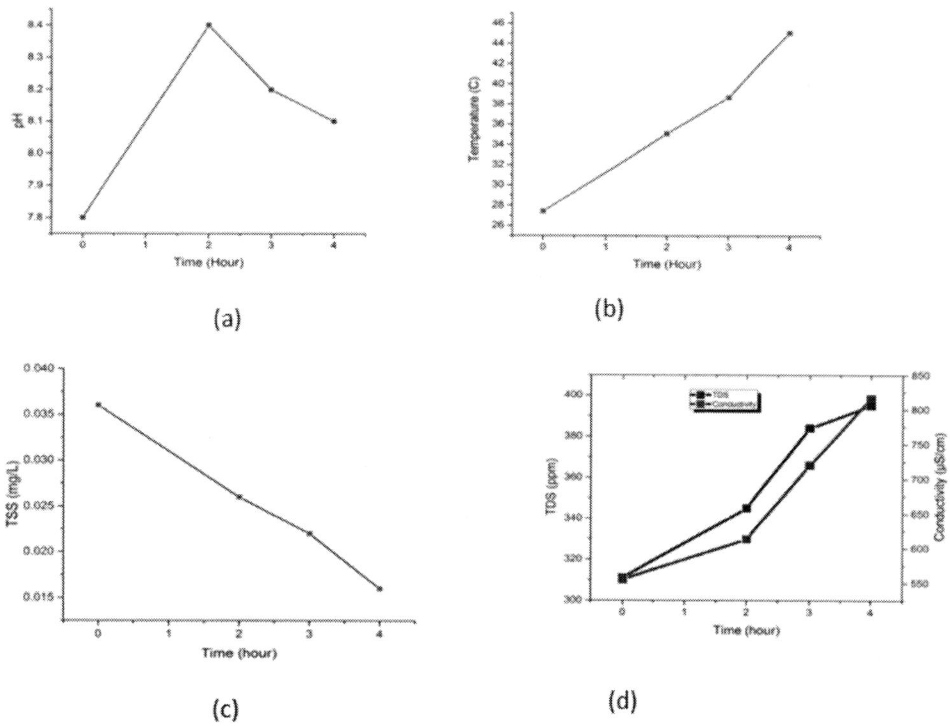

Figure 2. Physical characteristic of wastewater including (a) pH, (b) temperature, (c) TSS and (d) TDS and conductivity.

Figure 2(a) shows pH of wastewater was increase after 2 hours of plasma treatment. However, the pH was decreased for 3 and 4 hours of treatment. Carbonates dan bicarbonates were contained in Batik wastewater as glue the dyes to textile. Kommineni *et al.* reported that reaction between active species hydroxyl and carbonates will resulting OH^- as shown in Eq. 1 and 2 [6]. Thus it will makes the wastewater become alkaline and reduce the pH.

$$OH^{\bullet} + CO_3^{2-} \rightarrow CO3^{\bullet -} + OH^- \quad (1)$$

$$OH^{\bullet} + HCO_3^{2-} \rightarrow HCO3^{\bullet -} + OH^- \quad (2)$$

In Figure 2(b), temperature of wastewater was increase proportionally with treatment time. In the plasma, species actives have collides each other many times and resulting heat. As consequently, the temperature will increase for longer time of treatment. The high temperature will accelerate the formation of free radicals which play a role to degrade dyes [7].

Total suspended solid (TSS) indicate solids in the wastewater that can be trapped by a filter [8-10]. In Figure 2(c) it can be seen that the TSS value is relatively reduced after plasma treatment. With longer treatment time, the more number of active species were produced. This active species plays an important role to degrade organic compounds in waste. Therefore, the more active species were formed, the more effective to remove the organic and inorganic substances in wastewater. Total disolved solid (TDS) are a measure of the combined content of all inorganic and organic substances contained in a liquid in molecular, ionized or micro-granular (colloidal sol) suspended form [8-10]. TDS values increase with increasing treatment time because plasma produces new ionic species. This ionic compound will oxidize the textile waste. Furthermore, another indicator which have a close relationship with TDS is electrical conductivity. It will increase proportionally with TDS value as shown in Figure 2(d).

4. Conclusion

Based on the result, it can be concluded that the plasma corona discharge can be used as promising method to reduce concentration of batik wastewater. The percentage of wastewater degradation increase proportionally with treatment time. Continuous flow system can be used as simple application procedure in the wastewater treatment, even with longer treatment time compare to batch system. Moreover, from measurement of physical characteristics, the quality of wastewater become better after plasma treament.

Acknowledgments

Authors wish to thank PNBP Universitas Sebelas Maret No. 543/UN27.21/PP/2018 for financial support through Penelitian Unggulan UNS 2018.

References

[1] Nadeem K, Tezcanli-Guyer G and Dizge N 2017 *J. Water Process Eng.* **20** 9-39
[2] Lin C Y, Nguyen M L T and Lay C H 2017 *J. Cleaner Prod.* **168** 331-337
[3] Zhang J, Zheng-zheng and Yinni Z 2007 Low-Temperature Plasma- Induced Degradation of Aqueous 2,4 dinitrophenol *School of The Enviroment* (Nanjing University)
[4] Jiang B, Zheng J, Qiu S, Wu M, Zhang Q, Yan Z and Xue Q 2014 *Chem. Eng. J.* **236** 348-368
[5] Anggraini R A N, Kusumandari and Saraswati T E 2018 *AIP Conf. Proc.* **2014** 020037
[6] Kommineni S, Zoeckler J, Stocking A, Liang S, Flores A and Kavanaugh M 2007 *Advance Oxidation Process* Nwri (United State of America)
[7] Raheem R A, Al-gubury H Y, Aljeboree A M, Alkaim A F 2016 *J. Chem. Pharm. Sci.* **9** 1134-1138
[8] Rinawati, Hidayat D, Suprianto R and Dewi, S P 2016 *Analit: Analytical and Env. Chem.* **1** 36-45.
[9] Bisschops and Spanjers H *Environmental Technology* **24** 1399-1411
[10] Lin S H and Peng C F 1994 *Water Res.* **28** 277-283

IOP Conf. Series: Materials Science and Engineering **578** (2019) 012017 doi:10.1088/1757-899X/578/1/012017

Synthesis and characterization of vanadium doped $Bi_4Ti_3O_{12}$ as photocatalyst material

R Handayani[1], W N Safitri[1], N Aini[1], A Hardian[2], A Prasetyo[1,*]

[1]Department of Chemistry, Faculty of Science and Technology, Universitas Islam Negeri Maulana Malik Ibrahim Malang, Jl. Gajayana 50, Malang, Indonesia, 65144.
[2]Department of Chemistry, Universitas Jenderal Ahmad Yani, Cimahi, Indonesia.

E-mail: antoniaprasetyo@gmail.com

Abstract. One of interesting properties of $Bi_4Ti_3O_{12}$ is photocatalyst. The band gap of $Bi_4Ti_3O_{12}$ is 2.9 eV as results only work at ultraviolet wavelength range. Metal doped to $Bi_4Ti_3O_{12}$ can reduce band gap energy. In this work, $Bi_4Ti_3O_{12}$ was doped vanadium ($Bi_4Ti_{3-x}V_xO_{12}$ with x= 0.05; 0.1; 0.15; 0.2) using solid state reaction method. X-ray diffraction data show that $Bi_4Ti_{3-x}V_xO_{12}$ (x= 0.05; 0.1; 0.15; 0.2) were successfully formed with space group B2cb, however there is small impurities of VO_2 phase. Scanning electron microscopy (SEM) images showed that the particle of all obtained $Bi_4Ti_{3-x}VxO_{12}$ (x= 0.05; 0.1; 0.15; 0.2) have shape platelet-like. Diffuse Reflectance Spectroscopy data showed that the energy band gap of $Bi_4Ti_{3-x}V_xO_{12}$ decreased by increasing vanadium dopant.

1. Introduction

Aurivillius compound has widely studied due to the interesting properties of ferroelectric [1]. Aurivillius compound have general formula: $A_{n-1}Bi_2B_nO_{3n+3}$ (n= 1, 2, 3, 4,...) with cations A = Ca^{2+}, Sr^{2+}, Ba^{2+}, Pb^{2+}, Bi^{+3}, Na^{+1}, and cations B = Ti^{4+}, Nb^{5+}, Ta^{5+}, W^{6+} or Mo^{6+}. n is an integer that representing the amount of BO_6 layers perovskite blocks [2,3]. Recently, many researchers explored to properties of aurivillius compound besides ferroelectric such as photocatalyst, luminescence, and thermoelectric properties [4,5]. As well known that photocatalyst technology have potentially can be applied in many fields such as (a) waste treatment technology especially for organic dye waste, (b) to produce hydrogen gas by water splitting process technology. The photocatalyst properties of Aurivillius compound was reported by many researchers such as $Bi_4V_2O_{11}$, Bi_2WO_6, $Bi_4Ti_3O_{12}$ [6-8].

$Bi_4Ti_3O_{12}$ (BIT) is well known as three layered Aurivillius compound family and reported potentially applied as photocatalyst material. The band gap energy of BIT 3.1 eV which indicates that the work function of $Bi_4Ti_3O_{12}$ at ultra violet (UV) light range [9]. To applied it in tropical country like Indonesia area, this fact is not good, because mostly sun light in visible light range therefore need the technique as purpose to reduce the band gap energy. Metal doped to photocatalyst material is technique was used to reduce band gap energy. There are many metal can be used as dopant material photocatalyst such as, V, Mn, Fe, Cr. Chen, et al (2016) synthesized chromium doped BIT and reported the enhancing of photocatalytic activity as well as the photoabsorption in visible light range [8]. Meanwhile, Dumrongrojthanath, et al (2017) reported that Br doped $BiWO_6$ have higher photocatalytic activity and can work in visible light range [10]. It indicates that the metal doped to Aurivillius compound can solve the high band gap energy problem. Previous researches reported that vanadium metal can be used to improve the photocatalytic activity and optical properties of TiO_2 [11]. Hence, in this research, we synthesized vanadium doped the BIT to improve the optical properties using solid state reaction.

2. Experimental

Bismuth Oxide (Bi_2O_3), Titanium (IV) Oxide (TiO_2), Vanadium (V) Oxide (V_2O_5) were grinded for 1 hour in acetone and was calcined at 700 °C for 24 hours. The sample was put from furnace and then regrind for 1 hour and was calcined at 800 °C for 24 hours. The similar method have conducted for calcined at 850, 900, 900 and 1000 °C. Sample product was characterized using (a) powder X-ray diffraction (XRD) to identity the phase of sample, (b) scanning electron microscopy-energy dispersive spectroscopy (SEM-EDS) to identify the morphology as well as element of sample, and (c) diffuse spectroscopy reflectance to identify the band gap energy of sample.

3. Results and Discussion

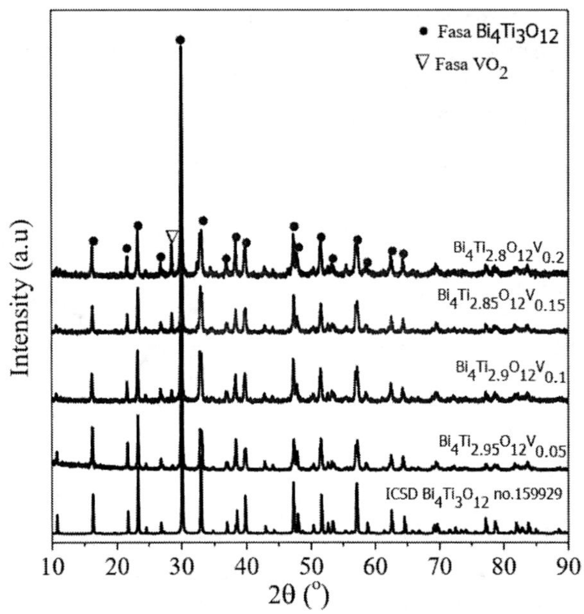

Figure 1. XRD data of $Bi_4Ti_{3-x}V_xO_{12}$ (x= 0.05; 0.1; 0.15; 0.2)

Figure 1 showed XRD pattern of $Bi_4Ti_{3-x}V_xO_{12}$ (x= 0.05; 0.1; 0.15; 0.2). All peaks of XRD indexed to International Crystallography Standard Data (ICSD) no 159929 with space group *B2cb* and as can be seen that compound $Bi_4Ti_{3-x}V_xO_{12}$ (x= 0.05; 0.1; 0.15; 0.2) was successfully synthesized but still found the impurities compound VO_2. The intensity of VO_2 peaks increases and it correspond to increases in %weight of vanadium. Figure 2 depicted the peak position of XRD data at around 2θ (°): 30 and it can been seen that the peak position is different which indicates the crystallite size of samples is different.

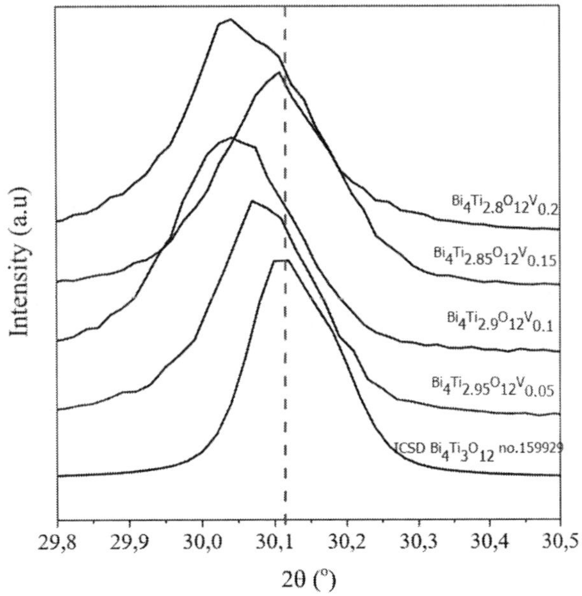

Figure 2. Peak position of XRD data of $Bi_4Ti_{3-x}V_xO_{12}$ (x= 0.05; 0.1; 0.15; 0.2) at around 2θ: 30°

Figure 3. Image of $Bi_4Ti_{3-x}V_xO_{12}$ (x= 0.05; 0.1; 0.15; 0.2) particle: (a) x= 0.05, (b) x= 0.05, (c) x= 0.15, and (d) x= 0.2

Table 1. The results of EDS analysis

Compound	Weight (%)			
	Bi	Ti	V	O
$Bi_4Ti_{2.95}V_{0.05}O_{12}$	66.45	9.44	0,48	23.62
$Bi_4Ti_{2.9}V_{0.1}O_{12}$	66.74	9.42	1.27	22.56
$Bi_4Ti_{2.85}V_{0.15}O_{12}$	63.42	9.55	1.12	25.91
$Bi_4Ti_{2.8}V_{0.2}O_{12}$	65.62	9.59	0.56	24.23

As can be seen at figure 3, the particle shape of sample is platelet-like and it similar to previous report by other researchers about the particle shape of BIT [12]. The distribution size of particle is not uniform as caused the agglomeration and it will influence to photocatalytic activity because the shape as well as distribution size of particle plays important role in photocatalytic activity [13,14]. The result of elemental analysis was tabulated at Table 1 and shown that vanadium was succesfully introduced to sample. The weight of vanadium which obtained from EDS analysis is different to theoretical calculation and it indicates that the sample is not homogen especially for the impurities VO_2.

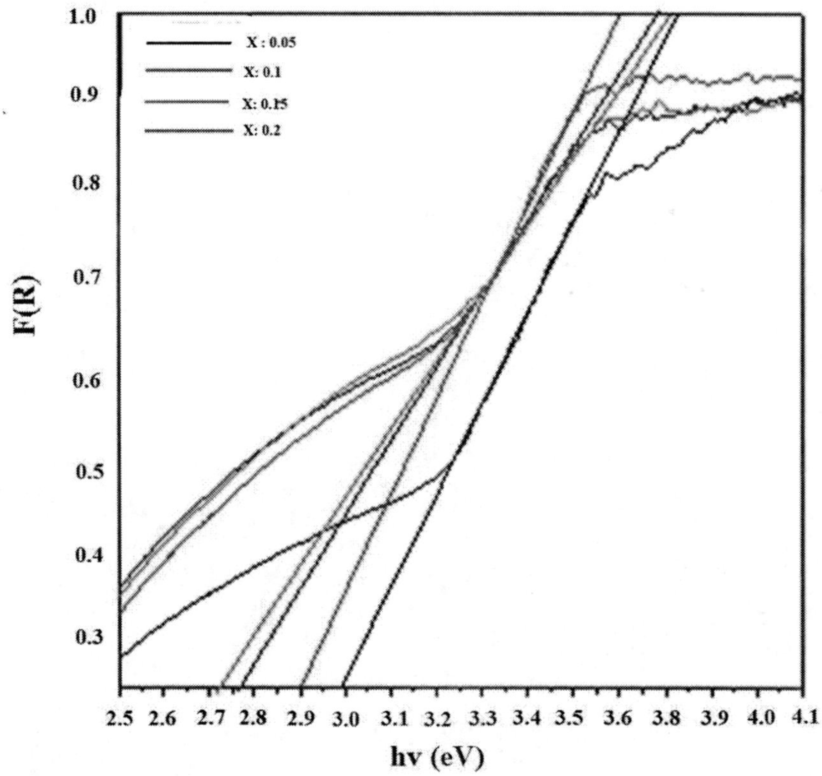

Figure 4. Diffuse reflectance spectroscopy of $Bi_4Ti_{3-x}V_xO_{12}$ (x= 0.05; 0.1; 0.15; 0.2)

Tabel 2. The band gap energy of Vanadium doped BIT

Compound	Band gap energy (eV)
$Bi_4Ti_{2.95}V_{0.05}O_{12}$	3.0
$Bi_4Ti_{2.9}V_{0.1}O_{12}$	2.9
$Bi_4Ti_{2.85}V_{0.15}O_{12}$	2.75
$Bi_4Ti_{2.8}V_{0.2}O_{12}$	2.71

Figure showed the DRS spectra of $Bi_4Ti_{3-x}V_xO_{12}$ (x= 0.05; 0.1; 0.15; 0.2) and was analyzed using Kubelka-Munk equation to calculate the band gap energy. The value of band gap energy $Bi_4Ti_{3-x}V_xO_{12}$ (x= 0.05; 0.1; 0.15; 0.2) as tabulated at Table 2. As can be seen that the band gap energy decreases and as consequence the work function of sample shifted to visible light range (at around 430 nm). The decreases of band gap energy was caused the new state between conduction band and valence band as result the band gap was formed is lower. As result, it give a advantage to BIT as photocatalytic material.

4. Conclusion

$Bi_4Ti_{3-x}V_xO_{12}$ (x= 0.05; 0.1; 0.15; 0.2) was synthesized using solid state reaction and found the impurities VO_2. The work function of $Bi_4Ti_{3-x}V_xO_{12}$ (x= 0.05; 0.1; 0.15; 0.2) is lower than $Bi_4Ti_3O_{12}$ as results the sampel product can work in visible light range.

Acknowledgements
The corresponding author thank's to Faculty Science and Technology, Universitas Islam Negeri Maulana Malik Ibrahim Malang for financial support by scheme Bantuan Publikasi Ilmiah Tahun 2018.

References

[1] Smolenskii G A, Isupov V A, Agranovskaya, A I 1961 *Fiz. Tverd. Tela (Leningrad)* **3** 899-901
[2] Aurivillius, B, 1949 *Arkiv Kemi Band I* **54** 463-480
[3] Aurivillius, B, 1949 *Arkiv Kemi Band I* 1949 **58** 499-512
[4] Liu X, Xu L, Huang Y, Qin C, Qin L, and Seo H J 2017 *Ceram. Int.* **43** 12372–12380
[5] Yu L, Hao J, Xua Z Li W, Chu R, and Li G 2016 *Ceram. Int.* 42 14849–14854
[6] Lu Y, Pu Y, Wang J, Qin C, Chen C, and Seo H J 2015, *Appl. Surf. Sci.* **347** 719-726
[7] Phuruangrat A, Dumrongrojthanath P, Thongtem S, and Thongtem T, 2018 *Mater. Lett.* **224** 67-70
[8] Chen Z, Jiang X, Zhu C, and Shi C 2016 *Appl. Catal. B* **199** 241-251
[9] Singh, J D, Seo, S S A, and Lee, H D 2010, *Phys.Rev.B* **82** 180103
[10] Dumrongrojthanath P, Phuruangrat A, Thongtem S, and Thongtem T 2017 *Mater. Lett.* **209** 501-504
[11] Wang T, and Xu T 2017 *Ceram. Int.* **43** 558-1564
[12] Zhao Z, Li X, Ji H, and Deng, M 2014. *Integr. Ferroelectr.* **154** 1 154-158
[13] Zhao W, Jia Z, Lei E, Wang, L, Li Z, and Dai, Y 2013 *J Phys.Chem. Solids.* **74** 111604-1607
[14] Faradis R, Azizah E N, Marella S D, Aini N, and Prasetyo A 2018 *IOP Conf. Ser.:Mater. Sci. Eng.* **333** 012019

IOP Conf. Series: Materials Science and Engineering **578** (2019) 012018 doi:10.1088/1757-899X/578/1/012018

Synthesis Of Imprinted Ionic Material ($SiO_{2(BGS/RHA)}$-TMPDT-Im-Au) For Adsorption Of Ion Au(III)

S Hastuti[1], S Wahyuningsih[1], T Martini[1] and I N Hayati[1]

[1] Department of Chemistry, Faculty of Mathematics and Natural Sciences, Sebelas Maret University, Surakarta, 57126, Indonesia

Email : hastuti.uns@gmail.com

Abstract. Synthesis of imprinted ionic material $SiO_{2(BGS/RHA)}$-TMPDT-Im-Au) was successfully studied. Sodium silicate from a mixture of ash bagasse and rice husk ($Na_2SiO_{3(BGS/RHA)}$) were used as the precursors. Furthermore, $Na_2SiO_{3(BGS/RHA}$ is modified with the N^1-(3-Trimethoxysilylpropyl)diethylenetriamine (TMPDT) in which Au(III) ions are used as templates. The synthesis results were characterized by Fourier-transform infrared spectroscopy (FTIR), Surface area analyzers (SAA), Thermo Gravimetric analysis (TGA), Scanning Electron Microscope with Energy Dispersive X-ray Spectroscopy (SEM-EDX). The adsorption study was carried out using a batch system conducted in various pH and contact time. The adsorption process achieved a maximum pH of 3 in a 60-minute equilibrium time. Au(III) adsorption capacity of $SiO_{2(BGS/RHA)}$-TMPDT-Im-Au was 10.44 mg/g. The kinetic studies show that the adsorption of Au(III) on $SiO_{2(BGS/RHA)}$-TMPDT-Im-Au follows a pseudo-second-order with the reaction rate constant (k) of 76.7331 g mol^{-1} min^{-1}.

1. Introduction

Gold is widely used in various fields, such as jewellery, catalysts and electronic fields. In the electronics field, gold is used for components such as ICs, contacts and cables. Mobile phones contain gold around 300-350 g/ton [1], while the ore is only around 5-30 g/ton [2]. Electronic waste will increase every year. To overcome this waste is to recover gold metal from waste. This also helps in providing gold raw materials.

Adsorption is one of the methods developed for gold extraction [1][3-6]. The adsorption method is the most suitable for extracting gold (III) metal ions at low concentrations. This is because of the low cost and high efficiency. Silica gel is a material developed for adsorbents. This material has several advantages such as good mechanical properties, high porosity, large surface area, no swelling, resistance to microbial attack and thermal stability [7-11]. One way to increase the ability of silica as an adsorbent is by surface modification using organic functional groups [11-14]. Imprinted ionic is a better method, because it combines modifications as well as the formation of templates that match the target metal ions [15-16]. Ionic imprinted material will increase capacity and selectivity of adsorption [16].

2. Experimental

2.1. Materials

The materials used are rice husk ash (RHA) and bagasse ash (BGS). In addition, chemicals are also used such as sodium hydroxide (NaOH), hydrochloric acid (HCl), HNO_3, thiourea from Merck. As a modifier is N1-(3-Trimethoxysilylpropyl)diethylenetriamine (TMPDT) from Aldrich. Whereas as a template is the solution of Au(III) 1000 ppm from analytical laboratory of UGM.

2.2. Physical measurements

Identification of functional groups was analyzed using FTIR (Fourier Transform Infra-Red Spectrophotometry) Shimadzu IR prestige 21. SEM-EDX (Scanning Electron Microscopy with

Energy Dispersive X-ray Spectroscopy) was used to analyze the surface morphology and elemental composition of a material. Thermal stability of a material was analyzed using TGA (Thermo Gravimetric Analyzer) STA-PT-1600 TG-DSC / DTA linseis thermal analyzer. Porosity of material was analyzed using SAA (Surface Area Analyzer) Quantachrome Nova Analyzer 1100. AAS (Atomic Absorption Spectrophotometry) Shimadzu AA 6650 was used to determine the concentration of metal ions.

2.3. Synthesis of imprinted ionic Au(III) material

Synthesis of sodium silicate from rice husk ash $Na_2SiO_{3(RHA)}$ and bagasse ash $Na_2SiO_{3(BGS)}$ refers to research [16] with several changes. In the synthesis of imprinted ionic material ($SiO_{2(BGS\text{-}RHA)}$-TMPDT-Au-Imp), 100 ml of a mixture of $Na_2SiO_{3(BGS)}/Na_2SiO_{3(RHA)}$ with a volume ratio of 3: 7 was added with 6 M HCl to almost forming a gel. After that the mixture was added with a mixture of 10 mL TMPDT and 10 mL Au(III) 1000 mgL^{-1} which had been stirred for 1 hour, then added 6 M HCl to form a gel. The gel that formed is left for 24 hours, then dried at 70 ° C until dry. Results in the form of dry solids of $SiO_{2(BGS\text{-}RHA)}$-TMPDT-Au was mashed up to 200 mesh size. The Au(III) in $SiO_{2(BGS\text{-}RHA)}$-TMPDT-Au was released with 7% thiourea in 1 M HCl. The result is imprinted ionic material ($SiO_{2(BGS\text{-}RHA)}$-TMPDT-Au-Imp). As a control, it was also synthesized of material $SiO_{2(BGS\text{-}RHA)}$-TMPDT by adding a mixture ($Na_2SiO_{3(BGS)}/Na_2SiO_{3(RHA)}$ =3:7) with TMPDT.

2.4. Determination of the optimum pH of adsorption

A 10 mL solution of Au(III) 15 mgL^{-1} in a variety of pH 1- 6 was contacted at 10 mg $SiO_{2(BGS\text{-}RHA)}$-TMPDT-Au-Imp for 1 hour. The mixture was stirred for 1 hour then filtered. Au(III) in filtrate was analyzed by AAS. Eq. 1 was used to calculate the percentage of Au(III) adsorbed.

$$P = 100 \times (C_0\text{-}C_e)/C_0 \qquad (1)$$

where P represents the amount of the metal ions adsorbed (%); C_0 and C_e represent the initial and the final concentration of the metal ions (mgL^{-1}).

2.5. The adsorption contact time

As much as 10 mL of Au(III) 15 mgL^{-1} at pH 3 was contacted with 10 mg $SiO_{2(BGS\text{-}RHA)}$-TMPDT-Au-Imp at a time variation of 0, 5, 10, 20, 30, 40, 50, 60, 90, 120 minutes. The mixture was stirred for 1 hour then filtered. The same treatment was carried out on $SiO_{2(BGS\text{-}RHA)}$-TMPDT.

3. Result and Discussion

3.1. Synthesis material imprinted ionic Au(III) ($SiO_{2(BGS\text{-}RHA)}$-TMPDT-Au-Imp)

The mechanism for the formation of imprinted ionic Au(III) material ($SiO_{2(BGS\text{-}RHA)}$-TMPDT-Au-Imp) can be explained in Figure 1. Reaction I (Figure 1.) shows the formation of orthosilicic acid (compound 1). Reaction II (Figure 1.) is the binding of $AuCl_4^-$ by TMPDT (compound 2). The release of the methoxy group from TMPDT is shown in reaction III (Figure 1). Then compound 1 reacts with compound 2 to form polymerization as shown in reaction IV (Figure 1). The reaction of V (Figure 1) is the release of $AuCl_4^-$ ions to produce $SiO_{2(BGS\text{-}RHA)}$-TMPDT-Au-Imp (compound 5).

In the release process $AuCl_4^-$, It was used thiourea 7% in 1M HCl solution. The release $AuCl_4^-$ was carried out 3 times then washed with distilled water (Figure 2). Thiourea is used for desorption of Au(III) because thiourea will reduce to Au(I) and then Au(I) forms a stable complex [16]:

$$Au^+(aq) + 2\ SC(NH_2)_2(aq) \longleftrightarrow Au[SC(NH_2)_2]_2^+(aq)$$

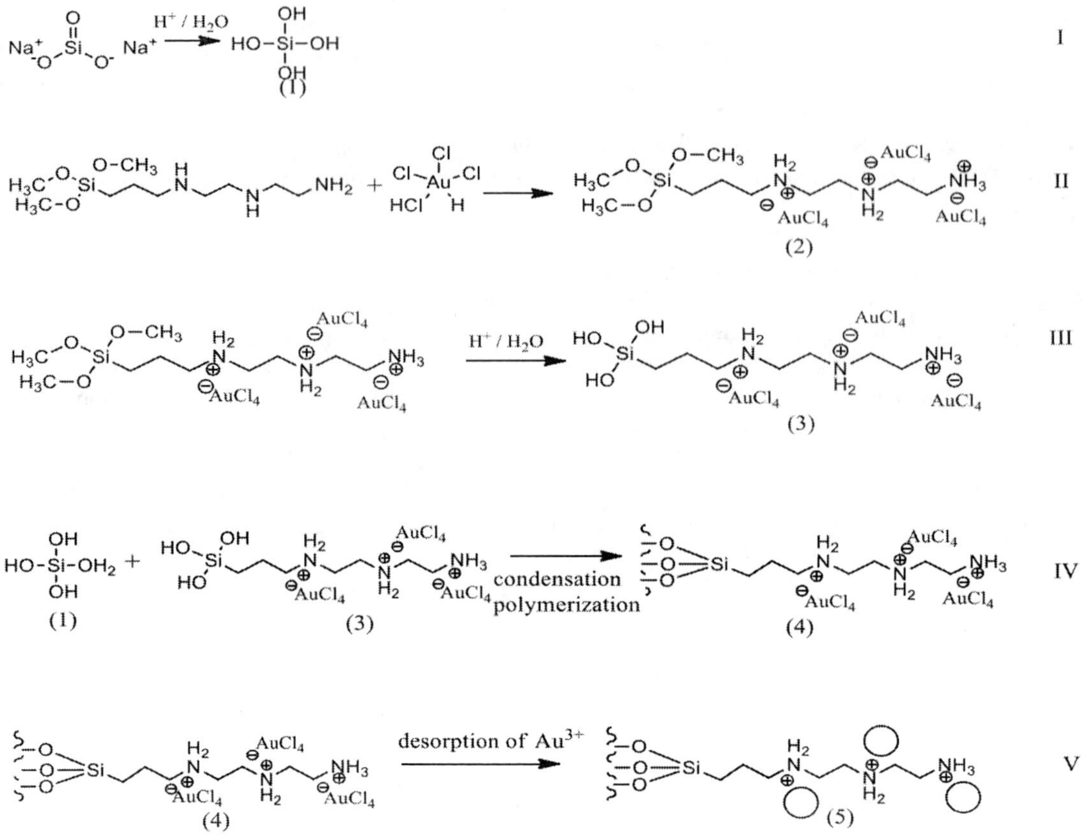

Figure 1. Mechanism of formation of $SiO_{2(BGS\text{-}RHA)}$-TMPDT-Au-Imp

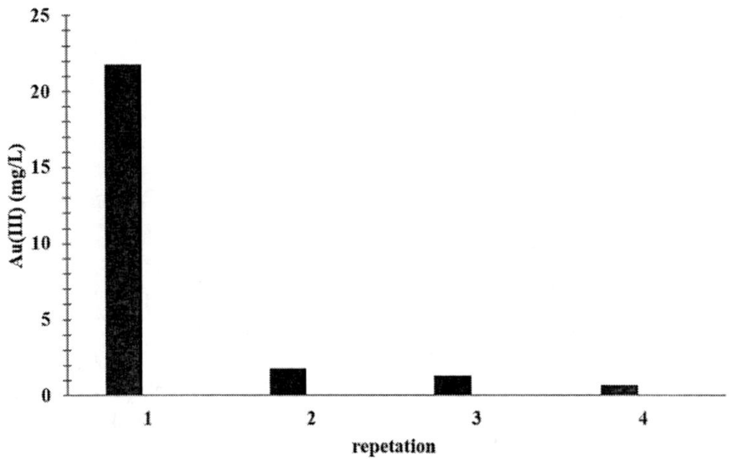

Figure 2. Au(III) ions released in each wash (mgL^{-1})

3.2. Characterization of Material

3.2.1. FTIR Analysis. Functional groups contained in the material was analyzed using FTIR. Based on Figure 3 that the $SiO_{2(BGS\text{-}RHA)}$-TMPDT, $SiO_{2(BGS\text{-}RHA)}$-TMPDT-Au and $SiO_{2(BGS\text{-}RHA)}$-TMPDT-Au-Imp appeared a peak of around 1480 cm^{-1} indicating a bond = C-NH- [17]. This proves that the

TMPDT modifier has been bound to silica. The Au(III) bound to the N atom will reduce the vibration intensity of the N-H bond in the compound [18].

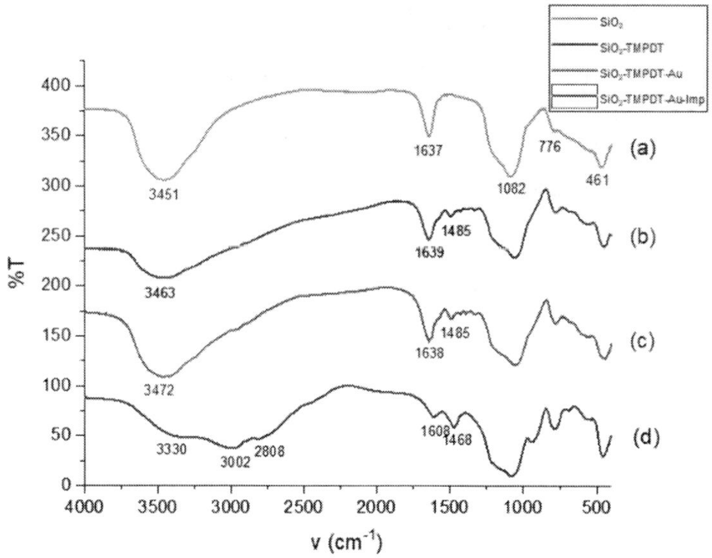

Figure 3. FTIR Spectra of $SiO_{2(BGS\text{-}RHA)}$ (a), $SiO_{2(BGS\text{-}RHA)}$TMPDT (b), $SiO_{2(BGS\text{-}RHA)}$-TMPDT-Au (c) and $SiO_{2(BGS\text{-}RHA)}$-TMPDT-Au-Imp (d)

3.2.2. TGA Analysis. Thermal stability of $SiO_{2(BGS\text{-}RHA)}$-TMPDT-Au-Imp was analyzed by TGA shown in Figure 4.The results show that there are two stages of weight loss. The first stage, the loss of H_2O molecules is 11.9% at a temperature of 35,4^{0}C - 128,8^{0}C. The second stage, lost the organic group of TMPDT by 39.4% in 187,4^{0}C -899,9^{0}C.

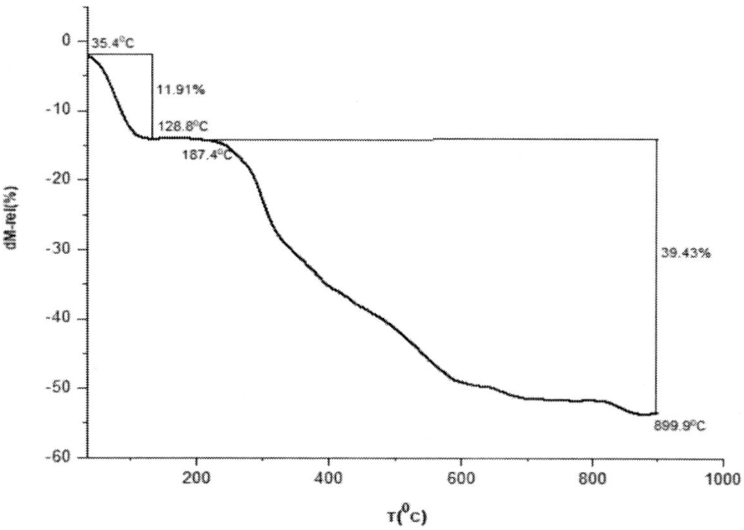

Figure 4. Curve of TGA of $SiO_{2(BGS\text{-}RHA)}$-TMPDT-Au-Imp

3.2.3. SEM-EDX analysis. To determine surface morphology and elements in $SiO_{2(BGS\text{-}RHA)}$-TMPDT-Au-Imp was analyzed using SEM-EDX. The results are shown in Figure 5 and Table 1. The SEM image indicate that $SiO_{2(BGS\text{-}RHA)}$-TMPDT-Au-Imp have irregular shapes [16].

IOP Conf. Series: Materials Science and Engineering **578** (2019) 012018 doi:10.1088/1757-899X/578/1/012018

Figure 5. Morphology of $SiO_{2(BGS\text{-}RHA)}$-TMPDT-Au-Imp

Tabel 1. Elements of the material $SiO_{2(BGS\text{-}RHA)}$-TMPDT-Au-Imp

Element	% mass
C	18.68
N	14.02
O	39.73
Si	18.52
Cl	8.25
Cu	0.08
Au	-

The presence of N and C elements in $SiO_{2(BGS\text{-}RHA)}$-TMPDT-Au-Imp (Table 1) indicates the presence of TMPDT in the material. Total 0% Au showed the release of Au (III) from the adsorbent to form Au (III) template.

3.2.4. SAA analysis. The porosity of adsorbent such as surface area and pore volume in $SiO_{2(BGS\text{-}RHA)}$-TMPDT-Au-Imp were 16,361 m^2/g, and 0,050 cc/g, respectively. While the surface area and pore volume for $SiO_{2(RHA)}$ are 296.387 m^2/g and 0,833 cc/g, respectively. Surface area of $SiO_{2(BGS\text{-}RHA)}$-TMPDT-Au-Imp is smaller than SiO_2. This is due to pore closure by TMPDT compounds [8].

3.3. Adsorption of Au(III) on $SiO_{2(BGS\text{-}RHA)}$-TMPDT-Au-Imp

3.3.1. The effect of pH. The effect of pH on the adsorption of Au (III) in $SiO_{2(BGS\text{-}RHA)}$-TMPDT-Au-Imp is shown in Figure 6. The optimum conditions were achieved at pH 3 of $SiO_{2(BGS\text{-}RHA)}$-TMPDT and $SiO_{2(BGS\text{-}RHA)}$-TMPDT-Au-Imp. In general at pH 1-6 the adsorption ability of $SiO_{2(BGS\text{-}RHA)}$-TMPDT-Au-Imp is greater than $SiO_{2(BGS\text{-}RHA)}$-TMPDT. The existence of templates on imprinted material allows Au(III) to be adsorbed more. At pH 1-2, percent adsorption is decreases. This is because $AuCl_4^-$ ions at low pH tend to form $HAuCl_4$ which is difficult to interact with adsorbents which positively charged [12]. Similarly at pH 4-6, it also tends to decrease. It is possible to occur the deposition of Au(III) ions into gold hydroxide ($Au(OH)_3$)[16].

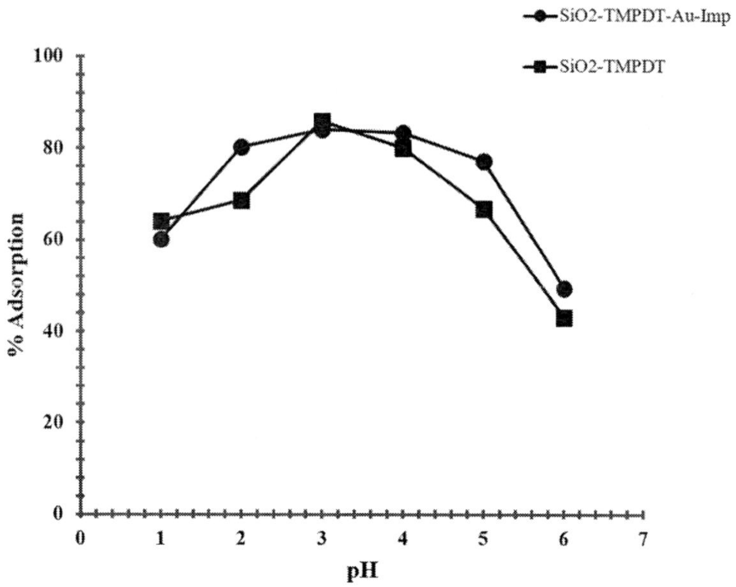

Figure 6. Effect of pH on Au(III) adsorption on SiO_2-TMPDT and SiO_2-TMPDT-Au-Imp

3.3.2. Kinetics of Adsorption. Figure 7. shows the effect of contact time on adsorption of Au(III) in $SiO_{2(BGS-RHA)}$-TMPDT and $SiO_{2(BGS-RHA)}$-TMPDT-Au-Imp. The balance time is achieved starting at 60 minutes for $SiO_{2(BGS-RHA)}$-TMPDT and $SiO_{2(BGS-RHA)}$-TMPDT-Au-Imp. The value of q_e on $SiO_{2(BGS-RHA)}$-TMPDT and $SiO_{2(BGS-RHA)}$-TMPDT-Au-Imp are 10.05 mg/g; 10.44 mg/g, respectively.

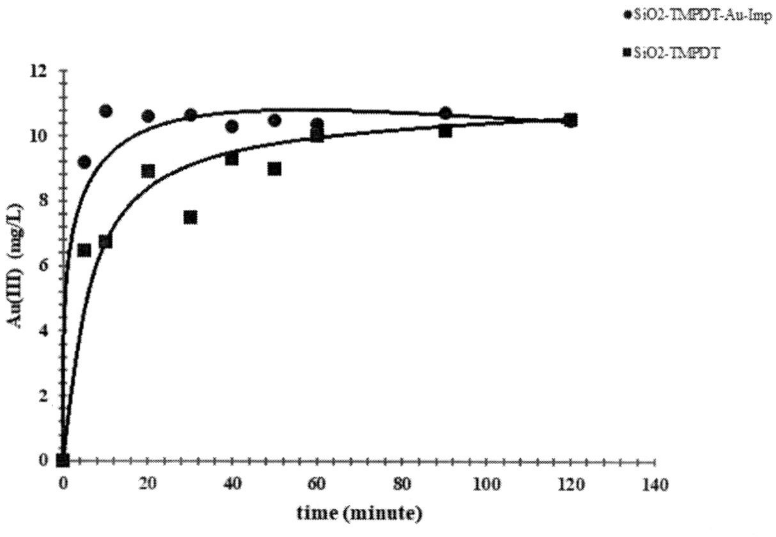

Figure 7. Adsorption contact time of Au(III) on SiO_2-TMPDT and SiO_2-TMPDT-Au-Imp

To determine the reaction kinetics that control the adsorption process, the reaction kinetics were tested for First order (Eq. 2) [19], second order (Eq. 3), pseudo-first order (Eq. 4) [20] and pseudo-second order (Eq. 5) [21]. They were applied to evaluate the experimental data of contact time. Kinetic model and parameters for Au(III) adsorption are presented in Table 2.

$$\ln C_e = -k_1 t + \ln C_0 \quad (2)$$

$$\frac{1}{C_e} = k_2 t + \frac{1}{C_0} \quad (3)$$

$$\log(q_e - q_t) = \log q_e - \frac{k_3}{2.303} t \quad (4)$$

$$\frac{t}{q_t} = \frac{1}{k_4 q_e^2} + \frac{t}{q_e} \quad (5)$$

where C_e represents of Au(III) concentration at equilibrium (mM), C_0 represents of Au(III) initial concentration (mM); q_e and q_t represent the amounts of Au(III) adsorbed at equilibrium (mol g^{-1}) and at any time t (mol g^{-1}); k_1, k_2, k_3, and k_4 represent the adsorption rate constant of first order (min^{-1}), second order ($mM^{-1}min^{-1}$), pseudo-first order (min^{-1}), and pseudo-second order (g mol^{-1} min^{-1}), respectively. The results of the calculation of k and R^2 values are shown in the Table 2.

Table 2. Constant rate and linear coefficient of Au(III) adsorption in various kinetics models on (A).$SiO_{2(BGS\text{-}RHA)}$-TMPDT and (B). $SiO_{2(BGS\text{-}RHA)}$-TMPDT-Au-Imp

Kinetic model	Constant rate (k)		R^2	
	A	B	A	B
First order	0.0082 min^{-1}	0.0047 min^{-1}	0.6643	0.2000
Second order	0.2639 mM^{-1} min^{-1}	0.1396 mM^{-1} min^{-1}	0.8052	0.2166
Pseudo-first order	0.0408 min^{-1}	0.0249 min^{-1}	0.7519	0.1144
Pseudo-second order	3.3961 g mol^{-1} min^{-1}	76.7331 g mol^{-1} min^{-1}	0.9911	0.9996

Based on the R^2 value, the adsorption kinetics of Au(III) on $SiO_{2(BGS\text{-}RHA)}$-TMPDT and $SiO_{2(BGS\text{-}RHA)}$-TMPDT-Au-Imp tend to follow pseudo-second order kinetics model. This is because the values for both adsorbents in pseudo-second order are close to 1. The pseudo-second order kinetics model illustrates that adsorption occurs chemically [3-4][12][22].

4. Conclusion

Synthesis of $SiO_{2(BGS\text{-}RHA)}$-TMPDT-Au-Imp can be done using $SiO_{2(BGS\text{-}RHA)}$ as a precursor, TMPDT as a modifier and Au(III) as a template. Formation of the template was done by desorption of Au(III) using 7% thiourea solution in 1M HCl for 3 repetitions. The process of maximum adsorption of Au(III) by $SiO_{2(BGS\text{-}RHA)}$-TMPDT and $SiO_{2(BGS\text{-}RHA)}$-TMPDT-Au-Imp occur at pH 3. Adsorption capacity (q_e) on $SiO_{2(BGS\text{-}RHA)}$-TMPDT and $SiO_{2(BGS\text{-}RHA)}$-TMPDT-Au-Imp are 10.05 mg/g; 10.44 mg/g, respectively. Constant rate (k) for $SiO_{2(BGS\text{-}RHA)}$-TMPDT and $SiO_{2(BGS\text{-}RHA)}$-TMPDT-Au-Imp are 3.3961; 76.7331 g mol^{-1} min^{-1}, respectively. Adsorption kinetics follow pseudo-second order kinetics model.

Acknowledgement

The authors gratefully express acknowledgments to Ministry of Research, Technology and Higher Education (RISTEKDIKTI) of the Republic of Indonesia.

References

[1] Hagelüken C and Corti C W 2010 *Gold Bull.* **43** 209.
[2] Ogata T and Nakano Y 2005 *Water Res.* **39** 4281.
[3] Fujiwara K, Ramesh A, Maki T, Hasegawa H and Ueda K 2007 *J. Hazard. Mater.* **146** 39.
[4] Ramesh A, Hasegawa H, Sugimoto W, Maki T and Ueda K 2008 *Bioresour. Technol.* **99** 3801.
[5] Shamspur T and Mostafavi A 2009 *J. Hazard. Mater.* **168** 1548.
[6] Bozkurt S and Merdivan M 2009 *Environ. Monit. Assess.* **158** 15.
[7] Xie F, Lin X, Wu X and Xie Z 2008 *Talanta* **74** 836.

[8] Jal PK, Patel S and Mishra BK 2004 *Talanta* **62** 1005.
[9] Zougagh M, Pavon J M and de Torrres AG 2005 *Anal Bioanal Chem.* **381** 1103.
[10] Mahmoud M, Maximous N and Masoud M 2004 *Microchim Acta* **147** 111.
[11] Tong A, Akama Y and Tanaka S 1990 *Anal. Chim. Act.* **230** 179.
[12] Hastuti S, Nuryono and Kuncaka A 2015 *Indones. J. Chem. 17(2)*108.
[13] Istiningrum R B, Tiwow C D, Nuryono and Narsito 2015 *Procedia Chemistry* **17** 132.
[14] Rubcumintara T 2015 *Inter. J. Chem Eng and Appl*, **6**(2) 95.
[15] Buhani, Narsito, Nuryono and Kunarti, E S 2010 *Desalin* **251** 83.
[16] Sakti S C W, Siswanta D and Nuryono 2013 *Pure Appl. Chem.* **85** 211.
[17] Chan C C P, Choudhurya N R and Peter P 2011 *Colloids and Surfaces A: Physicochem. Eng. Aspects* 377 20
[18] Wu Y, Zuo F , Lin Y, Zhou Y, Zheng Z And Ding X 2014 *Journal of Macromolecular Science*, Part A: *Pure Appl. Chem*. 51, 441
[19] Singh S, Verma L K, Sambi S S and Sharma S K 2008 *Proceedings of the World Congress on Engineering and Computer Science,* San Francisco, USA, 112.
[20] Lagergren S 1898 *Kungliga Svenska Vetenskapsakademiens Handlingar* **24** 1.
[21] Ho Y U S H 2004 *Scientometrics* **59** 171.
[22] Wang H, Bao C, Li F, Kong X and Xu J 2010 *Microchim. Acta,* **168** 99.

IOP Conf. Series: Materials Science and Engineering **578** (2019) 012019 doi:10.1088/1757-899X/578/1/012019

Patterned Porous Silicon Prepared by Reactive Ion Etching Technique

R Suryana[1], N D Pratiwi[1], M Handayani[1], M Santika[1] and O Nakatsuka[2]
[1]Department of Physics, Faculty of Mathematics and Sciences, Universitas Sebelas Maret, Jl. Ir. Sutami 36A Kentingan Surakarta 57126, INDONESIA
[2]Department of Materials Physics, Graduate School of Engineering, Nagoya University, Furo-cho, Chikusa-ku, Nagoya 464-8603, JAPAN

E-mail: rsuryana@staff.uns.ac.id

Abstract. A previous study, porous silicon (PSi) have been fabricated by electrochemical anodization method. The PSi surfaces have an inhomogeneous pores size in the range of 95–1450 nm. In this study, we have prepared patterned PSi using RIE technique in CF_4 gas which is activated by RF plasma. Before the RIE process, the surfaces Si was patterned using the photolithography method. Then, the Si surfaces were etched in various etching time to form patterned PSi. The patterned PSi samples were characterized by SEM, Dektak profilometer, and four-point probe equipment. We found the pore sizes and the pore depths for each the etching time were about 5.30 μm and 12.6 nm; 5.19 μm and 19.6 nm; 4.30 μm and 23.8 nm; 3.40 μm and 28.0 nm for 10 min, 20 min, 30 min, and 40 min, respectively. The pores depths increased with increasing the etching time. Meanwhile, the resistivity of patterned PSi increased with increasing the etching time i.e., 1.14×10^{-4} Ω.cm, 1.79×10^{-4} Ω.cm, 2.36×10^{-4} Ω.cm, and 2.60×10^{-4} Ω.cm for 10 min, 20 min, 30 min, and 40 min, respectively. This indicated that the pore could be considered as defects or dislocation inside material which resist electrons diffuse toward the electrode.

1. Introduction

Due to porous silicon (PSi) has good physical, chemical, and optical properties that can be applied in electronic and optic devices, research on PSi from year to year is increasing [1-7]. The previous study, we have fabricated PSi using the electrochemical anodization method on Si(100) surfaces [8]. The pores have an inhomogeneous pores size in the range of 95 nm–1450 nm. In addition, there were two groups of pores, i.e., the spherically shaped pores and the irregular shaped pores. The inhomogeneous pores size and the irregular shaped pores can be difficult in the device applications because the PSi fabrication is unreproducible. Therefore, in this research, the regular PSi fabrication has been conducted by a combination of photolithography and reactive ion etching (RIE) techniques. These techniques have good control of homogenous size and shape, porosity and depth pore of the Si surfaces [9]. The patterned PSi samples were characterized by the scanning electron microscopy (SEM) to observe the pore shapes and sizes, the Dektak profilometer to measure the diameter and the depth of pores, and the four-point probe (4PP) equipment to calculate the resistivity of pores.

2. Experimental

A 3-inch n-type Si (111) with resistivity 0.1–10 Ω.cm was used as substrate. It was put on a sample holder in the spin coater. The ZPN1150-90 photoresist was dropped on a Si(111) surface then was spun at 500-5000 rpm for 33 s. The sample was baked at 90°C for 90 s. The sample was placed below a photomask 5μm/5μm in the photolithography then exposed to UV light for 8 s. Finally, the sample was baked again at 110°C for 60 s then immersed into the MD3 developer and the DI-water. The pattern on Si surface consisting of the closed or opened area was observed using a digital microscope.

The samples were placed in the reactive ion etching (RIE) chamber. The etching time, the flowing CF_4 gas and the radio frequency (RF) power was controlled at 10 min, 10 sccm and 10 watts, respectively. After certain pressure was attained, CF_4 gas flowed into the chamber automatically. CF_4 gas plasma has a gray color which indicated that CF_4 ion become reactive can be seen through the chamber window. The procedures above were repeated for other samples for different the etching time i.e., 20 min, 30 min, and 40 min. Finally, each sample was immersed into the boiled-H_2SO_4 and H_2O_2 with ratio 3:1 for 1 min, overflowed into the DI-water for 1 min and then dried using N_2 gas. The samples were characterized using the SEM, the Dektak profilometer and the 4PP.

3. Results and discussion

Figure 1 shows the SEM images of patterned Si surfaces after the RIE process for 10 min (Fig. 1a), 20 min (Fig. 1b), 30 min (Fig. 1c), and 40 min (Fig. 1d), while the images below are magnification of one of the pores for each the etching time as shown in Fig. 1e-h for 10 min–40 min, respectively. Pore diameter for each the etching time is about 5.30 μm for 10 min (Fig. 1a), 5.19 μm for 20 min (Fig. 1b), 4.30 μm for 30 min (Fig. 1c) and 3.40 μm for 40 min (Fig. 1d). Based on the Dektak profilometer measurement for each the etching time, the pore depth is about 12.6 nm for 10 min, 19.6 nm for 20 min, 23.8 nm for 30 min, and 28.0 nm for 40 min. For ease discussion, the pores diameter and the pores depth versus the etching times is conducted in the graph as shown in Fig. 2. The pore diameter decreases with increasing the etching time while the pore depth increases with the etching time. The RIE is anisotropy etching having the lateral and vertical etching rates. Figure 2 conclude that the etching rate in the vertical direction is faster than that in the lateral direction. This result agreed to Tanaka et al., [10], they proposed that the lateral etching rate is faster than that in the lateral etching rate regardless of Si crystal orientation.

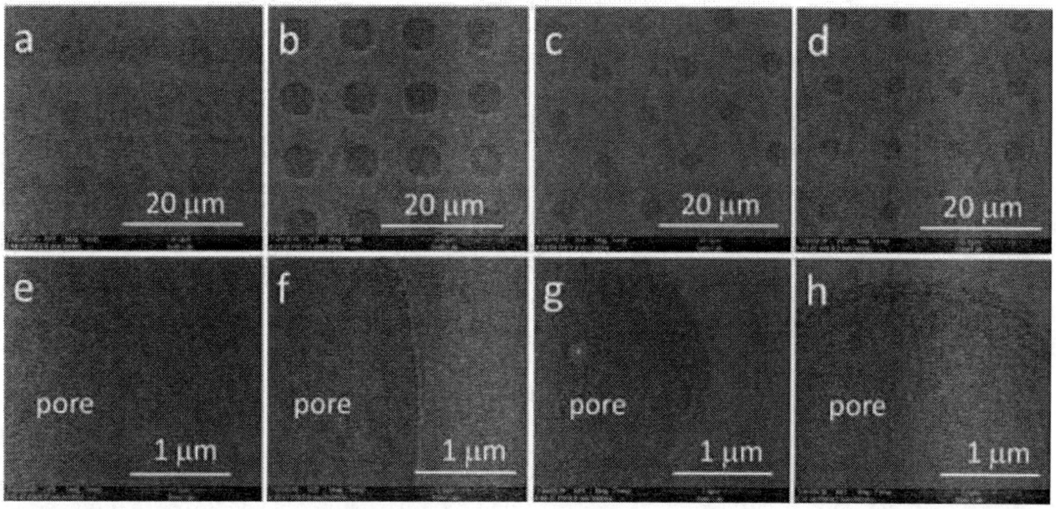

Figure 1. SEM images of patterned Si after the reactive ion etching for 10 min (a), 20 min (b), 30 min (c), and 40 min (d) while the images below are magnification of one of the pores for each the etching time i.e., 10 min (e), 20 min (f), 30 min (g), and 40 min (h).

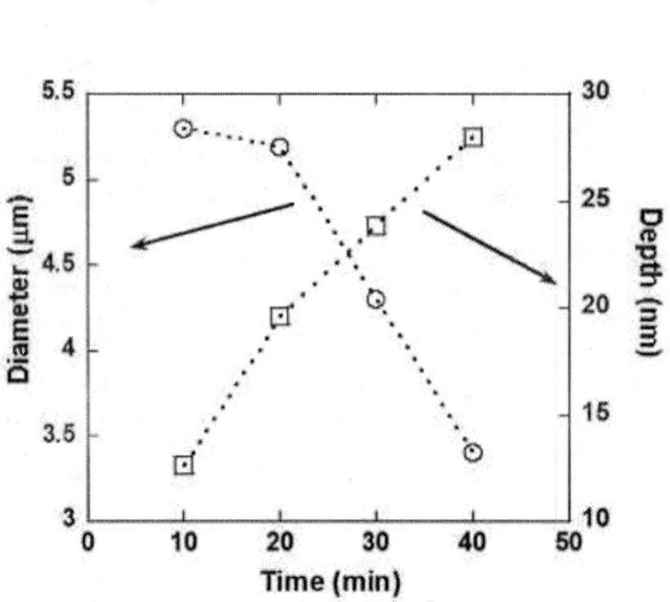

Figure 2. Measurement of pores diameter and pores depth on Si(111) surfaces after the reactive ion etching for the etching time of 10 min, 20 min, 30 min and 40 min using the Dektak profilometer

The unmasked portion (opened area) of the surface is exposed to etching plasma with high kinetic energy which the bottom surface obtained from vertical etching, and the lateral undercut spread from the sidewall of the vertical etching. However, in the (111) orientation case, the etch-steps start within the mask opening, traveling along the (111) planes and stop at the out-of-planed-oriented (111)-planes defined by the mask geometry [11]. Figures 1g and 1h show a double line or a shadow line, it indicated that the etch-steps stop on an out-of-plane which has angle 70.5° to the in-plane oriented planes [11].

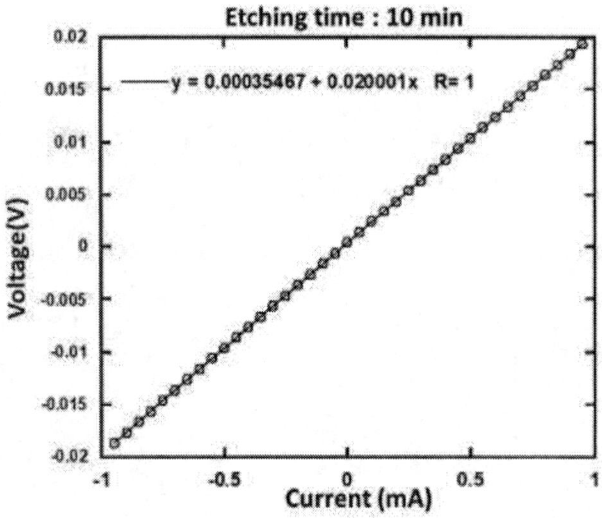

Figure 3. Current-voltage curve of the PSi at etching time for 10 min obtained from the four-point probe measurement.

The resistivity of the porous silicon can be calculated from equation $\rho = (V/I)\ w\ (\pi/\ln 2)\ F(w/s)$ [12] where (V/I) can be obtained from the I-V graph as result the measurement using the 4PP, w is thickness or the pore depth, $F(w/s)$ is a correction factor approaching unity as w approaches zero. In this research, w is in nanometer order. Now, the resistivity equation become $\rho = (V/I)\ w\ (4.53)$. For example, the PSi at etching time for 10 min as shown in Fig. 3 has (V/I) = 20.0 Ω which is obtained from linear equation of second term that is 0.020001×10^3. Therefore, resistivity value for etching time of 10 min is 1.14×10^{-4} Ω.cm. In the same way, it is obtained (V/I) and ρ for etching time of 20 min, 30 min and 40 min as summarized in Table 1. The resistivity increases with increasing the etching time. The resistivity of a semiconductor is affected by a defect or dislocation inside the material [13] due to the electrons difficult to diffuse and the defect or the dislocation can act as a scatter. In this research, the pores can be considered as the defects or the dislocation inside the material.

Table 1. Summarized the calculation of resistivity of PSi for each the etching time

Etching time (min)	V/I (Ω)	w ($\times10^{-6}$ cm)	ρ ($\times10^{-4}$ Ω.cm)
10	20.0	12.6	1.14
20	20.1	19.6	1.79
30	21.9	23.8	2.36
40	20.1	28.0	2.60

4. Conclusion

The patterned PSi has been successfully formed by a combination of the photolithography and the reactive ion etching technique. The pores size decreases with increasing the etching time while the pores depth increases with increasing the etching time. Furthermore, the resistivity of PSi increases with increasing the etching time. Due to the pores are considered as the defects or the dislocation in the material, the electrons facing an obstacle to flow toward the electrode. It is causing the PSi has low conductivity or high resistivity.

Acknowledgment

RS wish to thank Universitas Sebelas Maret 2018 for financial support through World Class University for Lecturer Exchange Program. NDP, MH, and MS would like to thank Universitas Sebelas Maret for financial support through a Global Challenge Scholarship 2018 Batch I during the Internship Program at Nagoya University, Japan. RS would like to thank Zaima/Nakatsuka Laboratory, Department of Materials Physics, Graduate School of Engineering, Nagoya University, Japan for providing all materials and facilities during the research in Nagoya University and also to thank Saito Laboratory, Department of Applied Physics, Graduate School of Engineering, Nagoya University, Japan for supporting SEM instrument.

References

[1] Nayef U M 2013 *Int. J. Basic & Appl. Sci. (IJBAS-IJENS)* **13(02)** 61
[2] Singh P, Sharma SN and Ravindra NM 2010 *JOM* **62** 15
[3] Kadhim RG, Ismail RA and Abdulridha WA 2015 *Int. J. Fil. Sci. Tec.* **4(3)** 199
[4] Mohammed IM and Shnieshil AH 2013 *Int. J. Application in Engineering and Management* **2(9)** 77
[5] Kumar P, Lemmens P, Ghosh M, Ludwi F and Schilling M 2009 *J. Nanomater.* **18** 7
[6] Russo L, Colangelo F, Cioffi R, Rea I and Stefano LD 2011 *Mater.* **4(6)** 1023
[7] Burham N, Hamzah AA and Majlis BY 2017 Self-Adjusting Electrochemical Technique for Producing Nanoporous Silicon *New Research on Silicon-Structure, Properties, Technology* (Materials Science) ed V I Talanin (Croatia: INTECH) chapter 6 pp 125-154
[8] Suryana R, Sandi DK and Nakatsuka O 2018 *IOP Conf. Series: Mater. Sci. Eng.* **333** 012034
[9] Pease F and Chou SY 2008 *Proceeding of the IEEE* **96(2)** 248

[10] Tanaka S, Sonoda K, Kasai K, Kanda K, Fujita T, Higuchi K and Maenaka K 2011 *Crystal Orientation Dependent Etching in RIE and Its Application* (IEEE MEMS) 217
[11] Oosterbroek RE, Berenschot JW, Jansen H V, Nijdam AJ, Pandraud G, van den Berg A and Elwenspoek MC 2000 *J. Microelectromech. Syst.* **9(3)** 390
[12] Smits FM 1958 *Bell Syst. Tech. J.* **37(3)** 711
[13] Sambles JR 1983 *Thin Solid Films* **106 (4)** 321

IOP Conf. Series: Materials Science and Engineering **578** (2019) 012020 doi:10.1088/1757-899X/578/1/012020

The effectiveness of the activated carbon from coconut shell and corn cob to adsorb Pb(II) ion and it's analysis using solid-phase spectrophotometry

S Saputro[1], L Mahardiani[1], M Masykuri[1], and A Z Jazuli[1]

Chemistry Education, Faculty of Teacher Training and Education, Sebelas Maret University, Jl. Ir. Sutami 36A Surakarta, Central Java-57126, Indonesia

E-mail: sulistyo_s@staff.uns.ac.id

Abstract. This research aimed to know the mass ratio of adsorbent combination of activated carbon from coconut shell and corn cob to adsorb Pb(II) ion and it's analysis using the solid-phase spectrophotometry (SPS). The technique of data collection is carried out by several stages, which are: (1) the production of carbonated coconut shell and corn cob using muffle furnace at a temperature of 350 °C for 60 minutes; (2) the activation of the coconut shell carbon using $ZnCl_2$ 15% activator and the corn cob carbon using HCl 1M, (3) contacting this adsorbent combination between activated carbon from coconut shell and corn cob for 20 mL of liquid simulations waste Pb(II) ion on variations ratio of 1:0; 0:1; 1:1; 2:1; 1:2; (4) the analysis levels of Pb(II) using SPS; (5) the characterization of adsorbents in the most optimum combination between activated carbon of coconut shell and corn cob using FTIR. The results showed that the adsorbent can increase the capabilities and the effectiveness of the absorption to Pb(II) ion and the optimum ratio from the comparison between coconut shell and corn cob is 1:2 with the absorption ability to adsorb Pb(II) ion in the simulations waste is 86.78%. SPS is an effective method with high sensitivity in the level of μg/L and the limit of detection (LoD) of 0.06 μg/L.

1. Introduction

We know that Indonesia is located in the tropical region and is traversed by the equator, besides Indonesia is also known as one of the agricultural countries, where the agricultural sector is very abundant. The tropical region is known to have unlimited potential for natural resources for the future of life. Based on statistical data, coconut plantation results in Indonesia reach 176,575.82 tons/year, while corn reaches 3,212,391 tons/year [1], agricultural products in the form of coconut and corn have byproducts in the form of coconut shells and corn cobs which are relatively more than the main product. These byproducts need to be processed so it does not become useless waste. Today, many resolutions have been found for the processing of solid waste, ranging from making souvenirs, biobriquettes, fertilizers, and even making environmentally friendly alternative energy.

On the other hand, the development of an increasingly advanced era has caused industry in Indonesia to also develop rapidly, as a result new problems arise in the form of environmental pollution, such as water pollution. Water pollution according to the Decree of the Minister of Population and Environment No: KEP-02/MENKLH/1988 concerning the Determination of

[1] Sebelas Maret University, Jl. Ir. Sutami 36A Surakarta, Central Java-57126, Indonesia.

Environmental Quality Standards are: the entry or inclusion of living things, energy substances, and/or other components into the water or changes in the water order by human activities or by natural processes, so that the quality of water falls to a certain level which causes water to become deficient or no longer functioning according to its designation [2]. Water pollution is mostly caused by industrial and household waste that is dumped carelessly without being processed first. The main sources of water pollutions come from industry sector. Some type of industry that contributes greatly are the mining and construction, manufacturing, power-generating, and food processing industries. One of the impacts from the mining industry is heavy metal contamination in the water [3]. Heavy metals contained in water can cause various problems. Starting from the problem of damage to the environment and ecosystems, to the health and safety problems of living things. Water pollution could be minimized among others; by recycling, by reusing, waste minimization, by mitigating, by preventing, and by compost [4]. Lead (Pb) is one of the heavy metals that can cause these problems.

Lead is a metal with a bluish gray color with high density (11.48 g ml-1 at room temperature) [5]. Acute lead toxicity in natural waters causes severe damage to the kidneys, reproductive system, liver, and brain, and the central nervous system, and can cause death [2]. Most Lead wastes are produced by Industrial activities and other produced by natural processes. The high content of Pb in the waters could be due to weathering and leaching of lead from waste rocks dumps [6]. Therefore, absorbents that can reduce lead content in polluted water are now being developed. Absorbent made activated carbon from coconut shell combined corn cobs is possible to be more effective in adsorbing Pb^{2+} ion in water, besides economical, both materials are also easy to obtain, and can be a useful waste treatment.

Solid-phase spectrophotometry is one method that can be used to determine the adsorption capacity of an adsorbent with better sensitivity (the ability to read concentration to the level of ppb or μg/L than other methods. In this research, the effectiveness of activated carbon from coconut shell and activated carbon corn cobs is tested to be used for Pb^{2+} ion adsorption using solid-phase spectrophotometry.

2. Experimental

The research method applied was laboratory experiment method. The basis of this research was a decrease in Pb^{2+} ion levels by using adsorbents from a combination of activated carbon from coconut shell with corncob and analysis conducted using solid-phase spectrophotometry. Then the analysis of sample characteristics using Fourier Transform Infrared Spectroscopy (FTIR).

2.1. Materials, Tools and Instrument.

Materials used are coconut shell, corn cobs, $Pb(NO_3)_2$, $ZnCl_2$ 15%, HCl 1M, H_2SO_4 0.5M, resin AG Muromac 50W-X2 H^+ form (100-200 mesh), distilled water, chloroform, and dithizone (diphenylthiocarbazone) 0.005%. The used tools were a muffle furnace, pan, oven, analytical balance, volumetric flask, volume pipette, drop pipette, beaker glass, measuring cup flask, whatman filter paper, blue litmus, erlenmeyer, watch glass, stirrer glass, aliquoting devices which are assembled using a syringe, blender, mortar and pestle, a 100 mesh sieve, porcelain bowls, stirrer bar, and a magnetic stirrer. The used Instruments were a UV-visible spectrophotometer from K-MAC and FTIR spectrophotometer from Shimadzu [7].

2.2. Research procedure

2.2.1. Production of adsorbent. Coconut shells and corn cobs are removed first and then washed clean to remove impurities with destilled water. Then dried in an oven for 60 minutes then put into muffel furnace at 350 °C for 60 minutes to become carbon. Mash coconut shell carbon and corn cobs separately then sieve with a size of 100 mesh [8].

2.2.2. Adsorbent activation. Coconut shells carbon soaked in a solution of $ZnCl_2$ 15% and corn cobs carbon soaked in a solution of HCl 1M at room temperature for 24 hours. Filter and rinse the residue to make the filtrate become neutral with distilled water, and then dried in an oven at 110°C for 24 hours. Adsorbent was tested by FTIR spectrophotometer before and after activation [7].

2.2.3. Resin preparation. In this research used Muromac resin AG Muromac 50W-X2 H^+ solid form (100-200 mesh size) before that is used, the resin dissolved in distilled water and silence a few moments to get more fluffy resin.

2.2.4. Determination of Pb(II) calibration curve. Pb (II) solution was 0 μg/L, 2 μg/ L, 4 μg/L, and 8 μg/L each taken 20 mL. Then added 2 mL of dithizon solution, 1 mL of H_2SO_4 0.5M solution, and 0.06 mL of fluffy resin. Then it was stired for 20 minutes and analyzed using solid-phase spectrophotometry (SPS) with a wavelength of 483 nm and 558 nm. Then the absorbance difference from the two wavelengths is taken, namely $\Delta A = A_{483nm} - A_{558nm}$. Where the ΔA obtained will be made a standard curve Pb(II) (ΔA vs. concentration).

2.2.5. Determination of Pb(II) species in simulation liquid waste. Simulation liquid waste solution was taken 20 mL, then added 1 mL of H_2SO_4 0.5M solution, 2 mL of dithizon solution, and 0.06 mL of resin. Then it was stired for 20 minutes and analyzed using a UV-visible spectrophotometer with a wavelength of 483 nm and 558 nm. Then the absorbance difference is calculated from both wavelengths, namely $\Delta A = A_{483nm} - A_{558nm}$. Where the ΔA obtained will be substituted in the equation of the calibration curve Pb(II) (ΔA vs. concentration), so that the Pb (II) species in the simulation waste are known.

2.2.6. Determination of the most effective comparison of adsorbents from coconut shells and corn cobs. Combine activated coconut shell carbon and activated corncob carbon in a row with a ratio of 1:0; 0:1; 1:1; 2:1; and 1:2 into a beaker glass containing 25 mL of Pb(II) simulated solution then input in the stirrer (stired for 30 minutes). The solution obtained was filtered with Whatman filter paper. The resulted filtrate was taken as much as 5 mL and then diluted it until 50 mL with distilled water in the volumetric flask. Furthermore, taken 20 mL from the resulting filtrate diluted and added with 1 mL of H_2SO_4 0.5M, 2 mL of dithizone, and 0.06 mL resin then stir it for 20 minutes. The mixture than analysed using a UV-vis spectrophotometer with a wavelength of 483 nm and 558 nm. ΔA obtained will be substituted in equation of Pb(II) calibration curve (ΔA vs concentration). Based on, the concentration of Pb(II) in a simulated liquid waste after the Pb(II) adsorption is known. Lastly the most effective adsorbent comparisons using Fourier Transform Infra Red (FTIR).

2.2.7. Determination limit of detection (LoD). The blank solution is 20 mL. Then added 1 mL of 0.5M H_2SO_4 solution, 2 mL of dithizon solution, 0.06 mL of resin and then stirred for 20 minutes and then analyzed using solid-phase spectrophotometry with a wavelength of 483 nm - 558 nm. The absorbance difference of the two wavelengths is calculated, namely $\Delta A = A_{483nm} - A_{558nm}$. Where the ΔA obtained will be substituted in the equation of the calibration curve Pb (II) (ΔA vs. concentration), so that the blank content is obtained. The determination of Relative Standard Deviation (RSD) was carried out to determine the limit of detection which was formulated as LoD = 3RSD. In this experiment data retrieval was repeated five times to obtain accurate results.

3. Results and Discussion

3.1. Production of activated carbon from coconut shell and corn cobs.

The earliest process is dehydration, both coconut shell and corn cobs, both of which have to go through this dehydration process, this process begins with the cleaning of impurities that are still in the coconut shell and corn cobs, such impurities, sand, coconut fiber and epidermis on coconut shells, corn hair, etc. After all the ingredients are free from impurities, the next stage is washing with distilled water. Furthermore, the dehydration stage is done by heating all the ingredients in the oven to a temperature of 100°C for 60 minutes, the purpose of the dehydration process is to remove the water content contained in the sample. After the dehydration process, the sample then enters the carbonization process. This carbonization process uses a muffle furnace, which is a device that functions to fabricate a sample with a certain temperature. Both coconut shell and corn cobs were put into the furnace at 350°C for 60 minutes. The temperature selection is based on several experiments carried out, and obtained the perfect results of drying at 350°C, because if the temperature is too high the sample coconut shell and corn cobs will run out and become ash, and vice versa when the temperature is too low, carbon will produce not perfect. After the carbonation process using this muffle furnace, the carbon obtained is ready to be activated. In this experiment activators used in both carbon were different. This is based on research journals that use different activators on different ingredients. In this research coconut shell carbon activator is $ZnCl_2$ 15%, and corn cobs used HCl 1M. The activation process is carried out by soaking each carbons into each activator for 24 hours.

Chemical activation in making activated carbon that is often used are KOH, $ZnCl_2$, and H_3PO_4. Some of these activators produce activated carbon that has a fairly large surface area with large pores. Activators that can be used in the manufacture of activated carbons, among others, strong acids, strong bases and salts of strong acids [9]. Activated carbon from coconut shell is activated using $ZnCl_2$. $ZnCl_2$ activator is a salt from strong acid namely HCl. Before activation of coconut shell carbon contains a lot of water, this is due to the hydroscopic nature and also was caused from water molecules trapped inside the hexagonal structure of carbon. Good quality activated carbons is which has lowest moisture content. The presence of water content in activated carbon can affect decrease of absorption ability. $ZnCl_2$ activator is a substance that is hydroscopic, so the use of $ZnCl_2$ as an activator will produce better activated carbon. $ZnCl_2$ 15% is used because $ZnCl_2$ in levels 15% was able to produce the highest yield with activated carbon, more effective than KOH and H_3PO_4 [10]. The process of activating corn cobs carbon is by using HCl 1M solution. Hydrochloric acid (HCl) as a chemical activator that has hydroscopic characters, so it can reduce water content in activated carbon. compared with other strong acid activators such as H_2SO_4 dan HNO_3, carbon was activated with HCl has better iodine absorption because HCl is better to reduce impurities so that more pores are formed and the absorption process becomes more maximal [11]. The activation stage in an materials that will be made as an adsorbent is very important to clean impurities in the pores and increase the surface area [12].

3.2. Determination of Pb(II) calibration curve and Pb(II) levels in simulated liquid waste

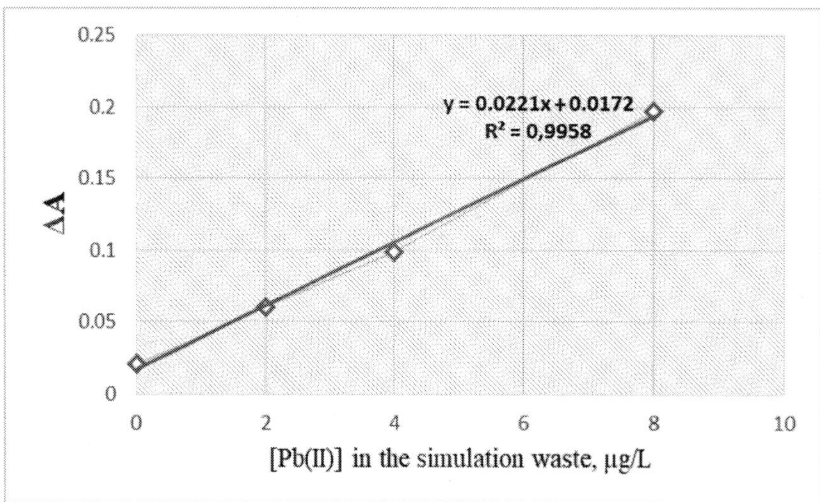

Figure 1. Calibration curve of Pb(II) solution

Based on the experimental results obtained a standard calibration curve with the equation $y = 0.021x + 0.017$. From this equation y is ΔA and x is concentration. The equation is used to find the concentration of simulated liquid waste.

Table 1. Determination results of Pb(II) levels in simulated liquid waste

Sample	[Pb(II)] calculation (ppb)	ΔA	[Pb(II)] actually (ppb)
Sample Pb(II)	50	0.127	48.87

From the table, we know that actually concentration of simulated waste is 48.87 part per billion. Its mean there is 48.87 μg of ions Pb^{2+} in the 1 Liter simulated waste. In the natural water, mean Lead metals maximum permissible concentration in water samples by NAFDAC, SON (2007), and WHO (2006) standards is 0.01 mg/L [13]

3.3. Determination the optimum combination of adsorbent for coconut shell and corn cob in a decrease levels of Pb(II) ions.

In the **Table 2**. shows the results of the contact obtained from the experiment and has been analyzed using solid-phase spectrophotometry (SPS). SPS is one type of UV-Vis instrument sepctrophotometry which has the ability to analyze up to μg/L levels. The basic principle of using UV-Vis sepctrophotometry is to measure the spectrum of light absorbed by a solution. Therefore, the solution to be analyzed must be sensitive to light. In simulated liquid waste samples containing heavy metal ion Pb^{2+}, the color of the solution is less sensitive. Therefore, to increase the sensitivity of the solution complexing is used. The complex that is formed can later increase the sensitivity of the solution and usually form a certain color, each compound or ion has a certain complex that matches. In this experiment, diphenyltiocarbazon (dithizon) complexing was used.

Table 2. The influence of the composition of the mass to ion adsorption of Pb (II)

Compotition of the mass	**Final [Pb(II)] (μg/L)**	**Adsorbed [Pb(II)] (μg/L)**	**Adsorbed [Pb(II)] (%)**
0 CS : 1 CC (1)	6.86	42.00	85.94
1 CS : 0 CC (2)	8.14	40.72	83.32
1 CS : 1 CC (3)	8.13	40.73	83.34
1 CS : 2 CC (4)	6.45	42.41	86.78
2 CS : 1 CC (5)	7.06	41.80	85.53

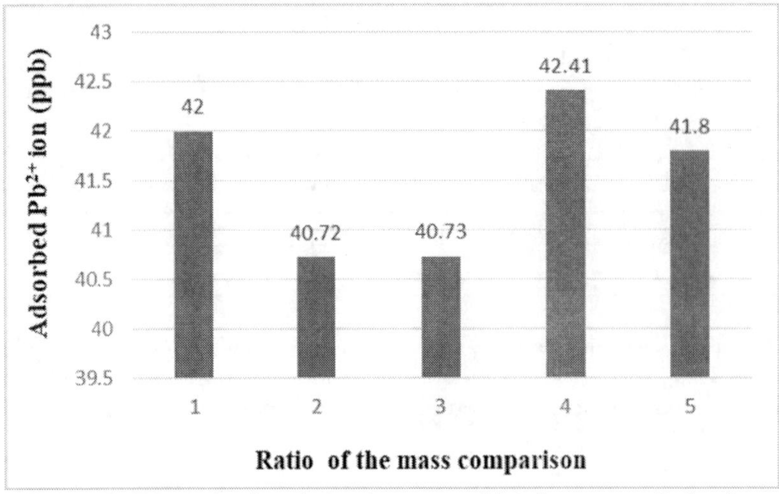

Figure 2. Determination the optimum combination mass of adsorbent

Based on the data research in **Table 2** and the graph in **Figure 2**, the optimum combination mass of adsorbent to adsorb Pb(II) metal ions in the simulated liquid waste was 1:2 with adsorbed concentration was 42.41 μg/L, and the percentage of adsorbed as 86.78 %. Beside on the data shown in the ratio of comparasion 0:1 and 1:0, we known the activated carbon from corn cobs is better than activated carbon from coconut shell. But after being combined it can increase adsorption ability.

3.4. Fourier transform infrared spectra analysis.

Analysis of structure and functional groups of activated carbon can be one of the reinforcement results of the research. In this research, the functional group analysis of the adsorbent on the most optimum combination of activated carbon are before activation, after activation, and after going through the contacting process with simulation waste containing Pb^{2+} ions. The analysis was performed using Fourier Transform Infra Red (FTIR) instrument. Fourier transform infra red (FTIR) is an infrared spectrophotometer which is an instrument used to measure infrared radiation absorption at various wavelengths [14]. The working principle of FTIR is to recognize the functional group of a compound from the infrared absorbance carried out on the compound.

Table 3. Interpretation of FTIR spectra adsorbent combination of coconut shell activated Carbon with Corn Cob.

wavelength (cm^{-1})			Interpretation
Before Activation	After Activation	After Contacting	
1600.02	1603.88	1604.84	There is C=C aromatic
1701.29	1698.40	1694.54	The existence of C=O carbonyl group
1357.94	-	-	The existence of C-O group
2835.48	2837.41	2836.45	There are C-H aldehyde
2876.95	2900.10	2890.45	
-	3326.39 3456.59	3328.31 3432.80	Stretching vibration O-H

In the interpretation of IR spectra. it is known that the adsorbent before activation has several differences with the adsorbent after activation. especially the presence of the -OH group formed after activation. Meanwhile. the results of the adsorbent activated after and before contacting have spectra readings that are not too much difference. In general. the three results of analysis using FTIR showed similarities in groups C=O carbonyl, C=C aromatic, and C-H aldehyde.

Based on the interpretation guide of IR spectra results. if there is a sharp absorption between 1820-1600 cm^{-1}. it can be ascertained that the sample contains a C=O carbonyl group. In the adsorbent sample before activation. this uptake appeared in spectra 1701.29. while the adsorbent after absorption activation appeared in spectra 1698.40. then on the adsorbent after absorption absorption appeared at 1694.54. Between the spectra after activation with the spectra after contacting. there is an absorption that occurs at almost the same number. This shows that there is no significant change in the C = O carbonyl group after contacting simulations.

3.5. Determination of detecting limit

An analytical instrument has a detection limit value. otherwise known as a limit of detection (LoD). which means the smallest number of analytes in a sample that can still be detected and still provide a more significant response than blank. In this research. The limit of detection (LoD) for each element was achieved as follows

$$\text{LoD} = \frac{3 \times S}{m}$$

Where S is the standard deviation of the blank readings and m represents the gradient of the calibration [6]. Limit of Detection (LoD) from the instrument (solid-phase spectrophotometry) was 0.06 μg/L. so it can be said that the sample at a concentration of 0.06 μg/L can be read. In the another instrument for examples. at atomic absorption spectroscopy (AAS) instruments had LoD of 0.290 μg/L and UV-Vis (HACH DR 5000 UV-Vis Spectrophotometer) is 3-300 μg /L [15] [16].

So that the solid-phase spectrophotometry is a sensitive and effective method to use in the analysis the decreased levels of Pb(II) adsorption results of activated carbon from coconut shell and corn cob in levels μg/L. Moreover on the previous research. solid-phase spectrophotometry had a limit of detection (LoD) until 0.014 μg/L [17].

4. Conclusion

The results showed that the adsorbent combination between activated carbon from coconut shell and corn cob can increase the capabilities and effectiveness of the absorption to adsorb Pb(II) ion in the simulations waste and the most optimum ratio from the comparison between coconut shell and corn cob is a ratio of 1:2 with the absorption ability to adsorb Pb(II) ion in the simulations waste is 86.78%. Solid-phase spectrophotometry is an effective method with high sensitivity in the level of detection up to μg/L. be marked from the limit of detection (LoD) of this research until 0.06 μg/L.

Aknowledgement

Authors wishing to acknowledge assistance from Head of Chemistry Laboratory, Chemistry Study Program. Faculty of Teacher Training and Education, and Head of Central Chemistry Laboratory, Sebelas Maret University for permission and used the facilities of the laboratory. This research was partially supported by the Research Grant to Increase The Capacity of Research Group 2018.

References

[1] Badan Pusat Statistika 2015 *Statistika Indonesia Tahun 2015* (Bandung: PBS).
[2] Rukaesih A 2004 *Kimia Lingkungan* (Yogyakarta: Adi Yogyakarta) p 92-93.
[3] Gambhir RS, Kapoor V, Nirola A, Sohi R and Bansal V. 2012 *J. Hum. Ecol.* **37**(02) 103.
[4] Owa FW. 2014 *Int. Lett. Nat. Sci.* **03** 1.
[5] Vogel AI 1990 *Buku Teks Analisis Anorganik Kualitatif Makro dan Semimikro* (Jakarta: PT Kalman Media Pustaka) p 207.
[6] Cobbina SJ, Duwiejuah AB, Quansah R, Obiri S and Bakobie N 2015 *Int. J. Environ. Res.*

Public Health **12** 10620.

[7] Saputro S, Mahardiani L and Wulandari DA 2018 *IOP Conf. Ser.: Mater. Sci. Eng.* **333** 012055.

[8] Rohmah DN, Saputro S, Masykuri M and Mahardiani L 2018 *IOP Conf. Ser.: Mater. Sci. Eng.* **333** 012056.

[9] Olivia ME and Netti H 2015 *Jurnal Teknik Kimia USU* **4**(1) 47-52.

[10] Esterlita and Herlina 2015 *Jurnal Teknik Kimia USU* **4**(1) 47-52.

[11] Alfiany H, Bahri S and Nurakhirawati 2013 *J. Nat. Sci.* **2**(3) 75-86.

[12] Mohd MN, Iqbaldi, Khudzir I, Azlan MIM, *et. al.* 2013 *J. Trop. For. Sci.* **25**(4) 503.

[13] Jenyo OA and Oladele AH 2016 *Eur. Sci. J.* **12**(24) 339.

[14] Fessenden RJ and Fessenden JS 1982 *Kimia Organik Jilid 1* (Jakarta: Erlangga).

[15] Kojuncu Y, Juzica MB, Umit A, *et. al.* 2004 *Sep. Sci. Technol.* **39**(11) 2751.

[16] Ali. MF and Sharkrani SA 2014 *Int. J. Civil Environ. Eng.* **14**(01) 8.

[17] Saputro S, Yoshimura K, Matsuoka S, Takehara K and Narsito 2009 *Anal. Sci.* **25** 1445.

IOP Conf. Series: Materials Science and Engineering **578** (2019) 012021 doi:10.1088/1757-899X/578/1/012021

Properties of biosurfactant produced by *Pseudomonas putida* grown in Crude Palm Oil (CPO)

V Suryanti[1,*], D S Handayani[1], A Masykur[1] and Lindasari[1]
[1]Department of Chemistry, Faculty of Mathematics and Natural Sciences, Universitas Sebelas Maret, Jl. Ir. Sutami 36A Surakarta 57126 Indonesia

[*]E-mail: venty@mipa.uns.ac.id

Abstract. Biosurfactants produced by *Pseudomonas putida* grown on crude palm oil (CPO) as carbon source have been characterized. Biosurfactant was produced in nutrient broth medium containing NaCl and 10% v/v of crude palm oil (CPO) with 5 days of incubation time. Biosurfactants are able to reduce water surface tension from 72 to 54 mN/m and has a CCM value of 798 mg/L. Biosurfactant has a water in oil (w/o) emulsion system. Biosurfactants are able to form emulsions between water and several hydrocarbons, including benzene, lubricating oil, palm oil, soybean oil, sunflower oil and olive oil. The largest emulsion index of 83% is obtained from the water emulsion with lubricating oil and can last up to 30 days.

1. Introduction

Surfactants are active compounds that have two functional groups in their molecules, namely hydrophilic groups and hydrophobic groups. Surfactants are classified into two, i.e. synthetic surfactants and biosurfactants. Synthetic surfactants are conventionally produced from chemical reactions in the form of petroleum derivatives, while biosurfactants are produced from cellular activity of microorganisms, such as bacteria and fungi [1-2]. Biosurfactants have several advantages compared to synthetic surfactants, including biodegradable, low toxicity and non-irritating. Biosurfactants have wide applications such as for bioremediation of oil-contaminated soils/pollutants and emulsifier in food, cosmetic, pharmaceutical and agricultural industries [3-5]. Biosurfactants have broad molecular coverage such as lipopeptides, glycolipids, phospholipids and polymeric biosurfactants. The molecular structures of biosurfactants are resulting from cellular activity of the microorganism which is influenced by the type of microorganisms and nutrients in the fermentation medium, such as carbon and nitrogen sources [6-7].

Genus Pseudomonas, especially the *Pseudomonas aeruginosa* are known to have great potential as biosurfactant producers [8-14]. Other species of Pseudomonas, such as *P. putida* have the potential to produce biosurfactants. These bacteria have been reported to produce a glycipipid class biosurfactant [15-17]. *P. putida* is a cylindrical gram-negative bacteria that are non-pathogenic. Another advantage of this species is its high resistance to extreme environmental conditions, such as extreme temperatures, extreme pH and high salinity [18].

Microorganism-based surfactants can be produced from renewable materials such as fatty acids, carbohydrates, proteins and vegetable oils. Crude palm oil (CPO) which contains fatty acids and its abundant presence makes CPO potentially as a carbon source in biosurfactant production. Fatty acid compounds in CPO are palmitic acid (45%), stearic acid (3%), myristic acid (1%), oleic acid (40%) and linoleic acid (10%) [19]. Palm oil has been used as a carbon source in the biosurfactant biosynthesis of glycolipids by *Pseudomonas* [20-21]. We have described the production of biosurfactants by *P. putida* using CPO as substrate on our previous work [14]. In this study, the properties of the biosurfactant produced by *P. putida* grown on CPO are discussed.

2. Experimental

2.1. Materials

All chemical were used are analytical grade from e-Merck. The crude palm oil (CPO) was obtained from Centre for Chemical and Packaging, Jakarta, Indonesia. The strain used in this work, *P. putida* FNCC 0071, was purchased from IUC Food and Nutrition, Universitas Gadjah Mada, Indonesia.

2.2. Media Used and Growth Condition
Cultures of bacteria were maintained on nutrient agar. The media were used for biosurfactants production composed of nutrient broth (8 g/L), NaCl (5 g/L) and CPO (5%v/v). Media mixture was sterilized prior used. Fermentation was carried out at 150 rpm for 5 days at room temperature.

2.3. Biosurfactant recovery
Culture liquid of *P. putida* was filtered by Buchner filtrations using Whatman filtration paper grade 42. The supernatant was then acidified to pH 2.0 with HCl 6 N and leaved overnight at 4 °C. The supernatant was extracted twice with *n*-hexane for removing the remaining CPO. The aqueous solution was then extracted twice using ethyl acetate. Sodium sulphate anhydrous was then added to the organic layer. The organic layer which was free form water was then evaporated.

2.4. Biosurfactant Characterization
The surface tension and interfacial tensions were evaluated by capillary rise method. The CMC value was achieved by dissolving of biosurfactant in distilled water and the surface tension was calculated for various concentrations of the biosurfactant. The CMC was shown by sudden changes in the surface tension, which was obtained by plotting the surface tension as a function of biosurfactant concentration.

The conductivity test was applied for investigating the emulsion type of biosurfactants. Sodium chloride as an electrolyte was added to the emulsion and the conductivity was measured. If the conductance increases with increasing of concentration of sodium chloride, the emulsion is oil in water type. On the other hand, it is water in oil type if there is no significant changes in its conductance.

The E24 of the biosurfactants were examined by addition of 0.1 mg of biosurfactant to a crew-capped tube containing 1 mL of distilled water and 1 mL of hydrocarbons. The mixtures were then vortexed for 2 mins and leaved to stand for 24 h. The emulsification index (E24) is known as percentage of height of emulsified layer (mm) divided by total height of the liquid column (mm). The E24 of the formed emulsions were observed for 30 days.

3. Results and Discussion
The FT-IR spectrum of the produced biosurfactant suggested that rhamnolipids, glycolipids type biosurfactants, were produced by *P. putida* as reported previously [14]. Increased biosurfactant concentration in the medium resulted in lower surface tension of the medium. The occurrence of a sudden change in the value of the surface value of the medium or surface tension which tends to be constant at the addition of certain concentrations of biosurfactants, indicates that the CMC is achieved [10]. The CMC value of biosurfactant produced by *P. putida* is obtained for 798 mg/L. Janek *et al.* (2013) reported that biosurfactant from *P. putida* with glucose as carbon has a CMC value of 130 mg/L [7]. Martinez-Toledo and Rodriguez-Vazquez (2011) obtained a CMC value of 430 mg/L [15]. Although the biosurfactant produced in this study had a higher CMC value compared to previous studies, the CMC value of biosurfactant was lower when compared to SDS which had a CMC value of 2,100 mg/L [2]. The biosurfactant produced by *P. putida* was able to reduce the water surface tension by 18 mN / m (72 mN/m to 54 mN/m) at the CMC value. These results are not much different from the decrease in surface tension by biosurfactants produced in the study of Martinez-Toledo & Rodriguez-Vazquez (2011), which is 18.70 mN/m [15].

The emulsion system consists of two substances with different polarity which when mixed, one of these substances will be dispersed into other substances. This allows an o/w or w/o emulsion system. Conductivity measurement of oil and sample mixture before and after sodium chloride addition can be used to determine the emulsion system. Conductivity measurement of water emulsion with palm oil which has been added with biosurfactant before and after the addition of sodium chloride was performed to determine the biosurfactant emulsion system (Table 1).

As presented in Table 1, the addition of sodium chloride to distilled water increased the conductivity value. In the water, sodium chloride will be ionized so that the more sodium chloride added the more the amount of ions moves in the water causes more increase in the conductivity value. Palm oil has a conductivity value of 0 (zero), both before and after the addition of sodium chloride. Sodium chloride is not soluble in oil so the value of conductivity remains zero. Conductivity measurements of the oil mixture and biosurfactant solution produced a constant conductivity value of 0.10 μs both before and after the addition of sodium chloride. The constant value of conductivity after adding sodium chloride indicates that biosurfactant forms an emulsion between water and palm oil with a w/o system.

Table 1. Biosurfactant emulsion type investigation by addition of NaCl.

NaCl	**Conductivity (μs)**		
(mg)	A	B	C
0	10.00	0	0.10
5	0.61×10^3	0	0.10
10	1.03×10^3	0	0.10
15	1.34×10^3	0	0.10
25	1.90×10^3	0	0.10
50	2.94×10^3	0	0.10
100	4.05×10^3	0	0.10

A : aquadest; B : palm oil; C : aquadest, palm oil and biosurfactants (800 mg/L)

The produced biosurfactant was characterized for main surfactant properties which is its ability to lower the surface tension of water and immiscible compounds mixtures (Table 2). The immiscible compounds which were used in this study were olive oil, sunflower oil, soybean oil, palm oil, lubricant oil, benzene, hexane, diesel oil, petroleum oil and kerosene. The emulsion stabilities were observed for 30 days by monitoring the E24.

One characteristic of biosurfactants is as an emulsifying agent. Biosurfactants produced in this study were analyzed for their ability as emulsifying agents and their stability using several hydrocarbons both with vegetable oil (such as: olive oil, sunflower oil, soybean oil, palm oil) and with petroleum derivatives (such as: lubricant oil, benzene, hexane, diesel oil, petroleum oil and kerosene). The ability of biosurfactants in emulsifying water with several hydrocarbons is summarized in Table 2.

The biosurfactants produced by *P. putida* have emulsification capabilities for both water emulsions with vegetable oils and petroleum-derived hydrocarbons. The highest emulsion index was obtained between water and lubricating oil with an emulsion index value of 83% and stable for 30 days. The water emulsion index with vegetable oil used in this study provides an emulsion index of about 42-50% and with benzene an emulsion index obtained was 26%. Water emulsions with olive oil can last up to 25 days with an emulsion index value of 30%. Water emulsions with sunflower oil and soybean oil are stable up to 20 days. These results indicated that the biosurfactant could be useful as an emulsifying and emulsion-stabilizing agent for vegetable oils. Water emulsions with other hydrocarbons such as hexane, diesel oil, petroleum ether and kerosene have a low emulsion index which is around 4-11%.

The ability of commercial surfactant such as SDS (Sodium Dodecyl Sulphate) in emulsifying water with several hydrocarbons was performed for positive control (Table 3). SDS have emulsion index values between 55-76% for water emulsions with all hydrocarbons used, and stable for 30 day. Based on these results, the average emulsification ability of biosurfactants produced by P. *putida* is still lower when compared to SDS. Nevertheless, the water emulsion index with lubricating oil produced by biosurfactants is better when compared to SDS, it shows that the biosurfactant produced by *P. putida* using CPO as a carbon source has the potential to be applied in the environment.

Table 2. The emulsification index of emulsion between water and immiscible compounds with the presence of biosurfactants.

Day	Emulsification Index (E24) (%)									
	Olive oil	Sunflower oil	Soybean oil	Palm oil	Lubricant oil	Benzene	Hexane	Diesel oil	Petroleum oil	Kerosene
1	48	48	50	42	83	26	7	4	11	11
5	44	44	44	37	79	22	7	4	7	7
10	42	36	41	31	75	11	0	0	0	6
15	33	31	39	27	75	4	0	0	0	0
20	30	29	31	19	71	4	0	0	0	0
25	30	14	27	15	71	4	0	0	0	0
30	11	11	7	4	67	4	0	0	0	0

Table 3. The emulsification index of emulsion between water and immiscible compounds with the presence of SDS (Sodium Dodecyl Sulphate).

Day	Emulsification Index (E24) (%)									
	Olive oil	Sunflower oil	Soybean oil	Palm oil	Lubricant oil	Benzene	Hexane	Diesel oil	Petroleum oil	Kerosene
1	67	71	67	63	55	76	72	65	60	67
5	57	54	58	54	48	61	64	60	59	60
10	56	54	56	54	46	59	62	58	57	60
15	56	54	56	54	45	58	55	58	53	60
20	56	54	56	54	45	58	53	58	53	59
25	56	54	56	51	45	56	52	58	47	59
30	56	54	56	51	45	48	50	58	47	59

4. Conclusions

Biosurfactant produced by *P. putida* has a CCM value of 798.46 mg/L, capable of reducing water surface tension from 72 to 54 mN/m. It has an w/o (water in oil) emulsion system. The produced biosurfactants capable of stabilizing the emulsion between water and benzene, lubricating oil, palm oil, soybean oil, sunflower oil and olive oil.

Acknowledgments

The authors acknowledge their thanks to Universitas Sebelas Maret and Ministry of Research, Technology and Higher Education Indonesia for the financial support to this research.

References

[1] Baker S C and Chen C 2010 *Biosurfactants.* United Kingdom: Landes Bioscience and Springer Science, Busines Media.

[2] Pornsunthorntawee O, Wongpanit P, Chavadej S, Abe M and Rujiravanit R 2008 *Bioresour. technol.* **99** 1589-1595.

[3] Kosaric N, Cairns W L and Gray N C C 1987 Biosurfactants and Biotechnology. New York, Marcel Dekker, Inc.
[4] Mulligan C N 2005 *Environ. pollut.* **13** 183-198.
[5] Nitschke M and Costa S G V A O 2007 Biosurfactants in food industry. *Trends Food Sci. Technol.* **18** 252-259.
[6] Kosaric N and Vardar-Sukan F 2015 *Biosurfactans Production and Utilization Processes, Technologies and Economics.* New York, CRC Press.
[7] Janek T, Łukaszewicz M and Krasowska A 2013 *Colloids Surf B: Biointerfaces.* **110** 379-386.
[8] Aparna A, Srinikethan G and Smitha H 2012 *Colloids Surf B: Biointerfaces.* **95** 23-29.
[9] Raza Z A, Cox M, Khan M S and Khalid Z M 2007 *Process Biochem.* **42** 686-692.
[10] Abdel-Mawgoud A M, Aboulwafa M M and Hassouna N A 2009 *Appl. Biochem. Biotechnol.* **157** 329-345.
[11] Kaskatepe B, Yildiz S, Gumustas M and Ozkan S A 2015 *Braz. J. Microbiol.* **46** 855-859.
[12] Suryanti V, Marliyana S D, Handayani D S and Ratnaningrum D 2013 *Indo. J. Chem.* **13 3** 229-235.
[13] Suryanti V, Marliyana S D and Wulandari A 2015 *AIP Conf. Proc.* **1699** 040003
[14] V Suryanti, D S Handayani, S D Marliyana dan S Suratmi 2017 *IOP Conf. Ser. Mater. Sci. Eng.* 176, 012003
[15] Martinez-Toledo A and Rodriguez-Vazquez R 2011 *Ann. Microbiol.* **61** 605-613.
[16] Hatef Z N and Khudeir S H 2017 *IOSR J Pharm. Biol. Sci.* **2** 13-18.
[17] V Suryanti, D S Handayani, A Masykur and Lindasari 2017 *IOP Conf. Ser. Mater. Sci. Eng.* 75, 012013
[18] Poblete-Castro I, Becker J, Dohnt K, Santos V M and Wittmann C 2012 *Appl. Microbiol. Biotechnol.* **93** 2279-2290.
[19] Man Y B C, Haryati T, Ghazali H M and Asbi B A 1999 *J. Am. Oil Chem. Soc.* **76** 237-242.
[20] George S and Jayachandran K 2012 *J. Appl. Microbiol.* **114** 373-383
[21] Hamzah A, Sabturani N and Radiman S 2013 *Sains Malays.* **42** 615-623.

IOP Conf. Series: Materials Science and Engineering **578** (2019) 012022 doi:10.1088/1757-899X/578/1/012022

Optimization of the use of fly ash and natural zeolite activated as composite for an adsorbent of lead heavy metal (Pb)

Pranoto[1], T Martini[1], B M W Sunardi[1] and E P Anandita[1]
[1]Departement of Chemistry, Faculty of Mathematics and Natural Sciences, Sebelas Maret University, Surakarta, Indonesia

E-mail: pakpranotomipa@staff.uns.ac.id

Abstract. Coal is an alternative source of energy besides petroleum and natural gas which are widely used so they can produce a large amount of waste such as fly ash. Natural zeolite is a natural material that is widely spread in the territory of Indonesia. Fly ash and natural zeolite contain a SiO_2 and Al_2O_3 that can be used as an adsorbent. This study aims to create a composite of fly ash/natural zeolite activated as adsorbent to optimize the performance of Pb heavy metal adsorption. Fly ash and natural zeolite were activated using NaOH 3M, and HCl 1M. The adsorption process was carried out with variations of composition 100:00, 75:25, 50:50, 25:75, 00:100 (w/w), variations in pH of 1, 2, 3, 4, 5, 6, and contact time variations 15, 30, 45, 60, 75, 90 min. The adsorbent characterization was performed with FTIR, XRD, SAA, and total specific acidity as well as the final measurement of Pb metal concentration determined by AAS spectrophotometer. The activation process can be shown on the FTIR spectra with the shift of wave numbers, XRD diffractogram showed the decrease of intensity in fly ash and natural zeolite, the increase of acidity value and surface area on each material shows the availability of more active groups. The adsorption of metal Pb reached the optimum condition on fly ash:natural zeolite composition 75:25 (w/w), pH condition of 5, and contact time 60 min, and adsorption capacity of 0,577 mg/g and 97% adsorption percentage following Langmuir isotherm. The adsorption result obtained is supported by characterization of adsorbent after and before adsorption using SEM EDX indicating the presence of Pb element on adsorbent after adsorption and the surface morphology became rougher due to effect of Pb metal absorption.

1. Introduction

Coal is an alternative energy source that has abundance in Indonesia so that it has a relatively cheaper price. According to the World Coal Association (2012) Indonesia ranks the 5th largest coal producing country in the world in 2011 [1]. Sumatra and Kalimantan are Indonesian regions which contain a lot of coal. Based on the National Energy Policy the use of coal in Indonesia in 2005 is equal to 15% of the total national energy needs, and will continue to experience up to 2025 reaching 34% as a form of reduction in fuel use. The amount of coal usage is also balanced with the amount of residual waste using coal such as fly ash [2]. The use of fly ash by knowing the elements and minerals in it is often used as raw material for cement production, construction materials, concrete reinforcement, and heavy metal adsorbent [3,4]. The use fly ash to remove heavy metals has two advantages, namely to protect environment of heavy metal contamination and to reduce fly ash waste in the environment [5].

Fly ash contains as much as 90% consisting of SiO_2 and Al_2O_3 [6]. Utilisation of fly ash for removal of some pollutants such as heavy metal so that fly ash can be used as an adsorbent [7] which has low cost benefits. Natural zeolite is a hydrated mineral alumino silicate from porous structures with chemical physics such as cation exchange, molecular sifting, catalysis and adsorption. Natural zeolite is used as an adsorbent because it has a surface area large, low cost, and easy to get the material [8]. Zeolite exchangeable ions makes particularly suitable for removing heavy metal ions from industrial effluent waters [9].

Heavy metal is a metal known for its biodegradability and accumulation in the living system, which causes serious illnesses and disorders [10] and can also spurring environmental pollution both water, soil and air pollution. Expose this heavy metal is very risky for humans, which can damage human physiology and systems other biology if the level received is exceeded. Heavy metal cannot be degraded or destroyed [11]. Highly toxic heavy metal, can cause death or interference health that does not recover in a short period of time, such as lead (Pb).

This study aims to determine the performance of the manufacture of fly ash/natural zeolite adsorbents that has been given an activation treatment which is expected to increase the number of active site (-OH, -Al-OH, Si-OH) with high adsorption capacity, so it can adsorb Pb heavy metals optimally and effectively.

2. Experimental

2.1 Materials and Tools

The raw material of composite used in this study is fly ash derived from industry food at Mojosongo, Central Java, and natural zeolite from Klaten, Central Java. Fly ash and natural zeolite were initially ground to a 150 mesh screen. The initial material is washed and activated, fly ash uses NaOH 3M, natural zeolite activated uses HCl 1M . The material is neutralized pH using aquades and made variation of composite composition and a performance test was carried out using model of solution from Pb 1000 ppm. The tools used are buchner, FTIR, XRD, SEM-EDX, AAS, SAA (BET), Desiccator.

2.2 Methods

The adsorption process is carried out by determining the optimum conditions for each variation in the adsorption process of Pb 6 ppm. Variation in composite composition is carried out on comparison of fly ash:natural zeolite 100: 00, 75:25, 50:50, 25:75, 00: 100 (w/w), pH variation solution from pH 1-6, and variation of contact time at 15, 30, 45, 60, 75, 90 min, as well determine the type of adsorption isotherm on the fly ash:natural zeolite composite adsorption Pb heavy metals with various concentrations of 2, 4, 6, 8, 10, 12 ppm.

2.3 Determination of the type of adsorption isotherm

Determination of adsorption isotherms by simple linear regression with equations Langmuir and Freundlich. The adsorption isotherm and its type can be known from the R value approaching one [12,13].

2.3.1 Langmuir Isotherm

$$\frac{Ce}{qe} = \frac{1}{K_L q_m} + \frac{1}{q_m} Ce \tag{1}$$

where Ce is equilibrium concentration (mg/L), qe is the amount adsorbed at equilibrium (mg/g), qm is the maximum adsorption capacity that corresponds to the monolayer coverage which is complete (mg/g), and K_L is the Langmuir constant which is related to the adsorption energy Langmuir isotherm adsorption model.

2.3.2 Freundlich Isotherm

$$\ln qe = \ln K_f + \frac{1}{n} \ln Ce \tag{2}$$

where K_f and 1/n are Freundlich constants associated with adsorption capacity and adsorption intensity. In the Langmuir equation a graph of Ce qe vs Ce is made and the Freundlich equation is made graph equation lnqe vs lnCe.

3. Results and Discussion

3.1 Activation of Fly ash and Natural Zeolite

The activation process can dissolve impurities that are on the surface of fly ash with damaging the outer layer, removing water that is still trapped in the pores of the zeolite crystals later can increase surface area, causing loss of amorphous and impurity organic so that the acidic site which is initially covered by impurities becomes activated. This matter can be seen in FTIR spectra of fly ash and natural zeolites in Figures 1 and 2, that there are differences in wave and increase in transmittance intensity after the activation process, because of the increase intensity of the vibration energy.

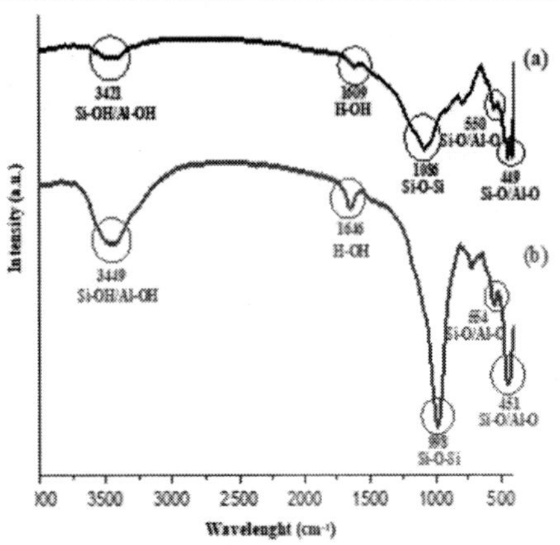

Figure 1. FTIR spectra of *fly ash* material before activation process

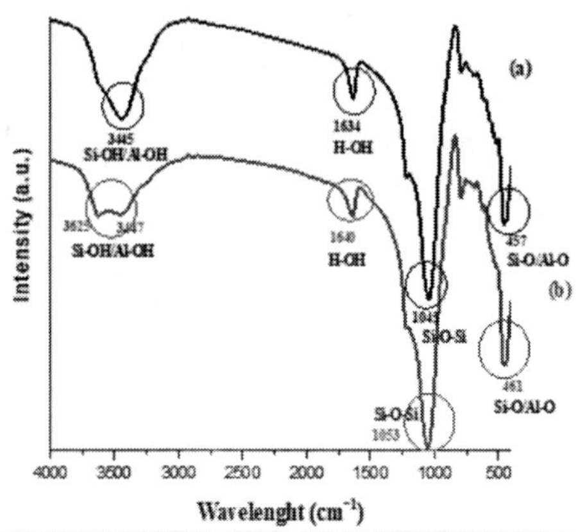

Figure 2. FTIR spectra of *fly ash* material after activation process

The presence of minerals in the fly ash and natural zeolite can also be seen in Figures 3 and 4, where on fly ash there is hematite, quartz low, mulite, and corundum, while at natural zeolite contains mordenite and clinoptilolite. The activation process also causes the intensity of the diffraction pattern increases and decreases due to loss impurity and damage to materials due to high temperatures.

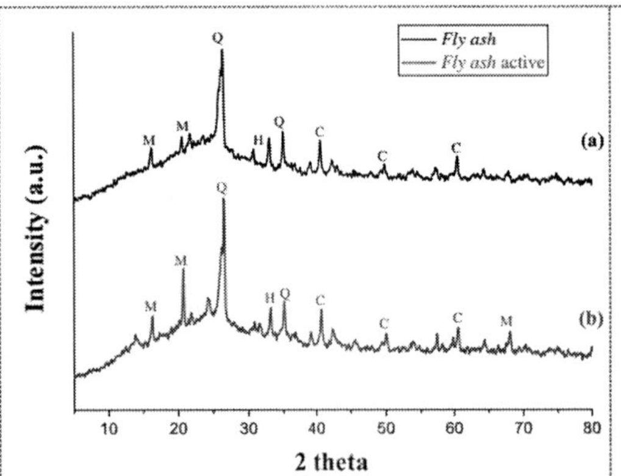

Figure 3. Diffraction patterns of *fly ash* material before activation process

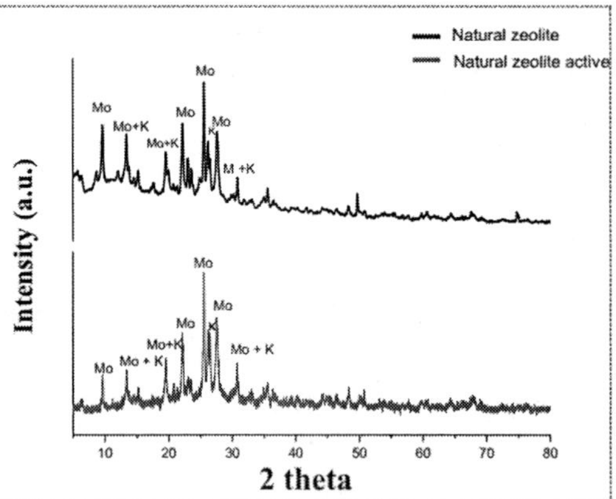

Figure 4. Diffraction patterns of natural zeolite material after activation process

3.2 Specific Total Acidity Analysis

Determination of specific total acidity was carried out using gravimetric method, with ammonia base was used as its adsorbate. Ammonia is a strong base that has a small molecular size so that it can be adsorbed and enter the active site pore of the material. Based on Table 1, the total specific acidity value of each adsorbent material increased after activation. This activation process can remove impurities that are on the surface of the material, thus causing the acidic site which was originally covered by impurities to be opened and appear. The large number of acid sites shows that the material has a cluster active which will bind more metal ions.

Table 1. Results of acidity analysis

No	Sample	Acidity (mmol/g)
1	Fly ash	0.824
2	Fly ash active	1.412
3	Natural zeolite	0.253
4	Natural zeolite active	1.056

3.3 Performance Test of Fly ash / Natural Zeolite Composite to Lead Metal (Pb)

Composite performance is carried out to determine the optimization adsorption process of heavy metals Pb. This optimization is carried out on variations in composition, pH of the solution, and contact time.

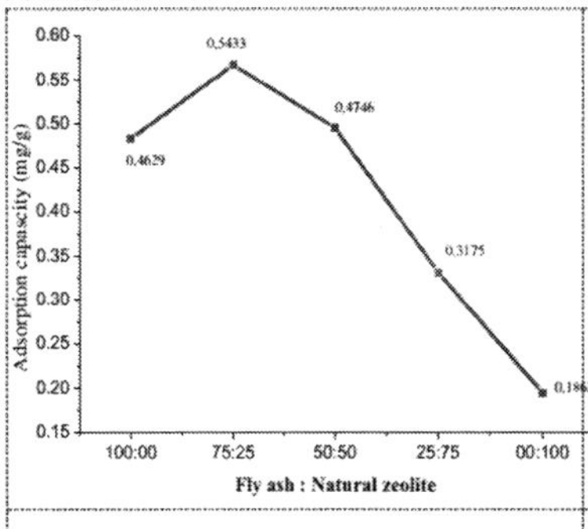

Figure 5. Graph performance of variation composition of composite

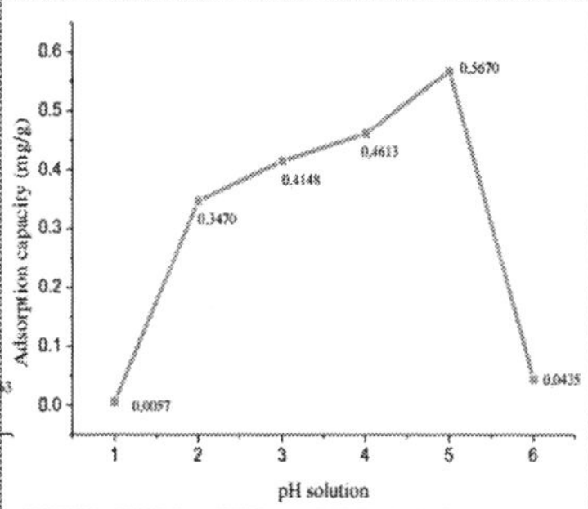

Figure 6. Graph performance of variation pH of the solution

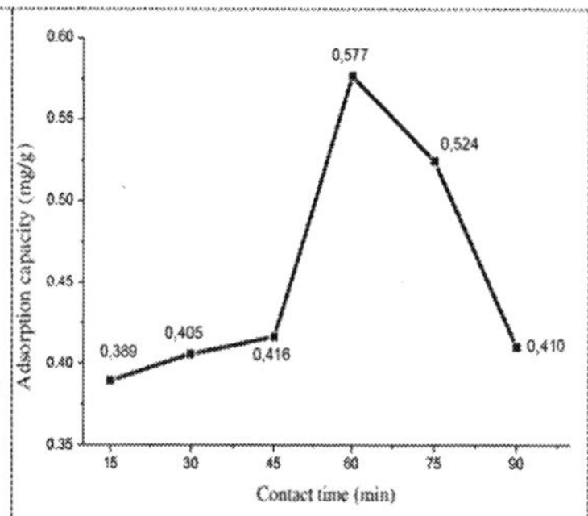

Figure 7. Graph performance of variation of contact time

It can be seen in Figures 5, 6 and 7. The optimization of performance of adsorbents on heavy metals Pb is the variation composition of fly ash:natural zeolite 75:25 (w/w) with adsorption capacity 0.5433 mg/g, the optimum composite shows that the two materials stick together to the pore wall so that the material surface area increases and supports the heavy metal adsorption process of lead (Pb). PH solution is conditioned using a buffer so that the pH of the solution does not change much when the adsorption process takes place. At pH <5 there is a decrease in adsorption capacity due to the acidic condition of the solution, the surface of the adsorbent positive which is surrounded by H^+ ion because the functional group in the adsorbent is protonated and will cause a competition with the Pb^{2+}. The conditions of pH 6 decreases the adsorption capacity and made Pb metal had saturated and formed $Pb(OH)_2$, the optimum variation in pH of the solution at pH 5 with adsorption capacity of 0.5670 mg/g, and variation optimum contact time at 60 minutes with an adsorption capacity of 0.5770 mg/g, shows that the adsorption capacity increases with increasing contact time, this happens because the the greater the interaction between adsorbent and Pb metal, but then it will experience a decrease in adsorption capacity if it has passed the optimum contact time, which occurs because more Pb metal adsorbed will reduce the adsorbent surface area so that the adsorbent can no longer adsorb Pb metal and cause desorption or the release of Pb metal which has been bonded to the adsorbent into the solution.

IOP Conf. Series: Materials Science and Engineering **578** (2019) 012022 doi:10.1088/1757-899X/578/1/012022

3.4 SEM EDX Analysis

The SEM results in Figures 8 and 9 show that the adsorbent after adsorption shows cracks in the adsorbent which is the place where metal ions are adsorbed, other than that it can be seen that the surface of the adsorbent has an increased level of roughness. This occurs because of the effect of adsorbing metal ions on the adsorbent, because of the type of adsorbate the watery interacts with the adsorbent. Pb attached to the surface of the adsorbent is proven by EDX results as shown in Table 2.

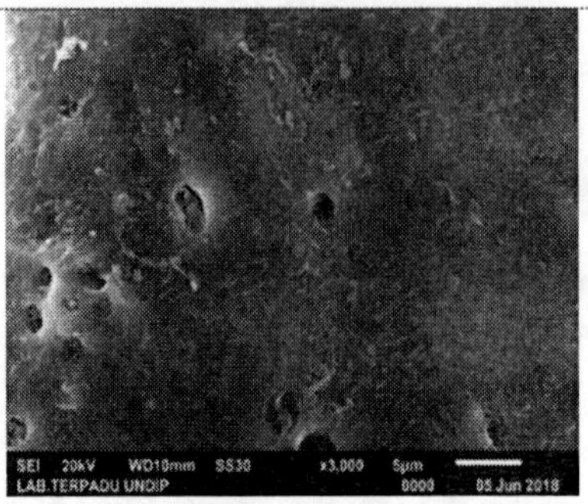

Figure 8. Morphology of the fly ash composite : natural zeolite 75:25 (w/w) before adsorption process

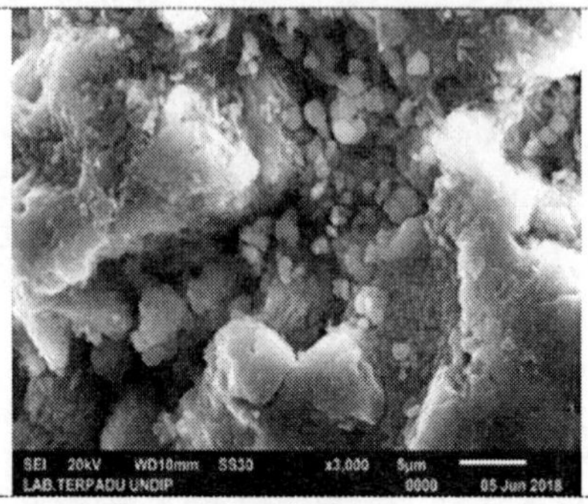

Figure 9. Morphology of the fly ash composite : natural zeolite 75:25 (w/w) after adsorption process

Table 2. Comparison of composite adsorbent elements before and after adsorption

Adsorben	**Kandungan Unsur (%)**		
	Si	**Al**	**Pb**
Fly ash/natural zeolite 75:25 (b/b) before adsorption	59.08	30.88	-
Fly ash/natural zeolite 75:25 (b/b) after adsorption	64.90	12.60	0.12

EDX results as shown in Table 2 show that Pb metal has been adsorbed on the adsorbent, this is shown from the shooting on the surface of a particular sample and the presence of Pb content in the composite after adsorption of 12%.

3.5 Specific Surface Area Analysis

Analysis of the surface area aims to find out how much the surface area of the adsorbent used in the adsorption process with Pb metal. The surface area can also show that the composite has been successfully formed by knowing the value of the measured surface area between the active fly ash forming material and the active natural zeolite with a new material fly ash/natural zeolite composite. The surface area of the adsorbent increases after the composting process which can be seen in Table 3, which is caused by the incorporation of the material that has been active who have many active sites and open pores.

Table 3. Result of specific surface area analysis

No	Sample	Specific Surface Area (m^2/g)
1	Fly ash active	27.940
2	Natural zeolite active	67.264
3	Composite	74.745

The incorporation of two active materials which has an active site that opens on the exposed surface and pores, causing fly ash particles to fill natural zeolite particles that will be interconnected each other so that it will form a space between layers that can increase the pore thickness and increase the surface area the fly ash/natural zeolite composite material so that it can be said that the composite material has been successfully formed because it has produced different characteristics from the initial material. The larger the area the surface of the active site on the material will be more which is wide this surface will provide space in the adsorption process for Pb metal.

3.6 Determination of Adsorption Isotherms

Determination of the adsorption isotherm aims to know what type of adsorption occurs in the adsorption process between adsorbent fly ash/natural zeolite to Pb metal. The coefficient of determination (R^2) of each of these isotherms can explain that the type of adsorption of fly ash/natural zeolite adsorbents (75:25) on Pb metal occurs chemistry ionic bonds and physically through van der walls forces.

Table 4. Parameter of isotherm adsorption

Adsorbent	Langmuir			Freundlich		
	qm	K_L	R^2	K_f	n	R^2
fly ash/natural zeolite composite 75:25	0.983	4.053	0.9661	0.689	3.571	0.8873

However, adsorption tend to follow the Langmuir isotherm which has an R^2 value which is closer to 1 means that on the surface of the adsorbent there is an active site that is proportional to the extent surface, which can adsorb only one molecule and adsorption is limited to formation of monolayer single layer, with adsorption capacity value of 0.983 mg/g.

4. Conclusion

Activation process on fly ash material, natural zeolite can dissolve the impurities there are a material that can be shown by characterization, where the FTIR spectra is fly ash, natural zeolites after activation experience a width and depth of transmittance bands, shift in wave numbers, diffraction patterns XRD shows a decrease and increase in intensity, an increase in acidity and surface area on fly ash, natural zeolite after activation shows availability active groups of Si-OH, Al-OH, Si-O-Si, Al-O-Al, Si-O, Al-O, and more adequate surface area in Pb heavy metal adsorption process. Adsorption of Pb metal reaches optimum conditions at fly ash:natural zeolite composition 75:25 (w/w), the pH condition of the solution 5, and the contact time of 60 minutes, as well as the adsorption capacity of 0.577 mg/g and adsorption percentage 97%. Adsorption of natural fly ash/zeolite adsorbents 75:25 (w/w) on Pb metal occurs chemically through ionic bonds and physically through van der waals forces, but adsorption tends following the Langmuir isotherm with adsorption capacity of 0.983 mg/g.

References

[1] World Coal Association 2012 *Coal statistics. Retrieved May* **1** p. 2013.
[2] Chindaprasirt P, Jaturapitakkul C, Chalee W, and Rattanasak U 2009 *Waste Manage. (Oxford)* **29**(2) 539-543.
[3] Ahmed Z T dan Hand D W 2014 *Industrial & Engineering Chemistry Research* **53(17)** 6985-6989.
[4] Cho H, Oh D and Kim K 2005 *J. Hazart. Mater.* **127**(1-3) 187-195.
[5] Kuncoro E P and Fahmi M Z 2013 *Procedia Earth Planet. Sci.* **6** 377-382.
[6] Jha B, Koshy N and Singh D N 2015 *Front. Environ. Sci. Eng. China* **9**(2) 216-221.
[7] Wang S, Ma Q and Zhu Z H 2008 *Fuel* **87**(15-16) 3469-3473.
[8] Samarghandi M R, Al-Musawi T J, Mohseni-Bandpi A and Zarrabi M 2015 *J. Mol. Liq.* **211** 431-441.
[9] Erdem E, Karapinar N and Donat R 2004 *J. Colloid Interface Sci.* **280**(2) 309-314.
[10] Mahmoud M E, Osman M M, Hafez O F and Elmelegy E 2010 *J. Hazart. Mater* **173**(1-3) 349-357.
[11] Karnib M, Kabbani A, Holail H and Olama Z 2014 *Energy Procedia* **50** 113-120.

[12] Shojaeipoor F, Elhamifar D, Moshkelgosha R. and Masoumia B 2016 *J. Taiwan Inst. Chem. Eng*. **67** 166-173.
[13] Huang X and Pan M 2016 *J. Mol. Liq*. **215** 410-416.

IOP Conf. Series: Materials Science and Engineering **578** (2019) 012023 doi:10.1088/1757-899X/578/1/012023

I-V characteristics on G4 DNA molecule: Electric field and current dependent hopping amplitude model

E N Sinurat[1], E Yudiarsah[1]

[1]Department of Physics, FMIPA Universitas Indonesia, Depok 16424, Indonesia.

E-mail : e.yudiarsah@sci.ui.ac.id

Abstract. We study the charge transport property of G4 DNA molecule by calculating the I-V characteristics at several temperatures. The G4 DNA we used consists of 32 G-tetrads built from four guanine bases and is contacted to metallic electrodes at both ends. The molecule is represented mathematically by using tight binding Hamiltonian model. Miller-abrahams scheme is used in calculating electron hopping constant under electric field which is applied along the molecular axis of DNA. Temperature is considered in the model by assuming twisting angles are distributed following Gaussian distribution. The current as voltage function is calculated using Landauer-Buttiker formalism from transmission probabilities. The result show that twisting motion due thermal agitation weakens the effect of electric field.

1. Introduction

The measurement of charge transport in DNA molecules [1] gave controversial results, DNA molecule has various conductivity properties ranging from insulators to proximity-induced superconductor [2]. Many factors can be the cause of these various results such as the influence of environment and molecule itself. Environmental effects may come from factors such as DNA-electrode contact, electric field, magnetic field [3], temperature [4], etc. The physical nature molecule itself may come to play in the form of DNA base sequence, DNA molecular length, internal DNA movement, etc. [5].

Recently the charge transport on G4 DNA has attracted the attention of researchers to study it. These attentions stems from its unique properties [6] and its long size so that more overlapping orbital which makes better conductivity [7]. In many charge transport experiments, a DNA molecule is directly connected to metallic electrode at both ends in order to obtain characteristics of electric current as voltage function [8]. In this study, we take into consideration the environmental influence on the G4 DNA molecules. The environmental influences brought into equation are electric field and temperature. The electric field effect of DNA molecule, which in turn affecting the transport process, in our knowledge is absent in the theoretical study of I-V characteristics of DNA molecule. In the model, the electric field causes significant change in electron hopping constant from a site to other site. The temperature causes structural fluctuations, such as vibration of DNA chain so that there is a change in the rotating angle between the nearest base pair and distance between one base to another base [9]. The structural vibration, in this study model with the twisting motion of G-quarted, can be seen as phonon excitation in DNA. The I-V characteristic of G4 DNA were studied using tight binding Hamiltonian model [5]. The electron transmission probability on the DNA molecule calculated using Green's function method was used in calculating I-V characteristics by using the Landauer-Buttiker formula.

IOP Conf. Series: Materials Science and Engineering **578** (2019) 012023 doi:10.1088/1757-899X/578/1/012023

2. Experimental

2.1. Theoretical model

In this study, we use tight binding approach in modelling single electron system in G4 DNA molecule. In this model, we assume that electrons can move from one base to neighboring base, both on the same strand (intrastrand hopping) and different strands (interstrand hopping diagonally and interstrand hopping vertically) but electrons cannot move from one backbone to backbone other [10].

G4-DNA is a structure constructed from square planar structures called G-quadruplex or G-quartet which consists of four guanine bases structure that are piled on top of each other and connected via Hoogsteen hydrogen bond. In G-quadruplex there is a cation between each pair of tetrad making G4 DNA structure more stable. Two adjacent G-quartets in the stack is separated by a distance of 3.4 Å makes a twiting angle of 30 ° [11]. In this study, the length of DNA molecule is 32 G-quarteds.

The single electron tight binding Hamiltonian model of G4 DNA molecule maybe written mathematically as follows.

$$H = \sum_{n=1}^{N}\sum_{\tau=1}^{4}\{(\varepsilon_n^{\tau} + \emptyset_n^{\tau})|n,\tau\rangle\langle n,\tau| + t_{n,n+1}^{\tau,\tau}|n,\tau\rangle\langle n+1,\tau|\}$$
$$+ \sum_{\tau=1}^{4}\{t_{n,n}^{\tau,\tau+1}|n,\tau\rangle\langle n,\tau+1| + t_{n,n+1}^{\tau,\tau+1}|n,\tau\rangle\langle n+1,\tau+1|\}$$
$$+ \sum_{\tau=1}^{4}\{B_n^{\tau_{\tau},\tau_{\tau}}|n,\tau_{\tau}\rangle\langle n,\tau_{\tau}| + t_{n,n}^{\tau,\tau_{\tau}}|n,\tau\rangle\langle n,\tau_{\tau}|\} + \mathrm{h.c} \tag{1}$$

The parameters in equation (1) have been explained in other report [12]. Based on Miller-Abrahams formula, effect of electric field which is applied parallel to the symmetrical axis of DNA on the electron hopping constant is modelled as follows [13]:

$$t_{i,i+1} = t_0 \exp\left\{-eE\left|\frac{Z(i+1) - Z(i)}{K_B T}\right|\right\} \tag{2}$$

In equation (2), $t_{i,i+1}$ is electric field affected electron hopping constant from initial site $(i+1)$ to final site (i). t_0 is electron hopping constant in the absence of electric field, k_B is Boltzmann constant and T is temperature, e is electron charge, $Z(i)$ is site-i position in the z axis of DNA. The effect of temperature in the form of G-quarted twisting motion is considered in this study following the model discussed in ref [9].

2.2. *Electric current calculation*

Transmission probability as explained in ref [14] is used to calculate electric current as voltage function using Landauer-Buttiker formula [11] :

$$I = \frac{2e}{h}\int_{-\infty}^{+\infty} T(E)[f_L(E) - f_R(E)]dE \tag{3}$$

,

where is $f_{L(R)}(E)$ is Fermi-Dirac distribution function

IOP Conf. Series: Materials Science and Engineering **578** (2019) 012023 doi:10.1088/1757-899X/578/1/012023

$$f_{L(R)}(E) = \{1 + \exp\left[\frac{E - \mu_{L(R)}}{k_B T}\right]\}^{-1} \tag{4}$$

Under bias voltage V, the chemical potential $\mu_{L(R)}$ of left (right) electrode can be written as

$$\mu_L = E_f + 1/2\, eV \text{ dan } \mu_R = E_f - 1/2\, eV \tag{5}$$

In this study, Fermi energy E_f and electron hopping constant on metallic electrode respectively are set to 8.45 eV and 0.75 eV following ref. [4].

3. Results and Discussion

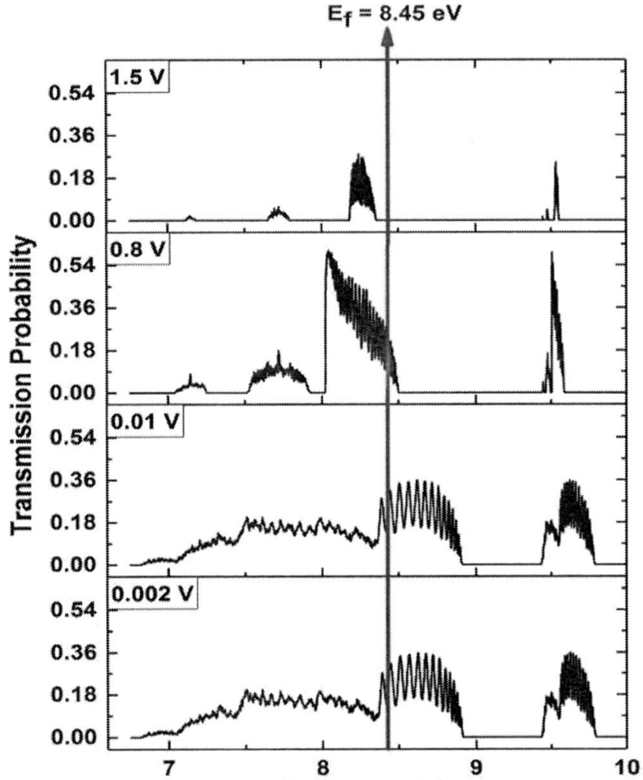

Figure 1. Electron transmission probability on 32 G-quarted stack of G4 DNA molecule calculated at temperature of 77.2 K and G-quarted twisting motion frequency of 2.00 meV for several voltages.

In this study of the properties of charge transport on G4 DNA molecule, the I-V charcteristics of this molecule has been calculated on 32-stack of guanine tetrad at G-quarted twisting motion frequency of 2.0 meV, temperatures of 77.2 K, 150 K, and 300 K. I-V Characteristics is calculated using Landauer-Buttiker formula from transmission probability. The transmission probability at temperature of 77.2 K shows that the increase of electric field causes transimission probability spectrum change in the form the energy gap becoms wider and energy bands become narrower. Howerver, the maximum transmission probability at first increases with the electric field, and after reaching a particular value its start decreasing. Figure 1 shows transmission probability calculated at temperature of 77.2 K and oscillation frequency of 2.0 meV for several voltages. The red vertical line is for showing the change in transmission probability spectrum around the Fermi energy. The states involve in the transport process are the states around Fermi energy of 8.45 eV in the range $E_f \pm 1/2\ eV$ (chemical potensial of left/right electrod). Since the transmission probability is used in calculating the I-V characteristics, the change in its spectrum with the electric field manifest in the I-V characteristics curve.

At temperature 77.2 K and low voltage of 2.0 mV, the charge transport process occurs in energy range of 8.45 eV ± 1/2 (0.002), which is around 8.449 eV – 8.451 eV. Even though the transmission probabilities are quite high, see Figure 1, since the number of state involved in the process is small, the current obtained is very small as can seen in Figure 2. As the voltages increases, the number of states involved in the transport process in increases since the Fermi energy still inside the band of states with high transmission probability. Therefor current increases considerably with voltages. At voltage of 0.8 V the charge transport process occurs in energy range of 8.45 eV ± ½ (0.8) which is around 8.05 eV - 8.85 eV. In this range a large number of states involved in the process and have high transmission probabilities. Despite of Fermi energy is located near band edge in energy band, the current through the molecule is very large. In contrast, at voltage of 1.5 V the charge transport process occurs in energy range of 8.45 eV ± ½ (1.5) which is around 7.7 eV - 9.2 eV below the Fermi energy. Other than low transmission probability, at this voltage less states involve in the transport process since the Fermi energy is located in energy gap, so the current flow smaller than the one at 0.8 V. As we increase the voltage even more, the current keep decreases, since the transmission probability becomes small and the energy band becomes farther away from Fermi energy. Similar trend is observe at current calculated for temperatures of 150 K and 300 K. The current to the voltage function tends to decrease when temperature is increased, this is because the structural irregularity in DNA molecule increases thus disrupting the electron transport procecess [4]. As the temperature increases, the voltage at which the current starts to decreases becomes higher, and at 300 K the current keep increasing with voltage up to voltage around 2.5 V, and then it decreases. This model shows that the change in the internal of G4 DNA molecule which affected the electron hopping constant may change the I-V characteristics curve of the molecule.

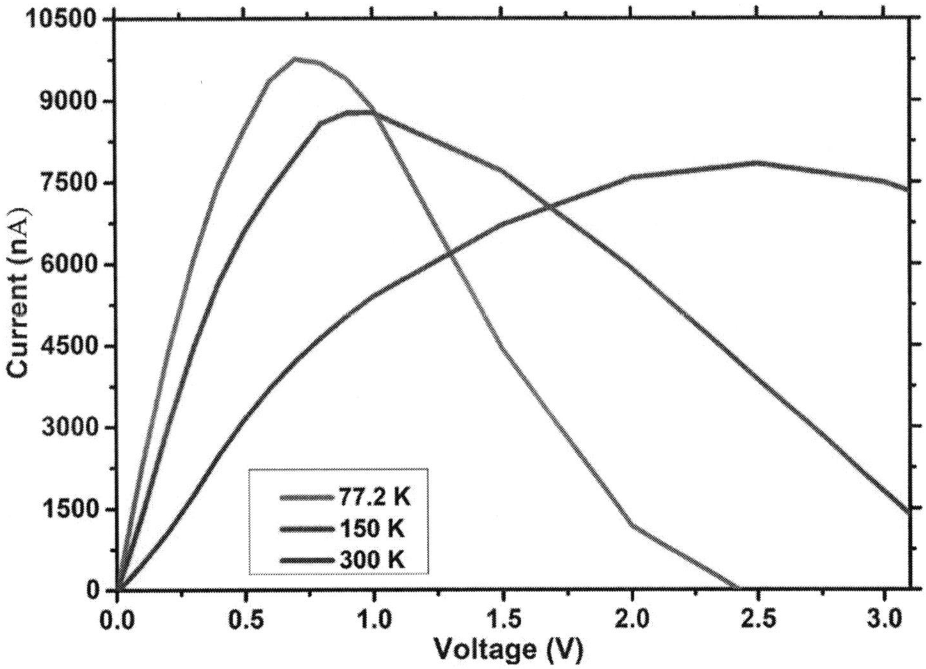

Figure 2. I-V characteristics calculated on 32 G-quarted stack of G4 DNA molecule for G-quarted twisting motion frequency of 2.00 meV and temperature of 77.2 K, 150 K, and 300 K

4. Conclusion

The I-V characteristics of 32 G-quarted stacked G4 DNA molecule have been studied under the influence of temperatures of 77.2 K, 150 K and 300 K. The I-V characteristic are calculated by taking into consideration the effect of electric field on the molecule in the form of electric field depedent electron hopping constant at G-quarted twisting frequency of 2.0 meV. The results show that the current increases with the electric field up to a certain value and then start to decreases. This trend is observed at all frequency used in this report. The increase of temperature causes electric current decreases because thermal fluctuations makes the structure of DNA molecule becomes irregular so that it hinder the transport process. In addition to that, the increase in temperture also results in the voltage where the current start to decreases becomes higher.

Acknowledgments

This project cannot be accomplished without assistance in the form of moral support, facilities, and material from various parties. Therefore, on this occasion the authors would like to express their deepest gratitude to Daniel K Suhendro for parameters used in calculation, CISCO Computer Laboratory Departemen Fisika for the facilities. The authors also thanks to Universitas Indonesia for funding the work through Hibah PITTA Nomor: 2259/UN2.R3.1/HKP.05.00/2018.

References

[1] Slinker J D, Muren N B, Renfrew S E and Barton J K 2011 *Nat. Chem.* **3** 230-235
[2] Simchi H, Esmaeilzadeh M and Mazidabadi H 2013 *J. Appl. Phys.* **113** 074701
[3] Endres R G, Cox D L and Sing R R P 2004 *Rev. Mod. Phys.* **76** 195-214
[4] Suhendro D K, Yudiarsah E and Saleh R 2010 *Phys. B Phys. Condens. Matter* **405** 4806–4811
[5] Schmidt B B, Hettler M H and Schön G 2007 *Phys. Rev. B* **75** 115125
[6] Guo A and Xiong S 2009 *Phys. Rev. B - Condens. Matter* **80** 035115

[7] Woiczikowski P B, Kuba T, Gutírrez R, Cuniberti G and Elstner M 2010 *J. Chem. Phys.* **133** 035103
[8] Kang D, Sun M, Zuo Z, Wang H, Lv S, Li X and Li L 2016 *Phys. Lett. A* **380** 977–982
[9] Yudiarsah E, Suhendro D K and Saleh R 2014 *AIP Conf. Proc.* **1617** 35
[10] Sinurat E N and Yudiarsah E 2017 *AIP Conf. Proc.* **1862** 030038
[11] Kang D, Jiang H, Sun Z, Qu Z and Xie S 2011 *J. Phys. Condens. Matter* **23** 55302
[12] Wiliyanti V and Yudiarsah E 2016 *AIP Conf. Proc.* **1729** 020033
[13] Ling L, Meller G, and Kosina H 2007 *Microelectronics J.* **38** 47-51
[14] Qi J, Edirisinghe N, Rabbani M G, and Anantram M P 2013 *Phys. Rev. B* **87** 85404

IOP Conf. Series: Materials Science and Engineering **578** (2019) 012024 doi:10.1088/1757-899X/578/1/012024

Influence of magnetic and electric field on I-V characteristic of the G4 DNA molecule

L Apriyanti [1], E Yudiarsah[1]

[1]Department of Physics, FMIPA Universitas Indonesia, Kampus UI Depok, Depok 16424, Indonesia

E-mail: e.yudiarsah@sci.ui.ac.id

Abstract. The I-V characteristics of G4 DNA molecule in applied magnetic and electric fields has been calculated. We use DNA molecule in the form of stacked of 32 guanine quadruplex (G4) structures. The DNA molecule is contacted to the metallic electrode at both ends. The system under investigation is modeled using tight-binding Hamiltonian approach. Magnetic and electric field affects electron hopping constant following Peirels Phase Factor and Miller-Abraham, respectively. The result shows slight current increment with the magnetic field and current increment with electric fields after certain voltage value.

1. Introduction

At this moment, science and technology are developing rapidly, particularly for nanotechnology and molecular electronic [1, 2, 3]. Various nanotechnology materials have been studied by researchers, one of which is biomolecular material. Biomolecular materials have the potential to support the development of nanoscale electronic devices.One of the biomolecular materials studied is deoxyribonucleic acid (DNA). The understanding of the electronic properties of DNA molecule has attracted a lot of attention of experimentalist and theorist [4]. DNA can be synthesized and has unique characteristics which are self-recognizing and self-assembly, in which scientists use DNA as an active element in a bottom-up nanotechnology approach [3, 5].

DNA molecule is composed of nucleotides, which are made from four unique nucleotides: adenine (A), thymine (T), guanine (G), and cytosine [6]. Four types of bases are for storing the genetic information concerning the structure and function of all living organisms [7]. A special kind of DNA molecule is composed solely of guanine nucleotides (G), G4DNA [8]. G4 DNA consists of a G-quartet arrangement, which is a planar structure of four guanines that are connected by hydrogen bonds [1].

During the past few decades, physics, chemistry, and biology communities studies focused on the nature of the transport of double-stranded DNA (ds-DNA), as one of the leading candidates for molecular electronics [8]. In this research, we studied the G4 DNA molecule. G4 DNA molecule is placed in an applied electric and magnetic field, which are set parallel to DNA symmetrical axis [9, 10]. This research aims to study electric and magnetic field influence on the electron transmission probabilityin G4 DNA molecule. The effect of electric field on G4 DNA molecule, which in turn will affect the transport properties the molecule,israrely, if ever, studied.

The parameters considered in this paper is including frequency, temperature, electric .field, and magnetic field. Green's function method along with tight-binding Hamiltonian of DNA molecule model is used in calculating the transmission probability. The molecule model and the method for calculating electron transmission probabilitywill be discussed in the theoretical method section. The

IOP Conf. Series: Materials Science and Engineering **578** (2019) 012024 doi:10.1088/1757-899X/578/1/012024

result of the calculation will be described and discussed in detail in the results and discussion. Then the results will be summarized.

2. Experimental

2.1 Theoretical Method

DNA model molecule used in this study isG4DNA modelof 32 base pairs. The unique structure of G4-DNA consists of a stacked of G quartet. Each G quartetis a set of four G bases put in a squarestructure with one G at its corner and is connected with hydrogen bonds. Schematic diagram of G4 DNA is shown in Figure 1. The system can be effectively described with a tight-binding Hamiltonian model.Tight-binding Hamiltonian model of G4-DNA is as follows [10, 11],

$$H = \sum_{i=1}^{L} \Big[\sum_{\tau=1}^{4} \{ (\varepsilon_i^{\tau} + \phi_i^{\tau}) |i,\tau\rangle\langle i,\tau| + t_{i,i+1}^{\tau,\tau} |i,\tau\rangle\langle i+1,\tau| \}$$
$$+ \sum_{\tau=1}^{4} \{ B_i^{\tau,q} |i,\tau\rangle\langle i,q| + t_{i,i}^{\tau,q} |i,\tau\rangle\langle i,q| \}$$
$$+ \sum_{\tau=1}^{4} \{ t_{i,i}^{\tau,\tau+1} |i,\tau\rangle\langle i,\tau+1| + t_{i,i+1}^{\tau,\tau+1} |i,\tau\rangle\langle i+1,\tau+1| \} \Big] + h.c. \quad (1)$$

Symbol in equation 1 can be explain as follow. L is the length of the DNA chain which is 32 G quartet. Indices i, , and $q$are consecutively for site index, strand index, and backbone index. Base's onsite energy and backbone's onsite energy are represented consecutively by parameters and . Electron's hopping constant, which depend on electric and magnetic fields, is represented by parameter$t$ .All parameter used in this work can be found in ref. [12].

Externalelectric field which is applied in perpendicular to the planar structure of G quartet, affect the electron hopping constant according to Miller-Abrahams[9,13].Meanwhile, the magnetic field which is also applied perpendicular to the planar structure of G quartet, change the phase of electron hopping constant. The change in phase is called *Peierls phase factor*[14]. Considering these two factors, electron hopping constant from site j to site i in external magnetic and electric fields can be stated as follows

$$t_{i,j} = t_o \exp - \left(i \frac{e}{\hbar} \int_j^i \vec{A} . \overrightarrow{dl +} eE \left| \frac{z(j)-z(}{K_B T} \right., \right. \quad (2)$$

where$\vec{A}$ is potential vector and E electric field.$z(i)$isz coordinate ofsite i.$t_0$is electronhopping constant without both electric field and magnetic fields. T, e, $\hbar$, and $k_B$consecutively are temperature, electron charge, Planck constant, and Boltzmann constant.

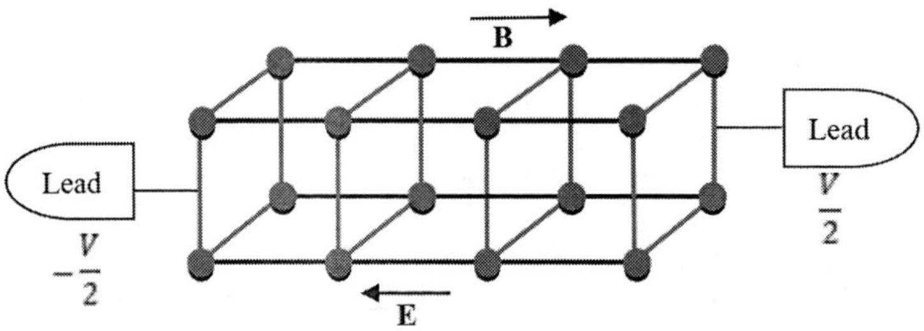

Figure 1. Scheme diagram of G4 DNA in the electric and magnetic fields.
The black color is the strand, the blue color is the site, and the red color is guanine bases.

Green's function methodis used in calculatingthe transmission probability of G4 DNA model. Green's function, advance/retarded, for single particle system represented with hamiltonian H can be stated

$$G^{R/A}(E) = \lim_{\eta \to 0^+}[E \pm i\eta - H^{\cdot}_{,} \quad (3)$$

where E is the energy of electron.Once theGreen's function is known, the transmission at a given electron energy can be calculated using the Fisher-Lee relation:

$$T = Tr[\Gamma_l G^R \Gamma_r G \quad (4)$$

represents the coupling of the molecule to the left electrode and represents the coupling of the molecule the right electrode. Function I can be expressed in terms of self-energy Σdue to left/right electrode as

$$\Gamma_{l/r} = i[\Sigma_{l/r} - \Sigma_{l}^{\cdot} \quad (5)$$

3. Results and Discussion

In this paper, we report the transmission probability calculated at two cases. In the first case, we want to study how the transmission probability change if the voltage is varied and the temperature and external magnetic field are kept constant. In the other case, we consider, instead of kept constant the magnetic field is varied and electric field and temperatureare constant.

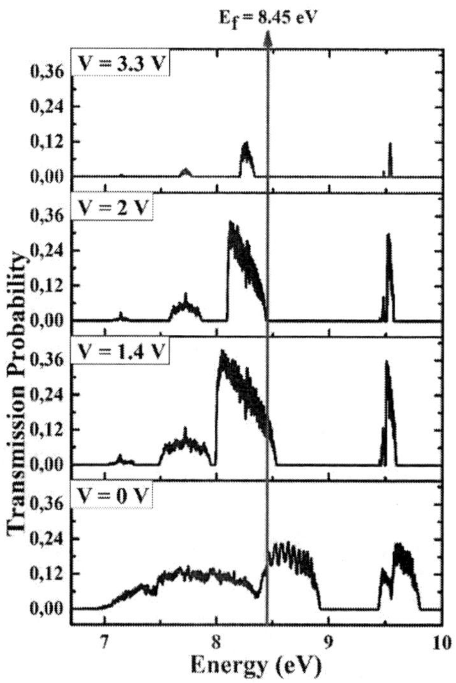

(a)

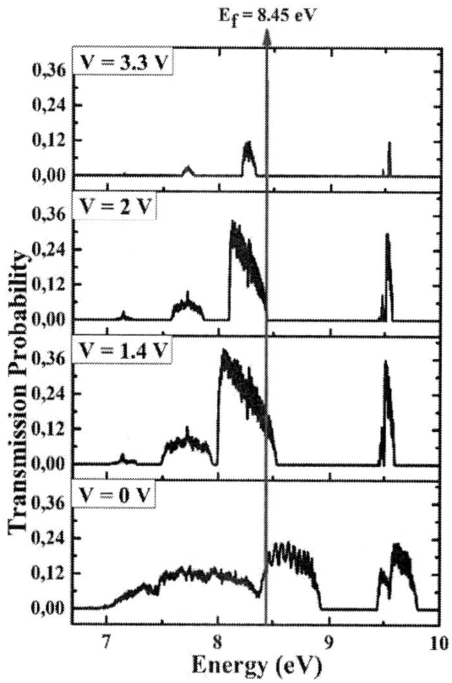

(b)

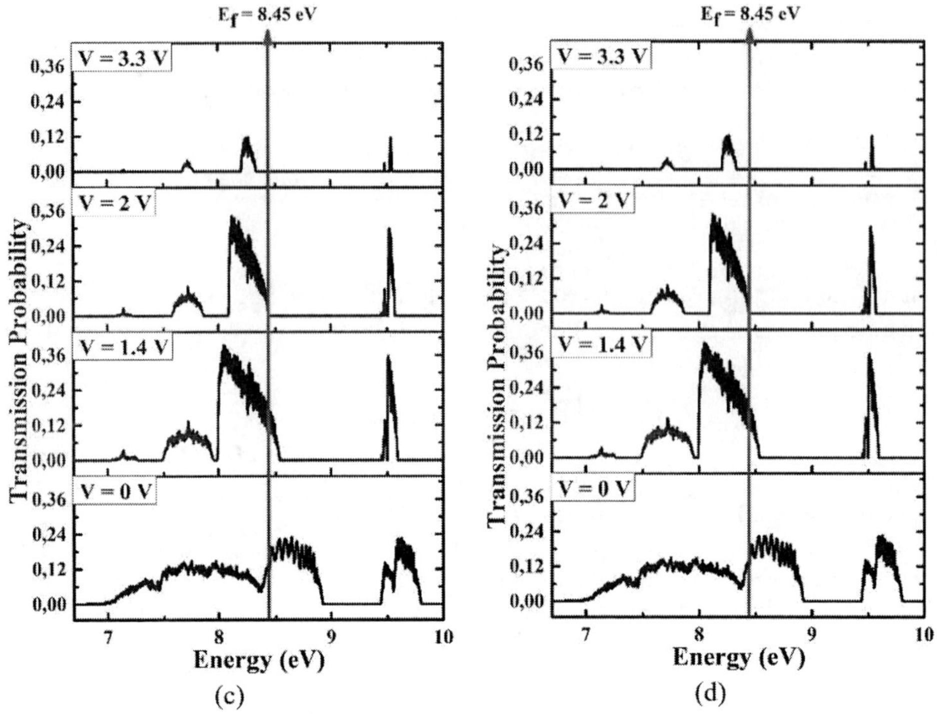

Figure 2. Electron transmission probability on 32 G quartet long G4DNA molecule calculated at temperature of 77.2 Kelvin, twisting motion frequency of 2 meV and external magnetic field of 3 Tesla for several voltages.

The first case is carried out by setting temperature, twisting motion frequency, and external magnetic field, respectively, to 77.2 K, 2 meV, and 3 T. The calculation results in this case are shown in Figure 2. As can be seen in figure, at zero voltage, the transmission probability spectrum shows two region with nonzero transmission probabilities separated by a gap, one broad band around energy range of 6.7 – 8.9 eV and one narrower band around energy of 9.4 – 9.8 eV. The gap is the region from 8.9 eV to 9.4 eV which have zero transmission probabilities. The red vertical line in the figure indicates the position of Fermi Increasing the voltage across DNA molecules, affecting the molecule manifested by the changes in electron hopping constants which are, in our model, becoming smaller. The increment of voltage is widening the gap between two bands and narrowing the band of states with nonzero transmission probability. Even though the band becomes narrower, its maximum transmission probability gets higher. Not only the maximum transmission probability gets higher, but also several other states experiences the increase in magnitude. As the center high energy band moves toward lower energy, the center of wide band moves toward opposite direction making the band edge gets closer to Fermi energy. Since the change in the transmission probability will be reflected in I-V characteristic, the electric effect on the molecule property modelled used in this study shows that the electric field can change the electrical properties. Within our model, since the band edge gets closer to the Fermi energy, at higher voltage less states with nonzero probability transmission are around the Fermi energy. Therefore less states are available for transport process.

IOP Conf. Series: Materials Science and Engineering **578** (2019) 012024 doi:10.1088/1757-899X/578/1/012024

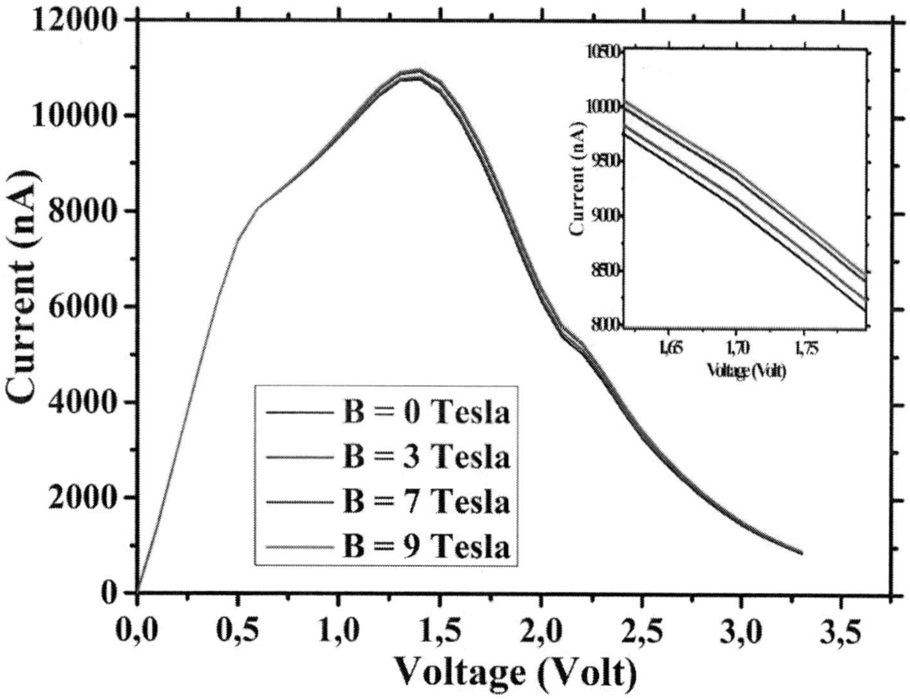

Figure 3. Electron transmission probability on 32 G quartet long G4DNA molecule calculated at temperature of 77.2 Kelvin, twisting motion frequency of 2 meV and external electric field of voltage of 0.4 V for several value of external magnetic field.

The result of the calculation of transmission probability at constant electric field for various values of magnetic fields, the second case, are shown in Figure 3. This calculation is carried out by setting temperature, twisting motion frequency, and external electric field, respectively, to 72.2 K, 2 meV, and the voltage of 0.4 V. As can be seen in the figure, the spectrum of transmission probability at no external magnetic field is very similar to the calculation result at low voltages discussed in the previous part. The presence of external magnetic field seems to not change much the transmission probability. The presence of external magnetic field only change the phase of the electron hopping constant not its magnitude. Therefore, the external magnetic field changes the phase of electron wave function which in turn changes the magnitude of electron wave function at one sites due to interferences of several wave function which come to that site via different route. Since the phase changes due to external magnetic field is small for small distance travel, the phase different of several waves arrive at one sites is also small. This result in very small changes in the electron transmission probability with the increment of external magnetic field.

4. Conclusion

The aims of this research are to study how the magnetic and electric field influence I-V characteristic on G4 DNA molecule. The transmission probabilitycalculated in the presence of external electric field shows that the higher the voltage, the wider the energy gap at the transmission probability spectrum. This is probably due to the electric field in the model makes the value of the electron hopping constants decreases with voltage.At theend, the effect of external magnetic field on the molecule included in the model, causes less states can be participated in the transport process. The calculation result of electron transmission probability under the influence of external magnetic field shows that the transmission probability in the range of applied magnetic field of 0 - 9 Tesla tends to be constant when the magnetic field increases. Therefore, the magnetic field influence on the transmission probability is not significant.

Acknowledgements
The author would like to thank all those who contributed to this study, both in the form of moral support, facilities, and material. Thanks to Cisco Laboratory of Department of Physic Universitas Indonesia for permission to use the laboratory. Thanks also to Universitas Indonesia for funding this study through Hibah PITTA DRPM 2018 contract number: 2259/UN2.R3.1/HKP.05.00/2018.

References

[1] Kang D, Sun M, Zuo Z, Wang H, Lv S, Li X and Li L 2016 *Phys. Lett. A.* **380** 977-982
[2] Klotsa D, Romer R A and Turner M S 2005 *Biophys. J.* **89** 2187-2198
[3] Yudiarsah E,Suhendro D K and Saleh R 2014 *AIP Conf. Proc.***1617** 35-38
[4] Apriyanti L and Yudiarsah E 2017 *AIP Conf. Proc.***1862** 030043
[5] Woiczikowski P B, Kuba T, Gutierrez R, Cuniberti G and Elstner M 2010 *J. Chem. Phys.* **133**, 035103
[6] Guo A and Xu H 2007 *Phys. Lett. A.* **364** 48-53
[7] Alvarez J R, Skachkov D, Massey S E, Lu J, Kalitsov A and Velev J P 2014 *Phys. Rev. Applied.* **1** 034001
[8] Guo A and Xiong S 2009 *Phys. Rev. B.* **80** 035115
[9] Jadhav P J, Limketkai B N and Baldo M A 2010 *Adv. Polym. Sci.* **223** 29-44
[10] Kang D, Jiang H, Sun Z, Qu Z and Xie S 2011 *J. Phys. Condens. Matter.* **23** 055302
[11] Guo A-M, Yang Z, Zhu H J and Xiong S J 2010 *J. Phys. Condens. Matter.* **22** 065102
[12] Wiliyanti V and Yudiarsah E 2016 *AIP Conf. Proc.***1729** 1–6
[13] Ling L, Meller G and Kosina H 2007 *Microelectronics J.* **38** 47-51
[14] Peierls R 1933 *Zeitschrift fur Phys.* **80** 763–791

IOP Conf. Series: Materials Science and Engineering **578** (2019) 012025 doi:10.1088/1757-899X/578/1/012025

Modelling electric field influence on density of states (dos) and localization length of poly(da)-poly(dt) dna

F Rahmani[1] and E Yudiarsah[1]
[1]Department of Physics, FMIPA Universitas Indonesia, Depok 16424, Indonesia.

E-mail : e.yudiarsah@sci.ui.ac.id

Abstract. DNA based microelectronics technology have been developed in the last couple of years, but researchers found that its stability can be volatile to the change in its environment. In this research, electron transport in DNA poly(dA)-(dT) have been analysed by observing its density of states (DOS) and localization length. Retarded green function is used to calculate DOS, whereas transfer-matrix method is employed in calculating localization length in order to determine the electronic structure of the DNA. Along the z-axis of DNA electric field is applied, it will alter hopping constant between sites in DNA. Tight binding model is incorporated with Miller-Abraham expression to yield electric field and also temperature influence in DOS. Effect of medium is considered within the calculation. The results tell us that as the electric field increases, the value of DOS decreases, so that the localization length. The increment of temperature also shows similar trend.

1. Introduction

Charge transport properties in DNA molecule system has received interest in scientific community over the last two decades. The fact that there are similarity between the arrangement of π orbital in DNA bases stacks and π orbital in stacked metallic aromatic organic crystals, like Bechgaard salts [1], could motivate scientist to be assured of DNA conductive ability [2]. Barton et al. reported that charge transfer in DNA which is intercalated by transition metal complex is distance-independent [3]. The mechanism of charge transport in DNA can be explained by two major processes, that are superexchange and hopping mechanism, both have been verified experimentally [4]. Variety of results has been occured and studied extensively both theoritically and experimentally [5-9]. *Ab initio* calculations predict dry poly(dG)-poly(dC) DNA to be a wide-gap semiconductor [5], and it is in agreement with experimental result that show it has 2.0 V voltage gap under low temperature [6]. Experiment performed by Yoo et al. concludes that poly(dA)-poly(dT) could behave as an n-type semiconductor, while poly(dG)-poly(dC) has a p-type semiconductor characteristic [7]. This behaviour owing to base-pair-sequence-dependence of the observed molecule.

The environment impacts (i.e electric field [7], magnetic field [8], and temperature [9]) also play important role in altering DNA charge properties, due to its sensitivity. Research conducted by Wong et al indicates higher efficiency in magnetic field induced radical cation propagation which leads to more electron conductivity in DNA duplexes [8]. Phonon excitation in DNA due to temperature effect can be treated as base pair twisting and stretching modes, and it affected on its localization length decrement. This outcome caused by the widening of twist angle distribution width [9]. Electric field can also modify DNA charge transport process. Experimental research by Chunda Zhou et al shows that when double-stranded DNA is conducted with AC electrical fields at frequencies of a few hundred Hertz, it exhibit breakage of the DNA conformation [10]. Study by Malyshev argue that perpendicular gating electric field could generate potential harmonic modulation on the base on-site energy along helical axis, so in consequence it leads to effective increase of transmission and voltage gaps [11]. Theoritical work by Li et al. reckon when electric field exists, it could transform hopping constant amplitude of an organic semiconductor [12], which in this study becomes a major influence of our DNA energy band calculation.

Numerical calculation in understanding qualitative properties of the poly(dA)-poly(dT) DNA in electric field influence has been done in this research. We studied its electron transport by examining electron density of states (DOS) and also localization length. Temperature effect is seen as collective phonon vibration excitation, wherein leads to fluctuation in twist angle and distance between base pair [9]. Miller-abraham formula is used in modelling the electric field influence on the molecule system [12]. Both incorporated into our Hamiltonian tight-binding expression [13]. In this work we account contribution of backbone disorder, watson-crick hopping constant amplitude, and homogeneous interelectrode field.

2. Theoritical Method

Periodic nucleotide chains consist of adenine (A) and its complementary base, thymine (T), is sandwiched between two electrodes to provide different potential in both ends. Charge can propagate along DNA base pairs and its backbone (sugar-phosphate). The pathway of charge propagation described as interstrand, intrastrand, and in addition we also consider in watson-crick pairs hopping. Electron charge also can hop from bases to the nearest backbone or vice versa, and the backbones are interconnected. Each base and each backbone is viewed as one site. Parameter $t_{i,j}^{\tau,\sigma}$ describes electron hopping between each site. Environment contact related effect is not considered in our calculation, since we treat our molecule as dry DNA system. The tight-binding Hamiltonian model is given by [13]

$$H = \sum_{i=1}^{L}\left[\sum_{\tau,\sigma=1}^{4}(\varepsilon_i^{\tau} + \phi_i^{\tau})|i,\tau\rangle\langle i,\tau| + t_{i,i}^{\tau,\sigma}|i,\tau\rangle\langle i,\sigma| + t_{i,i+1}^{\tau,\sigma}|i,\tau\rangle\langle i+1,\sigma|\right] + h.c \quad (1)$$

The length of the molecule is represented by L (pairs). Electron position in molecule is represented by i (sites), τ (base strands), and σ (backbone strands) symbols. Onsite energy at each site in a particular strand is denoted as ε_i^{τ} (8.22 eV for A, 9.06 eV for T, and 9.36 eV for backbone). Phonon vibration excitation is given by $\emptyset_i^{\tau}$ parameter, namely twisting mode effect on the onsite energy, where it is assumed to decrease with twist angle increment. This parameter contains not only electron-phonon coupling, but also the change of twist angle of the i-th base in strand τ [9]. Helical structure of DNA is not neglected in this work and preserved in the form of relative twist angle between neighboring base pairs in equilibrium, and the value is constant for all base pairs (36°). Gaussian distribution is used to randomly distribute the twisting angle of bases on temperature with zero mean and standard deviation proportional to $\sqrt{\frac{T}{f}}$; temperature (T) used in this study are 30 K and 77 K. f is proportional to the square of twisting frequency (ω) [9]. Onsite energy value for both metallic electrode is 8.45 eV, while molecule-electrode hopping constant is 0.75 eV.

The molecule is influenced by gated electric field along its helix axis. This perturbation will modify electron hopping constant (t) from site i to j in molecule site, according to Miller-Abraham formula [14]. So, the electron hopping constant will become

$$t_{ij} = \begin{cases} t_0 \exp\left(\frac{qE}{\kappa k_B T}(z_i - z_j)\right); & if\ (z_i - z_j) < 0 \\ t_0 & ; otherwise \end{cases} \quad (2)$$

where z_i is z coordinate of site i, q is electron charge, E is electric field amplitude which represented by voltage difference (V) across molecule, κ is the DNA in medium dielectric constant and k_BT is thermal energy at temperature T.

Transfer matrix method is employed to calculate localization length. However, this method will produce large number of matrices multiplication, where multiplication of exponentially small number with exponentially large number is inevitable. In consequence, there would be numerical error. To avoid this problem we use Gram-schmidt orthonormalization method to obtain coefficient value that is used to calculate Lyapunov exponents (δ). Then the smallest value of δ is taken to yield localization length (λ) [12].

$$\lambda = \frac{1}{\delta} \tag{3}$$

DOS of the molecule will be generated numerically by Green's function method. Furthermore, metallic electrodes at both end of DNA molecule is evaluated within calculation through the self-energy (Σ) matrices that will be subtituted into DOS formula. The DOS formula is given by

$$DOS(\varepsilon) = -\frac{Im}{\pi} Tr[G(\varepsilon, \vec{k})] \tag{4}$$

3. Results and Discussion

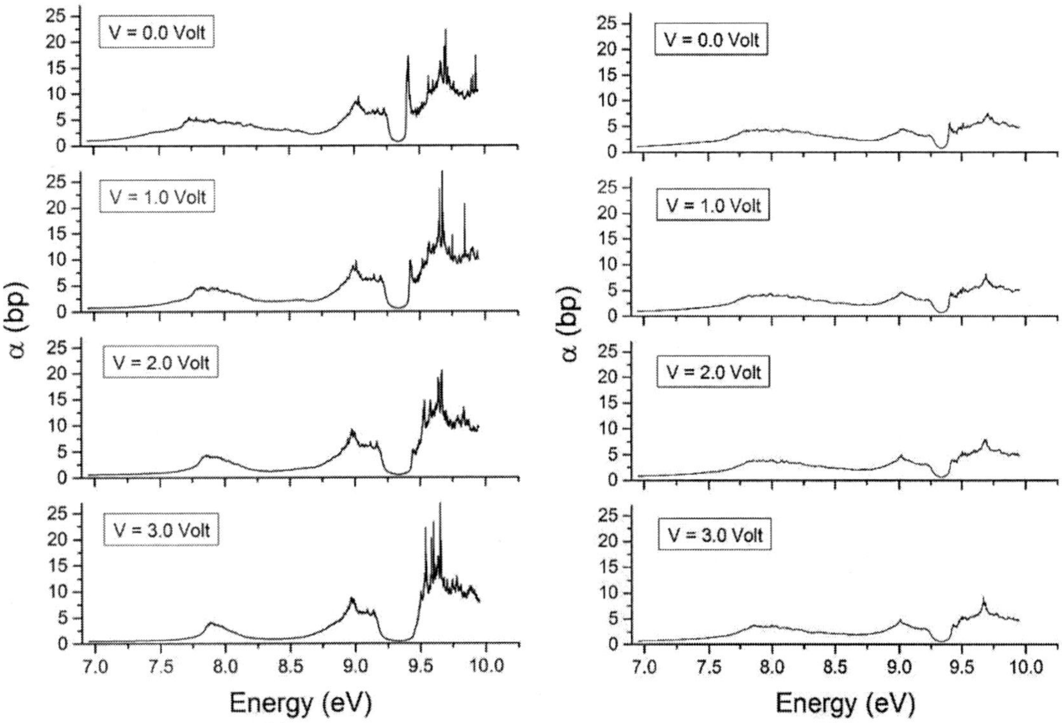

(a) (b)

Figure 1. Electron localization length of poly(dA)-poly(dT) DNA in voltage of 0 – 3 volt, calculated at temperature of (a) 30 K and (b) 77 K

Equation (2) shows hopping constant amplitude $t_{i,j}^{\tau,\sigma}$ are modified by electric field, so that later will change electron energetic movement that have been incorporated into the tight-binding Hamiltonian used in this work. This change appears in Figure 1 and Figure 2 through the widening of band gap, region in which DOS and localization length are almost zero, by the presence of electric field. We can also see the energy bands, where both aforementioned values are larger than zero, are narrowing. In Figure 1 the localization length do not show extended states, valued less than N (102 pairs), that being said electron can not reach the end of electrode. Work by Suhendro et al. argue if localized states emerge by disorder in backbone onsite energy together with thermal fluctuation, and it caused disorder in effective bases onsite energy, so that the difference between backbone onsite energy and bases onsite energy in few electronic energy states are quite minuscule to yield extended states [9]. The widening of band gap in Figure 1a imply that charge trasport in DNA poly(dA)-(dT) is compromised by the influence of electric field. A similar trend also occurs for DOS as seen in Figure 2a, where the peaks of energy band near gaps are degrading, in consequences it widen the band gaps. It interprets that within the framework used in this research, electric field disturbs the DNA molecule, defying the conventional assumption that electric field merely drop at molecule and electrode contact.

The electric field effect is visibly more prominent for 30 K temperature in either localization length (Figure 1a) or DOS (Figure 2a), but moment when temperature increase to 77 K (Figure 1b and 2b) the effect is no longer seen. From Figure 1 we can see that along with the increasing magnitude of temperature, localization length drops quite significantly. In high temperature standar deviation of twist angle distribution will be affected, and makes it wider. In accordance, it causes variation in both intrastrand and interstrand hopping amplitudes between sites, so it provide onsite energy disorder which hinder charge transport in the molecule [14].

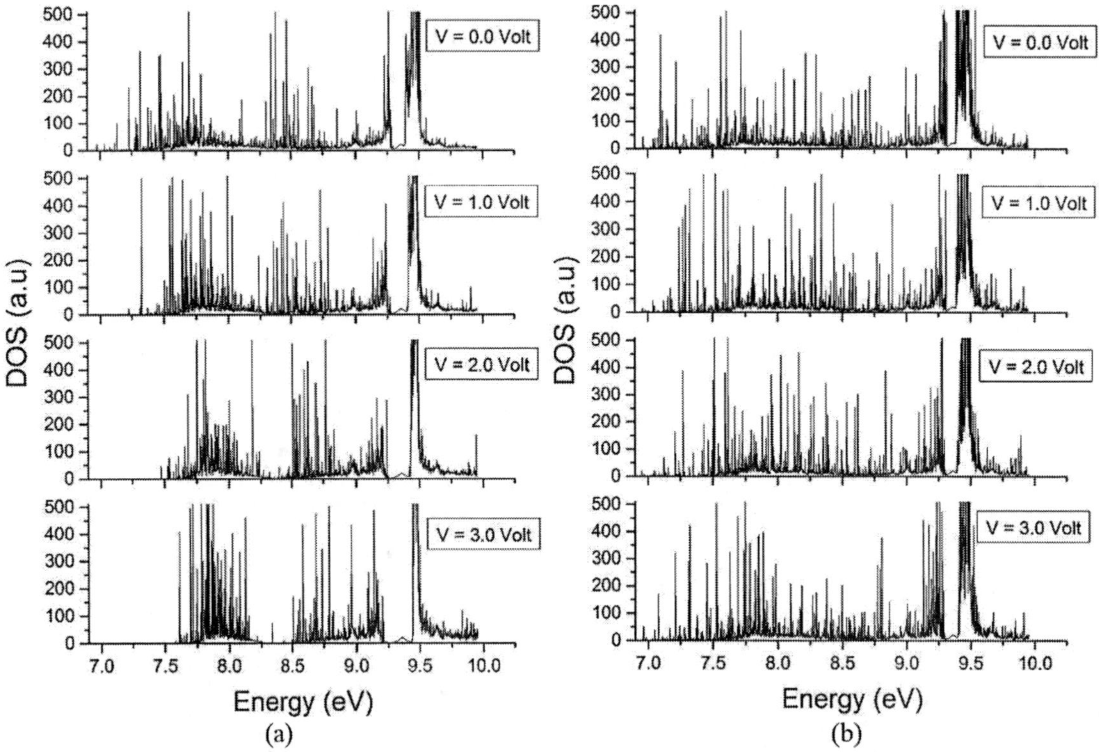

Figure 2. Electron density of states of poly(dA)-poly(dT) DNA in voltage of 0 – 3 volt, calculated at temperature of (a) 30 K and (b) 77 K

Thermal energy transforms the onsite energy of bases that would obtain negative impact on molecule energy structure. Thermal fluctuation have primary effect in localization length decrement than electric field. This could be seen in Figure 1b where voltage increment makes insignificant change to the localization length. As in DOS when the temperature is 77 K (Figure 2b) the earlier gap created by electric field now have been filled by several new states. This new states is analysed to be phonon excitation states which emerge by force of thermal energy. The electric field effect cannot also be seen affecting DOS in Figure 2b in terms of widening or narrowing band region. In addition, as the temperature rise up the DOS shows the molecule to be more disordered.

4. Conclusion

Theoritical work in calculating electric field influence on DOS and localization length of poly(dA)-poly(dT) DNA had been done. In 30 K temperature, higher electric field results in the widening of localization length and DOS band gap, but when the molecule is in 77 K the effect is no longer visible. The field effect occurs because electric field could modify hopping constant amplitude according to Miller-Abraham formula [12]. However, phonon excitation activated by thermal energy is more prominent than field effect.

Acknowledgments

We would like to thank Daniel K. Suhendro for giving us calculation parameters. We would also give our gratitude to CISCO Laboratory, Physics Department, Universitas Indonesia for letting us to use their computational facility. This research is supported by Hibah PITTA DRPM UI 2018 Number: 2259/UN2.R3.1/HKP.05.00/2018.

References

[1] Roth S R and David C 2015 *One-Dimensional Metals: Conjugated Polymers, Organic Crystals, Carbon Nanotubes and Graphene* (Germany: VCH Verlagsgesellschaft, Weinheim) pp 31-34
[2] Eley D D and Spivey D I 1962 *Trans. Farad. Soc.* **58** 411–415
[3] Murphy C J, Arkin M R, Jenkins Y, Ghatlia N D, Bossmann S H, Turro N J, and Barton J K 1993 *Science* **262** 1025–1029
[4] Jortner J, Bixon M, Langenbacher T, and Michel-Beyerle M E 1998 *Proc. Natl. Acad. Sci. USA.* ***95*** 12759–12765
[5] Mertig M, Kirsch R, Pompe W, and Engelhardt H 1999 *Eur. Phys. J. D.* **9** 45-48
[6] Storm AJ, van Noort J, de Vries S, and Dekker C 2001 *Appl. Phys. Lett.* **79** 3881
[7] Yoo K H, Ha D H, Lee J O, Park J W, Kim J, Kim J J, Lee H Y, Kawai T, and Choi H Y 2001 *Phys. Rev. Lett.* **87**, 198102
[8] Wong J R, Lee K J, Shu J J, and Shao F 2015 *Biochemistry* **54**, 3392–3399
[9] Suhendro D K, Yudiarsah E, and Saleh R 2010 *Phys. B Cond. Matt.* **405** 4806-4811
[10] Zhou C, Reisner W W, Staunton R J, Ashan A, Austin R H, and Riehn R 2011 *Phys. Rev. Lett.* vol. **106** pp. 248103
[11] Malyshev A V 2007 *Phys. Rev. Lett.* **98** 096801
[12] Ling L, Meller G, and Kosina H 2007 *Microelec. Jour.* **38** 47–51
[13] Rahmi K A and Yudiarsah E 2016 *AIP Conf. Proc.* **1729** 020037
[14] Yudiarsah E, Suhendro D K, and Saleh R 2014 *AIP Conf. Proc.* **1617** 35–38

IOP Conf. Series: Materials Science and Engineering **578** (2019) 012026 doi:10.1088/1757-899X/578/1/012026

Utilization of Yogyakarta natural zeolites to reduce NH_4 and NO_2 levels in shrimp pond water and its kinetic rate study

D D Anggoro[1], L Buchori[1], I Sumantri[1], Ivan[1], D A B Sejati[1], H Oktavianty[1]

[1] Department of Chemical Engineering, Diponegoro University, Semarang, 50275, Indonesia

E-mail: anggorophd@gmail.com

Abstract. Natural zeolite is an unique crystal that has a very small and uniform pore size when compared to other adsorbents, such as activated carbon and silica gel. This study aims to determine the potential of natural zeolite as an adsorbent, to absorb toxic gases in pond water, especially shrimp ponds and to study its kinetics rate of NH_4 and NO_2 adsorption. This study consisted of two stages: activated zeolite using NaOH, and tested the absorption power of NH_4 and NO_2 gases. Zeolite was activated using NaOH where 25 grams of zeolite were mixed with 0.5N NaOH and heated for 2 hours at 70°C, then stirred and dried in an oven at 110°C for 4 hours. Then zeolite was cooled in a desiccator. The results of NH_4 and NO_2 gas adsorption showed that natural zeolites can reduce NH_4 and NO_2 gas levels in shrimp pond water. The rate constant of pseudo-second-order, and the initial adsorption rate at any given time t, can be predicted from the equation which is full zeolite obtained the best perform of NH_4 and NO_2 adsorption.

1. Introduction

Development of cultivation systems from traditional to intensive with the majority shrimp ponds have potential towards increasing environmental pollution. Less optimal utilization of excessive feed will cause accumulation of organic matter. Decomposition of organic matter requires oxygen in the process, so that the availability of oxygen for the biota in it is reduced. If this happens continuously it will cause death for shrimp and other biota. Pollution materials that are difficult to decompose by microorganisms also cause hoarding and result in damage to the environment which will directly disrupt organisms that live in these environments. Organic pollution materials that function as fertilizers are actually detrimental due to algae blooms and aquatic plantscausing oxygen competition in the waters. The above factors are the cause of the decline in the body's resistance to the attack of the disease because of the poor quality of the environment, if this is left continuously then mass death will occur so that the population will decline [1].

Traditional methods for nitrogen removal from wastewater are denitrification, nitrification, chemical coagulation, adsorption, selective ion exchange, ammonia stripping, electrodialysis, filtration, reverse osmosis [2, 3, 4, 5]. Removal processes of dissolved nitrogen compounds, can vary. For intensive RAS (Return Activated Sludge), the most common practice involves utilizing nitrifying bacteria to convert ammonia-nitrogen to nitrate-nitrogen [6]. The rate limiting step of this process is the oxidation of ammonia [7]. Incomplete nitrification occurs when a lack of NOB (Nitrite Oxidizing Bacteria) productivity is present, leading to increased concentrations of nitrite. The adsorption process is more widely used in industry because it has several advantages, which are more economical and also does not cause toxic side effects and is able to eliminate organic materials. Adsorption is a process that occurs when molecules from substances liquid or gas accumulates on a solid / liquid surface, forming a thin layer formed from molecules or atoms. The most important thing in the adsorption process is the selection of a good type of adsorbent [6].

The adsorption process of solutions theoretically generally lasts longer than the adsorption process on gas, steam or pure liquid [8]. This is due to the adsorption of the solution involving competition between the solution components and the adsorption site. The solution adsorption process can be estimated qualitatively from the polarity of the adsorbent and the constituent components of the solution. The tendency of polar adsorbents to more strongly absorb polar adsorbates than non-polar adsorbates, and vice versa. The solubility of the adsorbate in the solvent is the determining factor in the adsorption process, generally the hydrophilic substance is difficult to adsorb in a dilute solution. Basically, an adsorbent must have a high specific surface area, which has small diameter pores so that the adsorbate retention process by the adsorbent takes place more effectively. Specifically, the pore size also determines the adsorption of a particular compound in solution. If the adsorbent's pore size gets smaller, the adsorption capacity is greater, assuming that the adsorbed component can enter the porous cavity. The greater number of adsorbents will provide a larger surface area for the adsorbate to be desorbed. In addition, the more the amount of adsorbent will also provide greater opportunity for contact with the adsorbate molecules [9].

One of the most potential adsorbents is zeolite. Zeolite has several properties such as having dehydration properties, high cation exchange, a good catalyst, and as an agent for other compounds [10]. Increased efficiency or optimization of zeolites as adsorbents can be done through activation. The activation process aims to cleanse the pore surface, remove disturbing compounds and increase the specific surface area [11]. At the activation stage, the adsorbent is first soaked using an activating material, including HCl, HNO_3, H_2SO_4 and H_3PO_4 [12].

This study aims to improve the quality of water using natural zeolite as an adsorbent so that the levels of NH_3 and NO_2 contained in water can be reduced and safe for shrimp breeding and also to aims kinetics rate of NO_2 and NH_3 adsorption using zeolite catalyst.. In this study, filters were made from PVC pipes with a pipe length of 100 cm and a diameter of 6 inches (15.24 cm). The results of this study are expected to reduce the level of Fe to 96% and Mn decreased to 84.3%.

2. Experimental

2.1 Zeolite Activation

Zeolite is activated using NaOH where 25 grams of zeolite are mixed with 0.5N NaOH and heated for 2 hours at 70°C while stirring and then dried in an oven at 110°C for 4 hours. The zeolite is then cooled in a desiccator. After drying, zeolite is ready for use. This study used 3 variations of zeolite, volume zeolite 1/3 part of pipe , volume zeolite 2/3 part of pipe and full voulme of zeolite that were filled in the pipe of filter.

2.2 Nitrite Levels Analysis

Analysis of nitrite levels was carried out using Diazotasi-Spectrofotometry method. In this study, samples with sulfanilic acid and N- (1-Nepthyl ethylene diamin) dihydrachloride in an acidic atmosphere (pH 2 - 2.5) formed a complex compound, then placed on a spectrophotometer to be analyzed at a wavelength of 420 nm.

2.3 Ammonium Levels Analysis

Analysis of ammonium levels was carried out by the Nessler-Spectrophotometry method. In this analysis, the sample in an alkaline atmosphere is reacted with Nessler reactants to form complex compounds that are yellow to brown. The intensity of the color that occurs is measured its absorbance at a wavelength of 420 nm.

IOP Conf. Series: Materials Science and Engineering **578** (2019) 012026 doi:10.1088/1757-899X/578/1/012026

3. Results and Discussion

3.1. NH_4 and NO_2 decreasing results

Figure 1 shows that the NH_4 concentration in shrimp pond water decreased from 300 minutes in all the amount of zeolite used. In the use of 25 grams a decrease in NH_4 concentration from 4.276 ppm to 3.879 ppm or around 9.284%. On the use of 50 grams and 100 grams there was a decrease in NH_4 concentration of around 15.084% and 18.569% respectively. This shows that the more amount of zeolite used, the greater the amount of NH_4 gas adsorbed.

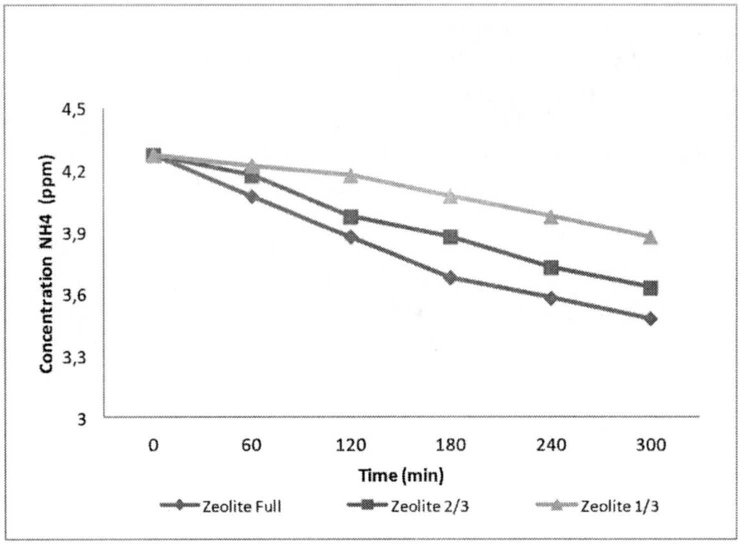

Figure 1. Decrease in NH_4 concentration with different amounts of zeolite

While Figure 2 shows that the NO_2 concentration in shrimp pond water decreased from 300 minutes in all the amount of zeolite used. In the use of 25 grams a decrease in NO_2 concentration from 0.457 ppm to 0.354 ppm or around 22.538%. On the use of 50 grams and 100 grams there was a decrease in NO_2 concentration of around 33.917% and 40.700% respectively. This shows that the more amount of zeolite used, the greater the amount of NO_2 gas adsorbed.

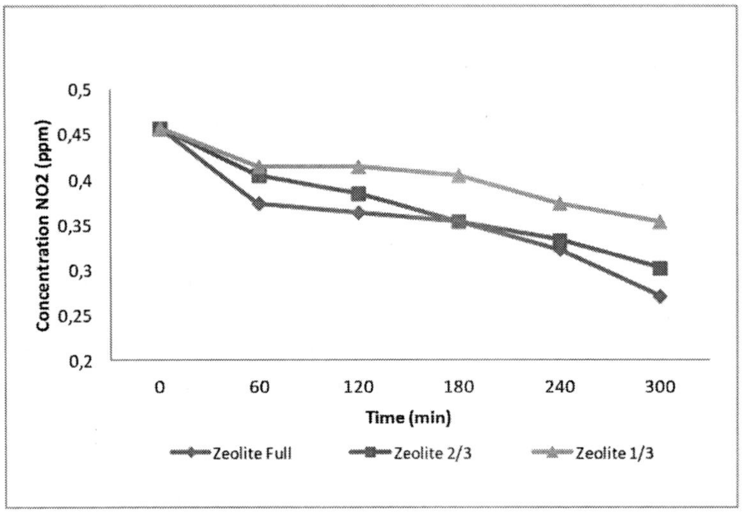

Figure 2. Decrease in NO_2 concentration with different amounts of zeolite

3..2. Kinetics rate of NH_4 and NO_2 adsorption

There are many mathematical models are reported in the literature, all attempting to describe quantitatively in the kinetic behaviour during the adsorption process. Among these models, the pseudo-first-order, pseudo-second-order and intraparticle diffusion model were used to test kinetic experimental data.

3.2.1. Pseudo-second-order model

The rate constant of pseudo-second-order adsorption is determined from the following equation given by Lagergren [14, 15]:

$$\frac{1}{C_A} - \frac{1}{C_{A0}} = k\,t \tag{1}$$

where, C_A and C_{A0} are the amounts of NH_4 and NO_2 adsorbed (mg/g) at equilibrium and at time t (min), respectively, and k (min^{-1}) is the rate constant of the pseudo-second-order adsorption.

Results of k value for NH_4 adsorption can be shown in table 1 below.

Table 1. Adsorption kinetic parameters of NH_4

Zeolite	NH_4 before	NH_4 after	K	R^2
Full	4.276	3.435	0.000188	0.98717
2/3	4.276	3.581	0.000138	0.99194
1/3	4.276	3.824	0.000072	0.98273

Results of k value for NO_2 adsorption can be shown in table 1 below.

Table 2. Adsorption kinetic parameters of NH_4

Zeolite	NO_2 before	NO_2 after	K	R^2
Full	0.457	0.222	0.004441	0.89637
2/3	0.457	0.250	0.003601	0.98858
1/3	0.457	0.297	0.002010	0.98002

Table 1 and table 2 showed that the utilization of full zeolite obtained the higher adsorption of NH_4 and NO_2 gases. It means that the more zeolites are used, the more NH_4 and NO_2 gases are adsorbed. It is because of the activated zeolite has a surface area and acidity that is easily modified so it can absorb more NH_4 and NO_2 gases because the framework of the zeolite has been substituted with ions from the gas.

4. Conclusion

Full zeolite utilization obtained the best perform of NH_4 and NO_2 adsorption, respectively gave a decrease in the concentration of 18.569% NH_4 and 40% NO_2. Adsorption kinetic data were analysed using the pseudo-second-order model. The advantage of using pseudo-second-order model is that the adsorption capacity, the rate constant of pseudo-second-order, and the initial adsorption rate at any given time t, can be predicted from the equation that respectively obtained 0.000188 and 0.004441 of k value for NH_4 and NO_2.

Acknowledgements

The authors gratefully acknowledge the financial support received in the form of a research grant from in addition to APBN LPPM RKAP Diponegoro University Fiscal Year 2018.

References

[1] Kilawati Y and Maimunah Y 2015 *Research Journal of Life Science* **2**(1) 50–59
[2] Tchobanoglous G 1979 *Wastewater Engineering: Treatment Disposal Reuse* (New York: McGraw-Hill, Inc.) 2nd ed. p. 88
[3] Henze M 1995 *Nutrient removal from wastewater* In New World Water (London: Sterling Publications Limited) p.114
[4] Bae B, Juang Y, Han W and Shin H 2002 *Water Res.* **36** 3330-3340
[5] Lin S H and Wu C L 1996 *Water Res.* **30**(8) 1851–1857
[6] Crab R, Defoirdt T, Avnimelech Y, and Verstraete W 2005 *Aquaculture* **270** 1-14
[7] Vadivelu V, Keller J, and Yuan Z 2007 *Water Science & Techno.* **56**(7) 89-97
[8] Rezaei F, and Webley P 2010 *Sep. Purif. Technol.* **70** 243-256
[9] Gonzales M D, Cesteros Y, and Salagre P 2011 *Micro. Meso. Mater.* **144** 162-170
[10] Pratomo S W, Mahatmanti F W, and Sulistyaningsih T 2017 *Indo. J. Chem. Sci.* **6**(2) 104-109
[11] Kundari N A, Susanto A and Prihatiningsih M A 2010 *Prosiding Seminar nasional VI* SDM Teknologi nuklir Yogyakarta, 18 November 2010
[12] Emelda, L, Putri, S M, and Ginting, S 2013 *Jurnal Rekayasa Kimia dan Lingkungan* **9**(4) 166-172
[13] Lagergren, S 1898 *K. Sven. vetensk.akad. handl.* **24**(4) 1–39
[14] Ho, Y S 2004 *Scientometrics* **59**(1) 171–177

Dealumination of ZSM-5 as catalyst to convert glycerol to glycerol monostearate

D D Anggoro[1], S B Sasongko[1], L Buchori[1], K C Sulistyani[1], A Oktavijaya[1], H Oktavianty[1]
[1] Department of Chemical Engineering, Diponegoro University, Semarang, 50275, Indonesia

E-mail: anggorophd@gmail.com

Abstract. This study aims to improve the acidity of the ZSM-5 catalyst as a catalyst that can produce the highest yield of glycerol monostearate from conversion of glycerol to glycerol monostearate. This study consisted of three stages: dealumination ZSM-5 Zeolite using H_2SO_4, catalyst characterization using ammonia absorption for acidity analysis and pyridine analysis for basicity catalyst, and GC-MS for Glycerol Monostearate molecular weight and purity analysis. Zeolite was dealuminated and then dried at a temperature of 110^{0}C for 1 hour, then calcined at a temperature of 550^{0}C for 4 hours. The dealumination variables used are acid concentration, temperature and time of dealumination. The test results obtained the optimum condition to produce the highest total acidity and the highest surface acidity at 7M of acid concentration, temperature of 60^{0}C, and 2 hours of dealumination time. The optimum conditions of the dealumination variables to produce monoglycerides as product respectively, 3.5-6.5 M of acid concentration, temperature of dealumination at 40-60^{0}C and 0.85-2 hours of dealumination time. This study also found other product namely glycerin, with high % yield on two catalyst samples which have the lowest total acidity.

1. Introduction

Glycerol monostearate is an ester compound produced from the esterification reaction between glycerol and stearic acid. The manufacture of glycerol stearic is carried out in a temperature range of 140^0-190^{0}C with the time used for the experiment is 8 hours, the optimal conditions are produced at a temperature of 180^{0}C with a reaction time of 8 hours which is obtained 94.58% ester, catalyst used acid catalyst (HCl) and base (KOH) with a concentration of 0.75% [1]. The use of this GMS (glycerol monostearate) product is for non-ionic surfactants in the oleochemical industry, this GMS is used in shampoos as pearlizing agents, emulsifiers and lotions, and in the food industry (ice cream, butter, etc.) as an opacifier.

Zeolite is defined as a group of crystalline crystalline aluminosilicate materials (less than 2 nm), composed of tetrahedral units of SiO_4 and AlO_4 which are interconnected with each other [2]. Zeolite is a type of aluminosilicate where the aluminosilicate type material has three main characteristics, namely a good and regular crystal structure, the ion in the framework structure can be easily exchanged with other ions, and ion exchange in the framework greatly affects the properties of zeolite. Zeolites have a unique structure, because they have cavities or pores with molecular dimensions which are part of the structure of the crystalline. Having a microscopic cavity, zeolites can be compared to enzymes, which have certain catalytic groups [3].

Zeolite has been widely used in the reaction of important chemical industries, such as for cracking (hydrocracking) and isomerization [4]. Kadja et al. [2] research zeolite that has been used in various fields, such as sorbent [5; 6; 7], cation exchange agents [8; 9; 10], as well as catalysts [11; 12; 13, 14; 15]. Further use of zeolite has also been rampant, such as optical-electronic materials [16; 17], and membranes [18; 19; 20]. However, until now the most commonly used zeolite utilization is as a catalyst.

Gu et al [21] conducted synthesis of Glycerol Monosterate from stearic acid and glycerol as the main material with the help of N, N-dimethylformamide (DMF) / toluene as a solvent, and $H_3PW_{12}O_{40}$/MCM-41 with weight ratio of $H_3PW_{12}O_{40}$ are 10% ,30%, and 50% as a catalyst. A mixture of glycerol and stearic acid with DMF solvent was stirred at 800 rpm and heated using oil. N_2 gas with steady flow is passed over the mixture, then the catalyst is added to the reactant when the desired temperature has been reached. In the results, it was found that 30% $H_3PW_{12}O_{40}$/MCM-41 was more active than 10%, 50% of $H_3PW_{12}O_{40}$/MCM-41. Stearic acid conversion at 30% $H_3PW_{12}O_{40}$ /MCM-41 as catalyst increased from 80% to 98% with the selectivity of glycerol monostearate 91%. This is related to higher catalytic activity due to acid strength and amount of acid. When the amount of $H_3PW_{12}O_{40}$ acid is too small, the activity of the catalyst is weak so it cannot direct the synthesis of diglycerides or triglycerides, whereas if the amount of $H_3PW_{12}O_{40}$ acid is too large then the large granules of $H_3PW_{12}O_{40}$ will form, the granules will block the pores, where monoglycerides will form more in pores diglycerides and triglycerides. The Si/Al ratio of the ZSM-5 Zeolite (obtained from International Zeolite in NH_4 form) is 30. Increased Si /Al ratio has an impact on catalytic activity due to acid strength and the amount of acid. The dealumination process can increase the Si/Al comparison ratio so that the catalyst is more acidic, and can increase zeolite activity and can produce high Glycerol Monostearate conversion.

This study was used three stages, namely the process of chemical treatment/dealumination, washing process and calcination process. Variables used in the dealumination process are temperature variables, time variables and acid solution concentration variables. This research is expected to get the optimum operating conditions of the dealumination process on zeolite catalysts.

2. Experimental

The experiment was held in Process Laboratory, Chemical Engineering, Diponegoro University. ZSM-5 catalyst came from Zeolit International, was dealings using H_2SO_4 solution (Merck). The H_2SO_4 solution was done by making a solution of H_2SO_4 (3-8 M) which is dissolved in 25 ml of distilled water then 10 grams of ZSM-5 is put into the mixture. The dealumination process is carried out over time variation (2-5 hours) with temperature of dealumination range of 40^0-60^0C. The mixture is filtered with a suction filter. The filtered sediment was dried in an oven at 110^0C for 4 hours. Modified catalysts, calcined with 550^0C for 3 hours. Monostearate synthesis is carried out by inserting glycerol and stearic acid into a three-neck flask, then adding 20% tetradecane w/w, and ZSM-5 catalyst. Heat and set the stirring speed to 800 rpm. Flow nitrogen gas into a three-neck flask, set the flow rate to 10 ml / minute and do the synthesis for 6 hours.

Catalyst acidity test is done by smoothing the calcined catalyst first. Weigh the initial weight of each sample. Pour ammonia and pyridine into the container and place it into a different desiccator. Enter the sample into each desiccator. Weigh the weight of sample on the first day, second day, and so on until it reaches 3 times constant. The results of the ZSM-5 dealumination catalyst process were used for glycerol synthesis with stearic acid. GC-MS analysis will be carried out to determine the molecular weight of the compounds produced, and show the purity of the compounds produced using the dealuminated ZSM-5 catalyst [22].

3. Results and Discussion

3.1 Effect Variables of Dealumination Process on Monoglycerides Yield

In this study, ZSM-5 which has been dehydrated with H_2SO_4 solution was used as a catalyst to synthesize glycerol and stearic acid into monoglycerides. The mechanism of monoglyceride formation reaction is as follows:

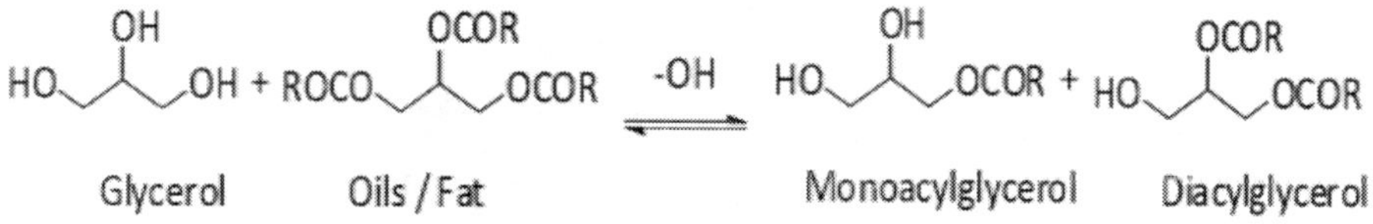

Figure 1. Mechanism of Monoglycerides Formation Reaction

The reaction between glycerol and fat will produce monoglycerides and diglycerides. Adding a base catalyst to the glycerolysis reaction will help break the OH bond in glycerol to form monoglycerides and diglycerides. Table 1 shows GCMS analysis of monoglycerides yield for 16 runs and also shows the glycerin product that obtained.

Table 1. GCMS Analysis of % Yield of Monoglycerides

Run	Dealumination Variables			% Monoglycerides	% Glycerin
	Acid Concentration (M)	Temp (°C)	Time (hour)		
1	3	40	2	7.23	7.24
2	3	40	5	8.62	10.82
3	3	60	2	20.97	26.23
4	3	60	5	4.42	7.52
5	7	40	2	21.07	12.05
6	7	40	5	3.85	2.78
7	7	60	2	4.08	6.27
8	7	60	5	4.12	7.04
9	5	50	3.5	18.7	0
10	1.4723	50	3.5	1.81	1.81
11	8.5276	50	3.5	0	0
12	5	32.3616	3.5	0	100
13	5	67.6383	3.5	0	5.34
14	5	50	0.854249	0	99.15
15	5	50	6.145751	10.09	0
16	5	50	3.5	0	0

3.2 Effect of Acid Concentration and Temperature of Dealumination on Monoglycerides Yield

Figure 2 and Figure 3 show the interaction of H_2SO_4 concentration and the dealumination temperature on the % area or % yield of Monoglyceride produced. Along with the increasing concentration of acid, it is followed by the greater% yield of the monoglyceride produced. In the temperature range of 40^0 - 60°C, along with the increasing concentration of acid it will give a higher% yield of Monoglyceride as well. In the temperature range of 40^0- 60°C, the use of 7 M concentration in the dealumination process gives a greater % yield (in the dark red area) than the use of 3 M acid concentration (in the red area). This is because the greater the concentration of H_2SO_4, the more aluminum metal released from zeolite will cause an increase in the Si/Al ratio. The high ratio between silica and aluminum causes increased adsorption capability. The acidity of a catalyst is defined as the ability of a catalyst to adsorb bases. So that the greater the ability of catalyst adsorption, the greater the acidity site, this happens because of the large number of zeolite pores that are exposed and the surface of the solid being wider due to Al being released from the zeolite. The increase in surface area of the dealuminated ZSM-5 shows a greater level of acidity thus increasing the ability of the catalyst in the synthesis of glycerol to become Monoglycerides.

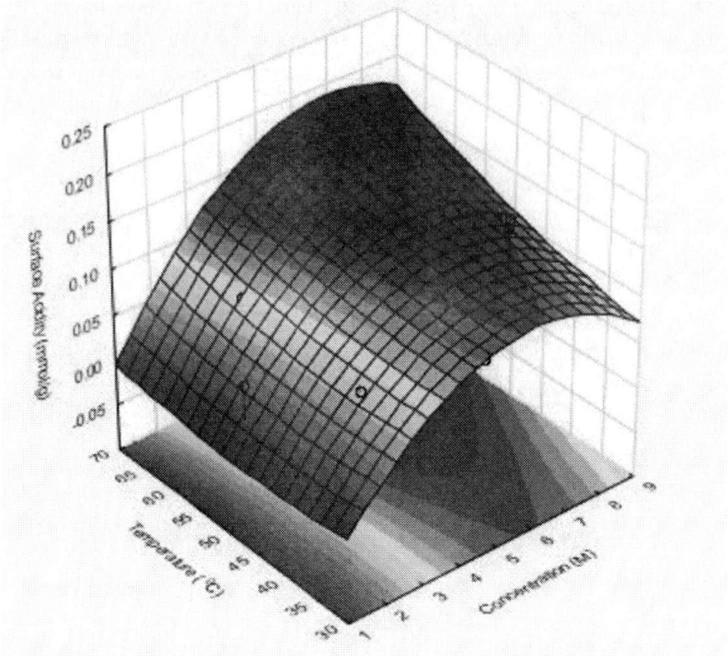

Figure 2. 3D Surface of Acid Concentration vs Dealumination Temperature

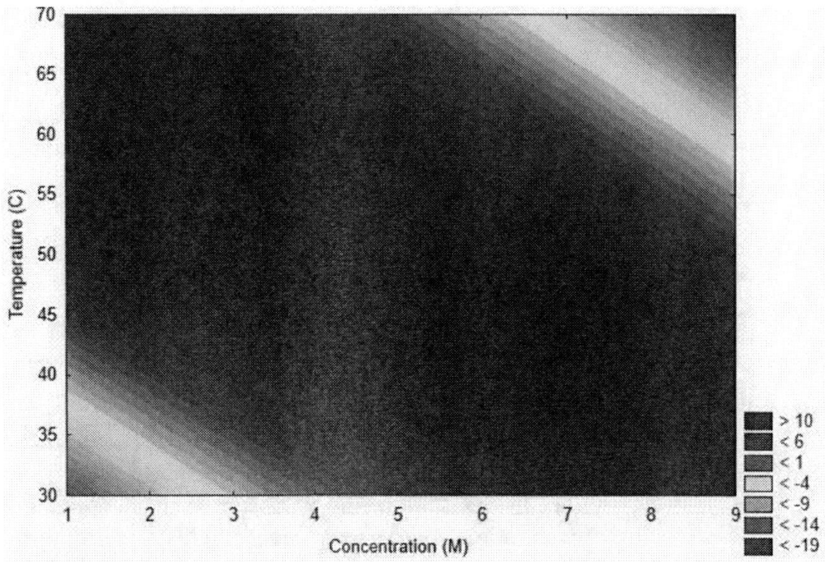

Figure 3. Contour of Acid Concentration vs Dealumination Temperature

3.3 Effect of Temperature and Time of Dealumination on Monoglycerides Yield

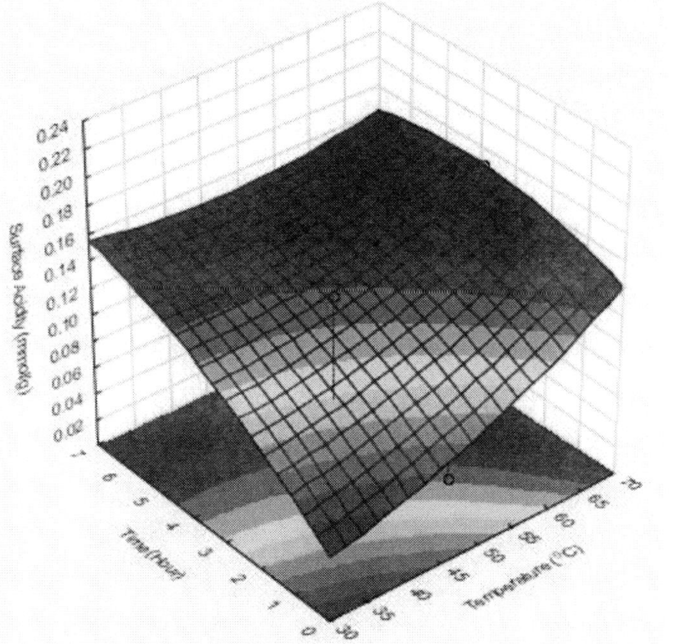

Figure 4. 3D Surface of Temperature vs Time of Dealumination

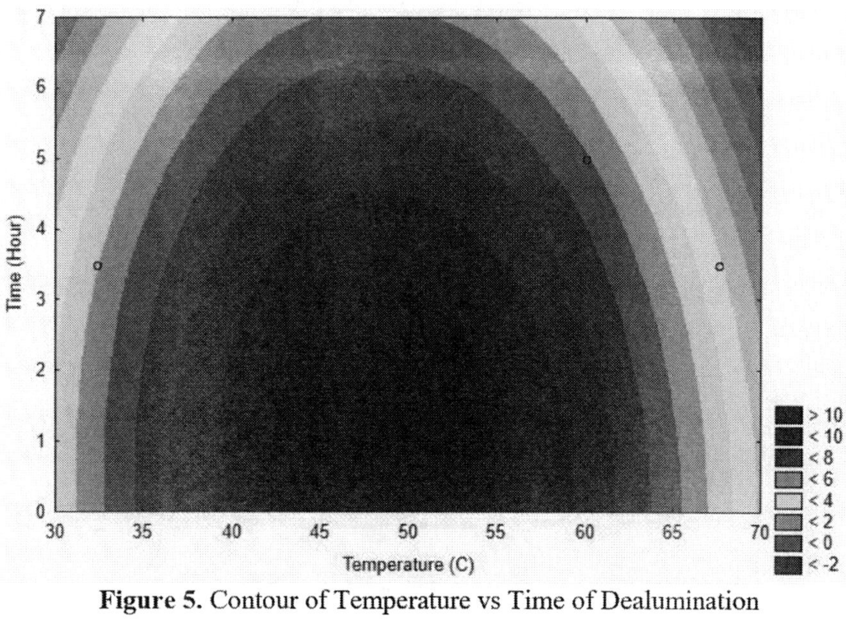

Figure 5. Contour of Temperature vs Time of Dealumination

Figure 4 and Figure 5 how the interaction between temperature and time of dealumination on the% yield of Monoglyceride produced. In the temperature range between 40^0-50°C and dealumination time range 0.85-2 hours, the dealumination temperature increase will increase the yield% formed (Monoglyceride formed). Where the dealumination temperature is increasing, the acidity of the resulting dealuminated ZSM-5 catalyst will also increase. This is because the longer the dealumination process causes the Si/Al ratio to rise. The increase in the Si/Al ratio causes a change in the skeleton of the zeolite. Changes in the zeolite framework will cause changes in the pores of the zeolite which will also cause changes in surface area. The increase in surface area of the dealuminated ZSM-5 shows a greater level of acidity thus increasing the ability of the catalyst in the synthesis of glycerol to become Monoglycerides.

But at temperatures above 60^{0}C, a rise in temperature will reduce the% yield of the Monoglyceride produced. This is caused by damage to some of the pores and catalyst framework so that it will decrease its ability to synthesize glycerol to Monoglyceride. Therefore the optimum dealumination temperature of this process is in the range of 40^0-50°C with optimum dealumination time of 0.85 to 2 hours.

3.4 Effect of Acid Concentration and Time of Dealumination on Monoglycerides Yield

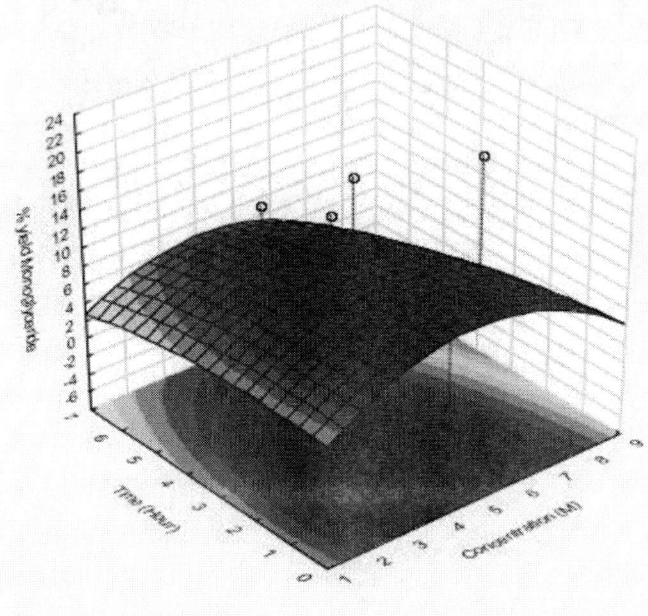

Figure 6. 3D Surface of Acid Concentration vs Time of Dealumination

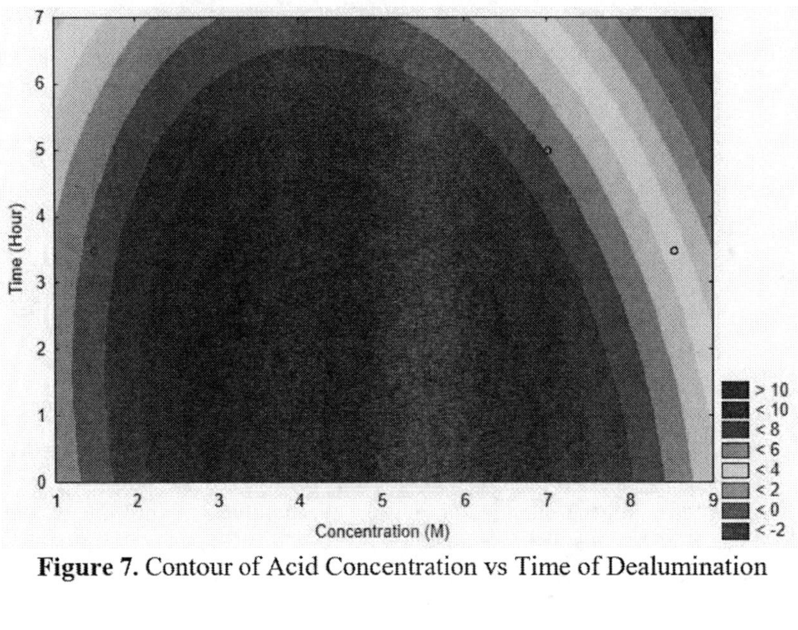

Figure 7. Contour of Acid Concentration vs Time of Dealumination

Figure 6 and Figure 7 show the interaction of concentration and time of dealumination on the% yield of the Monoglyceride produced. In the range of H_2SO_4 concentrations between 3.5-6 M and the dealumination time range for up to 3 hours, an increase in H_2SO_4 concentration will increase the% yield formed (Monoglyceride formed). Where the longer the dealumination time is, the acidity of the resulting dealuminated ZSM-5 catalyst will also increase. This is because the longer the dealumination process causes the Si/Al ratio to rise. The increase in the Si/Al ratio causes a change in the skeleton of the zeolite. Changes in the zeolite framework will cause changes in the pores of the zeolite which will also cause changes in surface area. The increase in surface area of the dealuminated ZSM-5 shows a greater level of acidity thus increasing the ability of the catalyst in the synthesis of glycerol to become Monoglyceride.

But at concentrations of H_2SO_4 above 7 M, the increase in H_2SO_4 concentration will reduce the % yield of Monoglyceride produced. This is caused by damage to some of the pores and catalyst framework so that it will decrease its ability to synthesize glycerol to Monoglyceride. Therefore, the optimum H_2SO_4 concentration of this process is in the range of 3.5-6 M with optimum dealumination time a maximum of 3 hours.

In this study, another product was found besides monoglyceride, glycerin. The effect of the detected ZSM-5 acidity on% monoglyceride yield can be seen from Figure 8. From Figure 8, it can be concluded that the higher the acidity of zeolite, the lower % yield of Glycerin can be obtained. Acidity increasing is caused by an increase in the Si/Al ratio and dissolution of CaO impurities. Si/Al ratio increasing can cause increased activity of the catalyst. The activity increasing of catalysts is useful in reducing the activation energy of the reaction of a compound [23]. This is what causes the conversion of glycerol to Glycerine decreases with increasing acidity because glycerol is converted into other compounds.

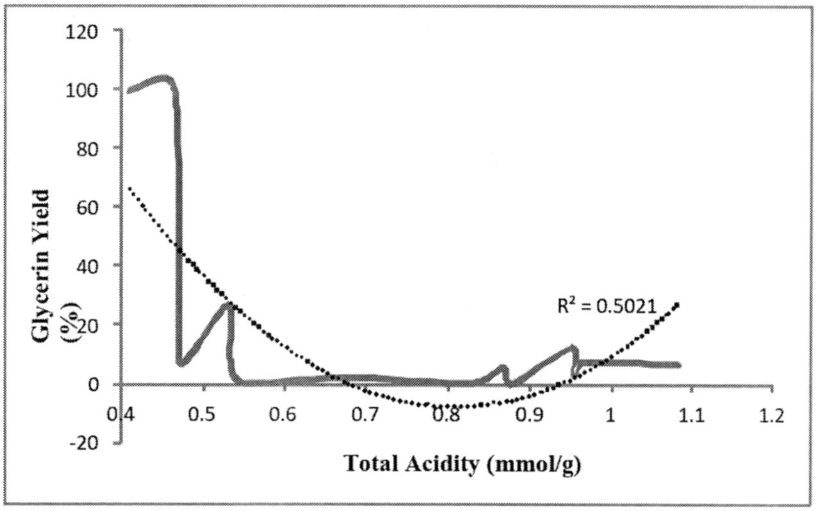

Figure 8. Effect of Acidity of Dealuminated ZSM-5 on Glycerin Yield

4. Conclusion

Acid concentration, temperature, and time in the ZSM-5 catalyst dealumination process affected the acidity generated from the dealuminated ZSM-5 catalyst. The total acidity of the ZSM-5 catalyst which has the most optimum results was found in sample 7 with a variable acid concentration of 7 M, temperature of 60 °C, and the processing time of catalyst for 2 hours. The surface acidity of the ZSM-5 catalyst which has the most optimum results was found in sample 7 with 7 M acid concentration variable, temperature 60 °C, and dealumination time of catalyst for 2 hours. The optimum conditions for producing Monoglyceride products are when the acid concentration variable is 3.5-6.5 M with dealumination temperature 40-60 °C and dealumination processing time for 0.85 to 2 hours. Glycerin products were found with% yield of 100% and 99.15% in two catalyst samples which had the lowest total acidity.

Acknowledgements
The authors gratefully acknowledge the financial support received in the form of a research grant from KEMENRISTEKDIKTI Indonesia No: 101-76/UN7.P4.3/PP/2018.

References

[1] Rosdiana I and Atun S *J. Penelitian Saintek* **20**:61-66
[2] Kadja G T M, Rilyanti M, Mukti R, Marsih I N, dan Ismunandar 2013 *Jurnal Matematika dan Sains* **18**(3).
[3] Gates B C 1997 *Catalytic Chemistry* New York: John Wiley & Sons, Inc.
[4] Muller J M, Mesquita G C, Franco S M, Borges L D, deMacedo J L, Dias J A, and Dias S C L 2015 *Micro. & Meso. Mat.* **204** 50-57 DOI: 10.1016/j.micromeso.2014.11.002
[5] Damjanovic L, Rakic V, Rac D S, and Aurox A 2010 *J. Hazard. Mater* **184** 477-484
[6] Lee T P, Saad B, Ng E P, and Salleh B 2012 *J. Chromatogr. A* **1237** 46-54
[7] Chang X, Lu G, Guo Y, Wang Y, and Guo Y 2013 *Micro. Meso. Mat.* **165** 113- 120
[8] Biskup B and Subotic B 2004 *Sep. Purif. Technol* **37** 17-31
[9] Wang Y, Lin F, and Pang W 2008 *J. Hazard. Mater* **160** 371-375
[10] Hou J, Yuan J J, and Shang R 2012 *Powder Technol* **226** 222-224
[11] Tukur N M and Al-Khattar S 2011 *Chem. Eng. J.* **166** 348-357
[12] Hu S, Shan J, Zhang Q, Wang Y, Liu Y, Gong Y, Wu Z and Dou T 2012 *Appl. Catal. A* **445-446** 215-220
[13] Wu L, Degirmeci V, Magusin P C M M, Lousberg N J H G M, and Hensen E J N 2013 *J. Catal.* **298** 27-40
[14] Corma, A, Mengual J, and Miguel P J 2013 *Appl. Catal A.* **460-461** 106-115
[15] Ordonez, E G F, Salbidegoitia J A, Ayastuy J L, Ortiz M A G, Marcos, M P G and Velasco J R G 2013 *Catal. Today* **227** 163-170
[16] Sauer J, Marlow F, and Schuth F 2001 Chapter 5 – Nanoporous materials for optical applications, *Handbook of Advanced Electronic and Photonic Materials and Devices,* Elsevier.
[17] Thang H V, Grajciar L, Nachtigall P, Bludsky O, Arean C O, Vrydova E, and Bulanek R 2014 *Catal. Today* **227** 50-56
[18] Lee H and Dutta P K 2002 *J. Phys. Chem. B* **106** 11898-11904
[19] Huang A, Liu Q, Wang N, Tong X, Huang B, Wang M, and Caro J 2013 *J. Membr. Sci.* **437** 57-64
[20] Wang Y, Otomo R, Tatsumi T, and Yokoi T 2016 *Micro. & Meso. Mat.* **220** 275-281
[21] Gu, Yi., Hong, Mei., Xiao, Guomin., dan Zhou, Lu., 2014. *Asian Journal of Chemistry*, **26**(22) 7564-7658
[22] Chen X, Wu D, He Y, and Liu S, 2009. *Anal. Chim. Acta.* **638**(1)16-22
[23] Machado, M and Pariente J P S 2000 *App. Cat. A: General* **203** 321-328

IOP Conf. Series: Materials Science and Engineering **578** (2019) 012028 doi:10.1088/1757-899X/578/1/012028

A microfluidic chip-based assay outperforms common methods for quantification of cell-free DNA and RNA

T N Susilawati[1,2]
[1] Department of Microbiology, Faculty of Medicine, Universitas Sebelas Maret, Surakarta, Indonesia
[2] College of Medicine and Dentistry, James Cook University, PO Box 902, Cairns Hospital, Cairns, Australia

E-mail: tri.susilawati@staff.uns.ac.id

Abstract. Plasma and serum contain cell-free DNA and RNA and are common specimens used for pathology investigation. Circulating nucleic acids in plasma and serum (CNAPS) represent an important target for analyzing several conditions including prenatal diagnostics, tumors, infections, and inflammatory disorders. It is highly expected that plasma and serum contain potential biomarkers for the detection of cancers and other pathological conditions. The major challenges are that CNAPS are present in small quantities and are highly degraded, hampering their downstream analysis. For example, obtaining acceptable quality and quantity of CNAPS remains a bottleneck for advanced studies such as those using next-generation sequencing (NGS) platforms as the currently available platforms require nanogram to microgram levels of deoxyribonucleic acid or ribonucleic acid (DNA/RNA) input. This paper examines the use of various methods to quantify CNAPS, including quantitative real-time polymerase chain reaction (qRT-PCR), digital PCR (dPCR), fluorometric assay, gel electrophoresis, and Bioanalyzer. In particular, the results of CNAPS quantification using a fluorometer, a NanoDrop spectrophotometer, and a microfluidic chip-based assay were compared. It is asserted that spectrophotometry, fluorometric assay and gel electrophoresis lack sensitivity for detecting CNAPS; thus, highly sensitive methods such as qRT-PCR, dPCR, and Bioanalyzer are required.

1. Introduction

First discovered in 1869, nucleic acids were called "nuclein" consistent with their localization in the nucleus [1]. Plasma DNA was firstly reported 70 years ago by Mandel and Métais in 1948 [2] and in the last decade, researchers have been actively reporting small amounts of extracellular DNA and RNA fractions. Technical advancements in molecular biology have facilitated the characterization of circulating nucleic acids in plasma and serum (CNAPS) and the investigation of their roles in numerous pathological conditions. Previous studies reported the increased levels of circulating DNA in pregnancy, various types of malignancies, inflammatory disorders, trauma, and stroke [3,4]. It has also been shown that specific DNA/RNA fragments can be used as biomarkers for myocardial infarction, stroke, diabetes, and cancer; thus facilitating diagnosis, prognosis and monitoring of treatments [3].

The major difficulties when working with CNAPS are the low yield and poor quality, thus requiring highly efficient methods for isolation and quantification. The levels of circulating DNA in healthy subjects varied from undetectable to several hundreds of nanogram per milliliter of a sample

[5-8]. The poor quality of CNAPS is indicated by the high level of DNA and RNA degradation. Two studies [9,10] reported a prevalence of short DNA fragments (mostly <200 nt) in the plasma of healthy subjects whilst the circulating RNA detected is mostly small RNAs (up to 200 nt long) and miRNAs (~20 nt long) [11].

As the initial assessment in most of the laboratory experiments, quantification of CNAPS provides valuable information prior to downstream analysis. Therefore, the method for quantification should be simple, fast and accurate. Spectrophotometry based on ultraviolet absorption is the most common method used in the quantification of DNA and RNA yield after isolation procedures. Other methods of nucleic acids quantification include fluorometric assay, gel electrophoresis, quantitative real-time polymerase chain reaction (qRT-PCR), digital PCR (dPCR), and a microfluidic chip-based assay using Bioanalyzer. This study aimed to compare the results of CNAPS quantification using fluorometry, spectrophotometry, and microfluidic chip-based assays.

2. Experimental

There were two steps involved in this study in which the blood samples were donated by healthy volunteers and from patients recruited for a prospective study investigating infectious agents causing fever [12]. Plasma and serum were prepared in two different types of Vacuette® tubes (Greiner Bio-One); i.e. the purple-top tubes containing K2EDTA and the red-top tubes containing clot activator and were processed within three hours after collection by centrifugation for 15 minutes at 2,000 g.

In the first experiment, six healthy volunteers were recruited into the study. Samples were collected using various methods; i.e., by using a vacuum system or syringe and needle with or without the application of a tourniquet. The High Pure Viral Nucleic Acid kit (Roche Applied Science) was used to isolate the nucleic acids from the samples according to the manufacturer's instructions. The quantity of double-stranded DNA (dsDNA) was determined using SYBR Green I dye (Life Technologies) in a microplate fluorescence assay (MFA). The GeneRuler Express DNA ladder (Fermentas) was serially diluted with EB buffer (Qiagen) to make the following concentration of DNA standards: 0.25, 0.0125, 0.063, 0.031, 0.016 and 0,008 ng/μL. The buffer alone was served as a "blank" to give 0 ng/μL of DNA standard. Then, 25 μL of each DNA standards were prepared in the 384 well plates (Greiner) and mixed with an equal volume (25 μL) of SYBR Green I diluted in 1:1250 in EB buffer to give 50 μL per well. For the assay of test samples, 5 μL of each DNA tests was diluted in 20 μL EB buffer in the microplate well and mixed with 25 μL SYBR Green I solution as above. The plates were incubated for 10 min in the dark at room temperature. Following incubation, the fluorescence in each well was measured at room temperature using a spectrofluorometer (POLARStar Omega) fitted with 485/520 nm excitation/emission filter set. The experiment was conducted in duplicate and the average reading of fluorescence intensity was used for developing the standard curve and for calculating the level of DNA in test samples. The concentrations of dsDNA in the samples were compared by using Wilcoxon Signed-Rank test.

In the second experiment, plasma and serum were collected from patients with fever. The DNA Mini Kit (Qiagen) and TRIzol LS reagent (Life Technologies) were used to isolate plasma/serum DNA and RNA, respectively, with the addition of glycogen during RNA precipitation. Following this, the isolated DNA and RNA were randomly selected for quantification using methods other than fluorometry assay. A total of 41 DNA samples originated from plasma or serum of 21 participants and 63 RNA samples from 33 participants were quantified using a Nanodrop 2000 spectrophotometer (Thermo Scientific). All DNA and RNA samples were assessed using a microfluidic chip-based platform (the Agilent 2100 Bioanalyzer) with suitable kits; i.e., Agilent High Sensitivity DNA kit and Agilent RNA 6000 Pico kit (Agilent Technologies). Data were presented by using descriptive statistics and the representative results of the bioanalyzer were presented. The study protocol was approved by the Human Research Ethics Committee of James Cook University and Cairns Hospital (ID: H4456 and HREC/12/QCH/7, respectively).

3. Results and Discussions

The comparisons of the CNAPS quantity in healthy volunteers and in patients with fever measured by using various methods are presented as follows. It can be seen in Figure 1 that the concentrations of plasma/serum DNA were below 1500 ng/mL (or less than 1.5 ng/µL) whilst those in the whole blood are tripled. In particular, the levels of dsDNA in serum were higher than that in plasma ($p < 0.05$). The method of blood collection itself did not affect the DNA concentrations in the samples as it can be observed from the figure that there was no difference in DNA levels regardless the application of a tourniquet or vacuum system.

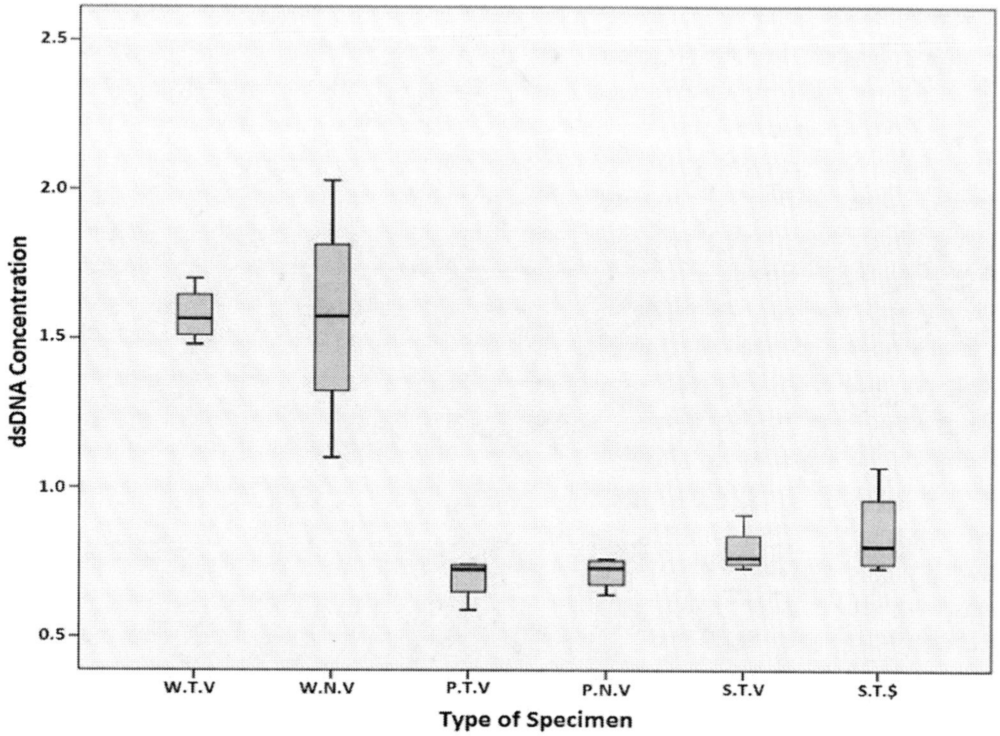

Figure 1. dsDNA concentration (ng/µl) in various blood specimens taken from healthy volunteers. W.T.V: whole blood, drawn by using a tourniquet and vacuum system; W.N.V: whole blood, drawn into a vacuum system without the use of tourniquet; P.T.V: plasma, collected with the use of tourniquet and vacuum system; P.N.V: plasma, collected into a vacuum system without the use of tourniquet; S.T.V: serum, collected with the use of tourniquet and vacuum system; S.T.$: serum, collected by using a syringe and needle with the use of tourniquet

The DNA levels in the plasma or serum of patients with fever measured by the Nanodrop 2000 spectrophotometry ranging from 1.4 until 12.4 ng/µL and the average concentration is 5.19 ± 3.57 ng/µL (data not shown). Among those, only 12 out of 41 (29%) samples are of acceptable quality, as indicated by the value of A 260/280, which is ≥ 1.8. As for the RNA levels, the concentrations are widely varied, ranging from 0.4 until 114.7 ng/µL and the average concentration is 12.79 ± 26.38 ng/µL (data not shown). Although the average of circulating RNA levels was generally detected higher than DNA, the quality was lower, only 8 out of 63 (12.7%) samples had the value of A 260/280 ≥ 1.8. In all DNA/RNA samples, the value of A 260/230 was less than 1, indicating contamination by phenolate ion, thiocyanates, and other organic compounds used during the nucleic acids isolation.

IOP Conf. Series: Materials Science and Engineering **578** (2019) 012028 doi:10.1088/1757-899X/578/1/012028

Figure 2 shows the results of CNAPS quantification using a microfluidic chip-based assay; i.e., using the Agilent 2100 Bioanalyzer. Data are presented in an easy-to-read format consisting of electropherograms, a gel-like image, and tabular results. In the bioanalyzer electropherograms, one cannot expect to see the 18S and 28S subunits as two distinct bands that show an acceptable quality of RNA. Instead, the whole degraded RNAs were grouped around the marker band and small peaks are usually present after the marker at 24–29 seconds. These peaks represented small RNA fragments about 100 bp that are especially noted when traditional RNA extraction methods using phenol are performed.

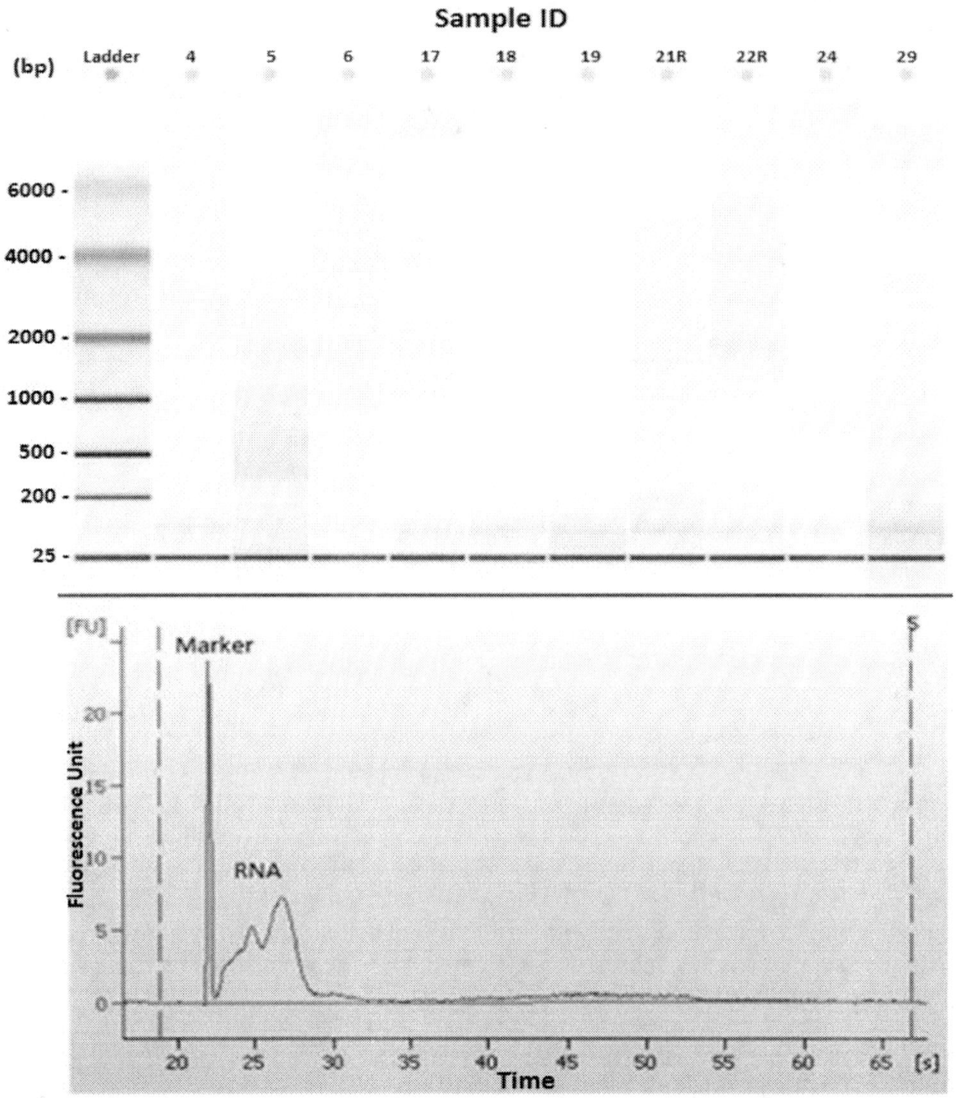

Figure 2. Typical quantity and quality of circulating nucleic acids, measured using Bioanalyzer with a detection limit of 200 pg per band. The low-molecular-weight smears in the gel-like image (above) indicate a low quantity (usually at picogram levels) of highly fragmented CNAPS. An electropherogram (below) from the plasma of a representative subject (ID# 005) indicates RNA of low quality in which the typical 18S and 28S subunits were not detectable, but all of the RNA was grouped around the marker band. In this sample, the RNA integrity number (RIN) is 2.4 (data not shown).

The origin of CNAPS is debatable but it is assumed that they probably derive from cells dead for apoptosis or necrosis as well as the results of the active release of newly synthesized DNA/RNA into circulation [13-16]. The low quantity and quality of CNAPS hinder their accurate quantification. Various methods of DNA and RNA quantification have been established and are discussed below.

Fluorometric assay utilizing dyes (fluorophores) that bind to a certain type of DNA or RNA has been used for quantification of nucleic acids. It is a simple and quick method, yet more sensitive than the spectrophotometric assay that measures nucleic acids' absorbance. A previous fluorometric study using PicoGreen reported that the total of circulating DNA was around 10 times higher than the levels of amplifiable DNA measured by qRT-PCR [7]. The study indicated that the fluorometry assay is sensitive enough to measure the levels of CNAPS, especially with the use of a highly sensitive dye such as PicoGreen. Indeed, PicoGreen and RiboGreen are recommended dyes for circulating cell-free DNA and RNA assay, respectively, due to their high sensitivity of detection [17,18]. SYBR Green I can be used as an alternative for quantification of circulating DNA since it has acceptable sensitivity but less expensive than PicoGreen. It is important to note that the yield of CNAPS is extremely variable from sample to sample (anywhere from less than 1 ng to more than 100 ng) so that some samples may fall below the limit of detection of fluorometry instruments, which is around 20 ng. This can produce false or non-consistent readings.

Spectrophotometry has poor sensitivity and the analysis is affected by the presence of contaminants (e.g., proteins, carbohydrates, salts, phenol, free nucleotides), contributing to the total absorbance of the sample and leading to overestimation of the concentration of CNAPS. Another problem is that purification of nucleic acid samples is required prior to spectrophotometry measurement in order to ensure accurate results. The purification of CNAPS can be problematic due to their short size. These small fragments of CNAPS pass through membrane filter used during column purification method, leading to significant loss of the isolated DNA/RNA. Thus, the spectrophotometry method is not applicable to analyze CNAPS as this method requires highly purified and adequate amount of nucleic acid samples.

The Agilent 2100 Bioanalyzer (Agilent Technologies) is a microfluidic chip-based assay that provides a more comprehensive analysis of circulating DNA and RNA including quantity, fragment size, and RNA integrity. Although the extent of CNAPS usually appears as faint smears, the results of bioanalyzer are far more accurate and more detail than fluorometry and spectrophotometry assays. In a situation when the CNAPS have been amplified, spectrophotometry or electrophoresis of DNA/RNA through agarose is a cheap and easy method to visualize the presence of DNA/RNA and often to estimate the quantity as well. Higher concentrations of agarose help to separate small DNA/RNA molecules; thus, visualization of CNAPS fragments can be conducted on a 1.5–2% (w/v) agarose gel.

Quantification of CNAPS using qRT-PCR usually involves the use of two fluorescent dyes (probes) and primers that are designed to amplify the particular gene of interest such as human telomerase reverse transcriptase (hTERT) [5], β-actin [19], or β-globin gene [20]. The presence of miRNA can be assessed using miR-21 gene [21], which is highly abundant and thus a good choice for control. Alternatively, RNU6 and miR-520d-5p [22] or random CNAPS such as glyceraldehyde-3-phosphate dehydrogenase (GAPDH) genes [23] could be used as an internal control for preventing false negative results when the study aims to look for pathogens' DNA/RNA in human plasma or serum using qRT-PCR. Although qRT-PCR is a fast method for gene quantification, its use is limited for targeting known genes so the total number of CNAPS within a sample is underestimated. In addition, the method requires standardization as the data generated by qRT-PCR are highly influenced by the instrument, user and experimental variability [24] so that comparison of the results cannot be made amongst different settings. It is also important to note that heparin can inhibit the detection of miRNAs in PCR-based assay so that other methods of miRNAs quantification should be considered in patients treated with heparin [25].

Digital PCR (dPCR) is an alternative method for quantification of a scarce amount of CNAPS. Unlike qRT-PCR which measures only amplifiable DNA, dPCR performs absolute measurement of CNAPS by single molecule counting. The method has been used for quantification of CNAPS in the field of cancer research [26-29], prenatal testing [30-32], transplantation medicine [33] and virology [34-36]. The clinical utility of dPCR should be studied further, however, since the method can produce false positive results due to cross-reactivity [37] in addition to lower throughput and sensitivity compared with qRT-PCR [36,38].

4. Conclusion

The extent of low quantity and highly degraded CNAPS necessitates an optimal sample processing for the success of downstream analysis. Spectrophotometry, fluorometric assay and gel electrophoresis lack sensitivity for detecting CNAPS; thus, highly sensitive methods such as qRT-PCR, dPCR and bioanalyzer are required. The advantages and limitations of each method have been discussed.

Acknowledgements

The study was conducted in the facility of Queensland Tropical Health Alliance (QTHA) laboratory, James Cook University, Australia. The author thanks AusAID, Far North Queensland Hospital Foundation and James Cook University for the funding support. The author acknowledges assistance and encouragement from QTHA colleagues. This article was based on a doctoral thesis written by the author, entitled "Evaluation of next-generation sequencing technology in determining infectious causes of fever" and contains excerpts thereof.

References

[1] Dahm R 2010 *EMBO Rep.* **11** 153–60
[2] Mandel P and Métais P 1948 *C. R. Acad. Sci. Paris* 1948 **142** 241–3
[3] Beck J, Urnovitz H B, Riggert J, Clerici M and Schutz E 2009 *Clin. Chem.* **55** 730–8
[4] Elshimali Y I, Khaddour H, Sarkissyan M, Wu Y and Vadgama J V 2013 *Int. J. Mol. Sci.* **14** 18925–58
[5] Ha T T et al 2011 *PLoS One* **6** e25969
[6] Catarino R, Ferreira M M, Rodrigues H, Coelho A, Nogal A, Sousa A and Medeiros R 2008 *DNA Cell Biol.* **27** 415–21
[7] Szpechcinski A, Struniawska R, Zaleska J, Chabowski M, Orlowski T, Roszkowski K and Chorostowska-Wynimko J 2008 *J. Physiol. Pharmacol.* **59** 675–81
[8] Horlitz M, Lucas A and Sprenger-Haussels M 2009 *PLoS One* **4** e7207
[9] Pinzani P, Salvianti F, Zaccara S, Massi D, De Giorgi V, Pazzagli M and Orlando C 2011 *Clin. Chim. Acta* **412** 2141–5
[10] Kirsch C, Weickmann S, Schmidt B and Fleischhacker M 2008 *Ann. N. Y. Acad. Sci.* **1137** 135–9
[11] Kishikawa T, Otsuka M, Ohno M, Yoshikawa T, Takata A and Koike K 2015 *World J. Gastroenterol.* **21** 8527–40
[12] Susilawati T N, Jex A R, Cantacessi C, Pearson M, Navarro S, Susianto A, Loukas A C, McBride W J H 2016 *Eur. J. Clin. Microbiol. Infect. Dis.* **35** 1137–49
[13] Beck J, Urnovitz H B, Riggert J, Clerici M and Schutz E 2009 *Clin. Chem.* **55** 730–8
[14] Elshimali Y I, Khaddour H, Sarkissyan M, Wu Y and Vadgama J V 2013 *Int. J. Mol. Sci.* **14** 18925–58
[15] van der Vaart M and Pretorius P J 2008 *Ann. N. Y. Acad. Sci.* **1137** 18–26
[16] Xue X, Teare M D, Holen I, Zhu Y M and Woll P J 2009 *Clin. Chim. Acta* **404** 100–4
[17] Silvers R, Keller H, Schwalbe H and Hengesbach M 2015 *Chembiochem* **16** 1109–14
[18] Li X, Wu Y, Zhang L, Cao Y, Li Y, Li J, Zhu L and Wu G . 2014 *Anal. Biochem* **451** 18–24
[19] Wang Q, Cai Y, Brady P and Vermeesch J R 2015 *Genet. Mol. Res.* **14** 12797–804
[20] Khordadpoor-Deilamani F and Akbari M T 2015 *Bratislavske Lekarske Listy* **116** 662–5
[21] Kumarswamy R, Volkmann I and Thum T 2011 *RNA Biol.* 2011 **8** 706–13
[22] Rice J, Roberts H, Rai S N and Galandiuk S 2015 *Surgery* **158** 1345–51

[23] Karatayli E, Altunoglu Y C, Karatayli S C, Yurdaydin C and Bozdayi A M 2014 *J. Virol. Methods* **207** 133–7
[24] Smith C J and Osborn A M 2009 *FEMS Microbiol. Ecol.* **67** 6-20
[25] Boeckel J N, Thome C E, Leistner D, Zeiher A M, Fichtlscherer S and Dimmeler S 2013 *Clin.* Chem **59** 1125–27
[26] Beaver J A et al 2014 *Clin. Cancer Res.* **20** 2643–50
[27] Bettegowda C et al Detection of circulating tumor DNA in early- and late-stage human malignancies 2014 *Sci. Transl. Med.* **6** 224ra224
[28] Yung T K, Chan K C, Mok T S, Tong J, To K F and Lo Y M 2009 *Clin. Cancer Res.* **15** 2076–84
[29] Taly V et al 2013 *Clin. Chem.* **59** 1722–31
[30] Tsui N B, Kadir R A, Chan K C, Chi C, Mellars G, Tuddenham E G, Leung T Y, Lau T K, Chiu R W and Lo YM 2011 *Blood* **117** 3684–91
[31] Barrett A N, McDonnell T C, Chan K C and Chitty L S 2012 *Clin. Chem.* **58** 1026–32
[32] Tsui N B, Hyland C A, Gardener G J, Danon D, Fisk N M, Millard G, Flower R L and Lo YM 2013 *Prenat. Diagn.* **33** 1214–16
[33] Beck J et al 2013 *Clin. Chem.* **59** 1732–41
[34] Brunetto G S, Massoud R, Leibovitch E C, Caruso B, Johnson K, Ohayon J, Fenton K, Cortese I and Jacobson S 2014 *J Neurovirol* **20** 341–51
[35] Strain M C, Lada S M, Luong T, Rought S E, Gianella S, Terry V H, Spina C A, Woelk C H and Richman D D 2013 *PLoS One* **8** e55943
[36] Sedlak R H, Cook L, Cheng A, Magaret A and Jerome K R 2014 *J. Clin. Microbiol.* **52** 2844-8
[37] Hudecova I 2015 *Clin. Biochem.* **48** 948-56
[38] Hayden R T, Gu Z, Ingersoll J, Abdul-Ali D, Shi L, Pounds S and Caliendo A M 2013 *J. Clin. Microbiol.* **51** 540–6

IOP Conf. Series: Materials Science and Engineering **578** (2019) 012029 doi:10.1088/1757-899X/578/1/012029

Improving method of TiO_2-Fe_2O_3 composite materials for self-cleaning glass preparation

J P Ramadhani[1], S Wahyuningsih[1], A H Ramelan[1]
[1]Inorganic Material Research Group, Faculty of Mathematics and Natural Sciences, Universitas Sebelas Maret, Jalan Ir. Sutami No.36A Kentingan Surakarta

E-mail: sayekti@mipa.uns.ac.id

Abstract. Fe_2O_3-TiO_2 composite have been synthesized as a self-cleaning material on glass. TiO_2 was prepared from titanium tetraisopropoxide (TTIP) by sol gel method. Meanwhile, Fe_2O_3-TiO_2 composite was prepared from TiO_2 synthesized with additional precursor $Fe(NO_3)_3{\cdot}9H_2O$. TiO_2 synthesized powder was added to Fe^{3+} solution in aquadest so it formed crystalline layer on TiO_2 powder. Core shell composite of Fe_2O_3-TiO_2 was obtained after calcination on temperature 600 °C. Identification and characterization of composite by X-Ray Difraction (XRD) showed that there are difractogram peaks of TiO_2 and Fe_2O_3. Another analysis by X-Ray Fluoroscence (XRF) showed that composition of Fe_2O_3-TiO_2 composite are 1:1, 1:1.5, 1:2, and 1:3 (w/w). The morphology of composite shown by Scanning Electron Microscope (SEM) that explained coverage character of TiO_2 with Fe_2O_3 formed core shell type. The electronic spectrum of Fe_2O_3-TiO_2 core shell result showed an increasing response in visible light so it can be aplicated as a coating material in manufacturing technology of self cleaning glass.

1. Introduction

Glass is a transparent material that also has some disadvantages, such as being contaminated with dirt if it is not gradually cleaned which will cause lighting reduction in the room and can even damage the surface of the material. The use of glass remains an option for buildings in the 21st century [1]. In addition to the growing knowledge of the nature of transparent materials [2], glass also beautifies property and becomes an alternative to save energy because of its benefits in relation to room lighting and heating of a building. The development of glass production processes has led to the widespread use of this material in almost all parts of the building [3].

In recently years, the research about photocatalyst has been developed. One of the important application of TiO_2. Photo-oxidation using TiO_2 photocatalysts is widely studied as a new alternative to pollution reduction techniques such as dust, mold, bacteria, and other organic compounds [4]. However, the practical application seems limited. Some reasons include low photon utilization efficiency. Another reason is the need for ultraviolet (UV) use as a source of excitation. To overcome this problem, modification of this catalyst has also been tried by doping with various other metal oxides, including Fe_2O_3.

Iron metal oxide (Fe_2O_3) has been shown to increase the catalytic activity of TiO_2 against oxidation of organic pollutants. This result is quite satisfactory because the nature of TiO_2 photocatalyst is very suitable for the construction of photochemical energy conversion systems and treatment of pollutants [5]. Several studies have claimed that TiO_2 as anatase coexists with metallic iron at acidic pH or in situations where iron doping levels are less than 1 % (m/m) which increases the photocatalytic properties of TiO_2. The higher the doping level or the higher pH, the α-Fe_2O_3 phase has a greater probability to form, the cause is TiO_2 transformation from anatase to rutile [6]. Therefore, to overcome some of these problems, it is necessary to modify the surface of the TiO_2 thin layer by adding Fe_2O_3 metal oxide as a dopant.

2. Experimental

2.1. Materials

The materials used in this study were titanium tetraisopropoxide (TTIP) (E Merck), $Fe_2O_3.9H_2O$ (E merck), glacial acetic acid, aquades, and 98% ethanol (E Merck). Calcination of TiO_2 and TiO_2-Fe_2O_3 was performed by Furnace Thermo. Meanwhile, the analysis and characterization of materials were done by XRD (Bruker D8 Advantage), XRF (Bruker S2 Ranger), SEM-EDX (SEM 250 Quanta). The degradation result of rhodamine B was calculated using UV Vis Spectrophotometer Pelkin Elmer Lambda 25.

2.2. Methods

The titanium (IV) isopropoxide solution was hydrolyzed using a glacial acetic acid solution (temperature 15 °C) at a ratio of 1: 10 (v/v) and then stirred continuously using a magnetic stirrer until a white sol gel was obtained, then stirred and heated at 90 °C until obtained white gel [7]. Then Fe^{3+} solution from precursor $Fe_2O_3.9H_2O$ was added into white gel. The mass variation of $Fe_2O_3.9H_2O$ precursor are 0.625 gram; 1.25 gram; 2.5 gram; and 5 gram(s). The mixing solution was stirred using a magnetic stirer and was heated at temperature 100 °C for ± 2 hours to improve crystal phase of α-Fe_2O_3 until obtained red-copper gel. Then red-copper gel has been calcined in oven at temperature 150 °C for 12 hours followed by roasted in furnace at temperature 600 °C for 2 hours [8]. The resulting TiO_2-Fe_2O_3 crystalline powder was then analyzed by XRD, XRF, spectrophotometer and SEM.

TiO_2-Fe_2O_3 composites then has been coated on a glass to prove the self-celaning ability. The test is using Rhodamine B that will be degradated by coated-glass under visible light. The degradation result of Rhodamin B was calculated using UV-Vis spectrophotometer.

3. Results and Discussion

3.1. Characterization of TiO_2-Fe_2O_3 by XRD and XRF

Synthesis of TiO_2 with the addition of Fe_2O_3 was carried out using the sol-gel method according to Wahyuningsih et al. [7]. The addition of metal is expected to replace several positions of Ti atoms in TiO_2. The calcination process was carried out at 400 °C for 2 hours to obtain the anatase phase TiO_2 crystal. X-ray diffractograms of TiO_2 and TiO_2-Fe_2O_3 are shown in Figure 1.

The XRD analysis results for TiO_2-Fe_2O_3 composites can be shown in Figure 1. (b)-(e), in this figure shows the diffractogram comparison of TiO_2-Fe_2O_3 composites with variations in Fe_2O_3 addition. In Figure 1 (b) overall XRD spectra results have not shown significant differences with pure TiO_2 XRD results spectra, this is because the addition of Fe_2O_3 is slightly so that the TiO_2 is still dominant. From Figure 1 (c) - (e) it is known that new peaks appear indicating the presence of Fe_2O_3 which is indicated by a peak of 2θ = 35° (d_{110} = 2.5176 Å) and 2θ = 48° (d_{104} = 2.6995 Å) in accordance with the JCPDS No.871-165 standard. In addition to the appearance of Fe_2O_3, FeOOH is produced which usually does appear together with Fe_2O_3 due to alkaline conditions at the time of synthesis. The peak that characterizes the FeOOH phase is the peak of 2θ = 21° (d_{110} = 4.1896 Å) and 2θ = 35° (d_{101} = 2.5297 Å) according to JCPDS No. 810-462.

The results of the analysis using XRD can also be used as a semi-quantitative analysis, and can be applied to calculate the phase ratio material contained. Identified TiO_2 material is the phase anatase and rutile. Each variation of the anatase and rutile ratio of TiO_2- Fe_2O_3 composites can be compared by looking at the bar diagram Figure 1. In other words, the addition of Fe_2O_3 dopant turned out to affect TiO_2 crystal structure changes. The addition of dopant was able to improve anatase transformation to rutile as shown in Figure 8. The occurrence of these structural changes will change the photoactivity of TiO_2 and further explained in the activation of TiO_2- Fe_2O_3 composites in degrading Rhodamine B.

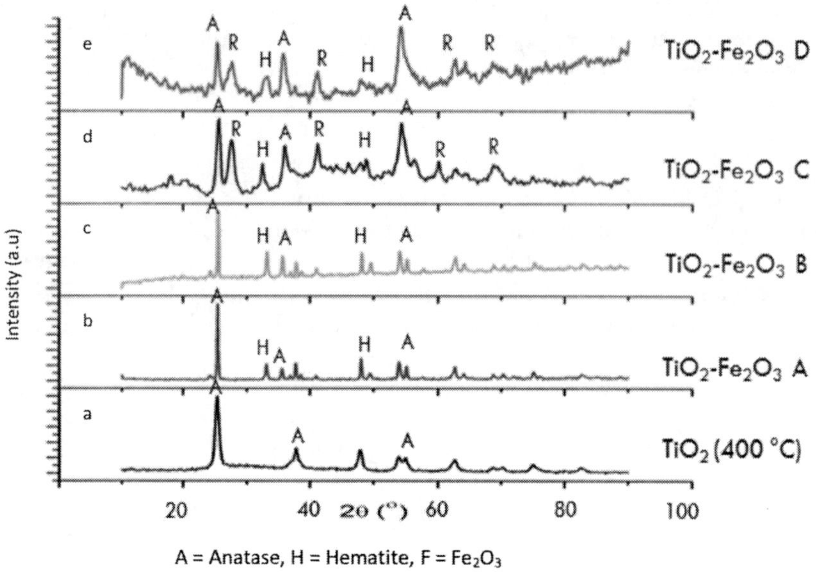

Figure 1. Diffractogram of (a) TiO_2, (b) TiO_2-Fe_2O_3 A (1:1), (c) TiO_2-Fe_2O_3 B (1:1,5), (d) TiO_2-Fe_2O_3 C (1:2), (e) TiO_2-Fe_2O_3 D (1:3)

Table 1. Relative Percentages of Fe_2O_3 in TiO_2 Synthesis Composite Results of TiO_2- Fe_2O_3 A (1:1), B (1:1,5), C (1:2), and D (1:3) based on XRF result

Compound	**Percentage**			
	A	**B**	**C**	**D**
TiO_2	49.22%	41.38%	33.62%	23.96%
Fe_2O_3	48.71%	56.67%	65.11%	74.95%
P_2O_5	0.53%	0.47%	0.42%	0.37%
SO_3	0.33%	0.28%	0.27%	0.25%
K_2O	0.23%	0.29%	0.26%	0.23%
Others	0.98%	0.91%	0.32%	0.24%

From the XRF data (Table 1) shows that the composite material TiO_2-Fe_2O_3 contains the main components of TiO_2 and Fe_2O_3. Fe_2O_3 produced in the XRF analysis (Table 1) also included FeOOH which was characterized by XRD. XRF analysis shows that the main composition of the composites is Ti, O, and Fe. The presence of Fe ensures the formation of metal oxides on the surface of TiO_2 and the appearance of O elements as chemical bonds between elements to form oxide compounds. This indicates that the composite material TiO_2-Fe_2O_3 has been successfully synthesized correctly.

Synthesis of TiO_2-Fe_2O_3 composites also contains other components which are impurities such as Cl, P_2O_5, SO_3, and K_2O with relatively small levels of less than 1%. The impurity can come from the remainder of the solvent which has not completely disappeared during calcination, and can also be due to contact with air.

3.2. Photocatalytic Activity Test

Activity test of TiO_2- Fe_2O_3 composites for degradation of Rhodamine B with time variation was carried out on compositions of TiO_2- Fe_2O_3 variation 1 to variation 4 and pure TiO_2. The results of degradation of Rhodamine B are shown in Figure 3-7. A total of 0.02 grams of composite was put into 10 mL of 10 ppm Rhodamine B which was conditioned illuminated by halogen lamps in a reactor with various time. The time variation used are: 0 minutes, 5 minutes, 10 minutes, 20 minutes and 40 minutes.

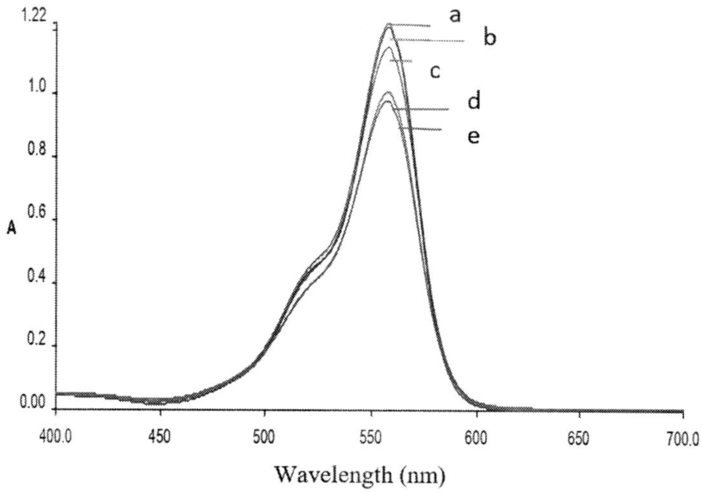

Figure 2. Rhodamine B Degradation Curve by TiO_2 photocatalyst for a period (a) 0 minutes, (b) 5 minutes, (c) 10 minutes, (d) 20 minutes, (e) 40 minutes

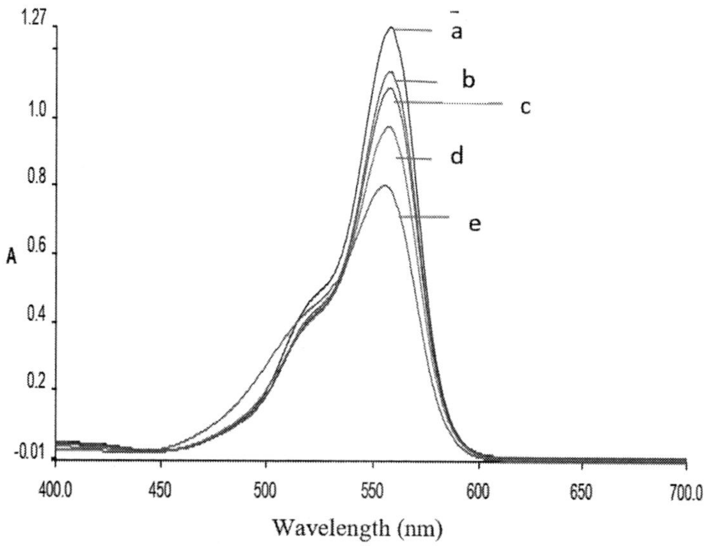

Figure 3. Rhodamine B Degradation Curve by TiO_2- Fe_2O_3 A (1:1) (a) 0 minutes, (b) 5 minutes, (c) 10 minutes, (d) 20 minutes, (e) 40 minutes

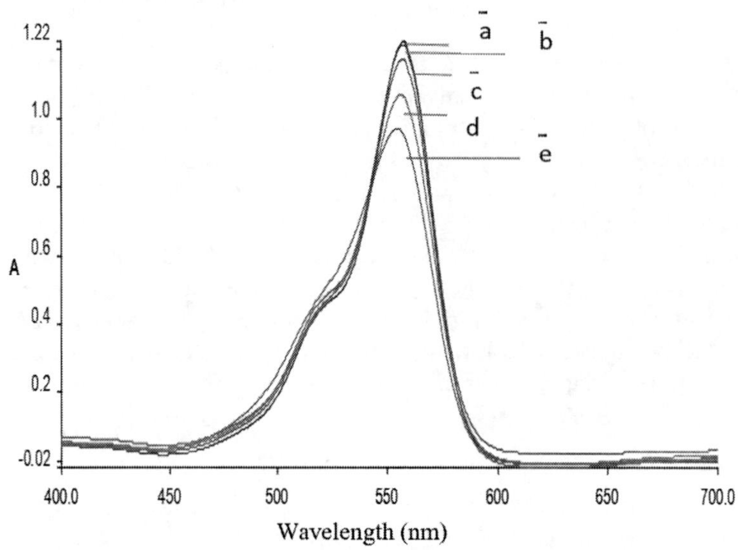

Figure 4. Rhodamin B Degradation Curve by TiO_2-Fe_2O_3 B (1:1,5) for a period (a) 0 minutes, (b) 5 minutes, (c) 10 minutes, (d) 20 minutes, (e) 40 minutes

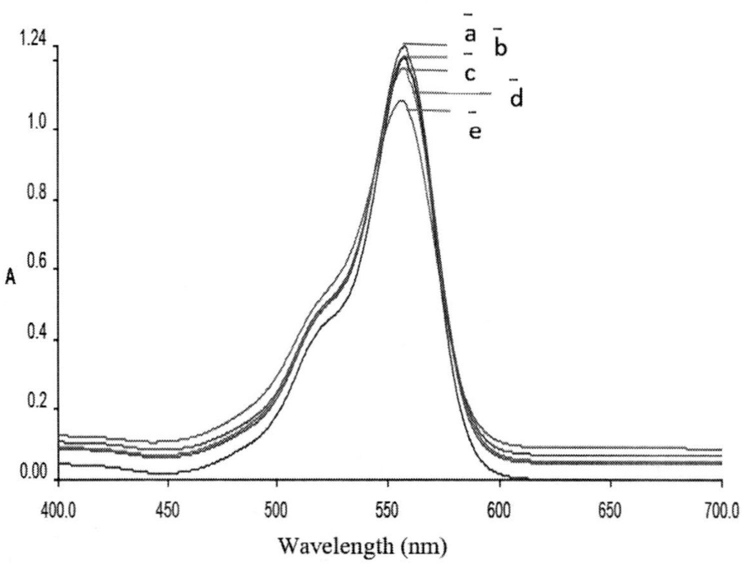

Figure 5. Rhodamin B Degradation Curve by TiO_2- Fe_2O_3 C (1:2) for a periode (a) 0 minutes, (b) 5 minutes, (c) 10 minutes, (d) 20 minutes, (e) 40 minutes

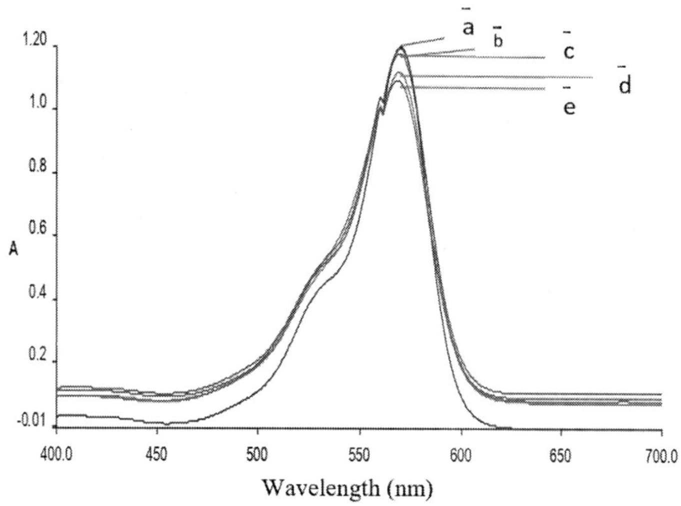

Figure 6. Rhodamin B Degradation Curve by TiO_2- Fe_2O_3 D (1:3) for a period (a) 0 minutes, (b) 5 minutes, (c) 10 minutes, (d) 20 minutes, (e) 40 minutes

Figure 2-6 shows a decrease in absorbance of Rhodamin B with variations in irradiation time for: 0 minutes, 5 minutes, 10 minutes, 20 minutes and 40 minutes by each variation of TiO_2-Fe_2O_3 composites. From the figure it can be seen that the decrease in absorbance of Rhodamin B is greater as the irradiation time increases from 0 minutes to 40 minutes. These results indicate that the longer the visible irradiation time, the greater the decrease in absorbance of rhodamine B.

This result is consistent with the research conducted by [9] and [10] which proves that the longer the visible light irradiation time will have an increasing effect on composite photocatalytic activity. This is because the longer the visible light irradiation, the more electrons will continue to be excited and play a role in the photodegradation activity of Rhodamin B. From the absorbance reduction data obtained (Figure 6) the effectiveness percentage can be determined. The percentage of degradation can show optimization of photocatalytic-based self-cleaning activities of TiO_2-Fe_2O_3 composites. The results of the percent degradation calculation can be shown in Table 2.

Table 2. Self-cleaning Rhodamin B uses a glass test sample at the time of contact exposure of 40 minutes

Sample	**η (%)**
TiO_2	38.45
Composite TiO_2-Fe_2O_3 A (1:1)	43.86
Composite TiO_2-Fe_2O_3 B (1:1,5)	42.72
Composite TiO_2-Fe_2O_3 C (1:2)	31.45
Composite TiO_2-Fe_2O_3 D (1:3)	20.99

The data obtained in Table 2 shows that the optimal results were obtained in TiO_2-Fe_2O_3 A (1:1) composites which were able to degrade Rhodamin B as much as 43.86%. These results indicate that the composition of Fe_2O_3 will affect composite self-cleaning activities. In a composite variation with a higher ratio of Fe_2O_3, it will reduce self-cleaning activity. This is due to the presence of the FeOOH crystalline phase which normally appears with Fe_2O_3 because of the synthesis process under alkaline conditions, where FeOOH has a photocatalytic activity that is not good at degrading Rhodamin B.

3.3. Hydrophobicity Test

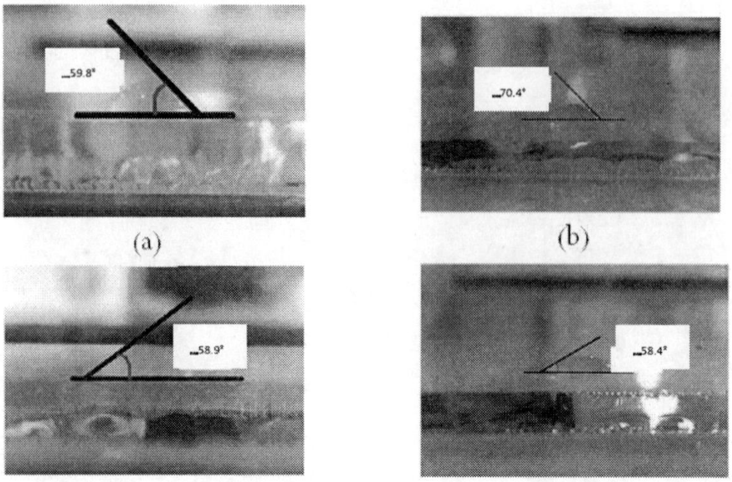

(c) Hydrophobicity test : (d)

Figure 7. Contact angle before sun-drenched of (a) Non-coated glass, (c) TiO_2- Fe_2O_3-coated glass, and after sun-drenched (b) non-coated glass, (d) TiO_2- Fe_2O_3-coated glass

The contact angle measurement was performed by using water droplets. Small contact angle could be achieved when a stable form hydroxyl grouped on the surface. Hence, when the water appeared, it could be directly bound to these hydroxyl groups. The results obtained from this process are shown in Figure 5. The contact angle of coated glass shows a decrease after it's irradiated by visible light means that self-cleaning process based on photocatalytic is occured so the characteristic of glass surface change from hydrophobic into hydrophilic.

4. Conclusion

TiO_2-Fe_2O_3 composites was successfully synthesized and applicated for self-cleaning glass. The longer irradiation time, the more decrease of concentration of Rhodamine B, which shows that the photocatalytic activity of the composite TiO_2-Fe_2O_3 is increasing. The maximum degradation of Rhodamin B reached 43.86% for 40 minutes by TiO_2-Fe_2O_3 A (1:1). In the hydrophobicity test, the contact angle of TiO_2- Fe_2O_3 coated glass after UV irradiation is 58.4°.

References

[1] Brauer D S, Gentleman E, Farrar D F, Stevens M M and Hill R G *Biomed. Mater.* **6** 045007
[2] Cedillo-González E I, Riccò R, Montorsi M, Montorsi M, Falcaro P and Siligardi C 2014 *Build. Environ.* **71** 7-14
[3] Zahedi F, Behpour M, Ghoreishi S M and Khalilian H 2015 *Energy.* **120** 287-295
[4] Gherardi F, Colombo A, D Arienzo M, Di Credico B, Goidanich S, Morazzoni F, Simonutti R and Toniolo L 2015 *Microchem. J.* **126** 54-62
[5] Wang H, Zhang R, Yuan Z, Shu X, Liu E and Han Z 2015 *Ceram. Int.* **41** 11844-11851
[6] Kazemi M, Mohammadizadeh M R 2011 *Thin Solid Films.* **519** 6432-6437
[7] Wahyuningsih S, C Purnawan, T E Saraswati, E Pramono, A H Ramelan, A Wisnugroho and S. Pramono 2014 *Journal of Enviromental Protection.* **5** 1630-1640
[8] Kazemi M and Mohammadizadeh M R 2011 *Appl. Surf. Sci.* **257** 3780-3785
[9] Liu H and Gao L 2006 *Journal of the American Ceramic Society.* **89** 370–373
[10] Xing, P., Sugimoto, T., Astsuhi, M 2009 *Journal of Colloid Interface Sci.* 259

Improving preparation method of TiO_2-NiO composite materials for self-cleaning glass

I Farichah[1], S Wahyuningsih[1], A H Ramelan[1]

[1]Inorganic Material Research Group, Faculty of Mathematics and Natural Sciences, Universitas Sebelas Maret, Jalan Ir. Sutami No.36A Kentingan Surakarta

E-mail: sayekti@mipa.uns.ac.id

Abstract. TiO_2-NiO Composite have been synthesized as self-cleaning material on glass. TiO_2 was prepared using titanium tetraisopropoxide (TTIP) by sol gel method, while TiO_2-NiO composite was prepared from TiO_2 synthesized with additional $Ni(NO_3)_2.6H_2O$ precursor. TiO_2 synthesized powder was added to Ni^{2+} concentrated solution in aquadest so it formed crystalline layer on TiO_2 powder. Core shell composite of TiO_2-NiO was obtained through the drying process at 110 °C and calcined at 700 °C. Identification and characterization of NiO-TiO_2 composite were carried out using X-Ray Diffraction (XRD), and X-Ray Fluoresecence (XRF). XRD diffractograms show the appearance of TiO_2 and NiO peaks. Another analysis with XRF showed that composition of TiO_2-NiO composite are 2:1; 1:1 and 1:2 (w/w). The electronic spectrum from TiO_2-NiO core shell result showed an increasing response in visible light so it can be applied as coating material in manufacturing technology of self cleaning glass.

1. Introduction

Glass is a non-crystalline solid object that is usually transparent and has a variety of practical, technological and decorative uses in things like window panels, cutting tools, optoelectronics and others. Scientifically, the term "glass" is often defined in a broader sense, including every solid object. which has a non-crystalline structure (amorphous) on the atomic scale and shows the transition from glass when heated to a liquid state [1]. Along with the development of the times, there have been many researches on the modification of glass substrate to have several advantages, including anti-bacterial, anti-fungal, anti-UV, not easily dirty and easy to clean. Glass is often installed in locations that are difficult to reach so there are obstacles to clean glass. In skyscraper buildings, there is a high risk of accidents in the process of cleaning the outer glass of the building. Therefore, self-cleaning glass is needed in order to reduce the risk of workplace accidents during glass cleaning.

Titanium dioxide (TiO_2) has been widely used in various fields of application especially as photocatalytic for self-cleaning [2], water treatment [3], antibacterial [4], and air purification [5]. In addition TiO_2 has been used for sensors [6], solar energy conversion [7], and as an antiultraviolet (UV) agent [8].TiO_2 also has the ability of photocatalysts to absorb UV light (200-380 nm) with a band gap energy of 3-3.2 eV but the effectiveness of TiO_2 in photocatalysts is still low, this is due to the high band gap energy that can only absorb ultra violet light, 4 % of all sunlight energy reaches the earth's surface [9].

TiO_2 photocatalytic activity can be increased through the process of doping metal ions. The use of doping methods is more beneficial than the impregnation method. This is due to the use of impregnation method on dopant insertion does not change the electronic structure of photocatalysts. Dopants function as electron trapping which can increase photocatalytic activity [10].

Nickel oxide (NiO) is one of the active catalysts for photocatalytic reactions. The addition of NiO into the photocatalyst can increase the decomposition of water, where hydrogen gas is formed on the surface of nickel oxide while oxygen gas is released from the photocatalyst surface [11]. In the study, Nickel metal was tested as a dopant to increase the efficiency of TiO_2 photocatalysts in visible light regions and reduction of band gap on TiO_2[12]. The anatase transformation to rutile due to the addition of dopants depends not only on the nature of the metal dopant (ionic radius), but also by the concentration of dopants [13]. In this paper, the addition of NiO to TiO_2 is expected to increase TiO_2 photocatalytic activity so that it can be better for self-cleaning applications.

2. Experimental

2.1. Materials

The materials used in this study were titanium tetra isopropoxide (TTIP) (E Merck), $Ni(NO_3)_2.6H_2O$ (E Merck). Calcination of TiO_2 and TiO_2-NiO was performed by Furnace Thermo. Meanwhile, the analysis and characterization of materials were done by XRD (Bruker D8 Advantage), XRF (Bruker S2 Ranger), and UV Vis Spectrophotometer (Pelkin Elmer Lambda 25).

2.2. Methods

2.2.1. Synthesis of TiO_2

The titanium (IV) isopropoxide solution was hydrolyzed using a glacial acetic acid solution (temperature 10-14 °C) at a ratio of 1: 10 (v / v) and then stirred continuously using a magnetic stirrer until a white sol gel was obtained, then stirred and heated at 90 °C until white gel (sol gel TiO_2) obtained [14]. And then, sol gel TiO_2 was roasted at 150 °C for 24 hours until formed the xerogel TiO_2 Xerogel TiO2 was further treated by calcined at 400 °C for 2 hours with a speed of 10 °C / minute. The resulting TiO_2 crystalline powder was analyzed by XRD.

2.2.2. Synthesis of TiO_2 – NiO

1.145 grams of xerogel TiO_2 was added to Ni^{2+} concentrated solution from $Ni(NO_3)_2.6H_2O$ precusore until obtained a stoichiometric ratio of TiO_2: NiO = 2:1; 1:1; and 1:2 (mol/mol), then stirred for 60 minutes and heated at 90 °C until obtained green gel (sol gel TiO_2-NiO). The resulting sol gel of TiO_2-NiO was roasted at 110 °C for 3 hours. The obtained xerogel TiO_2-NiO was calcined at 700 °C for 2 hours [15]. The resulting TiO_2-NiO crystalline powder was analyzed by XRD, XRF, and UV-Vis spectrophotometer.

3. Results and Discussion

Synthesis of TiO_2 with the addition of NiO metal was carried out using the sol-gel method [14]. The addition of metal is expected to replace several positions of Ti atoms in TiO_2. The calcination process was carried out at 700 °C for 2 hours. X-ray diffractograms of TiO_2 and TiO_2-NiO are shown in Figure 1. Figure 1. (A) shows the diffraction pattern of the synthesized TiO_2 is indicated by the presence of characteristic peaks that appear at a certain diffraction angle (2θ). The peaks include (2θ) = 25.46° (d_{101} = 3.498 Å), (2θ) = 37.432° (d_{004} = 2.402 Å), (2θ) = 48.196° (d_{200} = 1.888 Å), (2θ) = 54° (d_{105} = 1.698 Å), and (2θ) = 62.807° (d_{211} = 1.478 Å) which is the characterization area of anatase TiO_2 in accordance with JCPDS standard No.78-2486.

XRD analysis of TiO_2-NiO composite be shown in Figure 1. (B)–(D). Figure 1. (B)–(D) shows a comparison of the TiO_2-NiO composite diffractogram with variations in the addition of NiO from the XRD result. The data has shown the significant differences with the TiO_2 yield spectra and changes in crystalline phase occurred due to changes in the increasing mass of NiO. From the Figure 1. (B) – (D) it is known that new peaks appear indicating the presence of NiO indicated by the peak (2θ) = 43° (d_{200} = 2.08Å) in accordance with the JCPDS standard No.471-049. Besides the appearance of NiO,

$NiTiO_3$ also appeared in the region $(2\theta) = 24°$ ($d_{012} = 3.682Å$), $(2\theta) = 33°$ ($d_{110} = 2.697Å$), $(2\theta) = 49°$ ($d_{024} = 1.838Å$), and $(2\theta) = 57°$ ($d_{018} = 1.599Å$) in accordance with JCPDS standard No. 753-757.

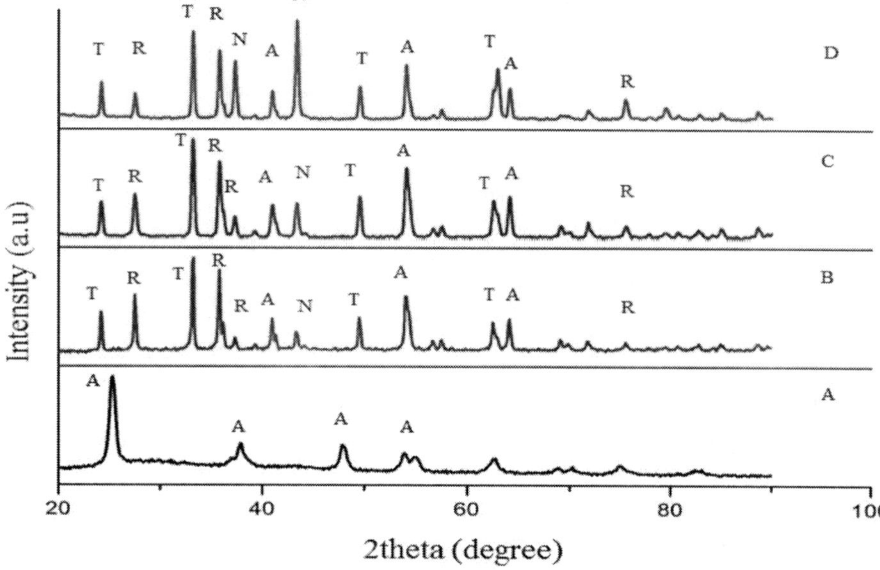

Figure 1. Diffractogram of (A) TiO_2 (B) TiO_2-NiO (2:1) (C) TiO_2-NiO (1:1) (D) TiO_2:NiO (1:2)

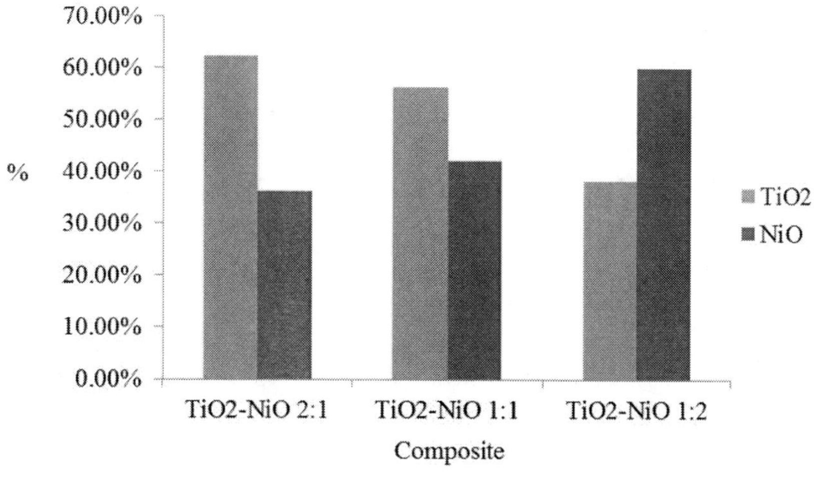

Figure 2. Composition of TiO_2 and NiO in TiO_2 – NiO Composite

The result of XRF analysis shown the ratio of TiO_2:NiO is 1:1 ; 1:1.5 ; and 1:2. Photocatalytic activity test of TiO_2-NiO composites was carried out by applying it to degradation of Methylene Blue for a period time. About 0.02 g of composite was put into 10 mL of 10 ppm Methylene Blue under halogen lamps illuminations in a reactor with varying time. Variations in the time used are: 0 minutes, 5 minutes, 10 minutes, 15 minutes and 20 minutes.

IOP Conf. Series: Materials Science and Engineering **578** (2019) 012030 doi:10.1088/1757-899X/578/1/012030

Methylene Blue concentration decreased due to photocatalytic degradation reaction by TiO_2-NiO composite. The results can be seen in Figure 3 by using composites (A) TiO_2 (B) TiO_2-NiO (2:1) (C) TiO_2-NiO (1:1) (D) TiO_2:NiO (1:2).

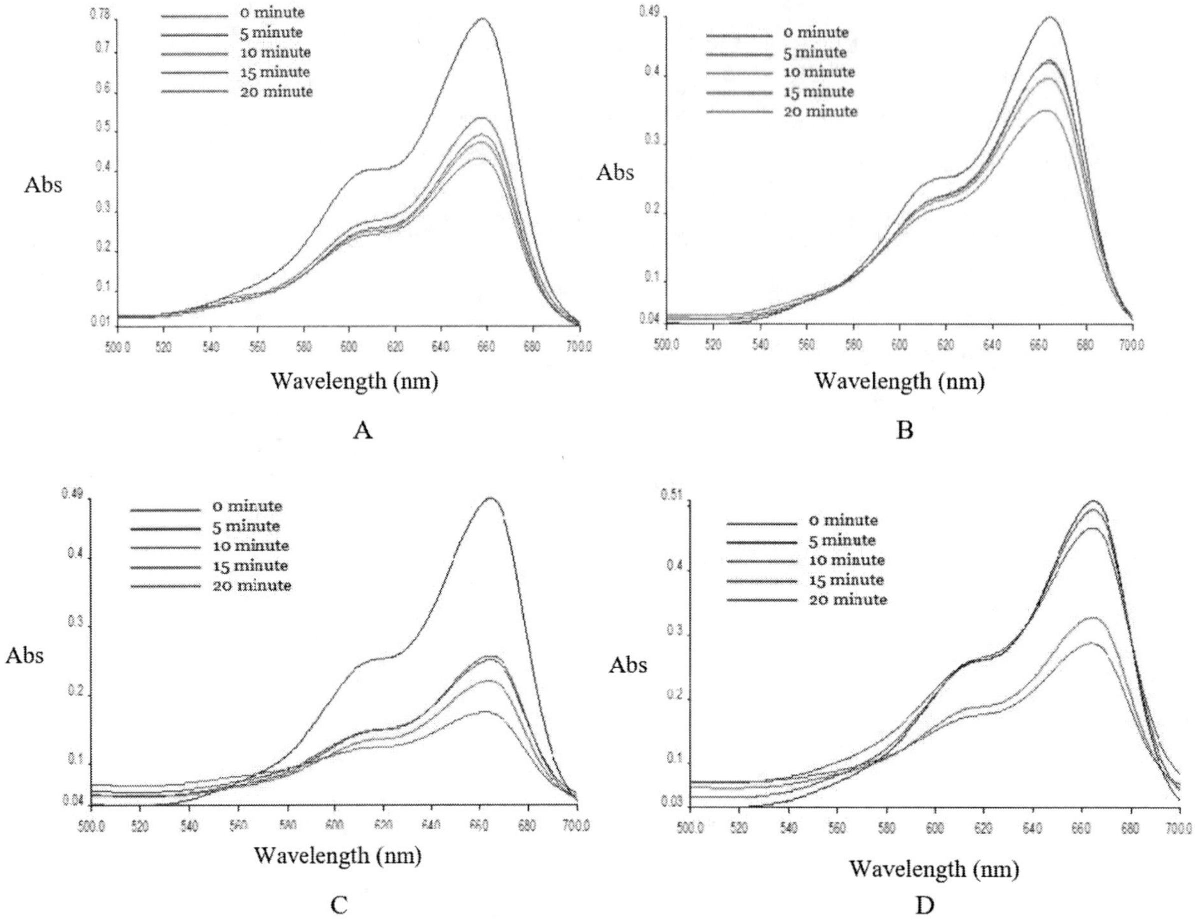

Figure 3. Methylene Blue photocatalytic Degradation by TiO_2-NiO composites (A) TiO_2 (B) TiO_2-NiO (2:1) (C) TiO_2-NiO (1:1) (D) TiO_2:NiO (1:2)

Based on the Figure 3 (A) – (D), generally the longer irradiation time, the more of Methylene Blue have degraded. From the figure 3 it can be seen that the decrease in absorbance of Methylene Blue is greater as the irradiation time increases from 0 minutes to 20 minutes. These results indicate that the longer the visible irradiation time, the greater the decrease in absorbance of Methylene Blue.

From the table 1 shows the percentage of degradation of Methylene Blue, in TiO_2 the degradation of Methylene Blue in the 5 minutes was 31.76962% while in the 20 minute it was 45.44908%. In TiO_2-NiO composite the highest degradation of methylene blue was 64.01827% within 20 minutes on TiO_2-NiO composite (1:1). This is consistent with research [16] which explains the duration of irradiation affects the degradation of Methylene Blue which proves that the longer the visible light irradiation time will have an increasing effect on composite photocatalytic activity. This is because the longer the visible light radiation, the more electrons continue to be excited and play a role in the photodegradation activity of Methylene Blue.

Table 1. Degradation of Methylene Blue on TiO_2 and TiO_2 – NiO Composite

Composite	Percentage of Degradation of Methylene Blue Based on Time Variation (%)			
	5 Minutes	10 Minutes	15 Minutes	20 Minutes
TiO_2	31.76962	37.39551	40.05956	45.44908
TiO_2-NiO (2:1)	13.27231	18.04149	53.98603	57.59598
TiO_2-NiO (1:1)	47.04096	48.40889	54.53982	64.01827
TiO_2-NiO (1:2)	33.96763	37.55842	56.21077	61.61098

(A) (B) (C

Figure 4. The image of contact angle calculation (A) No-coated glass before UV irradiation; (B) glass/TiO_2 before UV irradiation; (C) glass/TiO_2-NiO before UV irradiation

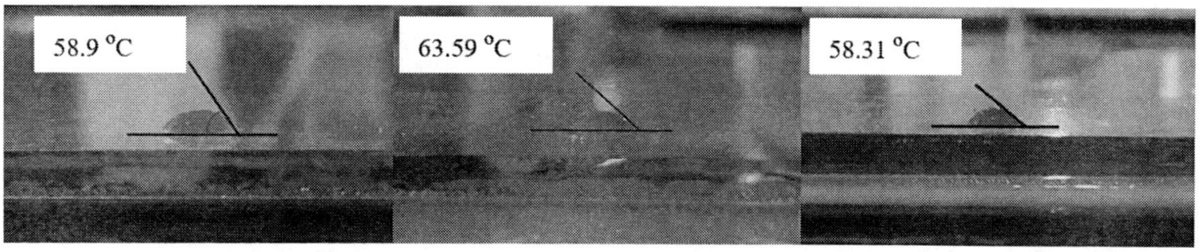

(A) (B) (C)

Figure 5. The image of contact angle calculation (A) No-coated glass after UV irradiation; (B) glass/TiO_2 after UV irradiation; (C) glass/TiO_2-NiO after UV irradiation

The contact angle measurements were carried out by using water droplets. The contact angle between water and the material was analyzed using ImageJ software. Small contact angles can reach stable hydroxyl forms grouped on the surface. Therefore, it can be done with this hydroxyl group. The results obtained from this process are shown in Figure 4-5. Based on measurements of TiO_2-NiO composite contact angles obtained contact angle on glass without coating before irradiation was obtained 59.8° (Figure 4A) and after irradiation 58.9° (Figure 5A) almost no change in this indicates the absence of photocatalytic activity on glass without coating so that the glass does not have self-cleaning properties. In glass/TiO_2, the contact angle before irradiation was 81.56° (Figure 4B) and after irradiation 63.59° (Figure 5B), whereas in glass/TiO_2-NiO the contact angle before irradiation was 68.65° (Figure 4C) and after irradiation 58.31° (Figure 5C). In glass/TiO and glass/TiO_2-NiO the decrease in contact angle increases the hydrophilic properties of the glass which indicates photocatalytic activity so that the glass has self-cleaning properties.

4. Conclusion

TiO_2-NiO composite was successfully synthesized and applied for self-cleaning glass. The longer irradiation time, so that the greatest of decrease in absorbance of Methylene Blue, which shows that the photocatalytic activity of the composite TiO_2-NiO is increasing. The maximum dye degradation of Methylene blue achieved was 64.01827% for 20 minutes. In the hydrophobicity test, the contact angle of glass/TiO_2 after UV irradiation was obtained 63.59 °C and glass/TiO_2 - NiO 58.31 °C.

References

[1] Karasu B and Ay N2000 *Ministry of Education Publication 3525*. Ankara (in Turkish).
[2] Carneiro J O, Teixeira V and Portinha A 2007 *Materials Science and Engineering B*. **138** 144–150
[3] Lachheb H, Puzenat E and Houas A 2002 *Applied Catalysis B*. **39** 75–90
[4] Zhang H, Liu H, Mu C, Qiu C and Wu D 2006 *Proceedings of 1st IEEE International Conference on Nano Micro Engineered and Molecular Systems*. 955–958
[5] Yu H, Zhang K and Rossi C 2007 *Indoor and Built Environment*. **16** 529–537
[6] Rella R, Spadavecchia J and Manera M G 2007 *Sensors and Actuators B*. **127** 426–431
[7] Jung H S, Lee J K and Nastasi M 2005 *Langmuir*. **21** 10332–10335
[8] Hoffmann M R, Martin S T, Choi W and Bahnemann D W 1995 *Chemical Reviews*. **95** 69–96
[9] Qi K, Cheng B, Yu J and Ho W 2017 *Chinese Journal of Catalysis*. **38** 1936–1955
[10] Liu H and Gao L 2006 *Journal of the American Ceramic Society*. **89** 370–373
[11] Motahari F, Mozdianfard M R, Soofivand F and Niasari M S 2013 *RSC advances*. **4** 27654–27660
[12] Takashi H, Y Sunagawa, S Myagmarjav, K Yamamoto, N Sato and A Muramatsu 2003. *Materials Transactions*. **44**
[13] Choi J, Hyunwoong P and Michael R 2010 J. *Phys.Chem.***114** 783-792
[14] Wahyuningsih S, C Purnawan, T E Saraswati, E Pramono, A H Ramelan, A Wisnugroho and S Pramono 2014 *Journal of Enviromental Protection*. **5** 1630-1640
[15] Riyas S, Yasir V A and Das P N 2002 *Bull. Mater. Sci*. **25** 267-273
[16] Alfina Bulan Tahta 2015 *Chemistry Student Journal*. **1** 599-605

A control of TiO_2 nanostructures by hydrothermal condition and their application: a short review

R Hidayat[1,*], G Fadillah[2,*] and S Wahyuningsih[3]
[1]Master Student of Chemistry, Universitas Sebelas Maret, Surakarta, Indonesia
[2]Chemistry Department, Universitas Islam Indonesia, Yogyakarta 55584, Indonesia
[3]Chemistry Department, Universitas Sebelas Maret, Surakarta, Indonesia

[*]Email: rahmathidayat@student.uns.ac.id, ganjar.fadillah@uii.ac.id

Abstract. TiO_2 (titanium dioxide) is an inorganic material that has many uses in various fields such as environment, sensor devices and solar energy conversion. Various method approaches have been developed for the synthesis of nanostructured TiO_2, but the hydrothermal method is still considered the most effective method and provides many advantages in nanostructure formation because the methods can be controlled. In the present study, we will discuss the formation of TiO_2 with various conditions on hydrothermal methods in TiO_2 nanostructure formation. Generally, TiO_2 nanoparticles can be obtained by hydrothermal methods. TiO_2 precursors, pH, reaction time and also temperature are factors that influence the formation of TiO_2 structures. The methods are commonly used because hydrothermal methods can easily control conditions to obtain a homogenous structure. The difference of TiO_2 structure is very influential in its properties and activity performance.

1. Introduction

Titanium dioxide (TiO_2) is an inorganic material which have high stability, low cost and safety for humans and the environment. Because of its properties, the materials has many uses in various fields such as for photocatalysis process, solar energy conversation and sensor devices. TiO_2 as a semiconductor is able to absorb photons to produce electrons so the material can be widely used in various fields. However, the limitation of the system shows that TiO_2 requires high-energy for producing an electron. Therefore, this material is often modified to reach the nanostructure. So far, there are many approach ways to the synthesis of TiO_2 nanostructured, i.e. via templating, chemical vapor deposition, solvothermal methods, direct oxidation, microwave methods and hydrothermal synthesis methods. However, hydrothermal synthesis is a common method widely used to prepare TiO_2 nanostructure because the methods have a simple route for synthesis.

Hydrothermal methods were firstly introduced in the 1990s by controlling morphology and shape of materials. Hydrothermal method is one of synthesizing nanoparticle materials that are easy to use. The Hydrothermal method can be used to obtain 0D, 1D, 2D and 3D nanostructures TiO_2. These methods are usually used at high temperature (>100 °C), a high pressure and an aqueous medium with a presence of catalyst [1]. Some literature reported that this method showed effective methods for synthesized nanomaterials with high crystallinity, purity and also a homogenous structure [2].

Currently, the development of hydrothermal methods for synthesized nanostructures is still an interisting topic to discuss. Some kinds of literature have reported that the development of these

IOP Conf. Series: Materials Science and Engineering **578** (2019) 012031 doi:10.1088/1757-899X/578/1/012031

methods not only focusing to obtain a homogenous nanoparticle but also but also to produce nanocomposites or nano-hybrid materials [3-5].

Wahyuningsih et al. (2016) firstly reported that TiO_2 powder could be synthesized by mixing a TiO_2 powder and NaOH solution at high temperature [28]. The results showed that the structure of TiO_2 was strongly influenced by several factors which previously explained [6]. In this study, we focused on discussing the production of hydrothermal TiO_2 nanostructures with regard to several factors such as time, precursor, pH condition, also temperature and its application.

2. Experimental

2.1 Synthesis of TiO_2 nanostructures

Some kinds of literature have reported that the TiO_2 nanostructures could be synthesized by hydrothermal methods in various conditions such as precursors, time and temperatures during the reaction process, and pH solution. Huang et al. (2016) reported that TiO_2 nanorods have been successfully synthesized from titanium oxalate as precursors in NaOH solution. Generally, titanium tetraisopropoxide (TTIP) was selected and reacted with oxalate. The reaction would produce a titanium oxalate which used as precursors for the synthesis of TiO_2 nanostructures. Beside, Wahyuningsih et al. (2016) reported that TiO_2 nanorods could be synthesized from TiO_2 powder. TiO_2 that was used in the hydrothermal reaction was brookite phase that resulted from mechanical milling of anatase TiO_2. TiO_2 was reacted in basic solution (NaOH) with various concentration of NaOH. TiO_2 nanostructures was not only reacted in basic solution but also in acidic solution like chloric acid or fluroric acid. Nanoparticles were characterized by spectroscopy techniques such as X-ray diffraction (XRD), X-ray photoelectron spectroscopy (XPS) and EDX (Energy Dispersive X-Ray).

3. Results and discussions

3.1. Precursor Effect

TiO_2 nanostructures could be synthesized in various conditions. The difference of hydrothermal condition gives different result especially for homogeneity and crytanillity of materials. As mentioned earlier, these methods can produce nanostructures, however, all of these are influenced by the selected precursors. Huang et al (2016) reported that TiO_2 nanorods (1D) could be synthesized from titanium oxalate in basic solution [7]. Beside, Fadillah et al. (2017) also reported that TiO_2 1D can be synthesized as directly using TiO_2 powder at a high base concentration [8]. However, the use of TiO_2 powder as a precursor resulted in poor homogeneous crystals. Generally, to ensure the shape of the structure, the material was characterized by X-ray photoelectron spectroscopy (XPS), EDX (Energy Dispersive X-Ray) and X-ray diffraction (XRD). The characterization result as shown on Figure 1.

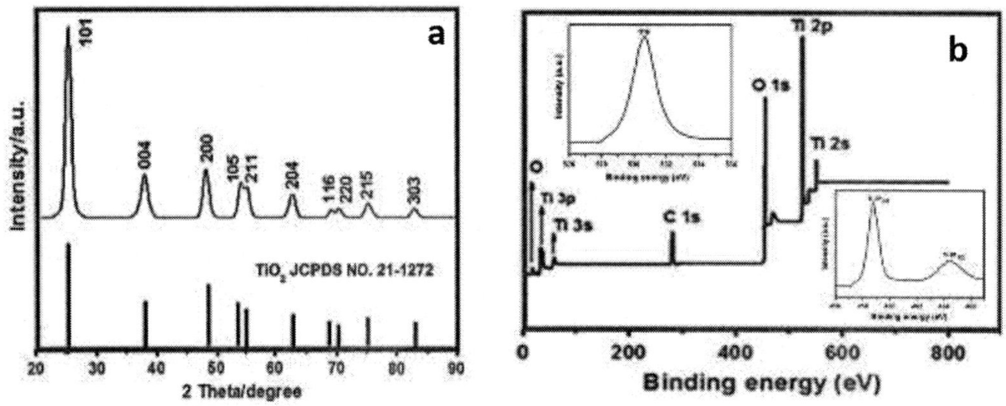

IOP Conf. Series: Materials Science and Engineering **578** (2019) 012031 doi:10.1088/1757-899X/578/1/012031

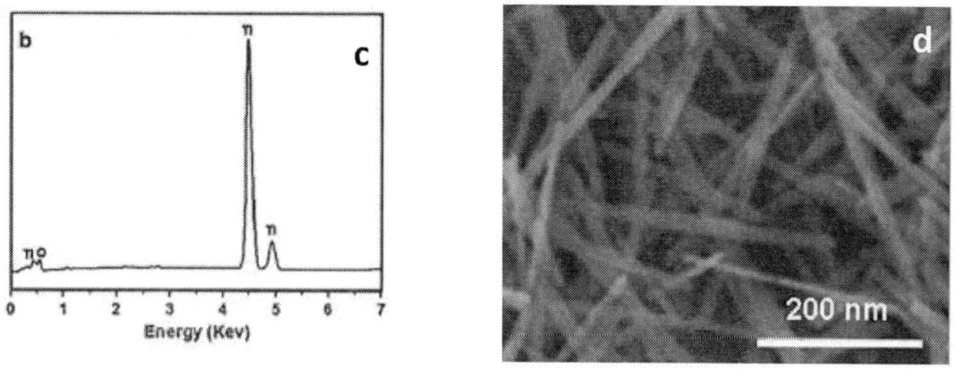

Figure 1. Characterization of TiO_2 nanostructures (a) XRD, (b) XPS, (c) EDX and (d) SEM [7].

Based on The XRD spectra (Figure 1a) shows that TiO_2 has a good crytallinity because the diffractogram shows a clear and sharp shape. The pattern of diffraction matches as well with the standard (following Number of JCPDS 21-1272). To ensure the materials have been successfully formed, the XPS data was used to support the data. XPS spectra in Figure 1b shows not only the elemental compositions but also the chemical status of TiO_2. The results indicated that the composition of materials only contains Ti and O. In addition, the TiO_2 was also confirmed using EDX data. The data reported that the sample contains only Ti and O elements like XPS results.

The SEM characterization showed that the TiO_2 nanorods have been successfully synthesized with average diameters between 10-20 nm. The results confirmed that the 1D materials could be synthesized from TTIP as precursors [9]. The other studies reported that TiO_2 nanoparticles could be formed by hydrothermal methods using TTIP as a precursor in ethanolic HNO_3 solution [10, 11]. The XRD characterizations results the XRD patterns of synthesized TiO_2 nanorods using hydrothermal methods. The results indicated that the diffraction of TiO_2 has changed to rutile phase (101 and 002). The surface morphology characterization both SEM and TEM show that TiO_2 nanorods have a distinctive shape that is rod-shape. The TiO_2 nanorods grows up regularly on the surface of FTO substrate. Whereas the image shows a single crystalline of TiO_2 nanorods and it has [001] growth direction based on SAD (selected area diffraction) pattern. The obtained TiO_2 nanorods have a uniform diameter that is around 50 nm. Tan et al. (2015) have synthesized layered nanostructured TiO_2 by using $TiCl_4$ as a precursor and tetramethylammonium hydroxide (TMAH) [12]. The 3D TiO_2 nanostructures can be obtained from TTIP in HNO_3 solution hydrothermally. TEM characterization of two TiO_2 nanoparticles as shown in Figure 2.

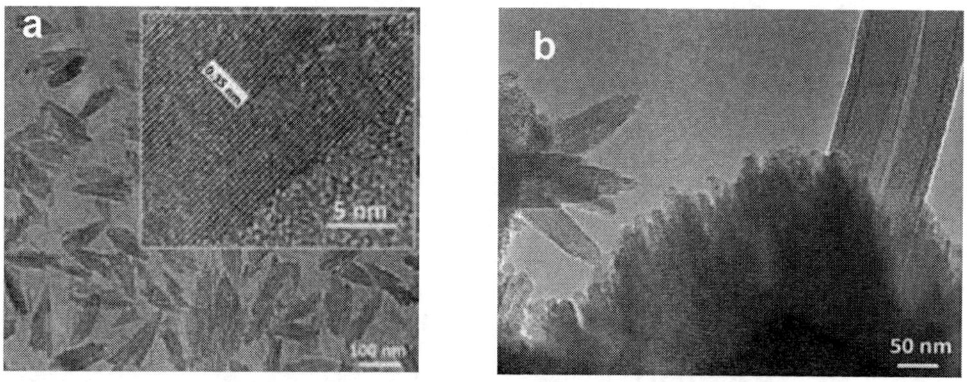

Figure 2. TEM images of TiO_2 nanoparticles (a) 2D [12] and (b) 3D respectively [13].

3.2. Temperature and Time

The effect of time and temperature during the reaction process show a significant influence on the formation of TiO_2 nanostructure. Erdogan et al. (2016) reported that the increasing temperature affected to intensity of the peak as shown in Figure 3 because of the change of TiO_2 crystallinity [13,14].

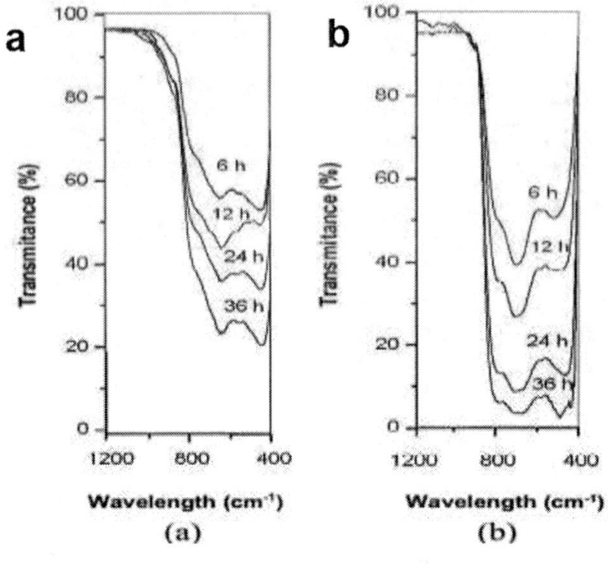

Figure 3. FT-IR spectra of TiO_2 in different temperatures (a) 150 °C and (b) 200 °C at different times [15].

Based on Figure 3, a reaction time of hydrothermal effects on crystal formation of TiO_2. It can be known from XRD patterns of TiO_2 nanorods in Figure 4. There is a different ratio of Rutile (101) and (002) that indicates in lower temperature, the nanorods have a random orientation then a higher temperature occurs well-aligned arrays of the vertical nanorods [15-18]. The orientation change is supported by morphology data from SEM analysis. It revealed that the difference of a time reaction is not only effects on a dimensional size but also a orientation of crystal growth on the surface of the substrate [19]. The rate of TiO_2 nanorods growth in the other literature is shown in Figure 5a.

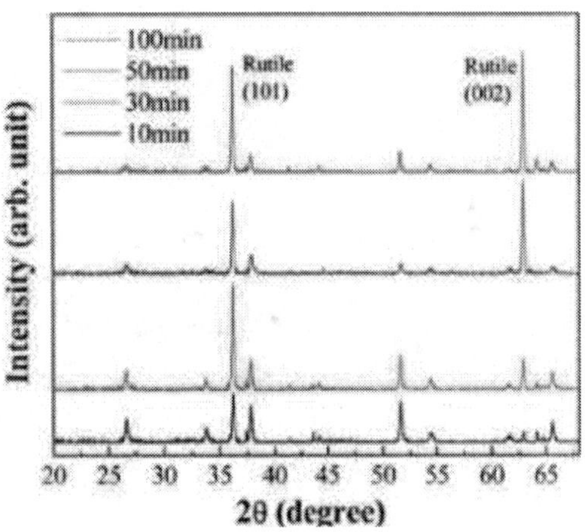

Figure 4. The effect of reaction time of hydrothermal on TiO_2 nanorods diffractogram [10].

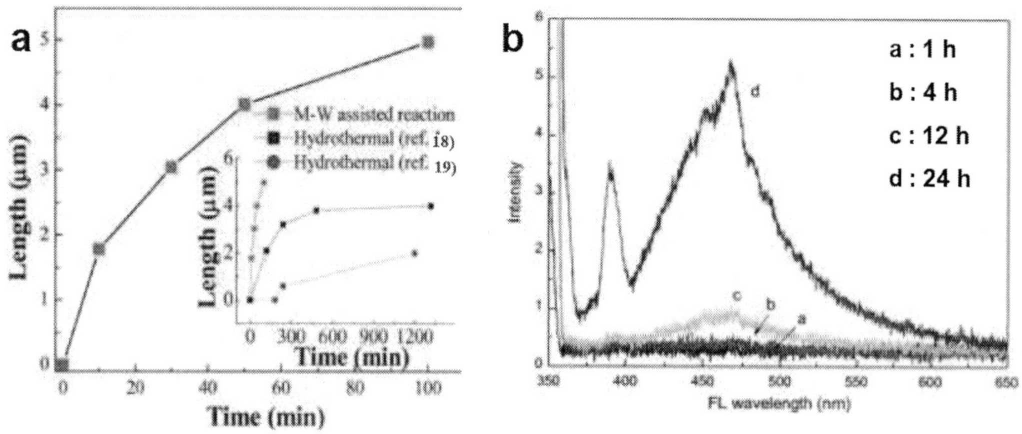

Figure 5. (a) The rate of TiO_2 nanorods growth during a microwave-assisted hydrothermal reaction compare with a conventional hydrothermal and (b) Luminescence spectra of TiO_2 influenced by reaction time duration [12].

The influence of reaction time to an optical characteristic of TiO_2 nanoparticles is shown in Figure 5b. This spectroscopy technique is used for knowing the crystalline TiO_2 formation [23]. TiO_2 nanoparticles are radiated with UV-Visible lamp so it caused excite TiO_2 molecules. The 1 h reaction time does not show emission peak, it is a different result in the 4 h treated that it shows a very weak emission at 470 nm. The maximum emission peak is obtained for 24 h. The result indicates that anatase TiO_2 nanoparticles are produced in a large amount during a long time reaction.

3.3. *pH Condition*

The pH is one of the factors to the formation of TiO_2 nanostructure. Nian et al. (2006) reported that different pH values give a different resulted shape of TiO_2 nanorods. The pH during reaction could be controlled by adding base solution such as NaOH into TiO_2, then by adding acid solution such as HNO_3 until pH values were achieved [20]. Some literatures reported that TiO_2 ID like nanotubes is a good precursor for the synthesis of TiO_2 nanorods due to it has a specific shape. So, the hydrothermal reaction for these nanotubes under acidic or basic condition impacts on bond cleavage and then the destruction of its [21]. The results of TiO_2 nanorods synthesized from nanotubes mention in Figure 6.

Figure 6a shows the influence of pH condition on XRD pattern of TiO_2 nanorods. There is a reduction of XRD peak intensity for TiO_2 anatase phase. It means that the higher pH values effect in the stability of TiO_2 anatase phase. This is confirmed by a morphology showed of TiO_2 nanorods in Figure 6b. The measurement of TiO_2 nanorods average size under highly acidic condition (pH 2.2) are about 10 nm. The pH condition has an effect on the crystalline sizes of TiO_2 nanostructure [22,23]. The average thickness of crystal face can be known from XRD data while the true lengths of the long and short particles can be determined from TEM images. Because of a high hydrothermal temperature, nucleation and crystallization of TiO_2 occur fastly after the nanotube destruction. Based on this phenomenon, the TiO_2 growth is controlled by kinetics, not thermodynamics, so there is obtained metastable anatase [24,25].

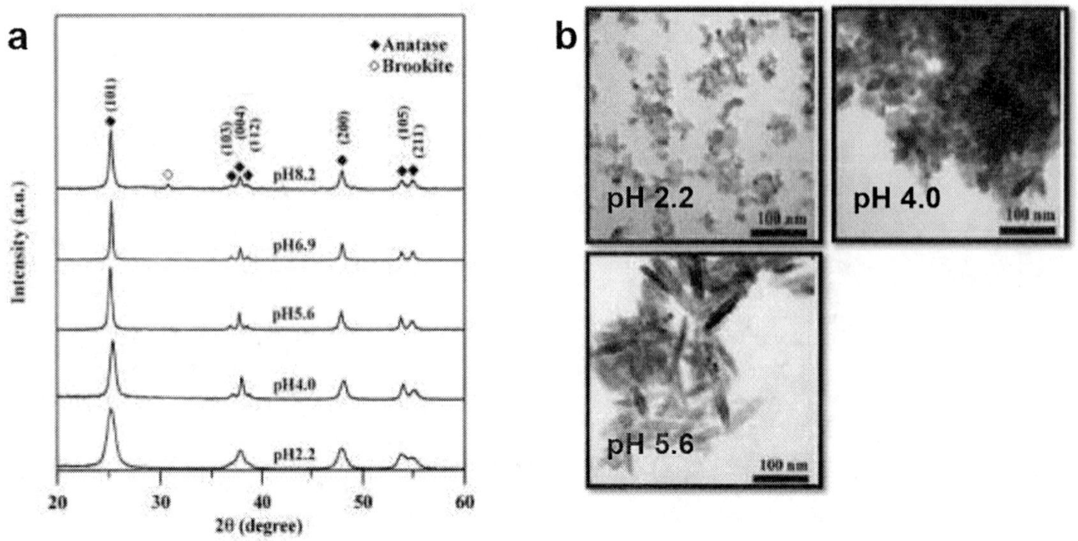

Figure 6. pH values dependence of TiO_2 nanorods (a) XRD spectra and (b) TEM images [21].

3.4. Energy application

The one of huge applications of TiO_2 nanoparticles is dye sensitized solar cells (DSSC). TiO_2 is commonly used as a semiconductor in photovoltaics. The materials are worked as harvesting light so the materials have an important function. Tacchini et al. (2012) reported that TiO_2 nanostructure have successfully synthesized by hydrothermal for DSSC semiconductor [26,28]. The results were obtained with the highest lighting conversion (3.2%). The determination of DSSC efficiency uses I-V measurement and IPCE analyzer as in Figure 7a.

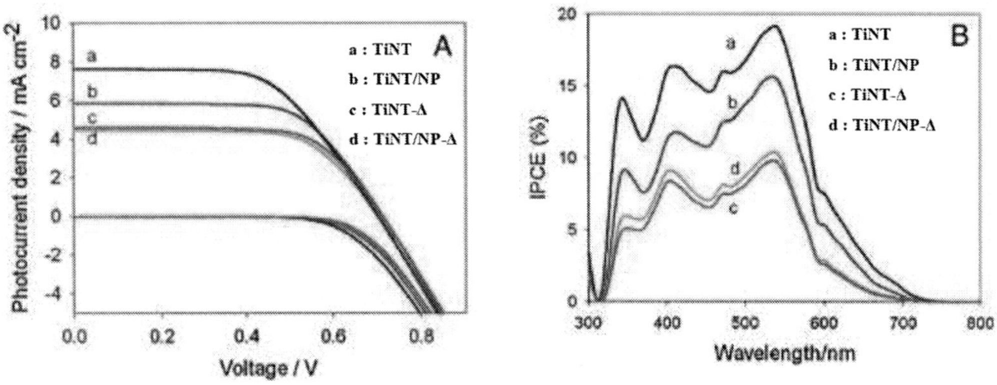

Figure 7. (a) I-V curves and (b) IPCE analysis of TiO_2 nanostructures [26].

Based on Figure 7b, The results revealed that the highest %IPCE is showed by TiO_2 nanotubes. The IPCE spectra confirm that TiO_2 nanotubes give the highest efficiency. Another research, TiO_2 nanorods can be synthesized with TiO_2 powder as precursor via the hydrothermal process for DSSC material [27]. The efficiency test shows that TiO_2 nanotubes have a higher efficiency that commercial TiO_2. It shows that the shape of nanostructures is one of the factors for electron mobility on the system.

3.5. Environmental application

The another application of TiO_2 nanoparticles as a semiconductor is used widely for wastewater treatment such as photochemical reaction for degradation of synthetic dye. Huang et al. (2016) reported that a photocatalytic performance of TiO_2 nanorods for Rhodamine-B (RhB) degradation [24]. Figure 8a shows the absorption spectra of RhB after treated with TiO_2 nanoparticles as catalyst. The results show that the presence of TiO_2 as photocatalyst could decrease a dye color as well. The optimum results were found at maximum degradation of dye more than 90%. The comparison study for 3 types of TiO_2 has reported in Figure 8b. The results show that TiO_2 nanorods showed significantly performance for decreasing a dye color. TiO_2 nanorods have a rod-shape and it makes electron mobility become more effective for producing hydroxyl radical to react with dye molecules.

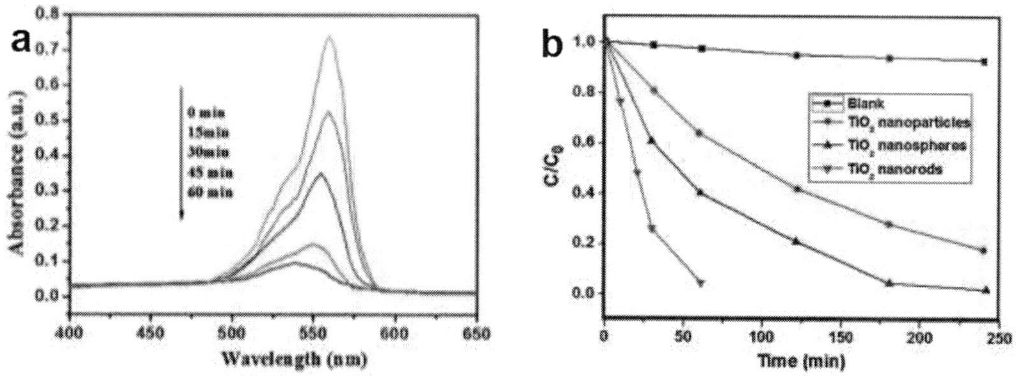

Figure 8. Photocatalytic activity of TiO_2 nanostructures (a) UV-Vis absorption spectra of RhB and (b) photodegradation rate of RhB [7].

4. Conclusion

Hydrothermal is one of methods for producing a TiO_2 nanostructure. The structure TiO_2 depends on several factors that are pH, time and temperature during the hydrothermal reaction and also precursors. Controlling the shape and size of the structure of TiO_2 greatly influences its chemical properties so that it can be widely used in several applications. In addition, the nanorods and nanotubes structure of TiO_2 shows a good performance both as a seminconductor and photocatalyst.

References

[1] Somiya S, Roy R and Bull 2000 *Mater. Sci.* **23** 453
[2] Zaman A C, Ustundag C B, Kaya F and Kaya K 2012 *Mater. Lett.* **66** 179-181
[3] Byrappa K and Adschiri T 2007 *Prog. Crys. Growth Charac. Mater.* **53** 117
[4] Kasuga T, Hiramatsu M, Hoson A, Sekino T and Nihara K 1998 *Langmuir* **14** 3160-3163
[5] Chen X, Liu L, Yu P Y and Mao S S 2011 *Science* **331** 746-750
[6] Dinh C T, Nguyen T D, Kleitz F, Do T O 2009 *ACS Nano* **3** 3737-3743
[7] Huang X, Meng L, Du M, Li Y J 2016 *Mater. Sci.: Mater Electron* **27** 7222-7226
[8] Fadillah G, Wahyuningsih S and Ramelan A H 2017 *IOP Conf. Series* **75** 012005
[9] Zhang H, Wang G, Chen D, Lv X and Li J 2008 *Chem. Mater.* **20** 6543
[10] Yang M, Ding B, Lee S and Lee J K 2011 *J. Phys. Chem. C* **115** 14534-14541
[11] Tompsett G A, Cooney R P, Metson J B, Rodgers K A and Seakins J M J 1995 *Spectrosc.* **26** 57
[12] Tan Z, Sato K and Ohara S 2015 *Adv. Powd. Tech.* **26** 296-302
[13] Erdogan N, Ozturk A and Park 2016 *J. Ceram. Int.* **42** 5985-5994
[14] Habib A, Haubner R and Stelzer N 2008 *Mater. Sci. Engineer. B.* **152** 60-65

[15] Collazzo G C, Jahm S L, Carreno N L V and Foletto E L 2011 *Brazilian J. Chem. Engineer*. **28**, 265-272
[16] Ayers M R and Hunt A J 1998 *J. Mater. Lett* **34** 292-293
[17] Ivanova T and Harizanova A 2001 *Solid State Ionics* **138** 228-230
[18] Karuppuchamy S and Jeong J M 2006 *J. Oleo. Sci.* **55** 264-266
[19] Feng X J, Shankar K, Varghese O K, Paulose M, Latempa T J and Grimes C A 2008 *Nano Lett.* **8** 3781
[20] Liu B and Aydil E S 2009 *J. Am. Chem. Soc.* **131** 3985
[21] Nian J N and Teng H J 2006 *J. Phys. Chem.* B **110** 4193-4198
[22] Li Y, Fan Y and Chen Y J 2002 *Mater. Chem.* **12** 1387
[23] Bischoff B L and Anderson M A 1995 *Chem. Mater.* **7** 1772
[24] Seok S I, Kim M S and Suh T S 2002 *J. Am. Ceram. Soc.* **85** 1888–1890
[25] Jiang Z, Wei W, Mao D, Chen C, Shi Y, Lv X and Xie J 2015 *Nanoscale* **7** 784
[26] Tacchini I, Anson-Casaos A, Yu Y, Martinez M T and Lira-Cantu M 2012 *Mater. Sci. Engineer. B* **177** 19-26
[27] Jeng M J, Wung Y L, Chang L B and Chow L 2013 *Intl. J. Photoenergy* **13** 1-8
[28] Wahyuningsih S, Fadillah G, Hidayat R and Ramelan A H 2016 *Procedia Chem.* **19** 632-637

IOP Conf. Series: Materials Science and Engineering **578** (2019) 012032 doi:10.1088/1757-899X/578/1/012032

One pot and sonically assisted synthesis of Fe_3O_4 and $SiO_2@Fe_3O_4$

OA Saputra[1], **WW Lestari** [1,2], **FR Wibowo**[1,2]
[1]Chemistry Department, Graduate School of Sebelas Maret University Surakarta, Indonesia, Jl. Ir. Sutami 36A Surakarta, Indonesia.
[2]Chemistry Department, Faculty of Mathematic and Natural Sciences, Sebelas Maret University Surakarta, Jl. Ir. Sutami 36A Surakarta, Indonesia.

E-mail: fajar.wibowo@gmail.com

Abstract. Magnetite (Fe_3O_4) and core-shell $SiO_2@Fe_3O_4$ nanoparticles were successfully synthesized through a one pot assisted by sonication method. The products were morphologically, physically and chemically characterized using Transmission Electron Microscopy (TEM), N_2 adsorption-desorption, and Fourier Transform Infra-red (FTIR). In this research, concentration of tetraethyl orthosilicate (TEOS), a precursor of SiO_2, was varied by 1, 3 and 5 mmol in order to control the SiO_2 shell thickness. The FTIR measurements showed that the increasing of TEOS concentration was parallel with the increasing of Si-O stretching vibration peak intensity revealing an enlargement of SiO_2 shell thickness. It was proved by morphological study where the addition of 1 mmol of TEOS has approximately 2 nm of shell thickness while by using 5 mmol was ~10 nm of shell thickness. Magnetite nanoparticle has 80.43 $m^2 g^{-1}$ of BET surface area but dramatically decreased to 55.74 $m^2 g^{-1}$ after incorporating of SiO_2 (TEOS 5 mmol) caused by the presence of CTAB in pore channel.

1. Introduction

Magnetite (Fe_3O_4), in nature, is deposited in rocks and naturally arising material contained 1:2 stoichiometric ratio of Fe^{2+} and Fe^{3+} ions. Generally, magnetite is a crystalline material with bravis crystal structure of face-centered cubic (FCC) close-packing of oxide ions, where Fe^{2+} occupied octahedral site and Fe^{3+} engaged half of octahedral and tetrahedral sites [1]. The synthesis of magnetite nanoparticle was firstly introduced by Massart [2] by a simplest method called co-precipitation. Briefly, the co-precipitation of magnetite nanoparticle involves Fe^{2+} and Fe^{3+} ions with molar ratio by 1:2 catalysed by alkaline (OH^-). The formation of Fe_3O_4 is relatively fast if the solution reached pH between 8 and 14. This method, however, have a minor obstacles, i.e. difficult to control the size of Fe_3O_4 nanoparticles and produce an aggregate of particles [3]. To control the particle size of magnetite nanoparticle, researchers usually introduce coating or template agent. For example, Petcharoen and Sirivat [4] synthesized magnetite nanoparticle through co-precipitation method and added oleic acid and hexanoic acid resulting a good particle size distribution. Si et al. [5] controlled the particle size of magnetite nanoparticle by incorporating of polyelectrolyte. Sun et al. [6] synthesized of biocompatible Fe_3O_4 nanoparticle in addition of polyethylene glycol (PEG-6000) and oleic acid, a class of surfactant, to control the dispersion and size of magnetite nanoparticle. In this study, we used cationic surfactant, hexadecyl trimethyl ammonium bromide (also known as cetyl trimethyl ammonium bromide, CTAB) to synthesize magnetite nanoparticle. This surfactant is often used as pore structure directing agent in synthesis of mesoporous silica nanoparticle.

In order to protect the magnetite nanoparticle from instability in acid environment, the formation of SiO_2 shell is introduced. The SiO_2 shell was often post-fabricated after obtaining Fe_3O_4 nanoparticle [7,8]. However, there are limited article reporting a one pot synthesis of core-shell $SiO_2@Fe_3O_4$ nanoparticle. Thus, in this paper, we reported a facile synthesis of core-shell $SiO_2@Fe_3O_4$ through one pot method.

2. Experimental

2.1. Materials

Iron(II) sulphate heptahydrate ($FeSO_4.7H_2O$), iron(III) chloride hexahydrate ($FeCl_3.6H_2O$), hexadecyltrimethylammonium bromide (CTAB), ethanol, were purchased from Merck. Tetraethyl orthosilicate (TEOS) and ammonium hydroxide (NH_4OH, 25-28%) were commercially available in Sigma-Aldrich. Ultrapure distilled water was obtained from MIPA Terpadu laboratory. All chemicals were used as received without further purification.

2.2. Synthesis of Magnetite Nanoparticles (MNPs)

Iron(III) chloride hexahydrate (2 mmol) and Iron (II) sulphate heptahydrate (1 mmol) was dissolved in 50 mL of distilled water with and without CTAB. After completely dissolved, ammonium hydroxide (32%, 5 mL) was wisely dropped to the solution. The dark brown precipitate was observed after first drop of ammonium hydroxide addition and turned to be black precipitate after several minutes. The precipitate was stirred for 1 h for equilibrium, separated using external magnetic field, washed several times using distilled water and ethanol until reaching neutral pH, and dried using freeze drying machine (freeze for 1 h at -49 °C, freeze drying 12 h). The obtained MNPs with and without CTAB (with CTAB noted as MNPs+CTAB) were characterized using FTIR, TEM and N_2 adsorption desorption.

2.3. Synthesis of $SiO_2@Fe_3O_4$ Nanoparticle

Synthesis of $SiO_2@Fe_3O_4$ nanoparticles were conducted in one pot approached assisted by sonication techniques. After obtaining MNPs by following the previous procedure in the presence of CTAB, amount of TEOS (1, 3, and 5 mmol) was added and sonicated for 30 minutes. Afterward, the mixture was stirred for 1 h, separated by external magnetic field, washed with distilled water and ethanol, and dried using freeze drying machine. The final products marked as $1SiO_2@Fe_3O_4$, $3SiO_2@Fe_3O_4$ and $5SiO_2@Fe_3O_4$ were characterized using FTIR, TEM, and N_2 adsorption-desorption.

3. Result and Discussion

3.1. Synthesis of Magnetite and $SiO_2@Fe_3O_4$ Nanoparticles

Precipitation of Fe^{3+}:Fe^{2+} 2:1 in basic condition (pH ~11) is a conventional method to synthesize MNPs. This method is relatively easy to perform and yielding high amount of precipitate of MNPs. Briefly, MNPs is constructed by mixing Fe^{2+} ion and Fe^{3+} ion with 1:2 molar ratio catalyzed by base. In this research, we used $FeSO_4.7H_2O$ and $FeCl_3.6H_2O$ as Fe^{2+} and Fe^{3+} ions sources, respectively. According to Petcharoen and Sirivat [4], Fe^{2+} ion forms $Fe(OH)_2$ after addition of base (OH^-) with molar ratio 2:1 to Fe^{2+} ion. However, Fe^{3+} ion forms $Fe(OH)_3$ in basic solution (3 mol of OH^-). Magnetite (Fe_3O_4) is a result of $Fe(OH)_2$ and $2Fe(OH)_3$ reaction. The base (ammonium hydroxide) plays important role in the formation of MNPs, especially in the size of MNPs. Sun et al. [6] reported that the addition of ammonium hydroxide until reaching pH ~11 produced MNPs with average particle size below 10 nm, but it turn to bigger than by increasing or decreasing the pH value (~ 10 nm at pH 10 or 12, but >20 nm at pH 9 or 13). The size of MNPs influences the magnetic properties where superparamagnetic is obtained if the size of MNPs below 20 nm, but the size 20-100 nm is categorized as stable single-domain ferrimagnetic [1]. In our investigation, we adjusted the pH of solution around 10 concerning the green chemistry principle, use as small amount chemical as possible.

Synthesis of $SiO_2@Fe_3O_4$ was performed by directly adding TEOS as silica precursor into a same pot of MNPs with the presence of CTAB. The concentration of TEOS was varied in order to evaluate the shell thickness of SiO_2 toward Fe_3O_4 as core. The role of CTAB is to direct the pore structure of SiO_2 shell by forming a micelle structure in proper concentration. In our study, the concentration of CTAB was slightly more than its critical micelle concentration. According to Figure 1, the TEOS as SiO_2 precursor, covered the Fe_3O_4. The proposed mechanism is as follow: TEOS is adsorbed in the surface of Fe_3O_4 then hydrolyzed to form silanol group, and finally polymerized to form SiO_2 while the pore structure directed by CTAB.

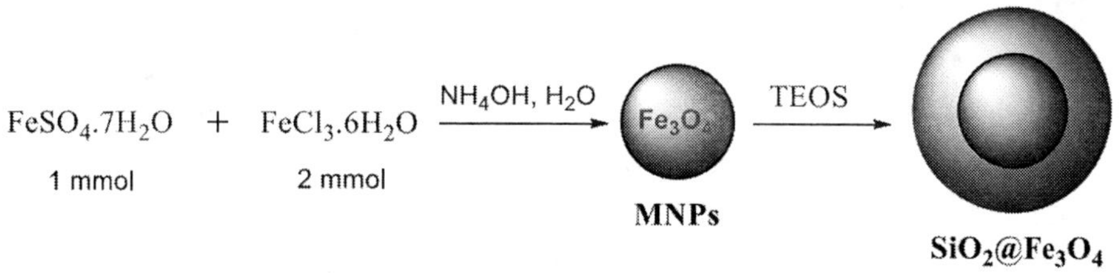

Figure 1. Schematic illustration of synthesis of MNPs and $SiO_2@Fe_3O_4$

3.2. Functional Group Analysis

Functional group analysis is one of the simple techniques to chemically characterize the materials in term of their functional group using FTIR spectroscopy. Figure 2 shows the FTIR spectrums of MNPs, $1SiO_2@Fe_3O_4$, $3SiO_2@Fe_3O_4$, and $5SiO_2@Fe_3O_4$. Briefly, the MNPs have characteristic peaks around 3400 cm^{-1} and 570 cm^{-1} assigned as stretching vibration of O-H and Fe-O, respectively. Coating of Fe_3O_4 by SiO_2 shell was evaluated by FTIR showing an increasing of Si-O absorbance band at around 1052 cm^{-1} following an increasing of TEOS concentration. This indicates that the shell thickness of SiO_2 becomes thicker.

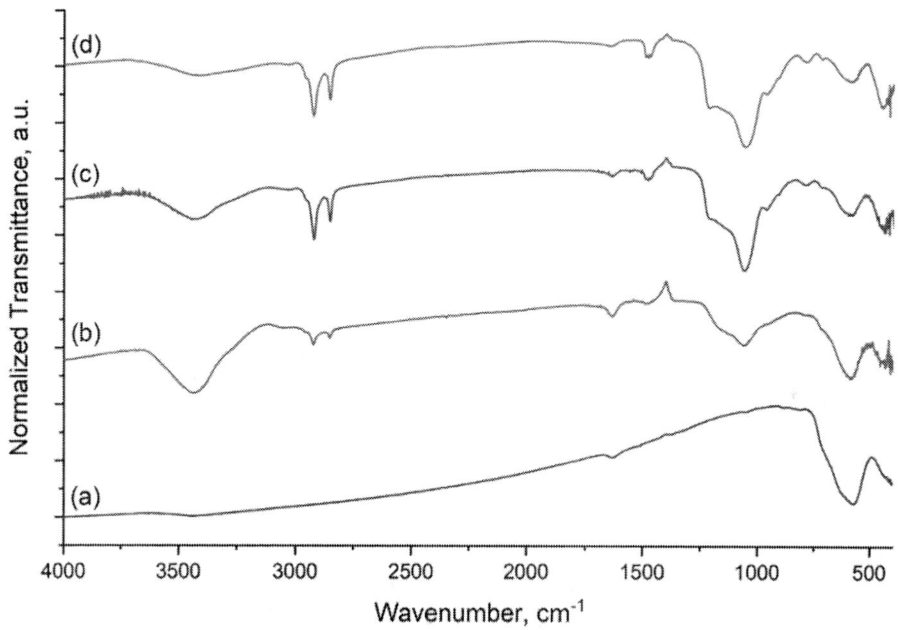

Figure 2. FTIR spectrum of (a) MNPs, (b) $1SiO_2@Fe_3O_4$, (c) $3SiO_2@Fe_3O_4$ and (d) $5SiO_2@Fe_3O_4$

3.3. Morphological Analysis

The morphology of MNPs was observed by TEM measurement. Figure 3 reveals the TEM pictures of MNP and MNP in the presence of CTAB. MNP have irregular shapes and tend to sticking together to form agglomerate. However, after incorporating CTAB, the particles of MNP was dispersed but still in irregular form. The role of CTAB in the formation of MNP is to prevent agglomeration through coating the surface of MNP thus they are not attached each other.

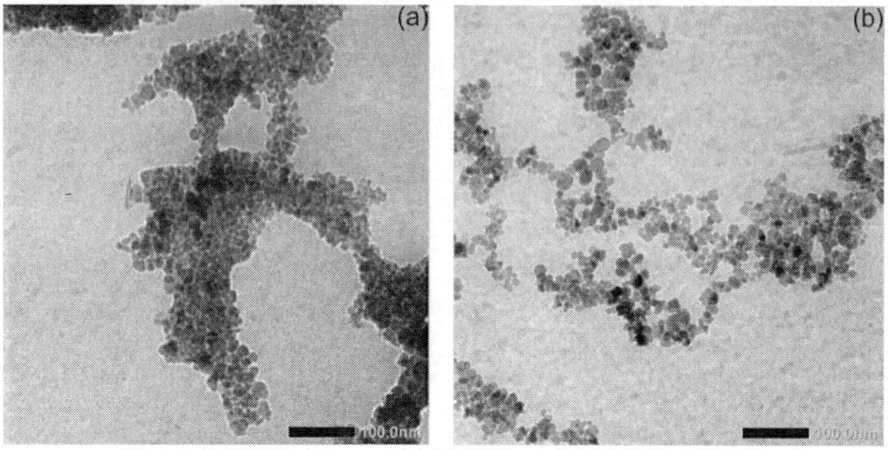

Figure 3. TEM pictures of (a) MNPs and (b) MNPs+CTAB

Figure 4 displays TEM images of $SiO_2@Fe_3O_4$ nanoparticles. It was mentioned before that the addition of high concentration of TEOS as SiO_2 precursor caused increasing of SiO_2 shell thickness. Figure 4a reveals the TEM image of $1SiO_2@Fe_3O_4$ (1 mmol TEOS) having very thin SiO_2 shell approximately ~2 nm. However, by incorporating high concentration of TEOS (5 mmol), the SiO_2 shell thickness increased to around 10 nm as seen in Figure 4b. The SiO_2 shell in both $1SiO_2@Fe_3O_4$ and $5SiO_2@Fe_3O_4$ not covered a single particle of MNP but group particles of MNP. This is due to the MNP tend to stick together preventing the SiO_2 shell covering a single nanoparticle.

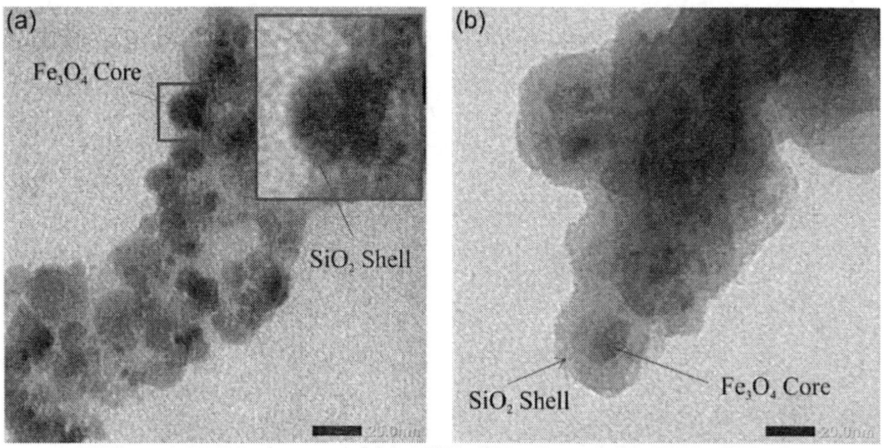

Figure 4. TEM pictures of (a) $1SiO_2@Fe_3O_4$ and (b) $5SiO_2@Fe_3O_4$

3.4. N_2 Adsorption-Desorption

The surface area of MNP and $SiO_2@Fe_3O_4$ was evaluated using N_2 adsorption-desorption. Figure 5 tells hysteresis loop of both MNP and $SiO_2@Fe_3O_4$. According to IUPAC classification [9], the porous structure type of both was type IV, a class of mesoporous having a pore diameter between 2-50 nm.

The pore size distribution was calculated using Barrett-Joyner-Halenda (BJH) method [10], and found that the average pore size of MNP was 8.34 nm while $5SiO_2@Fe_3O_4$ was 1.69 nm. The decreasing of pore size of MNP after covering by SiO_2 shell due to there are still CTAB remaining in the pore channel proved by the FTIR data (the presence of C-H stretching absorbance band). The blocking of pore channel by CTAB was not only decrease the average pore size but also decrease the surface are of the nanoparticle calculated by Brunauer-Emment- Teller (BET) method. The MNP has 80.43 m^2 g^{-1} of surface area while $5SiO_2@Fe_3O_4$ was 55.74 m^2 g^{-1}.

The decreasing of surface area of $5SiO_2@Fe_3O_4$, in fact, is due to the presence of CTAB in pore system of SiO_2 pore channels. In this research, we also measured the N_2 adsorption-desorption of Fe_3O_4 with the presence of CTAB and found that the BET surface area is 43.68 m^2 g^{-1}. Comparing to the Fe_3O_4 with the presence of CTAB, the $5SiO_2@Fe_3O_4$ has higher BET surface area by 27.61%.

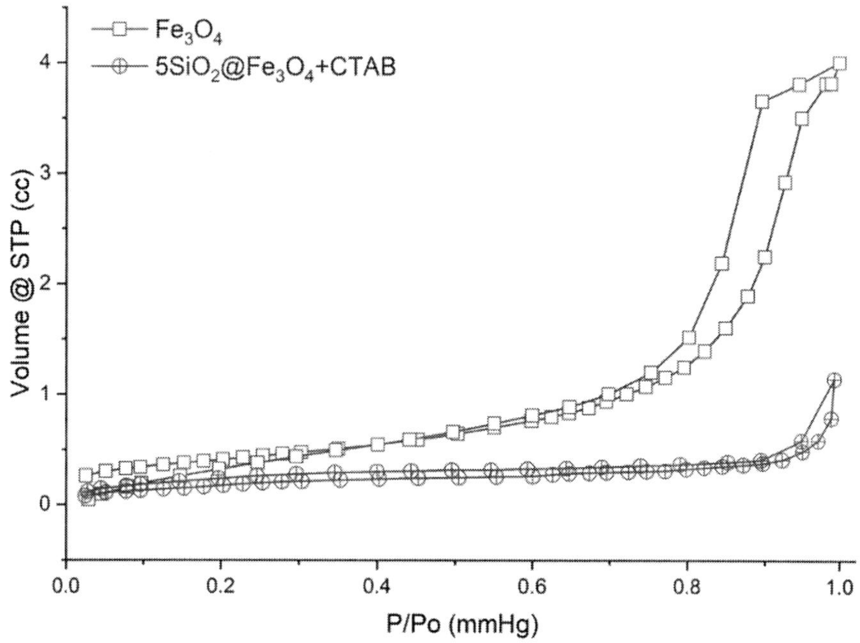

Figure 5. Hysteresis loop of N_2 adsorption-desorption of MNPs and $5SiO_2@Fe_3O_4$

4. Conclusion

The MNP and $SiO_2@Fe_3O_4$ were successfully synthesized through one pot assisted by sonication treatment. Functional group analysis revealed that the increasing of TEOS concentration induced the increasing of Si-O intensity showing an increasing of SiO_2 shell thickness. It was supported by TEM measurement, where $1SiO_2@Fe_3O_4$ has around 2 nm of SiO_2 thickness while $5SiO_2@Fe_3O_4$ was 10 nm of SiO_2 thickness. The MNP has 8.34 nm of average pore size and 80.43 m^2 g^{-1} of surface area. However, after covering by SiO_2, the pore size and surface area decreased.

Acknowledgement
Authors would like to acknowledge to Universitas Sebelas Maret for support funding through Hibah Pascasarajana (No. 543/UN27.21/PP/2018).

References

[1] Mirabello G, Lenders JJM, Sommerdijk NAJM 2016 *Chem. Soc. Rev.* **45** 5085
[2] Massart R 1981 *IEEE Trans. Magn.* **17** 1247
[3] Laurent S, Forge D, Port M, Roch A, Robic C, Vander Elst L, et al. 2008 *Chem. Rev.* **108** 2064

[4] Petcharoen K, Sirivat A 2012 *Mater Sci Eng B* **177** 421
[5] Si S, Kotal A, Mandal TK, Giri S, Nakamura H, Kohara T 2004 *Chem. Mater.* **16** 3489
[6] Sun J, Zhou S, Hou P, Yang Y, Weng J, Li X, et al. 2007 *J. Biomed. Mater. Res. Part A* **80** 333
[7] Gai S, Yang P, Li C, Wang W, Dai Y, Niu N, et al. 2010 *Adv. Funct. Mater.* **20** 1166
[8] Kang X, Cheng Z, Yang D, Ma P, Shang M, Peng C, et al. 2012 *Adv. Funct. Mater.* **22** 1470
[9] Rouquerolt J, Avnir D, Fairbridge CW, Everett DH, Haynes JH, Pernicone N, et al. 1994 *Pure Appl. Chem.* **66** 1739
[10] Jin X, Wang Q, Sun J, Panezai H, Bai S, Wu X. 2017 *Microporous Mesoporous Mater.* **254** 77

IOP Conf. Series: Materials Science and Engineering **578** (2019) 012033 doi:10.1088/1757-899X/578/1/012033

Electrical Properties of Barium Titanate Doped Lanthanum ($Ba_{1-x}La_xTiO_3$) Fabricated Using Co-precipitation Method

M D Noviastuti[1], Y Iriani[1,*] and K Kusumandari[1]

[1]Department of Physics, Faculty of Mathematics and Natural Sciences, Universitas Sebelas Maret, Jl. Ir. Sutami 36A Kentingan Surakarta 57126, Indonesia

*E-mail: yofent_iriani@staff.uns.ac.id

Abstract. Barium Titanate doped Lanthanum ($Ba_{1-x}La_xTiO_3$) with variations in the concentration of mole dopants Lanthanum (x) 0; 0.01; 0.02; 0.03; 0.04 and 0.05 by co-precipitation method. Bulk $BaTiO_3$ and $Ba_{1-x}La_xTiO_3$ were sintered at 1400 °C for 4 hours. The influence of variations concentration mole dopants of Lanthanum was analyzed through microstructure and electrical properties. The results of X-Ray Diffraction (XRD) analysis showed that the influence of variations concentration mole dopants of Lanthanum caused the diffraction angle to shift to the right. The influence of the variation concentration dopants mole of Lanthanum cause the dielectric constant to increase. The highest dielectric constant of 257.81 is obtained when the concentration of the La dopant mole is 5% (x=0.05) at a frequency of 40 kHz. The influence of Lanthanum dopant concentration variation on the morphology of the sample is that the more mole of dopant La added to $BaTiO_3$ the smaller the size of the sample

1. Introduction

Ferroelectric material has important role in the field of electronics. The ferroelectric phenomenon can be characterized by the presence of a hysterisis curve which is a graph of the relation between electric polarization and an electric field [1].

Commonly ferroelectric materials are used for energy storage for example are Barium Titanate ($BaTiO_3$), Titaninate Zirconium Bismuth (BZT) and Barium Strontium Titanate ($BaSrTiO_3$) [2]. In this study the ferroelectric material was used Barium Titanate because it has a high dielectric constant (> 1000) at room temperature [3]. Barium Titanate has ABO_3 perovskite structure [4]. The characteristic of ferroelectric can be modified using A or B ion doping. The selection of ions doping is adjusted with atomic radius at the position of atom A and B.

Barium Titanate can be synthesed with several method such as sol-gel method [5], Chemical Solution Deposition (CSD) [6], co-precipitation, soft chemical process and hydrotermal method [7]. This study use co-precipitation method because it has easier process. The co-precipitation method is carried out by constant stirring, so the solution will be homogeneous and it produces precipitate [8]. Barium Titanate was modified with Lanthanum doping. The addition of a little Lanthanum in a perovskite lattice (0.05 per unit formula) increase the permittivity value and decrease the curie temperature [3]. In this research Lanthanum replaces Ba^{2+} ion because La^{3+} and Ba^{2+} ions have almost same radius, La^{3+} (1.15 Å) and Ba^{2+} (1.35 Å) [9].

2. Experimental

Fabrication of Barium Titanate doped Lanthanum use co-precipitation method. Concentration of mole Lanthanum (x) was varied at 0; 0.01; 0.02; 0.03; 0.04 and 0.05. Barium Hydroxide [$Ba(OH)_2$] (Sigma Aldrich, 95%), Lanthanum Nitrate [La ($NO_3)_3$], (Sigma Aldrich, <99%), Titanium Tetrabutoxide [Ti (C_4H_9O)] (Sigma Aldrich, 97%), Oxalic Acid ($H_2C_2O_4$) (Sigma Aldrich> 99%), Isopropanol [$(CH_3)_2(CHOH)$] are raw materials which use for synthesis Barium Titanate doped Lanthanum. Oxalotitanic acid (HTO) solution was made with mixed Titanium Tetrabutoxide in Isopropanol (IPA) and Titration Oxalic acid in isopropanol (IPA). $Ba(OH)_2$ and La (x = 0; 0.01; 0.02; 0.03; 0.04 and 0.05) were mixed in HTO solutions. Aquades added in that solution with tritation. The final solution was precipitated for 24 hours at room temperature. The solution was washed by ethanol and aquades until it has PH > 5, then the solution was filtered with filter paper.

$BaTiO_3$ and $BaTiO_3$ doped La powder were heated at T= 100 °C for 5 hours and then powder crushed for 1 hour. $BaTiO_3$ and $BaTiO_3$ doped La powder were pressed become bulk. The bulk was sintered for T=1400°C for 4 hours. The bulk BLT was characterized with X-Ray Diffraction (XRD), Scanning Electron Miscroscopy (SEM), Sawyer Tower method and RCL meter.

3. Result and Discussion

The result of X-Ray Diffraction (X-RD) characterization obtained the diffraction pattern peaks which formed from the graph of the relationship between Intensity (I) and the diffraction angle (2θ). The peaks of diffraction pattern in the $Ba_{1-x}La_xTiO_3$ sample can be seen as. Figure 1. The peaks of the diffraction pattern matched using the International Center for Diffraction Data (ICDD). ICCD is used to determine the angle shift that occurs between experimental data and theory and it was used to determine the lattice parameters between values a, b and c.

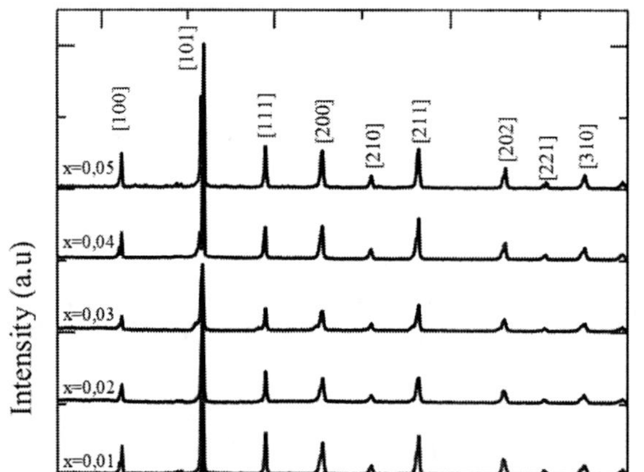

Figure 1. The diffraction pattern of the $Ba_{1-x}La_xTiO_3$ sample with variations in mole and La concentrations.

The peaks of $Ba_{1-x}La_xTiO_3$ sample diffraction pattern shown in Figure 1. It was matched with the ICDD database pdf # 831880 have Barium Titanat ($BaTiO_3$). The addition concentration of mole La dopant in Barium Titanate showed diffraction angle shift. $Ba_{1-x}La_xTiO_3$ sample shifts to the right as the concentration of La dopant mole is added in Figure 2.

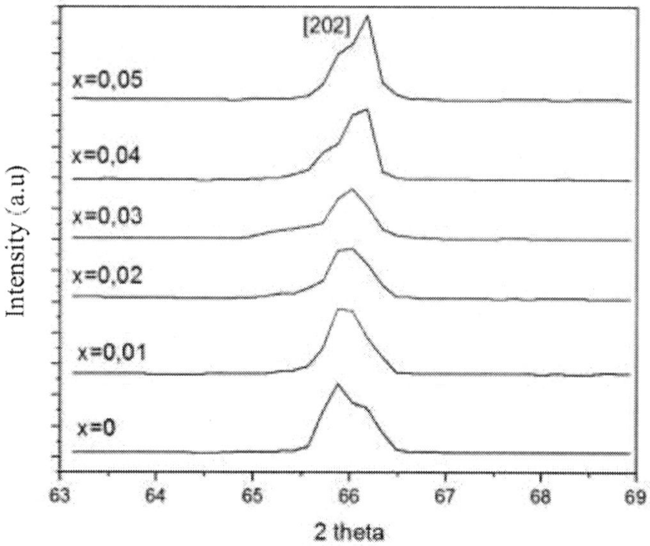

Figure 2. Magnification of the diffraction angle peaks [202] of the $Ba_{1-x}La_xTiO_3$.

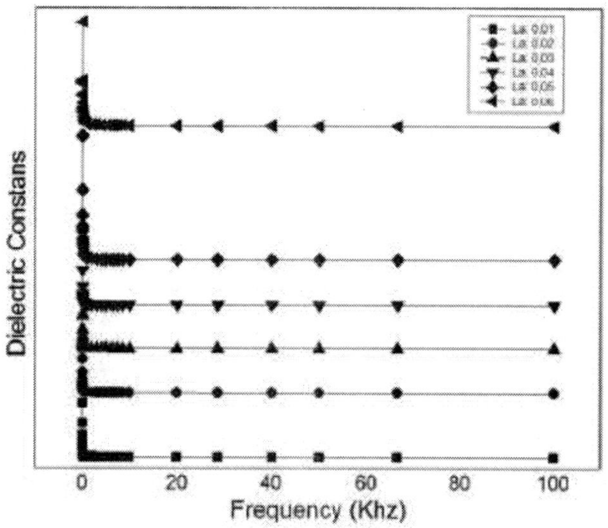

Figure 3. The dielectric constant for each frequency at mole doping variation La (x).

The dielectric constant of $Ba_{1-x}La_xTiO_3$ sample was measured by the RCL meters. RCL meter measurements obtained capacitance values. Measurement use *Vrms* = 1 Volt and the frequency of 0.01 kHz - 100 kHz. The relations between frequency and dielectric constant of $Ba_{1-x}La_xTiO_3$sample can be seen in Figure 3.The dielectric constant decrease when frequency 0.01 kHz to 10 kHz and the dielectric constant were more stable at frequency of 10 kHz to 100 kHz. When the frequency is low, dielectric constant is high because the dipole follows the shape of the field and the dielectric permitivity value equals in the static field [10].

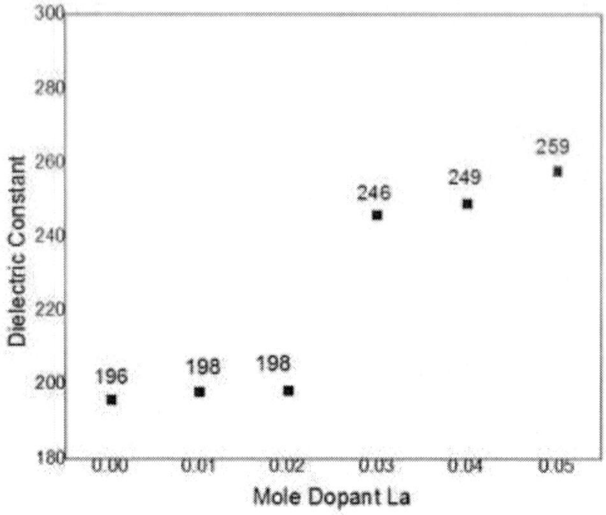

Figure 4. Dielectric constant of $BaTiO_3$ doped La.

The dielectric constant increase as the number of mole La dopant was added increase (Figure 4), this is because the radius of the La atom (1.15 Å) is smaller than the Ba atom (1.35 Å) so that the grain size to be small. The small grain size cause unit cell volume become small. The volume of small cell units cause large polarization, because the volume of cell units is inversely proportional to polarization. The large polarization on sample cause a large dipole moment so that the stored charge gets bigger, because the dipole moment and charge are directly propotional. The more stored charge in a sample, the capacitance becomes more large so that the dielectric constant is large because of the capacitance with the dielectric constant is directly proportional.

The results of characterization using SEM on Barium Titanate doped Lanthanum ($Ba_{1-x}La_xTiO_3$) sample such as morphology, roughness and grain size. The morphology of $Ba_{1-x}La_xTiO_3$ can be seen in Figure 5.

(a)

(b)

Figure 5. Photo SEM $Ba_{1-x}La_xTiO_3$ (a) x=0 and (b) x=0.01.

(c) (d)

(e) (f)

Figure 5. Photo SEM $Ba_{1-x}La_xTiO_3$ (c) x=0.02 (d) x=0.03 (e) x=0.04 and (f) x=0.05.

The influence of variations in the number mole La dopant caused small grain size, because the La atom will be replaced Ba atom position which the atomic radius of La is smaller than the Ba atom. The roughness of BLT sample was calculated using WSxM 4.0 Beta 9.0 software. The results of roughness calculations with software can be see Table 1. The influence of variations concentration mole La dopant causes the sample roughness level to be smaller. The grain size and Roughness of the $Ba_{1-x}La_xTiO_3$ can be seen in Table 1.

Table 1. Grain size in $Ba_{1-x}La_xTiO_3$ samples.

La mole in $Ba_{1-x}La_xTiO_3$	Grain Size (μm)	Roughness
x= 0	5.17	597.367
x= 0.01	3.03	509.140
x= 0.02	2.60	383.633
x= 0.03	1.84	367.202
x= 0.04	1.50	286.348
x= 0.05	0.41	256.064

4. Conclusion

The influence of the variation concentration dopant mole of Lanthanum in $BaTiO_3$ causes the diffraction angle shifts to the right and crystal size to smaller. The dielectric constant increase as the concentration dopant mole of Lanthanum was added in $BaTiO_3$ increase. The grain size and roughness to smaller as the concentration of mole La dopant was added in Barium Titanate there more.

Acknowledgement
The authors would like to thank PUPT UNS Grant of 2018.

References

[1] Takayama R Y, Tomita K, Ijima and Ueda I 1996 *J. Appl. Phys.* **61** 4011-15
[2] Jamaluddin A, Susilowati E, Budiawanti S and Iriani Y 2014 *Adv. Mater. Res.* **896** 358362
[3] Morrison F D, Derek C S and Anthony R W 2001 *Int. J. Inorg. Mater.* **3** 1205-10
[4] Pradhan S and Roy G S 2013 *Researcher* **5** 63
[5] Iriani Y, Saputri D F, Hadiati S, Ramelan A H and Variani V I 2014 *Adv. Mater. Res.* **91** 205-9
[6] Setyadhani R T, Iriani Y and Jamaluddin A 2014 *Adv. Mater. Res.* **896** 229-32
[7] Vijayalakshmi R and Rajendran V 2010 *J. Nanomater. Biostruct.* **5** 511-17
[8] Khollam Y B, Bhoraskar S V, Deshpande S B, Potdar H S, Pavaskar N R, Sainkar S R and Date S K 2002 *Mater. Lett.* **57** 1871-79
[9] Wang Y, Cui B, Liu Y, Zhao X T, Hu Z Y, Yan Q Q, Wu T, Zhao L L and Wang Y Y 2014 *Scripta. Mater.* **90-91** 49-52
[10] Sonia, Chandrasekhar M and Kumar 2016 *J. Ceram. Int.* **42** 10587-92

IOP Conf. Series: Materials Science and Engineering **578** (2019) 012034 doi:10.1088/1757-899X/578/1/012034

The effect of holding time variation on the efficiency of DSSC TiO_2 transparent with dye *Ipomoea aquatica* Forsk

Khairuddin[1], C. Cari, A. Supriyanto, T. Y. Septiawan[2]
Physics Department, Graduate Program, Universitas Sebelas Maret,
Jl. Ir. Sutami No. 36A, Kentingan, Surakarta, 57126, INDONESIA

E-mail: [1] khairuddin_70@yahoo.com, [2] trioyudas88@gmail.com

Abstract. TiO_2 semiconductor material is commonly used as a working electrode in dye-sensitized solar cell (DSSC). There are two types of TiO_2 used, namely the form of nanopowder and transparent paste. In this study, the fabrication of DSSC has been successfully carried out using TiO_2 with a transparent type of paste. DSSC fabrication was carried out by placing transparent TiO_2 paste 18 NR-T using three variations of holding time in the heating process using furnace Carbolite 1100. Variations of holding time used were 10 minutes, 20 minutes and 30 minutes at 500 °C. Natural dye that is used comes from *Ipomoea aquatica* Forsk. For DSSC characterization, I-V test was done using Keithley 2620A and absorbance test using UV-Vis Spectrophotometer Lambda 25. The results of the characterization obtained the highest efficiency 5.5 x 10^{-2}%. While the absorbance test obtained absorption peaks in the wavelength range for each watercolor dye 400-440 nm and 640-680 nm.

1. Introduction

Dye-sensitized solar cells (DSSC) are promising devices for using solar energy because of low production costs compared to conventional semiconductor solar cells, and also in high light for electrical conversion efficiency [1]. Compared to amorphous silicon cells, DSSCs have been widely researched for their low-cost potential, assembling simple technology and good efficiency. There are four main factors that influence the power conversion efficiency (η) of a solar cell device: dye sensitizer structure, photoanode, opposing electrode and electrolyte. Most importantly, photo morphology that plays an important role in achieving high power conversion efficiency. Therefore, it is still important and critical to developing new structures from photoanodes for DSSCs [2].

The research on DSSC carried out has attracted much attention as a substitute for potential alternative energy for Si-based photoelectric cells mainly due to the relatively high conversion efficiency and cost effectiveness. DSSC uses fast electron injection from molecular light absorbers for large band gap of nanocrystalline TiO_2 semiconductor films on the conduction of transparent glass substrates (photodiodes) during light illumination [3]. DSSC as a type of cheap and renewable energy device has been considered a prospective alternative for silicon solar cells due to high energy conversion efficiency and ease of manufacture because O'Regan and Grätzel first showed efficient conversion > 7% to facilitate titanium mesoporous nanoparticle dioxide (TiO_2) . Since then, much interest has been developed for the fundamental aspects and application of technology, and the efficiency of photovoltaic conversion, now exceeding 12%. The DSSC is usually assembled with an electron transport layer made of 10-15 μm nanoparticulate films as thick as photo-codes, organic dye molecules as photosensitizers, redox electrolytes consisting of redox pair iodine, and opposite electrodes Pt. The four simple components and their interfaces have been widely surveyed to minimize photoelectron loss, which leads to high power conversion efficiency [4].

Because of its flexible and exotic nature, TiO_2 nanomaterials have been used as photocatalysts, photovoltaic materials, gas sensors, etc. The efficiency of DSSC based on TiO_2 nanoporous and perovskite photoelectrode, was 20.1%. It was reported that the structure of TiO_2 nanoparticles, which showed the best efficiency in DSSCs, had trapped sites in the contact area between TiO_2 nanoparticles and this disrupted electron transport in thin films of TiO_2. One-dimensional (1D) TiO_2 nanomaterials such as nanotubes (NTs), nanorods (NRs), and nanowires (NWs) have shown a reduction in recombination rates for electron-hole pair excitation and display optical and electrical uniqueness.. In particular, vertical growth NRS TiO_2 allows shorter and less undisturbed electrical path directions for photogenerated operators and increases load separation and transportation cost properties in many photoelectrochemical devices such as DSSC [5].

In nature a lot of organic materials have coloring molecules that are very promising to be used as ingredients for solar cells. Terms dye can be used as an active material (sensitizer), the material must be able to become a medium of transfer of an electric charge when absorbing photon energy. Plant pigmentation occurs because the structure of the pigment interacts with sunlight and changes the wavelength to be transmitted or reflected by plant tissue [6]. The weakness of DSSC with dye type ruthenium is small in nature and not environmentally friendly because it is toxic, so it is a consideration to be applied to large-scale DSSC. Dye is environmentally friendly and abundant in nature, the dye from the leaves, seeds, fruit, stems, and roots of plants becomes an alternative choice as a sensitizer in DSSC [7]. Substances such as chlorophyll, β-carotene, anthocyanin, tannin, curcumin, etc. in plants can be applied as a sensitizer.

Chlorophyll is widely found in green plants that function as givers of green pigments in plants. This compound plays a role in the process of photosynthesis of plants by absorbing and converting light energy into chemical energy. Chlorophyll is the main pigment in photosynthesis, absorbs more blue and red light, where supplementary pigments such as carotenoids and phycobilin can increase the absorption of green-blue and yellow spectra. The attractive properties of photosynthetic pigments are applied such as sensitizers in solar cells [8]. Sensitizers function to absorb light energy and use them to excite electrons, so that the more light energy absorbed, the more electrons are expected to be generated to increase the solar cells efficiency.

2. Experimental

2.1. The TiO_2 18 NR-T Preparation

The transparent TiO_2 paste made from a mixture of TiO_2 18NR-T paste with ethanol stirred with a magnetic stirrer for an hour at 350 rpm.

2.2. The.Preparation of Natural Dye Ipomoea aquatica Forsk

The Chlorophyll natural dye is prepared by dissolving 20 grams of *Ipomoea aquatica* Forsk and 60 ml of ethanol then stirring with 450 rpm speed for an hour. Dye is kept for 24 hours at room temperature and then filtered.

2.3. The Deposition of TiO_2 (Working Electrode)

The process of TiO_2 deposition on *Flouro Tin Oxide* (FTO) glass was carried out by pouring transparent TiO_2 paste using the spin coating method with active areas of 1 x 1 cm. After the TiO_2 paste is dripped then rotated at a speed of 1000 rpm for 15 seconds. Then the deposition results are heated using the Furnace (Carbolite 1100) at a temperature of 500 °C. At this stage, three variations of the process of holding time are made which are 10 minutes, 30 minutes, and 60 minutes. Then immersed into the dye solution for 24 hours.

2.4. The Deposition of Platinum (Counter Electrode)
The platinum solution is made from a mixture of *Hexaclhoroplatinic* (IV) Acid 10% with Isopropanol. The platinum deposition process is carried out by heating the FTO glass using a hot plate with a temperature of 250 °C for 15 minutes. Then the platinum solution is dripped appropriately until a platinum layer forms on the FTO glass.

2.5. The Fabrication of DSSC
The DSSC fabrication is made by assembling work electrode + Dye, electrolyte, and the opposing electrode. The arrangement of the circuit resembles a sandwich.

3. Results and Discussion

3.1. Performance of DSSC
The characterization of DSSC performance was tested by using I-V meter Keithley 2620A with a light source of 1000 watt illumination lamp. This is done because the energy given is almost the same as the solar photon radiation energy which is equal to 1360 W/m^2 [9]. From the measurement results obtained the efficiency shown in Table 1.

Table 1. Performance of DSSC

Variation	I_{sc} (mA)	V_{oc} (mV)	I_{max} (mA)	V_{max} (mV)	η (%) x 10^{-2}
10 minutes	0.00019	0.50020	0.00012	0.30313	3.5
30 minutes	0.00023	0.59113	0.00013	0.31832	5.0
60 minutes	0.00018	0.54554	0.00015	0.33333	5.1

From Table 1. It can be seen that during the process of holding time for 10 minutes produces a DSSC efficiency of 3.5 x 10^{-2}%. When the holding time process is raised to 30 minutes DSSC efficiency also increases to 5.1 x 10^{-2}%. In the 60 minutes of holding time process, the efficiency of DSSC the increase is small.

The curve of DSSC performance efficiency will be shown in Figure 1.

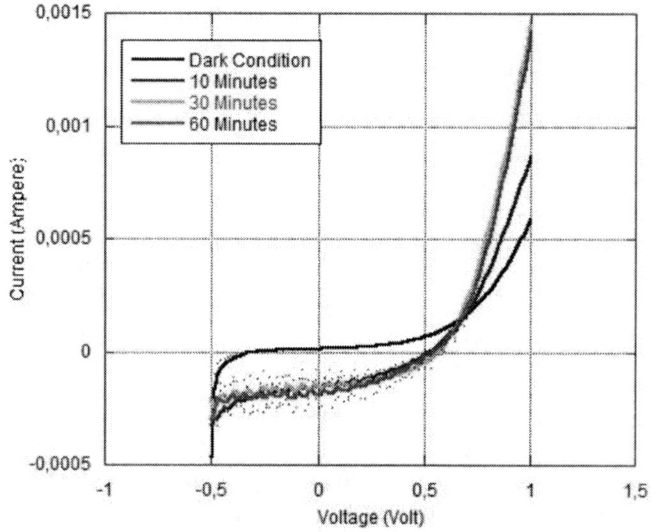

Figure 1. The DSSC Performance Curve

3.2. The Absorbance of Dye Ipomoea aquatica Forsk

The characterization of wavelength absorption was done by using UV-Vis Spectrophotometer Lambda 25. The test result is shown in Figure 2.

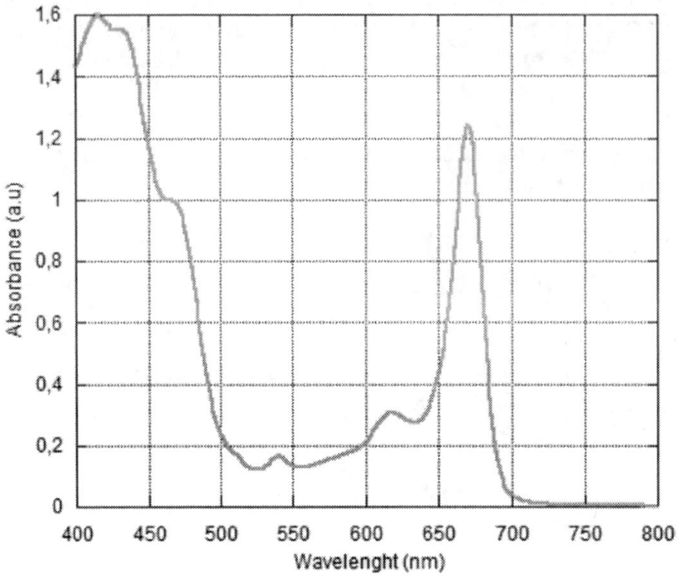

Figure 2. The Dye *Ipomoea aquatica* Forsk Curve Absorbance

From Figure 2. It can be seen that the peak of the absorbance at *Ipomoea aquatica* Forsk dye is 416 nm and 670 nm. The graph shows that the Ipomoea aquatica Forsk dye is a chlorophyll-a. The result of preparation of natural dye shows that absorbance spectrum of dye *Ipomoea aquatica* Forsk according to the references [10].

4. Conclusion

The DSSC fabrication with paste TiO_2 18 NR-T transparent and dye *Ipomoea aquatica* Forsk has been succesfully performed. The test results showed an increase in the efficiency of DSSC in the process of holding time 10 minutes, 30 minutes, and 60 minutes. The biggest efficiency increase is the increase from the holding time of 10 minutes, 30 minutes, and 60 minutes. The highest efficiency obtained at 60 minutes holding time is 5.1×10^{-2} %. The peak absorbance of *Ipomoea aquatica* Forsk dye is at a wavelength of 416 nm and 670 nm. From the results of the research on *Ipomoea aquatica* Forsk dye according to chlorophyll-a.

Acknowledgment

This Research was partly supported by the Higher Education Project Grant with contract no. 543/UN27.21/PP/2018.

References

[1] Meng L, Chen H, Li C, & Dos Santos, M P 2015 *Thin Solid Films* **577** 103-108.
[2] Lu J, Liu Z, Zhu C, Zhang M, & Wan M 2015 *Mater. Lett.* **159** 61-63.
[3] Saji V S, & Py M 2010 *Thin Solid Films* **518**(22) 6542-6546.
[4] Kim H S, Kim Y J, Lee W, & Kang S H 2013 *Appl. Surf. Sci.* **273** 226-232.
[5] Pawar U T, Pawar S A, Kim J H, & Patil P S 2016 *Ceram. Int.* **42**(7) 8038-8043.
[6] Narayan M R. 2012 *Renew. Sus. Energ. Rev.* 16(1), 208-215.
[7] Kimpa M I, Momoh M, Isah K U, Yahya H N, & Ndamitso M M 2012 *Mater. Sci. Appl.* **3**(5), 281-286.

[8] Nygren K 2010 *Solar cells based on synthesized nanocrystalline ZnO thin films sensitized by chlorophyll a and photopigments isolated from spinach* (Linköping: Linköping University).
[9] Chen C J 2011 *Physics of solar energy* (New Jersey: John Wiley & Sons).
[10] Syafinar R, Gomesh N, Irwanto M, Fareq M, & Irwan Y M 2015 *Energy Procedia* **79** 896-902.

IOP Conf. Series: Materials Science and Engineering **578** (2019) 012035 doi:10.1088/1757-899X/578/1/012035

Size and aspect ratio dependency of sensitivity of ellipsoidal metal nanoparticle based liquid sensor

M Muldarisnur[1], **E Oktorina**[1], **N Fridayanti**[1], **E Zeni**[1], **E Elvaswer**[1] and **S Syukri**[2]
[1] Department of Physics, Faculty of Mathematics and Natural Sciences, Andalas University, West Sumatera, Indonesia
[2] Department of Chemistry, Faculty of Mathematics and Natural Sciences, Andalas University, West Sumatera, Indonesia

E-mail: muldarisnur@sci.unand.ac.id

Abstract. Detection of bio chemical species in a liquid is very important for research and technical purposes. In the last decade, metallic nanoparticles have been shown to be potential for such measurement. Resonance wavelength (or frequency) of free electron oscillation on metallic nanoparticle surface is very sensitive to tiny changes of its surrounding environment. Change in surrounding medium shifts the resonance wavelength with a slope determined mainly by the size and geometry of nanoparticle. We numerically optimized size and aspect ratio of silver and gold nanoparticles to obtain the highest sensing sensitivity. Here, sensitivity is defined as the slope of resonance wavelength on refractive index of liquid. We found that size, aspect ratio, and material of metal affect sensing sensitivity of metal nanoparticle based sensor significantly.

1. Introduction

Detection of biological and chemical substances is a routine and an important step in scientific research, environmental hazard monitoring, and food and process control [1]. Surface plasmon resonance (SPR) sensing is now widely investigated to detect biological and chemical species dissolved in liquids or gasses [2,3]. SPR sensing offers high sensitivity and measurement accuracy due to strong dependency of plasmonic resonance on local change. Surface plasmons occur when free electrons on metal oscillate coherently upon illumination by electromagnetic wave having appropriate wavelength. There are two types of SPR, namely propagating surface pasmon polaritons (SPP) and non-propagating localized SPR (LSPR) [4]. SPP occurs at the interface of planar (thin film) metal and has been intensively investigated to realize nanoscale optical devices and to study interactions and binding mechanism of proteins and other biological structures. LSPR, on the other hand, occurs at the surface of metalic nanoparticles.

Compared to SPP, LSPR based sensing is often chosen as it can be easily excited without a complex optical set up using prism, grating, or scatterer. Furthermore, the wavelength at which coherent oscillation occurs (resonance wavelength) depends dielectric constants of metal and surrounding medium, the size, and the geometry of metal nanoparticles [5,6]. Dependency of the resonance intensity and wavelength on local dielectric environment has been used to develop label-free optical sensing. Adsorption of biological or chemical species on the surface of metal nanoparticles changes refractive index of surrounding locally. The changes can even be monitored in a real time [7,8].

IOP Conf. Series: Materials Science and Engineering **578** (2019) 012035 doi:10.1088/1757-899X/578/1/012035

Significant progress of SPR based sensing is mainly facilitated by the availability of many chemical synthesis recipes to obtain gold and silver nanoparticles. Gold and silver nanoparticles have been synthesized in forms of sphere [9], cube [10], rod [11], bar [12], rice [12], pyramid [9], and triangle [13]. Gold and silver nanoparticles are vastly used because of their high free electron density and low loss.

We report numerical study on dependency of sensing sensitivity of ellipsoidal metal nanoparticles on its size and aspect ratio. Ellipsoidal particles are promising for biochemical sensing because their resonance exhibits polarization dependent spectra due to anisotropy in particle dimension. We found that size, aspect ratio, and type of metal influences the sensing sensitivity of ellipsoidal nanoparticles significantly.

2. Simulation details

2.1 Numerical method

Extinction spectra of gold and silver nanoparticles were calculated using boundary element method [14,15]. Full Maxwell's equations were solved for a metal nanoparticle in a homogeneous dielectric medium. Metal nanoparticle having frequency-dependent dielectric constant is separated abruptly with its surrounding dielectric medium. Relative magnetic permeability of metal and dielectric is assumed to be unity. When metal particle is excited by external field (e.g. plane wave), electrons in metal will be polarized and resulting in induced electromagnetic fields. The surface of metal is discretized and Maxwell's equations are solved at the boundary using appropriate boundary conditions.

2.2 Nanoparticle Parameters

Calculations were performed for gold and silver nanoparticles with three different aspect ratios (0.25, 0.50, and 0.75) and five different particle diameters (10 nm, 30 nm, 50 nm, 70 nm, and 100 nm). Here, aspect ratio is the ratio of minor axis to major axis. The nanoparticles were excited using plane wave at wavelength of 200 – 1000 nm polarized along ellipsoidal minor axis and propagate along ellipsoidal major axis. Refractive index of medium surrounding particles was varied from 1.0 to 2.0. The variation of medium refractive index is aimed at mimicking refractive indices of liquids or gases containing biological and chemical species. Refractive indices of many gasses lie between 1 – 1.1, while those of chemicals are between 1.25 – 1.70.

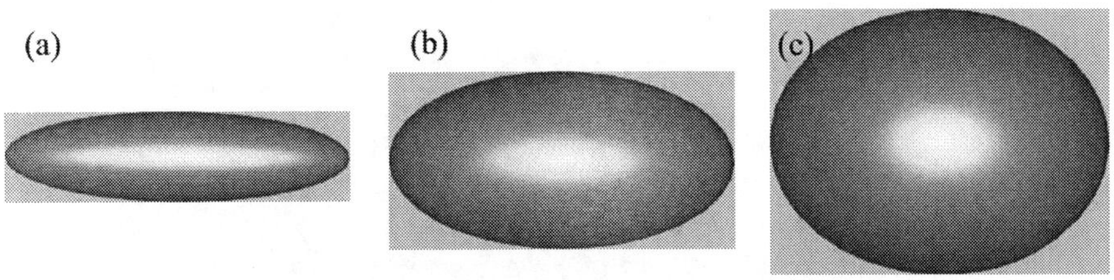

Figure 1. Ellipsoidal nanoparticles with different aspect ratio, (a) 0.25, (b) 0.50, and (c) 0.75.

The key parameter of SPR based sensing is sensitivity. Sensitivity of metal nanoparticle were calculated from extinction spectra. The sensitivity describes the shift of extinction spectra when refractive index of surrounding medium increases by one, therefore it has a unit of nm/RIU where RIU refers to refractive index unit. Adsorption of biological or chemical species may increase effective refractive index of medium resulting in a stronger screening of the Coulombic restoring force.

3. Results and Discussion

3.1 Effect of aspect ratio

The dependency of extinction (absorption + scattering) spectra of 20-nm silver nanoparticle on aspect ratio is plotted in figure 2. The intensity of extinction peaks increases monotonously with aspect ratio, while resonance wavelength decreases non-linearly. The resonance wavelength blue-shift with aspect ratio because of an increase in Coulombic restoring force when particle size becomes smaller. It is clearly visible that the shift of the resonance wavelength is much stronger for aspect ratio less than one. Larger shift means higher sensing sensitivity. Therefore, in this work aspect ratio was assumed to be 0.25, 0.50, and 0.75.

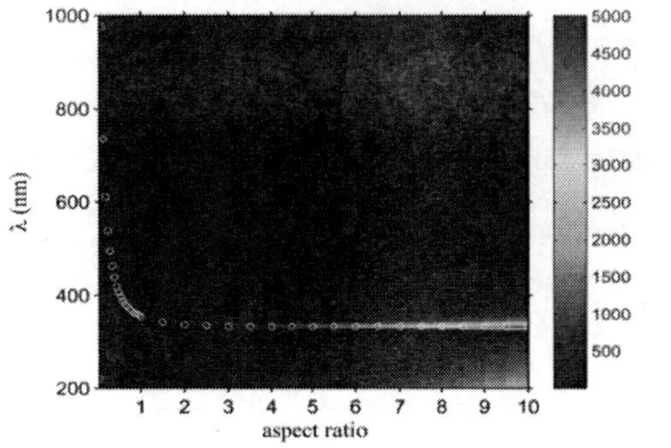

Figure 2. Surface plot of extinction spectra versus aspect ratio of ellipsoidal silver nanoparticle embedded in air. Diameter of nanoparticle is assumed to be 20 nm. Intensity of extinction spectra is given in colorbar. Resonance wavelengths are indicated as white circles.

3.2 Effect of refractive index of medium

Example of normalized extinction spectra of 30 nm gold and silver nanoparticles for three different aspect ratio are given in figure 3. Similar calculation was performed for particles with a diameter of 10 nm, 50 nm, 70 nm, and 100 nm.

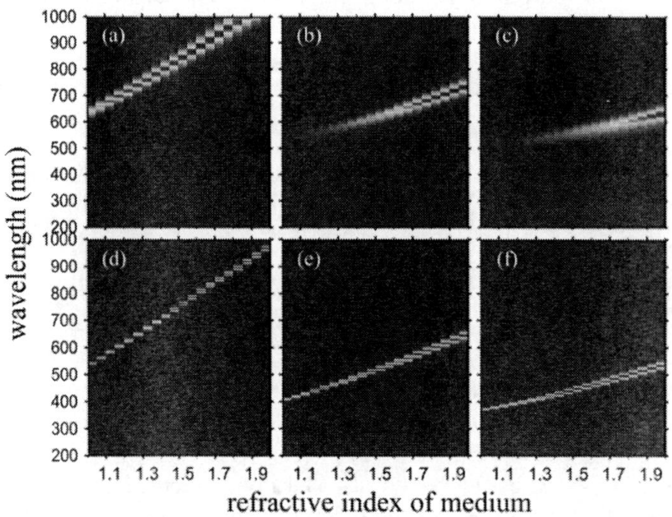

Figure 3. Intensity normalized extinction spectra of 30 nm nanoparticle for different aspect ratio as a function of refractive index of medium, (a) Au aspect ratio 0.25, (b) Au aspect ratio 0.50, (c) Au aspect ratio 0.75, (d) Ag aspect ratio 0.25, (e) Ag aspect ratio 0.50, and (f) Ag aspect ratio 0.75.

IOP Conf. Series: Materials Science and Engineering **578** (2019) 012035 doi:10.1088/1757-899X/578/1/012035

Red-shift of extinction peak with refractive index of surrounding medium can be observed for all particle diameter and aspect ratio. The shift of extinction peak does not vary linearly with refractive index of medium. The slope of the shift seems to depend on aspect ratio and particle size. Larger red-shift with refractive index means a higher sensing sensitivity. Resonance wavelength of silver is shorter than that of gold nanoparticles. Red-shift of extinction peak with refractive index of medium is also clearly visible. Broadening of extinction peak for silver nanoparticle is not as significant as that of gold nanoparticles.

Red shift of the extinction peak with refractive index of medium is caused by a lowering in Coulombic restoring force as a result of an increase in polarization charge at the surface of nanoparticles on the dielectric medium side. Broadening of extinction peak is also clearly visible when nanoparticle diameter is larger than 50 nm. The broadening is caused by retardation effect where oscillations of free electrons are delayed and damped [16]. Radiation losses and damping is dominant for particles larger than 100 nm, therefore affecting sensing sensitivity.

Imaginary part of dielectric constant of silver is much smaller than that of gold nanoparticles. As a consequence, damping and broadening of silver extinction peak is much lower than those of gold nanoparticles. Small peak width is advantageous for sensing because it allows for precise assignment of resonance wavelength and thus refractive index of investigated sample.

3.3 The shift of resonance wavelength

Red shift of extinction peak of gold and silver particles as a function of refractive index of surrounding medium, diameter, and aspect ratio is displayed on figure 4.

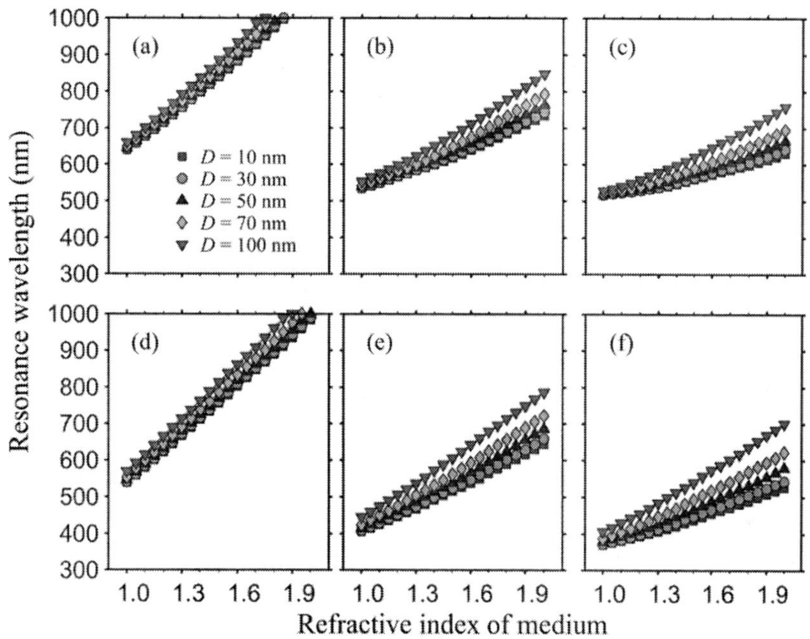

Figure 4. Dependency of red shift of extinction spectra of ellipsoidal gold nanoparticle with aspect ratio of (a) 0.25, (b) 0.50, and (c) 0.75, and ellipsoidal silver nanoparticle with aspect ratio of (d) 0.25, (e) 0.50, and (f) 0.75. Diameters of nanoparticles are given in the legend of Figure (a). The color and type of data marker are the same for all sub-plots.

One can clearly see that resonance wavelength red shifts with particle diameter and refractive index of medium, but blue shift with aspect ratio. The red shift shows nicely quadratic dependency on refractive index of medium within the simulation range. The slope of red shift of silver nanoparticle spectra (figure 4 (d), (e), and (f)) is larger than that of gold nanoparticles (Figure 4 (a), (b), and (c)). In the work reported by Anker, et al. [17], the red shift is assumed to be linear with refractive index of medium. The linear trend seems to fit quite nicely with their data where refractive index of solvent less than 1.6. However, as can be seen from figure 4, linear trend is just a rough estimation because quadratic dependency is valid over a larger range of medium refractive index. Quadratic dependency is even more clearly visible for larger particles. Larger slope of silver nanoparticle red shift corresponds to a steeper wavelength dependency of real part of dielectric constant of silver.

3.4 Sensing Sensitivity

Sensitivity of SPR based sensor is determined by the slope of resonance wavelength to the refractive index of medium. The obtained sensitivity has a unit of nm/RIU. Sensitivity of ellipsoidal nanoparticles is shown in figure 5. Sensitivity increases linearly with refractive index as expected from quadratic dependency of red-shift. In addition, sensitivity of ellipsoidal nanoparticles decreases with aspect ratio and increases with particle diameter. The increase of sensitivity with refractive index can be explained as a result of polarization charge build up in the dielectric medium. Polarization reduces oscillation restoring force that results in red shift of extinction peak. High sensitivity at higher refractive index is related to a strong light localization at the interface of metal and dielectric. Linear variation of sensitivity with refractive index was also predicted by Miller and Lazarides [18].

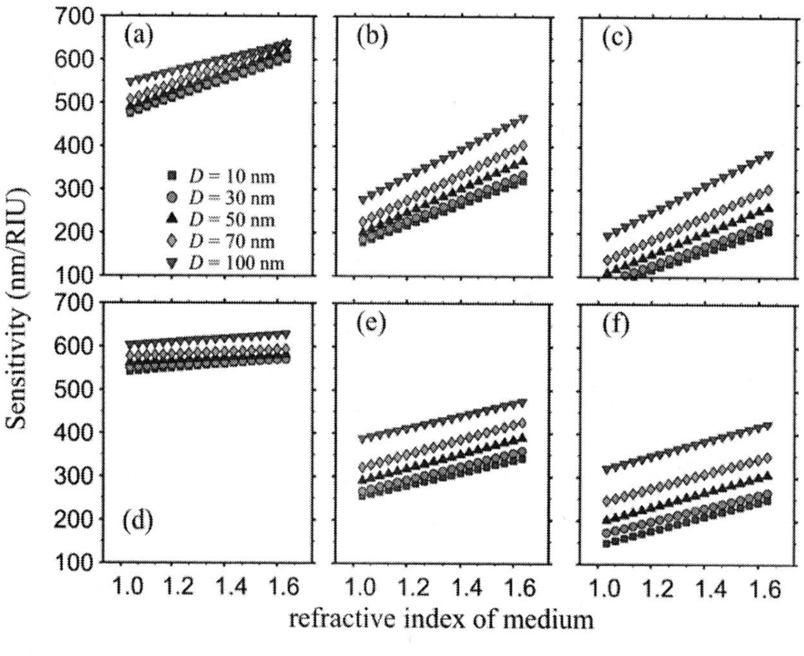

Figure 5. Dependency of sensitivity of ellipsoidal gold nanoparticles with aspect ratio of (a) 0.25, (b) 0.50, and (c) 0.75, and ellipsoidal silver nanoparticles with aspect ratio of (d) 0.25, (e) 0.50, and (f) 0.75. Diameters of nanoparticles are given in the legend of Figure (a). The color and type of data marker are the same for all sub-plots.

The increase of sensitivity with particle diameter is caused by larger variation of real part of dielectric constant of metal nanoparticles at longer wavelength. Real part of dielectric constant of metal determines resonance wavelength of surface plasmon. Silver has a steeper variation of real part of dielectric constant with wavelength than that of gold, therefore the sensitivity of silver nanoparticles to refractive index of medium is higher. However, gold has important advantages compared to silver, namely biocompatibility and chemical inertness. Silver can easily be oxidized and corroded in a humid environment.

Lowering sensitivity with aspect ratio is likely due to reduction of particle anisotropy when aspect ratio increases close to unity. Ellipsoidal nanoparticles shows incident light polarization dependency where longitudinal and transversal plasmonic mode may occurs. The resonance wavelength of longitudinal mode depends sensitively on particle size and refractive index of medium. When aspect ratio reach unity, the modes overlap and then reduces sensitivity.

4. Conclusions

Sensitivity of SPR sensing using ellipsoidal metal nanoparticle depends on ellipsoid aspect ratio, diameter, type of metal, refractive index of surrounding medium. Smaller ellipsoidal aspect ratio is more promising for sensing due to its much higher sensitivity. Larger particles consistently show a higher sensitivity than the smaller one. Sensitivity increases linearly with refractive index of medium. Silver nanoparticles show a higher sensitivity than gold. The obtained sensitivity for gas (refractive index slightly above 1) and chemical species (refractive index 1.3 – 1.6) sensing is relatively high that are well above 200 nm/RIU and in best case up to 600 nm/RIU. For sensing application, small aspect ratio and large particle diameter are recommended.

Acknowledgement

The authors would like to thank the Ministry of Research and Technology and Higher Education of the Republic of Indonesia through the 2017-Hibah Fundamental Research Grant No. 059/SP2H/LT/DRPM/2017 and the 2018-PDUPT Research Grant No. 050/SP2H/LT/DRPM/2018.

References

[1] Andreescu S and Sadik O A 2004 *Pure Appl. Chem.* **76** 861.
[2] Huber A, Demartis S, and Neri D 1999 *Journal of Molecular Recognition* **12** 198.
[3] Cheng C S, Chen C Q, and Lu C J 2007 *Talanta* **73** 358.
[4] Hutter E and Fendler J H 2004 *Adv. Mater.* **16** 1685.
[5] Stewart M E, Anderton C R, Thompson L B, Maria J, Gray S K, Rogers J A, and Nuzzo R G 2008 *Chem. Rev.* **108** 494.
[6] Ghosh S K, Nath S, Kundu S, Esumi K, and Pal T 2004 *J. Phys. Chem. B* **108** 13963.
[7] Lee K-S and El-Sayed M A 2006 *J. Phys. Chem. B* **110** 19220.
[8] Ly N, Foley K, and Tao N 2007 *Anal. Chem.* **79** 2546.
[9] Wiley B, Im S, Li Z-Y, McLellan J, Siekkinen A, and Xia Y 2006 *J. Phys. Chem. B* **110** 15666.
[10] Sherry L, Chang S, Schatz G, Duyne R V, Wiley B, and Xia Y 2005 *Nano Lett.* **5** 2034.
[11] Bakshi M S, Sachar S, Kaur G, Bhandari P, Kaur G, Biesinger M C, Possmayer F, and Petersen N O 2008 *Crystal Growth & Design* **8** 1713.
[12] Wiley B, Chen Y, McLellan J, Xiong Y, Li Z-Y, Ginger D, and Xia Y 2007 *Nano Lett.* **7** 1032.
[13] Washio I, Xiong Y, Yin Y, and Xia Y 2006 *Adv. Mater.* **18** 1745.
[14] Hohenester U and Trügler A 2012 *Computer Physics Communications* **183** 370.
[15] Abajo F J G d and Howie A 2002 *Phys. Rev. B* **65** 115418.
[16] Wokaun A, Gordon J P, and Liao P F 1982 *Phys. Rev. Lett.* **48** 957.
[17] Anker J N, Hall W P, Lyandres O, Shah N C, Zhao J, and Duyne R P V 2008 *Nat. Mater.* **7** 442.
[18] Miller M M and Lazarides A A 2005 *J. Phys. Chem. B* **109** 21556.

Effect of nanoparticle geometry on sensitivity of metal nanoparticle based sensor

M Muldarisnur[1], N Fridayanti[1], E Oktorina[1], E Zeni[1], E Elvaswer[1] and S Syukri[2]
[1] Department of Physics, Faculty of Mathematics and Natural Sciences, Andalas University, West Sumatera, Indonesia
[2] Department of Chemistry, Faculty of Mathematics and Natural Sciences, Andalas University, West Sumatera, Indonesia

E-mail: muldarisnur@sci.unand.ac.id

Abstract. Bio chemical detection using metal nanoparticles has been intensively investigated due to availability of robust nanoparticle synthesis methods whether chemically or using lithographic approach. SPR based sensor employs the shift of plasmonic resonance wavelength when local surrounding medium of metal nanoparticles change even slightly. The shift of the resonance wavelength depends sensitively on geometry and size of nanoparticle. The resonance wavelength and its shift were calculated numerically using boundary element method. Here, we compare the sensitivity of gold and silver nanoparticles to sense small change in refractive index of medium. Four different geometries (spherical, ellipsoidal, cylindrical, and donut-like nanoparticles) of different size were investigated to find the highest sensor sensitivity.

1. Introduction

Currently, we witness an increasing demand for robust and versatile biochemical sensing techniques for industrial, environmental, medical, and chemical analysis purposes [1]. Among many other approaches, surface plasmon resonance (SPR) sensing is now considered as a promising approach for detecting biological and chemical species in liquid or gaseous samples [2,3]. SPR sensing allows for real-time and label-free measurement by using only a small amount of sample. Surface plasmons are coherent oscillations of conducting electrons of a metal at the boundary with dielectric medium when illuminated with electromagnetic wave of appropriate wavelength [4]. The oscillations cause local field enhancement at the very surface of metal, therefore any small changes on the metal surface can be detected. Surface plasmon polariton (SPP) sensing on the surface of planar metal has been well studied and used for investigating biological structures for example binding and releasing mechanism of proteins. The excitation of SPP needs special optical set up such as prism, grating, or scatterer which then inhibits miniaturization. Consequently, metal nanoparticle based sensor using localized surface plasmonic resonance (LSPR) is often more preferred because of simplicity in its excitation and strong dependency of its sensitivity on shape and size of nanoparticles [5,6].

SPR based sensing is performed based on the shift of plasmonic resonance peak when local refractive index of medium that is in contact with particle changes. The wavelength at which coherent oscillation occurs (resonance wavelength) is determined by the dielectric constants of metal and surrounding dielectric medium, size, and shape of nanoparticles. SPR sensing make use of local change of medium refractive index upon adsorption of biological or chemical species on the surface of metal nanoparticles [7,8]. Dependency of the resonance wavelength on local dielectric environment has been used to develop label-free optical sensing. Localized SPR based sensing are progressing very rapidly due to the possibility to control size and shape of metal nanoparticle precisely using either lithographic technique or wet chemical synthesis. Gold and silver nanoparticles are two of most investigated materials. The reasons are still their high free electron density and low loss (negative part of dielectric constant). Gold and silver nanoparticles can be synthesized into many different geometries such as spherical, cylindrical, bar, ellipsoidal, and triangular [9,10,11,12].

Here, we report the effect of nanoparticle geometry on the sensitivity of SPR-based sensor. Four different shape of nanoparticles namely spherical, cylindrical, ellipsoidal, and donut-like will be compared to obtain the highest sensitivity. Shape and size of nanoparticle determine resonance wavelength and consequently affects the sensitivity of SPR based sensor.

2. Experimental

2.1 Calculation method

Boundary element method [13,14] was used to calculate extinction spectra of gold and silver nanoparticles. This method solves full Maxwell's equations for a metal nanoparticle embedded in a homogeneous dielectric medium. The metal nanoparticle having frequency-dependent dielectric constant is separated by an abrupt boundary with the surrounding dielectric medium. The metal and the dielectric medium are non-magnetic, therefore their relative magnetic permeability was assumed to be one. Free electrons in metal nanoparticle are polarized when illuminated with a plane wave. The polarization induces Coulombic restoring force that is proportional to particle size. Periodic change of the restoring force results in periodic coherent oscillation of free electrons. For calculation, the surface of metal is discretized and Maxwell's equations are solved using appropriate boundary conditions.

2.2 Geometry of nanoparticles

Calculations were performed for gold and silver nanoparticles with four different shapes and different particle diameters (see Figure 1). The nanoparticles were excited using a monochromatic plane wave with a frequency that varies between 200 – 1000 nm. Refractive index of medium surrounding particles was varied from 1.00 to 2.00 with a step of 0.05 to match the refractive indices of liquids or gases containing biological and chemical species. Refractive index of gasses is a little bit above that of vacuum (i.e. 1.0 – 1.1), while refractive index of most of chemicals are 1.3 – 1.6.

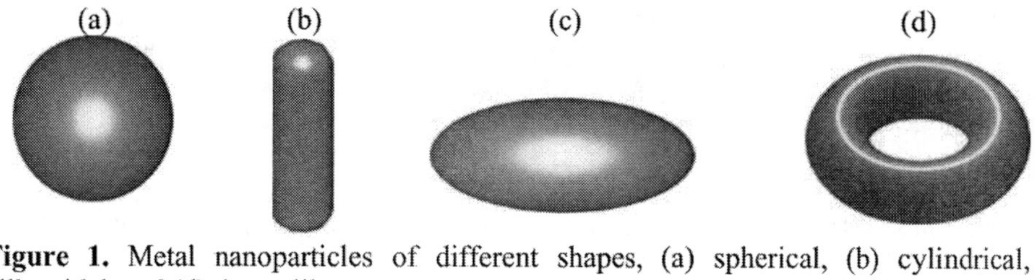

Figure 1. Metal nanoparticles of different shapes, (a) spherical, (b) cylindrical, (c) ellipsoidal, and (d) donut-like.

The sensing sensitivity of metal nanoparticle was calculated from its extinction spectra. The sensitivity describes how big the shift of extinction peak when refractive index of medium surrounding nanoparticle increases by one. The sensitivity has a unit of nm/RIU, where RIU refers to refractive index unit. A good sensor should have a high sensitivity so that it can detect any tiny change surrounding medium.

3. Results and Discussion

3.1. Extinction Spectra

Sample of extinction spectra of gold and silver nanoparticles are shown in figure 2. Here, extinction spectra are defined as surface plot of extinction cross section versus wavelength of incoming light and refractive index of surrounding dielectric medium. figure 2(a), (b), (c), and (d) are spectra of gold nanoparticles, while figure 2(e), (f), (g) and (h) are spectra of silver nanoparticles.

Maximum of extinction spectra of spherical nanoparticles increases significantly with the increase of medium refractive index and particle diameter. In addition, the extinction peak position (resonance wavelength) red shifts in a non-linear manner when refractive index of medium increases. The slope of the red shift depends on particle size and type of metal. Broadening of extinction spectra of silver nanoparticles is less than the broadening of gold nanoparticles. A narrow secondary extinction peak at a shorter wavelength than the primary one is only visible in the extinction spectra of silver nanoparticle.

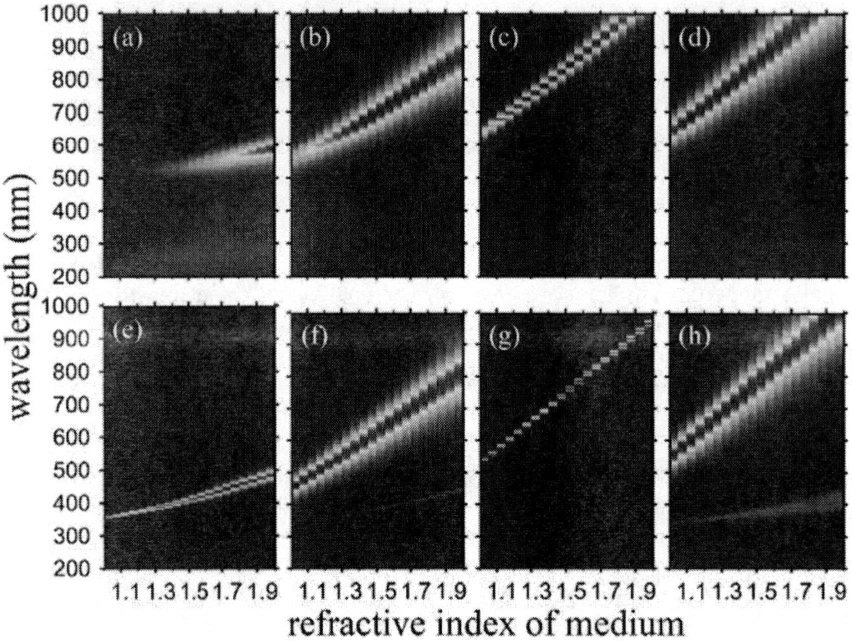

Figure 2. Extinction spectra of nanoparticle as a function of refractive index of dielectric medium, (a) Au sphere diameter 30 nm, (b) Au cylinder diameter 50 nm and height 100 nm, (c) Au ellipsoid diameter 30 nm and aspect ratio 0.25, (d) Au donut diameter 100 nm and ring radius 20 nm, (e) Ag sphere diameter 30 nm, (f) Ag cylinder diameter 50 nm and height 100 nm, (g) Ag ellipsoid diameter 30 nm and aspect ratio 0.25, (h) Ag donut diameter 100 nm and ring radius 20 nm. The extinction spectra for each plot has been normalized by its maximum, therefore the blue color scale refers to zero whereas red refers to one.

The red shift of the extinction peak with refractive index of medium is caused by an increase of polarization charge at the surface of metal. The polarization reduces Coulombic restoring force determining resonance wavelength. The red shift of extinction peak with particle size is caused by retardation of the exciting field. The separation distance of free electrons on the opposite surface of nanoparticle increases with particle size. The increase of separation distance lowers the restoring force appearing as red shift of resonance wavelength. The broadening of extinction peak with particle size is caused by retardation of electron oscillation for large particle. Lower imaginary part of dielectric constant of silver results in much narrower extinction peak than that of gold nanoparticle. The red shift of extinction peak with refractive index of medium and particle size is also obvious for non-spherical nanoparticles like cylindrical, ellipsoidal, and donut-like geometries. The explanation for the red shift is also similar with that of spherical particles. An increase of medium refractive index causes an increase in polarization charge at the medium part of the boundary which then decreases restoring force of oscillation and thus increases resonance wavelength. The slope of red shift depends on particle shape and size. The ellipsoidal nanoparticle seems to have the narrowest peak width. The broadening of extinction spectra for large particle size is caused by excitation of higher multipole radiation. The contribution of quadrupole mode is indicated by non Lorentzian extinction peak for big particles.

A quadrupolar peak at shorter wavelength than the primary (dipolar) peak is visible for silver nanoparticles. The quadrupolar peak is much narrower than that of the dipole. The quadrupole peak does not red shift significantly with refractive index of medium, therefore shows a lower sensitivity.

3.2. Resonance Wavelength

Resonance wavelengths of gold and silver particles for different refractive index of medium for different particle size and shape are shown in figure 3 and figure 4, respectively. The resonance wavelength shifts to a longer wavelength as refractive index of medium increases. The red shift fits quadratic dependency with refractive index. Except for donut-like particle, an increase of particle dimension (diameter or height) results in a red shift with a slope that depends on particle shape. There is no much difference between the shift of resonance wavelength between gold and silver nanoparticles.

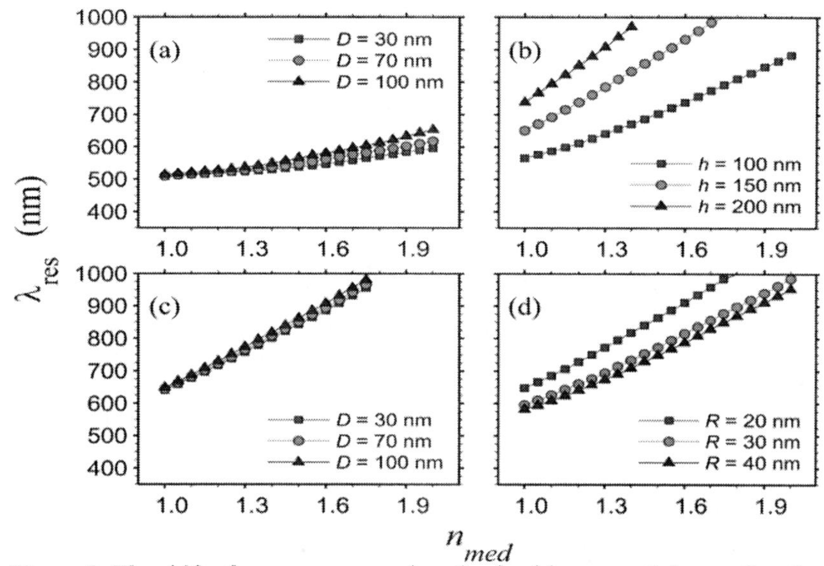

Figure 3. The shift of resonance wavelength of gold nanoparticle as a function of refractive index of dielectric medium for different particle shape and size, (a) spherical, (b) cylindrical with a diameter of 50 nm, (c) ellipsoidal with an aspect ratio of 0.25, and (d) donut-like with a diameter of 100 nm. Other dimensions of nanoparticles are given in the corresponding plot legend.

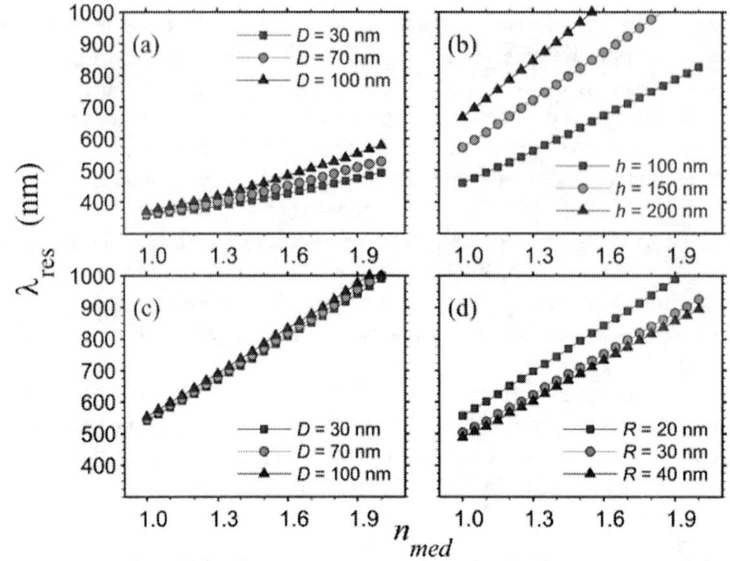

Figure 4. The shift of resonance wavelength of silver nanoparticles as a function of refractive index of dielectric medium for different particle shape and size, (a) spherical, (b) cylindrical with a diameter of 50 nm, (c) ellipsoidal with an aspect ratio of 0.25, and (d) donut-like with a diameter of 100 nm. Other dimensions of nanoparticles are given in the corresponding plot legend.

The dependency of resonance wavelength on particle dimension is stronger for cylindrical and donut-like nanoparticles. The shift of resonance wavelength of silver nanoparticle is larger than that of gold nanoparticle. Larger shift of resonance wavelength of silver particle is caused by stronger wavelength dependency of real part of silver dielectric constant.

3.3. Sensing Sensitivity

The slope of resonance wavelength versus refractive index medium indicates the sensitivity of SPR sensor. Larger slope is demanded to be able to discriminate tiny local change of medium refractive index. The change can be caused by the presence of biological or chemical species. The sensitivity has a unit of nanometer per refractive index unit (nm/RIU). The sensing sensitivity of spherical, cylindrical, ellipsoidal, and donut-like gold and silver particles are shown in figure 5 and figure 6, respectively. The sensitivity is linearly dependent on refractive index medium as expected from quadratic dependency of resonance peak shift.

In general, sensitivity increases with refractive index of medium. It means that higher refractive index chemical or biological substances are much easier to detect. However, care must be taken because extinction peak tends to broaden at high medium refractive index due to retardation effect. On contrary to spherical and cylindrical particles, the sensitivity of ellipsoidal and donut-like particles do not always increase with refractive index of medium. The slope of sensitivity of decreases gold ellipsoidal particle with a diameter of 100 nm and donut-like particle with a 20 nm ring radius.

IOP Conf. Series: Materials Science and Engineering **578** (2019) 012036 doi:10.1088/1757-899X/578/1/012036

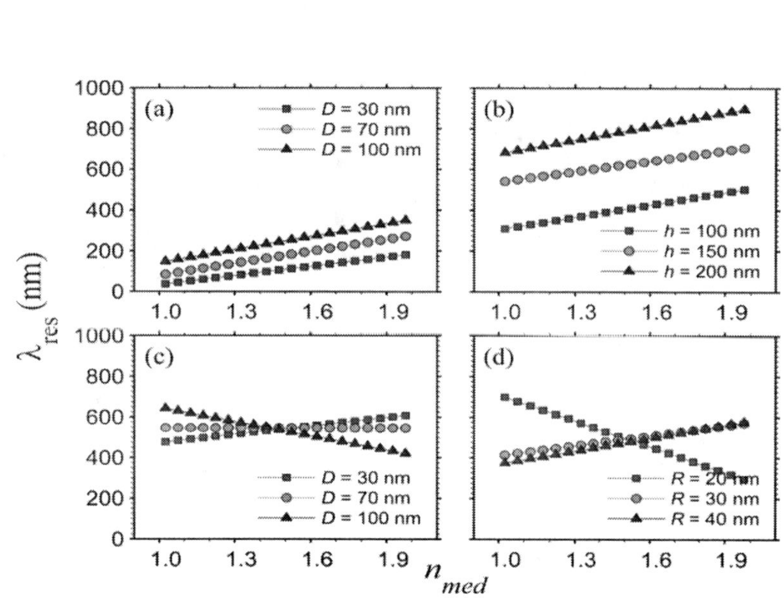

Figure 5. The sensitivity of gold nanoparticles as a function of refractive index of dielectric medium for different particle shape and size, (a) spherical, (b) cylindrical with diameter of 50 nm, (c) ellipsoidal with aspect ratio of 0.25, and (d) donut-like with diameter of 100 nm. Other dimensions of nanoparticles are given in the corresponding legend.

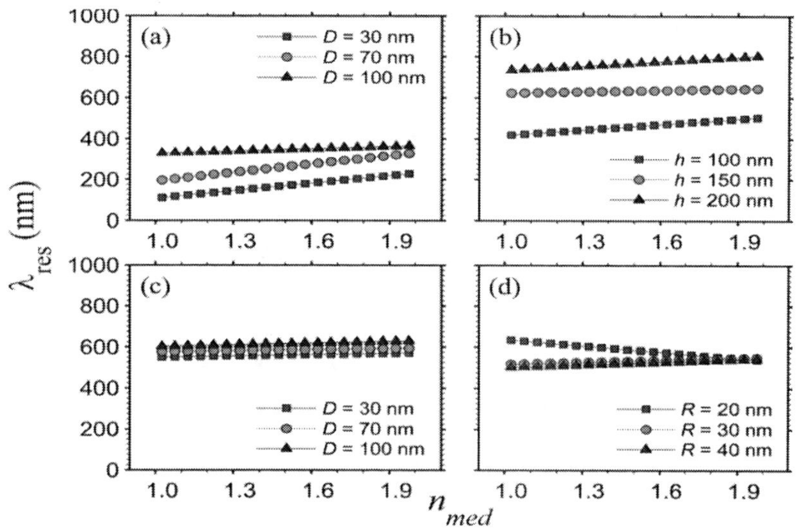

Figure 6. The sensitivity of silver nanoparticles as a function of refractive index of dielectric medium for different particle shape and size, (a) spherical, (b) cylindrical with diameter of 50 nm, (c) ellipsoidal with aspect ratio of 0.25, and (d) donut-like with diameter of 100 nm. Other dimensions of nanoparticles are given in the corresponding legend.

From figure 5 and figure 6 we can see that cylindrical nanoparticle has the highest sensing sensitivity, followed by ellipsoidal particles, donut-like particle, and the last spherical particles. High sensitivity of cylindrical and ellipsoidal nanoparticles is likely due to a stronger shape anisotropy. Anisotropic nanoparticles shows longitudinal and transversal modes whose resonance wavelength depends sensitively on particle dimension. The highest sensitivity is obtained by using cylindrical nanoparticle with a high aspect ratio (ratio of height to diameter). The sensitivity of cylindrical nanoparticle is always higher than 400 nm/RIU. It is also clearly visible that spherical particles have the lowest sensitivity that lies in the range between 100 - 400 nm/RIU. The dependency of sensitivity on composition of metal is weaker than its dependency on particle shape and size.

4. Conclusion

Sensitivity of SPR sensor increases linearly with refractive index of surrounding medium. Particle size and shape affect sensitivity of SPR sensor significantly. The dependency of sensitivity on particle composition (e.g. gold and silver) is much weaker than on particle shape and size. Larger particle size is, in general, results in higher sensing sensitivity. The highest sensitivity is shown by cylindrical nanoparticle, while the lowest sensitivity is displayed by spherical one. The sensitivity of rod shape nanoparticle is always above 400 nm/RIU, while that of spherical particles is below 400 nm/RIU.

Acknowledgement

The authors would like to thank the Ministry of Research and Technology and Higher Education of the Republic of Indonesia through the 2017-Hibah Fundamental Research Grant No. 059/SP2H/LT/DRPM/2017 and the 2018-PDUPT Research Grant No. 050/SP2H/LT/DRPM/2018.

References

[1] Andreescu S and Sadik O A 2004 *Pure Appl. Chem.* **76** 861.
[2] Huber A, Demartis S, and Neri D 1999 *Journal of Molecular Recognition* **12** 198.
[3] Cheng C S, Chen C Q, and Lu C J 2007 *Talanta* **73** 358.
[4] Hutter E and Fendler J H 2004 *Adv. Mater.* **16** 1685.
[5] Stewart M E, Anderton C R, Thompson L B, Maria J, Gray S K, Rogers J A, and Nuzzo R G 2008 *Chem. Rev.* **108** 494.
[6] Ghosh S K, Nath S, Kundu S, Esumi K, and Pal T 2004 *J. Phys. Chem. B* **108** 13963.
[7] Lee K-S and El-Sayed M A 2006 *J. Phys. Chem. B* **110** 19220.
[8] Ly N, Foley K, and Tao N 2007 *Anal. Chem.* **79** 2546.
[9] Wiley B, Im S, Li Z-Y, McLellan J, Siekkinen A, and Xia Y 2006 *J. Phys. Chem. B* **110** 15666.
[10] Bakshi M S, Sachar S, Kaur G, Bhandari P, Kaur G, Biesinger M C, Possmayer F, and Petersen N O 2008 *Crystal Growth & Design* **8** 1713.
[11] Wiley B, Chen Y, McLellan J, Xiong Y, Li Z-Y, Ginger D, and Xia Y 2007 *Nano Lett.* **7** 1032.
[12] Washio I, Xiong Y, Yin Y, and Xia Y 2006 *Adv. Mater.* **18** 1745.
[13] Hohenester U and Trügler A 2012 *Computer Physics Communications* **183** 370.
[14] Abajo F J G d and Howie A 2002 *Phys. Rev. B* **65** 115418.

Ab Initio DFT Study of Band Gap of Armchair and Zigzag Graphenes

N U J Hauwali[1], I Syuhada[1], V E Ligasetiawan[1] and T Winata[1,*]
[1]Physics for Electronic Material Research Division, Physics Department
Faculty of Mathematics and Natural Sciences, Bandung Institute of Technology
Jl. Ganesa 10, Bandung 40132, West Java, Indonesia

[*]E-mail: toto@fi.itb.ac.id

Abstract. Graphene sheet is a single-atom thick material and attracts great interest due to its extraordinary physical properties, but it has zero bandgap. At the boundary of graphene sheet, two types of stable edges were the zigzag edge and armchair edge. The electronic states of graphene largely depend on the edge structures. In this research, we used Quantum ESPRESSO package to calculate the band gap of armchair and zigzag graphene due to the increase of their number of cells. The results show that the graphene band gaps are influenced by the number of cells in the armchair and zigzag configuration.

1. Introduction

Graphene is made from carbon atoms arranged on a honeycomb structured by hexagons form. The structure flexibility of graphene is reflected in its electronic properties. The sp^2 hybridization between planar structure with a formation of a ρ bond between carbon atoms are separated by 1.42 Angstrom [1]. The theory of graphene has been established in 1947 as a first step to understand the electronics properties of graphite 3D. But it has predicted theoretically that 2D material is impossible to be grown. In 2004, Andre Geim and Novoselov exfoliate the graphene for the first time and they have received the 2010 Nobel Prize in Physics for their work on graphene research. The tight binding calculation of graphene shows that its conduction and valence band touch at Dirac points in the Brillouin Zone where energy dispersion are linier with respect to momentum [2]. Graphene has a potential as a nanotechnology material and can change the soft matter and the 2D material. Graphene can be used as a transparent conducting layer, sensor, supercapacitor, solar cell, but it has zero bandgap [3].

Nanographene have an important edge geometry dependence on their electronic structures. The are two stable configurations of graphene, armchair and zigzag configuration. The armchair and zigzag configurations are fundamental edges configurations, and there are clear different between armchair and zigzag configuration. The gap can be tuned by changing the ribbon width in graphene nanoribbon [4].

In this research, we study the band gap of armchair and zigzag graphene. We study the influence of increase the number of cells in armchair and zigzag graphene to the band gap of graphene using Quantum ESPRESSO package. Quantum ESPRESSO is the open source code always used for electronic calculations based on density functional theory (DFT), or many-body perturbation theory, using plane-wave basis sets and pseudopotentials [5].

IOP Conf. Series: Materials Science and Engineering **578** (2019) 012037 doi:10.1088/1757-899X/578/1/012037

2. Experimental

Description of the Coulomb's interaction between nuclei and electron is needed to calculate the band gap and density of states. Researcher usually solve this problem by using Hartree-Fock method. Since 1970's, density functional theory (DFT) has been extensively used for the calculations of electronic structure of atoms, molecules, and solids. The formalism is based on the Hohenberg-Kohn theorem [6], which states that the ground-state energy of system can be expressed as a functional of the ground-state of a single-particle density.

The main problem for calculating the electronic structure is to find the energy, that can be written in Schrodinger equation for certain Hamiltonian:

$$\hat{H} = \hat{T} + \hat{V} \tag{1}$$

In solid case, we calculate the Coulomb interaction between nuclei and electron. We can write 3 types of interaction: nucleus-nucleus interaction, nucleus-electron interaction, and electron-electron interaction.

$$\hat{H} = \sum_{I}^{N_n} \frac{h^2}{2m_I} \nabla_i^{\,2} + \frac{h^2}{2m_e} \sum_{i}^{N_e} \nabla_i^{\,2} + \sum_{I \neq J}^{N_n} \frac{e^2}{8\pi\varepsilon_e} \frac{Z_I Z_J}{|\bar{R}_I - \bar{R}_J|} + \sum_{I}^{N_n} \sum_{i}^{N_e} \frac{e^2}{4\pi\varepsilon_e} \frac{Z_I}{|\bar{R}_I - \bar{r}_i|} + \sum_{i \neq j}^{N_e} \frac{e^2}{8\pi\varepsilon_e} \frac{1}{|r_i - r_j|} \tag{2}$$

Depend on Born-Oppenheimer approximation, we can have a nucleus with fixed position than we can neglect the kinetic energy.

$$\hat{H} = \frac{1}{2} \sum_{i}^{N_e} \nabla_i^{\,2} + \sum_{i \neq j}^{N_e} \frac{1}{2} \frac{1}{|r_i - r_j|} + \sum_{I}^{N_n} \sum_{i}^{N_e} \frac{e^2}{4\pi\varepsilon_e} \frac{Z_I}{|\bar{R}_I - \bar{r}_i|} \tag{3}$$

In 1929, Hartree proposed that the electron system can be characterized by the electron density, which is given by the sum of all electrons at a point **r**.

$$n(\boldsymbol{r}) = \sum_{i=1}^{N_e} \delta(\boldsymbol{r} - \boldsymbol{r}_i) \tag{4}$$

Hohenberg-Kohn theorem leads to energy

$$E[n] = \langle \psi[n] | \hat{H} | \psi[n] \rangle = F_{HK}[n] + \int d^3r \, V_{ext}(\boldsymbol{r}) n(\boldsymbol{r}) \tag{5}$$

where F_{HK} as Hohenberg-Kohn functional denoted as:

$$F_{HK}[n] = \langle \psi[n] | \hat{T} + \hat{V} | \psi[n] \rangle \tag{6}$$

In 1965, Kohn-Sham [7] assumed that the ground state of the original system is equal to the ground state density of some chosen non-interacting systems. Thus we can write

$$E_{KS}[n] = T_S + E_H[n] + E_{ext}[n] + E_{xc} \tag{7}$$

In this work, the first step to obtain the structure of graphene system is to generate a single layer graphene. We use graphite CIF file to generate an armchair and zigzag graphene through VESTA and ASE Python. The band gap and density of states (DOS) are obtained through planewave DFT calculations in Quantum ESPRESSO package. Exchange-correlation (XC) potential is approximated by PBE-GGA, and k-points are chosen to be "automatic" for self-consistent field (SCF) and non-self consistent field calculation (NSCF), and chosen to be "crystal_b" for band and DOS calculation, where the k-point were generated by using XCRYSDEN. The pseudopotential is taken from Kresse-Joubert PAW algorithm [8]. In this research we are not doing a relax calculation since we only need a band gap and DOS. To obtain these two results, the SCF calculation is considered more efficient. After calculating the SCF, we calculate the NSCF as a prequal to the projected DOS. After calculating the NSCF, we calculate the band gap and DOS of armchair and zigzag graphene.

3. Results and Discussion

3.1. Armchair graphene

3.1.1. Band gap and Fermi energy

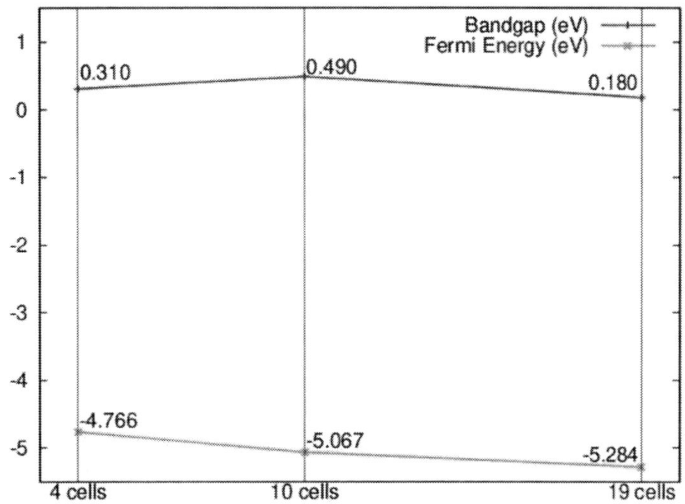

Figure 1. Band gap and Fermi energy.

For different number of cells in armchair graphene (Figure 1), the Fermi energy decrease when the number of cells increase, the Fermi energies are -4.766 eV, -5.067 eV and -5.284 eV, respectively for 4, 10 and 19 number of cells. For its band gap, the value is 0.310 eV for 4 cells, increase to 0.490 eV for 10 cells and decrease to 0.180 eV for 19 cells. We can see from this result that the band gap value is oscillating, but the value is between 0.0 and 0.5 eV.

3.1.2. Density of states

The density of states of the system describes the number of states per an interval of energy at each energy level available to occupied.

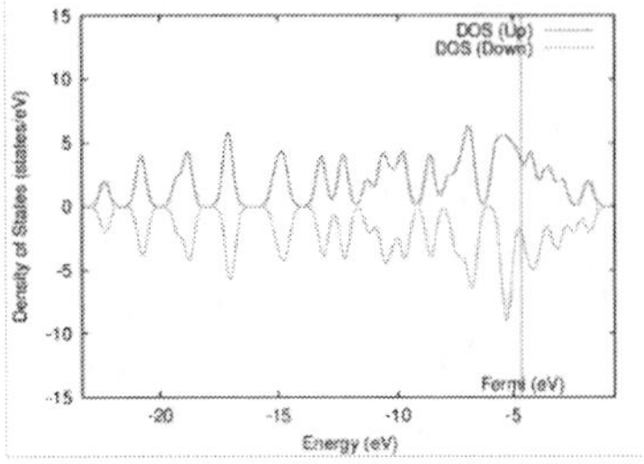

Figure 2. DOS for 4 cells.

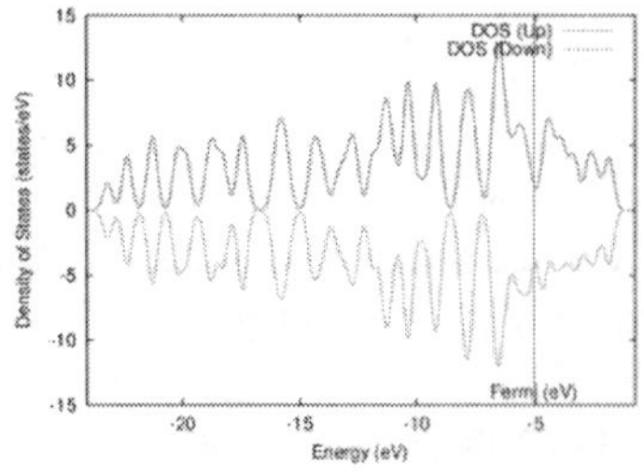

Figure 3. DOS for 10 cells.

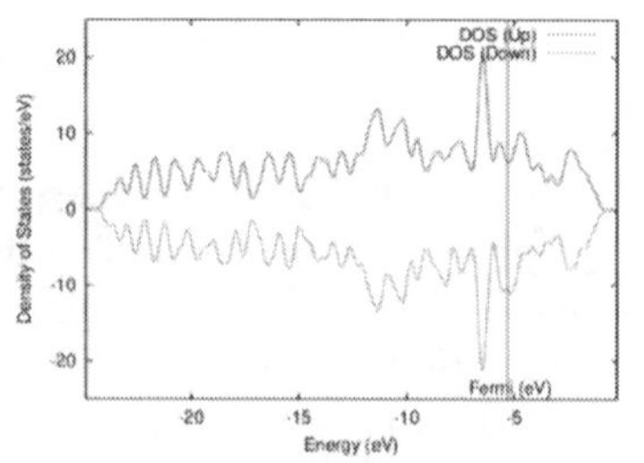

Figure 4. DOS for 19 cells.

Figure 2, 3 and 4 are all the total density of states of armchair graphene. The Fermi energy is shown by the vertical line in each curve. As we see from these density of states curves, we can say that the armchair graphene is metal-like, since there is no gap in each density of states curve. The total density of states curve shows the antisymmetric between spin up and spin down, indicates that the armchair graphene is paramagnetic-like [9,10].

3.2. Zigzag graphene

3.2.1. Band gap and Fermi energy

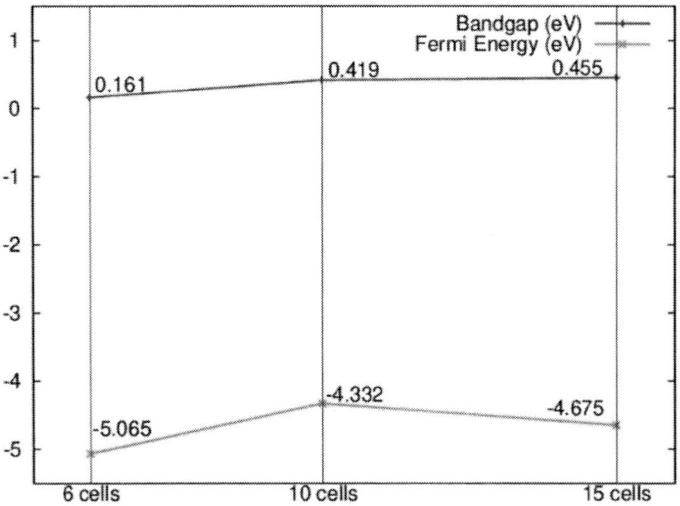

Figure 5. Band gap and Fermi energy.

For different number of cells in zigzag graphene, the Fermi energy increase from -5.065 eV to -4.332 eV for 6 to 10 number of cells but decrease to -4.675 eV when the number of cells increased to 15. For spin up band gap, the band gap increased when the number of cells is increased. When we use 6 cells, the gap is 0.161 eV, increase to 0.419 eV for 10 cells, and then 0.455 eV for 15 cells. We can see from these results that the band gap of zigzag graphene is between 0.0 eV and 0.5 eV.

IOP Conf. Series: Materials Science and Engineering **578** (2019) 012037 doi:10.1088/1757-899X/578/1/012037

3.2.2. *Density of states*

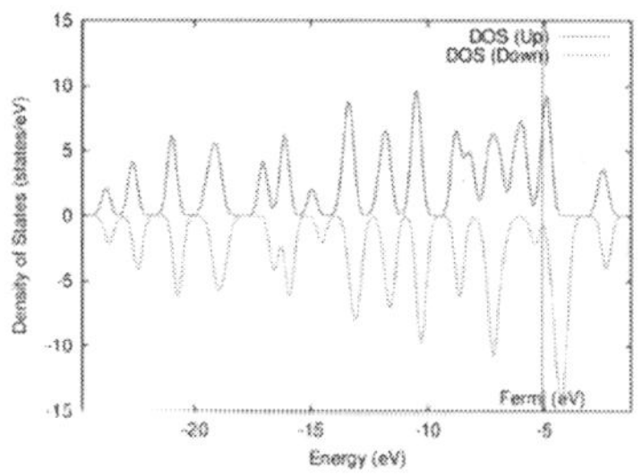

Figure 6. DOS for 6 cells.

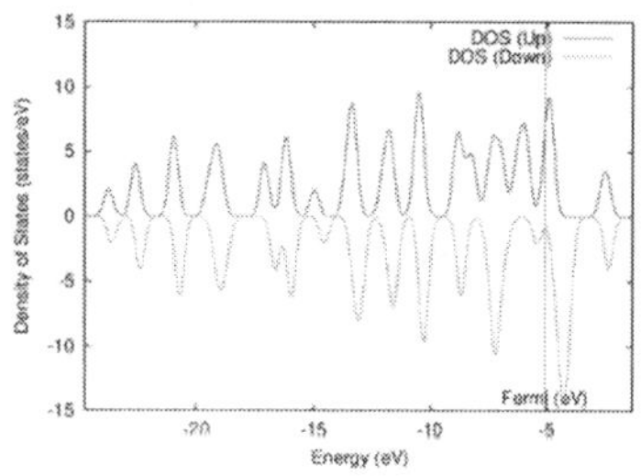

Figure 7. DOS for 15 cells.

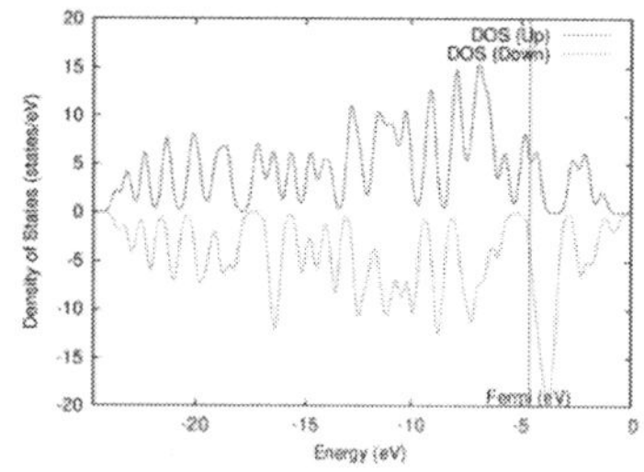

Figure 8. DOS for 15 cells.

Figure 6, 7, and 8 are the density of states of zigzag graphene. From the overall of the density of states curves, it is seen that the zigzag graphene is paramagnetic-like. Since there is no gap shown in these density of states curves, then the zigzag graphene is metal-like material.

4. Conclusion

Based on the results above, one can say that there is an influence between the number of cells and the band gap of armchair and zigzag graphene. For armchair graphene, the band gap is oscillating when we increase the number of cells, but the band gap is increase when we increase the number of cells for zigzag graphene. The value of armchair and zigzag band gap not more than 0.5 eV. From DOS calculation, one can say that these two types of graphene are paramagnetic-like.

Acknowledgement

This work is supported by Republic of Indonesia Ministry of Research, Technology and Higher Education (127/SP2H/PTNBH/DRPM/2018).

References

[1] Neto A H C, Guinea F, Peres N M R, Novoselov K S and Geim A K 2009 *Rev. Mod. Phys.* **81** 109
[2] Sahu S and Rout G C 2017 *Int. Nano Lett.* **7** 81
[3] Wadhwa P, Kumar S, Kumar T J D and Kumar R 2018 *J. Appl. Phys* **123** 16
[4] Enoki T and Kiguchi M 2018 *Phys. Rev. Applied* **9** 037001
[5] Giannozzi P *et al.* 2017 *J. Phys.: Condens. Matter.* **29** 46
[6] Hohenberg P and Kohn W 1964 *Phys. Rev.* **136** B864
[7] Kohn W and Sham L J 1965 *Phys. Rev.* **140** A1133
[8] Prandini G, Marrazzo A, Castlli I E, Mounet N and Marzari N 2018 *npj Comput. Mater.* **4** 72
[9] Ma S, Wurentuya B, Wu X, Jiang Y, Tegus O, Guan P and Narsu B 2017 *RMC Adv.* **7** 27454
[10] Bolandhemat N, Mahmudur Md, Zainuddin H, Chan K T and Shualbu A 2017 *J. Mater. Sci. Eng.* **6** 6

IOP Conf. Series: Materials Science and Engineering **578** (2019) 012038 doi:10.1088/1757-899X/578/1/012038

Fabrication of Nano-Calcium Carbonate Precipitate by Ultrasonic Milling Technique using Ethylene Glycol Media

E Sulistiyono[1*], M Handayani[1], M R Arwanda[2] and T Kusumaningsih[2]
[1]Research Center of Metallurgy and Materials, Indonesian Institute of Sciences
[2]Department of Chemistry, Faculty of Mathematics and Natural Sciences, SebelasMaretUniversity, Surakarta

[*]E-mail: eko221068@gmail.com

Abstract, Calcium Carbonate Precipitate is one of the filler ingredients which have the advantage of being neutral to all materials, light and harmless. Along with the widespread use of composite materials, it is necessary to develop smaller size nano calcium carbonate materials having nano-size properties that are superior to ordinary calcium carbonate precipitates. The purpose of this study is to obtain nano-carbonate precipitate materials with the particle size below 100 nm. In this study, the method used is starting from the preparation of calcium carbonate precipitate through a carbonation process, and then an ultrasonic milling process with ethylene glycol media is carried out. The process step is calcination, then the slaking process with water and then the carbonation process using carbon dioxide gas. Variables in the process of making calcium carbonate precipitates were not carried out, whereas in the ultrasonic milling process the variables in the process were ultrasonic time and the use of amplitude. The results of the experiment are then analysed by testing the sediment rate to obtain optimal grain size results, then the PSA Nano analysis in wet conditions and SEM analysis after the product is dried. The experimental results showed that the optimum yield of calcium carbonate precipitate granules with a size of 64 nm at 20 minutes of the ultrasonic time and the use of waves with an amplitude of 30%.

1. Introduction

Ethylene Glycol is one of the alcohol compounds that has more than one branch chain, so that it can form more than one (OH) group. With a number of more than one chain, ethylene glycol is easier to bind tightly to suspended particles in it so that it is easier to be attacked using microbubble due to ultrasonic wave effect. This can be proven from the results of research on the manufacture of ZnO nanoparticles which are capable of producing particle sizes below 100 nm [1]. Besides that, the use of ethylene glycol is also effective for the manufacture of other oxide materials such as TiO_2 [1]. As ultrasonic media, ethylene glycol has the advantage of having a very high boiling point of 197.3 °C. The boiling point is very beneficial in terms of resistance to microbuble attack caused by ultrasonic waves, able to increase the solvent temperature by more than 100 °C. However, the use of ethylene glycol media has the disadvantage of not being suitable for particles that cannot bear at high temperatures during the process of separating particles with ethylene glycol after treatment with ultrasonic waves [2].

The material to be processed into nano-size particles in this experiment is calcium carbonate which results from the process of re-precipitation of calcium hydroxide with carbon dioxide gas. The result of this deposition process is an amorphous calcium carbonate known as Light Calcium Carbonate Precipitate, which has a lighter density than calcium carbonate in the form of Aragonite Crystals and calcite [3]. Calcite and Aragonite are the two main constituents of $CaCO_3$, which are known to be very easy to form in large quantities in nature both biologically and with technology. Interactions that occur between CaO-CO_2-H_2O systems have been studied in several researches. Several theories have succeeded in developing the core of carbonate and the complex model of the surface has been successfully applied to the dissolution and precipitation of $CaCO_3$ in solution with free of Magnesium [4]. Natural calcite rocks are minerals formed through crystallization in shallow sea areas. Calcite belong to the carbonate group with the same chemical composition, namely carbonate ($CaCO_3$), but sometimes these minerals also compose rocks that have different compositions [5]. Limestone ($CaCO_3$) is one of the many industrial minerals that are needed and used directly or indirectly. Direct use of limestone, for example, is in the industries of cement and sugar, ceramics, ingredients added to the smelter process and building materials, while the indirect use of limestone is still needs to be processed in advance into light calcium carbonate or quicklime (CaO). Light calcium carbonate and quicklime or burning lime are as much needed for the needs of the paint industries, toothpaste, paper bleach, cosmetics, soil neutralization, building materials and others [6]. In this study, the process of making Light Calcium Carbonate Precipitate from limestone originating from Klaten, Central Java was carried out. The raw material is then dissolved in ethylene Glycol as a media and given ultrasonic waves. In this experiment, the experimental variables used were ultrasonic time and ultrasonic wave amplitude in ethylene glycol media.

The calcium carbonate precipitated by mechanical stirring had a particle size of about 20 μm. By contrast, the particle size of vaterite formed under ultrasonic irradiation was about 2 μm, with a specific surface area of 25-30 m^2/g. The major polymorph of calcium carbonate formed by ultrasonic irradiation was vaterite with some calcite present. For 40 kHz ultrasonic irradiation, the specific surface area of the calcium carbonate increased with increasing amplitude. The particle size of vaterite formed at this frequency was about 2 μm, and its distribution was sharper than that obtained at 20 kHz. The mode diameter of the synthesized vaterite was found to decrease with increasing amplitude at 40 kHz [7].

The de-agglomeration and charge nullification were achieved using surface modification of CaCO3 nanoparticles using triethoxy vinyl silane (TEVS) under ultrasound environment. The size of unmodified and modified $CaCO_3$ was found to be ≈ 50-80 and 50-90 nm, respectively, with square crystal structure morphology [8].

The result shows that calcium carbonate nano particles fulfill the requirements of calcium carbonate nano material with averages of the particle size obtained from PSA analysis is 109 nm, with the number of particles has reached 90.7% in a particle size range of 60 to 100 nm. From XRD analysis, the crystal structure of calcium carbonate nano particles suits very well with calcite pattern [9]. The characterization using XRF showed that CaO content of precipitated calcium carbonate from natural limestone Sukabumi had high purity of 99.16%. The particle distribution using scanning electron microscope (SEM) showed that precipitated calcium carbonate from natural limestone Sukabumi revealed 1.79-11.46 μm, meanwhile the particle distribution of precipitated calcium carbonate Merck showed larger particles with the size of 3.22-10.68 μm [10].

2. Experimental

2.1. Materials used in the experiments

2.1.1. Natural Materials

The ultrasonic process in this study was to use calcium carbonate precipitates obtained from the calcination of limestone at a temperature of 900 °C for approximately six hours in a size of approximately 1 cm. In the calcination process, 5,000 g of limestone in the form of calcium carbonate in Calcite form experienced the release of carbon dioxide gas to form 2,600 g of calcium oxide. Furthermore, calcined limestone was ground and continued with the slaking process which is mixing as much as 2,500 g of calcined limestone with 3,000 mL of distilled water and stirring for 4 hours. In this process, there is a reaction forming calcium hydroxide accompanied by heat which boils the remaining water of the reaction. The next process is carbonatazion, by taking 200 g of the slaking process and added 500 mL of distilled water. This process can occur by adding CO_2 gas to $Ca(OH)_2$ solution. The carbonatazion process is carried out with pH control until the pH becomes 6. Then proceed with filtration and drying of carbonatazion results with an oven at 150 °C for 1 day.The end result of the process is a mild white calcium carbonate compound known as Light Calcium Carbonate precipitate. The form of raw materials during the process can be seen in the Figure 1.

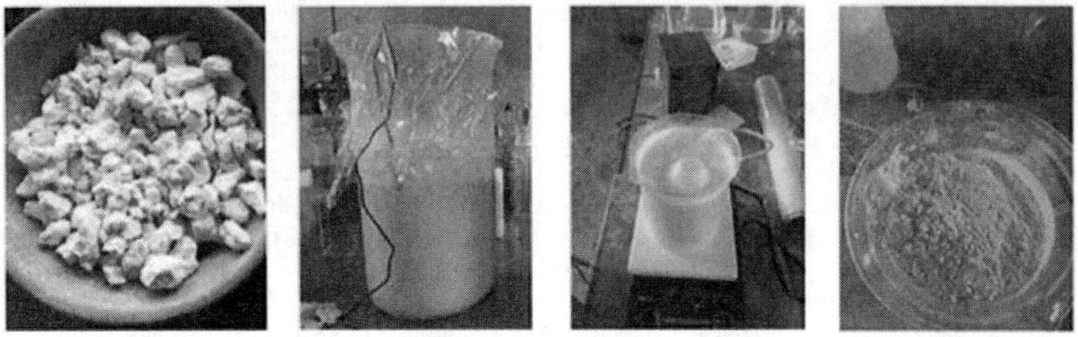

(a) (b) (c) (d)

Figure 1. Results of the manufacturing stage of Light Calcium Carbonate Precipitate for experimental materials: (a) calcination results, (b) slaking results, (c) carbonatization results and (d) drying results.

The results in the form of calcium carbonate precipitates are then carried out by characterization process using XRF analysis. XRF analysis results showed that light Calcium Carbonate was obtained with a high purity level of 99.16%, and only a small amount of phosphorus and sulfur impurities..

Table 1. XRF Analysis of Light Calcium Carbonate Precipitate

Component	Limestone (%w)	Process result (% w)	Comparison with Merck (%w)
CaO	90.93	99.16	99.40
MgO	6.54	Nd	Nd
SiO_2	0.74	Nd	Nd
Al_2O_3	0.47	Nd	Nd
P_2O_5	0.27	0.14	Nd
SO_3	0.05	0.14	0.16
Fe_2O_3	0.71	Nd	Nd

[a]laboratorium of Faculty of MIPA-UI. Nd = No Detector

2.1.2. Commercial Materials

In this experiment in addition to materials derived from nature calcium carbonate as the row material for the process used materials derived from commercial purchases. These materials include Ethylene Glycol with Merck's Pro Analysis spesificatrions with catalog number : 1.09621.2500, aquadest form the ion exchange process and Filter paper Whatman Catalog number 42, 150 mm diameters.

2.2. Analysis Instrument

In this experiment, the analysis of the experimental results was carried out in stages where the first step was observed by sedimentation test by looking at the deposition velocity of calcium carbonate in ethylene glycol media before and after the ultrasonic process was carried out. The second stage is testing particle distribution using PSA Nano equipment from Beckman Coulter with Delsa Nano C 812 type. The results of PSA Nano analysis are then compared with visual testing with SEM using SEM equipment from JEOL with Type JSM-639OA.

2.3. Experiment

In this experiment the process of reducing the size of calcium carbonate precipitates was reduced from the carbonatation process of limestone from Bayat, Klaten Region, Central Java. So that the process of reducing the size of the grain is obtained nano-size calcium carbonate powder. In this experiment the grain size reduction process is carried out by using ultrasonic waves in ethylene glycol media. The experiment was carried out in a 500 ml beaker glass using 10 g of calcium carbonate precipitate and a 200 mL ethylene glycol media. The experiment was carried out 10 times with variables changing ultrasonic amplitude 20%, 25%, 30%, 35% and 40%, with an ultrasonic processing time of 10 minutes and 20 minutes. The ultrasonic process is carried out in a maximum temperature range of 90 °C and without heating from outside. After administration of calcium carbonate ultrasonic waves contained in ethylene glicol media the sedimentation rate was tested by the stokes method to see the longest sediment rate. The best results from the experiments with the longest sediment rate were analyzed by analysis of nano PSA and compared with before the ultrasonic process was carried out. Samples with the best results from measurements with nano PSA were then dried so that nano-size calcium carbonate dry solids were obtained. Dry solids are then analyzed visually the granular shape and size of the granules with SEM equipment.

3. Results and Discussion

3.1. Sedimentation Rate

The results of the analysis using the Stokes Method showed that the higher the amplitude process, the sedimentation rate decreases.However, at the point of amplitude 35% shows a different tendency, where the amplitude is added, the sedimentation rate of calcium carbonate granules is higher.The effect of ultrasonic time on the process of sedimentation rate of calcium carbonate granules was seen to have the same tendency between the ultrasonic time of 10 minutes and 20 minutes. In the ultrasonic process in 20 minutes results in a precipitation rate of calcium carbonate granules longer than the ultrasonic process in 10 minutes.

From the graph above, it can be seen that the optimal process occurs at the amplitude of 30% both for ultrasonic time 10 minutes and 20 minutes. For further analysis, the PSA Nano analysis will be carried out at 30% amplitude and ultrasonic time in 10 minutes and 20 minutes.

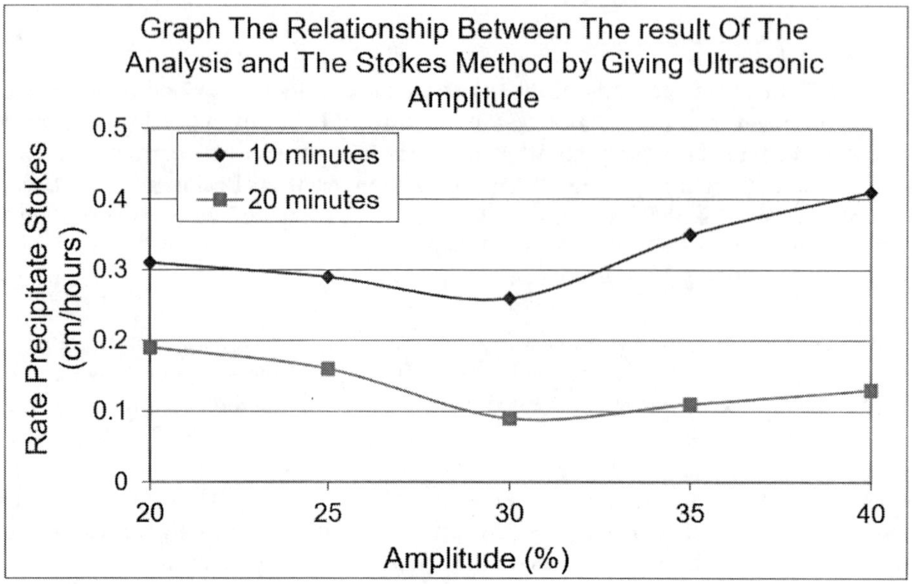

Figure 2. The results of measurements of sedimentation rate for amplitude during the ultrasonic process in 10 minutes and 20 minutes.

3.2. PSA Nano analysis

PSA Nano analysis has been carried out on three samples, namely a sample of calcium carbonate solids dissolved in ethylene glycol without ultrasonic waves, samples given ultrasonic waves in 10 minutes and 20 minutes at 30% amplitude.

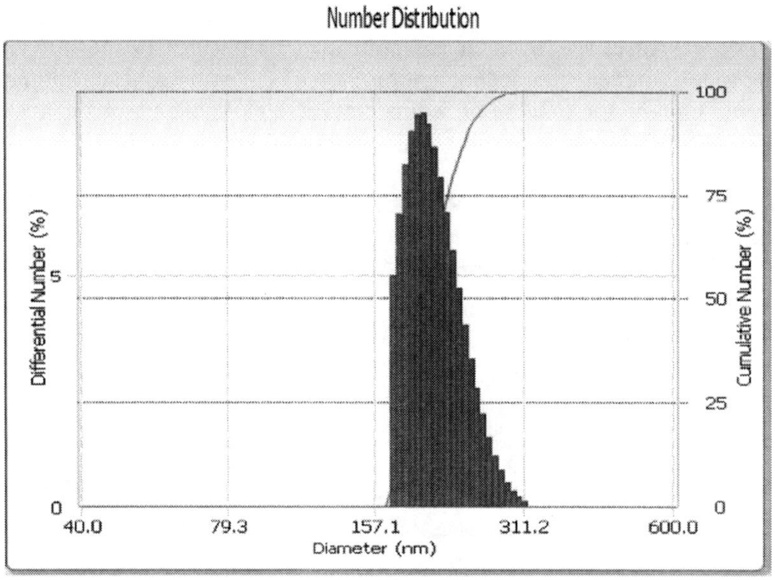

Figure 3. Results of measurements of the distribution of calcium carbonate particles before ultrasonic process using PSA Nano, The analysis conditions is in a solution with a concentration of 10 g/200 mL of ethylene glycol.

The results of the PSA nano analysis showed the distribution of particles size in the range of 170 nm to 311 nm and the highest number of particles in the range 185 nm to 212 nm as depicted in figure 3. The results of the analysis illustrate that the results of carbonatazion process of limestone from Bayat area, Klaten Regency, can produce very fine precipitate calcium carbonate granules, but the particle size has not reached the nanoparticle size. For this reason, it is necessary to do an ultrasonic process so that nano-size particles can be obtained.

The measurement of grain size was performed using the PSA Nano method on the optimum ultrasonic process. The results of the experiment using the Stokes Method obtained two optimum points:at the ultrasonic process in 10 minute with 30% amplitude and the ultrasonic process in 20 minutes with 30% amplitude.

The results of PSA Nano analysis showed that the particle size distribution resulted from the ultrasonic process in 10 minutes gave a particle size range of 57 nm to 193 nm. This result indicated that the cumulative particle size below 100 nm had reached 93.7%, this can be seen in the figure 4.

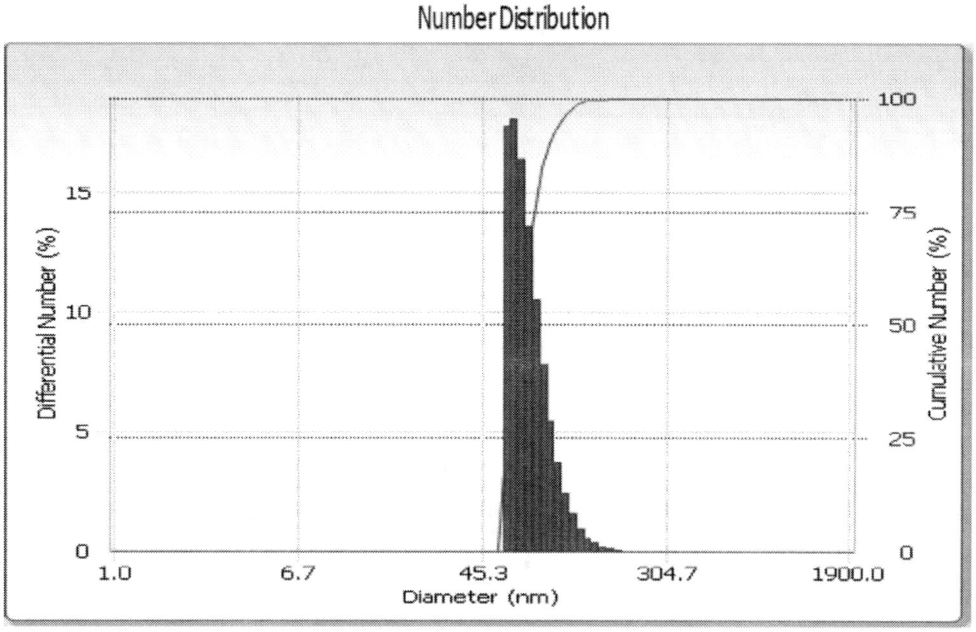

Figure 4. Results of measurements of calcium carbonate particle distribution using PSA Nano analysis after ultrasonic processin 10 minutes.The condition is in a solution with a concentration of 10 g/200 mL of ethylene glycol.

The ultrasonic process in 20 minutes gave the particle size in the range of 56 nm to 104 nm, where the cumulative particle size below 100 nm reached 99.9% as shown in figure 5. The results of the ultrasonic process with a duration of 20 minutes has produced almost perfect calcium carbonate nanoparticles, so it can be concluded that the ultrasonic process with 30% amplitude and 20 minutes of ultrasonic time can convert calcium carbonate material into nano calcium carbonate precipitate.

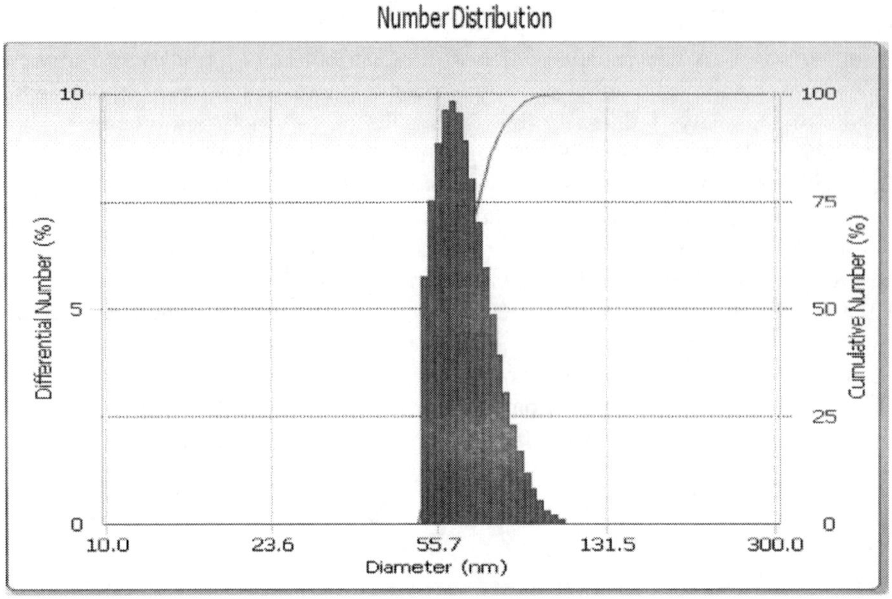

Figure 5. Results of measurements of calcium carbonate particle distribution using PSA Nano analysis after ultrasonic process in 20 minutes. The condition is in a solution with a concentration of 10 g/200 mL of ethylene glycol.

3.3. The result of SEM analysis of nano-calcium carbonate precipitate

After analyzing using PSA Nano, the next step is to visually see calcium carbonate particles using the SEM method. Unlike the PSA analysis, where the PSA analysis the sample does not go through the drying process, it is immediately analyzed in the solution. While in the SEM analysis, sample must go through the drying process first. The results of SEM analysis on calcium carbonate samples without ultrasonic process are depicted in Figure 6.

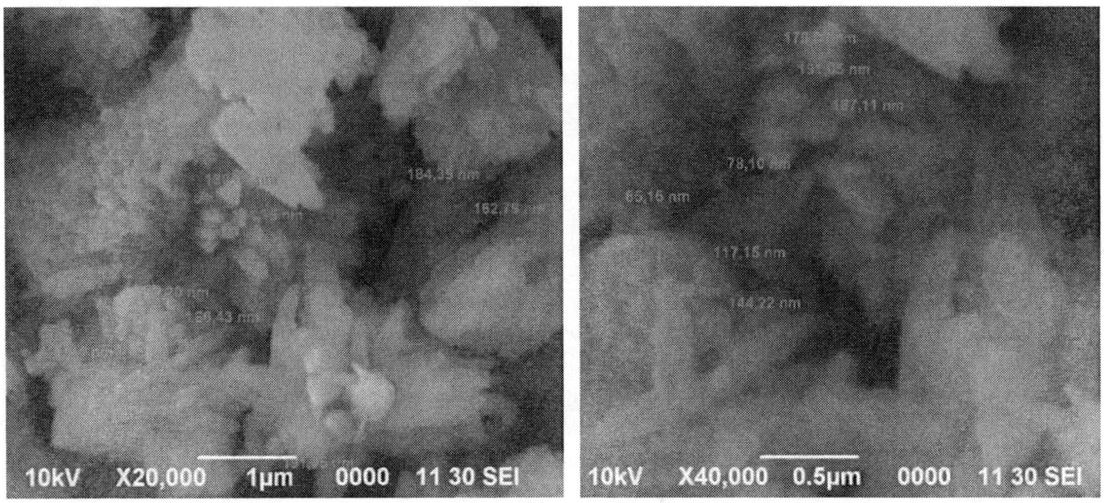

Figure 6. SEM Analysis result of calcium carbonate precipitate before ultrasonic process where (left) magnification 20,000 times and (right) magnification 40,000 times.

IOP Conf. Series: Materials Science and Engineering **578** (2019) 012038 doi:10.1088/1757-899X/578/1/012038

From the results of SEM analysis, it can be seen that in the sample with a magnification of 20,000 times and a magnification of 40,000 times, there are several particles with sizes below 100 nm but they are mainly have the particles size above of 100 nm. It match with the result of the PSA analysis, that the size of the granules in the raw material is not nano-sized, where the particle size is in the range of 170 nm to 311 nm. By looking at the grain shape, it appears that the shape of the grain is irregular but shows a rectangular shape, which gives an indication of calcite compounds.

SEM analysis measurements with a magnification of 40,000 times for the samples with ultrasonic process at 30% amplitude in 10 minutes and 20 minutes show the results of the average grain size are smaller than without ultrasonic process as shown in figure 8.

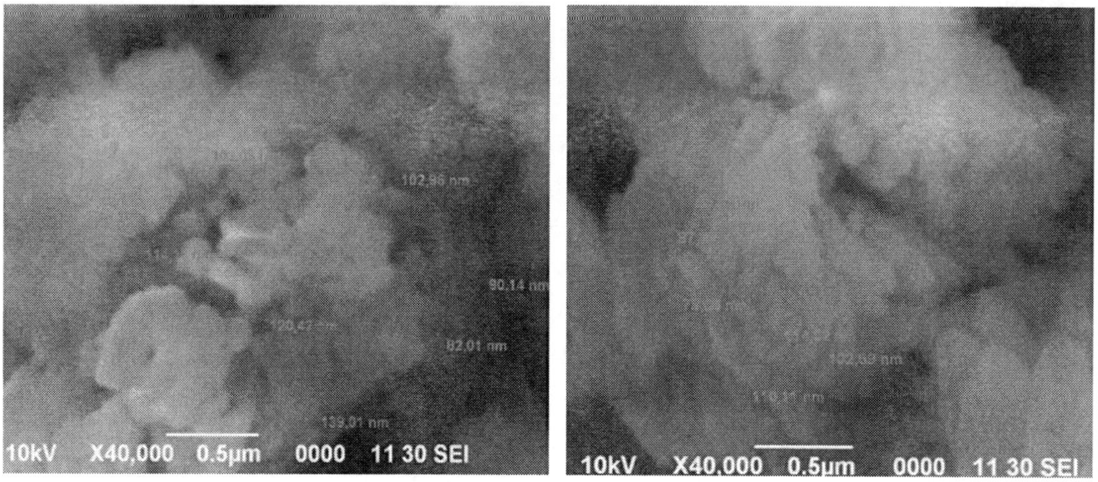

Figure 7. SEM Analysis result of calcium carbonate precipitate after ultrasonic process with magnification 40,000 times, (left) 10 minutes, and (right) 20 minutes.

From figure 7, it shows that the ultrasonic process for 20 minutes give better result than the 10 minute ultrasonic process. Based on SEM images, it can be seen that the size of the granules in the ultrasonic process of 10 minutes is between 82 nm to 140 nm with an average size of 112 nm. Whereas in the ultrsonic 20 minutes the grain size is between 79 and 114 with an average of 93 nm.

This is different from the PSA analysis which show the particle size of 57 nm to 193 nm and particles size below 100 nm has reached 93.7%. For the ultrasonic process in 20 minutes shows that there are 5 points of particles size below 100 nm and there are only 3 points above 100 nm. This is different from the PSA analysis which shows 99.9% nano-sized particles. The reason is that there is a coagulation process in drying process with ethylene glycol media. Another possibility is that there is also a little residual of ethylene glycol in the particle.This can be seen from the crystalline image which showsaccumulate large grains. Qualitatively, the measurement of particle size by SEM analysis shows the tendency of reduction forthe particle size of the samples when it is subjected to ultrasonic waves. Secondly there is a tendency to reduce grain size with the longer use of ultrasonic waves.

4. Conclusion

Qualitative measurement results with the Stoke method show that the optimum point of the ultrasonic process is at 30% amplitude, with a duration of 10 minutes and 20 minutes. The results of measurements with PSA nano at the optimum point show calcium carbonate precipitate in nano size as much as 93.7% for 10 minutes of ultrasonic process and 99.9% in 20 minutes of ultrasonic process. Analysis with SEM qualitatively showed that there is a tendency to reduce particle size by giving ultrasonic waves and in 20 minutes better than 10 minutes. During the drying process, the separation of ethylene glycol media with calcium carbonate precipitate results in a coagulation process which disrupts the observation process with SEM.

References

[1] Madhusree K and Dey T K 2012 *J. Thermochim. Acta*, **535** 58-65
[2] Terry A E and Liwei W 2008 *Powder Technol.* **182** 51-55
[3] Crocker M J 1997 Sonochemistry and Sonoluminescence *Enclycopedia of Accoustic* John Wiley and Sons, Inc) **1** chapter 26
[4] Pokrovsky O S 1998 *J. Cryst. Growth* **186** 233-39
[5] Oates J A H 1998 *Lime and Limestone Chem. Tech.* 130-53
[6] Lailiyah Q, Baqiya M and Darminto 2012 *J. Sci. Art ITS* **1** 69
[7] Yoshiyuki K, Kohei Y and Nobuki N 2010 *Ultrasononic Sonochem.* **17** 617-20
[8] Navichandra S, Ananda M, Hansora D P and Satyandra M 2015 *Nanosci. Nanoeng.* **3**
[9] Handayani M, Sulistiyono E, Firdiyono F and Fajariani E N 2018 *IOP. Conf. Ser.: Mater. Sci. Eng* **333** 01243
[10] Sulistiyono E, Handayani M, Firdiyono F and Fajariani E N 2018 *IOP. Conf. Ser.: Mater. Sci. Eng* **333** 012042

IOP Conf. Series: Materials Science and Engineering **578** (2019) 012039 doi:10.1088/1757-899X/578/1/012039

Antibacterial activity of copper oxide nanoparticles prepared by mechanical milling

M I Amal[1*], J T Wibowo[2], L Nuraini[1], G Senopati[1], M Y Hasbi[1], G Priyotomo[1]
[1] Research Centre for Metallurgy and Materials, Indonesian Institute of Sciences, Tangerang Selatan, Indonesia
[2] Research Centre for Oceanography, Indonesian Institute of Sciences, Jakarta, Indonesia

*E-mail: muha137@lipi.go.id

Abstract. In this study, Cu_2O (copper oxide) nanoparticles were synthesized by using mechanical milling and tested for antibacterial properties. The raw material for copper oxide powder (macroparticles) was refined in size using an instrument of shaker mill with a 1: 1 ball and powder ratio and a time variation of up to 16 hours. X-Ray Diffraction (XRD) diffraction data showed changes in crystallinity and size depending on the grinding time from the calculation of the crystallite size. The results of microstructure observation using an electronic microscope (Scanning Electron Microscope, SEM) also showed changes in particle size correlated with the length of the milling process. Antibacterial activity was carried out on the *Vibrio eltor*, *Bacillus subtilis*, *Staphylococcus aureus*, *Candida tropicalis*, and *Saccharomycetes cerevisiae* with the effect of particle size on antibacterial activity is discussed.

1. Introduction

The group of metal oxides compound is well-known to have wide range of applications in addition of their abundance in nature and easiness of synthesis. Copper oxide, in particular, has been studied for its potential as an antifouling agent, antibacterial active agent, pesticide, photocatalyst, solar cells, pigments, flame retardants, and chemotherapeutic agent [1−7]. The functional properties of materials are generally enhanced when it comes into nanoscale. Generally, two approaches have been introduced to prepare copper oxide nanoparticles, i.e. top-down and bottom-up that apply physical and chemical process or combination of both. Physical vapor deposition (PVD), chemical vapor deposition (CVD), sol-gel synthesis, sonochemistry, solvothermal, and mechanochemical synthesis are among the methods. One emerging technique that employs biological entity as natural biofactories for producing nanoparticles has also been introduced recently [8]. Aforementioned techniques to produce copper oxide nanoparticles have their own benefit and drawback, thus careful selection of appropriate technique to prepare copper oxide nanoparticles that meets designated functional properties is a must.

The application of copper oxide as antifouling agents has been introduced long time ago. However, there is a raised concern regarding the toxicity of copper oxide that can be harmful to other marine biota. Some countries such as Sweden and Netherlands have restricted the use of copper-containing antifouling paints, though in Indonesia it is still used widely. Some findings suggest that the risk of copper oxide harmful depends on the marine environment. One way to reduce the risk is by decreasing the amount of copper oxide in the marine coating while preserve its effectivity. It can be done by

IOP Conf. Series: Materials Science and Engineering **578** (2019) 012039 doi:10.1088/1757-899X/578/1/012039

refining the size into nanoscale, thus increasing the active surface area and functional antifouling activity. Other benefits of nanoparticle are high compatibility to various organic and inorganic polymer matrix as well as good stability at high temperature.

In this study, copper oxide nanoparticle was prepared by applying mechanical milling technique, a physical top-down approach. This method is known for its efficient cost and easiness to scale-up, while it can be detrimental to some materials, especially organic compounds. The structural properties, morphology and particle size of refined powders were analysed. Finally, antibacterial activity was assessed using agar diffusion method and its correlation of effectivity to particle size was studied.

2. Experimental

The type of mechanical milling applied in this experiment was a high energy ball mill designed by Research Center for Physics, Indonesian Institute of Sciences. The milling process was interrupted in 3 minutes interval and continued after 2 minutes stopping to avoid temperature elevation. The Ball-to-Powder Ratio (BPR) was 1:1 with the diameter of alumina ball was 1 mm. In order to investigate the effect of mechanical milling process to the particle size, the duration of milling was varied from 4 to 16 hours. The raw material was commercially available with initial particle size of 2–5 µm (Sigma-Aldrich).

The as-milled copper oxides were characterized for its structural properties by using X-Ray Diffraction (XRD; Miniflex Rigaku). The visual observation was assisted by using Scanning Electron Microscope (SEM) attached with Energy Dispersive X-ray Spectrophotometer (EDX) (JEOL JSM-6390A) to analyze its composition. Particle Size Analyzer (PSA; LS 200 Beckman Coulter) was used to determine the particle size distribution of as-milled copper oxide particles.

The qualitative antibacterial assay of the as-milled copper oxides powder was carried out using agar diffusion method against *Vibrio eltor*, *Bacillus subtilis*, and *Staphylococcus aureus* with methanol (Bratachem) as positive control. The detail of this qualitative antibacterial assay procedure is reported elsewhere [9].

The effect of particle size to antibacterial activity of refined copper oxides particle was investigated against *Candida tropicalis* and *Saccharomycetes cerevisiae* with *n*-hexane (Merck) as positive control. Firstly, extract of inorganic compound with a concentration of 2.4 mg/ ml using 20% n-hexane as a solvent was prepared. Culture of microbial targets, i.e. *Candida tropicalis* and *Saccharomycetes cerevisiae* was diluted 10 times. 100 µl of Yeast Mannitol Broth (YMB) (Sigma-Aldrich) media is inserted in the microplate of the first column to the third column. The first column was containing the media where 100 µl of 2.4 mg/ml concentration test was added, and then the mixture was homogenized by stirring using a micropipette. Following the homogenization, the dilution was inserted by the media mixture and the test extracts for the first column into the second column and then was homogenized. The dilution was processed until the third column, so that the desired final concentration is 1.2, 0.6, and 0.3 (µg/ml). Each column that has contained a mixture of media and extract with various concentrations was deposited with 10 µl of target microbial suspension (*C. tropicalis*) with a ten times dilution. Furthermore, the mixture was incubated using a shaker with room temperature for 24 hours. Yeast growth was observed by looking at the turbidity. The mixture was measured with a microplate reader at a wavelength of 595 nm. The same treatment was carried out for *S. cerevisiae* target microbes with similar ten times of dilution, and *n*-hexane solvent was used comparison (positive control).

3. Results and Discussions

Figure 1 shows the morphology of copper oxides after refinement process by high energy shaker mill up to 16 hours. All SEM images showed inhomogeneous distribution of particle from tens of micrometer to sub-micron of size. This finding is common for mechanomilling results, as the powders can be agglomerated in some extent. Yet, the pulverization resulted to powder refinement which increased as longer milling applied.

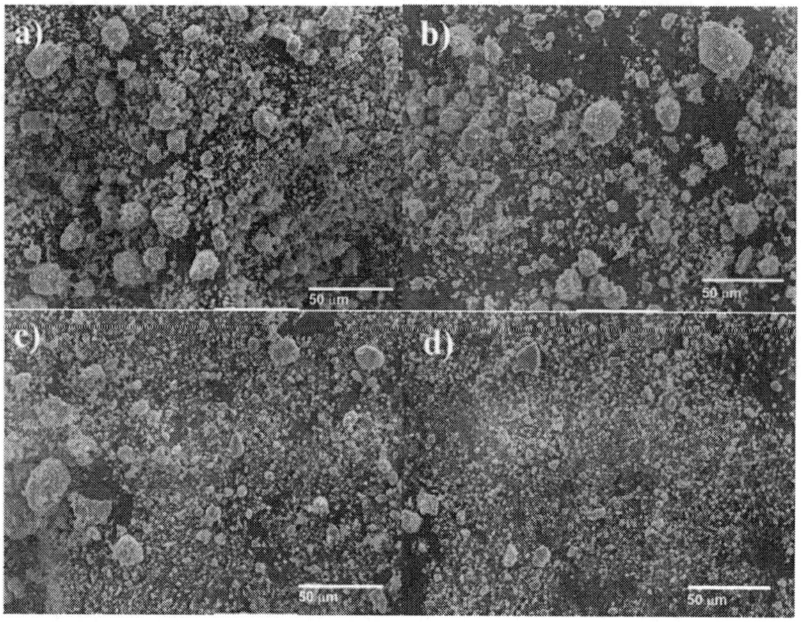

Figure 1. Scanning electron micrographs of as-milled copper oxides powder after 4, 8, 12, and 16 hours for a), b), c), and d) respectively.

The semi quantitative EDX analysis revealed the presence of copper and oxygen without any impurities as shown in Figure 2. The atomic percentage ratio indicated a Cu_2O phase except for samples of 12 and 16 hours milling, having a slight distortion due to the presence of CuO.

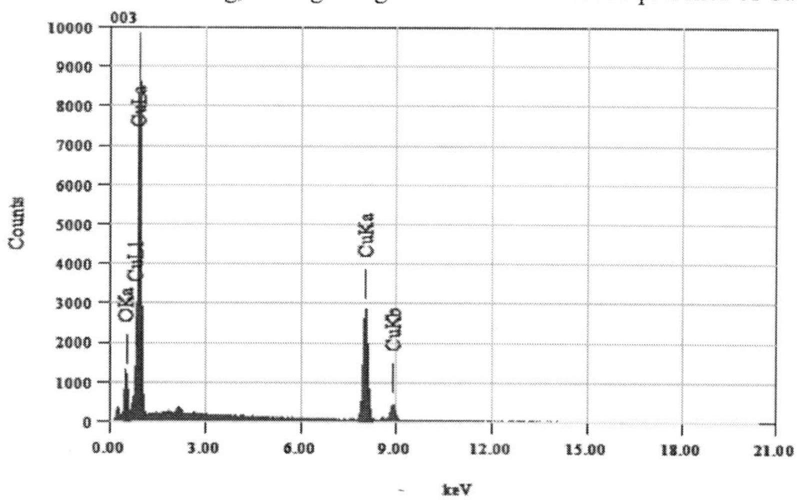

Figure 2. EDX spectra of as-milled copper oxides powder after 4 hours. The other samples also showed the similar result.

X-Ray diffractogram of samples in Figure 3 also confirmed the phase formation of these copper oxides compounds. The peaks at 29.5, 36.5, 42.3, 61.4, and 73.5° referred to Cu_2O phase in accordance to ICSD 00–005-0667 (copper oxide). The as-milled powders of 12 and 16 hours showed additional peak at 38.7° that fitted to copper dioxide peak of ICSD 00–045-0937. The evolution of some Cu_2O into CuO is possible due to extensive process of oxidation and more generated thermal energy after longer milling hours. XRD spectra also show that crystallinity of samples was decreased toward increasing of milling time. The change in crystallinity was affected by the refinement of powder as well as the deformation of crystal structure due to collision of ball milling.

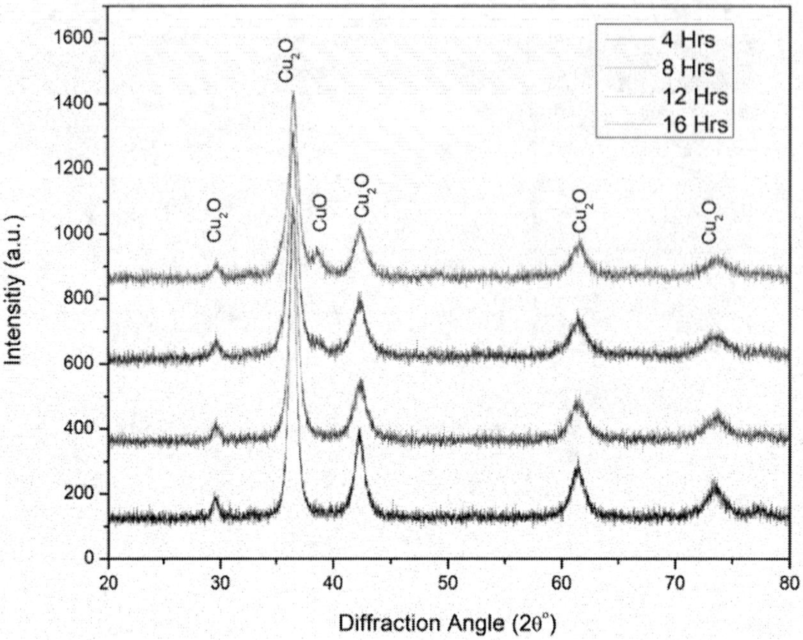

Figure 3. XRD spectra of Cu_2O powders with different milling hours. As-milled powders of 12 and 16 hours also contained CuO phase.

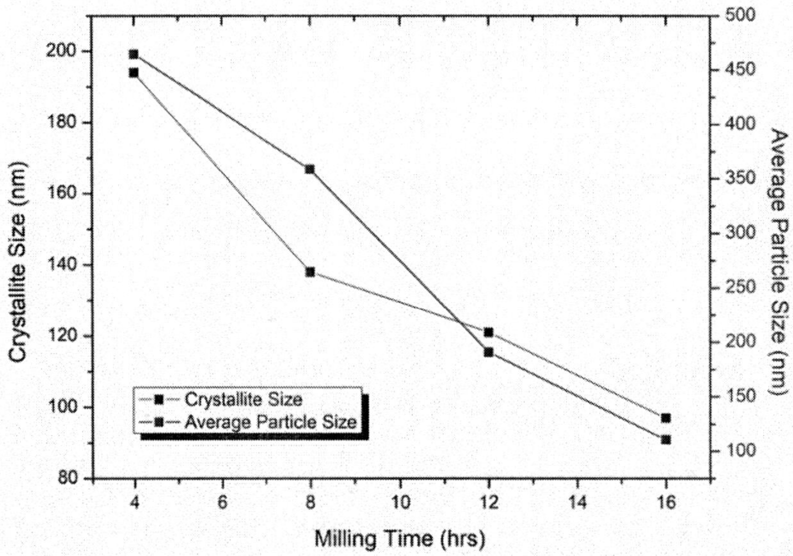

Figure 4. Crystallite size and average of particle size of samples.

Figure 4 shows the change of average particle size and crystallite size of samples after milling. The average particle size was obtained from particle size analysis result while crystallite size was calculated from XRD data by Scherrer equation. Both particle and crystallite size have straightforward correlation to the mechanical milling process, which longer milling time gave finer particle size. The discrepancy between average particle size and crystallite size became smaller as milling time increased. Eventually, a particle was equivalent to single crystallite after refinement process of 12 hours, while before that the particles were either agglomeration of small particles or single big particle. The average particle size was decreased from around one micrometre of original particle size before milling to around 110 nm after 16 hours milling. The physical refinement method of mechanical milling is difficult to reach particle size under 100 nm due to limitation of ball size. To obtain particle size below 100 nm, it is needed extremely small size *of* milling balls.

Table 1. Qualitative antibacterial assay using as-milled Cu_2O after 16 hours and methanol as control. (+) indicates activity, while (-) indicates inactivity.

Sample	*Vibrio eltor*	*Bacillus subtilis*	*Staphylococus aureus*
24 hours incubation with Cu_2O concentration of 1 mg/mL			
Cu_2O	+	+	+
Methanol	-	-	-
72 hours incubation with Cu_2O concentration of 1 mg/mL			
Cu_2O	+	+	+
Methanol	-	-	-

Generally, bacteria are distinguished based on the difference in the structure of bacterial wall. Gram-negative bacteria have an outer cell membrane in contrast of gram-positive bacteria. Yet, the cell wall of gram-positive bacteria is rich of peptidoglycan. To examine the antibacterial activity of as-milled copper oxides, gram-negative bacteria of *V. eltor* and gram-positive of *B. subtilis* and *S. aureus* were used. Table 1 shows the qualitative antibacterial activity assay using as-milled Cu_2O after 16 hrs and methanol as control. The inhibition zone was observed after 24 and 72 hours incubation. The antibacterial activity indicated that the as-milled have activity toward bacteria compared to methanol. The antibacterial mechanism depends on the bacteria type, i.e. gram-negative or gram-positive. The oxidation of microbial surface molecules due to the ionic interaction between positive charge of nanoparticles and negative charge of microbe, will lead to microbes death. While in the gram-positive bacteria, the antibacterial mechanism is related to cell wall biodestruction through degradative effects of DNA [10]. The qualitative antibacterial assay suggested the potency of copper oxide nanoparticle as antibacterial agent. Based on this, the quantitative analysis was carried out to study the antibacterial activity efficiency of copper oxides nanoparticles toward *C. tropicalis* and *S. cerevisiae*. *C. tropicalis* is a species of yeast and a common pathogen in neutropenic hosts. While, *S. cerevisiae* is the only yeast cell that has Berkeley bodies, a unique organelle that involves in particular secretory pathways.

Figure 5 shows the antibacterial activity examination against *C. tropicalis* and *S. cerevisiae*. It is obvious that particle size affected the surface area and the ability to continuously release copper ions in solution. Hence, the interaction of copper ions toward microbial cell wall would be more intense if the particle size is smaller. In addition, both Cu^+ and Cu^{2+} ions have antimicrobial effect that only differs in mechanism. Thus the presence of different copper oxides in samples 12 and 16 hours did not influence the antimicrobial activity efficiency. The results of antibacterial activity showed similar tendency in as-milled powders of 8, 12, and 16 hours. It hinted that to increase antibacterial activity, the particle size is needed to be below 100 nm.

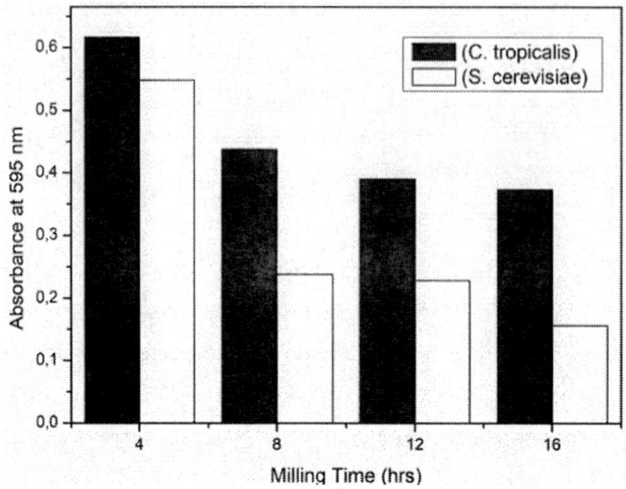

Figure 5. Antibacterial activity examination using as-milled Cu_2O with different milling hours against *C. tropicalis* and *S. cerevisie*.

4. Conclusion

The nanosized copper oxides were successfully prepared by cost-effective and simple method of mechanical milling. Through 16 hours of milling by using high-energy shaker mill, the particle refinement showed 90% reduction from initial particle size. Both the semi quantitative analysis of EDX and structural analysis of XRD depicted the presence of only copper oxides compound. Yet, in as-milled powders of 12 and 16 hours, the copper oxide phases were CuO and small presence of Cu_2O. The antibacterial activity was performed against *Vibrio eltor*, *Bacillus subtilis*, and *Staphyllococus aureus* with methanol as positive control for qualitative assay and *Candida tropicalis* and *Saccharomycetes cerevisiae* for quantitative assay. The result showed the effectivity of nanoparticle copper oxides as biocidal agent. Furthermore, it also indicated that smaller the particle size, the higher the efficiency of antibacterial activity.

References

[1] Anyaogu K C, Fedorov A V and Neckers D C 2008 *Langmuir* **24** 4340–4346
[2] Rahman A, Ismail A, Jumbianti D, Magdalena S and Sudrajat H 2009 *Indo. J. Chem.* **9(3)** 355–360
[3] Kent R D and Vikesland P J 2016 *Environ. Sci. Technol,* **50** 6772–6781
[4] Wellia D V, Kusumawaty Y, Diguna L J and Amal M I 2017 Introduction of Nanomaterials for Photocatalysis in Nanocomposites for Visible Light-induced Photocatalysis *Nanocomposites for Visible Light-induced Photocatalysis* ed M Khan, D Pradhan, Y Sohn (Cham: Springer) chapter 1 pp 1–17
[5] Pasquier A D , Duan Z, Pereira N and Lu Y 2010 *ECS Trans.* **28(4)** 179–190
[6] Baker I and Gibbs R S 1943 *Ind. Eng. Chem.Anal.* **15(8)** 505–508
[7] Chen M J, Lin Y C, Wang X N, Zhong L, Li Q L and Liu Z G 2015 *Ind. Eng. Chem.Res.* **54(51)** 12705–12713

[8] Khan M M, Ansari S A, Amal M I, Lee J and Cho M H 2013 *Nanoscale* **5(10)** 4427–4435
[9] Putra M Y, Wibowo J T, Murniasih T and Rasyid A 2016 *AIP Conf. Proceed.* **1744** 020039
[10] Azam A, Ahmed A S, Oves M, Khan M S and Memic A 2012 *Int. J. Nanomedicine* **7** 3527 3535

Study on optimum parameters of high content paraffin wax microemulsion

A Fitroturokhmah[1], R Kusumaningrum[2], G S T Sinaga[1], W B Widayatno[1], A S Wismogroho[1], A Noviyanto[2,3], M I Amal[4]*
[1] Research Centre for Physics, Indonesian Institute of Sciences, Tangerang Selatan, Indonesia
[2] Nano Center Indonesia, PUSPIPTEK, Tangerang Selatan, Indonesia
[3] Department of Mechanical Engineering, Mercu Buana University, Jl. Meruya Selatan, Kebun Jeruk, Jakarta Barat, Indonesia
[4] Research Centre for Metallurgy and Materials, Indonesian Institute of Sciences, Tangerang Selatan, Indonesia

E-mail: muha137@lipi.go.id

Abstract. As a by-product of crude oil processing, paraffin wax has high economic potential to be converted into microemulsion for various applications. Herein, paraffin wax microemulsion was prepared by emulsion inversion point (EIP) method in the presence of non-ionic surfactants. The focus of the study is to determine optimum preparation parameters for obtaining stable microemulsion. The initial results showed that paraffin wax microemulsion can be prepared when hydrophilic-lipophilic balance (HLB) value was 10 with a wax composition of 40 wt%. This HLB value is used as a reference for obtaining stable microemulsion by varying emulsification temperature and time. Emulsion stability was determined by observing phase change for 60 days, which then was quantized using particle size analyzer. The morphology of the microemulsion droplets was observed using optical microscope. The droplets showed a bright spot under polarized light owing to its solid phase in room temperature. Stable microemulsion with particle size average diameter of about 290.6 nm could be formed at emulsification temperature of 80 °C and emulsification time of 40 minutes.

1. Introduction

As by product of crude oil processing, paraffin wax is a saturated hydrocarbon with carbon numbers range C18-C90, soluble in crude oil, but it will begin to precipitate when the oil cools to the wax appearance temperature (WAT) [1]. Generally it is used as raw materials for making candles, but it has a low economic value. To increase its economic value, the paraffin wax is converted into another product such as emulsions. Emulsions are dispersions made up of two immiscible liquid phases, called the dispersed phase that dispersed in the continuous phase, which is mixed using mechanical shear and surfactant. It is found in many industries, such as food, pharmaceutical, petroleum, agrochemical, paint, coating, and so on [2, 3]. Paraffin wax emulsions are generally used for waterproof coatings as hydrophobic properties to various surfaces, such as ceramic materials, paper, cardboard, wooden furniture, etc.

Emulsions are not formed spontaneously and they also tend to be unstable after the formation. To stabilize paraffin wax emulsion, anionic, cationic or non-ionic surfactants are frequently used [4].

Nature of surfactant helps in deciding stability of microemulsion. A dipole and hydrogen bond interactions stabilizes non –ionic surfactant and electrical double layer stabilizes ionic surfactant (anionic and cationic). Surfactants are responsible for reducing the attractive forces that exists naturally in the form of surface tension. The choice of surfactant is based on the hydrophilic-lipophilic balance (HLB) value, help to develop desired emulsion. Surfactant with low HLB values are useful for forming water in oil (W/O) emulsions and high HLB values are used to form oil in water (O/W) emulsions [3]. HLB values between 10.25 and 10.75 were produce a stable wax emulsion solution because of its conductivity almost unchanged [5].

Previous study concluded that stable paraffin wax emulsions can be obtained by mixing all components in stages or adding aqueous surfactant to the oil, whereas the EIP method can produce emulsion with small droplet sizes and kinetically stable. EIP method as one of low-energy method was effectively used to prepare microemulsion. Low-energy emulsification method can utilize the chemical energy stored in the material and produce nano-emulsions almost spontaneously [2].

In the present study, we prepared paraffin wax microemulsions using the EIP method, and investigated the effect of surfactant concentrations, temperature of reaction, and stirring time on its stability and particles distribution. The properties of microemulsion were studied by optical microscope, particle size analyzer, and viscosity measurements.

2. Experimental

2.1. Materials

Paraffin wax was provided by PT. Kirana Mitra Abadi, Indonesia, with melting point is 57 °C and oil content about 20 wt%. The paraffin wax was analyzed by using Gas Chromatography (GC) to obtain the distribution of hydrocarbon molecular weight. The main components of this paraffin wax are normal alkanes with a range of carbon numbers 22-30 which are classified as long-chain hydrocarbons (Figure1). Non-ionic surfactants were sorbitan monooleate (Span 80) and polyethylene (20) sorbitan monooleate (Tween 80) were used without further purification. The water was deionized water.

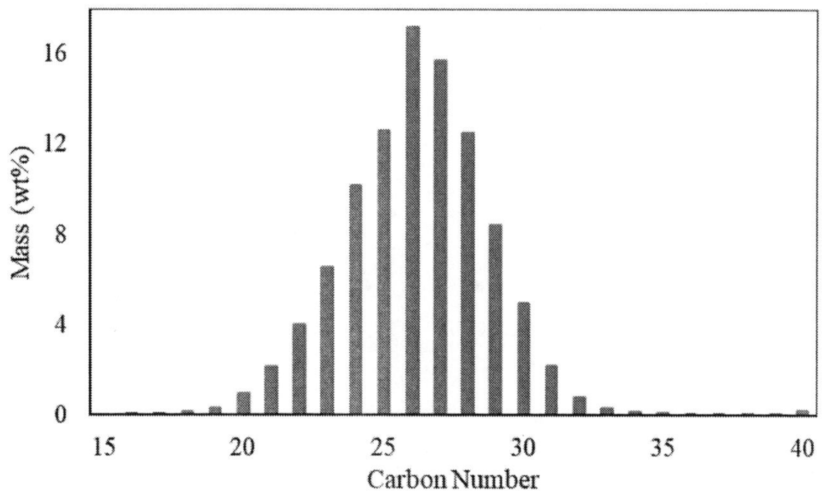

Figure 1. Carbon number distribution of paraffin wax.

2.2. Methods

2.2.1. Preparation of emulsion. Microemulsion was prepared initially from a mixture of paraffin wax and surfactant by adding water dropwise and stirring under various stirring time and heating temperature at 60-85 °C. The rate of water addition was kept constant at approximately 1.5 mL/min.

IOP Conf. Series: Materials Science and Engineering **578** (2019) 012040 doi:10.1088/1757-899X/578/1/012040

The amount of paraffin wax in all emulsions was 40 wt%, while the addition of surfactant was varied from 5 to 9 wt%.

2.2.2. Microemulsion Characterizations. Little emulsion sample was placed in the sample bottle and weighed then dried for 1 hour at 100 °C to observed its solid content. The dried sample was weighed again and the solid content shown by Eq. 1[2].

$$Solid\ content = \frac{sample\ mass\ after\ drying}{sample\ mass\ before\ drying} \times 100\% \tag{1}$$

Emulsion stability was determined by observing phase change for 60 days, which then the fraction of emulsion phase was calculated according to Li (2010) calculation. The morphology of paraffin wax microemulsion was observed using optical microscope and to see the particle distribution. The particle size of microemulsion was determined using particle size analyzer (PSA) of Nanoplus Particulate Systems version 5.1 / 3.00.

3. Results and Discussions

A Stable emulsion can be formed using the right combination of surfactants which has an HLB value close to that required by the oil phase. HLB value describes the affinity of surfactant to the water and oil phase, and their effect on emulsion stability depending on volumetric proportion of the phase [6]. Base on previous report, HLB value that suitable for making stable microemulsion wax is between 9.5 and 10.3, for which there was no creaming [2]. In this study, microemulsions were initially prepared with high paraffin wax content (i.e. 40 wt%) and the amount of surfactant mixture of Tween 80 and Span 80 were ranging from 5 to 9 wt% to obtain appropriate concentration surfactant in the making of microemulsion.

Paraffin wax microemulsion in this study has a total solid content about 50 wt%. The advantage of high solid content paraffin wax is to reduce the cost of transporting wax emulsion from production site to the place of use, concentrated formulation is prepared with the content of the dispersed phase (wax) being about 50-60 wt% of the emulsion [7].

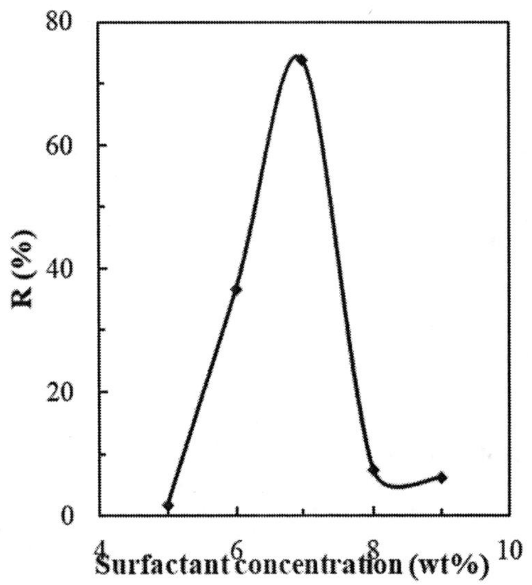

Figure 2. Effect of surfactant concentration to stability of paraffin wax microemulsion

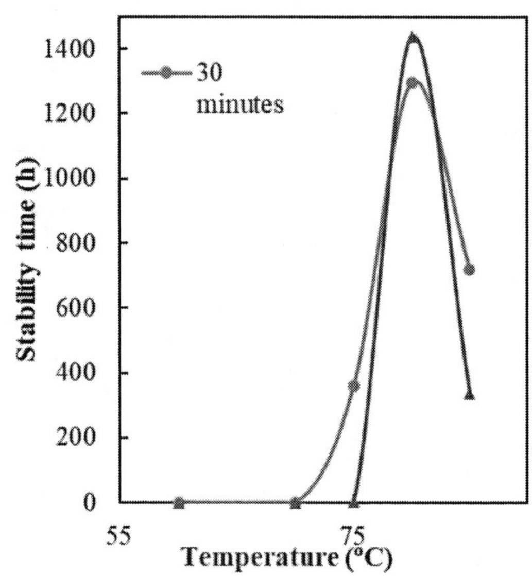

Figure 3. Effect of temperature and stiring time to stability of paraffin wax microemulsion

Figure 2 shows the stability of microemulsion as a function of surfactant concentration. Mixing Tween 80 and Span 80 as emulsifier at 5-10 wt% has lowest surface tension which plays an important role in mixing oil phase and water to form an emulsion. The highest microemulsion stability was obtained at a surfactant concentration of about 7 wt% and changed to be a solid emulsion phase at the concentration of more than 8 wt%. During the EIP process, the emulsion system changes from a W/O emulsion to an O/W emulsion and bicontinuous or lamellar structure can be formed during this process [2].

3.1 Effect of temperature

Paraffin wax used as the base material of microemulsion is solid at room temperature. During the emulsification process, paraffin wax has to be melted with increased of temperature for easier mixing with surfactant and water. Figure 3 shows microemulsion stability time as a function of temperature in different stirring time. Optimum parameter for achieving the highest microemulsion stability such as temperature and stirring time were obtained at 80 °C and 40 minutes. Making microemulsion at a temperature of 60 °C with both stirring times 30 and 40 minutes, microemulsion process can't be occurred perfectly, because the emulsification temperature is close to the paraffin wax melting point. At a temperature of 85 °C microemulsion becomes thicker and turns into a solid phase after several times. This result is different from previous studies which stated that the stability of emulsions can be increased by increasing the temperature ranging from 55 to 85 °C [2]. It is possibly due to differences in paraffin wax composition used. The increase of temperature makes non-ionic surfactants more hydrophobic (with appearance of cloud point) [8].

3.2 Particle distribution

Particle size distribution of paraffin wax microemulsion which prepared at temperature 80 °C with different stirring time shows that the effective emulsification time to produce a stable microemulsion is 40 minutes. When the emulsification time increased, particle size distribution is decrease. At the emulsification time of 30 minutes, particle size is distributed at about 10 μm (Figure 4) and when the emulsification time raised 40 minutes, particle size is distributed at about 5 μm (Figure 5). This may occur because longer emulsification time at certain temperature can provides enough energy to break up the particle size. That is break up of droplets caused by adsorption of surfactants and droplet collisions which may or may not cause coalescence to occur [5]. A stable microemulsion can be formed if the droplet breaking up process is more dominant than droplet re-coalescence [9].

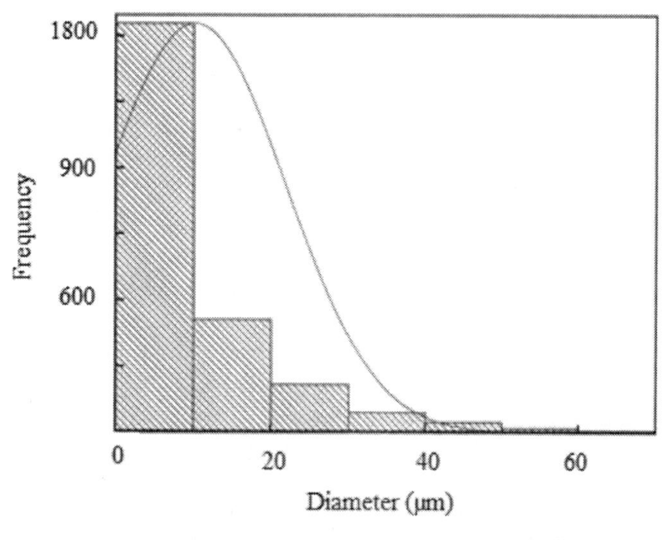

Figure 4. Particle size distribution of microemulsion at stirring time 30 minutes

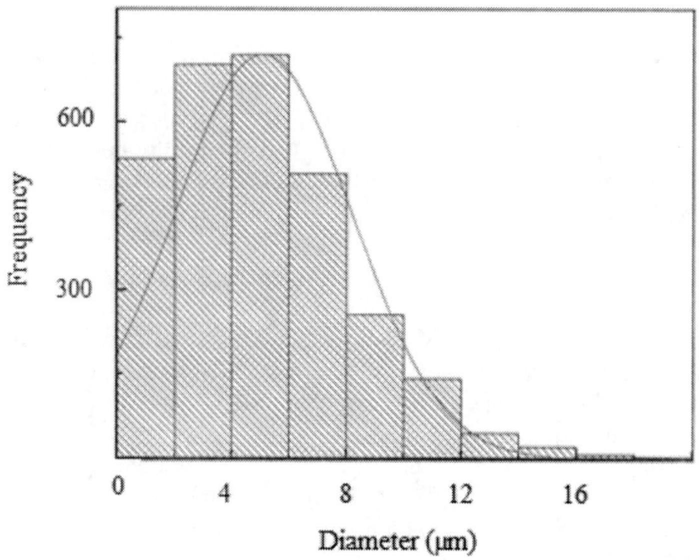

Figure 5. Particle size distribution of microemulsion at stirring time 40 minutes

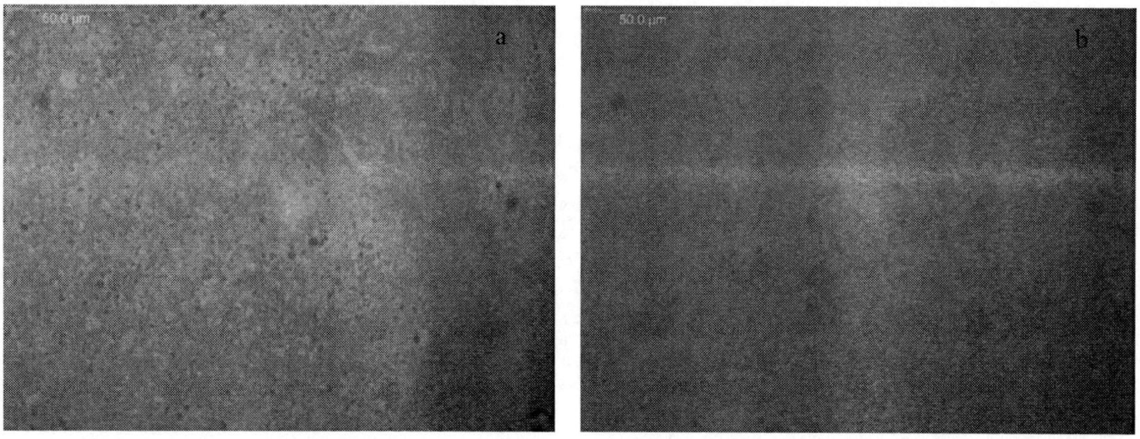

Figure 6. Microscopic image of paraffin wax microemulsion at 80 °C and emulsification time 30 minutes (a) and 40 minutes (b)

According to the particle size distribution data of PSA, paraffin wax microemulsions with good stability have average diameter about 290.6 nm. Particle size distribution data becomes supporting data to the emulsion stability test where emulsification time at 40 minutes produces the highest microemulsion stability. This can occur because of flocculation is prevented and allows the system to remain dispersed without separation. Examination of paraffin wax microemulsion under microscope are consistent with particle size distribution data, microemulsion at higher emulsification time has a smaller diameter appearance (Figure 6). Previous research has explained that coalescence can be prevented by small droplets, since these droplets cannot be deformed and therefore surface fluctuations can be prevented. Small droplet size causes a large decrease in the gravitational force and Brownian motion may be sufficient to overcome gravity [10].

4 Conclusion

The stability of microemulsion prepared using Tween 80 and Span 80 increases with increasing temperature and emulsification time to the optimum parameters, namely optimum stability at a temperature of 80 °C and emulsification time for 40 minutes, wherein the average particle size distribution 290.6 nm. Further studies to reduce particle size distribution to nano-emulsion will be carried out using nano bubble method.

Acknowledgment
This work was supported by Insinas Kemitraan contract number 058/P/RPL-LIPI/INSINAS-1/III/2018, "Development of Aluminium Metal Powder for Pigments and Hebel Additive". The authors are gratefully acknowledged the Laboratory of High Resistant Materials, Research Center for Physics-LIPI for providing research facilities.

References

[1] Ahn S, Wang K S, and Shuler P J 2005 *SPE Int. Symp. on Oilfild Chemistry* 2-4 February 2005 (Texas: Society of Petroleum Engineers Inc.) p 1–9
[2] Li C ,Mei Z, Liu Q, Wang J, Xu J and Sun D 2010 *Colloids Surfaces: A Physicochem. Eng. Asp.* **356** 71–77
[3] Kale S N and Deore S L 2017 *Sys Rev Pharm* **8** (1) 39–47
[4] Koroleva M Y, Gorbachevski O S and Yurtov E V 2017 *Theor. Found. Chem. Eng.* **51** (1) 125–32
[5] Shu L and Hongjun Y 2013 *Eur. Chem. Bull.* **2** (1) 15–7
[6] Morales C J, Riebel U, Guzman N M and Guerra M 2011 Lat. Am. Appl. Res. **41** 107
[7] Kwao-Boateng E, Kapustin V M and Makhin D Y 2017 *Chem. Process Eng. Res.* **54** 7–14
[8] Danghui D, Fengyan L and Tianbo Z 2012 *China Pet. Process. Petrochemical Technol.* **14** (1) 15–9
[9] Jadhav A J, Holkar C R, Karekar S E, Pinjari D V and Pandit A B 2015 *Ultrasonic Sonochemistry* **23** 201-7
[10] Tadros T, Izquierdo P, Esquena J and Solans C 2004 *Adv. Colloid Interface Sci.* **108–109** 303–18

Multi-walled Carbon Nanotubes Reinforced-Based Magnesium Metal Matrix Composites Prepared by Powder Metallurgy

M Handayani[1], M Ganta[2], N Darsono[1], E Sulistiyono[1], F P. Lestari[1], A Erryani[1], I N G P Astawa[1], Lusiana[1], S Azhari N[3].

[1] Research Centre for Metallurgy and Material, Indonesian Institute of Sciences, (LIPI), Tnagerang Selatan, Indonesia

[2] Faculty of Engineering and Technology, Sampoerna University, South Jakarta, Indonesia

[3] Department of Metallurgy Engineering, Faculty of Engineering, University of Jenderal Achmad Yani, Bandung, Indonesia

E-mail: murni.handayani@lipi.go.id

Abstract. Nanostructured carbons, for example, carbon nanotubes (CNTs) arise much attraction in the last decades because of its remarkable physical and mechanical properties. The unique physical, chemical and mechanical properties of CNTs make them attractive for biomedical applications. CNTs have been used to modify conventional biomedical materials to enhance mechanical properties, biocompatibility, or to impart other functionalities. One of the metal materials which attract high attention in the biomedical field for orthopaedics application is Magnesium. Recently, many researchers are studying the magnesium reinforced with a different kind of reinforcement. As a lightest metal structure material, magnesium matrix composites show many advantages over monolithic magnesium and magnesium alloys. In this study, Magnesium metal matrix was reinforced by multi-walled carbon nanotubes (MWCNTs) in order to observe the physical and the mechanical properties of the composites. The MWCNTs-Reinforced based magnesium MMC was performed using powder metallurgy process with MWCNTs loading are: 0, 0.1, 0.2, 0.3 and 0.5 %. The temperature of the sintering process was conducted at 600°C by flowing argon gas during the process with the holding time of 2 hours. The experiment result exhibited that the porosity of MMC was increased with the reinforcement by the increasing of the loading MWCNTs observed by metallography and SEM. The mechanical properties showed that the loading of MWCNTs 0.1 % increased the micro-hardness of MMC about 19% compared to pure magnesium, with the peaks at the loading of MWCNTs 0.3 % which increased the mechanical properties about 0.31% at highest hardness value of 46.5 HV.

1. Introduction

In recent year, the biomaterial implant attracts a lot of attention. Biomaterial implant expected to be an aid for the human body and its performance. Biomaterials can be made by metal, ceramics, polymers and composites. The fundamental of biomaterial is compatibility within human body, non-toxicity, and osteo-integration. Materials for implant was also can be classified by its degradation performance bio-inert material and bio degradable materials. Bio-inert means it can be lasting as long as needed and bio-degradable defines as the ability to degrade safely. The example of bio-inert is Metal such as Stainless Steel, Titanium Alloy, Cobalt-Chrome Alloy, and composites. Metal implants such as stainless steel, titanium are not degradable in human's body which require advanced action to remove implant after human's tissues healed. Bio-degradable its self can be classified as pure Metal, Metal - Alloy, and composites. Biodegradable metallic material took an important rule since traditional metal implant gives several notable problems. Some metals release toxic ions as a result of corrosion post-implantation, causing allergic reactions, local anaphylaxis and inflammation [1–4].

Magnesium (~ 1.74 g/cm^3) is the lightest materials among several structural materials such as Steel (~7,7 g/cm^3), Aluminium (~2.7 g/cm^3), and Titanium (4.506 g/cm^3). However, Magnesium has inferior physical and mechanical properties. Magnesium Tensile Strength is 160 MPa. Let's we compare with other such as general steel (~370 Mpa), Aluminium (~310 Mpa), Titanium Alloy Medical grade (~950 Mpa).

It is not so common using pure Magnesium as a biomaterial since the low mechanical properties and very fast break down on the human body. Several non-toxic metals have been used for the alloying metal such as Ca, Sr, Zn, Si, Sn and Zr. Other than binary alloy of Magnesium, Magnesium metal matrix composite can be used and the physical, mechanical, and corrosion behaviour of pure magnesium can be enhanced or manipulated using reinforcement. A lot of reinforcement material have been utilized from bioactive glass, zinc oxide, hydroxyl Apatite, Carbon [3,5,6]. Magnesium-based composite is known to have high specific mechanical properties [7], low density, improved thermal and dimensional stability and better damping properties. Commonly, magnesium MMC reinforced with ceramic powder due to ease fabrication and low-cost production [8]. The use SiC [9] and Ti [10] increased the tensile strength however the ductility are usually sacrificed relative to the monolithic Mg. Solution to this problem is to look for further advancement in the properties of Nano-size reinforcement. Multiwalled Carbon nanotubes (CNT) with its electrical, mechanical, optical and other superior properties appear to be one of important element to various this application. Nevertheless, conventional carbon formulation may not be good enough for several applications such biomaterial implant. Class of carbon allotropes provides unique properties, Multi-walled carbon nanotubes are similar to hollow graphite fiber, except that they have a much higher degree of structural perfection, which are having a diameter of 10-200 nm [11–13]. Magnesium based composites can be tailored by using powder metallurgy followed by sintering at vacuum or inert gas [8], casting technology [14], and friction stir processing [15].

Powder metallurgy process performed to produce MWCNTs-Reinforced based magnesium MMC. Powder metallurgy simply means a method of a manufacturing process which pieces was delivered by blending certain metal powder such as Mg and CNT, compaction and sintered below the melting point. Improper selection of material may result in failure of the manufacturer requirements [16]. Proper material selection results in a good performance with minimum cost [17].

The objectives of this study were to characterize the physical and mechanical properties of metal matrix composites with MWCNT as reinforcement. Powder metallurgy was used to fabricate, and the attempts were made to compare the effect of increasing MWCNTs ratio on physical and mechanical properties of sample obtained.

2. Experimental

The starting material was Magnesium and MWCNTs in the form of powder. Magnesium composite (Mg-MWCNTs) containing pure magnesium 100% (diameter 60 – 300 μm, Merck USA), with MWCNTs (diameter of 20 – 40 nm, 90wt% of purity, Cheap Tubes Inc. USA) ratio of 0.1%, 0.2%, 0.3% and 0.5% (w.r.t to Magnesium weight) was prepared using powder metallurgy. Magnesium powder was homogenously mixed with the MWCNTs by using Shaker mill and the respective weight percentages of MWCNTs w.r.t Mg was shown in Table 1. The ratio between the starting material and grind media (stainless steel ball) is 1: 2. Therefore, we prepared 1 gr of mixed powder and 2 gr of grind ball. The starting powder was mixed for 15 minutes.

Table 1. Designation of prepared sample

Sample (Code)	Magnesium [gr]	MWCNTs [mg]
Mg	1	0
Mg – 0.1% MWCNTs	1	1
Mg – 0.2% MWCNTs	1	2
Mg – 0.3% MWCNTs	1	3
Mg – 0.5% MWCNTs	1	5

The mixed samples were then compacted at a pressure of 350 Psi. Compacted powder separated from compaction machine for further treatment as shown in Figure 1(e). Compacted billets were then sintered at 600°C by using tube furnace with Argon gas as protective atmosphere for 2 hours as shown in Figure 1(f). The microstructure of the sintered sample was studied by using metallurgical optical (MT7100, Meiji Technology) and electron microscope (JEOL JSM 6390, Japan). Hardness was measured by using Microvicker (Mitutoyo, HM-200).

3. Results and Discussion

Morphologies of the sintered samples were observed on the images acquired by optical and electron microscopes, as well as chemical composition by using EDAX. Figure 1 compare the morphologies as a function of MWCNTs weight content. The observed samples are not etched because we would like to examine the homogeneity and the rules of the MWCNTs addition. From the image below we may have said that by Increasing the weight percent of MWCNTs increased the affinity of closed porosity formation. The number of porosity was decreased followed by the increasing of pore diameter, as the MWCNTs content increased.

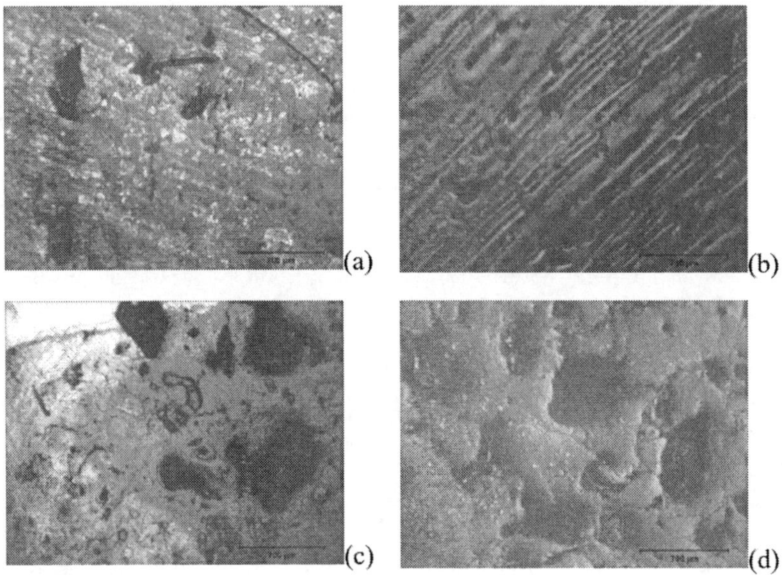

IOP Conf. Series: Materials Science and Engineering **578** (2019) 012041 doi:10.1088/1757-899X/578/1/012041

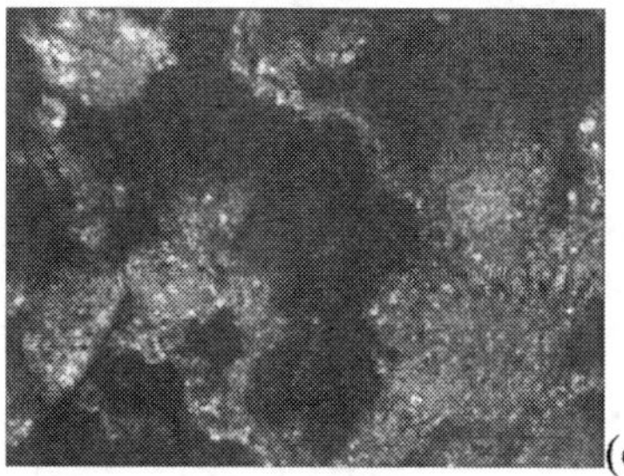
(e)

Figure 1 Surface observation by using optical microscope: (a) pure Mg (b) 0.1% (c) 0.2% (d) 50K 0.3% (e) 0.5% wt % of MWCNTs.

MWCNTs have large surface area. Therefore, it tends to interact each other MWCNTs to form MWCNTs aggregates rather than the interaction magnesium as matrix producing small clusters in the matrix. MWCNTs are stable materials as general carbon allotropy such as graphite, carbon active, and graphene. In order to make a good interface, MWCNTs needs to be decorated or functionalized with active functional group.

Figure 2 shows the electron micrograph of the sintered sample and the elemental spectra of the scanned area of the sample.

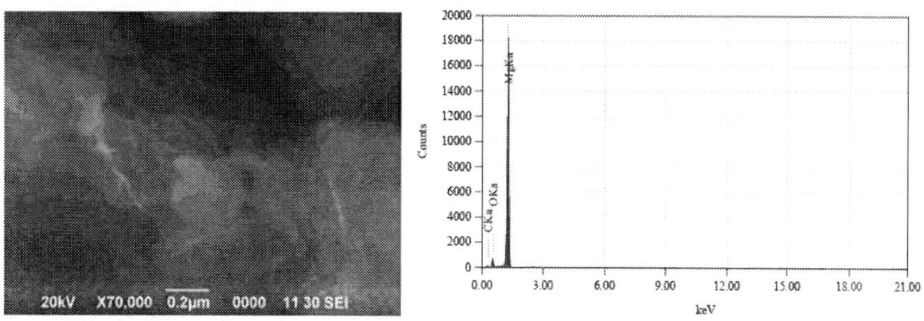

Figure 2. Electron micrograph of Sintered pellet showing clustering of MWCNTs

From the spectral element image derived from the EDAX, it shows that the MWCNTs are keep remain in the sintered sample. From the Figure 2, the fracture image of the sintered sample, MWCNTs exist in the form of agglomerates and generally placed in the grain boundary. As we mentioned before, lack of dispersion due to poor dispersibility of the MWCNTs in the magnesium matrix. That could be the reason as the amount of MWCNTs increased composite tends to be weaker, brittle and easily destroyed by an external force. Thus, it leads to poor mechanical properties of the nanocomposite.

Microhardness testing was used to determine the physical properties of the resulted sample. Microhardness of the measuring samples are shown as shown in Figure 3. Pure Mg has the lowest level of hardness compared to nanocomposite. It shows that hardness value increased as well as the weight ratio of MWCNTs. Magnesium matrix containing 0.3% of MWCNTs has the highest hardness value at the level of 46.5 HV compared to pure and other samples.

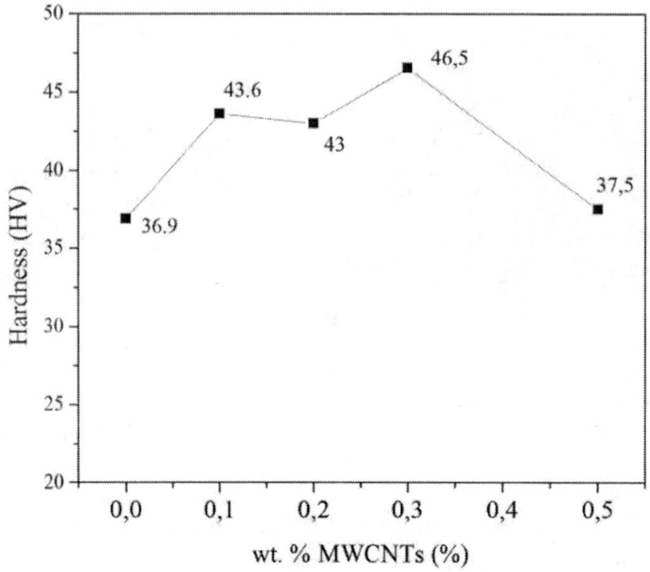

Figure 3. Microhardness of the sintered Mg Composites for 0-0.5 wt % MWCNTs addition

Compare to other typical experiment, we have a good agreement such as Meysam et al [5] for the work on the mechanical properties of magnesium matrix reinforced with hydroxyapatite and also N. Saikrishana et al. on their work on the investigation of the hardness of MWCNT/Mg Composites [18]. The mechanism of Strengthening on the Magnesium depends on the reinforced material. MWCNTs can be a grain refining agent during solidification, reducing interparticle spacing and small dimension and large surface area as mention in the work of [19].

4. Conclusion

Magnesium as degradable material is suitable for biodegradable implant application however pure magnesium provides low strength. The amount of MWCNTs in Mg matrix resulted the level of porosity increase. The pore size was also increased by increasing the MWCNTs content. Optimum strength of composite can be obtained at 0.3% of MWCNTs provides highest hardness value of 46.5 HV. Further treatment of MWCNTs in order to make sure the dispersion of MWCNTs in magnesium matric is good.

Acknowledgements

The authors gratefully acknowledge Research Center for Metallurgy and Materials (RCMM)-Indonesian Institute of Science (LIPI) for providing the facilities for this research, the authors wish to express gratitude also to all staffs of RCMM-LIPI for their kind continuous supports.

Reference

[1] Sezer N, Evis Z, Kayhan S M, Tahmasebifar A and Koç M 2018 *J. Magnesium Alloy.* **6** 23–43
[2] Chen Q and Thouas G A 2015 *Mater. Sci. Eng. R Reports* **87** 1–57
[3] Vojtech D, Cizova H, Volenec K 2006 Kovove Mater. **44** 211-223
[4] Zheng Y F, Gu X N and Witte F 2014 *Mater. Sci. Eng. R Reports* **77** 1–34
[5] Haghshenas M 2017 *J. Magnesium Alloy.* **5** 189–201
[6] White A A, Best S M and Kinloch I A 2007 *Int. J. Appl. Ceram. Technol.* **4** 1–13
[7] Zheng M, Wu K and Yao C 2018 *Mater. Sci. Eng A* **318** 50-56
[8] Goh C S, Wei J, Lee L C and Gupta M 2005 *Nanotechnology* **17** 7-12
[9] Gupta M, Lai M O and Saravanaranganathan D 2000 *J. Mater. Sci.* **35**(9) 2155–2165

[10] Hassan S F and Gupta M 2002 *J. Alloy. Compd* **345** 246–251
[11] Ajayan P M 1999 *Chem. Rev* **99** (7) 1781-1800
[12] Jong K P D E and Geus J W 2007 *Catal. Rev.* **42** 481-510
[13] Ono-ogasawara M and Myojo T 2013 *Adv. Powder Technol.* **24** 263–9
[14] Uozumi H, Kobayashi K, Nakanishi K, Matsunaga T, Shinozaki K, Sakamoto H, Tsukada T, Masuda C and Yoshida M 2008 *Mater. Sci. Eng. A* **495** 282-287
[15] Sunil B R, Reddy G P K, Patle H and Dumpala R 2016 *J. Magnes. Alloy.***4** 52-61
[16] Karande P and Chakraborty S 2012 *J. Mater.* **37** 317–24
[17] Shanian A and Savadogo O 2006 *J. Power Sources* **159** 1095–104
[18] Saikrishna N, Reddy G P K, Munirathinam B, Dumpala R, Jagannatham M and Sunil B R 2018 *J. Magnes. Alloy.***6** 83-89
[19] Suárez S, Ramos-Moore E, Lechthaler B and Mücklich F 2014 *Carbon* **70** 173-178

IOP Conf. Series: Materials Science and Engineering **578** (2019) 012042 doi:10.1088/1757-899X/578/1/012042

Preparation of Fe_2O_3-TiO_2/GO Composites as a Visible Light Oxidation Photocatalyst Material

U R K Sari[1], S Wahyuningsih[1], A H Ramelan[1]
[1] Inorganic Material Research Group, Faculty of Mathematics & Natural Sciences, Sebelas Maret University
Jl. Ir. Sutami 36A, Kentingan, Jebres, Surakarta 57126, Indonesia

E-mail: sayekti@mipa.uns.ac.id

Abstract. Fe_2O_3-TiO_2/Graphene Oxide (Fe_2O_3-TiO_2/GO) has been synthesized through a step by step synthesized. Preparation of Fe_2O_3-TiO_2 composites from Sukabumi iron sand with high iron content. Meanwhile, preparation of GO separately from graphite on strong acid condition H_2SO_4 and addition oxidator $KMnO_4$ treatment has been carried out. First composition analyzed Sukabumi iron sand shown content of iron about 50% and content of Ti about 10%. Fe_2O_3-TiO_2 composites prepared with addition TiO_2 powder thus obtained a ratio of Fe_2O_3-TiO_2 =1:1(w/w). Fe_2O_3-TiO_2 composites was roasted at 400 °C, 500 °C, and 600°C shown optimalhematite formed at 600 °C. Formation of Fe_2O_3-TiO_2/GO composites were at Fe_2O_3-TiO_2:GO= 1:1, 1:2, and 1:3 (w/w), respectively. Fe_2O_3-TiO_2/GO composites (A, B, C) shown excellent photocatalyst activities. Fe_2O_3-TiO_2:GO (1:3) composites have the optimum ability for photocatalytic activity of Rhodamine B degradation achieved about 93.96% for 40 minutes illumination.

1. Introduction

Advanced oxidation processes (AOPs) have emerged as promising alternative strategies for water treatment, especially for persistent and nonbiodegradable contaminants. Oxidation process of organic pollutant will initiate by electron and hole production from photochemistry reaction in the photocatalytic oxidation reaction. Fe_2O_3, which is an abundant, cost-effective and environmentally benign n-type semiconductor with band-gap of 2.2 eV, has been extensively studied because of its peculiar and fascinating physicochemical properties and wide potential applications in diverse fields such as water splitting [1] and photocatalysis [2], [3].TiO_2 in iron sand has benefits in the photocatalyst field and low cost natural materials. Raw materials for TiO_2 production are actually widely available in nature such as ilmenite ($FeTiO_3$) [4] and iron sand [5]. Magnetic particles in the form of Fe_2O_3 and Fe_3O_4 in iron sand content can be separated by magnetic separation method. The magnetic separation process will separate magnetic particles from iron sand approximately 99% [6]. Composite material of TiO_2 and Fe_2O_3 can absorb up to 550 nm wavelength adsorption edge, due to lower band gap energy of Fe_2O_3 that TiO_2. Fe_2O_3 are abundant, cost-effective and environmentally benign n-type semiconductor with band-gap of 2.2 eV, compare to TiO_2 with band gap of 3.2 eV. Composite of TiO_2-Fe_2O_3 has been extensively studied because of its peculiar and fascinating physicochemical properties and wide potential applications prospect.

Recently, Fe_2O_3 has been used to form hybrid materials with graphitic nanostructures for lithium ion batteries [7] and electrochemical detection of chromium [8]. However, there is still rarely found regarding the application of GO–Fe_2O_3 heterogeneous catalyst for photodegradation of organic contaminants to date. [9] claim that a reduced GO/TiO_2 composite will significantly increase the photovoltaic response and significantly prolongs the mean lifetime of electron–hole pairs, compared with pure TiO_2. [10] suggests chemically bonded P25-graphene composite as a high performance photo-catalyst for degradation of methylene blue.

Graphene has unique and superior properties compared to other materials. In the visible energy range (1.5-3 eV) graphene is transparant [11]. Graphene compared to GO, GO sheets have more oxygen function groups due to exfoliation.

There is an increase in photocatalytic activity factors that affect them, including structural properties (area width, grain size, morphology), optical properties (energy gap, electron recombination and holes) [12-15] and magnetic properties of the material [16]. GO can increase photocatalytic activity because it has a large surface area and acts as an electron transfer [17]. therefore, it is necessary to modify the properties of Fe_2O_3-TiO_2 material by adding GO [18] as a photocatalyst material for degradation photocatalyst of Rhodamine B.

In this work, we report an easy and scalable way of synthesizing Fe_2O_3-TiO_2/GO composite through a step by step synthesized. Preparation of Fe_2O_3-TiO_2 composites was conducted from Sukabumi iron sand with high iron content. While, the preparation of GO was carried out from graphite on strong acid condition H_2SO_4 and addition oxidator $KMnO_4$. Fe_2O_3-TiO_2 composites prepared with addition TiO_2 powder to gain a ratio of Fe_2O_3-TiO_2 =1:1(w/w).

2. Experimental

2.1 Material and Equipment

The materials used was iron sand from Sukabumi, sulphuric acid 95-98% (Merck), ammonia (NH_4OH) 25%, TiO_2 anatase powder, graphite, $KMnO_4$, H_2O_2, aquadest, and *Rhodamine B.* The equipment that require in this research was ultrasonic cleaner, platenary ball miller, oven, sentrifuge, thermometer. The materials characterization both before and after treatment were carried out by *Scanning Electron Microscopy* (Quanta 250), *Fourier Transform Infra Red* (Shimadzu PC 8201), *X-Ray Fluoresence* (Bruker S2 Ranger), *X-Ray Difraction* andUv-Vis Spectrometry Double Beam (Shimadzu PC 1601.

2.2 Preparation of Iron Sand

The Sukabumi iron sand was analyzed by XRF (*X-Ray Fluoresence)* and XRD (*X-Ray Difraction).* The magnetic and non magnetic particles was separated by magnet field. Then, the separated particles were milled by platenarry ball milling (300 rpm) for 5 hours. Milling process was conducted in ratio w/w (10:1) (130 gram : 13 gram).

2.3 Fe_2O_3-TiO_2

10 gram of iron sand was leached with an addition 100 ml H_2SO_4 98% (9 M) for 5 hours under reflux 90 ^oC. Then, the liquor of iron sand was hydrolyzed using NH_4OH (6M) until solid precipitate was apparanced. The precipitate was analyzed by XRF (*X-Ray Fluoresence).*

2.4 GO

2.5 gram of Graphite oxidized with addition $KMnO_4$ (8 gram) and H_2SO_4 98% (150 ml). Then, added 200 ml aquadest and 5 ml of H_2O_2 and stirred for 30 minutes. After that, 200 ml of aquadest were slowly added until pH 7. Solid powder was dried at 80 ^oC for 8 hours.

2.5 Composites Fe_2O_3-TiO_2/GO

Composites Fe_2O_3-TiO_2/GO synthesized with ratio 1:1, 1:2, and 1:3 (w/w). GO (0.5 gram, 1 gram and 1.5 gram) was mixed with 60 ml aquadest and 0.5 gram of crystal composites Fe_2O_3-TiO_2 of iron sand. Then, the solution mixture was sonicated for 1 hours and hydrothermal process at 150 ^oC for 3 hours.

2.6 Photocatalytic activity of Fe_2O_3–TiO_2/GO composites

0.01 gram composites Fe_2O_3/TiO_2-GO (1:1; 1:2; and 1:3) added 10 ml Rhodamine B 10 ppm. Then, mixed solution is irradiated 10, 20 and 40 minutes. Finally, degradation results characterized by UV-Vis.

IOP Conf. Series: Materials Science and Engineering **578** (2019) 012042 doi:10.1088/1757-899X/578/1/012042

3. Results and Discussion

Firstly, the pretreatment Sukabumi iron sand was analysed by XRF. The result in Figure 1 and Table 1 shown that Sukabumi iron sand contains Fe 50.48%, Ti 8.65%, and other elements under 5%.

Table 1. Analysis of pretreatment Sukabumi sand

Element	Concentration (%)
Fe	50.48
Ti	8.65
Si	3.07
Al	1.16
Ca	0.78

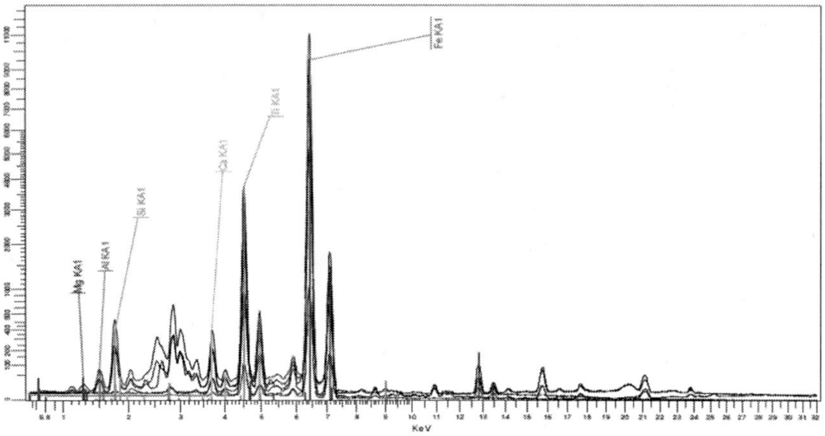

Figure 1. XRF analysis of pretreatment Sukabumi sand

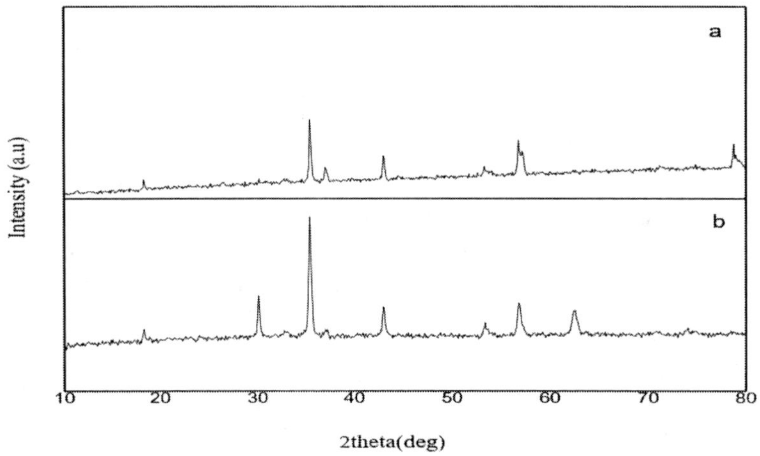

Figure 2. X-ray diffractogram a) Sukabumi iron sand b) Sukabumi iron sand (magnetic separation)

Sukabumi iron sand (magnetic separation) shown new peaks indicated that γ-Fe_2O_3 is starting to form and other peaks are indicated Fe_3O_4.

The process for extracted Fe_2O_3 and TiO_2 in iron sand used hydrometallurgy method, the dissolution process by using concentrated sulfuric acid solution to obtain Titanyl Sulphate ($TiOSO_4$) and Ferro Sulfate ($FeSO_4$).The materials was leached under reflux with H_2SO_4 9 M. The mechanism reaction that may occur during the leaching process is in equation 1 and equation 2.

$$FeTiO_{3(s)} + 2\ H_2SO_{4(aq)} \longrightarrow FeSO_{4(aq)} + TiOSO_{4(aq)} + 2H_2O_{(l)} \quad (1)$$

$$TiO_{2(s)} + Fe_2O_{3(s)} + 4\ H_2SO_{4(aq)} \longrightarrow TiOSO_{4(aq)} + Fe_2(SO_4)_{3(aq)} + 3H_2O_{(aq)} \quad (2)$$

The filtrate from the leaching was precipitation with the addition of NH_4OH 6 M to produced Fe_2O_3/TiO_2. This process could dissolve sediment Fe_2O_3-TiO_2obtained the ratio between the content of Fe_2O_3 and TiO_2 is 4.46:1

Table 2. XRF result precipitation

Compound	Concentration (%)	Element	Concentration (%)
Fe_2O_3	75.19	Fe	52.59
TiO_2	16.84	O	32.80
Al_2O_3	1.33	Ti	10.09
SiO_2	1.24	Al	0.71
P_2O_5	0.68	Si	0.58

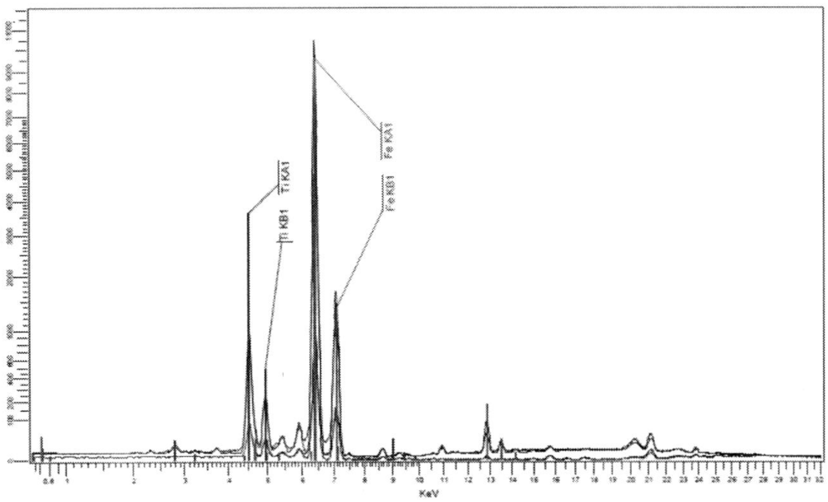

Figure 3. Analysis XRF of Fe_2O_3-TiO_2 sediment

Fe_2O_3/TiO_2 composites that have dried up and after adding TiO_2 anatase are carried out calcination variations of 400 ^{0}C, 500 ^{0}C, and 600 ^{0}C to obtain Fe_2O_3-TiO_2 composites crystals. From the results of X-Ray Diffraction (XRD) analysis shows specific differences in the diffractogram pattern of calcination results shown in Figure 4.

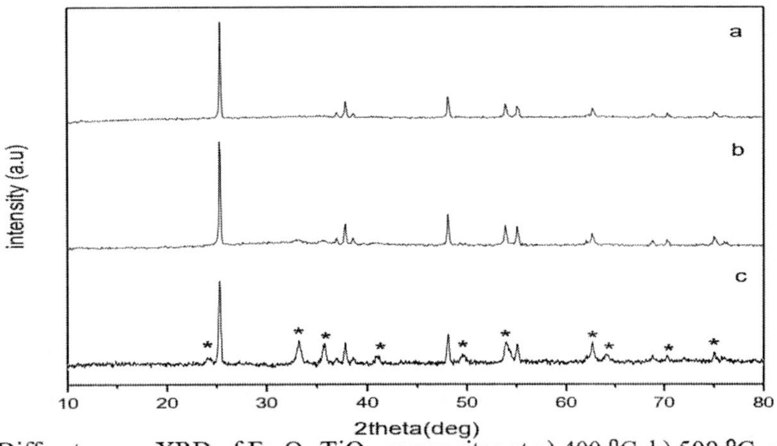

Figure 4. Diffractogram XRD of Fe_2O_3-TiO_2 composites at a) 400 ^{0}C, b) 500 ^{0}C, and c) 600 ^{0}C (*= α-Fe_2O_3)

Calcination 400 ^{0}C and 500 ^{0}C still form the same diffractogram that is the peaks that indicate TiO_2 anatase. XRD diffractogram at 600 ^{0}C calcination shows α-Fe_2O_3(ICSD No. 15840) pada

2θ= 24.2156°; 33.2436°; 35.7482°; 41.0043°; 49.6522°; 54.0104°; 62.7409°; 64.1241°; 70.3294°; 75.0640°, TiO_2 anatase (JCPDS No. 78-2486) at 2θ= 25.3113; 48.1137°; 55.1844°; 68.8008°, γ-Fe_2O_3 (ICSD No. 172905), Fe_3O_4 (ICSD No. 49549) at 2θ= 35.7482°; 70.3294°, Cr_2O_3 (ICSD No. 250078) at 2θ= 33.2436°, MnO (ICSD No. 657304) at 2θ= 38.6735° and 68.8008°.

Table 3. XRF results in hematite formation at 600 °C calcination temperature

Element	Concentration (%)
O	35.07
Fe	32.17
Ti	29.33
La	1.44
S	0.44

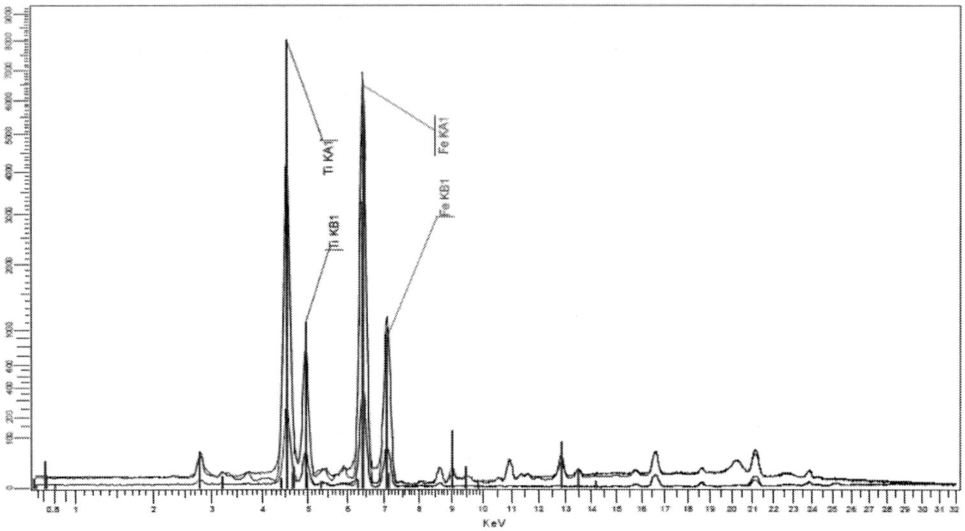

Figure 5. XRF of hematite calcination temperature at 600 °C

The results of the analysis of iron sand in the formation of Fe_2O_3-TiO_2 composites crystal shown a decreased in Fe and an increase in TiO_2 due to the addition of TiO_2 anatase powder, so it is obtained ratio of Fe_2O_3-TiO_2 is 1:1 (w/w)

Graphene Oxide was characterized by XRD and FTIR shown in Figure 6, 7, and 8. In Figure 6 it shown that the peak was intensed in graphite (ICSD No. 76767) 2θ= 26.42°, become lost and replaced with a peak on GO 2θ = 10.77° and 42.47°. Graphene Oxide has increased $d_{spacing}$(Å)= 8.22 Å from $d_{spacing}$ grafit 3.37 Å.

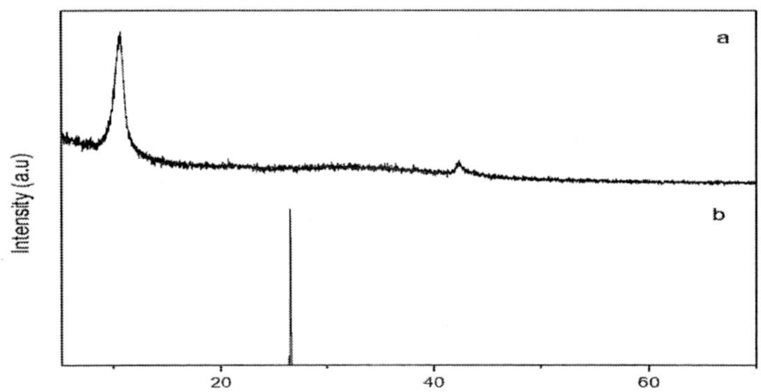

Figure 6. Diffactogram of a) Graphene Oxide b) Graphite

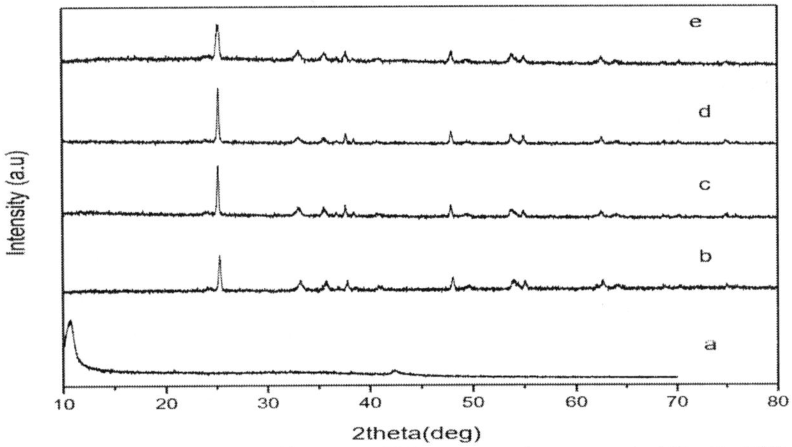

Figure 7. Diffractogram of a) GO b) Fe_2O_3-TiO_2 composites crystal c) Fe_2O_3-TiO_2 :GO (1:1) d) Fe_2O_3-TiO_2 :GO (1:2) and e) Fe_2O_3-TiO_2 :GO (1:3)

The peak intensity increases with the addition of GO in Fe_2O_3/TiO_2 composites crystal. However, the peak intensity on Fe_2O_3-TiO_2/GO (1:3) composites decreases and the width of the peak widen because the GO number increases. Thus, the composites formation is optimal in Fe_2O_3-TiO_2/GO (1:2). This was shown the interaction of GO in composites formation. Spectra (C) is formed uptake of –OH, C=C, C-OH and Fe-O which has been successfully formed on Fe_2O_3 - TiO_2/GO composites.

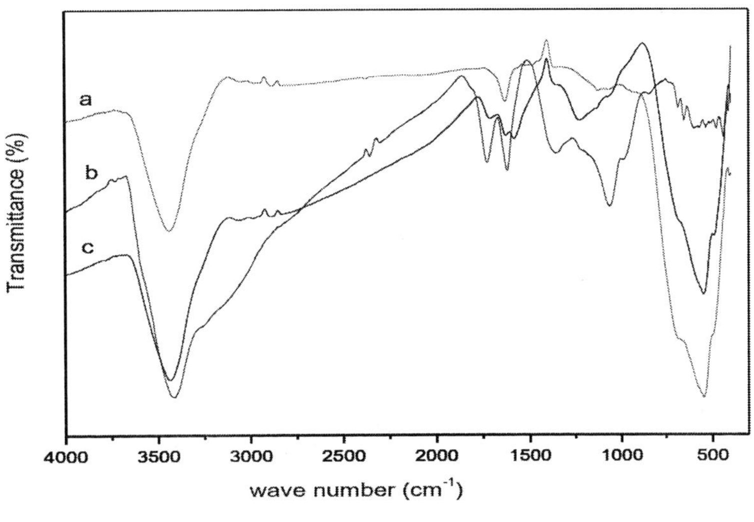

Figure 8. FTIR of a) Fe_2O_3-TiO_2 composites crystal b) GO c) Fe_2O_3-TiO_2 :GO (1:1; 1:2; and 1:3)

Decrease in concentration (Figure 9) shown the optimization of photocatalytic activity of Fe_2O_3-TiO_2/GO composites. The degradation of Fe_2O_3-TiO_2/GO composites increased significantly with increasing GO content in composites from 1:1 variations, 1:2 variations to 1:3 variations. Therefore, the composite Fe_2O_3-TiO_2/GO (1:3) shown the best photocatalytic activity.

IOP Conf. Series: Materials Science and Engineering **578** (2019) 012042 doi:10.1088/1757-899X/578/1/012042

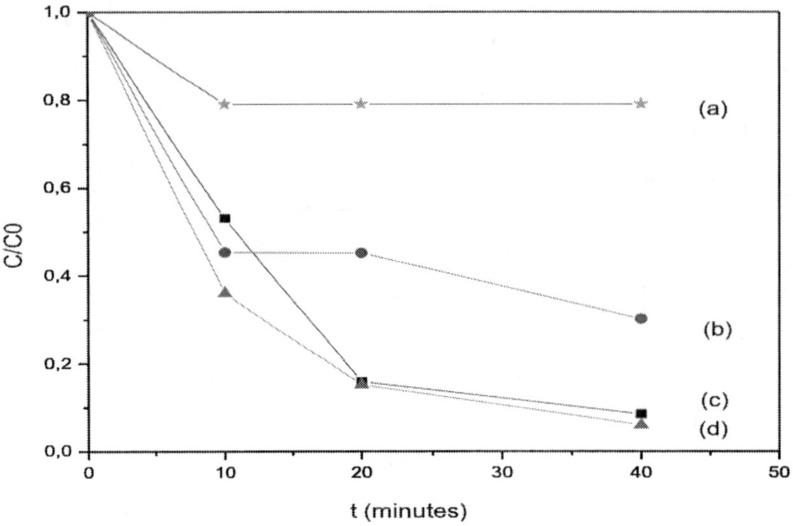

Figure 9. Results of photocatalytic degradation of Rhodamine B of (a) Fe_2O_3/TiO_2 (b) Fe_2O_3-TiO_2/GO (1:2) (c) Fe_2O_3-TiO_2/GO (1:1) and (d) Fe_2O_3-TiO_2/GO (1:3)

4. Conclusion

In conclution, Preparation of Fe_2O_3-TiO_2 composites through magnetic separation, hydrometallurgy and by addition of TiO_2 anatase powder to obtained a ratio of Fe_2O_3-TiO_2= 1:1 (w/w). The amount of Graphene Oxide added to the of on Fe_2O_3-TiO_2/GO composites crystal gives a large change in photocatalytic activity.

References

[1] Gondal M A, Hameed A, Yamani Z H, Suwaiyan A 2004 *Appl Catal A Gen* **268** (1–2) 159–67.
[2] Akhavan O, Azimirad R 2009 *Appl Catal A Gen* **369**(1–2) 77–82.
[3] Akhavan O 2010 *Appl Surf Sci* **257**(5) 1724–8.
[4] Wahyuningsih S, A H Ramelan, H P Pranata, Q A Hanif, Y A Ismoyo, and K F Ichsan 2016 *J. Phys. Conf. Ser* **776** 012026.
[5] Turney T W 1995 *Titanium Extraction*. United States Patent .
[6] Baba A A, Adekola F A, Arodola O A, Ibrahim L, Bale R B, Ghosh M K and Sheik A R 2012 *Journal Metallurgical Material Engineering* **18** 67-78.
[7] Zhu X J, Zhu Y W, Murali S, Stoller M D, Ruoff R S 2011 *ACS Nano* **5** (4) 3333–8.
[8] Prakash A, Chandra S, Bahadur D 2012 *Carbon* **50** (11) 4209–19.
[9] Wang P, Zhai Y, Wang D, Dong S 2011 *Nanoscale* **3** 1640–1645.
[10] Zhang H, Lv X, Li Y, Wang Y, Li J 2011 *ACS Nano* **4** 380–386.
[11] Novoselov K S, Geim A K, Morozov S V, Jiang D and Katsnelson M I V 2004 *Science* **306** 666.
[12] H J Lin, T S Yang, C S Hsi, M C Wang and K C Lee 2014 *Int. J. Therm. Sci.* **40** 10633-10640.
[13] Y Xu, M Zhang, Mi Zhang, J Lv, X Jiang, G He, X Song and Z Sun 2013 *Appl Surf Sci.* **315** 299-306.
[14] R Saravanan, V K Gupta, V Narayanan and A Stephen 2013 *J. Mol. Liq.* **181** 133-141.
[15] B Li and Y Wang 2010 *Superlattice and Microstructures* **47** 615-623.
[16] C Li, R Ounesi, Y Cai, Y Zhu, M Ma and L Zhu 2014 *Appl. Catal. B: Environ.* **156** 314-322.
[17] Jo and N 2015 *Royal Society of Chemistry*. pp 16024-16035.
[18] Zhang J J, Qi P, Li J, Zheng X C, Liu P, Guan X X and Zheng G P 2018 *J. Ind. Eng. Chem.* **61** 407-415.

Effect of TiO_2 Addition on Physical and Spectroscopic Properties of TeO_2-ZnO-Na_2O Glasses

Ahmad Marzuki*, Devara Ega Fausta
Department of Physics, Sebelas Maret University, Surakarta 57126, Indonesia

*E-mail: amarzuki@mipa.uns.ac.id

Abstract. This paper presents the effect of TiO_2 addition on the physical and optical properties of TeO_2-ZnO-Na_2O glasses. For this study, glasses with the compositions of 63.16TeO_2 - (31.57-x) ZnO – 5.26 Na_2O – xTiO_2 (where x = 1.06, 2.11, 3.16, and 4.22 mol%) were prepared with melt quenching technique. Glasses were characterized in order to obtained glass density, refractive index, and absorption spectra ranging from UV to IR regions. From these basic data, we have evaluated the values of molar volume, oxygen packed density, polaron radius, inter-ionic distance, field strength, and optical band gap. It was observed that incorporating TiO_2 into the glasses to substitute ZnO have resulted in decreasing the values of density, inter-ionic distance, and field strength; and yet increasing molar volume, oxygen packed density and polaron radius. The structural units building the glass networks have also been evaluated.

Keywords: Tellurite glasses, optical bandgap, refractive index, oxygen packed density, polaron.

1. Introduction

First tellurite glass was synthesized by Burger H., et al. at 1985 [1]. Since then, tellurite glasses have received more attention and researched for various applications by employing their properties such as: low meting point, good thermal stability, high refractive index, low phonon energy, wide transmission window, high rare-earth ions solubility and theoretically possess low minimum loss [2, 3, 4, 5, 6]. Several important applications employing these properties are intended to tailor optical devices such as optical amplifiers, fiber laser, photonic nanowires and sensors [7, 8].

TiO_2 has many unique optical and electrical characteristics and therefore has been researched for many different applications such as: sunscreen [9], self-cleaning coating [10] and solar cell [11]. It is well known that TiO_2 is a nucleating agent [12, 13]. Addition of TiO_2 into a glass can therefore enhance crystallization [14]. TiO_2 addition into a glass has also been studied because of its role in modifying glass refractive index [15]. Glass refractive index generally increase with increasing TiO_2 content. In addition of that, it was also shown that addition of TiO_2 can increase glass chemical durability and nonlinear refractive index which is useful for nonlinear optical devices.

This present work reports the effect of TiO_2 addition on the physical and spectroscopic properties of TeO_2-ZnO-Na_2O glasses. Physical studies are performed by evaluating glass density, molar volume, oxygen packing density (OPD), ionic concentration, polaron radius, inter ionic distances, and field strength. While the spectroscopic studies will include investigation glass absorption in UV-Vis and IR regions.

2. Experiment

Tellurite glasses with composition (in mol%): 63.16 TeO_2– (31.57-x) ZnO – 5.26 Na_2O – xTiO_2 (where x = 1.06, 2.11, 3.16 and 4.22) were synthesized using melt quenching technique. For this purpose, appropriate amount of each chemical was weighed and then mixed thoroughly in an agate mortar. The prepared mixture was then placed in platinum crucible and then melted in an electrical furnace at temperature of 900°C for about 35 minutes. Casting was carried out by pouring the molten glass into a pre-heated brass mould. In order to release any thermal and mechanical stress that might occur during fabrication process, the obtained glass was then annealed at 265°C for 6 hours and subsequently cooled to room temperature at cooling rate of 10^0C/minute. The obtained glasses were then cut into the size of 1.5 cm x 1.5 cm and then optically polished to meet the requirement for further optical characterization.

Density of all glasses were measured by applying Archimedes principles that is by immersing glass into distilled water contained in a pycnometer at room temperature. Using the obtained density, several other physical quantities are calculated. Among them are:

a. Molar volume (V_m)

$$V_m = \frac{\sum_i x_i M_i}{\rho_g} \qquad (1)$$

b. Oxygen Packed Density (OPD)

$$OPD = \frac{1000.N_O \rho_g}{M_g} \qquad (2)$$

c. Polaron radius (r_p)

$$r_p = \frac{1}{2}\left(\frac{\pi}{6N}\right)^{1/3} \qquad (3)$$

d. Field strength (F)

$$F = \frac{z}{{r_p}^2} \qquad (4)$$

In the above equation ρ_g is glass density, x_i and M_i are molar fraction and molecular weight of the ith component of glass, N_O is number of oxygen atom for each composition, N is number of site density and z is atomic number of Ti.

Refractive indices of all samples were measured using Brewster angle method at room temperature. Optical absorption spectra at UV-Vis-NIR range were measured at room temperature using UV-Vis Spectrometer Perkin Elmer Lamda 25 in the wavelength 200-1000 nm while thorecorded using Shimadzu Prectige IR 21 Spectroscopy by KBr disc technique.

3. Results and Discussion

3.1. Physical analysis

Figure 1 shows the change in density and molar volume as function of TiO_2 content and **Table 1** tabulates the value of density, molar volume, OPD, polaron radius, ionic concentration, inter-atomic distance and field strength. Decreasing glass density as TiO_2 content increase can be linked to atomic mass of Ti (47.87 sma) compared to that of Zn (65.39 sma) as well as atomic radius of Ti atom (2.0 A^0) as compared to that of Zn atom (1.53A^0). This substitution can therefore make the glass lighter while at the same time its volume is decreased. In other words the glass molar volume increases with the increases of TiO_2 content. This increase in molar volume can be linked to the increase in volume which also means an increase in the amount Non-Bridging Oxygen (NBO) inside the glass network [16, 17, 18]. It can also be seen from **Table 1** that the increase of TiO_2 content leads to increasing OPD, ionic concentration, and polaron radius, while decreasing the value of interatomic distance and field strength.

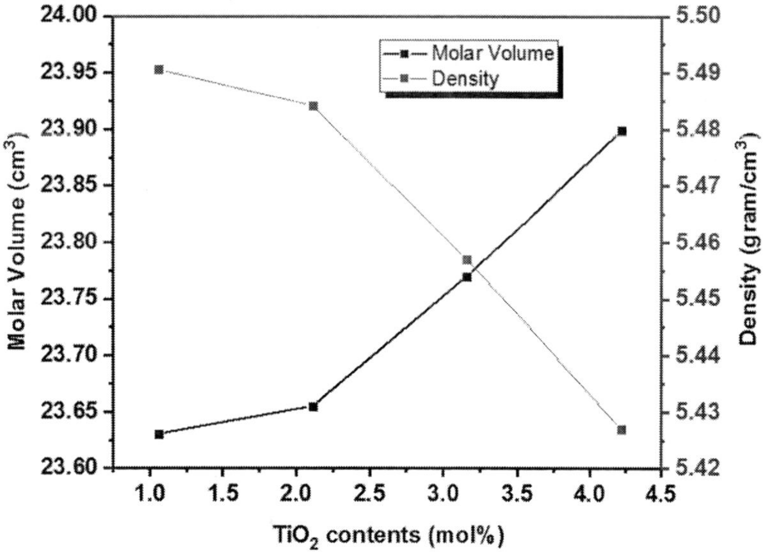

Figure 1. Variation of density and molar volume of all samples as function of TiO_2 content

Table 1. Density, Molar Volume, OPD, Ionic Concentration, Polaron radius, Inter Ionic Distances, Filed Strengh of all Glasses

Composition Code	Density (g/cm^3)	Molar Volume (cm^3)	OPD (mol/L)	Ionic Concentration ($x10^{22}$ $ions/cm^3$)	Polaron Radius (Å)	Inter Ionic Distance (Å)	Field Strength (10^{-15} cm^{-2})
TZNT1	5.4906	23.63	69.49	2.7017	2.356	0.0123	1.438
TZNT2	5.4842	23.65	69.86	5.3724	4.685	0.0062	0.368
TZNT3	5.4570	23.77	69.97	8.0069	6.983	0.0041	0.164
TZNT4	5.4269	23.89	70.03	10.635	9.276	0.0031	0.092

3.2. Optical Band Gap Energy and Refractive Index

Optical band gap energy is a measure of energy range separating valence band and conduction band energies and its value is sensitive to structural change. Optical band gap energy can be calculated using equation:

$$\alpha h\nu = B\frac{(h\nu - E_g)^n}{h\nu} \quad (5)$$

where α is absorption coefficient, $h\upsilon$ is photon energy, B is constant, E_g is the optical band gap energy and n is a constant (n equals 2 for indirect band gap and equals ½ for direct optical band gap). In the present work, the absorption edge indicates that it is indirect optical band gap [19] and therefore n is taken equals ½.

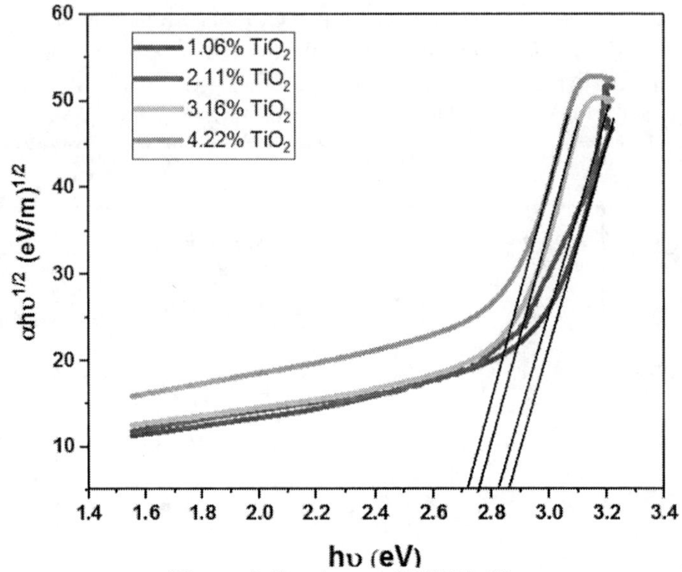

Figure 2. Band gap of TZNT Glass

Figure 2 shows a plot of $(\alpha h\nu)^{1/2}$ agains hv for all samples. The value of E_g was obtained by extrapolating the linear part of the curve to intercept the *hυ*-axis. The optical band gap obtained by extrapolating the linier region from the curve which $(\alpha h\nu)^{1/2} = 0$. The values of E_g along with refractive index measured using Brewster angle's method are presented in **Table 2**. While the Optical band gap energy decreases with the increase of the TiO_2 content, the glass refractive index increase with the increase of TiO_2 content. This result may be resulted from reconstruction of tellurite network cause by incorporating of Ti^{4+} ions that increases the number of NBO and thus increases polarizabity which in turn raise refractive index [19].

Table 2. *Eg* and *n* for all Glasses

Composition Code	E_g (eV)	n
TZNT1	2.86	1.911
TZNT2	2.82	1.961
TZNT3	2.75	1.976
TZNT4	2.71	2.005

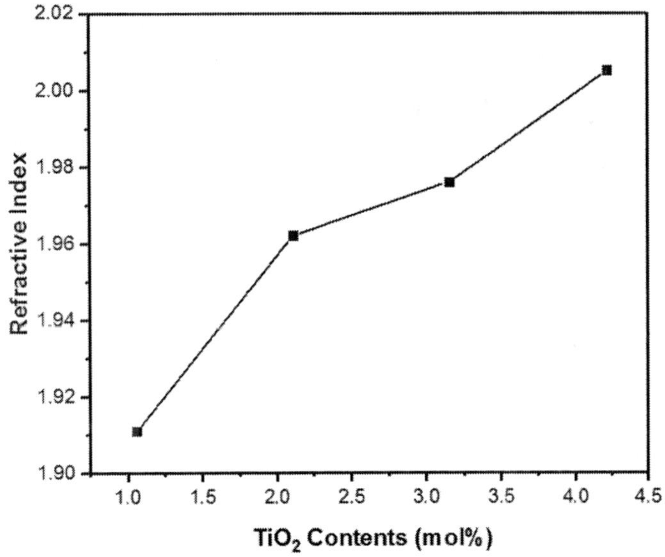

Figure 3. Refractive index of all Glasses as function of TiO_2 content

3.3. Absorption Spectra

Figure 4. shows optical absorption spectra of all the investigated glass recorded at room temperature at wavelength ranging from 400- 800 nm. It can be seen that the UV-edge is red-shifted as TiO_2 concentration is increased, so is the absorption coefficient. This red-shifting is explained in the same way of decreasing E_g with the increase of TiO_2.

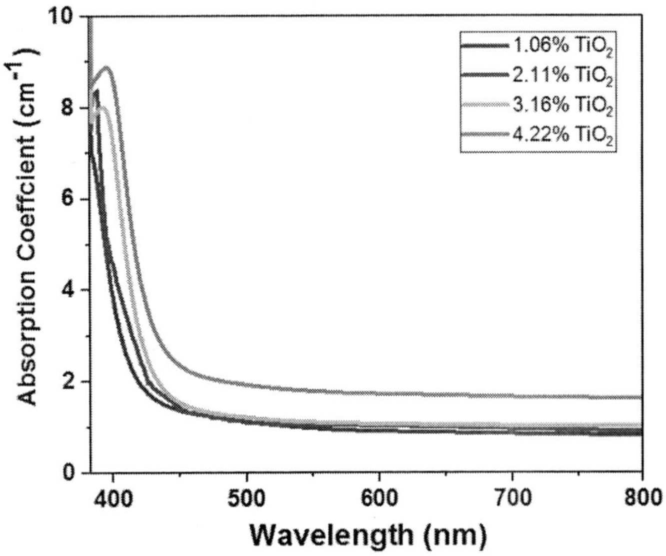

Figure 4. UV-Vis Absorption of TZNT Glass

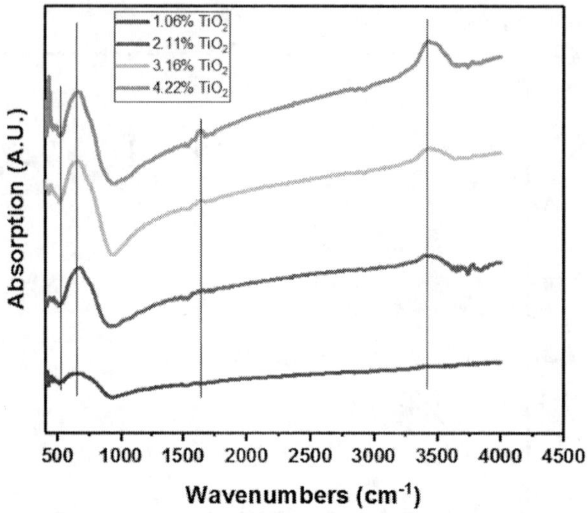

Figure 5. FT-IR Spectra of TZNT Glass

Figure 5 shows the FTIR absorption spectra of all glass samples. Several peaks within the range of 400-4000 cm^{-1} were observed. High similarity as can be seen for all spectrum indicates that all glass samples have similar glass network [20]. The peaks around 3426 cm^{-1} can be assigned as a peak originated from absorption of hydroxyl which might be arise from the fact that the absorption spectra were recorded in situ and, due to high humidity, re-adsorption of water might be occured [21[[22] [23]. The peaks at 1636 cm^{-1} is assigned to vibrations of hydroxyl group [24]. According to [25] the modifier cations like Na_2 possess vibrational sites below 500 cm^{-1}. Broad band which centered between 500 to 600 cm^{-1} are related to TiO_2 components [24]. In this work, that band slightly increased from 513 to 543 cm^{-1} as the higher addition of TiO_2 components.

Otherwise, the last strong band occurred between 647 to 670 cm^{-1}. According to [26] the shift of band, near 640 cm^{-1} towards higher frequencies (690 cm^{-1}) due to vibration between TeO_2-ZnO. It was shown that increase of TiO_2 composition does not strongly affect the characteristic vibrational band of pure TeO_2. It can be explained because both lengths of Te-O and Ti-O bonds are 1.9-2.0 Å. In this bond there is also vibration of Te-O bond in the TeO_3 pyramidal units which exhibiting strong infrared absorbtion [27]. For the further investigation, deconvolution was performed between 500 to 800 cm^{-1} in all glass samples. **Figure 6.** shows deconvoluted peak from TZNT Glasses. It was shown that, there are several peaks which constructed the strong peak between 500 to 800 cm^{-1}.

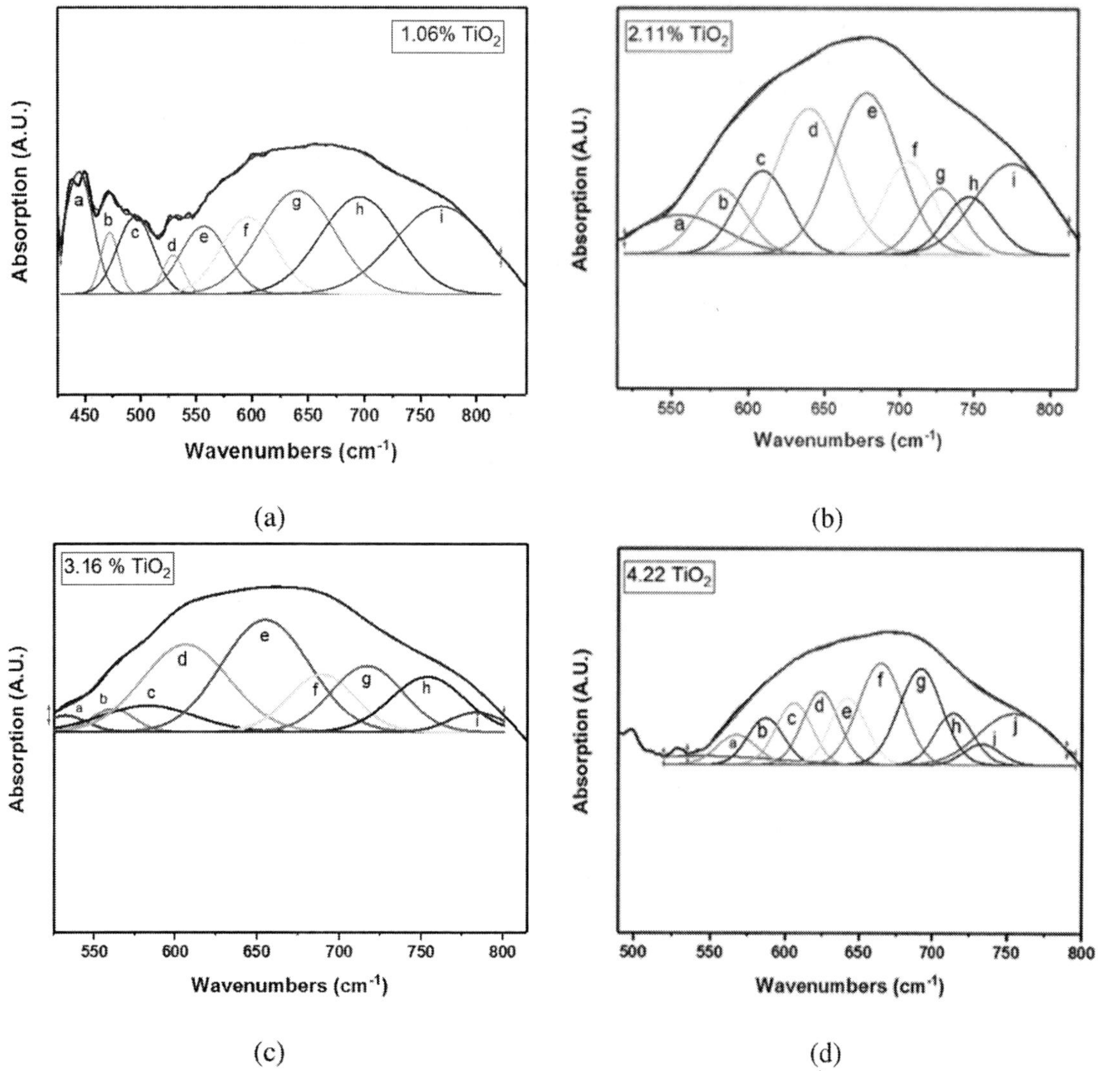

Figure 6. Deconvoluted peak at 500 to 700 cm^{-1}

4. Conclusion

TeO_2-ZnO-Na_2O glasses doped with various concentration of TiO_2 have been fabricated and characterized for physical and spectroscopies studies. It was shown that addition of TiO_2 causes decreasing in density, inter-ionic distance, field strength and optical band gap energy. In addition, the increase in TiO_2 content caused an increase in refractive index and shifts the UV-edge to longer wavelength.

References

[1] M Burger H, Vogel W and Kozhukharov V 1985 Infrared Phys. 25 395-409. (https://doi.org/10.1016/0020-0891(85)90114-9).
[2] Marzuki A, Fausta D E 2018 IOP Conf Ser Mater Sci Eng 333 (1) 012015. (DOI: 10.1088/1757-899X/333/1/012015).

[3] Marzuki A, Yunianto M, Riyatun, Purwanto H 2018 KEM 772 85-89. (DOI: 10.4028/www.scientific.net/KEM.772.85).
[4] Zhu Y, Shen X, Su X, Zhou M, Zhou Y 2019 Mater. Lett. 244 175-177 (DOI: https://doi.org/10.1016/j.matlet.2019.02.078).
[5] Fu W, Zhang C, Hou G, Xia J, Li G, Ping Y, 2019 Optik 182 308-313. (DOI: https://doi.org/10.1016/j.ijleo.2019.01.034).
[6] Yang Z, Wu Y, Yang K, Xu P, Zhang W, Dai S, Xu T 2017 Opt. Mat. 72 524-528. (DOI: https://doi.org/10.1016/j.optmat.2017.06.057).
[7] Tong L, Hu L, Zhang J, Qiu J, Yang Q, Lou J, Shen Y, He J, Ye Z 2006 Opt. Express 14 (1) 82-87. (DOI: **https://doi.org/10.1364/OPEX.14.000082**).
[8] Azlan M N and Halimah M K 2018 Results in Physics 11 58-64. (DOI: https://doi.org/10.1016/j.rinp.2018.08.017).
[9] Lewicka Z A, Yu W W, Oliva B L, Contreras E Q, Colvin V L 2013 J. Photochem. Photobiol, 263 24-33. (DOI: https://doi.org/10.1016/j.jphotochem.2013.04.019).
[10] Chen P, Wei B, Zhu X, Gao D, Gao Y, Cheng J, Liu Y 2019 Ceram. Int. 45 (5) 6111-6118. (DOI: https://doi.org/10.1016/j.ceramint.2018.12.085).
[11] Deng X, Wang Y, Cui Z, Li L, Shi C 2018 Superlattices Microst 117 283-287 (DOI: https://doi.org/10.1016/j.spmi.2018.03.051).
[12] Krzˇmanc M M, Dosˇler U, and Suvorov D 2012 J. Am. Ceram. Soc. 1–7 (DOI: 10.1111/j.1551-2916.2012.05216.x).
[13] Mukherjee D P, Das S K 2016 Journal of Asian Ceramic Societies 4 (1) 55-60 (DOI: https://doi.org/10.1016/j.jascer.2015.11.004).
[14] Zhang L, Liu S 2018 J. Non-Cryst. Solids 500 366-370 (DOI: https://doi.org/10.1016/j.jnoncrysol.2018.08.023).
[15] Karlsson S, Bäck L G, Kidkhunthod P, Lundstedt K, & Wondraczek L (2016 *Opt. Mater.Express, 6 (4) 1198. (*https://doi.org/10.1364/OME.6.001198).
[16] Altaf M and Chaundry M.A, 2010 *J. Mod. Phys.***1** 201-205. (DOI: https://doi.org/10.4236/jmp.2010.14030).
[17] Ruengsri S., Kaewkhao, J., Limsuwan, P, 2012 *Procedia Engineering***32** 772-779. (DOI: https://doi.org/10.1016/j.proeng.2012.02.011).
[18] Edukondaki A., Kavitha, B., Hameed, A., Kumar, K.S, 2012 *IOP Conf.* **73** 012127. (DOI: 10.1088/1757-899X/73/1/012127).
[19] Stambouli W., Elhouichet, H., Ferid, M, 2012 *J. Mol. Struc* **1028** 39-43. (DOI: https://doi.org/10.1016/j.molstruc.2012.06.027).
[20] Souri, D, 2010 *Mid. East J. Sci. Research***5** 44-48. (DOI: http://dx.doi.org/10.1016/j.molstruc.2016.05.052).
[21] Sahar, M R., Rohani M.S. and Kasim, A, 2009 *S. State Sci. and Tech.***1** 27-30. ISSN 0128-7389.
[22] Perumal, S., Sambandam, C.G., Prabu, K.M., Anathakumar S, 2014, IJRET **3** 651-657. (DOI: 10.15623/IJRET.2014.0304116).
[23] Nurbaisyatul, E.S., Azman, K., Azhan, H., Razali W.A., Noranizah, A, 2014 *Jurnal Teknologi* **69:2** 97-100.
[24] Beranek, R and Kisch, H, 2008, *Photochemical and Photobiological Science* **7** 40-48. (DOI: 10.1039/b711658f).
[25] ElBatal, F.H., Marzouk, M.A., ElBatal, H.A., 2016 *J. Mol. Struc.***1121** 54-58. (DOI: http://dx.doi.org/10.1016/j.molstruc.2016.05.052).
[26] Burger, H., Kneipp, K., Hobert, H., Vogel, W. 1992 *J Non-Cryst. Solid***151** 134-142. (DOI: https://doi.org/10.1016/0022-3093(92)90020-K).
[27] Heo, J., Lam, D., Sigel, H.S., Mendoza, E.A., Hensley, D.A, 1992 *J. Am. Ceram. Soc* **2** 277-281 (DOI: 10.1111/j. 1151-2916.1992. tb08176.x).

IOP Conf. Series: Materials Science and Engineering **578** (2019) 012044 doi:10.1088/1757-899X/578/1/012044

Antipyretic properties of carica (*Vasconcellea pubescens* A.DC.) fruit and seeds ethanolic extract in experimental animals

H Sasongko *, P Widiasih, and N L Putri
Department of Pharmacy, Faculty of Mathematics and Natural Sciences, Universitas Sebelas Maret. Jl. Ir. Sutami 36A Surakarta 57126, Central Java, Indonesia.

E-mail: heru_sasongko@staff.uns.ac.id

Abstract. Fever is the condition of body temperature which more than usual caused by an increase of central control temperature in the hypothalamus. An infection or virus usually cause this condition. Some of the active substances of plant secondary metabolites had been researched lower fever; they were flavonoids, saponins, and tannins. Carica (*Vasconcellea pubescens* A.DC) or mountain papaya is one of the fruits that had the compounds. This research purposed to know the antipyretic effect of ethanolic extracts of carica fruits and seeds towards Wistar rats. The study used pre-test – post-test only controls group design. Thirty Wistar rats were divided into six groups on each extract sample. Normal group was not given any treatment, negative control was given 2.5 mL aquadest, positive control was given acetaminophen syrup with the dosage 1.5 mL/kg body weight (b.w), and the dosages of carica fruit and seeds ethanolic extracts (EECF/ EECS), there were 200, 400, and 800 mg/kg b.w. The rats were s.c induced fever with 20 mL/kg b.w 20% yeast suspension. Temperature measurement was performed every 30 minutes for 6 hours. Data processed using One Way ANOVA. The results showed that ethanolic extracts of carica fruits and seeds with a dosage of 800 mg/kg b.w were able to lower the temperature significantly ($p < 0.05$) towards negative control in Wistar rats.

1. Introduction

The fever is definition a sign and symptoms of an underlying condition caused by an infection or viruses [1]. Fever is a significant feature of the infective and non-infective inflammatory processes, and it is one of the commonest physical signs found in patients [2]. Fever is a very common complaint in children accounting for as many as 20% of pediatric visits to doctors. Most children aged 0–36 months who have a fever have a focus of infection, which can be identified by careful history and examination. A viral upper respiratory tract infection is the most common focus [3]. Uncontrolled fever causes seizures [4], so it can reduce using antipyretic drugs [5].

Acetaminophen or paracetamol is nonsteroidal anti-inflammatory drugs for antipyretic which is widely used in many people because it's safe and easy to get [6]. The previous study, acetaminophen had adverse effect hepatotoxicity if the drugs used in a long time and overdose [3,4]. A case report from Chinese mention a 12-month-old boy who presented with hepatotoxicity disseminated intravascular coagulation and persistent renal insufficiency four days after repeated ingestion of a supratherapeutic dosage of paracetamol [9]. This adverse effect reason many people choose natural drugs as an alternative treatment.

The carica (*Vasconcellea pubescens* A.DC.) is one of the plants that come from Dieng plateau, Wonosobo, Central Java. Mountain papaya has to contain flavonoid, alkaloid, tannin, and phenol. Carica had been used in pharmacology test and prove that its fruit has antioxidant activity through lipid peroxidation inhibition [10) and analgesic activity [11]. In many studies doing antioxidant, analgesic, anti-inflammatory and antipyretic activity associated with secondary metabolites in some plant [7–9]. Recently, several plant-derived secondary metabolites have been described as interfering directly with molecules and mechanisms involved in inflammatory and fever processes [15–18]. However, to the best of our knowledge, the antipyretic effect of carica has not been explored in experimental fever models. This study aims to determine the antipyretic activity of carica fruit and seeds towards Wistar rats that had been induced fever.

2. Experimental

2.1. Chemical
Acetaminophen (Sanbe Farma), yeast (Na Kok Liong), WIDA NS 0.9% NaCl® (PT Bhakti Widatra), ethanol (pharmaceutical grade).

2.2. Animal
This research used 30 white male Wistar rats (*Rattus norvegicus*) and its 2-3 month age (150 – 200g). All animal handling procedures have been approved by the ethics committee of the Faculty of Medicine, Universitas Sebelas Maret with the number 502 / IV/ HREC / 2018.

2.3. Extraction
The fruit of carica (*Vasconcellea pubescens* A.DC) used in this study was collected from the Dieng Plateau, Central Java, Indonesia. This fruit and seeds are processed with maceration extraction methods using 70% ethanol for five days.

2.4. Phytochemical screening

2.4.1. Flavonoid screening. Flavonoid screening is done with *Bate-Smith Test*, the extract added by HCl then boiled in a water bath during 15 minutes. The positive reaction showed red color solution [19].

2.4.2. Saponin screening. Saponin screening is done with mixing the extract with 5 mL boiled water in test tubes, then shake until showed the positive reaction that is foam [20].

2.4.3. Polyphenol and Tannin screening. Polyphenol and Tannin screening did with add the extract with 70% ethanol. Then, add 2-3 drops 1% $FeCl_3$ solution. The positive reaction showed a bluish black or brownish green solution [21].

2.4.4. Alkaloid screening. This screening is done with dissolving 1 g extract. Two mL solution evaporated in porcelain bowl using a water bath. The residue is dissolved with 5 mL HCl 2 N then its divided into 2 test tubes. First tubes added by three drops of HCl 2 N and second tubes attached by three drops dragendorff reagent. Formation of sediment and reaction with dragendorff reagent will show orange residues which is a positive result of this screening [19].

2.4.5. Antipyretic test of mountain papaya fruit and seeds ethanolic extracts. Antipyretic test of carica fruit was determined using Ahmadiani modification method with yeast suspension induced in animal models [22]. The temperature was measured in rectal using a digital thermometer. Measurement of temperature was done before testing to knowing pretest temperature. All rats were given 20% yeast suspension with the dose 20 mL/kg body weight (b.w) as fever induced and injected subcutaneously (at the nape of neck). Fever temperature was checked after 3 hours since injection of yeast suspension. The rats divided into six groups on each extract sample which each group consists of five rats. Group I was given 2.5 mL aquadest as a negative control. Group II was given 1.5 mL/kg b.w acetaminophen syrup as a positive control. Group III – V was given carica fruit and seeds ethanolic extracts (EECF/ EECS) 200, 400, and 800 mg/kg b.w. Group VI not given any treatment as a normal group. Rectal temperature was then measured immediately after per oral granting as 0 hours. Antipyretic activity was measured during 6 hours, and the measurement of rectal temperature was checked every 30 minutes.

2.5. Statistical analysis
The data of rectal temperature processed in Area Under Curve (AUC). AUC was then measured using the following formula 1 :

$$AUC = \frac{(First\ temperature + Second\ temperature)}{2} \times interval\ of\ first\ and\ second\ temperature \quad (1)$$

The number of AUC in each group were then statistically analyzed using One Way ANOVA and followed by Post Hoc Test using Least Significant Difference (LSD).

3. Results and Discussion

3.1. Maceration and phytochemical screening of carica fruits and seeds
The phytochemical screening of carica fruits and seeds extract showed that it contains secondary metabolites consist of flavonoid, polyphenol, tannin and saponin that can be showed in table 1.

Table 1. Phytochemical screening of ethanolic extract of mountain papaya fruits and seeds

The phytochemical screening result by Simirgiotis explain the fruits of the *Vasconcellea pubescens*

No	Metabolite	Theory [19]	Test	Result	
				Fruits	Seeds
1.	Flavonoid	Red	Brownish red	Positive	Positive
2.	Saponin	Stable foam	Stable foam	Positive	Positive
3.	Alkaloid	The precipitate formed and with dragendorf reactant will form an orange precipitate	Black precipitate formed	Negative	Negativ e
4.	Polyphenol and Tannin	Bluish-black or green	Brownish green	Positive	Positive

have 19 phenolic compounds were identified [23]. Phytochemical analysis with the Wagner and Bladt method described that carica fruit extract showed the presence of flavonoids, tannins and phenolic [10]. The quantitative test results show that the contains total flavonoids quercetin equivalent in Dieng Plateau with value 633.35 mg / L, respectively [19].

3.2. Antipyretic Test
The antipyretic test using yeast suspension induction caused by S. cerevisiae could increase temperature by synthesis of prostaglandins [24]. The mechanism of yeast causes fever which is an increase in rat body temperature due to the presence of microorganisms in yeast that are considered as foreign objects by the immune system [25]. The S. cerevisiae and its cell wall components such as mannans induce fever by increasing plasma levels from IL-1b, interferon c, and TNF-α. This mediator works on the organum vasculosum lamina terminalis in the central nervous system. The mediator induces the formation of prostaglandins which play an important role in fever mediators [26]. The result of the antipyretic test can be seen in table 2 and 3.

Table 2. The average measurement in preliminary (t0) and fever temperature after 3 hours (ta) in the fruits extract sample

No	Groups	Rats Rectal Temperature (°C)	
		t0	ta
1	Negative Control	36.07±0.642	36.9±0.704
2	Positive Control	36.13±0.309	37.43±0.396
3	EECF 200 mg/kg b.w	35.63±0.323	37.13±0.532
4	EECF 400 mg/kg b.w	35.64±0.360	36.9±0.456
5	EECF 800 mg/kg b.w	35.92±0.271	37.42±0.463
6	Normal Control	35.38± 0.249	35.38± 0.249

Table 3. The average measurement in preliminary (t0) and fever temperature after 3 hours (ta) in the seeds extract sample

No	Groups	Rats Rectal Temperature (°C)	
		t0	ta
1	Negative Control	36.07±0.642	36.9±0.704
2	Positive Control	36.13±0.309	37.43±0.396
3	EECS 200 mg/kg b.w	35.49±0.563	36.93±0.292
4	EECS 400 mg/kg b.w	35.26±0.267	36.71±0.363
5	EECS800 mg/kg b.w	35.14±0.292	36.71±0.710
6	Normal Control	35.38± 0.249	35.38± 0.249

The result from Table 2 and 3 showed that increasing the temperature 0.83°C – 1.5°C of the rat. Increasing temperature above 0.5°C can be said that rats have a fever [27]. Peroral granting was given after checked fever temperature and directly measured rectal temperature as 0 minutes. The result of rectal temperature measurements during 6 hours was showed in figure 1 and 2.

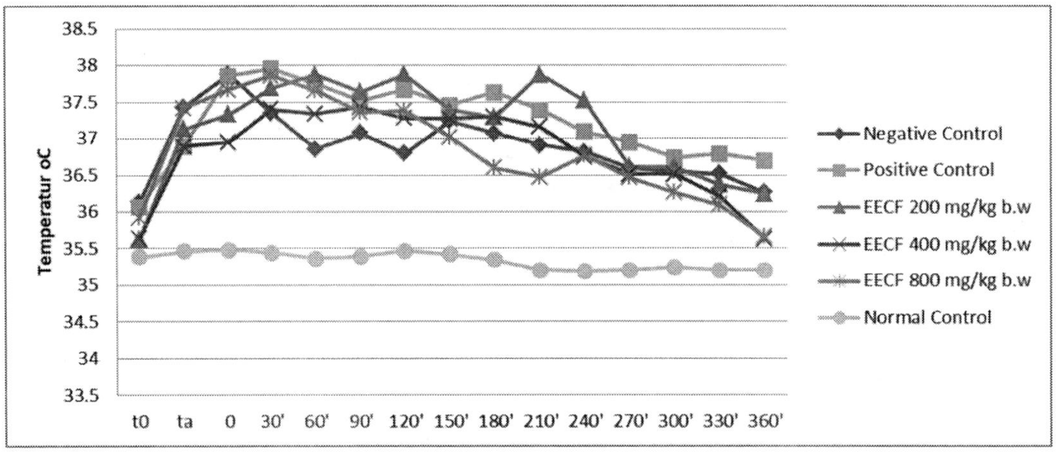

Figure 1. The graph of the decline in the average temperature of 6 groups. t0 = pre-test temperature, ta = post-test temperature, EECF =ethanolic extract of carica fruits

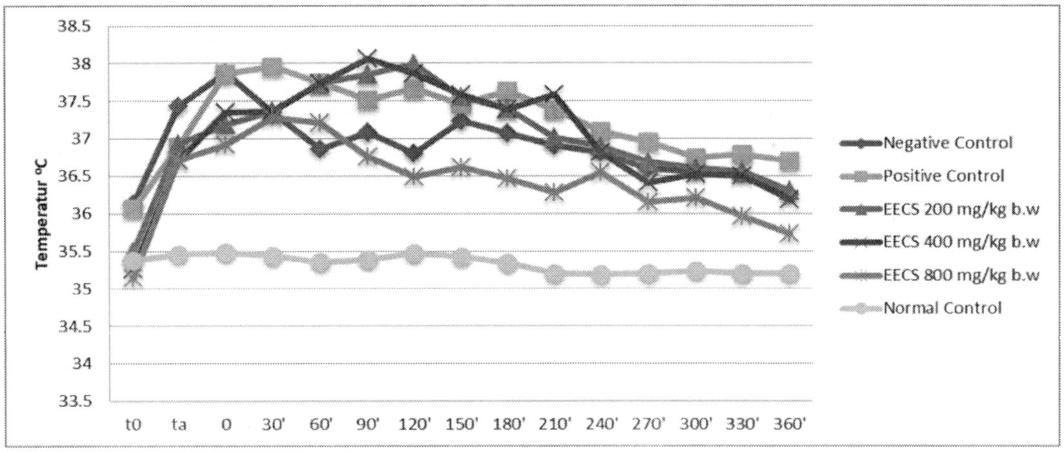

Figure 2. The graph of the decline in the average temperature of 6 groups. t0 = pre-test temperature, ta = post-test temperature, EECS =ethanolic extract of carica seeds

Figure 1 and 2 showed that negative control had more temperature than positive control and other groups of carica extract doses. It caused negative control was not given antipyretic agent treatment. Increased and decreased temperature in graph caused by fever induction that is still running. Yeast worked to raise the temperature during 9-18 hours [24]. The decrease of temperature in each group then statistically analyzed using One Way ANOVA and followed by Post Hoc LSD. Data that analyzed in statistical was AUC from all groups. AUC average in each group can be seen in table 4.

Table 4. Data area under the curve (AUC) of carica fruit (EECF) and seeds (EECS) ethanolic extracts

No	Groups	AUC ± CV	
		EECF	EECS
1	Positive Control	13286.1 ± 0.96*	13268.1 ± 0.96*
2	Negative Control	13446.9 ± 0.96	13446.9 ± 0.96
3	Dose 200 mg/kg b.w	13427.55 ± 0.38	13392.15± 0.34
4	Dose 400 mg/kg b.w	13303.5 ± 1.25	13397.55± 0.21
5	Dose 800 mg/kg b.w	13277.75 ±0.83*	13153.66 ± 0.26*
6	Normal Control	12727.65 ± 0.26*	12727.65 ± 0.73*

Description : *= significant difference between negative control groups ($p<0.05$)

The statistical analysis showed that normal control gave a significant difference in all groups because this control was not given any treatments. Negative and positive control gave significant differences because positive control contained acetaminophen as antipyretics agent [28]. Ethanolic extract of carica at the dose 800mg/kg b.w showed a significant difference ($p<0.05$) than other extract doses. Other carica extract had the effect of lowering temperature that can be seen in Table 4. Both AUC gave lowering temperature than negative control but still can not provide a significant difference. This result was due to the dose at 200, and 400 mg/kg b.w had not reached levels of maximum effective.

Mountain papaya fruits and seeds extract at the dose of 800 mg/kg b.w had more concentration than another dose. Furthermore, it is plausible to suggest that the antipyretic activities of plant involved, at least in part, modulation of the prostaglandin mechanism through direct inhibition of prostaglandin activity or indirect inhibition of prostaglandin synthesis, via inhibition of the enzyme cyclooxygenase [18,29]. Some study concludes that bioactive compounds namely flavonoids, tannins, and saponins have synergistic action, and thus confirmed the folklore of the plant for the treatment of various ailments [29,30].

4. Conclusion

The results showed that carica (*Vasconcellea pubescens* A.DC.) fruits and seeds ethanolic extracts with the dosage of 800 mg/kg b.w was significantly ($p < 0.05$) able to lower the temperature towards negative control in Wistar rats. At the dose possible had more concentration of metabolite than another dose. Further research is expected to conduct a quantification test of secondary metabolites of carica fruits and seeds extract as of the correlation between the dose, and the effect can be seen more clearly.

Acknowledgments

The authors would like to thank Universitas Sebelas Maret which funded this research with the PKLP PNBP 2018 Grants scheme and Pharmacology and Chlinical Pharmacy group research.

References

[1] Monto AS, Gravenstein S, Elliott M, Colopy M, Schweinle J. 2000 *Arch Intern Med.* **160(21)** 3243-3247.
[2] Circiumaru B, Baldock G, Cohen J. 1999 *Intensive Care Med.* **25** 668-673.
[3] El-Radhi AS, Carroll J, Klein N 2009 *Clinical Manual of Fever in Children* (Berlin : Springer) p. 1–24.
[4] Noebels J, Avoli M, Rogawski M, Olsen R, Delgado-Escueta A 2012 *Jasper's Basic Mechanisms of the Epilepsies* (Oxford University Press) 1258 p.
[5] Sullivan JE, Farrar HC. 2011 *Pediatrics.* **127** 580-5877.
[6] Allahyry E, Tarig SH, Chohedri AH, Fattahi ZS, Talei F 2006 *Armaghane Danesh.* **11(1)** 1–7.
[7] James LP, Mayeux PR, Hinson JA 2003 *Drug metabolism and disposition.* **31(12)** 1499-1506.
[8] Brune K, Renner B, Tiegs G 2015 *European Journal of Pain.* **19(7)** 953-965.
[9] Kwok, K. L., Fu, Y. M., and Ng, D. K. K 2004 *Hong Kong Medical Journal.* **10(1)** 61-64.
[10] Sasongko, H., Efendi, N. R., and Sugiyarto 2018 *AIP Conference Proceedings.* **2019(1)** 050001
[11] Sasongko H 2017 *In vivo analgesic effect of ethanolic extracts of exocarp, mesocarp, and seeds of Carica pubescens. In: Unity in Diversity and the Standardisation of Clinical Pharmacy Services: Proceedings of the 17th Asian Conference on Clinical Pharmacy (ACCP 2017), July 28-30, 2017, Yogyakarta, Indonesia.* (United Kingdom: CRC Press) 271.
[12] Saravanan, S., Arunachalam, K., and Parimelazhagan, T 2014 *Industrial Crops and Products.* **54** 272-280
[13] Eldahshan, O. A., and Abdel-Daim, M. M 2015 *Cytotechnology.* **67(5)** 831-844.
[14] Aouey, B., Samet, A. M., Fetoui, H., Simmonds, M. S., and Bouaziz, M 2016 *Biomedicine & Pharmacotherapy.* **84** 1088-1098.
[15] Haslam, E. 1996 *Journal of natural products.* **59(2)** 205-215.
[16] Calixto JB, Otuki MF, Santos ARS 2003 *Planta Med.* **69(11)** 973–83.
[17] Calixto JB, Campos MM, Otuki MF, Santos AR 2004 *Planta Med.* **70(02)** 93–103.
[18] Kanashiro, A., Machado, R. R., Malvar, D. D. C., Aguiar, F. A., & Souza, G. E 2008 *Journal of Pharmacy and Pharmacology.* **60(3)** 357-362.
[19] Minarno, E. B 2015 *el–Hayah.* **5(2)** 73-82.
[20] Yadav, R. N. S., and Agarwala, M. 2011 *Journal of phytology.* **3(12)**.
[21] Edeoga, H. O., Okwu, D. E., & Mbaebie, B. O 2005 *African journal of biotechnology.* **4(7)** 685-688.
[22] Ahmadiani A, Javan M, Semnanian S, Barat E, Kamalinejad M 2001 *Journal of ethnopharmacology. 75(2-3) 283-286.*
[23] Simirgiotis, M. J., Caligari, P. D., and Schmeda-Hirschmann, G 2009 *Food Chemistry.* **115(2)** 775-784.
[24] Ghauri MA, Khan GJ, Khan S, Javeed A, Naeem HS, Ashraf M 2017 *African Journal of Pharmacy and Pharmacology. 11(30) 355-361.*
[25] Swantara, I., Rachman, R., and Puspawati, N. 2017 *Jurnal Kimia.* 107-112.

[26] Ahmad, S., Rehman, T., and Abbasi, W. M. 2018 *Journal of Ayurveda and integrative medicine.* **9(3)** 177-182.

[27] Tesema, S. H., and Makonnen, E 2015 *Ethiopian journal of health sciences.* **25(2)** 139-146.

[28] Emire Z, Yabalak E, Görmez Ö, Gizir AM 2017 *J Serb Chem Soc.* **82(1)** 99–106.

[29] Zakaria ZA, Raden Mohd. Nor RNS, Hanan Kumar G, Abdul Ghani ZDF, Sulaiman MR, Rathna Devi G, et al. 2006 *Canadian journal of physiology and pharmacology.* **84(12)** 1291–1299.

[30] Sugiyarto, Novalina, D., Susilowati, A. and Sasongko, H 2018 AIP Conference Proceedings **2019(1)** 050005.

IOP Conf. Series: Materials Science and Engineering **578** (2019) 012045 doi:10.1088/1757-899X/578/1/012045

Optimization Combination of Suweg Starch (*Amorphophallus campanulatus Decne*) and Gembili Starch (*Dioscorea esculenta* (Lour.) Burk.) as Filler of Ibuprofen Tablet by Simplex Lattice Design Method

D E Ermawati [1], N N Hidayah[1], S Rohmani[1]

[1] Department of Pharmacy, Matematics and Natural Science Faculty, Universitas Sebelas Maret, Surakarta, Indonesia

E-mail: dianekaerma@gmail.com

Abstract: Tablet is pharmaceutical dosage form that simple and efficient in drug delivery system. Ibuprofen tablet is usually taken along with food or milk to reduce side effects on the gastrointestinal tract. Starch of Suweg and Gembili are not yet widely developed as excipients in tablet formulation. Based on Biopharmaceutics Classification Systems,ibuprofen includes in BCS class II with high permeability and low solubility. This study want to formulate ibuprofen tablets using combination of suweg and gembili starch with wet granulation method. However, the optimum proportion of starch composition is not yet known, so optimization is needed with the *simplex lattice design method* using *Design Expert software*. Obtained 8 formulas with physical response: granule flow time, tablet weight uniformity and disintegration time of tablets. The optimization shows that the optimum composition of suweg and gembili starch are 51.809%: 48.191%. The result of statistical analysis [t-test] physical parameters of granule and tablet of ibuprofen optimum formula were not significantly different (*p-value* > 0.05) with *Design Expert* prediction. Based on the requirements of Pharmacopeia Indonesia 4th Edition the results of optimum formula test provide good physical test with granule flow time 3.583±0.130 seconds; tablet weight uniformity 498.333±5.080 mg, hardness tablet 4.780±0.130 Kg, friability vvof tablets 0.904% and the disintegration time of tablets 6.833±0.750 minutes. The dissolution profile of the optimum formula showed %DE at 60 minutes is 76.975±0.037%. Consentration of ibuprofen in tablet is 191.755±5.847 mg and percentage of ibuprofen in tablet is 95.877±2.293%.

1. Introduction

Tablet is one of the most commonly used drug dosage forms because efficient, practical, and ideal for administering orally active substances [1] . Tablet is solid preparation, made in flat or double convex press, generally round, containing one type of drug with or more with or without additives. Additional substances used can function as filler substances, developer substances, and wetting agents [2]. The filler (diluent) functions to increase the volume of the mass to be easily made. The filler material is added if the active ingredient was small or difficult to press. For example lactose, starch, dibase calcium phosphate, and microcrystalline cellulose [3].

Various types of tubers are spread in all regions in Indonesia which have the potential to produce starch, including those that have not been widely developed and the utilization is not optimal is suweg tubers (*A. campanulatus Decne*) and gembili (*D. esculenta* (Lour.) Burk.). The results of previous studies by Richana and Sunarti [4] showed that *A. campanulatus Decne* tuber starch had low amylose (19%) and high peak viscosity (700 BU) so that it was well developed for thickening and filling materials. Gembili tubers with high starch yield (21.44%) are also very potential to be developed into starch products. The use of a combination of fillers in the tablet preparation formulation will affect the physical properties of tablets produced because the filler material has a large portion of the overall weight of the tablet.

Ibuprofen tablets are usually taken along with food or milk to reduce the side effects of the drug. Therefore, a new idea was developed to replace the components of tablet fillers with natural ingredients as well as food ingredients, namely suweg tubers (*A.campanulatus Decne*) and gembili (*D. esculenta* (Lour.) Burk.), the aim that Ibuprofen tablets are produced better and could helped minimize the side effects of the drug. Comparison of the filler composition in the formulation of a tablet Ibuprofen preparation can determine the physical properties of the tablets produced. Therefore, the research was optimized to obtain the optimum composition of tuber starch filler mixture with the *simplex lattice design* method using *Design Expert* software.

2. Experimental

2.1. Materials

GenesysTM 10s UV/VIS spectrophotometer, moisture analyzer, Guoming RC-1 paddle type dissolution tester, knife, grater, flannel, oven, mortar, 18 mesh sieve, 16 mesh sieve, 90 mesh sieve, 70 mesh sieve, drop pipette, beker glass, funnel, ruller, stopwatch, water bath, Prescia BJ 4501C and Sartorius BP 221S analytical balance sheet, Shanghai Tianhe Pharmaceutical Machinery TD punch single tablet printing machine, tablet friability tester C2, YD-1 hardness tester tablet, desintegration tester.

Suweg tubers (*A. campanulatus Decne*) and gembili tubers (*D. esculenta* (Lour.) Burk.) were obtained from Kaliwungu, Kendal, Central Java, Indonesia; Ibuprofen PT Konimex, Indonesia; monobasic potassium phosphate 0.2 M and 0.2 M sodium hydroxide obtained from Universitas Sebelas Maret Chemical Laboratory; and magnesium stearate 1%, talk 2%, amprotab 5%, gelatin 10%, and hydrochloric acid 0.1 N pharmaceutical grade from Bratachem.

2.2. Methods

2.2.1. Isolation of Tuber Suweg and Gembili

Determination of suweg and gembili tubers was carried out at the Biology Laboratory of the Mathematics and Natural Sciences Faculty, Universitas Sebelas Maret, Surakarta, Indonesia. The process of making starch from tubers is done by extracted with water solvent twice (1:3 and 5:3) for 15 minutes. After 15 minutes, the extract is filtered so that liquid starch is obtained, then dried using an oven with a temperature of 50±2 ^{0}C for 6 hours. Dry starch is ground and sieved using 70 and 90 mesh [4] then organoletics test.

The sample of 1.0 g of tuber starch was flattened above the pan of the Moisture Analyzer. The tool will heat the sample at 105 ^{0}C until it shows a constant readable water content value (± 3-5 minutes).

2.2.2. Wet Granulation and Evaluation

Ibuprofen tablets are made using wet granulation method. The ingredients are weighed according to the desired weight in each formula (Table 1). First, the binder (10% gelatine) is made by 10 g gelatine powder diluted with hot water ad 100 mL. Active substances (ibuprofen), crushers (amprotab) and fillers (suweg and gembili starch) are mixed and ground until homogeneous. Then the binder (10% gelatin) is added until a moist mass can be obtained. When the granule is passed on a 16 mesh sieve, the granule is dried in oven at 60 ^{0}C for 2 hours. Dry granules are sieved again with 18 mesh sieve, then mixed with lubricant (Mg stearate and talc) and stirred until homogeneous.

Granule evaluation includes : Flow time test of granule, weighed 50 g of granule and poured into a funnel with stem closed. The stem is opened and the granule is allowed to flow until it runs out. The time needed for the granule to flow until it is used up is recorded using a stopwatch. The test was repeated 3 times.

Table 1. Ibuprofen tablets formula based on *Simplex Lattice Design* Methods with variation proportion of suweg and gembili starch

Ingredients (mg)	F1	F2	F3	F4	F5	F6	F7	F8
Ibuprofen	200	200	200	200	200	200	200	200
Suweg starch	125	125	250	187.5	-	62.5	-	250
Gembili starch	125	125	-	62.5	250	187.5	250	-
Gelatine	10	10	10	10	10	10	10	10
Mg stratate	5	5	5	5	5	5	5	5
Talcum	10	10	10	10	10	10	10	10
Amprotab	25	25	25	25	25	25	25	25

2.2.3. Tableting and Evaluation

Tablet pressed with each tablet weigh 500 mg with a dose of 200 mg ibuprofen was performed using a single punch tablet pressed. Tablets that have been press then evaluation including :

2.2.3.1. Weight uniformity test

Weighed 20 tablets one at a time with an analytical scale, then the average weight and deviation of each tablet were weighed. The deviation of the weight of two tablets should not be more than 5% of the average weight and none of the tablets deviate more than 10% of the average.

2.2.3.2. Hardness test

The tablet is placed on the hardness tester with the tablet standing, then the tool lever is fully pressed. The scale achieved when a broken tablet is read. The test was repeated 3 times.

2.2.3.3. Fragility test

Weighed a total of 20 tablets, then put into the cylinder friabilator. The friabilator test equipment is run for 4 minutes or 100 rounds. Then the tablet is taken and released again, the tablet is weighed again and the difference in weight is calculated before and after testing.

$$\% \text{ Fragility} = (\text{initial weight-end weight})/(\text{initial weight}) \times 100\% \quad (1)$$

2.2.3.4. Disintegration time

Inserted 6 tablets one at a time into the tube of disintegration test equipment, then the device moved up and down regularly 30 times per minute in medium water temperature of 36-38^{0}C. Tablets are disintegrated if there are no tablet parts left behind.

2.3. Concentration of Ibuprofen in Tablet

2.3.1. Medium of Ibuprofen tablets

Medium of 0.1 N NaOH was made by dissolving 4.0 g of NaOH with distilled water in a 1.0 L. Maximum wavelength of ibuprofen into 0.1 N NaOH medium.

Liquid stock of Ibuprofen was prepared by dissolving 50.0 mg of ibuprofen ad 10.0 mL ethanol 96%, then put 1.0 mL solution was dissolved with 0.1 N NaOH solution ad 100.0 mL volumetric flask and obtained a solution with a concentration of 50.0 µg/mL. The solution was pipetted 12.0 mL put into a 25.0 mL volumetric flask dissolved with 0.1 N NaOH solution sufficiently, so that a solution with a concentration of 24.0 µg/mL was obtained. The maximum absorption of ibuprofen was measured at a wavelength of 200-400 nm.

2.3.2. Calibration curves of ibuprofen in the medium of 0.1 N NaOH

From stock liquids, dilution of 0.1 N ibuprofen in NaOH was performed with a concentration of 12.0; 16.0; 20.0; 24.0; 28.0; and 32.0 µg/mL ibuprofen.

2.3.3. Ibuprofen concentration in tablets

Taken 10.0 tablets randomly, weighed and determined the average weight. Ten tablets were ground to powder, then weighed as much as 100.0 mg of tablet weight. Then dissolved in 0.1 N NaOH into

100.0 mL flask, shaken until homogeneous, then filtered. The filtrate obtained was pipetted as much as 2.0 mL and put in a measuring flask ad 10.0 mL, then added 0.1 N NaOH. The absorbance was observed at the maximum wavelength using a UV-VIS spectrophotometer.

2.4. Dissolution Test

2.4.1. Dissolution media

Dissolution media used phosphate buffer pH 7.2. This buffer was prepared by mixing 50.0 mL of monobasic potassium phosphate 0.2 M and 22.4 mL of sodium hydroxide 0.2 M then diluted with distilled water ad 200.0 mL. Put of 100.0 mL solution and diluted with distilled water ad 1.0 L, then for pH adjustment 0.1 N HCl or 0.1 N NaOH was added.

2.4.2. Determination of the maximum wavelength of ibuprofen

Ibuprofen stock liquid was prepared by dissolving 50.0 mg of ibuprofen ad 10 mL ethanol 96%. Put of 1.0 mL ibuprofen solution then dissolved ad 100 mL of phosphate buffer pH 7.2 in a 100 mL volumetric flask and obtained with a concentration of 50 μg/mL. The solution was pipetted 12.0 mL into a 25.0 mL volumetric flask dissolved with phosphate buffer pH 7.2 sufficient until the flask boundary mark was obtained solution with a concentration of 24.0 μg/mL. The maximum absorption of ibuprofen in buffer solution was measured at a wavelength of 200-400 nm.

2.4.3. Calibration curves of ibuprofen in phosphate buffer pH pH 7.2

Ibuprofen liquid stock was then diluted in phosphate buffer pH 7.2 with a concentration of 8.0; 12.0; 16.0; 24.0; 28.0; and 32.0 μg/mL.

2.4.4. Tablet dissolution

Pour of 900 mL dissolution medium into a vessel, then heated at temperature of 37±0.5°C. The ibuprofen tablet was put into the dissolution vessel then rotated at a speed of 50.0 rpm. Samples were taken as much as 10.0 mL at intervals of 0, 2, 5, 10, 15, 20, 30, 45, and 60 minutes. Each sample taken is then replaced with 10.0 mL dissolution medium. The absorbed sample is measured at it's maximum wavelength. According to the Indonesian Pharmacopeia 5th Edition [5], within 60 minutes it must dissolve not less than 80% $C_{13}H_{18}O_2$ of the amount indicated on the label.

2.5. Data Analysis

Optimum formula was obtained by *Simplex Lattice Design* methods using Design Expert software. Data on the physical properties of tablets obtained by measurement of time until the granules flow out for flow time, calculation of the Coefficients of Variation (CV), balancing tablet weight of 5% and 10% for tablet weight uniformity, average scale value on hardness tester tools for tablet hardness, calculation of tablet fragility%, tablet disintegration time with disintegration test equipment.

Profiles of dissolution of ibuprofen tablet optimum formula for the combination composition of tuber starch fillers was obtained by comparing the dissolution test results with the requirements listed in Indonesian Pharmacopeia 5th Edition (2014).

The data obtained were analyzed statistically using a *T-Test* for one sample (one sample T-Test) with a 95 % confidence level. Test results are compared with predictive results.

3. Results and Discussion

The results of the determination in the Biology Laboratory of the Mathematics and Natural Sciences Faculty, Sebelas Maret University Surakarta, Indonesia based on Flora of Java book stated that the plants used in this study were Suweg (*A. campanulatus Decne*) and Gembili (*D. esculenta* (Lour.) Burk.).

High moisture content can increase the growth of microbes and enzymes that cause damage to starch. The limit of microbial water content can still grow is 14-15% . Water content of tuber starch with replications resulted in an average water content in suweg tuber starch (*A. campanulatus Decne*) which was 16.6% and gembili tuber starch (*D. esculenta* (Lour.) Burk.) was 18.7%. The test results of water content in tuber starch showed high values can be made possible due to environmental conditions when the research is still entering the rainy season so the air is more humid.

3.1. Flowing Time of Granules

The mathematical equation for the granule flow time is $Y_1 = 0.03$ (A) + 0.04 (B) + 2.79 (A) (B) – 1.23 (A) (B) (A-B) where A is suweg tuber starch and B is gembili tuber starch. The graph of mixing 2 components of suweg and gembili starch shows that the greatest response value is the combination of +2.79 (A) (B) which means that the mixture of suweg starch and gembili can increase the flow time of ibuprofen tablet granules (Figure 1).

3.2. Uniformity of Tablet Weight

Mathematical equation for granule flow time is $Y_2 = 5.03$ (A) + 5.03 (B) – 7.77 (A) (B) where A is suweg tuber starch and B is gembili tuber starch. The graph of mixing 2 components of suweg and gembili starch shows that the greatest response value is –7.77 (A) (B), which means that the mixture of suweg and gembili starch can reduce the uniformity of ibuprofen tablet weight (Figure 1).

3.3. Disintegration Time

The mathematical equation for granular flow time is $Y_1 = 7.39$ (A) + 9.61 (B) where A is suweg starch and B is gembili starch. The graph of mixing 2 components of suweg and gembili starch shows that the greatest response value is +9.61 (B) which means that the gembili starch may reduce the disintegration time of ibuprofen tablets (Figure 1).

The formula of the highest response is chosen as the optimum formula. From these optimization obtained the optimum formula proportion combination of suweg and gembili starch. The results of the prediction of Simplex Lattice Design with a optimum ratio of 51.809 % : 48.191 % has the highest total response value.

From the optimization results, it was made ibuprofen tablets with the same method with ibuprofen tablets made for optimization. Furthermore, ibuprofen tablets with the optimum formula composition combination of suweg starch and gembili starch were obtained from the optimization using Simplex Lattice Design method from the physical properties of granules and tablet.

The uniformity of ibuprofen tablet weight of the optimum formula test results were significantly different from the predictive results ($p>0.05$), but it was acceptable because it was still included in the requirements of tablet weight uniformity. The deviation of the weight of two tablets no more than 5 % of the average weight of 20 tablets and none of the tablets deviate more than 10 % from an average of 20 tablets.

Table 2. Physical response results of granule and tablets ibuprofen based on *Simplex Lattice Design* method

Formula	**%w/w**		**Flowing time (second)**	**Weight uniformity (mg)**	**Disintegration time (minute)**
	Suweg	**Gembili**			
1	50	50	3.3	500.1	6.7
2	50	50	2.9	500.2	7.3
3	100	0	3.3	502.7	5.8
4	75	25	3.0	501.9	7.7
5	0	100	4.2	502.8	9.0
6	25	75	5.6	501.8	7.2
7	0	100	3.2	502.5	8.8
8	100	0	3.1	502.0	5.3

Table 3. Point prediction value of physical respon of granule and tablet ibuprofen optimum formula based on Simplex Lattice Design method using Design Expert software

Suweg starch	Gembili starch	Flowing time of granul	Weigh uniformity	Disintegration time	Desirability value
51.81%	48.19%	3.50	500.66	7.08	0.86

The test results of determining the optimum formula levels of ibuprofen tablets using UV-VIS spectrophotometer show that on average each tablet contains of 191.76 ±5.85 mg ibuprofen with concentration of 95.88±2.29 % ibuprofen. According to the requirements listed [5], ibuprofen tablets are not less than 90% and not more than 110 %, so that the optimum level of the active ingredient of ibuprofen each tablet meets the requirements. From the results of the analysis showed that the optimum formula ibuprofen tablet had dissolution profiles which meet the requirements of Indonesian Pharmacopoeia 5th edition (2014) with %DE (Disolution Efficiency) which was not less than 80 % within 60 minutes which was 94.85±0.037 %.

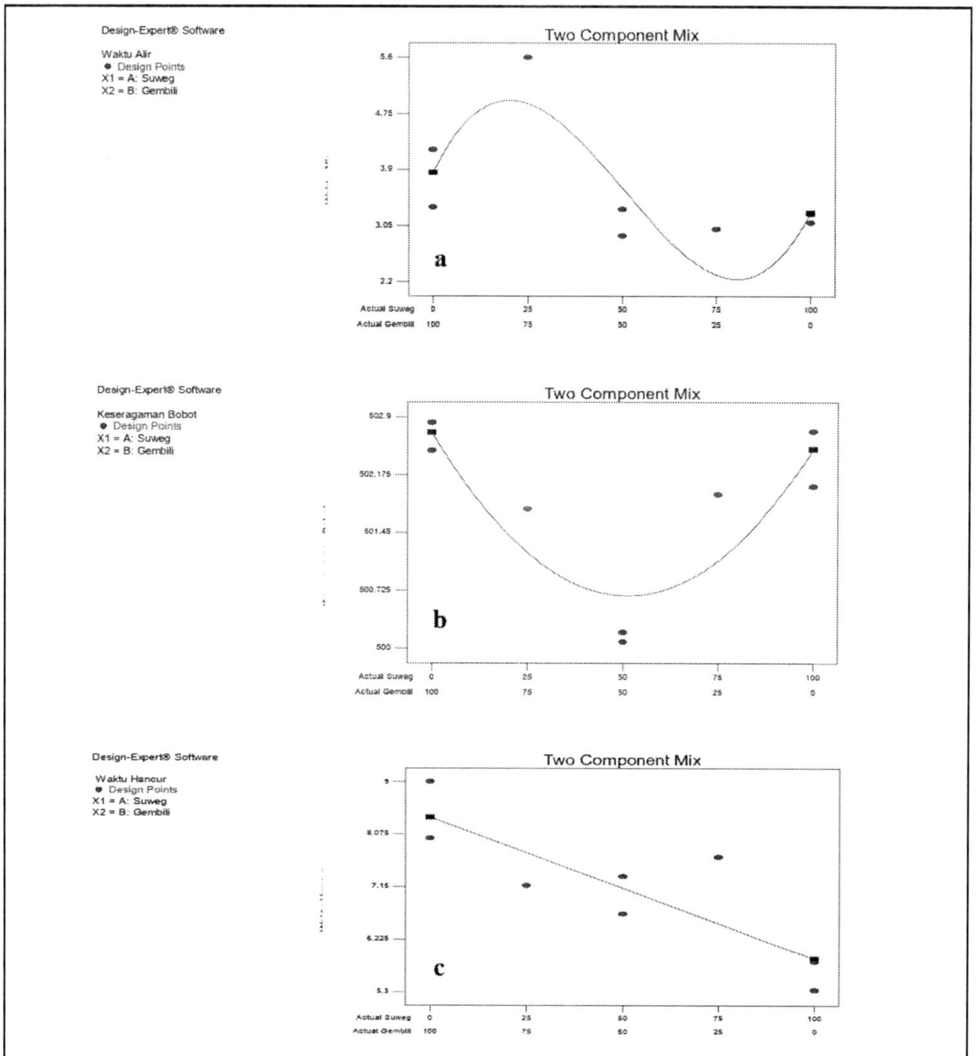

Figure 1. Diagram counter plot of filler combination suweg-gembili of ibuprofen tablets with phsical responses : flowing time of granule which means that the mixture of suweg starch and gembili can increase the flow time of granules **[a]**; uniformity of tablet weight, the mixture of suweg and gembili starch can reduce the uniformity of ibuprofen tablet weight **[b]**; and disintegration time the gembili starch may reduce the disintegration time of ibuprofen tablets **[c]**

Teble 4. Physical response value of granule and tablet ibuprofen optimum formula compare with Simplex Lattice Design prediction value

Physical responses	SLD	Optimum formula	p-value (>0.05)	Information
Flowing time of granul	3.50	3.58	0.370	Not significant
Weigh uniformity	500.66	498.33	0.004	Not significant
Disintegration time	7.08	6.83	0.490	significant

4. Conclusion

The results showed that the optimum formula ibuprofen tablet was obtained from the composition of the ingredient suweg starch (*A. campanulatus Decne*) and gembili starch (*D. esculenta* (Lour.) Burk.) with a ratio of 51.809 % : 48.191 %. The increase in the proportion of suweg starch as a filler on the ibuprofen tablet optimum formula gave better physical properties of granules and tablets : granule flow time 3.583±0.130 seconds; tablet weight uniformity 498.333±5.080 mg with CV value is 0.021 %, tablet hardness is 4.780±0.130 Kg, tablet fragility is 0.904 % and disintegration time is 6.833±0.750 minutes. The optimum dissolution profile of ibuprofen tablet showed that %DE at 60 minutes is meet the requirements of not less than 80 %, namely 94.85±0.037 %. While the results of the determination of the levels showed that on average each tablet contained 191.76±5.85 mg ibuprofen and the concentration of ibuprofen each tablets is 95.88±2.29 %.

References

[1] Ermawati D E, Sulaiman S, and Purwantini I 2017 *J. Pharm. Sci. Clin. Res.* **2**(1) 53-65.
[2] Fell J T and Newton J M 1970 *J. Pharm. Sci.* **59**(5) 688-691.
[3] Chebli C and Cartilier L 1998 *Int. J. Pharm.* **171**(1) 101-110.
[4] Richana N and Sunarti T C 2004 *J. Pascapanen* **1**(1) 29–37.
[5] Ditjen POM 1995 *Farmakope indonesia Edisi ke IV* (Jakarta: Departemen Kesehatan Republik Indonesia).

Optimization of Self-Nanoemulsifying Drug Delivery System (SNEDDS) of *Annona muricata* L. leaves chloroform extract using VCO (Virgin Coconut Oil) as an oil phase

A A R Sulkhan[1], A N Artanti[1], D E Ermawati[1] and F Prihapsara[1]
[1] Pharmacy Department, Faculty of Mathematics and Natural Sciences, Sebelas Maret University, Indonesia

Email: anif.apt@staff.uns.ac.id

Abstract. *Annona muricata* L. also called as soursop is one of medicinal plants that have a broad spectrum of biological activities and the most impressive of which is anticancer activity. Annonaceous acetogenins (ACGs) have been shown to successfully induce death in cancer cell. However, the solubility of ACGs is poor, so modification of formula of Self-Nanoemulsifying Drug Delivery System (SNEDDS) can be used for improving drug solubility. This study was used *Annona muricata* L. leaf chloroform extract then optimized to formula of SNEDDS using D-Optimal mixture design (Design Expert 9 Trial software). Sixteen formulas from the software consisting of oil phase (VCO), surfactant (tween 80-cremophor EL), and co-surfactant (propylene glycol) were tested of loading dose extract and characterized for transmittance, emulsification time, and phase separation. One sample t-test was used for verification, then the optimum SNEDDS was evaluated against particle size, polidispersity index, zeta potential, and TLC test was used to predict the active ingredient in the extract. The optimum SNEDDS consisting of 22.00% oil : 60.79% surfactant : 17.21% co-surfactant showed desirability value 0.608 and can loaded 25 mg/g extract. The optimum SNEDDS showed 84.4% (SE ± 1.64) of transmittance; 14.79 seconds (SE ± 1.81) of emulsification time; and 0.898 (SE ± 0.01) of phase separation. The verification test showed no significant difference ($p > 0.05$) between the observation and the prediction from the software. Evaluation results showed 42.73 nm (SE ± 0.49) of the particle size; polidispersity index 0.37 (SE ± 0.10); -27.43 mV (SE ± 0.26) of zeta potential, and TLC results showed similarity of active ingredient in the extract with the active compound of the fraction. Therefore, from these results suggest that the optimum SNEDDS achieved the nanoparticle size for SNEDDS formulation and may be improving drug solubility.

1. Introduction

Annona muricata L. also called as soursop is one of medicinal plants that have a broad spectrum of biological activities and the most impressive of which is anticancer activity. Annonaceous acetogenins (ACGs) have been shown to successfully induce death in cancer cell [1]. Besides that, the other compounds such as flavonoids also play an important role in induce death in cancer cell. According to Yang et al [2] inhibition of prostate tumor growth increased by synergistic interactions among ACGs and flavonoids in soursop leaves extract. Phytochemical study of soursop leaves extract that has been done shows that the level of flavonoids significantly higher concentration in chloroform extract than methanolic extract [3]. Soursop leaves also rich of lipid content compared to the other part of the plant due to the presence of ACGs which are derivates of long chain fatty acids (C32 or C34) [4]. But, ACGs are highly insoluble in water (<1 μL/mL) so their drug formulation have been limited [5].

However, these poorly water soluble drugs leads to low oral bioavailability. Yang et al [2] showed that the maximum plasma concentration of oral administration of 100 mg/kg soursop leaves ethanolic extract was less than 40 ng/mL in <1.0 hour [2]. Tablet dosage formulation of the soursop ethanolic leaves extract also has poor solubility when the concentration were tested [6]. Self-Nanoemulsifying Drug Delivery System (SNEDDS) is one of the lipid based drug delivery system that efficient for improving drug solubility. Hence, this study was aimed to optimized *Annona muricata* L. leaves chloroform extract to formula of SNEDDS that consist of oil phase (VCO), surfactant (tween 80-cremophor EL), and co-surfactant (propylene glycol) to achieve physical stability and further enhance drug solubility.

2. Experimental

2.1. Materials

Soursop fresh leaves (Karanganyar, Central Java), aquabides (IKA), analytical grade solvents like chloroform, methanol, ethyl acetate, and hexane, TLC silica gel 60 F_{254} (Merck), AGF, cremophor EL (Kolliphor® EL Sigma pH 6,0-8,0), tween 80, propylene glycol, glycerin, VCO, PKO, olive oil, aquades (Brataco). Oven, blender (Philips), analytical balance (Mettler Toledo), rotary evaporator (Stuart), waterbath (Grant), vortex mixer (Maxi Mix), sonicator (Branson), magnetic stirer (IKA), particle size analyzer (HORIBA SZ-100), micropippete (Gilson), pH meter (Eutech), sentrifuge (Mini Spin plus), UV/Vis spectrophotometer (Genesys™).

2.2. Method

2.2.1. Preparataion of Soursop Leaves Chloroform Extract

Fresh leaves were picked from the tree at Karanganyar, Central Java and identified in Biology Department, Faculty of Mathematics and Natural Sciences, Sebelas Maret University based on C.A. Backer & R.C. Bakhuizen van den Brink [7]. The leaves were separated from the stalk, washed, air-dried at room temperature (25 °C), then dried with oven at 45 °C, and crushed into coarse powder. Extract was prepared by percolating the dry powdered sample with chloroform for 2x24 hours. The liquid extract was thereafter concentrated using rotary evaporator and waterbath until giving some dark-green crude extract [8].

2.2.2. Solubility Study of The Extract in Oil Phase, Surfactant, and Co-Surfactant

A total of 1.0 mg extract was added into each eppendorf tube with 1.0 mL different component: oil (VCO, PKO, olive oil), surfactant (tween 80 (100%), cremophor EL (100%), tween 80:cremophor EL (85%:15%)), and co-surfactant (propylene glycol, glycerin). The mixture was vortexed on vortex mixer for 1.0 minute, incubated in waterbath at 45 °C for 10 minutes, sonicated for 15 minutes, and then left for two days at room temperature. The samples were then centrifuged at 3000 rpm for 20 minutes to separate the undissolved extract. Component that was able to dissolve more of the extract were selected visually for optimization [9].

2.2.3. Optimization SNEDDS Formula with D-Optimal Mixture Design

Three independent variables were studied: oil (A), surfactant (B), and co-surfactant (C) against three response variables: transmittance (%), emulsification time (second), and phase separation (F) using D-Optimal mixture design. Oil, surfactant, and co-surfactant that showing maximum solubility of the extract from the previous study were chosen as independent variables. Table 1 showed the independent variables and the levels selected for optimization.

Table 1. Levels of components used in SNEDDS formulation.

Code	Selected Component	Low	High
A	Oil	12.00%	22.00%
B	Surfactant	60.00%	70.00%
C	Co-Surfactant	12.00%	22.00%

2.2.4. Drug Loading Measurement
Samples of soursop leaves chloroform extract (100, 50, and 25 mg) were added into 1.0 g SNEDDS system. SNEDDS were then homogenized with a vortex for 1.0 minute, sonication for 10 minutes, and incubation at 45 °C for 15 minutes in waterbath. The extract that was able to dissolve more in that system and did not show any signs of instability were selected visually for the drug loading dose per 1.0 g SNEDDS [10].

2.2.5. Characterization of SNEDDS
2.2.5.1. Transmittance
A 1.0 mL SNEDDS was added with aquabides until volume of 50 mL. The mixture was homogenized with a vortex for 1.0 minute. Percentage of transmittance then was analyzed by UV-Vis spectrophotometer at 650 nm [11].

2.2.5.2. Emulsification Time
A 200 μL SNEDDS was dripped quickly into 50 mL of Artificial Gastric Fluid (AGF) medium pH 1.2 at 37±0,5 °C under 100 rpm continuous stirring. The emulsification time was taken as the time to form homogenous mixture spontaneously upon dilution [12].

2.2.5.3. Phase Separation
SNEDDS was added into eppendorf tube 1.5 mL then was centrifuged at 3500 rpm for 3x10 minutes [13]. The phase separation was visually observed by comparing the height of emulsion that remain stable with the total height of emulsion [14].

2.2.6. Determination of Optimum Formula of SNEDDS and Verification
D-Optimal mixture design was used to determine the optimum formula of SNEDDS. Three replicates of these optimum formula of SNEDDS were characterized for transmittance, emulsification time, and phase separation and thereafter were verified using one sample t-test.

2.2.7. Evaluation of Optimum Formula of SNEDDS
2.2.7.1. Particle Size and Zeta Potential
A 1.0 mL SNEDDS was diluted with aquabides until volume of 50 mL, then homogenized with a vortex for 1.0 minute, and 3.0 mL of sample was evaluted against particle size, polidispersity index, and zeta potential using particle size analyzer (Nanopartica SZ-100, HORIBA).

2.2.7.2. Thin Layer Chromatography (TLC) Test
Chloroform extract (E) that used in this study and active fraction (F) of soursop leaves from the previous study [15] were developed in TLC plate with the mobile phase chlorofom:ethyl acetate (7:3 v/v). Then, the spots were visualized by UV lamp at 254 and 366 nm [16].

2.2.8. Data Analysis
Design Expert 9 Trial (Stat-Ease Inc.) software was used to determined the optimum formula of SNEDDS. The effect of oil, surfactant, and co-surfactant were evaluated against the respond and statistically significant result if $p < 0.05$. Verification between predicted value and observation value was conducted using one sample t-test by analytical software and statistically no significant difference if $p > 0.05$.

3. Result and Discussion

3.1. Characterization of Soursop Leaves Chloroform Extract
Determination result (113/UN27.9.6.4/Lab/2017) showed that the sample used in this study was from family Annonaceae, genus *Annona*, and species *Annona muricata* L. Characteristics of soursop extract correspond to the study from Shofa [17] that the extract have dark-green colored, high consistency,

distinctive smell of leaves, and moisture content was 1.3%. This moisture content result can illustrate the residual solvent and water content of the extract. The amounts of chloroform solvent residue of the pharmaceutical product should be less than 0.0006% [18]. But, these moisture content result in this study not yet fully illustrated the chloroform solvent residue because the water and volatile oil content also affect.

3.2. Solubility Study of The Extract in Oil Phase, Surfactant, and Co-Surfactant

Selection of right component is important prerequisite for formulation of stable SNEDDS. The drug should have good solubility in each component so as the precipitation of drug during shelf life of formulation and after dilution in GI lumen can be avoided [19]. The solubility result of the extract was maximum in VCO, tween 80:cremophor EL (85%:15%), and propylene glycol. So, these were selected component for further formulation of SNEDDS.

The predominant fatty acid in VCO is lauric acid (>47%) which is medium chain fatty acid (MCFA) [20]. High percentage of MCFA in oil phase have been preferred for nanoemulsion formulation to achieve stability [21]. Saryanti [22] that surfactant combination of tween 80:cremophor EL (85%:15%) enhance SNEDDS formulation stability. In addition, long chain alcohols (C8-C12) such as propylene glycol are good co-surfactant for SNEDDS formulation [23].

3.3. Formula Composition of SNEDDS using D-Optimal Mixture Design

D-Optimal mixture design of Design Expert 9 Trial software showed sixteen formulas with various concentration of oil, surfactant, and co-surfactant with the total concentration is 100% (Tabel 2). These 11 formulas from the software have different composition with 5 formulas replicated. This is in order to reduce any possible errors that may occur.

Table 2. Formula composition from D-Optimal mixture design (Design Expert 9 Trial software) and the characteristics of SNEDDS.

Run	Percentage of Oil (A), Surfactant (B), and Co-Surfactant (C)			Characteristics of SNEDDS of Soursop Leaves Chloroform Extract		
	Oil (%)	Surfactant (%)	Co-Surfactant (%)	Transmittance (%)	Emulsification Time (second)	Phase Separation (F)
1	17.33	65.33	17.33	83.00	15.00	0.923
2	12.00	66.00	22.00	88.40	8.00	0.969
3	22.00	63.00	15.00	91.30	24.00	0.885
4	22.00	66.00	12.00	84.90	28.00	0.885
5	18.00	70.00	12.00	81.90	43.00	0.923
6	22.00	60.00	18.00	86.20	6.00	0.885
7	12.00	70.00	18.00	78.70	26.00	0.961
8	12.00	70.00	18.00	83.80	30.00	0.969
9	18.00	60.00	22.00	85.50	18.00	0.923
10	12.00	66.00	22.00	84.90	11.00	0.969
11	19.67	65.67	14.67	79.70	9.00	0.904
12	18.00	70.00	12.00	76.80	39.00	0.923
13	22.00	60.00	18.00	90.60	14.00	0.885
14	15.00	63.00	22.00	85.90	8.00	0.942
15	18.00	60.00	22.00	81.80	10.00	0.923
16	15.00	70.00	15.00	70.60	20.00	0.942

SNEDDS in 4 g system

3.4. Drug Loading

Drug loading measurement was aimed to identifying the maximum concentration of the extract that can dissolve homogeneously in the SNEDDS system. The test result obtained from drug loading at 25 mg/g indicated that the SNEDDS system is capable of dissolving extract. The solubility testing parameters were indicated by the clarity, homogenicity, and no precipitation in the SNEDDS system when visually observed.

3.5. Characteristics of SNEDDS

Characteristics of SNEDDS describing physical stability such as transmittance, emulsification time, and phase separation were used to determining the optimum formula in this study. Percentage transmittance was used to describe the clarity of SNEDDS system formed in aqueous media. Nanoemulsion have droplets with diameter on order of less than 100 nm, so nanoemulsion are often transparent in appearance as the droplet size is significantly smaller than the wavelength of visible light [24]. The transmittance value from the 16 run formulas ranged from 70.60% to 91.30% (Table 2).

Emulsification time is an important parameter to assess efficiency of SNEDDS to form nanoemulsion spontaneously in GI media under mild agitation. A perfect SNEDDS formulation will able to disperse completely and quickly which takes about less than 1.0 minute with a view clear nanoemulsion [11]. This study showed that 16 formulas of SNEDDS have emulsification time less than 1.0 minute with a clear appearance (Table 2).

Besides that, nanoemulsion also a kinetically stable system that takes quite long time to separate into different phase. Accelerated stability testing with centrifugation method was aimed to shorten the observation time by stiring at 3500 rpm for 20 minutes could be said equivalent to gravity for ±1,0 year [13]. The centrifugation force exerts pressure to the SNEDDS system thus destroying the interface layer which causes the oil to separate. Phase separation value (F) was obtained from the separation level of SNEDDS system. The higher the remining stable layer, the higher the phase separation value will be.

3.6. Optimum Formula of SNEDDS and Verification

The experimental mixture design technique is applied for optimization for independent varibles when response variables are dependent only on the proportions of the ingredients of mixture. D-Optimal mixture design is commonly used to reveal main effects and interaction effects between the independent variables against response variables. The interaction between independent variables against response variables in this study showed by polynomial equation from the Design Expert 9 Trial software (Table 3).

Table 3. Physical stability responses of SNEDDS based on mathematics model and ANOVA statistical analysis.

Response	Polynomial Equation	Mathematics Model	Lack of Fit [p >0.05]	p-value (ANOVA) [p <0.05]
Transmittance (%)	$Y = 13.45\,A + 0.34\,B + 3.17\,C - 0.14\,AB - 0.29\,AC + 0.01\,BC$	quadratic	0.3175	0.0197
Emulsification Time (second)	$Y = 20.93\,A + 6.12\,B + 33.14\,C - 0.43\,AB - 0.37\,AC - 0.64\,BC$	quadratic	0.0885	0.0031
Phase Separation (F)	$Y = 1.94\,A + 0.01\,B + 0.02\,C - 1.17\,AB - 9.77\,AC - 2.27\,BC$	quadratic	0.6690	<0.0001

All physical stability responses provide a quadratic model because the ANOVA statistical analysis have p <0.05 (Table 3). These result supported with Lack of Fit >0.05 which means there is no significant difference between the observation value with the prediction value from the model so that the quadratic model is precise. The polynomial equation of transmittance and emulsification time response showed that interaction between surfactant and co-surfactant in the mixture give the greatest influence in increasing percentage transmittance with the coefficient +0.01 and decreasing the emulsification time with the coefficient -0.064. This result correspond with the study from Saryanti [22] in which increased of surfactant combination of tween 80-cremophor EL and co-surfactant propylene glycol may increase trasmittance. Moreover, emulsification time is also mediated by the interaction of surfactant and co-surfactant by form the interface layer between oil and water [12]. Co-surfactant helps surfactant in reducing the interface tension between oil and water so that emulsification occurs spontaneously.

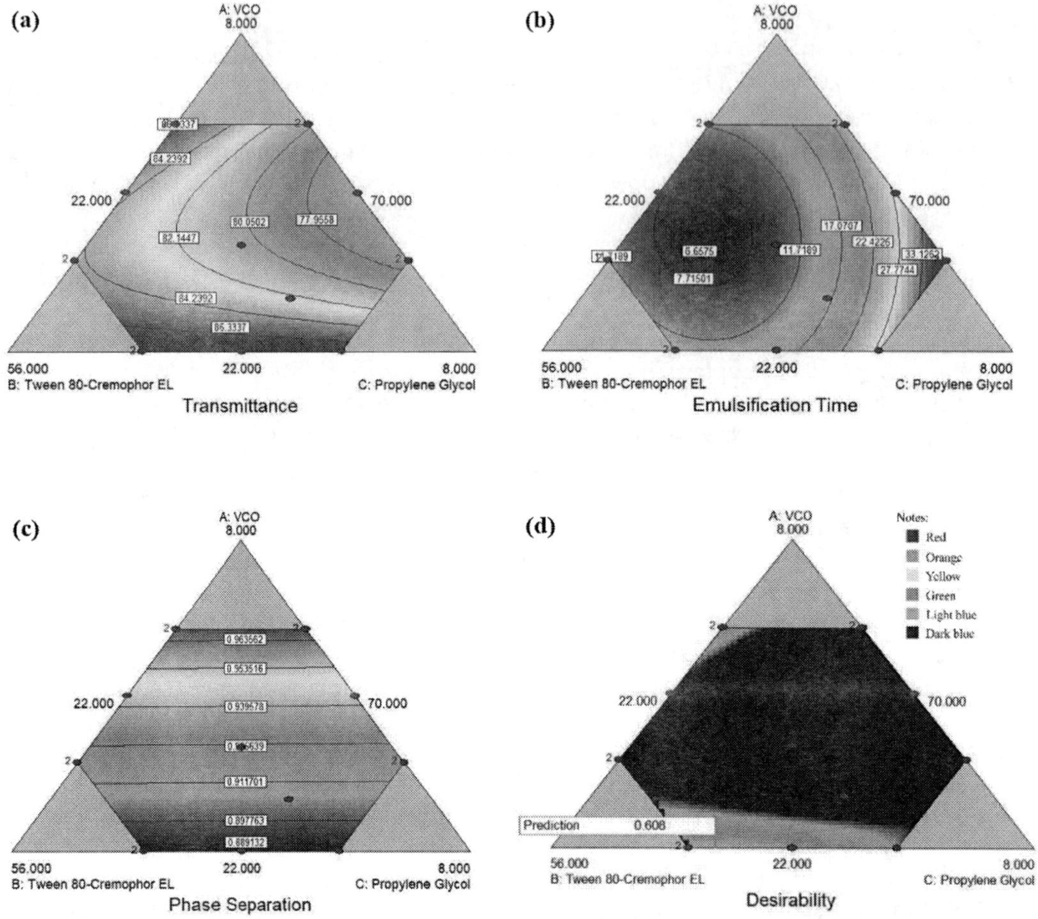

Figure 1. Contour plot diagram physical stability responses of SNEDDS: (a) transmittance, (b) emulsification time, and (c) phase separation. (d) Super impose diagram showed the optimum area of the oil : surfactant : co-surfactant component.

Contour plot diagram showed the interaction between oil (VCO), surfactant (tween 80- cremophor EL), and co-surfactant (propylene glycol) against each of the responses. The highest response indicated by the red area and the lower response indicated by the area of yellow, green, light blue, until dark blue (Figure 1). Based on optimization result, selected optimum formula of SNEDDS showed highest desirability value (0.608) in super impose diagram. This formula will produce optimum characteristic value that correspond with the desired value. The composition of the optimum formula is 22.00% oil : 60.79% surfactant : 17.21% co-surfactant.

Verification of optimum formula was aimed to compare between observation value with prediction value from the Design Expert 9 Trial software. Non-statistically significant results ($p > 0.05$) between predicted and observed value of transmittance, emulsification time, and phase separation on one sample t-test were shown in Table 4. As the result, it can be concluded that the optimum formula suggested by Design Expert 9 Trial software have no significant difference with the observation value so fulfill the verification.

Table 4. Verification of optimum formula of SNEDDS.

Response	Prediction	CI	Observation ± SE	p-value [p >0,05]
Transmittance (%)	88.39	84.09 - 92.71	84.64±1.64	0.150
Emulsification Time (second)	12.91	20.60 - 7.19	14.79±1.81	0.409
Phase Separation (F)	0.884	0.88 - 0.89	0.898±0.01	0.393

n=3 for each response

3.7. Evaluation of Optimum Formula of SNEDDS

3.7.1. Particle Size and Zeta Potential. Result of the particle measurement was 42,73 nm with polidispersity index (PI) value 0.37. The optimum SNEDDS showed uniform particle size distribution because the PI value closer to 0,0 [11]. Zeta potential was used to determine the value of repulsion between dispersion particles so the aggregation and phase separation can be avoided. The zeta potential in this research was high enough, that was -27.43 mV.

3.7.2. Thin Layer Chromatography (TLC). Active ingredient in the chloroform extract (E) was predicted using simple TLC method that use active fraction (F) of soursop leaves as positive control (Figure 2). These active fraction obtained from the previous study showed inhibition activity against HeLa cells with IC_{50} 96,00 µg/mL [15]. TLC spot at $UV_{366\ nm}$ with hRf value 95 was identified as active ingredient in comparison with active fraction that had the same hRf value. These result according to Pranatami [17] that hRf value of the most active fraction of soursop leaves was 95 with the same mobile phase in this study. Moreover, all of the plates showed a light-red alongside the spots. It is important to show that there was similarity of active ingredient in the extract with the active compound of the fraction.

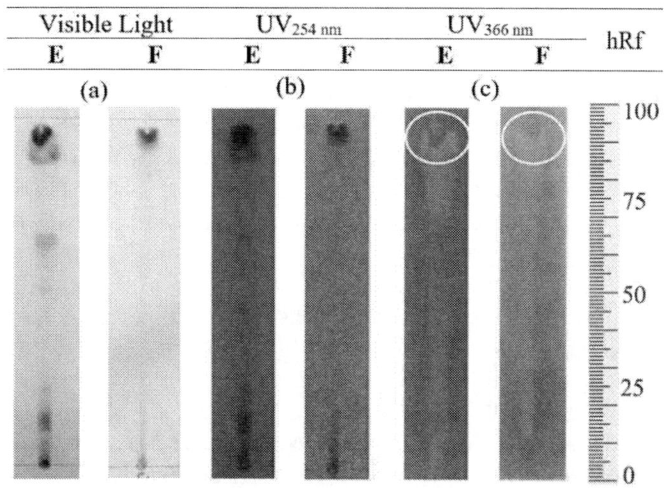

Figure 2. TLC profile of chloroform extract (E) and active fraction (F) of soursop leaves visualized by (a) visible light, (b) $UV_{254\ nm}$, and (c) $UV_{366\ nm}$. The stationary phase was TLC silica gel 60 F_{254} and the mobile phase was chloroform:ethyl acetate (7:3 v/v).

4. Conclusion

The optimum formula of SNEDDS of soursop leaves chloroform extract based on D-Optimal mixture design was 22.00% VCO, 60.79% tween 80-cremophor EL, and 17.21% propylene glycol with drug loading extract was 25 mg/g system. The verification test showed no significant difference ($p > 0.05$) between the observation and the prediction from the software. The optimum formula of SNEDDS fulfill the criteria of nanoemulsion with the 84.64% (SE ± 1.64) of transmittance; 14.79 seconds (SE ± 1.81) of emulsification time; and 0.898 (SE ± 0.01) of phase separation; 42.73 nm (SE ± 0.49) of the particle size; polidispersity index 0.37 (SE ± 0.10); and -27.43 mV (SE ± 0.26) of zeta potential. In addition, from the TLC results, we can predict that there was similarity of active ingredient in the extract with the active compound of the fraction.

Acknowledgement
This work is a collaboration with Natural Medicine Research Group and was supported by grant project from Sebelas Maret University.

References

[1] Tellez A V C, Gonzalez E M, Yahia E M and Vazquez E N O 2016 *Arab. J. Chem.* **11** 662–91
[2] Yang C, Gundala S R, Mukkavilli R, Vangala S, Reid M D and Aneja R 2015 *Carcinogenesis* **36** 656–65
[3] Rosdi M N M, Daud N N N N M, Zulkifli R M and Ya'akob H 2015 *J. App. Pharm. Sci.* **5** 45–8
[4] Agu K C and Okolie P N 2017 *Food. Sci. Nutr.* **5** 1029–36
[5] Hong J, Sun Z, Li Y, Guo Y, Liao Y, Liu M and Wang X 2017 *Int. J. Nanomedicine* **12** 5053–67
[6] Elisya Y, Kardono L B S and Simanjutak P 2014 *AJAS* **02** 323–9
[7] Bracker C A and Bakhuizen B R C 1963 *Flora of Java (Spermatophyte Only)* (Noordhoff-Groningen: The Netherlands)
[8] Setyorini H A, Kurniatri A A, Adelina R and Winarsih 2016 *BPK* **44** 279–86
[9] Pratiwi L, Fudholi A, Martien R and Pramono S 2017 *J. Young. Pharm.* **9** 341–6
[10] Aprian R F 2015 Formulasi SNEDDS (*Self-Nano Emulsifying Drug Delivery System*) Ekstrak Etanolik Akar Purwoceng Gunung (*Artemisia lactiforia* Wall. ex DC.) dengan VCO sebagai Fase Minyak, Tween 80 sebagai Surfaktan dan PEG 400 sebagai Kosurfaktan, Yogyakarta: UGM
[11] Pratiwi L, Fudholi A, Martien R and Pramono S 2016 *Int. J. PharmTech. Res.* **9** 380–7
[12] Winarti L, Martien R, Suwaldi and Hakim L 2016 *IJPCR* **8** 425–32
[13] Restu W K, Sampora Y, Meliana Y and Haryono A 2015 *Procedia Chem.* **16** 171–6
[14] Ermawati D E, Martodihardjo S and Sulaiman T N S 2017 *JPSCR* **02** 78-89
[15] Agpri K S 2018 Sinergisitas Efek Sitotoksik Kombinasi Fraksi Aktif Daun Sirsak (*Annona muricata* L.) dan Kayu Secang (*Caesalpinia sappan* L.) pada Sel Kanker Serviks HeLa, Surakarta: UNS
[16] Pranatami D A 2014 Uji Sitotoksisitas Fraksi Kloroform Ekstrak Daun Sirsak (*Annona muricata* L.) terhadap Sel Raji dan Profil Kandungan Kimia Fraksi Teraktif, Surakarta: UNS
[17] Shofa A S 2017 Optimasi Formula Sediaan SMEDDS (*Self-Microemulsifying Drug Delivery System*) Ekstrak Kloroform Daun Sirsak (*Annona muricata* L.) dengan Fase Minyak Kemiri, Surakarta: UNS
[18] Grodowska K and Parczewski A 2010 *Acta. Pol. Pharm,* **67** 3–12
[19] Venkatesh M and Mallesh K 2013 *J. Drug. Deliv. Ther.* **3** 131–42
[20] FAO 2006 *Virgin Coconut Oil* (Bangkok: FAO Regional Office for Asia and the Pacific)
[21] Huang J 2014 Composition of Nanoemulsion Delivery Systems *WO2015123631A1*
[22] Saryanti D 2016 Optimasi Self-Nanoemulsifying Drug Delivery System Ekstrak Akar Purwoceng Gunung (*Artemisia lactiflora* wall ex. DC) dan Uji Absorpsi secara In Vitro [Thesis], Yogyakarta: UGM
[23] Cerpnjak K, Zvonar A, Gasperl M and Vrecer F 2013 *Acta Pharm.* **63** 427–45
[24] Gupta A, Erai H B, Hatton T A and Doyle P S 2016 *Soft Matter* **12** 2826–41

IOP Conf. Series: Materials Science and Engineering **578** (2019) 012047 doi:10.1088/1757-899X/578/1/012047

Influence of CMC-Na Concentration to Physical Properties of SNEDDS Kersen leaves Ethanolic Extract Gel

D E Ermawati[1] and C Wulandari[1]

[1] Department of Pharmacy Diploma, Matematics and Natural Science Faculty Sebelas Maret University, Surakarta, Indonesia

E-mail: dianekaerma@gmail.com

Abstract: Kersen leaves (*Muntingia calabura* L.) consist of flavonoid, saponin and tannin that has bacteriostatic activity to bacteria caused acnes such as *Staphylococcus epidermidis*. In this research the extract is loaded into SNEDDS [*Self Nano Emulsifying Drugs Delivery System*] in order to optimalize the absorbtion of active substances because of the lipophilic character of stratum corneum. SNEDDS then dispersed into hydrogel matrix in order to help aplication on skin. Kersen leaves extracted using ethanol 96 % by maceration method. SNEDDS formula compotition are candlenut oil, tween 80 as surfactant and PEG 400 as co-surfactan in ratio 1:6:1, that contained 30 mg extracts each system. The hydrogel matrix made with carbopol with variation concentrations: 0.06 %; 0.13 % and 0.19 %, then physical property test: organoleptic, homogeneity, pH, viscosity, separation, adhesive and cycling test. Data were analyzed using One Way ANOVA to know the influence of variation of CMC-Na concentration. The results showed that SNEDDS Kersen leaves Ethanolic Extract with 0.06 % carbopol concentration indicated an optimum formula based on physical stability tests. Statistical analysis of the variation of carbopol concentration showed a significant difference on physical characteristics there are: viscosity, pH, homogeneity, and separation test.

1. Introduction

Acnes usually can be treatmen with antibiotics. According to Utami [1] the long term use of antibiotics causing microbial resistance, also cause organ damage and immunohypersensitivity, so that we need to develop acne treatment using traditional plants as active ingredients. One of the plants that can be used is kersen leaves. Based on Handayani's research [2], ethanolic extract of kersen leaves has bacteriostatic activity against *Staphylococcus epidermidis* which is a bacterium that causes acne at concentration of 9,0 ppm and has diameter of inhibitory 14.0 mm. Nanoemulsion is one of pharmaceutical dosage form that choosen to develope ethanolic extract of kersen leaves. Nanoemulsion has small particle size (200 - 300 nm) so can speed up the process of dissolving active substances of ethanolic extract of kersen leaves into skin [3], and also causes the drug to penetrate micron-sized bacterial cells more effectively. SNEDDS (Self Nano Emulsifying Drugs Delivery System) consist of the composition of oil, surfactants and cosurfactants [4]. Candlenut oil is chosen as an oil phase because it has better solubilization ability of hydrophobic drugs [5]. The selection of tween 80 as a surfactant because of its stable and have no toxicity and PEG 400 as a cosurfactant aims to create higher HLB (Hydrophilic Lipophilic Balance) values that meet the requirements in making

SNEDDS [6]. Gel preparations with extracts as active ingredients have been found, but have a less aesthetics appearance due to the brown color of the extract. That causes the preparation to be less pleasant when applied to the skin. SNEDDS ethanolic extract of kersen leaves that dispersed into hydrogel is one solution because it is able to produce clear appearance and the effectiveness of active substances to get the action target into skin. Carbopol is polymer that widely used as hydrogel matrix. Research of Kumar and Kumar, 2011 [7] shows that the physical properties of gel preparations that use carbopol as gelling agent are better than HPMC and CMC-Na. The basis for choosing carbopol concentration in this study refers to Djajadisastra, 2009 [8], which states that the optimal concentration of carbopol is 0.13-0.18 % with the concentration of propylenglycol as humectant between 5-9 % [6]. Based on the description above, SNEDDS ethanolic extract of kersen leaves (*Muntingia calabura* L.) were dispersed on hydrogel with variations carbopol concentration of 0.06 %; 0.13 %; and 0.19 %, and the physical properties of the preparation were tested which included organoleptic, homogeneity, pH, adhesion, dispersion, viscosity and stability test.

2. Experimental

2.1. Materials

GenesysTM 10s UV/VIS spectrophotometer, moisture analyzer, *rotary evaporator* (STUART RE300DB), pH meter (LUTRON PH-208), drop pipette, beker glass, *vortex* (MAXI MIX II), *sonicator* (DSA50–GL1-1.8L), *viscosimeter* (RION VT-04/03), stopwatch, water bath, Prescia BJ 4501C and Sartorius BP 221S analytical balance sheet, mikroskop (Nikon E100), centrifugation (HERAEUS FRESCO 17 CENTRIFUGE), *refrigerator* (LG).

Etanol 96 % (Bratachem), HCl (Merck), Mg powder (Merck), $FeCl_3$ (Merck). Candlenut oil (CV. Agung Jaya), Tween 80 (Merck), PEG 400 (Merck), Carbopol (Bratachem), Gliserin (Cusson), TEA (Bratachem), Propilenglikol (DOW), Metil Paraben (Bratachem).

2.2. Methods

2.2.1. Preparation ethanolic extract of kresen leaves

Kersen leaves taken from Cepu, Blora, Central Java, Indonesia. The selected leaves are old leaves because they contain more active compounds. Kresen leaves of 2.0 kg were washed with clean water, then dried into oven at temperature of 40 – 60 °C for 1 hour. The dried leaves then powdered into a smaller size by hand. Kersen powder of 300 grams, masetaed with 96 % ethanol until the surface of the powder was completely submerged, left for 3x24 hours with several times stirring. Maserate was filtered, then evaporated with a rotary evaporator and concentrated with waterbath.

2.2.2. Detection of active compounent kersen leaf ethanolic extract

The thickened extracts were tested for moisture content with Moisture analyzer, rendement was calculated, and phytochemical qualitative tests included flavonoid, saponin and tannin tests. Test for flavonoids : by dissolving the sample into 2.0 mL 96 % ethanol solvent, adding a few drops of HCl and Mg powder. A positive result if it shows radish/orange colour of extract. *Tanin Test* : by adding $FeCl_3$ 1 % reagent to the extract, positive results if formed in black or greenish brown. Saponin test : by dissolving the sample into aquadest, and shaking it. Positive results if a stable foam is formed which does not disappear after 10 seconds.

2.2.3. SNEDDS formulation and Hydrogel

SNEDDS formulas of Kersen Leaves Ethanol Extract (*Muntingia calabura* L.) consists of oil phase: surfactant: co-surfactant with optimum ratio of 1: 6: 1 and extract content of 30 mg for every 1.0 g of system. The ingredients are weighed, then homogenization using vortex for 1-2 minutes, then incubated at 45 °C for 10 minutes. The SNEDDS formed underwent transmittance tests and centrifuged at 3000 rpm for 15 minutes at room temperature. The Hydrogel formula of SNEDDS at Table 1.

Hydrogel formulation made of 100 g. The hydrogel base is made using carbopol that dispersed into warm aquadest, adding TEA while stirring homogeneously. Methyl paraben is dissolved in glycerin, then mixed in a hydrogel base. Propylene glycol is added, stirred homogeneously, added with distilled water ad 100 % and added 2 g of SNEDDS, stirred ad homogeneously.

Table 1. Hydrogel formula of SNEDDS Kersen ethanolic extract with variation concentration of CMC-Na

Ingredients (mg)	**% w/w**		
	F1	**F2**	**F3**
SNEDDS concist extract 30 mg/system	2	2	2
Carbopol	0.063	0.125	0.188
Gliserin	10	10	10
Propylene Glycole	15	15	15
Triethanoamine	1	1	1
Methyl Parabene	0.1	0.1	0.1
Aquadest	ad 100	ad 100	ad 100

2.2.4. Physical test of SNEDDS Kersen Leaf Ethanolic Extract (*Muntingia calabura* L.)
Physical test including organoleptic test, homogeneity, pH, adhesion, dispersion, viscosity and cycling test. Tests were carried out at 0, 14 and 28 day.

2.3 Data Analysis
Shapiro-Wilk test using SPSS 21 Software and data that normal distribution continue analysed with *One-Way ANOVA* to get significant information. Significant information data of One-Way ANOVA then continue analysed *Post HOC* test.

3. Results and Disscussion
The results of the determination in the Biology Laboratory of the Mathematics and Natural Sciences Faculty, Sebelas Maret University Surakarta, Indonesia based on Flora of Java book written by C.A Backer and R.C. Bakhuizen van den Brink, Jr. (1963, 1968) stated that the plants used in this study were kersen leaves (*Muntingia calabura* L.).

The result of kersen leaves ethanolic extract was 26.30 g with a yield value of 8.76 %. The results of the water content test showed that the thickened moisture content of the extract was 14.95 %, where the requirements of this water content had met the specified requirements, namely for thick extracts containing 5-30 % [9]. Organoleptic test results of kersen leaves ethanolic extract are dark green with the characteristic smell of kersen leaves and thick consistency. Phytochemical content test was carried out on extracts and SNEDDS and gel following results at Table 2.

Table 2. Phytochemical qualitative test of kersen leaves ethanolic extract and SNEDDS kersen leaves ethanolic extract

Sample test	**Phytochemical qualitative test**		
	Flavonoid	**Saponin**	**Tanin**
extract	+	+	+
SNEDDS	+	+	+

3.1. Transmitance Test and Centrifugation Test on SNEDDS
SNEDDS formula was selected based on a trial of several formula comparisons, and the formula was obtained by comparing the composition of candlenut oil: tween 80: PEG 400 was 1: 6: 1 with an extract content of 30 mg (every 1.0 g of SNEDDS). The amount of extract content has been in

accordance with the literature, and has been able to provide bacteriostatic effects on the bacteria *Staphylococcus epidermidis*. The transmittance test on the selected formula shows the transmittance rate is 92.19 %. The results of particle size test using PSA showed that the droplet extract size in SNEDDS was 15.2 nm. Centrifugation tests were carried out on SNEDDS. The result is that there is a phase separation with a separation ratio (F) value of 0.92. In several studies, an accelerated stability test was used to shorten observation time. The right centrifugation force will cause oil separation by damaging the emulgator layer absorbed around each grain. The more stable the emulsion is the greater the centrifugation force needed to damage the emulsion layer [10].

3.2. *Physical test of SNEDDS Hydrogel of Kersen Leaves Ethanolic Extract (Muntingia calabura* L.)

Observation of physical properties was carried out at 0 day, 14 day and 28 day. There are : Organoleptic testing of SNEDDS hydrogel aims to determine organoleptic changes during storage at during 28 days. Organoleptic testing includes observation of the color, smell, and consistency of the hydrogel. The results of organoleptic observations are presented in Table 3.

The test results showed that the hydrogel formulation in three formulas did not show organoleptic changes. This shows that the hydrogel formulation is stable at room temperature storage. The difference in carbopol concentration affects the level of consistency of gel preparations.

Homogeneity testing aims to find out that all ingredients are mixed homogeneously. SNEDDS hydrogel of kersen leaves ethanolic extract showed that all formulas were homogeneous and did not change during the storage period. All formulas show evenly dispersed particles. It can be concluded that the variation of carbopol gelling agent concentration on SNEDDS hydrogel formulation does not affect the homogeneity.

Table 3. Organoleptics test of SNEDDS Hydrogel of kersen leaves ethanolic extract during 28 days for Formula 1, 2 and 3

Formulas	Organoleptic test	Day to-		
		0	14	28
F1	Consistency	Viscous	Viscous	Viscous
carbopol	Colour	Yellow Brown	Yellow Brown	Yellow Brown
0.06 %	Odor	Kersen leaves	Kersen leaves	Kersen leaves
F2	Consistency	Thick	Thick	Thick
carbopol	Colour	Yellow	Yellow	Yellow
0.13 %	Odor	Kersen leaves	Kersen leaves	Kersen leaves
F3	Consistency	Solid	Solid	Solid
carbopol	Colour	Clear yellow	Clear yellow	Clear yellow
0.19 %	Odor	Kersen leaves	Kersen leaves	Kersen leaves

3.2.1. *Viscousity test*

Viscosity testing aims to determine the level of viscosity hydrogel that produced. A good gel is not too liquid or not too thick. Viscosity test results are presented in Figure 1. The viscosity values of optimum gel formulation are 2000-4000 cps [11]. The viscosity test results show that only formula I meets viscosity requirements. Based on observations it is known that the viscosities value of all formula have decreased. This can be caused by an increaseing temperature, where the heat obtained will increase the distance between moleculs so that the force will decrease and the distance becomes tenuous. Based on the normality test with the Shapiro-Wilk method during storage time showed that the viscosity data of formula I, formula II and formula III were not normally distributed (Sig <0.05). One Way Anova test cannot be done, because the data is not normally distributed. Based on the results of viscosity testing it is known that the higher concentration of carbopol influence the viscosity of the gel.

3.2.2 pH test

pH testing aims to determine the pH value of the hydrogel taht produced. The pH requirements that do not irritate the skin are between 5-9 [6]. The pH of an overly acidic preparation can irritate the skin, whereas if the pH is too alkaline it can make the skin dry. The pH test results are presented in Figure 1. In observations every 14 days it is known that the pH of each formula has decreased. Changes in pH values can indicate a reaction or damage to the constituent components in the formulation, it caused may the hydrogel that used carbopol, so that it can reduce the pH value [7].The results of ANOVA test between formulas showed that the pH of formula I, formula II, and formula III experienced a significant difference (Sig = 0,000). Post Hoc follow-up tests were conducted to find out which groups gave significant differences and obtained the results that the formula I, formula II and formula III each had a significant difference.

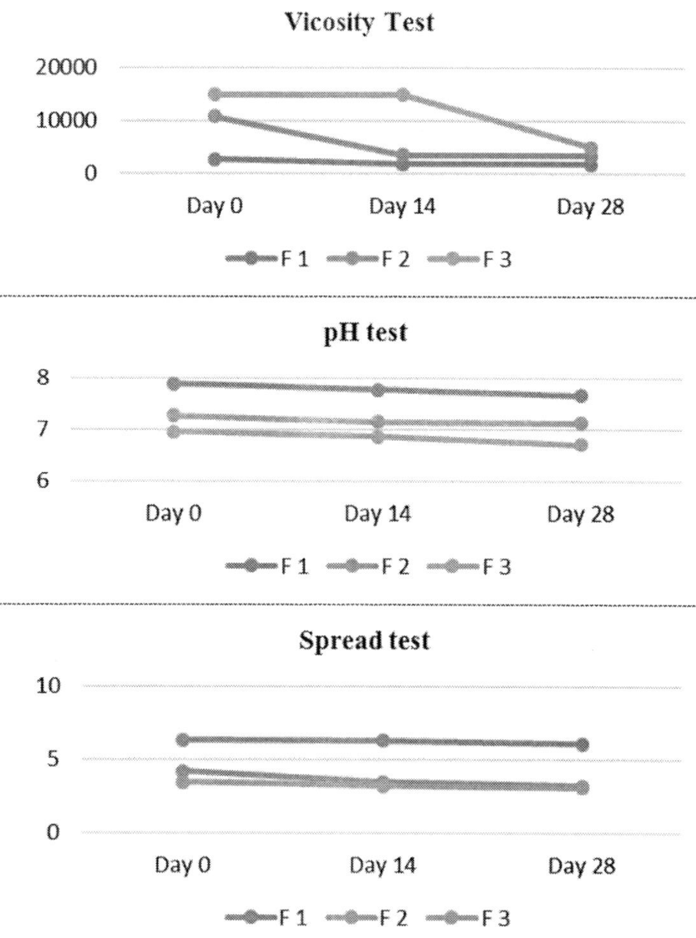

Figure 1. Diagram of physical properties test of SNEDDS hydrogel of kersen leaves ethanolis extract includes viscosity, pH and spread test during 28 days. Higest concentration of CMC-Na caused viscosity and pH decrease each weeks, and increase the spread test.

3.2.3 Cycling Test

Cycling test of hydrogel formula conducted for 6 cycles shows that formula I, formula II and formula III stable organoleptically and Hydrogel formula not separate likes syneresis process. Syneresis is the process where a liquid is separated from the gel, it cause when gel placed too long or

IOP Conf. Series: Materials Science and Engineering **578** (2019) 012047 doi:10.1088/1757-899X/578/1/012047

stored in the open air, and the temperature rises [8]. This results in contraction in the gel mass. The entangled liquid will come out and be above the surface of the gel· Based on the results of the cycling test, it was concluded that the gel formula of SNEDDS stable the active component of kersen leaves ethanolic extract when stored at cool temperatures (2-8 °C) and room temperature (25 ±2 °C).

3.2.4 Spread test

Spread power test aims to determine the ability of the gel formulation to spread when applied to the skin. Spread test results are presented in Figure 1. The optimum spread of gel formula is between 5-7 cm^3. Based on this literature, the spread value that meets the requirements only in formula I. Observations at day 0, day 14 and day 28 indicate that the dispersion power of each formula has decreased. During storage there can be a decrease in dispersion due to the retention of solvent which is absorbed by the gelling agent [5].

4 Conclusion

Kersen leaves ethanolic extract have flavonoid, tanin, and saponin compounend. Based on the evaluation of the physical properties test of SNEDDS hydrogel of kersen ethanolic extract that increasing the concentration of carbopol as a gelling agent causes an increase viscosity and a decrease in dispersion and pH values. Formula I with concentration of carbopol 0.06 % shows the best results, because it meets the main requirements as a optimum gel formula.

References

[1] Utami E R 2012 *Sainstis* **1**(1) 124-138.
[2] Handayani V 2015 *J. Fitofarmaka Indones.* **2**(1).
[3] Sadurní N, Solans C, Azemar N, and García-Celma M J 2005 *Eur. J. Pharm. Sci.* **26**(5) 438-445.
[4] Makadia H A, Bhatt A Y, Parmar R B, Paun J S and Tank H M 2013 *Asian J. Pharm. Res.* **3**(1) 21–27.
[5] Sulastri A and Chaerunisaa A Y 2017 *Farmaka* **14**(3) 17–26.
[6] Rowe R C, Sheskey P J, Cook W G, and Quinn M E 2013 *Handbook of Pharmaceutical Excipients–7th Edition*. Taylor & Francis.
[7] Kumar V and Kumar S 2011 *Int J Pharm Pharm. Sci.* **3** 55-57.
[8] Djajadisastra J, Mun'im A, and Np D 2009 *Jurnal Farmasi Indonesia* **4**(4).
[9] Evan E 1943 *Moisture contents of extract* University of Louisville. [Accessed: 29-Dec-2018].
[10] Ermawati D E 2017 *J. Food Pharm. Sci.* **4**(3).
[11] Garg A, Aggarwal D, Garg S, and Singla A K 2002 *Pharm. Technol.* **26**(9) 84-105.

IOP Conf. Series: Materials Science and Engineering **578** (2019) 012048 doi:10.1088/1757-899X/578/1/012048

Antibacterial Activity of Natural Paper from Banana Peel (*Musa paradisiaca* Linn.) with Additive Essential Oils

W A E Setyowati[1] and E Susanti V H [1]
[1]Department of Chemistry Education, Faculty of Teacher Training and Education,Universitas Sebelas Maret, Jl. Ir. Sutami 36A Kentingan Surakarta 57126 Indonesia

E-mail: widiastuti_aes@staff.uns.ac.id

Abstract. Several essential oils have been evaluated as antibacterial agent in the production of natural paper from banana peel (*Musa paradisiaca* Linn.), namely cinnamon, clove, lime and lemon oil. The purpose of the research was to examine the antibacterial activity of the papers with additive of those several essential oils againt *Staphylococcus aureus* and *Escherichia coli*. Paper preparation was successfully carried out by alkalization method, using 4% NaOH at a temperature of 100 °C for 1 h. The essential oils as antibacterial agent were added with concentration 2.5% to the pulp (1:40). The antibacterial activities were tested using Kirby-Bauer disc diffusion method. All of the paper produced meet the standard of ISO 6519:2016 about Basic Paper for Laminated Plastic Wrapping Paper within the parameter of pH and water content. The result of the research showed that all of the paper with essential oils additive has antibacterial activity againt *S. aureus*, but only paper with cinnamon oil additive that have antibacterial activity againts *E. coli*. The use of cinnamon oil gave the largest inhibition zone diameter both on *S. aureus* and *E. coli*. Paper with cinnamon oil additive had inhibition zone diameter 18 mm on *S. aureus* (strong category) and 9 mm on *E. coli* (medium category).

1. Introduction

Bacteria are the cause of many diseases suffered by the people of Indonesia, where one of the spreads is through food. Bacterial contamination in food can occur directly where contaminants are obtained from the place of life or origin of the food ingredients, or indirectly where contaminants are obtained through intermediaries. Wrapping paper unwittingly has become a bacterial contamination intermediary on food. So far, the raw material for making paper is wood cellulose. The more paper needs, the more wood is needed, so that over time more and more deforestation will occur. Banana peel has been reported to be an alternative raw material for making paper because of its high cellulose content. The resulting banana peel paper is reported to have fulfilled the SNI requirements on the parameters of pH, moisture content, grammage and brightness [1].

The way to weaken bacteria is by giving antibacterial compounds, a chemical agent that is able to inactivate bacteria by inhibiting bacterial growth (bacteriostatic) and killing bacteria (bactericidal) [2]. The mechanism of inhibition of bacteria by antibacterial substances can be done in 5 ways: 1) inhibiting microbial cell wall synthesis, 2) damaging the integrity of microbial cell walls, 3) inhibiting microbial cell protein synthesis, 4) inhibiting nucleic acid synthesis, and 5) damaging microbial cell nucleic acids [3].

Essential oil (also called volatile oil) is an aromatic oily liquid obtained from various plants (flowers, buds, seeds, leaves, twigs, bark, herbs, wood, fruit and roots). Essential oils are known to contain various chemical compounds, such as terpenes, aldehydes, alcohols, esters, phenolics, ethers, and ketones [4]. Essential oils are reported to have activity as antibacterial, antifungal, antiviral insecticides and antibiotics [5, 6]. Some essential oils, including cinnamon oil, clove oil, lime oil and orange oil, are reported to have antibacterial activity against *S. aureus, B. subtilis K. pneumoniae,P. vulgaris, P. aeruginosa* and *E. coli* [7]. Cinnamon oil has the main component of E-cinnamaldehyde with a percentage reaching 94.67% [8]. The main component of clove oil is eugenol which is used to reduce pain [9]. While the main components of lime and orange oil are limonene and β-pinena [10]. By adding essential oils as antibacterial agents in making natural paper with banana peels as raw material, antibacterial paper is produced to prevent bacterial contamination of food.

2. Experimental

2.1. Materials

The material for making paper was Kepok banana peel, NaOH, tapioca, essential oil consisting of cinnamon oil, lemon oil, clove oil and lime oil. Material for antibacterial activity assay was Mueller Hinton Broth, Mc. Farland, NaCl, potassium chloride and distilled water. A strain of gram-negative bacteria, *E. coli* ATCC 25922, and gram-positive bacteria, *S. aureus* ATCC 25923 was used. Distilled water are used as a negative control, tetracycline as a positive control for gram- positive bacteria, and chloramphenicol is positive control for gram-negative bacteria.

2.2. Method

2.2.1. Paper Production

Paper preparation carried out by alkalization method, using 4 % NaOH at a temperature of 100 °C for 1 h. Bleaching process with 10 % hydrogen peroxide solution. The essential oils as antibacterial are added with concentration 2.5 % to the pulp (1:40). Essential oils used in this research are cinnamon oil, clove oil, lime oil, and lemon oil. Tapioca is added at concentration of 10 % as sizing agent.

2.2.2. Water Content and pH Measurement

The natural paper produced are tested for pH based on SNI ISO 6588-1.2010 and water content by dry-oven method based on SNI ISO 287:2010. All of the paper produced meet the standard of ISO 6519:2016 about Basic Paper for Laminated Plastic Wrapping Paper within the parameter of pH and water content.

2.2.3. Antibacterial Activity Assay

The antibacterial activities are tested using Kirby-Bauer disc diffusion method. All bacterial cultures were diluted with 0.9 % NaCl to 0.5 MacFarland. A bacterial suspension was swab on Mueller-Hinton Broth using sterile cotton bud. The culture was incubated at 37 °C for 24 hours. The diameter of inhibition zone was measured using calipers.

Paper disc from natural banana peel paper in some variation essential oil and put the disc in a petri dish with medium content bacteria inside. The petri dish was incubated at 30 °C for 24 hours. Inhibition Zone Diameter (IZD) was measured as a middle line start from the clear spot around the disc.

3. Result and Discussion

3.1. Physical and Chemical Characteristic of paper

Natural paper from banana peel has been succesfully produced by alkalization method. The delignification process occurs with the addition of NaOH. Lignin contained in fiber sources (cellulose) will soften into strong fragments due to the transfer of hydrogen ions from hydroxyl groups in lignin to alkaline hydroxyl ions (Figure 1). Tapioca was added as sizing agent meanwhile essential oils was added as aroma agent as well as antibacterial agent. The appearance of paper produces as shown in Figure 2 are yellowish white with aroma according to the added essential oil. The physical and chemical properties of the paper produce shown in Table 1 and Table 2.

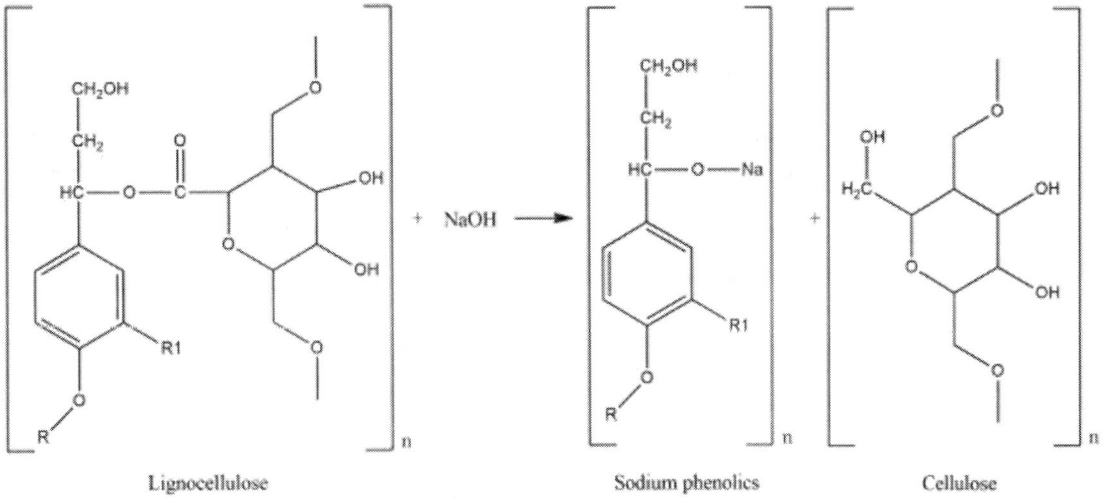

Figure 1. Delignification Reaction

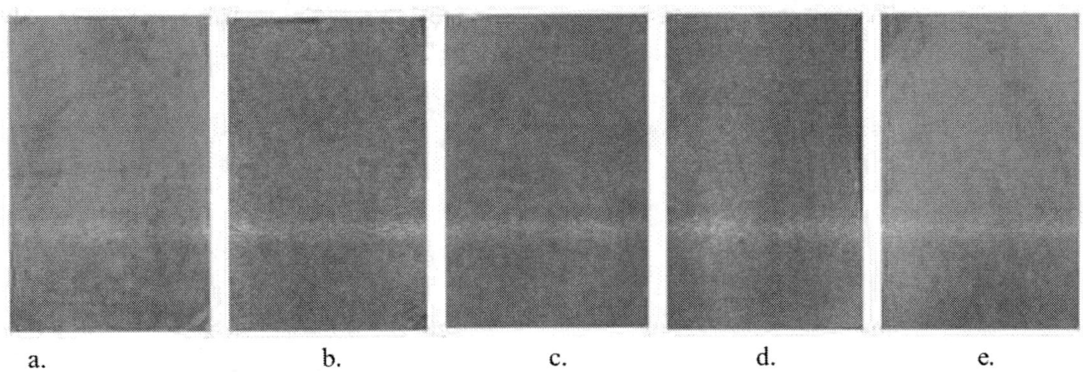

Figure 2. Banana peel natural paper: a. without essential oil, b. with lemon oil, c. with cinnamon oil, d. with lime oil, e. with clove oil

Table 1. Physical properties of banana peel natural paper

Banana peel natural paper	Grammage (g/m2)
without essential oil	115.8
with lemon oil	133.6
with cinnamon oil	166.4
with lime oil	158.4
with clove oil	142.2

According to SNI 8218:2015, food packaging paper and cardboard have two levels of grammage. High grammage has a value of 255-500 g/m^2, while the low grammage is <225g/m^2. The results showed that the paper produced meet the SNI standard with a low grammage paper category.

Table 2. Chemical properties of banana peel natural paper

Banana peel natural paper	Water Content (%)	pH
without essential oil	6.28	7.1
with lemon oil	7.04	6.7
with cinnamon oil	5.1	6.7
with lime oil	6.02	6.4
with clove oil	7.16	6.5

Based on Table 2, natural paper from banana peel (*Musa paradisiacal* Linn) produced has a range of pH 6-7 and the water content 5-7%. The standards of Basic Paper for Laminated Wrapping Paper on pH parameters ranging from 6-9 and water content parameter is a maximum of 8%. All of the paper produced meet the standard of ISO 6519:2016 about Basic Paper for Laminated Wrapping Paper within the parameter of pH and water content.

3.2. Antibacterial Activity

The antibacterial activity of the paper produced was indicated by the inhibition zone diameter (IZD) of *S. aureus ATCC 25923* and *E. coli ATCC 25922* (Figure 3).

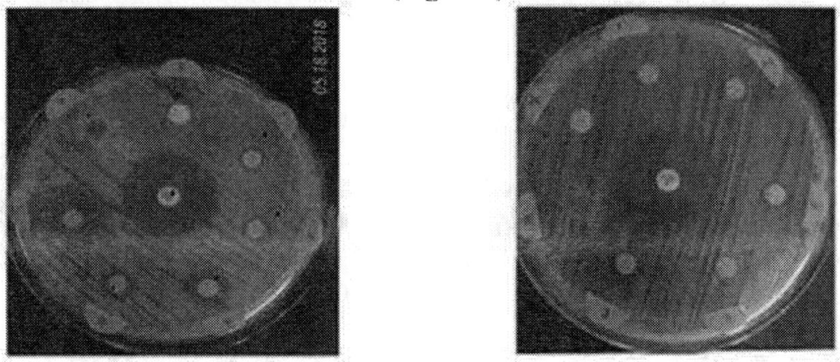

S. aureus ATCC 25923 *E.coli ATCC 25922*

Figure 3. Antibacterial activity of papers against *S. Aureus* and *E. coli*: 1. without essential oil, 2. with lemon oil, 3. with cinnamon oil, 4. with lime oil, 5. with clove oil, 6. commercial food paper

In vitro research in this study shows that all four essential oils added as antibacterial agents in making paper from banana peels can inhibit the growth of *S. aureus* with different effectiveness, from medium to strong categories. Whereas in *E. coli*, only paper with cinnamon oil additive has antibacterial activity in the medium category (Table 3). This is consistent with several previous studies which showed high antibacterial activity of cinnamon oil both against gram positive bacteria and gram negative bacteria [10], while clove oil did not show activity against *E. coli* [6]. No inhibitory activity was observed in distilled water as a negative control. Tetracycline as a positive control for gram-positive bacteria, and chloramphenicol as a positive control for gram-negative bacteria each has inhibition zone diameter greater than 20 mm which shows very strong antibacterial activity. In comparison, commercial food wrapping paper was tested for both bacteria, showed bacterial contamination on commercial food wrapping paper.

Table 3. Inhibition Zone Diameter of papers against *S. Aureus* and *E. coli*

Sample	*S. aureus* ATCC 25923		*E. coli* ATCC 25922	
	IZD (mm)	Category	IZD (mm)	Category
Paper without essential oil	6.84	Medium	0.0	Unactive
Paper with clove oil	6.67	Medium	0.0	Unactive
Paper with lime oil	11.48	Strong	0.0	Unactive
Paper with lemon oil	8.86	Medium	0.0	Unactive
Paper with cinnamon oil	18.0	Strong	9.0	Medium
Commersial food paper	0.0	Unactive	0.0	Unactive
Positive control	26.0	Very stong	23.0	Very Strong
Negative control	0.0	Unactive	0.0	Unactive

Antibacterial test results generally show that paper with essential oil additives has higher antibacterial activity against *S. aureus* than *E. coli* (Figure 4). This is related to the differences in the characteristics of the cell walls of the two bacteria. *S. aureus* and *E. coli* have different cell wall compositions. *S. aureus* is a gram positive bacterium whereas *E. coli* is a gram negative bacterium. About 90-95% of the cell wall of gram-positive bacteria consists of peptidoglycan, where other molecules, such as acids and proteins, are linked. Meanwhile gram-negative bacteria have a more complex cell wall with a peptidoglycan layer that is 2-3 nm thick [3]. An important characteristic of essential oils with their components is their hydrophobicity, which allows them to separate lipids from bacterial and mitochondrial cell membranes, disrupt the cell structure and make it more permeable [11, 12]. Because the gram-negative structure is more complex than gram-positive, hydrophobic essential oil molecules do not easily enter the gram-negative cell wall, so that essential oils are less able to influence the growth of gram-negative bacterial cells. This shows that essential oils on paper work as antibacterial agents by damaging the integrity of microbial cell walls.

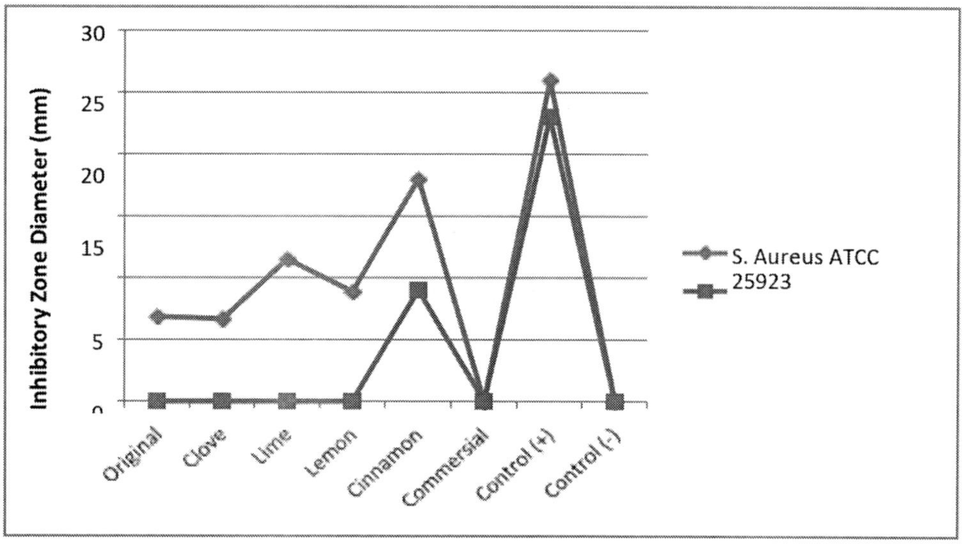

Figure 4. Comparison of antibacterial activity of natural paper with several essential oils against *S.aureus* and *E.coli*

Cinnamon essential oil as antibacterial agent on the paper has a greater inhibitory effect on *S. Aureus* and *E.coli*. Paper with cinnamon oil additive had inhibition zone diameter 18 mm on *S. aureus* (strong category) and 9 mm on *E. coli* (medium category). This shows that the chemical component in cinnamon oil is a source of antibacterial substances. Cinnamon oil is known to contain 38 components, with cinnamoneye as the main component [13,14], followed by benzaldehyde, benzoic acid and benzyl alcohol. The rest is found in very small quantities [10]. Antibacterial activity of products with cinnamon oil may be due to their main component, cinnamaldehyde, which has hydrophobic properties.

4. Conclusion

The result of the research showed that all of the paper with essential oils additive has antibacterial activity against *S.aureus*, but only paper with cinnamon oil additive that have antibacterial activity against *E. coli*. The use of cinnamon oil gave the largest inhibition diameter zone both on *S. aureus* and *E. coli*. Paper with cinnamon oil additive has inhibition diameter zone 18 mm on *S. aureus* (strong category) and 9 mm on *E. coli* (medium category). Cinnamon oil has a greater inhibitory effect on *S. aureus* and *E. coli*, because of its chemical component, namely cinnamaldehyde.

Acknowledgement

Thanks to Universitas Sebelas Maret for the research funding provided through the UNS 2018 PNBP fundamental research scheme.

Refferences

[1] Agustina E S W and Susanti V H2018 *J Phys: Conf. Ser.* **1022**(1)
[2] T D Brock, and M T Madigan 1994 *Biology of Microorganisms.* Sixth ed. Prentice Hall International, Inc
[3] F N F Fratianni, L D Martino, R Coppola and V D Feo 2013 *Pharmaceuticals* **6** 1451-1474
[4] Akhtar M S, D B Azam T. 2014 *Biol. Sci. Pharm. Res.* **2**(1) 1–7
[5] Burt S A. 2004 *Inter. J. Food Microbiol.* **94** 223-253
[6] S Kordali, R Kotan, A Mavi, A Cakir, A Ala, and A Yildirim 2005 *J. Agric. Food Chem.* **53** 9452-9458
[7] Prabuseenivasan S, M Jayakumar and S Ignacimuthu 2006 *BMC Complementary and Alternative Medicine* **6** 39
[8] S S M Ammar, K Mokhtaria, A A Amar, B B Tahar, D Moulay, H S Mohamed and B Laid 2017 *Asian J Anim Vet Adv.* **12**(1) 24-30
[9] J S Packyanathan and G Prakasam 2017 *J. Pharm. Sci. & Res.* **9**(7) 1203-1204
[10] Marie-Laurelota, D Eroccaserra, Fealixtomi, Camillejacquemond and J casanova 2002 *J. Agric. Food Chem.* **50** 796–805
[11] K Knobloch, H Weigand, N Weis, H M Schwar, and H Vigenschow 1986 *16th International Symposium on Essential Oils* Edited by: Brunke E J. De Gruyter, Berlin 429-445
[12] Sikkema J, De Bont J A M, and Poolman B 1994 *J. Biol. Chem.* **269** 8022-8028.
[13] Simic A, Sokovic M D, Ristic M, Grujic-jovanovic S, Vukojovic J, Marin P D 2004 *Phytother. Res.* **18** 713-717
[14] M T Baratta, H J D Dorman, S G Deans, C Figueiredo, J G Barroso, G Ruberto 1998 *Flav. Fragr. J.* **13** 235-244

IOP Conf. Series: Materials Science and Engineering **578** (2019) 012049 doi:10.1088/1757-899X/578/1/012049

Optimation Formula of SNEDDS Dosage from Ethanol Extract of Turmeric (*Curcuma domestica*) With Waste Oil of Eel (*Anguilla spp.*) as A Carrier

N Y Rahayu[1], A Budiharjo[2], A Pangastuti[2], A N Artanti[3], F Prihapsara[3] and M Harini[2]
[1] Bioscience Graduate Programme, Sebelas Maret University, Surakarta, Indonesia
[2] Department of Biology, Sebelas Maret University, Surakarta, Indonesia
[3] Department of Pharmacy, Sebelas Maret University, Surakarta, Indonesia

E-mail: anif.apt@staff.uns.ac.id

Abstract. Turmeric have a lot of advantages for us, usually for treatment a lot of desease, and it has efficiency as a drug. But turmeric extract have very low solubility and it cause minimal oral bioavailability. While nutrition of waste oil of Eel (*Anguilla spp.*) are higher than the other fish, because this waste oil have palmitic acid, linoleic acid, and oleic acid that usefull for treatment some disease. So, in this research, turmeric extract made by *Self-Nanoemulsifying Drug Delivery System* (SNEDDS) method with waste oil of Eel (*Anguilla spp.*) as a carrier to overcome solubility of turmeric extract. In this research, the turmeric was extracted with maceration method using ethanol as a solvent. This extracting showed high average rendemens are 6.62-8.19 %. Percentages of curcumin for this turmeric extract was calculated by TLC-densitometry, shows 2.99 % b/b. This extract had the optimization of the formulation of the composition of surfactant by Tween 80, cosurfactant by PEG 400, and waste oil of Eel (*Anguilla spp.*) as a oil carrier. After done some tests, include loading dose test on the selected optimal formulation, then we had determine that quality. It included by counting emulsification time, observations of the particle size, its distribution, and zeta potential of nanoemulsi using PSA. The results of SNEDDS optimal dosage was calculated by *Design Expert 7* had a mass (gram) ratio of waste oil of Eel (*Anguilla spp.*):Tween 80 (surfactant): PEG 400 (cosurfactant) = 0.55 :4.10:0.35 added with loading dose 0.5 g of Turmeric extract in 5 g system of SNEDDS. The result had a transmittance value of 91.48-92.17 % with emulsification time less than 30 seconds and it classified type A (Sakhti *et al*, 2013). Particle size of the chosen SNEDDS nanoemulsion turmeric extract that calculated by PSA, shows 11.9 nm and with polydispersity index are 0.118 droplet and zeta potential of -11.8 mV.

1. Introduction
The increasing of human population is the one problem in our life because it is not followed by increasing of human health [1]. So, we have to increase healthy service, availability drugs completely, efficiently drugs, safetly drugs with affordable price. The main problem of herbal medicine are poorly water soluble and the oral delivery [2] [3]. SNEDDS is one of the medical preparation method to increase the absorption of the lipophilic drugs [4], because SNEDDS are isotropic mixtures that consist of oil, surfactant, co-surfactant and drug in form oil in water emulsion. This forms is a good

mode to deliver poorly soluble drugs orally by increasing its bioavailability and stability [5] [3]. Turmeric is a rhizomes belonging to a ginger family (Zingiberaceae) that contain important bioactive component called curcumin [6] [7]. It has bioactive components with regard to potential as a herb medicine [6] [8] [1], but because of its low bioavailability, its usefulness, and attainable blood levels associated to its consumption is laced with doubt [9]. While fatty fish specially waste oil Eel (*Anguilla spp.*) are considered to be of great nutrition [10]. This waste oil have palmitic acid and oleic acid that can reduce cholesterol in blood [11]. This is due to their high natural component consist of essential n-3 polyunsaturated fatty acids (PUFAs) such as eicosapentaenoic acid and docosahexaenoic acid. These fatty acids have been shown to have potential benefits for human health [12][13]. Nanoemulsion form is a good choice that is expected to increase solubility and oral bioavailability [14][15][16] of turmeric extract ethanol fraction and waste oil of Eel (*Anguilla spp*). When they are combined into water phase in a slow agitation condition, spontaneously, it will form nanoemulsion of oil phase in water (O/W) [2]. The successfulness of this research is expected to be a nanoemulsion formula of ethanol fraction from turmeric extract with waste oil of Eel (*Anguilla spp*) as a carrier to be a good product as alternative one of herb medical with good bioavailability.

2. Experimental

2.1 Equipment and Materials

Turmeric rhizome and waste of Eel (*Anguilla spp*) were obtained from UNAGI Surakarta, Indonesia. Component of optimation of SNEDDS: aquabidest, aquadest, chloroform, ethanol, methanol, PEG 400, tween 80 (Brataco). Component of artificial gastric fluid (AGF) consist of NaCl, HCl 37 %, and aquades. Equipment which were used are blender, glass made tools (Pyrex), oven, stopwatch, waterbath (Grant), rotary evaporator (RVO 400 SD Boeco Germany), vortex (Maxi Mix II Thermolyne), magnetic stirrer (IKA C-MAG HS7), analytical balance (KERN ALJ), yellow tip (Kan Jian), sonicator (Branson 1510), flacon glasses, spectrophotometer UV Vis (Perkin Elmer Lambda 25), micropipet 50-1000 μL, PSA (Horiba SZ-100), and analytical scale (Metler Teledo).

2.2. Procedure

2.2.1. Maceration of Ethanol Fraction from Turmeric Extract and Waste Oil of Sidat (Anguilla spp).

Turmeric powder was extracted with ethanol solvent (1:5) for 5-7 days. Then turmeric's macerate was filtered to separate it from simplisia powder. Thus, the filtrat was evaporated by rotary evaporator for 1 hour until it became concentrated in 70 °C temperature and 3 rotary speeds. Then rendemens of turmeric extract was calculated. Percentages of curcumin was calculated by TLC-densitometry. Waste of Eel (*Anguilla spp*) was extracted by maceration method for 3-4 days with added 400 mL chloroform and 400 mL methanol, mixed until it became homogeneous mixture. Macerate was filtered to separate it from sludge. Thus, evaporate it by rotary evaporator in 60 °C temperature for 1 hour and 3 rotary speeds to be concentrated oil.

2.2.2. Optimization of surfactant, cosurfactant, and oil composition of SNEDDS-Turmeric Extract.

The higest transmittan value was choosen from formulas of oil: surfactant: cosurfactant are 1:1:1 until 1:8:1. The selected formula was optimized and selected by using *Design Expert 7* based on transmittan value and emulsification time in AGF, aquadest, and aquabidest.

Candidate of SNEDDS formula was taken 100 μL and added with aquades becomes 50 mL. Thus, it was homogenized by using vortex for average 30 minutes. The successfulness of making of nanoemulsion showed homogeneous emulsion and give clear visual. The transmittance of emulsifying formula of SNEDDS was measured by spectrophotometer on the length wave 650 nm with aquades blank to determine the clarity level.

Emulsification time was done by dropped quickly 1 mL SNEDDS formulation of turmeric extract in 500 mL medium AGF without pepsin, aquadest, and aquabidest, which were placed in 37 °C temperature on the magnetic stirrer with speed about 120 rpm. The observation was done by the time needed from the first drop until it forms emulsion using stopwatch. Nanoemulsion was formed which was dissolved in the medium perfectly [15].

2.2.3. *SNEDDS formula selection.*
Then, loading dose was done by added turmeric extract of ethanol fraction 250, 500, and 750 mg into 5 g optimal selected system component (carrier oil: waste oil of Eel (*Anguilla spp*), surfactant: tween 80, and consurfactant: PEG 400) which was appropriate with optimization results that had been done.

The mixture was sonicated for average 15 minutes to be homogeneous system, then it was placed in 45 °C temperature for 10 minutes. The loading dose was choosen by based on the higher dose of turmeric extract on the homogenous formulas. The result of the mixture was keep for 24 hours in the room temperature, freeze temperature, and 45 °C temperature to determine that homogenity. The homogeneous formula is the selected formula for SNEDDS [4].

Particle size, size distribution, and zeta potential was measured using Particle Size Analyzer (PSA). Every two drops of SNEDDS of turmeric extract formula were emulsified into 5 mL of aquades, then taken 3 mL into covet to be analyzed.

3. Results and Discussion

3.1. *Maceration and curcumin content of turmeric extract*
Turmeric extract was macerated by ethanol as a solvent (1:5). This extract had high rendemens, it showed between 6,62-8,19 %. It contained 2,99 % w/w of curcumin level. Curcumin is a non polar liposoluble compound. It dissolved into semi polar solvent, one is ethanol as a good extractant. The high concentration of ethanol showed that a lot of curcumin can be dissolved in this solvent [7].

Table 1. Rendemens of Turmeric Extract by Ethanol Fraction

No	Mass of Turmeric Powder (g)	Volume of Ethanol (mL)	Mass of Turmeric Extract (g)	Persentage of Rendemens (%)
1	151.00	755.00	10.00	6.62
2	67.79	338.95	4.75	7.01
3	150.57	752.85	11.05	7.34
4	111.8	559.00	8.5	7.60
5	118.5	592.5	9.7	8.19

3.2. *Optimization of surfactant and cosurfactant composition of SNEDDS-Turmeric Extract*
This aims to determine the right composition ratio of surfactant and co-surfactant which is able to produce a homogeneous formula after added with waste oil of Eel (*Anguilla spp.*) as a carrier. The results of surfactant and co-surfactant with waste oil of Eel (*Anguilla spp.*) (Table 2) showed that combination of waste oil of Eel (*Anguilla spp.*) : tween 80 : PEG 400 with ratio 1:6:1, 1:7:1, and 1:8:1 were able to produce a homogeneous mixture.

Table 2. Optimization of Waste oil of Eel (*Anguilla spp.*): Tween 80: PEG 400

Rasio composition	Composition (g)			Condition
	Oil	Tween 80	PEG 400	
1 : 1 : 1	1.67	1.67	1.67	not homogeneous
1 : 2 : 1	1.25	2.50	1.25	not homogeneous
1 : 3 : 1	1.00	3.00	1.00	not homogeneous
1 : 4 : 1	0.83	3.33	0.83	not homogeneous
1 : 5 : 1	0.71	3.50	0.71	not homogeneous
1 : 6 : 1	0.63	3.73	0.63	Homogeneous(F1)
1 : 7 : 1	0.55	3.88	0.55	Homogeneous(F2)
1 : 8 : 1	0.50	4.00	0.50	Homogeneous(F3)

The transmittance value of chosen formula (F1, F2, and F3) were measured using spectrophotometer to determine the highest value as a candidate for SNEDDS composition, showed that formula with the highest transmittance is formula F2 with average transmittance value is 96.966 % (Table 3).

Table 3. Transmittance Homogeneous Composition

Rasio composition of waste oil of Eel (*Anguilla spp*):Tween 80:PEG 400	Condition	Transmittance (%)
1 : 6 : 1	Homogeneous(F1)	65.697
1 : 7 : 1	Homogeneous(F2)	96.966
1 : 8 : 1	Homogeneous(F3)	96.647

The choosen formula was calculated by *Design Expert 7 Programe* to be eigth runs with 0.55 g flat waste oil of Eel (*Anguilla spp*). Every formula from eight run of choosen formula showed high transmittance (more than 90 %), while emulsification time in AGF, aquadest, and aquabidest are between 4-14 seconds (Table 4). It showed that the formulas was in nanoemulsion form type A [3], trusted as a good deliver for turmeric extract in formula SNEDDS.

Table 4. Transmittance Value and Emulsification Time from Eight Runs SNEDDS Formula

Run	Transmittance (%T)	Emulsification Time in AGF (s)	Emulsification Time in aquadest (s)	Emulsification Time in aquabidest (s)
1	91.85	7	10	12
2	91.97	5	10	9
3	91.83	6	10	8
4	92.23	6	9	8
5	91.48	9	8	12
6	91.75	4	11	6
7	92.17	6	9	14
8	91.91	12	8	9

The eight runs formulas were choosen would be calculated by *Design Expert 7 Programe* to decide the optimum SNEDDS with turmeric extract formula. Linier models graph showed that higher level of surfactant and lower level of cosurfactan increased transmittan value (Figure 1) and emulsification time in aquadest (Figure 3) and aquabidest (Figure 4), but decreased emulsification time in AGF (Figure 2).

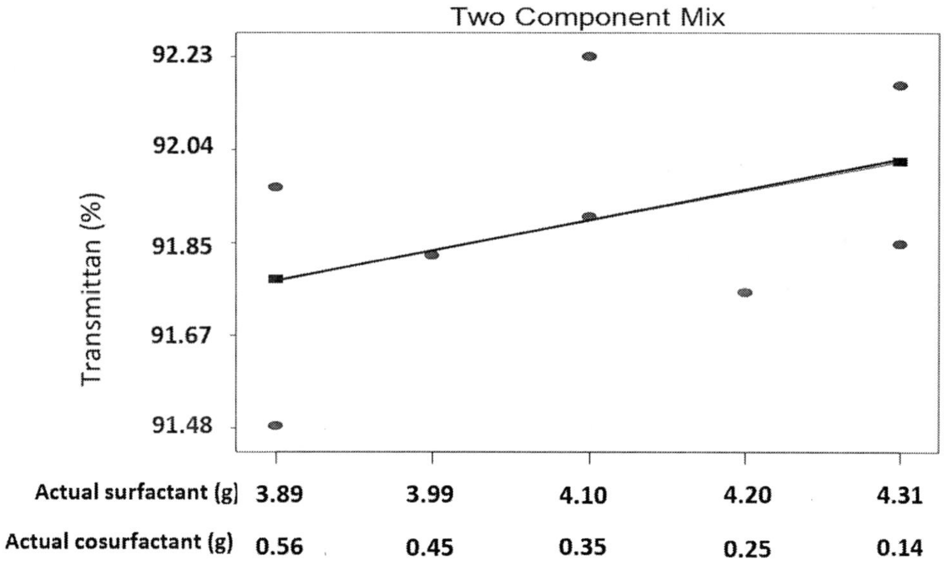

Figure 1. Linier Models of Two Component Mix of Transmittan Value

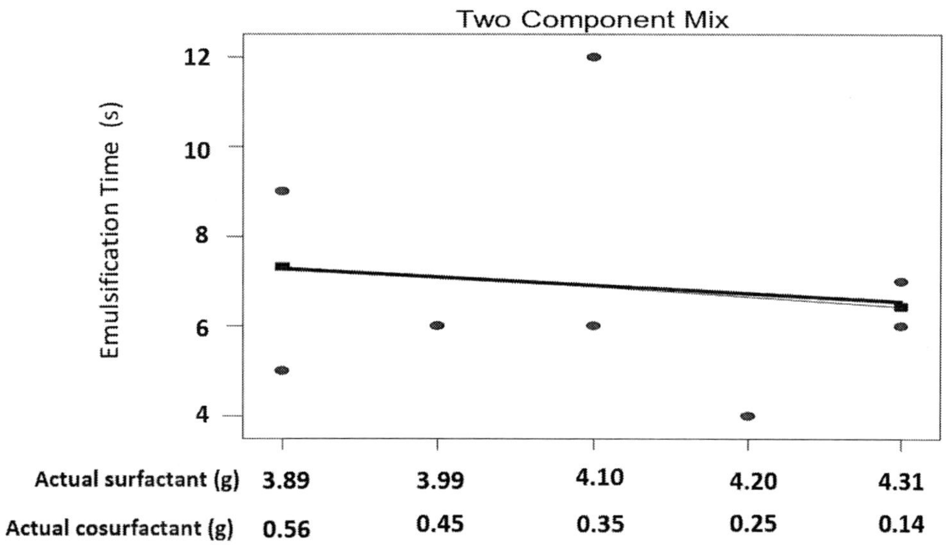

Figure 2. Linier Models of Two Component Mix of Emulsification Time in AGF

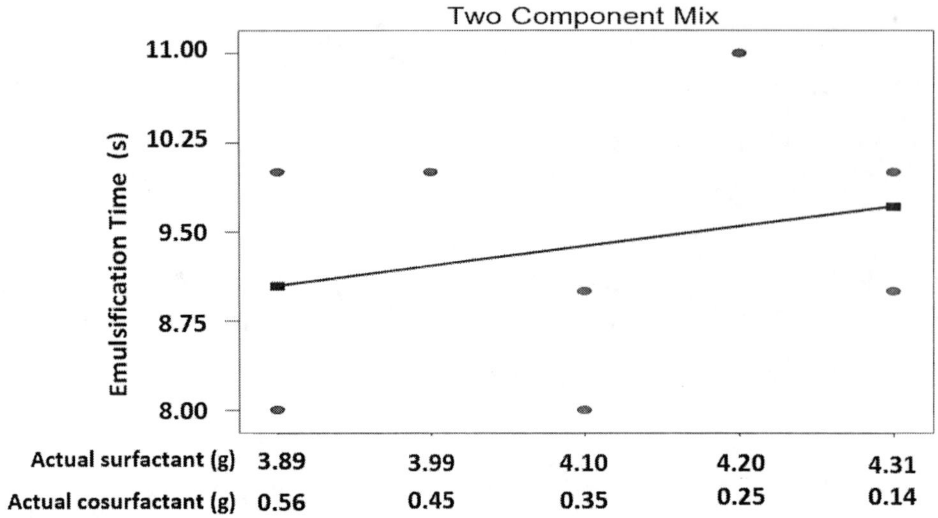

Figure 3. Linier Models of Two Component Mix of Emulsification Time in Aquadest

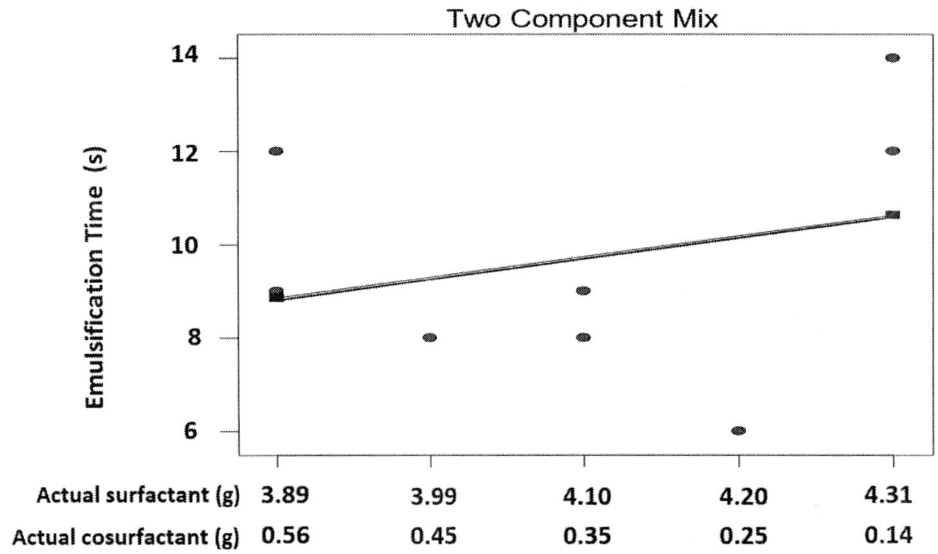

Figure 4. Linier Models of Two Component Mix of Emulsification Time in Aquabidest

The flat composition of waste oil of Eel which was mixed with ratio formulation of surfactant and cosurfactant lead to the formation of SNEDDS, so there is a relationship between the droplet size and the concentration of the surfactant and cosurfactant are being used. By increasing the cosurfactant concentration could lead to droplets with smaller mean droplet size [2], but the other hand the mean droplet size may bigger with increasing surfactant concentrations, so the linier models above showed the optimum composition surfactant and cosurfactant to be homogeneous emulsion of SNEDDS formulas. The stabilization of the oil droplets depend on of the localization of the surfactant molecules at the oil-water interface. The interfacial disruption elicited by enhanced water penetration into the oil droplets mediated by the increased surfactant concentration [2].

The highest desirability value of SNEDDS turmeric (*Curcuma domestica*) extract was 0.887 (Figure 5). This formula will produce optimum characteristic SNEDDS of turmeric extract. The composition of optimum formulation of SNEDDS were 4.10 g of surfactant Tween 80; 0.35 g of cosurfactant PEG 400, and added 0,55 g flat of waste oil of Eel (*Anguilla spp*) (Table 3).

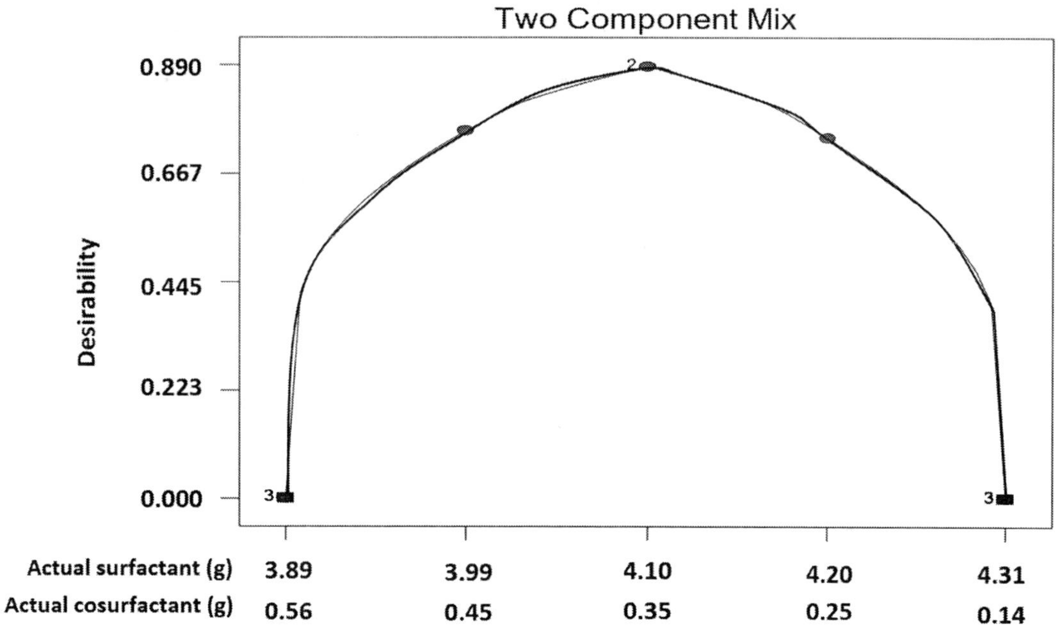

Figure 5. Two Component Mix of Desirability Solution of Optimation of Surfactant and Cosurfactant

Based on solution that calculating by *Design Expert 7 Programe,* this optimation of SNEDDS increased the high transmittan value and decreased emulsification time in AGF, aquadest, and aquabidest (Table 5). Of course, the optimum composition with the highest desirability provided the best result for formulation SNEDDS with 500 mg loading doses of turmeric extract.

Table 5. Solution of optimation SNEDDS turmeric extract (*Curcuma domestica*)

Waste oil of Eel (g)	Surfac. (g)	Cosurfac. (g)	Transmit value (%)	Emulsif. time in AGF (s)	Emulsif. time in aquadest (s)	Emulsif. time in aquabidest (s)	Desirability
0.55	4.10	0.35	91.90	6.88	9.38	9.75	0.887

Characteristic of SNEDDS was measured using PSA, showed that the result of particle size is 11.9 nm (Figure 6), distribution size was 0.118 (PI<1), and zeta potential was -11.8 mV (Figure 7).

It showed that formula has nanoemulsion particle size less than 100 nm and uniform particle size distribution (PI<1) [16]. Increasing composition of surfactant to cosurfactant ratio lead to decrease in mean droplet size [17]. Accordance with Savale's research, showed that fatty acid medium chain until long chain supported stability of SNEDDS emulsion [15], in this riset using waste oil of Eel which had medium until long chain of fatty acid, had stability formula SNEDDS emulsion.

Turmeric extract which has low bioavailibility oral need an exact formula which is able to be absorbed in lumen gut, so it can make a lot of advantages for human healty. This research showed that the optimation SNEDDS with oil waste of Eel (*Anguilla spp.*) as a carrier was the exactly formulation as drug delivery system for turmeric extract (*Curcuma domestica*).

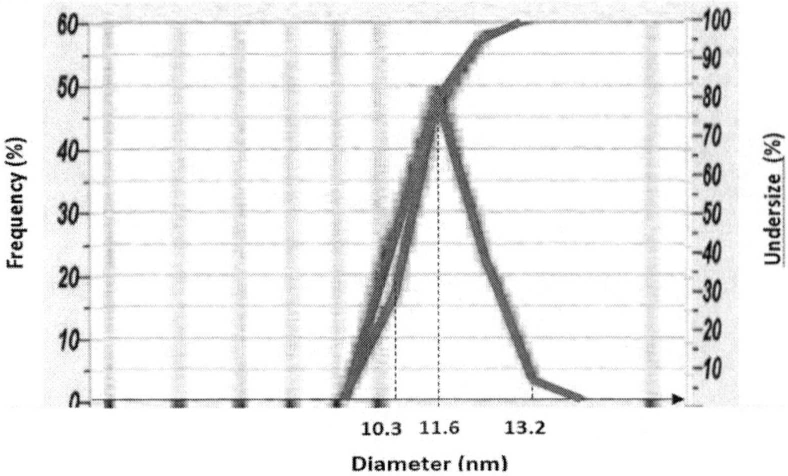

Figure 6. Particle Size of SNEDDS of Turmeric Extract

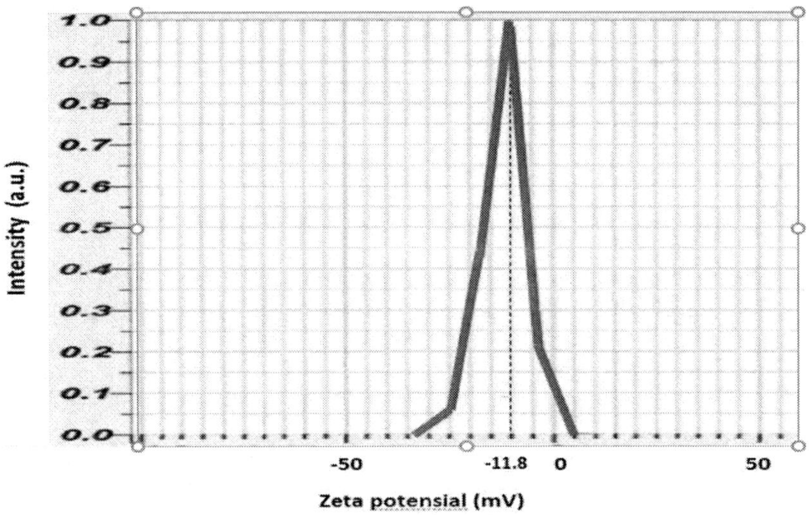

Figure 7. Zeta Potential of SNEDDS of Turmeric Extract

4. Conclusion

The results of this research showed that the optimum SNEDDS formula was waste oil of Eel (*Anguilla spp*) tween 80 : PEG 400 (0.55 : 4.10 : 0.35) added 0.5 g turmeric extract in 5 g system SNEDDS. This formula had emulsification time value less than 30 seconds, the particle size value 11.9 nm, and zeta potential value -11,8 mV.

Acknowledgment

We thank to Grant of Universitas Sebelas Maret for Ministry of Research and University Technology 2017 which has funded this research until the end.

References

[1] Ling J, Bo W, Guangping L, Hui J, Shaoping L 2012 *Food Chemistry Elsevier* **130** 229-235
[2] Amrutkar C, Salunkhe K S, Chaudari S R 2014 *Am. J. PharmTech Res* **3** (4) 43-61
[3] Sakthi U, Josephine R L F, Kiran B U 2013 *Biomed. & Pharmacol. J.* **6** (2) 355-362

[4] Khan A W, Sabna K, Shahid H A, Rakesh K S, Javed A. 2014 *Drug Deliv.* **22** (4) 552-561
[5] Martien R, Adyatmika, Irianto I D K, Farida V, Sari DP 2012 Majalah Farmaseutik **8** (1)133-144
[6] Ejaz A, Wu D, Kwan P, Meydani M 2009. *Journal of Nutrition* 919-925
[7] Popuri A K, Pagala B 2013 *International Journal of Innovative Research & Studies* **2** (5) 289-299
[8] Hadi M A, Hameedi E H, Kadhum N J,AzizD Z, Al-Saddi A H, Zaidan H K 2016 *Journal of Chemical and Pharmaceutical Science* **9** (3) 1098-1106
[9] Alrawaiq N S, Abdullah A 2014. *Int.J.PharmTech Res* **6** (1) 280-289
[10] Gunes F E 2013 *Journal of Marmara University Institute of Health Sciences* **3** (4) 220-228
[11] Sandita A, Maulana I P, Syafnir L 2015 *Prosiding Penelitian Spesia Unisba.* 388-396 ISSN 2460-6472
[12] Ngo D H, Vo T S, Ngo D N, Kim S K 2012 *International Journal of Biological Macromolecules* **51** 378-383
[13] Syahbana I, Suseno S H 2014 Processing Industry *Pakistan Journal of Biotechnology* **11** (1) 25-31
[14] Makadia H A, Bhatt AY, Parmar R B, Paun J S, Tank H M 2013 Future Aspects *Asian J. Pharm.* Res **3** (1) 20–26
[15] Savale S K 2015 *International Journal of Research in Pharmaceutical and Nano Sciences* **4** (6) 385-397
[16] Zhao T 2015 *Self-nanoemulsifying drug delivery system (SNEEDS) for the oral delivery of lipophilic drugs* (University of Trento Department of Industrial Engineering: Italy)
[17] Vanitasagar S, Subhashini N J P. 2013 *Int J Pharm Bio Sci* **4** (2) 511-521
[18] Reis C P, Neufeld R J, Ribeiro A J, Veiga F 2006 *Nanotechnology, Biology and Medicine* **2** 8-21

Optimization of self nanoemulsifying drug delivery system with ethanolic extract of ginger (*zingiber officinale*) and eel bone oil (*anguilla* spp.) extracted by maceration as carrier

II Wijayanti[1], A Budiharjo[1,2], A Pangastuti[1,2], F Prihapsara[3], A N Artanti[3], M Harini[2]

[1] Bioscience Graduate Program, Universitas Sebelas Maret. Jl. Ir. Sutami 36A Surakarta 57 126, Central Java, Indonesia. Tel./fax.: +62-856-786686168/+61-857-43210550,

[2] Department of Biology, Faculty of Mathematics and Natural Sciences, Universitas Sebelas Maret. Jl. Ir. Sutami 36A Surakarta 57 126, Central Java, Indonesia

[3] Department of Pharmacy, Faculty of Mathematics and Natural Sciences, Universitas Sebelas Maret. Jl. Ir. Sutami 36A Surakarta 57 126, Central Java, Indonesia.

E-mail: fea.prihapsara@staff.uns.ac.id

Abstract. SNEDDS is a drug delivery method by mixing oil, surfactant, cosurfactant and drug that can form spontaneous nanoemulsion oil in water in the digestive tract and produce nanometer-sized droplet. Aim of this study was to get optimation for SNEDDS (Self Nanoemulsifying Drug Delivery System) with ethanolic extract of ginger, eel fish bone oil extracted by maceration as carrier, tween 80 as surfactant and PEG 400 as cosurfactant that applicable as a nanoemulsion formula. Optimation formula was measured based on transmittance and emulsification time of mixture of eel fish oil, ginger extract, tween 80, and PEG 400 into 8 different compositions. The Transmittance values were measured using UV/VIS Spectrophotometric at 650 nm and emulsification times were measured using stopwatch by diluting 1 mL of SNEDDS into AGF (Artificial Gastric Fluid), aquadest, and aquabidest. The optimum SNEDDS formula was determined based on the optimization results using *Design Expert 7 program* with the independent variable was composition of tween 80 and composition of PEG 400 and the dependent variable was transmittance and emulsification time in AGF, aquadest, and aquabidest. Result of *Design Expert 7 program* optimization formula showed that SNEDDS of ginger extract will be optimized in 4.2 g of tween 80 as surfactant; 0.25 g of PEG 400 as a cosurfactant; 500 mg of ethanolic ginger extract and 0.55 g of maceration extract of eel fish bone oil with transmittance value was 92 ± 0.034 %; emulsification time in AGF was 10 second, in Aquadest was 10 second, and Aquabidest was 9 second. The result of Analysis of Variance indicated not significant differences between treatment groups and optimization by *design expert 7 program* (($P = 0.1359$ ($P>0.05$)).

1. Introduction

Nanoparticles are currently used widely as an alternative to targeted treatments and carrier agents. Interdependence between the shape, size, and biochemical properties of the surface of the nanoparticles gives effect to biodistribution, cellular internalization, and intracellular traficking from micro and nanoparticles [4]. SNEDDS (Self Nanoemulsifying Drug Delivery System) can be defined as a multi isotropic component consisting of oil, surfactants, co-surfactants which have the ability to form oil molecules as micro and nano-sized drug delivery through mild stirring with dilution (soluble) in a liquid medium such as gastrointestinal fluid. Inside the gastrointestinal tract, SNEDDS can form oil droplet less than 5 μm which carries drug compounds and increases drug absorption and drug molecular stability [8].

Oil can be absorbed properly as an oil droplet in the form of oil emulsion in water or in the form of digestion. Nanoemulsion is a thermodynamically stable system with a droplet size of less than 100 nm. Nanoemulsion has been widely studied to improve the bioavailability of drugs that are difficult to dissolve in water [11]. The presence of surfactants and cosurfactants is known to have effect on the size of particles formed because some of the materials used as surfactant and cosurfactant have hydrocarbon chain structures with hydrophobic chains. PEG 400 is known to have low toxic levels and has been widely used in various fields such as pharmaceuticals, biomedicine, and dispersing agents [7][14].The type of cosurfactant and the relative length of the hydrophobic chains affect the area of microemulsion and nanoemulsion [9].

Ginger (*Zingiber officinale*) has been widely used throughout the world as a food additive that causes a spicy taste and traditional medicine as a deterrent to treating hypertension and some other cardiovascular diseases. Active ingredients found in ginger include volatile compounds and phenol compounds known as gingerol, sesquiterpenoid and shogaol [1].

The eel known as 'unagi' in Japan is very expensive because it has a protein content of 16.4% and high vitamin A of 4700IU [5]. Based on the results of the analysis, eel oil obtained SFA as much as 19.87%, MUFA 25.84%, PUFA 13.84%. The main fatty acid composition of eel oil is palmitic acid as much as 13.58%, oleic acid 20.94%, linoleic acid 4.01%, EPA 1.57% and DHA 4.84% [12].

This research was done to get optimation formula for SNEDDS with ethanolic extract of ginger, eel fish bone oil as a carrier, tween 80 as a surfactant and PEG 400 as a cosurfactant that applicable as a nanoemulsion formula.

2. Experimental

2.1 Materials

Materials used in this research includes rotary evaporator (RVO 400 SD Boeco Germany), waterbath (Grant), vortex (Heidolph REAX top), sonicator (Branson 1510), magnetic stirrer (IKA C-MAG HS 7), flacon, analytic pair of scales (KERN ALJ), measuring glass (Pyrex), test tube (Pyrex), stopwatch, UV/Vis spectrophotometric (Perkin Elmer Lambda 25), yellow tip (Kan Jian), micropipet 50 – 200 μl, *Particle Size Analyzer* (PSA) (Horiba SZ – 100), ethanol p.a (Merck), PEG 400 (Brataco), Tween 80 (Brataco). aquades (General), NaCl, HCL 37% (Merck), Aquabides (Aldrich Yogyakarta).

2.2 Extraction of ginger

Ginger extract is obtained by macerating ginger powder with ethanol at ratio 1: 5 (w/v) for 2 x 24 hours and stir for 30 minutes every 24 hours. The ethanol solution was separated from the sediment by using a flannel cloth to be subsequently evaporated using a rotary evaporator at 70 °C to obtain a thick ginger extract.

2.3 Extraction of eel fish bone oil

Eel fish bone oil was obtained by maceration of eel fish bone with methanol and chlorofom in ratio of 1: 2 (v/v) for 3 days. Then methanol and chloroform solution was evaporated at temperature 55 °C using rotary evaporator to obtain oil extract.

2.4 SNEDDS formulation

SNEDDS formulation was done by mixing eel bone oil, tween 80 as surfactant, PEG 400 as cosurfactant, with 8 compositions, vortex for 2 minutes, sonication for 15 minutes and conditioned in a waterbath at 45 °C for 10 minutes. Then they were allowed to stand for 24 hours at room temperature.

2.5 Measurement of transmittance value

Transmittance values were observed using UV/Vis spectrophotometric by diluting 1 mL of SNEDDS without ginger extract into 50 mL of aquadest through transmittance measurement at 650 nm wavelength with aquadest as a blanko to determine the level of clarity of the emulsion. Transmittance values approaching to the aquadest (100%) are estimated to have reached the nanometer size of the emulsion.

The results of the comparison of eel bone oil, tween 80 and PEG 400 were selected based on the largest transmittance value (close to aquadest) then developed into 4 formulas by changing the composition of tween 80 and PEG 400. The four formulas were performed mathematical calculations to determine the upper and lower limits to put into the expert design program 7. In this study, the optimal formula selected was SNEDDS with ratio of oil: surfactant: cosurfactant as much as 1: 7: 1.

2.6 SNEDDS formulation with ginger extract

The addition of ginger extract as a drug compound was given at an optimal formula ratio of 500 mg extract. Then it was homogenized with vortex for 2 minutes, sonication for 15 minutes and conditioned in a waterbath at 45°C for 10 minutes. Then, the mixing results were allowed to stand for 24 hours at room temperature for later observation if there was sediment formed.

2.7 Emulsification Time measurement

Measurement of emulsification times were measured by diluting 1 mL of samples in artificial gastric fluid (AGF), aquadest, and aquabidest at temperature 37 °C while stirring at magnetic stirrer.

3. Results and Discussion

3.1 Result

Formula composition of oil: surfactant: cosurfactant then obtained into comparison with the highest transmittance value that is a ratio of 1: 7: 1 (96%). Then made into 4 composition groups to determine the upper and lower limits of the formula. The results of measurement of upper and lower limit of oil, surfactant, and cosurfactant are presented in Table 1.

Table 1 . Upper and lower limit of oil, Tween 80, and PEG 400

Oil : Tween 80 : PEG 400	**Oil (%)**	**Surfactant (%)**	**Cosurfactant (%)**
1 : 7 : 1	12.5 (upper limit)	87.5 (lower limit)	12.5 (upper limit)
0.75 : 7.25 : 0.75	8.57	91.43	8.57
0.5 : 7.5 : 0.5	5.88	94.12	5.88
0.25 : 7.75 : 0.25	3.13 (lower limit)	96.87 (upper limit)	3.13 (lower limit)

The four composition groups with the upper limit and lower limit, then made into 8 compositions to measure transmittance value and the emulsification time in AGF, aquadest and aquabidest. The results of measurement of transmittance value (%) are presented in table 2.

Table 2. Transmittance value of 8 runs of SNEDDS without ginger extract

Run	Transmittance
1	92.16
2	91.94
3	92.5
4	93.26
5	92
6	91.68
7	92.04
8	92.39

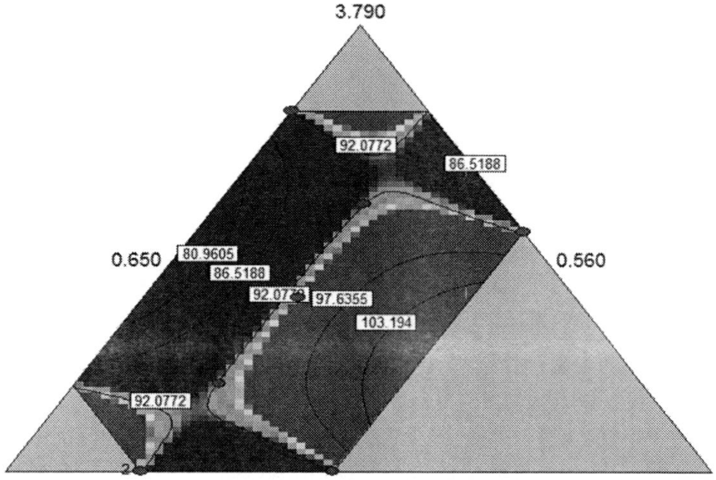

Figure 1. Pseudoternary phase diagram of transmittance value for oil, surfactant and cosurfactant compositions

The eight run compositions then measured to obtain emulsification time into AGF, aquadest, and aquabidest. The results of measurement of emulsification time (in seconds) presented in Table 4.

Table 3. Results of the experiment run 8 emulsification time

Run	AGF	Aquadest	Aquabidest
1	8	35	7
2	6	21	8
3	6	32	9
4	10	33	9
5	13	23	9
6	12	41	10
7	8	36	8
8	7	34	9

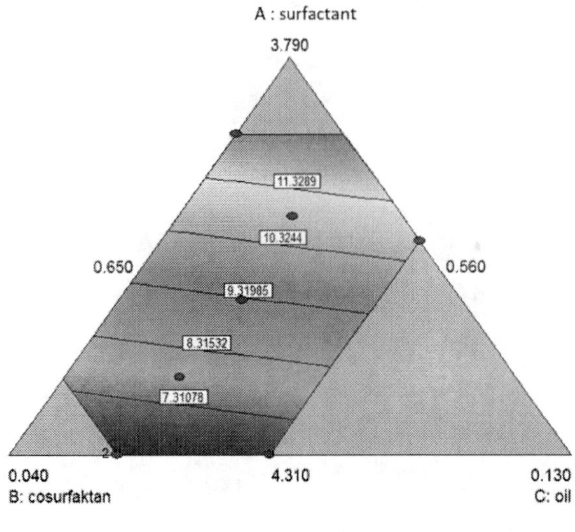

Figure 2. Pseudoternary Phase Diagram of Emulsification Time in Artificial Gastric Fluid

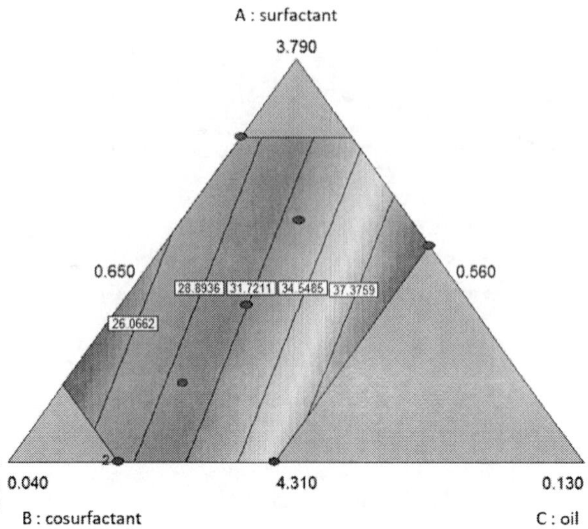

Figure 3. Pseudoternary Phase Diagram of Emulsification Time in Aquadest

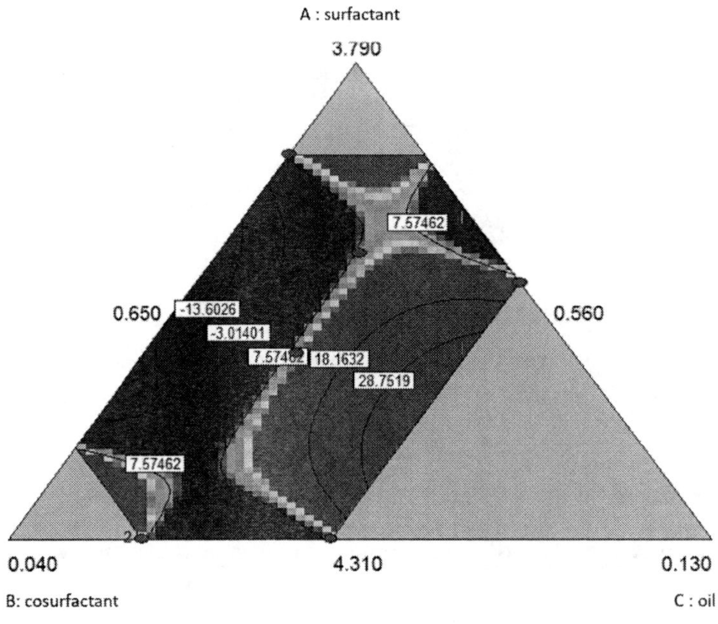

Figure 4. Pseudoternary Phase Diagram of Emulsification Time in Aquabidest

Transmittance and emulsification time data were then analyzed using *design expert program 7* to obtain optimal formula data for oil, surfactant and cosurfactant composition. The result of optimization formula based on *design expert program 7* was presented in figure 5.

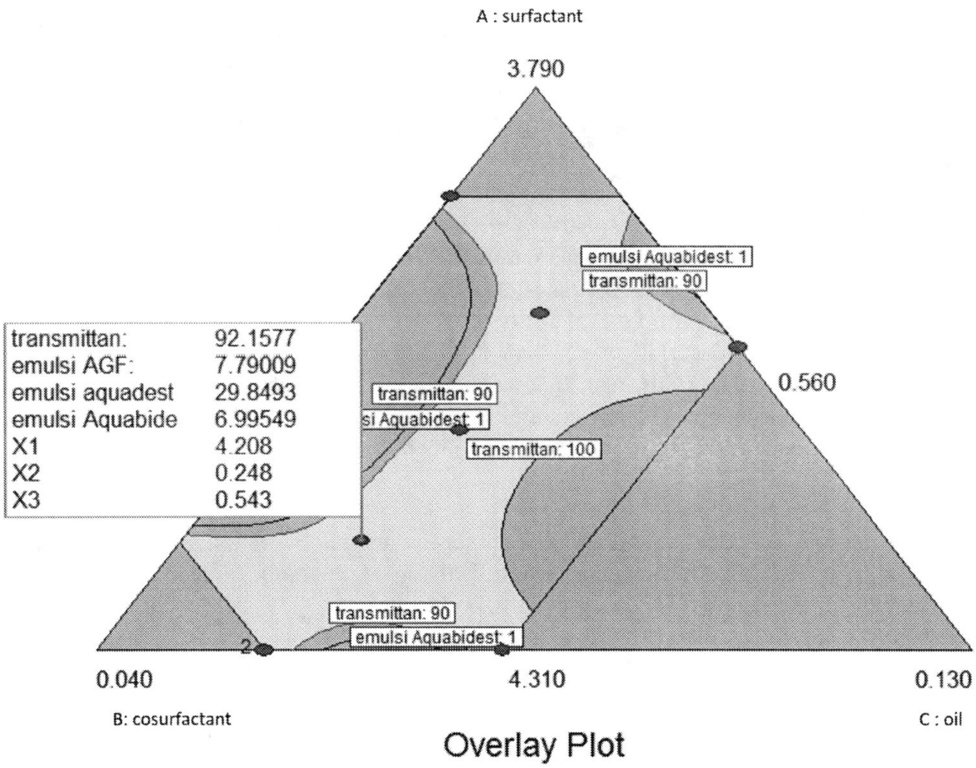

Figure 5. Pseudoternary phase diagram optimum comparison of oil, surfactant and cosurfactant compositions

The results of the optimal formula were then obtained into 5 g of SNEDDS with ginger extract and eel bone oil as carrier with calculations using the upper and lower limit of oil, surfactant and cosurfactant. So that the optimal formula was obtained with the composition of 0.54 g of eel bone oil, 4.2 g of surfactant (tween 80), 0.3 g of cosurfactant (PEG 400) and 500 mg of ginger extract. The final optimal formula then verified by transmittance and emulsification time measurement to observe if there are differences between programme and verification result. The results of measurements of transmittance value, emulsification time of SNEDDS of ginger extract with eel bone oil as carrier were 92% of transmittance value, 10 second of AGF emulsification time, 10 second of aquadest and 9 second of aquabidest. The result of Analysis of Variance indicated not significant differences between treatment groups and optimization by *design expert 7 program* (($P = 0.1359$ ($P>0.05$)).

3.2 Discussion

The results of measurements of transmittance value and emulsification time of SNEDDS of ginger extract with eel bone oil as carrier were 92% of transmittance value, 10 second of AGF emulsification time, 10 second of aquadest and 9 second of aquabidest.

Transmittance value of 92% indicated that SNEDDS transmittance was close to the aquadest transmittance value of 100%. The higher of transmittance value showed that the particle size was getting closer to the nanometer. SNEDDS is one of the drug formulations that can produce nanometer droplet size and increase solubility in liquid media and oral absorption of a type of lipophilic drug with oil phase as carrier. SNEDDS particle size is generally less than 100 nm and can increase oral bioavailability for drug compounds that have low solubility in water [15]. Measurement of emulsification time to determine the length of time needed by the system to be digested in some liquids such as aquadest, aquabidest, and the digestive tract which in this case is AGF (Artificial Gastric Fluid) at 36 °C, the faster an emulsified system, the faster the absorption in digestive tract, this showed that the SNEDDS system emulsification time below 1 minute was only 10 seconds in AGF media, 10 second in aquadest, and 9 second in aquabidest can be fully emulsified.

Tween 80 used in TiO_2 nanoparticles is known to have antifungal and antibacterial abilities both gram-positive and gram-negative bacteria [10]. Surfactants can increase the solubility of hydrophobic compounds in water. Tween 80 is known to have a low level of toxicity [2][3].

Tween 80 was chosen as surfactant because it has several advantages such as being able to transfer hydrophobic upconversion nanoparticles into the liquid phase, low prices, biocompatible and has the ability to prevent absorption of non-specific proteins due to its unique molecular structure. Tween coated was formed from hydrophobic interactions between the tails of fatty acids in tween and the oleic acid layer on the surface of the nanoparticles. Tween that coated the nanoparticles maintains the shape of the nanoparticles, has good solubility in water, low toxicity and has the ability in drugs delivery [6][13][15]. Some cosurfactants such as PEG 400 can change the shape of nanoparticles, size, and other surface properties at different levels depending on the molecular structure, nature of head group, hydrophobic tail length and ion counter type. Termination of the growth of nanoparticles is controlled by diffusion lanes and surfactant attachments on the molecular surface [14].

4. Conclusion

optimization formula showed that SNEDDS of ginger extract will be optimized in 4.2 g of tween 80 as surfactant; 0.25 g of PEG 400 as cosurfactant; 500 mg of ethanolic ginger extract and 0.54 g of maceration extract of eel bone oil with transmittance value was 92 ± 0.034 %, emulsification time in AGF was 10 second, in aquadest was 10 second, and aquabidest was 9 second. The result of Analysis of Variance indicated not significant differences between treatment groups and optimization result by *design expert 7 program* (($P = 0.1359$ ($P>0.05$)).

Acknowledgments
This study was supported by the Directorate General for Strengthening Research and Development, Community Service, Ministry of Research and Higher Education of the Republic of Indonesia.

Reference

[1] Akinyemi AJ, Ademiluyi AO, and roboh G 2014 *J Med Food* **17**(3) 317-323
[2] Cheng M, Zeng G, Huang D, Yang C, Lai C, Zhang C, and Liu Y 2017 *J Chem eng* **314** 98-113
[3] Feng J, Zeng Y, Ma C, Cai X, Zhang Q, Tong M, Yu B, and Xu P 2006 *Appl environ microbiol* **72** 7390-7393
[4] Gratton SEA, Ropp PA, Pohlhaus PD, Luft JC, Madden VJ, Napier ME, and Desimone JM 2008 *PNAS* **105** 11613-11618
[5] Haryono 2008 *Fauna Indonesia* **8**(1) 22-26
[6] Hillgren A, Lindgren J, and Alden M 2002 *Int J Pharm* **237** 57–69
[7] Landfester K, Musyanovych A, and Mailander V 2009 *J Polym Sci Part A: Polym Chem* **48** 493-515.
[8] Mohd AB, Sanka K, Bandi S, Diwan PV, and Shastri N 2017 *Drug Deliv 2015* **22**(4) 499-508
[9] Parmar N, Singla N, Amin S, and Kohli K 2011 *Colloids and Surfaces B: Biointerfaces* **86** 327-338
[10] Rajendar V, Chakra CHS, Rajitha B, Rao V, and Park SH 2016 *Mater Sci: Mater Electron* **28**(4) 3272–3277
[11] Rao SV, Shao J, and Agarwal P 2008 *Int J Pharm* **362**(1-2) 5-10
[12] Syahbana I 2015 *Fish oil extraction from eel bone (anguilla bicolor bicolor, McClelland 1844) with wet rendering method* (Jakarta: UI)
[13] Vaisman L, Wagner HD, and Marom G 2007 *Adv in colloid interface Sci* **128** 37-46
[14] Vidyasagar CC and Naik YA 2012 *Arab J Chem* **9** 507-510
[15] Zhao Y, Wang C, Chow AHL, Ren K, Gong T, Zhang Z, and Zheng Y 2009 *Int J Pharm* **383** 170-177

Solidification of meloxicam self-nano emulsifying drug delivery system formulation incorporated into soluble and insoluble carriers using freeze drying method

I Kuncahyo[1,2], **S Choiri**[3,4], **A Fudholi**[2]
[1]Department of Pharmaceutical Science and Technology, Faculty of Pharmacy, Setiabudi University, Surakarta, Indonesia, 57127
[2]Faculty of Pharmacy, Universitas Gadjah Mada, Sekip Utara, Yogyakarta, Indonesia, 55281
[3]Department of Pharmacy, Universitas Sebelas Maret, Ir. Sutami 36A, Surakarta, Indonesia, 57126

E-mail: syaiful.apt@gmail.com

Abstract. Solidification of liquid self-nano emulsifying drug delivery system (SNEDDS) formulation is required to enhance the feasibility and flexibility for further formulation processes. This study purposed to compare the soluble and insoluble carriers namely mannitol and fumed silica, respectively on the nano-emulsion formation and physical properties. The optimized SNEDDS formulation comprising of virgin olive oil, Tween 80, and PEG 400; and meloxicam was impregnated into solid carriers e.g. mannitol and fumed silica using a freeze-drying method. The physical mixture was carried out to compare the aforementioned method. Vibrational spectroscopy, thermal analysis, and morphological characteristic, droplet size and distribution, and drug release were performed to characterize the solid SNEDDS (S-SNEDDS). The result showed that crystallization of meloxicam did not observe in the S-SNEDDS formulations, which confirmed by the vibrational spectroscopy and thermal analysis. The morphological characteristic of S-SNEDDS was similar to the native carriers. The soluble carrier did not affect the formation of the nano-emulsion compared to the insoluble carrier. In addition, the S-SNEDDS enhanced the drug release of meloxicam up to 3-4 folds increment.

1. Introduction

A poorly water-soluble drug has a big hurdle and limitation to be developed for further formulation owing to a low solubility. It has a consequence, in which a low bioavailability is achieved [1,2]. Several approaches have been introduced and purposed to enhance the bioavailability. One of them, a lipid formulation using a self-emulsification mechanism is still attractive and interesting manner to be developed to the date [3,4]. Self nano-emulsifying drug delivery system (SNEDDS) gains a wide attention owing to the simplicity in the formulation, cost efficiency, and effective for bioavailability enhancement. SNEDDS comprising of oil, surfactant, and co-surfactant as an isotropic mixture when it is introduced into the medium, the nano-droplet formation is occurred spontaneously [5]. However, the liquid lipid formulation has numerous limitations e.g. flexibility, feasibility, and stability in the packaging and further formulation [4,6].

Furthermore, the solidification of SNEDDS is addressed to solve these issues. Solid SNEDDS can be applied to the further formulation, stability enhancement, and flexibility in packaging [7,8]. Moreover, a new concept of dry nano-emulsion is achieved, when the solid SNEDDS solubilized in

To whom any correspondence should be addressed

the medium [9]. However, the solidification process involves soluble and insoluble carriers, in which it differs from their characteristics [7,10]. Mannitol, a soluble material is a non-hygroscopic material and can absorb the liquid lipid at a high loading into the mesoporous of mannitol crystal structure [11]. Fumed silica, a non-soluble and generally known as a drying material had a high adsorbent capacity of liquid or viscous materials. It is usually used to solidify the liquid SNEDDS formulation using physical mixture and spray drying methods [8]. Thus, this study purposed to characterize the solid SNEDDS formulation using soluble and insoluble carriers namely mannitol and fumed silica on the nano-droplet formation and self-emulsification characteristics and drug release behavior. The vibrational, thermal, and morphological characteristics were also evaluated. Meloxicam as a poorly water-soluble drug was selected as drug model and loaded to the SNEDDS formulation.

2. Experimental

2.1. Material

Meloxicam (MLX) was obtained from Dexa Medica (Palembang, Indonesia). Olive oil was purchased from Brataco (Surakarta, Indonesia), Tween 80 and poly ethylene glycol (PEG) 400 were obtained from Merck (Darmstadt, Germany), mannitol was obtained from Roquette (Lestrem, France), and fumed silica was obtained from Cabot (Tokyo, Japan).

2.2. Preparation of SNEDDS formulation

Optimized SNEDDS formulation comprising of 26% olive oil as an oil, and 61.8% Tween 80 as a surfactant, 12.2% PEG 400 as a co-surfactant was weighed separately and mixed together using ultrasonicator and stirrer subsequently for 15 min. A 0.015% MLX was incorporated into SNEDDS formulation until a homogenous and clear solution was achieved under stirring condition. SNEDDS formulation was characterized by emulsification time, transmittance, and droplet size and distribution.

2.3. Solidification of SNEDDS

Hydrophilic soluble and insoluble carriers namely mannitol and fumed silica were used to solidify the SNEDDS formulation using a freeze-drying technique. Briefly, a one part of SNEDDS was diluted 10 times and added by 5 parts of mannitol and 1 part of fumed silica separately under gently agitation until either a clear or bluish solution was achieved. Prior the drying process, the solution was stored at -20°C overnight. Thereafter, the solution was dried using a Thermo PowerDry LL1500 freeze dryer (Waltham, MA) at the temperature of -60°C, the pressure of 0.75 mBar for 24 h. The solid SNEDDS mannitol (S-SNEDDS M) and FS (S-SNEDDS FS) were collected and stored in the desiccator until further characterization was performed.

2.4. Characterization of solid SNEDDS

The S-SNEDDS was characterized by a vibrational spectroscopy using an attenuated total reflectance-Fourier transform infrared (ATR-FTIR) spectrophotometry Thermo Nicolet iS50 (Waltham, MA), differential thermal-thermogravimetric analysis (DT-TGA) Shimadzu DT-TGA 60 (Tokyo, Japan), Morphology using a JEOL JSM-6510LA scanning electron microscopy (SEM) (Tokyo, Japan) equipped with a JEOL JEC-3000PC auto fine coater (Tokyo, Japan), and the drug release using an Erweka DT-820 dissolution tester (Heusenntam, Germany).

2.5. Drug release characterization

An equivalent to 15 mg of MLX for each solid SNEDDS was weighed accurately and filled in a gelatin capsule. The drug release characterization was carried out using an Erweka DT-820 dissolution tester (Heusenntam, Germany) equipped with a type II model USP. A 900 mL of phosphate buffer 0.05M pH 6.8 was used as the medium at 37±0.5°C. The sample was withdrawn at predetermined sampling time and analyzed spectrophotometrically using a validated analytical method.

2.6. Data analysis

The obtained data was compared statistically with confidence level of either 95% (p=0.05) or 99% (p=0.01).

3. Results and Discussion

MLX has been successfully incorporated into the SNEDDS formulation. Furthermore, the solidification process was carried out by using a water-soluble carrier, mannitol, and water-insoluble carrier fumed silica. The ET and %T of SNEDDS and S-SNEDDS formulation is presented in figure 1a. The liquid SNEDDS formulation was the lowest ET among solid formulations ($p<0.05$) owing to spontaneous emulsification without releasing from the carrier. SNEDDS produced nano-droplet formation when it introduced into water [3]. However, the emulsification process of S-SNEDDS M subsequently occurred after the solubilization of the water-soluble carrier [7]. Therefore, the emulsification time of S-SNEDDS M was lower than that of S-SNEDDS FS formulation and higher than that of SNEDDS. The ET of S-SNEDDS FS proved the highest value owing to the interaction and retardant action of the hydrophilic group in silica. Naturally, FS is not soluble but it has a hydrophilic characteristic [12]. However, it has a unique interaction by means of chemical interaction between Si-OH groups and hydrophilic groups in the SNEDDS formulation [8,12]. Hence, it produced the longest ET. The shorter ET, the more appropriate formulation was. The ET contributes to the critical quality attributes of the formation of nano-emulsion during dilution [13,14].

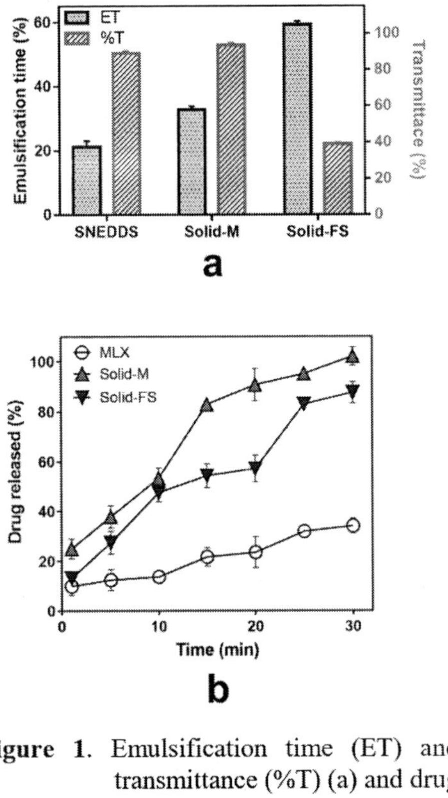

Figure 1. Emulsification time (ET) and transmittance (%T) (a) and drug release profile (b) of meloxicam and solid SNEDDS meloxicam

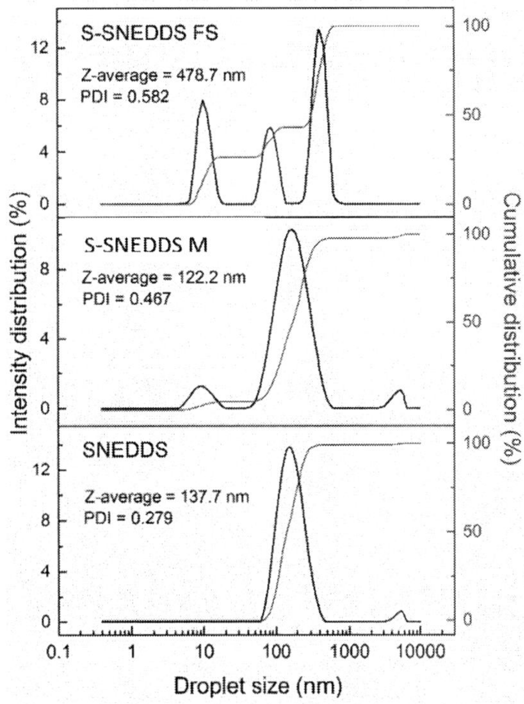

Figure 2. Droplet size distribution of SNEDDS, solid SNEDDS using mannitol (S-SNEDDS M) and fumed silica (S-SNEDDS FS)

The transmittance (%T) depicts the particle size owing to its color. Generally, clear to bluish nano-emulsion when it diluted with the medium had a nano-size range of droplet size. Thus, the higher the

%T, the more acceptable formulation was [15]. The %T of SNEDDS and S-SNEDDS M formulations was not a significant difference (p>0.01). Although, the %T of S-SNEDDS FS was reduced significantly because fumed silica did not soluble in the medium thus produced the turbid solution and reduced the %T. The drug released of MLX and S-SNEDDS formulations is presented in figure 1b. Both of S-SNEDDS formulations increased the drug release about 3.0-4.0 folds compared to the MLX drug release profile. This result proved that significant increasing of drug released had a great probability to enhance the bioavailability [8,10]. The drug release of S-SNEDDS M was higher than that of S-SNEDDS FS. It was affected by the interaction between a hydrophilic component in the SNEDDS formation and non-soluble FS, thus it reduced the drug release rate. Chemical interaction between the droplet (hydroxyl in fatty acid side chain) and the non-soluble fumed silica, a hydrogen bonding has significantly reduced the drug released. The S-SNEDDS M proved that the drug release occurred subsequently by the solubilization of carrier and the lipid formulation was emulsified. The drug released figured out the release in term of "formation of nano-droplet size from S-SNEDDS" [10].

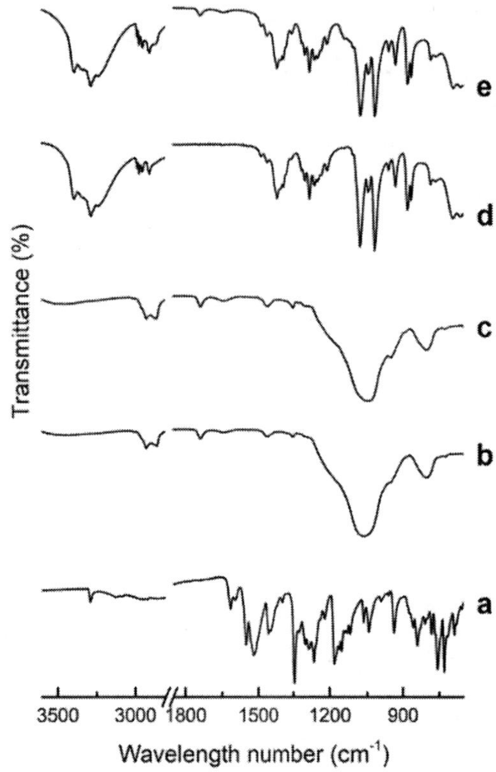

Figure 3. FTIR Characterization of meloxicam (a), mannitol (b), fumed silica (c), solid SNEDDS mannitol (d) and solid SNEDDS fumed silica (e)

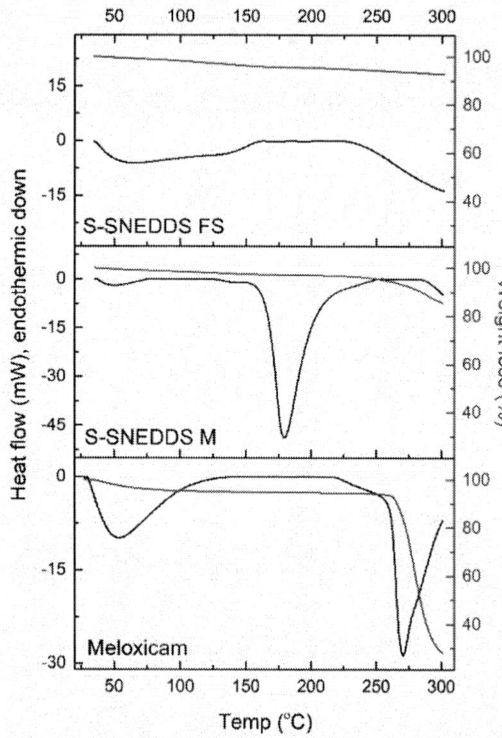

Figure 4. Thermogram of DSC and TGA of Meloxicam and Solid SNEDDS mannitol (S-SNEDDS M) and fumed silica (S-SNEDDS FS)

In order to investigate the nano-droplet formation, the particle size distribution after solidification was carried out. It is presented in figure 2. The SNEDDS produced monodisperse system and had the particle size of 137.7 nm and PDI of 0.279. This result suggested that the optimal formulation for nano-emulsion delivery was achieved. This system had a narrow distribution of particle under nano-droplet size [5,16]. Solidification altered the formation of nano-droplet size after dilution. In S-SNEDDS M, a new peak was observed around the droplet range of 3-11 nm owing to the micellar

IOP Conf. Series: Materials Science and Engineering **578** (2019) 012051 doi:10.1088/1757-899X/578/1/012051

formation. The solidification not only affected the emulsification process, but also it had interaction between the lipid formulation and solid carrier, thus it altered the phase behavior of nano-droplet formation. The nano-droplet formation was observed at peak of 50-400 nm. The nano-emulsion peak of S-SNEDDS M had a lower droplet size than SNEDDS, but it had a wider droplet size distribution. The obtained data of PDI was 0.461 owing to the formation of the micellar formation thus it increased the PDI. On contrary, the poly disperses system was observed in the S-SNEDDS FS after dilution. It was affected by the micellar formation at the range particle from 8-20 nm, nano-emulsion formation at the range of 30-80 nm and non-soluble fumed silica at the range of particle of 200-500 nm. The blatant alteration was observed owing to a high intensity of a peak was observed in the micellar region. This interaction between SNEDDS formulation and hydrophilic side of FS altered the phase behavior and induced formation of the micelle. The intensity of micelle was higher than that of the nano-emulsion owing to the separation of the isotropic mixture. In particular, the surfactant and co-surfactant induced the formation of the micellar solution. The micelle has different transport mechanism compared to the nano-emulsion passed through p-gp efflux and tick junction opening [5,14].

In order to characterize the S-SNEDDS formulation, vibrational spectroscopy, thermal behavior, and morphological characteristics were evaluated. The vibrational spectra of MLX and S-SNEDDS formulation are presented in figure 3. MLX (figure 3a) had specific peaks at 3305 cm^{-1} according to the N-H stretching vibration and 1627 and 1162 cm^{-1} according to the C=N and S=O stretching vibrations, respectively. A similar report of MLX spectra has been reported by a previous research [17]. There was no significant and specific peak shifting between fumed silica and S-SNEDDS FS owing to a low fraction of MLX in the S-SNEDDS FS. As similar results as to the S-SNEDDS M, there was no significant peak shifting between the mannitol and S-SNEDDS M. Hence, it indicated there was no interaction between the solidified material and SNEDDS formulation.

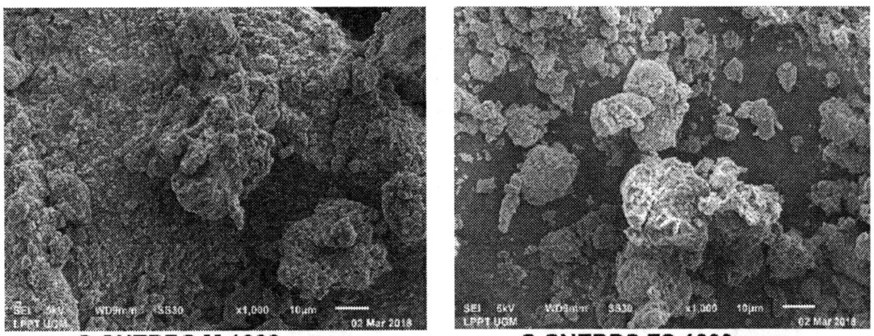

S-SNEDDS M 1000x mag. **S-SNEDDS FS 1000x mag.**

Figure 5. SEM photograph of solid SNEDDS mannitol and fumed silica using freeze drying method

The thermal analyses are presented in figure 4. A board endothermic peak was observed on the MLX thermogram around 40-100°C according to the loss of water adsorption (4.98%) in the solid MLX. Moreover, the sharp endothermic peak at 270.6°C according to the melting point of MLX followed by decomposition which was confirmed by loss of its weight about 65.4% from 260 to 300°C. The S-SNEDDS M had a weak and board endothermic peak owing to the evaporation of water adsorption. The sharp endothermic peak was observed at 179.8°C owing to the mannitol melting point. There was no peak observed around of MLX melting point, thus it was indicated there was no crystallization of MLX in the S-SNEDDS M. The slowly decreasing the gravimetric line from 260 to 300°C indicated that there was a decomposition of MLX followed by the mannitol decomposition. The thermogram of S-SNEDDS FS showed a broad endothermic peak was observed around 50-150°C owing to the water absorption and adsorption evaporation. In addition, the similar pattern of decreasing the heat flow around 255-300°C and confirmed by a slow decreasing of gravimetric line confirmed the amorphous form in the surface of the fumed silica. Owing to the adsorption phenomena of MLX in the surface the fumed silica, there was a possibility of formation of the amorphous state of

MLX during drying. According to the morphological characteristics on the surface of S-SNEDDS formulation (figure 5), there was no crystallization in the surface of the S-SNEDDS formulation. The S-SNEDDS M had an irregular characteristic and the agglomerate had a spherical characteristic and rough at the surface without a cracking after the freeze-drying process. The absorption phenomena of the soluble carrier promoted a different characteristic in the surface of insoluble carrier, especially under the freeze-drying treatment. The agglomerate of fumed silica was observed owing to the presence of SNEDDS formulation. The SNEDDS as a liquid form increased the cohesiveness of fumed silica particle thus the agglomerate was formed.

4. Conclusion

The solidification using mannitol and fumed silica as soluble and insoluble carriers, respectively promoted different characteristics by which it was affected by a different mechanism of solidification. The soluble carrier had a positive effect on the drug release and formation of the nano-emulsion. On contrary, the insoluble carrier had an advantage on the powder appearance. Although, it disturbed the nano-droplet and self-emulsification formations. For further development, the soluble carrier could be developed to enhance the flowability and loading capacity on solidifying the SNEDDS formulation.

Acknowledgement

This research was supported by “Lembaga Pengelola Dana Pendidikan” (Indonesian Endowment Fund for Education).

References

[1] Kalepu S and Nekkanti V 2016 *Drug Deliv. Transl. Res.* **6** 319–32
[2] Mu H, Holm R and Müllertz A 2013 *Int. J. Pharm.* **453** 215–24
[3] Date A A and Nagarsenker M S 2007 *Int. J. Pharm.* **329** 166–72
[4] Garg V, Kaur P, Singh S K, Kumar B, Bawa P, Gulati M and Yadav A K 2017 *Eur. J. Pharm. Sci.* **109** 297–315
[5] Rehman F U, Shah K U, Shah S U, Khan I U, Khan G M and Khan A 2017 *Expert Opin. Drug Deliv.* **14** 1325–40
[6] Agrawal A G, Kumar A and Gide P S 2015 *Drug Dev. Ind. Pharm.* **41** 594–604
[7] Chavan R B, Modi S R and Bansal A K 2015 *Int. J. Pharm.* **495** 374–84
[8] Kim K S, Yang E S, Kim D S, Kim D W, Yoo H H, Yong C S, Youn Y S, Oh K T, Jee J-P, Kim J O, Jin S G and Choi H G 2017 *Drug Deliv.* **24** 1018–25
[9] Rahman M A, Mujahid M, Rahman M A and Mujahid M 2018 *Braz. J. Pharm. Sci.* **54** 1-14
[10] Inugala S, Eedara B B, Sunkavalli S, Dhurke R, Kandadi P, Jukanti R and Bandari S 2015 *Eur. J. Pharm. Sci.* **74** 1–10
[11] Peng T, Zhang X, Huang Y, Zhao Z, Liao Q, Xu J, Huang Z, Zhang J, Wu C, Pan X and Wu C 2017 *Sci. Rep.* **7** 1-12
[12] Oh D H, Kang J H, Kim D W, Lee B-J, Kim J O, Yong C S and Choi H-G 2011 *Int. J. Pharm.* **420** 412–8
[13] Badawy M E I, Saad A-F S A, Tayeb E-S H M, Mohammed S A and Abd-Elnabi A D 2017 *J. Environ. Sci. Health Part B* **52** 896–911
[14] Singh Y, Meher J G, Raval K, Khan F A, Chaurasia M, Jain N K and Chourasia M K 2017 *J. Controlled Release* **252** 28–49
[15] Bali V, Ali M and Ali J 2010 *Colloids Surf. B Biointerfaces* **76** 410–20
[16] Parmar K, Patel J and Sheth N 2015 *Asian J. Pharm. Sci.* **10** 396–404
[17] Ainurofiq A and Choiri S 2018 *Pharm. Dev. Technol.* **23** 464–75

IOP Conf. Series: Materials Science and Engineering **578** (2019) 012052 doi:10.1088/1757-899X/578/1/012052

Formulation and the release of eugenol from cream using glycerin base

S Rohmani, N Miararani, A Yugatama, D E Ermawati and F Prihapsara
Pharmacy Department, Faculty of Mathematics and Natural Sciences, Sebelas Maret University, Indonesia

Email: licha.apt@gmail.com

Abstract. Eugenol consist of phenol, which have antibacterial effect. The aims of this research were to observed the characteristics of cream dosage form, eugenol released profile, and determined the effect of glycerin as water base through in vitro evaluation. Cream was made with glycerin as enhancer and water base. Various concentration of glycerin (10%, 15%, and 20%) were added to eugenol cream. Cream evaluation included organoleptic, homogenity, pH, stickiness, spreadability, viscosity, and eugenol released test from cream base of each formula. Drug released test was carried out eith Erweka Dissolution tester using apparatus 5 paddle overdisk. Phosphate buffer 7.4±0.05 was using as medium, with 32°C and 200 rpm rotation. The resul showed that eugenol can be formulated into cream, because of the evaluation has met the reqruitments of test quality. Those quality were organoleptic, homogenitu, pH, stickiness, viscosity and spreading test. Concentration analysis of eugenol cream in glcerin various concentration were evaluated using UV-Vis spectrophotometer. The result showed that total concentration of eugenol in cream were 107.11%; 99.91%; 105.63%. The result of eugenol released test were 11.53%; 11.38%; 10.96%. Eugenol cream with 10% of glycerin concentration has the highest perfomance as enhancer and water base.

1. Introduction

Indonesia is one of the largest producer of essential oil in the world. It can produce 50-150 types of essential oils available [16]. Essential oil has been proven that it gave benefits for health. One of the type of essential oil is eugenol. It were stated that eugenol has anti-inflamatory activity for acute lung injury (ALI) and antioxidant activity [17]. According to other research, it was revealed that eugenol has anticancer activity, antidepressant activity and anthelmintic activity [1]. Due to eugenol activities that have been proven before, it was important foe eugenol to be developed as a dossage form. In order to give more benefits of eugenol for health.

One way to utilize eugenol potential is by developin it into a high quality of pharmaceutical product, whic can be used widely by society. The requirement of pharmaceutical product to qualified are safe to use, effective for medicine, comfortable for patient, and having both physical and chemical stability. It can quarantee the quality of production and distribution process until it can be consumed by patient. Pharmaceutical preparation has big potential to developed such as topical products, one of them is cream. Topical product is use on skin surface to give it local effect. It has been proven that eugenol have pharmacologic activity which could be developed as topical product for antibacterial, anti-inflammatory, anthelmintic, antioxidant, and anticancer. Morover, eugenol has antidepressan ctivity which became an additional value.

The cream preparation, excipient or base is often to use as additional material usually to repair its concistency. It also can be used as cosolvent by increasing the solubility of the drug ingredients. Increasing solubility of cream will make drug easy to release from their base and influence the activity [5]. A base is suitable for cream if it were not strongly bind with other ingredients, since the drug must release immedietly from base before it penetrated to skin [9]. There are two steps for topical drug before it gives effect. Firstly the drug must be able to released from their base and go to to skin surface. Secondly they can penetrate through the skin membrane to achieve its action[6]. This research aims to find out the correlation between physical stability and realeasing profile of cream eugenol with various concentration of glycerin (10%, 15%, and 20%). These amount of optimal glycerin will increase eugenol penetration into the skin, so the optimal therapy effect can be reached.

2. Experimental

2.1 Materials

Eugenol, Glycerin (Bratako), Sodium Lauril Sulphate (Bratako), Vaseline Album (Bratako), Stearic Acid (Bratako), Setil Alcohol (Bratako), KH2PO4 (Merck), NaOH (Merck), Ethanol p.a (Merck), cellophane membrane (Merck), Dissolution tester, Viscometer (Rion vt-04E Co, TLD) , pH meter (Lutron), Spektrohotometer UV-Vis (Genesys ™).

2.2 Methods

2.2. 1. Cream Formulation

Cream was made by melting oil base including cetyl alcohol, vaselin album, stearic acid above the waterbath. Then heat the water base and disosolving glycerin, sodium lauryl sulphate, and aquadest. The melting oil base was put into a warmer mortar, then eugenol was added. After that the oil phase was mixture with water base, and stirred until cream emulsion was formed.

Table 1. Formulation of cream eugenol

Material	Amount (%)		
	F1	F2	F3
Eugenol	1.25	1.25	1.25
Glycerin	10	15	20
Sodium lauril sulphate	2	2	2
Vaseline Album	25	20	15
Stearic acid	10	10	10
Cetyl Alcohol	3	3	3
Aquadest	Ad 100	Ad 100	Ad 100

2.3. pH Test

pH evaluation was determined by using pH meter (Lutron). Cream was taken at a certain amount and it was mixed with aquadest. A mixture of cream and aquadest was analysed using pH meter. pH meter was placed in the mixture. The change on the scale of pH meter can be seen. The number displayed on pH meter was an exact pH from the product [14].

2.4. Stickiness Test

0.25 gram of sample was put on two glass objects. Then pressed the glass with 1 kg load or above, and left it for 5 minutes. After that, the glass objects were place on the tool that was given load 80 gram. The time was recorded until the two glass objects were separated [11].

2.5. Spreadability Test

0.5 gram of cream was placed on the midle of circular glass, and left it for 1 minute. Distribution diameter of cream was measured. After that it was added with 150 gram additional load and left it for 1 minute, then constant diameter of cream was measured.

2.6. Viscosity Test
The viscosity of cream was measured by using viscosimeter Rion vt-04E Co, TLD. The bowl was filled with a half of the cream sample to be tested. The rotor was put in the middle of the bowl containing the cream, it was turned on to starts rotating, the viscosity pointer will automatically move to the right. Once it is stable, it is displayed on the scale on the viscometer.

2.7. Determining Composition
1 gram of eugenol cream was diluted into 10 mL of ethanol, and then stirred for 15 minutes in 25 °C. Cream solution was distilled, and filtrate was generated. After then, it was diluted and analyzed using UV-vis spectrophotometer at a wavelength of 200-400 nm.

2.8. Eugenol Release from Base
Release test uses type 5 of dissolution test tool. Membrane used was cut and adjusted to the area of diffusion disc. Chamber was filled with 500 mL of phosphate buffer and the temperature was set at 32 ± 0.5 °C. Rotational speed was adjusted into 200 rpm. Medium fluid was taken as much as 10 mL at 0, 15, 30, 45, 60, 90, 120, 180, 240, 300, 360, 420, and 480 minutes. The sample liquid was analyzed using UV-vis spectrophotometer at wavelength of 283 nm[13].

3. Results and Discussion

3.1 pH Test
Potential for Hydrogen or pH test is aim to find out the safety and comfort of preparations when it used on skin. pH profile can determine the stability of an active ingredient when it is conditioned in an acidic or alkaline atmosphere[10]. According to SNI-16-4399-1996, the ideal pH should be in line with the skin pH; if the pH is too acidic, it will cause irritation to the skin. Meanwhile, if the pH is too alkaline, it can result in scaly skin[12]. Based on table 2, it is seen that the three formulas have been met the requirements of SNI.

3.2. Stickiness Test
The test aim to find out how long cream is able to attach on skin surface. The test conducted for 4 weeks shows that formula 1 has the longest stickiness, while the formula 3 has the fastest stickiness. It is because formula 1 has the highest oil-based concentration (Vaseline album) than other formula. Formula 3 has the lowest oil based concentration so it make formula 3 has the fastest time for cream to stay on skin surface.

3.3. Spread Test
Distribution test aims to find out cream ability to spread evenly while it is applied to the skin. A good cream is expected to spread easily on the skin surface without significant pressure. The easier preparation spread on the skin, the biggest the skin contact to the medicine will be. Becuase of that the absorption of the medicine will be more optimal. Good spread for semisolid preparation is 5-7 cm[15]. Data of spread test can be seen in table 4.

3.4. Viscosity Test
The result of visocsity test can be seen on table 5. Formula 1 has the highest viscosity while formula 3 has the lowest viscosity. It is due to highest concentration of glycerin will decrease cream viscosity. Decreasing viscosity can be caused by the increasing of globule. Thinner or softer cream will easy to move in dispersion medium and the globule will form a bigger particle. Biger size of globule will cause the reduction on the area of cream surface. Good cream viscosity according to SNI 16-4399-1996 is 2,000-50,000 cP or equivalent to 20-500 d.Pas. Only formula 1 with a base of 10% glycerin has viscosity according to SNI.

3.5. Determining Composition

Test for determining composition was presented in ratio between measured composition and theoretical composition. The result of eugenol composition can be seen in table 6. It showed that all formulas have met the requirements, which is in ratio of 80-11%[3].

3.6. Eugenol Release Test

The percentage of released eugenol can be seen in table 7. The value showed that eugenol composition in the buffer medium. Besides accumulated amount, diffused eugenol is left in the cellophane membrane. It can be concluded that the total amount of eugenol was able to diffused, has higher value than measured value in medium liquid. The most important factor for active substance released were preparation formulation, testing medium, membrane type, intensity of movement and temperature, selected excipient (Vaseline and glycerin), and viscosity of preparation which will influence the active substance release[15]. Eugenol has a phenol component so eugenol is easily soluble in glycerin. An active substance that is soluble in polar solvents in large quantities causes the drug to be difficult to release and penetrate through the membrane. For glycerin it is as a humectant. humectants work by forming a layer on the surface of the skin to protect. Membranes coated with glycerin make eugenol difficult to penetrate and diffuse. Therefore eugenol can be released from the formula with the highest concentration of glycerin.

Table 2. pH test of eugenol cream

pH	Week 1	Week 2	Week 3	Week 4
Formula 1	6.50	6.53	6.53	5.80
Formula 2	6.50	6.50	6.50	6.60
Formula 3	6.60	6.60	6.60	6.60

Table 3. Stickiness test of eugenol cream

Stickiness Test	Week 1	Week 2	Week 3	Week 4
Formula 1	2.67	4.34	3.33	3.58
Formula 2	1.56	3.85	2.25	2.56
Formula 3	1.51	3.61	2.41	2.51

Table 4. Spread test of eugenol cream

Spread Test (cm)	Week 1	Week 2	Week 3	Week 4
Formula 1	4.50	4.23	4.67	5.40
Formula 2	6.40	5.52	6.40	6.22
Formula 3	6.83	5.90	6.60	6.40

Table 5. Viscosity test of eugenol cream

Viscisity(dPas)	Week 1	Week 2	Week 3	Week 4
Formula 1	80	80	80	80
Formula 2	40	35	35	35
Formula 3	30	30	25	25

Table 6. Determining composition of eugenol

Formula	Replication	Sample weight (mg)	Eugenol teoritical concentration (ppm)	Research eugenol concentration	Average %recovery
	1	1000	1250	1338.75	
1	2	1000	1250	1342.76	107.11±0.37
	3	1000	1250	1333.88	
	1	1000	1250	1206.75	
2	2	1000	1250	1267.75	99.91 ± 2.93
	3	1000	1250	1272.63	
	1	1000	1250	1270.85	
3	2	1000	1250	1367.11	105.63±3.80
	3	1000	1250	1322.95	

Table 7. Amount of eugenol that penetrated through in vitro test (release profile)

Formula	% eugenol penetrated	SD	CV (%)
1	11.290	0.298	2.639
2	11.773	0.299	2.541
3	9.503	0.256	2.695

4. Conclusion

The difference in base concentration will affect the stability of the cream such as viscosity, spreadability, and eugenol release profile. The higher the concentration of glycerin, the smaller the viscosity, the wider the spread of power, and eugenol is difficult to release.

Acknowledgment

Thank you to Sebelas Maret University for the financial support.

References

[1] Anita Y, Widiyarti G and Abbas J 2014 *J. App. Pharm. Sci.* **4** 62-65
[2] Ansel 1989 *Pengantar Bentuk Sediaan Farmasi Edisi Keempat* (Jakarta: UI Press)
[3] AOAC 1998 *AOAC Peer Review Methods Program, Manual and Policies and Procedures.* (USA: AOAC International)
[4] Astuti I D H 2010 *Majalah Obat Tradisional* **15** 94-99
[5] Barry B W 2001 *Eur. J. Pharm. Sci* **14** 101-114
[6] Boylan J and Chopper Z 1994 *Handbook of Pharmaceuticals Excipients 2nd ed* (Washington: American Pharmaceutical Association)
[7] Chan C C, Lee Y C, Lam H and Zhang X M 2004 *Analytical Method Validation and Instrument Performance Verification* (New Jersey: Wiley)
[8] Ditjen POM 1985 *Formularium Kosmetika Indonesia* (Jakarta: Departemen Kesehatan Republik Indonesia)
[9] Idson B 1975 *J. Pharm. Sci* **64** 915-916
[10] Lachman L, Lieberman H A and Kanig J L 1986 *The Theory and Practice of Industrial Pharmacy 3rd ed* (Philadelphia: Lea & Febiger)
[11] Miranti L 2009 *Pengaruh Konsentrasi Minyak Atsiri Kencur (Kaemferia galangal) dengan Basis Salep Larut Air terhadap Sifat Fisik Salep dan Daya Hambat Bakteri Staphylococcus aureus Secara In Vitro* [Thesis] (Surakarta: Fakultas Farmasi Universitas Muhammadiyah Surakarta)
[12] Naibaho H, Olivia , Yamelan Y, Paulina and Wiyono 2013 *Jurnal Ilmiah Farmasi* **2** 27-33

[13] Qisti B W 2017 *Optimasi Propilenglikol dan Etanol sebagai Peningkat Penetrasi Ibuprofen dalam Sediaan Gel dengan Metode Simplex Lattice Design* [Thesis] (Jember: Fakultas Farmasi Universitas Jember)
[14] Tranggono, Iswari R, Latifah and Fatmah 2007 *Buku Pegangan Ilmu Pengetahuan Kosmetik* (Jakarta: PT. Gramedia Pustaka Utama)
[15] Voight R 1994 *Buku Pelajaran Teknologi Farmasi Edisi V, ed. Noerono S, Soewandi, Widianto, Mathilda* (Yogyakarta: Universitas Gajah Mada Press)
[16] Widi A and Putra N N 2014 *Jurnal Bahan Alam Terbarukan* **3** 12-16
[17] Yogalakshmi B, Viswanathan P and Anuradha C 2010 *Toxicology Journal* **268** 204–212

IOP Conf. Series: Materials Science and Engineering **578** (2019) 012053 doi:10.1088/1757-899X/578/1/012053

Renal Function Parameter on Acute Toxicity Test of Kapulaga (*Amomum cardamom*) Seed Extract in Rat

R D Yudhani[1*], R N Pesik[2], S Azzahro[3], A F Anisa[3] and R Hendriyani[3]
[1] Pharmacology Departement, Medical Faculty, Universitas Sebelas Maret (UNS) Surakarta, Indonesia
[2]Pathology Anatomy Departement, Medical Faculty, Universitas Sebelas Maret (UNS) Surakarta, Indonesia
[3]Students of Medical Faculty, Universitas Sebelas Maret (UNS) Surakarta, Indonesia

[*]Email: rdyudhani@gmail.com

Abstract. Kapulaga (*Amomum cardamom*) is one of herb which is commonly used for spices or medicine, especially the seeds. Previous studies showed that the essential oil of cardamom have antimicrobial, antiinflamatory, analgesic and antispasmodic activities. There was also evidence that *A.cardamom* has antiartherogenic and antidiabetic activity in rat. Despite the medicinal benefits of *A.cardamom*, this herb is not yet standardize for herb medicine. For its standardization, *A.cardamom* has to pass the preclinical and clinical studies to ensure its efficacy and safety profile. The aims of this study was to examine the safety profile of *A.cardamom* seed extract based on renal parameter function (ureum and creatine levels). Acute toxicity test was conducted based on the OECD 420 *Fixed Dose Prosedure* guideline that consists of two test steps, preliminary and main tests. For preliminary test, initially with 300 mg/kg BW dose of *A.cardamom* seed extract, followed by 2000 mg/kg BW. The main test consists of control and treatment group and each group used 5 rats. The rats in both groups were given 2000 mg/kg BW in a single dose of *A.cardamom* seed extract. The ureum and cretainine levels were assessed at day 14th using an enzymatic-photometric method. The data were analyzed by independent sample test. The results revealed that the ureum and creatinine levels in control and treatment groups were not statistically different. The mean of ureum levels in the control and tretment groups were 47.24±6.18 and 43.98±6.78 ($p=0.45$) and the mean of creatinine levels were 0.43±0.04 and 0.36±0.12 ($p=0.31$) respectively. These results show that a high single dose of *A.cardamom* seed extract (2000 mg/kgBW) did not toxic in rat based on renal function parameters.

1. Introduction

The health-promoting properties of many herbs and spices are numerous and well recognized. Moreover, herbs and spices have a traditional history of use with strong roles in cultural heritage [1,2]. One of herb which is commonly used for spices or medicine is *cardamom*. *Cardamom* is a famous aromatic spice and widely used in Eastern, Arab, and Scandinavian culinary. Moreover, this herb is also common as ingredients of Indian ayurveda and Chinese traditional medicine [3]. In Indonesia, *cardamom* not only use for culinary purposes but also commonly use as medicinal herb [4]. *Cardamom* belongs to the ginger family *Zingiberaceae* and consist of two genus, *Elletaria* and *Amomum*. Both forms are small spindle-like seedpod, but they are difference in size and appearance.

Elettaria is smaller and light green in colour but *Amomum* is bigger and its appearance is blackish grey [3].

Many studies have shown that *cardamom* plays an important wide range of health promoting roles against various pathologic conditions such as constipation, colic, diarrhea, dyspepsia, vomiting, headache, epilepsy, and cardiovascular diseases [1]. Previous studies revealed that *Amomum cardamom* leaf extract have an activity to decrease the atherogenic index and blood sugar levels in diabetic rats model [5]. Whilst, *Elletaria cardamom* was reported have many properties such as antibacterial, antiviral and antifungal [6]. Others studies reported that the essential oil of *cardamom* have antimicrobial, antiinflamatory, analgesic and antispasmodic activities [7]. All this collected evidences support the potential development of *A.cardamom* as standard herb medicine. Despite the medicinal benefits properties of *A.cardamom*, this herb is not yet standardize for herb medicine. For its standardization, *A.cardamom* has to pass the preclinical and clinical studies to ensure its efficacy and safety profile.

The kidney (renal) is a major target for drug-induced toxicity due to its function as the major organ of excretion. Kidney are naturally exposed to a greater proportion of circulating drug and chemicals compound and plays an important role as the primary eliminator of exogenous drugs and toxin. It makes the kidney is vulnerable to develop various form of injury. Renal toxicity has been reported for various agents including a large number of drugs and leads to acute kidney injury which is associated with high morbidity and mortality [8,9]. An important and unregulated source of potentially renal toxic substances is the alternative/complementary products, include herbal remidies, natural products and nutritional supplements that are commonly used in community and widely available at most health food stores. Moreover, interaction of herbal products with conventional drug is also potential source of renal toxicity [9]. The development of drug with minimal potential renal toxic effect would be one strategy to prevent these problem [8].

The uses of *A. cardamom* as medicinal plant are common, but there was lack of a proven studies on the toxicity of these treatment, to our knowledge. This study aims to examine the safety profile of *A.cardamom* seed extract based on renal parameter function (ureum and creatine levels). This study was conducted in order to obtain scientific evidence related to *A.cardamom* toxicity, needed for its development as standardized herbal medicine as well as for developing herbal medicines with less potential of renal toxic effect.

2. Experimental

2.1 Acute Toxicity Test

Pharmacological and toxicological evaluations of medicinal plants are essential for drug safety and development. Toxicity study of medicinal plants is important in predicting their safety. The main types of toxicological evaluations include: acute toxicity, subacute toxicity, subchronic toxicity, and chronic toxicity studies. Acute toxicity of medicinal plant extract refers to those adverse effects occurring following oral or dermal (topical) administration of a single dose of a substance or multiple doses given within 24 hour. There are some methods in acute toxicity test, and one of them is fixed dose procedure[10].

The acute toxicity test in this study was conducted based on the OECD 420 fixed dosed procedure that consist of two test steps, preliminary and main test. OECD 420 fixed dose procedure is one of OECD guidelines for testing chemical compound which avoided death of animal as endpoint to asses the acute toxicity and relied instead on the observation of clear signs of toxicity at one of a series of fixed dose levels. The preliminary test used initial dose of *cardamom* seed extract 300 mg/kg body weight because based on this procedure, if there was no information from any in vitro or in vivo study for standard dose which expected to show toxicity evidence of the testing compound, the preliminary dose use 300 mg/kg body weight as initial dose [11]. Using this initial dose, the rat survived, so after 24 hours from the first treatment, the dose was increased to 2000 mg/kg body weight as maximal dose in OECD 420 fixed dose procedure [11]. The main test consists of two groups, control group (5 rats) and treatment group (4 rat and 1 rat from preliminary test). Treatment group was given high single dose of *cardomom* seed extract baased on preliminary test (2000 mg/kg BW).

2.2 Renal Function Parameters

In the 14th day, the renal function parameters (ureum and creatinine) were assessed. Blood samples were collected from rat's orbital vein (3 cc) of the control and treatment groups. Ureum and creatinine level were measured using enzymatic-photometric method. All the collected data were statistically analyzed with shapiro wilk and independent sample t-test.

3. Results and Discussion

In recent times, plant origin drugs play a vital role in the management of various chronic disease and have received a great attention by researcher as alternative source to pharmaceutical drugs. Moreover, a wide range of natural products isolated from herbs and spices have been shown to posses immunomodulatory effect that potentially have beneficial against many diseases. Whilst, there is a lack of proven scientific studies on the toxicity and adverse effect of these treatments [1,12].

As mentioned before, *cardamom* is one of the common ingredients and well-known aromatic spice that widely used as herbal medicine for many condition such as for improving digestion, stimulating metabolism and exhibit antioxidant and antiinflamatorry effects [3,13]. The kidney is routinely assessed during preclinical safety evaluation due to its important role as a central detoxification organ. It leads an extraordinary exposure of renal tissues to drugs, reactive metabolites or environmental compounds [14,15]. In this present study, we conducted an acute toxicity test to observe the safety profile of *cardamom* seed extract based on renal function parameter (ureum and creatinine).

Table 1. showed that *A.cardamom* seed extract which was given to rats as a high sigle dose (2000 mg/kgBB) may not be toxic based on renal function parameter. This was assessed from ureum and creatinine levels between the control and treatment group which were not statistically different ($p > 0.05$) and mortality was not observed in any of the experimental groups.

Table 1. Renal function parameters in rats which were given high single dose of *A.cardamomum* seed extract. Data are expressed as mean ± SD.

Renal Function Parameters	*A.cardamomum* seed extract		
	Control Group (0 mg/kgBW)	Treament Group (2000 mg/kgBW)	P
Ureum(U/L)	47.24 ± 6.18	43.98 ± 6.78	0.45
Creatinine (U/L)	0.43 ± 0.04	0.36 ± 0.12	0.31

Results are expressed as the mean ± SD (n=5)

Ureum and creatinine are two important chemical substances in blood which can be evaluated to help assessing the Glomerular Filtration Rate (GFR) and renal function. The elevated level of ureum and creatinine in blood can be used to assess the progression of kidney damaged. Ureum is an organic compound and plays a vital role in the metabolism of nitrogen-containing compounds. Urea is filtered by the kidneys into urine as a waste product from dietary protein. Healthy kidney remove this compound from blood and its level will increase when the kidney failure occurs [14]. Creatinine is the breakdown product of creatinine phosphate in muscle and is excreted through the kidney along with other waste products. The balance of creatinine concentration between its generation and excretion is maintained by the kidneys at a fairly constant rate. The quantity of creatinine in serum depends on their generation, glomerular filtration and tubular secretion of this compound. Both serum ureum and creatinine are widely accepted as important biomarkers to assess renal function due to its pivotal role in the diagnosis and follow up of kidney failure [14,16,17].

Increased creatinine levels and blood urea nitrogen simultaneously indicate impaired renal filtration [18]. Blood urea nitrogen levels indicate urea levels in the blood. As mentioned above, urea is one of the body's disposal products. Urea is produced when the liver metabolizes proteins and eliminates through the kidneys. The body will maintain urea levels in the blood to stay normal so that the liver and kidneys should function normally [19].

Table 1 demonstated that, ureum level between control (47.24 ± 6.18) and treatment group (43.98 ± 6.78) were not statistically different (p=0.45). Moreover the creatinine level in both control (0.43 ± 0.04) and treatment (0.36 ± 0.12) groups were still in normal range based on study by Derelanko[18] which reported that the normal creatinine level in rat is in range 0.3-0.8 mg/dl. Cretinine is considered as one of the most reliable indicators of renal function and the increasing of blood creatinine is strongly related with renal damaged [20]. These evidence supported that acute high single dose of *A.cardamom* did not alter the renal function, since no changed observed in ureum level and the creatinine level is still in normal range.

This study revealed that acute high single dose of *A.cardamom* did not alter the blood ureum and creatinine level compared to control. Whist, Table 1 showed that both ureum and creatinine level in the treatment group (43.98 ± 6.78 and 0.36 ± 0.12) were slighty lower than those in control group (47.24 ± 6.18 and 0.43 ± 0.04), although there were not statistically difference. It might be due to the antioxidant properties of *A.cardamom*. Chemical analysis of *cardamom* showed, it contains antioxidant flavonoid which observed have effect to induce *glutathione S-transferase* (GST) and to reduce the lipid peroxidation in Swiss Albino Mice [21]. In accordance with this, Lim et al.[22]. reported that ethyl acetate fraction of *A.cardamom* has a high content of polyphenol and flavonoid compounds, which was proven plays a role in decreasing lipid peroxidation in liver tissues.

In addition, several studies demonstrated that most of the components of *cardamom* act as antioxidants such as limonene, cineole, linalool, pinene and borneol. *Cardamom* also contains Cu and Mn that are required for the activation of *superoxide dismutase* (SOD) enzyme which play an important role in the reducing of lipid peroxidation process [3]. This antioxidant activities of *cardamom* is also likely to occur in the kidneys and protect them from damage.

4. Conclusion

This present study suggested the safety profile of a high single dose of *A.cardamom* seed extract (2000 mg/kgBW) in rat based on renal function parameters.

References

[1] Majdalawieh A F and Carr R I 2010 *J. Med. Food* **13** 371-81
[2] Tapsell L C, Hemphill I, Cobiac L, Sullivan D R, Fenech M, Patch C S, Roodenrys S, Keogh J B, Clifton P M, Williams P G, Fazio V A and Inge K E 2006 *Med. J. Aust.* **185** S4-S24
[3] Das I, Acharya A, Berry D B, Sen S, Williams E, Permaul E, Sengupta A, Bhattacharya S and Saha T 2012 *Br. J. Nutr.* **108** 984-97
[4] Tjitrosoepomo 2015 *Taksonomi tumbuhan obat-obatan* (Yogyakarta: UGM Press)
[5] Winarsi H, Sasongko, N D, Purwanto A and Nuraeni I 2013 *Agritech.* **33** 273-80
[6] Gilani A H, Jabeen Q, Khan A and Shah A J 2008 *J. Ethnopharmacol.* **115** 463-72
[7] Jamal A, Javed K, Aslama M and Jafri M A 2006 *J. Ethnopharmacol.* **103** 149-53
[8] Tiong H Y, Huang P, Xiong S, Li Y, Vathsala A and Zink D 2014 *Mol. Pharm.* **11** 1933-48
[9] Perazella M A 2009 *Clin. J. Am. Soc. Nephrol.* **4** 1275-83
[10] Teke G N and Kuete V 2014 *Toxicological Survey of African Medicinal Plants* (Elsevier) pp 63-98
[11] Organization for Economic Co-operation and Development (OECD) 2001 *OECD Guidlines for Testing of Chemicals, Test No. 420: Acute Oral Toxicity Fixed Dose Procedure* pp 1-14
[12] Kifayatullah M, Mustafa M S, Sagupta P, Sarker M R, Das A and Das S K 2015 *J. Acute Dis.* **4** 309-15

[13] Goyal S N, Sharma C, Mahajan U B, Patil C R, Agrawal Y O, Kumari S, Arya D S and Ojha S 2015 *Int. J. Mol. Sci* **16** 27457-69
[14] Nisha R, Srinivasa K S R, Thanga M K and Jagatha P 2017 *J. Clin. Path. Lab. Med.* **1** 1-5
[15] Fuchs T C, Frick K, Emde B, Czasch S, von Landenberg F and Hewitt P 2012 *Toxicol. Pathol.* **40** 1533-48
[16] Kunitoshi I, Yoshiharu I and Koshiro F 1997 *Kidney Int.* **51** 850-54
[17] Venkateswarlu K, Heerasingh T, Babu C N, Triveni S, Manasa S and Babu T N B 2017 *Braz. J. Pharm. Sci.* **53** 1-8
[18] Derelanko M J 2008 *The toxicologist's pocket handbook-second edition* (New York: CRC Press)
[19] Purwaningsih S, Handharyan E and Lestari I R 2015 *Jurnal Akuatika* **6** 30-40
[20] Alarifi S, Al-Doaiss A, Alkahtani S, Al-Farraj S A, Al-Eissa M A, Al-Dahmash B and Al-Yahya M M 2012 *Saudi J. Biol. Sci.* **19** 103-10
[21] Bhattacharjee S, Rana T and Sengupta A 2007 *Asian Pac. J. Cancer Prev.* **8** 578-82.
[22] Lim D W, Kim H, Park J Y, Kim J E, Moon J Y, Park S D and Park W H 2016 *BMC Complement Altern. Med.* **16** 1-10

IOP Conf. Series: Materials Science and Engineering **578** (2019) 012054 doi:10.1088/1757-899X/578/1/012054

Permeability of Piroxicam with Sodium Lauryl Sulfate as Surfactant

F Prihapsara[1*], W A Saputra[2], A N Artanti[1], D E Ermawati[1], S Rohmani[1] and A Yugatama[1]
[1]Pharmacy Department, Faculty of Mathematics and Natural Science, Universitas Sebelas Maret
[2]Pharmacy Science and Technology Department, Faculty of Pharmacy, Universitas Gadjah Mada

[*]E-mail: fea.prihapsara@staff.uns.ac.id

Abstract. Piroxicam is a nonsteroidal anti-inflammatory drug (NSAID) which have characteristic insoluble in water and poorly fluidity. The profile absorption from piroxicam was depend on its dissolution rate in the gastrointestinal tract (GI). To improve the solubility in the water, it can used Sodium Lauryl Sulfate (SLS) as a surfactant which have mechanism by forming a micelle. The aim of this study was to measure the influence gradiet concentration of SLS with Piroxicam absorption that can observe by inverted intestinal sac method. The concentration of piroxicam solution that added by SLS are 1, 2, and 3% respectively. The preparation method using Crane and Wilson tube containing mucosal fluid. After preparation, it was taken at 37^{0}C in waterbath and added with serosal solution to intestinal sac by turned upside down and tied to a cannula. The sample was placed in the tube containing mucosal fluid and constantly flowing by oxygen gas. To measure the absorsion of piroxicam that used serosal solution which every 15 minutes was taken on tube. After this, it was diluted using $Ba(OH)_2$ and $ZnSO_4$ then continuous with separating process. The absorbant of supernatant was measured using Spectrophotometer UV-Vis. Data analyse was calculated by one-way ANOVA. According to P_m, the values 0.25x10^{-3} cm/minute (0% SLS), 0.39x10^{-3} cm/minute (1% SLS), 1.45x10^{-3} cm/minute (2% SLS) and 1.26x10^{-3} cm/minute (3% SLS) respectively. In conclusion, 2 % SLS significantly give the highest absorption of piroxicam by altering the membrane permeability.

1. Introduction

There are several factors that can affect the absorption process, one of them is drug solubility. Drug which have low solubility in water, dissolution rate is the rate limiting step of drug bioavailability [1]. Absorption of a drug can be defined as a process of transferring drug through biological barrier into the blood and lymphatics system. Drug absorption can determined by in vitro, in situ and in vivo method. Absorption in vitro through the intestine is based for determining the cumulative absorption of drug from intestinal lumen. This method is used for study various influential factors against permeability of the intestinal wall. Development for design of drugs to optimize the bioavailability especially for the drug with poor solubility and very slow absorption rate [2].

Physicochemical profile of drug is important to determine the bioavailability profile. It is not only important to optimize the formulation but also it can used to evaluation of medicinal products [3]. Piroxicam is a crystalline powder colorless, odorless, bitter taste, in the form of yellow monohydrate. Insoluble in water, in dilute acids and cyclohexane, slightly soluble in methanol, ethanol, isopropanol, in dimethyl formamide (1:10), dimethylsulfoxide (1:50), acetone (1:50), ethyl acetate (1:80), chloroform (1:20), and solubility in alkaline solution (1:100) (Florey, 1986). The principle of piroxicam solubility is stable at pH 7.5 with pKa 6.3. Factors that influence absorption rate i.e. pH, buffer, temperature and the presence of additional ingredients such as surfactant [4]. Sodium lauryl sulfate (SLS) can reduce surface tension between drug and medium at once form micelles so the drug molecules will carried by the micelle dissolves into the medium [5]. Addition of surfactant was assumed to be able to interact complex with drugs then affect the permeability of the membrane [6]. The permeability is affected by the diffusion coefficient, surface area, partition coefficient and a thick membrane.

Sodium Lauryl Sulfate (SLS) is anionic surfactant, stable in many medium, easy to mix with other components, non-toxic, no irritating, and effective in a wide pH range [7]. One of the important thing is the ability of surfactant to improve drug solubilitation which insoluble or small dissolved in a medium. Surfactant on low concentration, reduce surface tension and increase the absorption rate of drugs [8]. Surfactant in high concentration will gather to form aggregates called micelles [9].

2. Experimental

2.1. Materials and Instrument

Piroxicam (Nantong General Pharmaceutical Factory), SLS (Merck), NaH_2PO_4 (Merck), Na_2HPO_4 (Merck), NaOH (Merck), $Ba(OH)_2$ dan $ZnSO_4$ (Merck), dan NaCl (Merck), aquadestilata and male *Wistar* rat. Instrument which were used in this research are Crane and Wilson tube were modified by Yuwono from Institut Teknologi Bandung (ITB), Spectrophotometer Genesys 10, Analytical balance (Shimadzu Scientific Ltd.), Digital pH meter Omega models 5003, Centrifuges MLW T51.1, Sonicator Elma T 570, Waterbath (Shimadzu Scientific Ltd.), tools for surgery, glassware,volume pipette, test tube and flask.

2.2. Procedure

2.2.1. Calibration Curve of Piroxicam. Stock solution was prepared from 100 mg of Piroxicam were dissolved in a sonicator and than added with phosphate buffer (pH 7.4). The Series of stock solution i.e 10, 12, 14, 16, 18 µg/ml, then takes 1.0 ml of each series add 4.0 ml SLS (1% w/v) in phosphate buffer solution pH 7.4. The solution was measured at a maximum absorbance wavelength (200 -400 nm).

2.2.2. Permeability Test. Male wistar rats are adapted in the experimental room for a week before test. Male wistar rats are observed health and behavior. Male wistar rats which used in the experiment were healthy, showed no behavioral abnormalities and deviations from normal circumstances, and no weight loss exceeded 10% [10]. Principles of laboratory animal care guidelines were followed and prior permission for conducting the study (Ref. No. FO-UGM-FA-13-03).

Determination of the membrane permeability was observed with intestinal inverse sac method. Pylorus partion from male wistar was removed and then collected the intestine part. It was cleaned with NaCl 0.9% w/v and then divided into 2 equal length. In this study was using the top part intestine to placed sample treatment (0%, 1%, 2% and 3% SLS) and the bottom part as a negative control (without piroxicam).

Mucosal solutions are made from 20 mg% piroxicam in phosphate buffer pH 7.4 with SLS at level 0, 1, 2, and 3% (w / v). Mucosal solution put in a waterbath at a temperature of 37 °C. 1.5 ml of serosal solution incorporated into the intestinal sac that has been turned upside down and tied to the cannula, then inserted into the tube which already contains the mucosal fluid and constantly flowing O_2 gas at a rate of approximately 100 bubbles per minute to maintain the structure of the intestinal wall. During the trial maintained that all parts of the intestinal mucosal submerged in the liquid. Samples were taken every 15 minutes until 75th minutes.

Serosal fluid's sample was taken from intestinal sac then added with $ZnSO_4$ 5% and 0.3 N $Ba(OH)_2$ than centrifuged for 25 minutes. Supernatan was taken in tube and measured in the maximum wavelength of absorbance by spectrophotometre UV-Vis.

Total cumulative piroxicam was calculated from calibration curve of piroxicam, absorbance of sample reduced with negative control. Absorption rate, permeability of membrane, and lag time were calculated from the regression equation curve cumulative amount of piroxicam absorbed after steady state conditions in each replication. The data were analyzed with one way ANOVA test to compare the value of absorption rate, permeability of membrane and lag time between any various concentration of surfactant.

3. Result and Discussions

3.1. Calibration Curve of Piroxicam

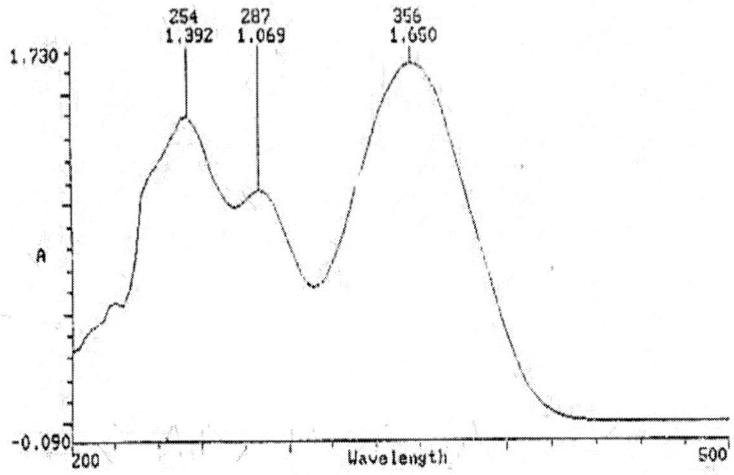

Figure 1. Wavelenght Maximum of Piroxicam.

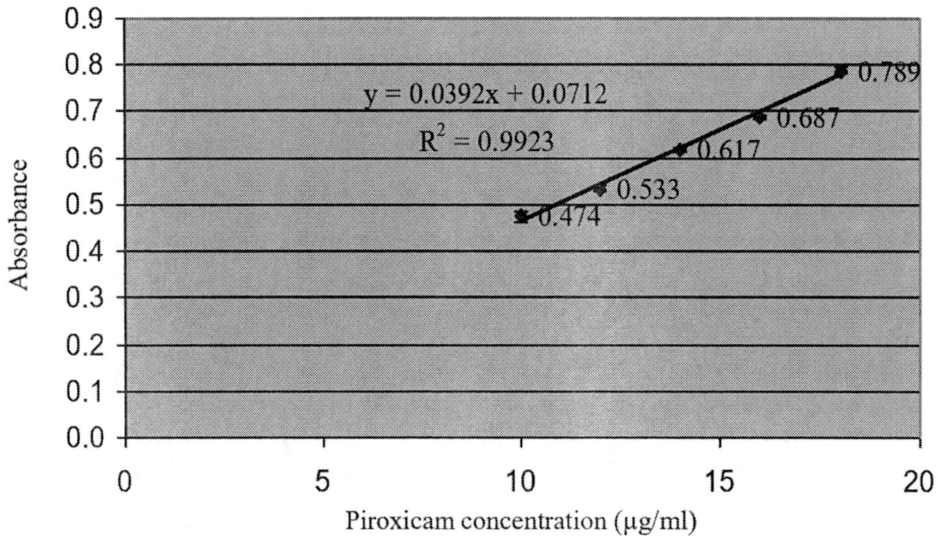

Figure 2. Calibration Curve of Piroxicam.

Wavelength maximum of Piroxicam was 356 nm. Drug absorption from gastrointestinal tract into the blood generally occurs after it dissolved in the absorption site. Solubility related with absorption rate which has a linear curve. SLS as a material that can increase the solubility of drug and it was expected to increase the piroxicam absorption rate.

3.2 Permeability Test

Table 1. Total cumulative piroxicam was absorbed (x 10^{-2} mg).

Time (minute)	SLS 0%	SLS 1%	SLS 2%	SLS 3%
15	0	0.540±0.431	2.268±1.773	1.512±1.292
30	0.144±0.125	1.320±1.238	5.655±3.589	4.998±3.434
45	0.548±0.250	2.355±1.003	9.579±4.179	8.232±4.106
60	1.469±0.173	3.243±1.222	14.086±6.661	12.983±6.071
75	2.358±0.350	5.156±1.894	19.727±8.324	16.165±8.105

Table 2. The value of absorption rate (K), permeability of membrane (Pm), and the lag time in various concentration of SLS.

Groups	Absorption rate (μg/minute)	Permeability of membrane (x10^{-3} cm/minute)	*Lag time* (minutes)
SLS 0%	0.050±0.009	0.250±0.043	29.480±4.400
SLS 1%	0.077±0.017	0.390±0.087	13.874±10.474
SLS 2%	0.289±0.107	1.440±0.538	10.309±3.407
SLS 3%	0.252±0.103	1.260±0.516	12.054±6.416

The value of Absorption rate, permeability of the membrane, and the lag time was linier with total cumulative piroxicam curve after steady state conditions. Table II showed the absorption rate, permeability of membrane, and lag time related with increased levels of SLS. The results showed that 2% SLS give the highest absorption. SLS which have highly polarity was easily soluble in water. Interaction between piroxicam-SLS formed a complex which is more soluble in water and intestinal fluid compared with piroxicam.

4. Conclusions

The highest absorption in solution piroxicam is addition 2% SLS and then followed by addition 3%, 1% and 0% SLS seen from the Pm, which is 1.44; 1.26; 0.39; and 0.25 (x10^{-3}cm / minute).

Acknowledgments

This work was supported by the grant of Mandatory Research Group Universitas Sebelas Maret 2018.

References

[1] Takano R, Furumoto K, Shiraki K, Takata N, Hayashi Y, Aso Y and Yamashita S 2018 *Pharm. Res.* **25** 2334-44
[2] Savjani K T, Gajjar A K and Savjani J K 2012 *ISRN Pharm.* **2012** 195727
[3] Shargel L, Yu A and Wu S 2005 *Apllied Biopharmaceutics and Pharmacokinetics*, 2nd Edition (Surabaya: Airlangga University Press)
[4] Chillistone S and Hardman J G 2017 *Anaesthesia Intensive Care Med.* **18** 335-39

[5] Esteves R, Dikici B, Lehman M, Mazumder Q and Onukwuba N 2016 *Beyond* **1** 27-35
[6] Choudhary S, Talele P and Kishore N 2015 *Colloids Surf. B. Biointerfaces* **132** 313-21
[7] Salager J L 2002 *Surfactants Types and Uses Laboratory of Formulation, Interfaces, Rheology and Process* (Colombia: Universidad De Los Andes)
[8] Tseng W C, Chu S P, Kong P H, Huang C K, Chen J H, Chen P S and Huang S D 2014 *J. Agric. Food Chem.* **62** 9059-65
[9] Lotya M, King P J, Khan U, De S and Coleman J N 2010 *ACS Nano* **4** 55-62
[10] Manzoor M, Raza S and Chaudhry B 2013 *IOSR JPBS* **6** 21-33

IOP Conf. Series: Materials Science and Engineering **578** (2019) 012055 doi:10.1088/1757-899X/578/1/012055

Effect of Temperature and Length of Storage to Chloramphenicol Eye Drop's Concentration

A Yugatama[1], R Nurmalinda[1], S Rohmani[1], D E Ermawati[1], and F Prihapsara[1]
[1]Pharmacy Department, Faculty of Mathematics and Natural Sciences, Sebelas Maret University, Indonesia

Email: adiyugatama.apt@gmail.com; adi.yugatama@staff.uns.ac.id

Abstract. Chloramphenicol is a broad spectrum antibiotic that acts by inhibiting protein synthesize. One of chloramphenicol dosage form which widely found in the marketplace is eyedrop. Drug storage will affect drug stability. This research aimed to know the effect of temperature and length of storage to chloramphenicol eye drop's concentration. Analysis of chloramphenicol concentration in eyedrops was done by UV spectrophotometry. This analysis was done to eyedrops after stored 0; 7; 21; and 30 days unsealed with various storage condition: under the sunlight, room temperature, and refrigerated. Data obtained was analyzed statistically using one-way ANOVA with 95% CI by SPSS for windows software. The result showed that the concentration of chloramphenicol eyedrop before stored was to meet the requirements which were 95.96%. The result of chloramphenicol concentration in eyedrop after stored 30 days in 3 storage condition was reduced significantly. While sunlight and storage temperature did not affect the reduction of chloramphenicol concentration significantly.

1. Introduction

Chloramphenicol is a broad spectrum antibiotic that acts by inhibit protein synthesis. Chloramphenicol having a bacterostatic activity against almost all gram positive and negative bacteria, and against Spirochaeta, Chlamydia trachomatis and Mycoplasma. Chloramphenicol having a good bacterisid activity against Str. pneumonia, Neiss. meningitides, and H. Influenzae [1].

One of chloramphenicol dosage form available in market is eyedrops, in 0.25–1% concentration. Eyedrops is a sterile solution which combined and packaged for eyes purposes. Eyedrops have to be sterile, isohydrile, and isotonic. This dosage form could not be used six weeks after being unsealed [2].

Drug storage affect drug stability utmost. Drug stability could be seen by concentration reduction during storage [3]. Drugs have to be stored in room temperature and avoid direct sunlight to ensure drug stability [4]. Temperature, light, moist and air condition are factors that induced drug degradation. Although reduction 10% of prior concentration of active ingredients is acceptable [5].

There are some theories stated about relation of drug stability to temperature, light and length of storage. Thus, it need to be studied about the effect of temperature and length of storage to chloramphenicol concentration in eyedrops.

2. Experimental

2.1. Materials

Equipments used were UV-Vis Spectrophotometer (Genesys 10S UV-Vis), analytical balance (Ohaus), micropipet (Joanlab), and glassware. Materials used were chloramphenicol standard, chloramphenicol eyedrops brand X, aquadest and ethanol.

2.2. Methods

2.2.1. Standard Solution. Standard solution was made by reconciling 100 mg chloramphenicol in 10 mL ethanol then diluted in aquadest until reached the concentration of 1000 ppm. From chloramphenicol 1000 ppm solution, was taken 0.1; 0.5; 1; 1.5; 2; 2.5; 3 mL and added to aquadest until 100 mL to generate standard solution of 1, 5, 10, 15, 20, 25, and 30 ppm chloramphenicol.

2.2.2. Sample Treatment. Samples were tested by stored in refrigerator temperature (3-7°C), room temperature (27-30°C), and exposed to direct sunlight. The drugs were unsealed before stored. Concentration of chloramphenicol in eyedrops then analyzed at day 0; 7; 21; and 30 after storage.

2.2.3. Determination of Chloramphenicol Concentration in Sample. As much as 5 mL samples were added by 25 mL aquadest. The mixture was taken 1 mL and added by aquadest until 100 mL. The solutions were measured its absorbance using spectrophotometer at 278 nm wavelength.

2.3. Data Analysis

Absorbance of samples were analyzed by linear regression until obtained chloramphenicol concentration in eyedrops in various storage condition. The data then analyzed by one way Anova to know the effect of temperature and length of storage to chloramphenicol concentration in eyedrops.

3. Results and Discussion

3.1. Determination of Maximum Wavelength

The result of 10 ppm chloramphenicol standard solution scanning in figure 1 showed that maximum wavelength is 278 nm. Based on other research, maximum wavelength of chloramphenicol is 278 nm [6].

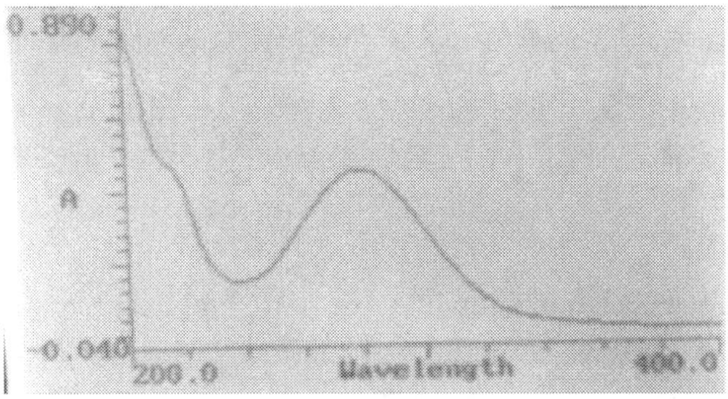

Figure 1. Maximum wavelength of chloramphenicol

3.2. Determination of Chloramphenicol Standard Curve

Standard curve equation obtained was $y= 0.0441x + 0.0115$ and the correlation coefficient (r) was 0.9999. Based on requirement, $r > 0.997$ showed that the curve is having a good linearity [7]. This can be concluded that standard curve equation obtained was having a good linearity. The chloramphenicol standard curve could be seen in figure 2.

IOP Conf. Series: Materials Science and Engineering **578** (2019) 012055 doi:10.1088/1757-899X/578/1/012055

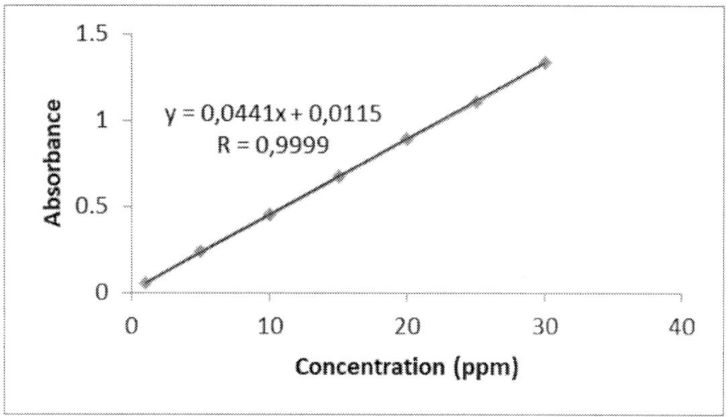

Figure 2. Chloramphenicol standard curve

3.3. Determination of Chloramphenicol Concentration in Sample

The result of chloramphenicol concentration in eyedrops determination after storage showed that there were reductions of chloramphenicol concentration both in refrigerator temperature, in room temperature, and exposed to direct sunlight. The prior concentration of chloramphenicol in eyedrops was 95.96%. This is in accordance with the requirements stated in USP standard that the concentration of chloramphenicol in eyedrops is not less than 90% and not over 130%. In addition, the deadline for using sterile preparation after unsealed is 28 days [8]. At 30 days length of storage after unsealed in refrigerator temperature, room temperature, and direct sunlight exposure, the chloramphenicol concentration in eyedrops were above 90%. This showed that chloramphenicol eyedrops brand X is fulfill the standard and still be able to use until 30 days after unsealing. Chloramphenicol concentration in various storage condition could be seen in figure 3.

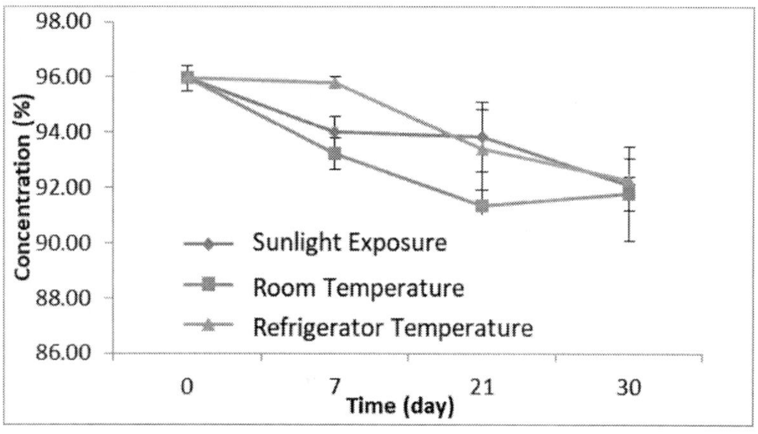

Figure 3. Curve of chloramphenicol concentration in various storage condition

During storage, concentration of chloramphenicol in brand X eyedrops were reduced. The longer the storage period, the lesser concentration of chloramphenicol found in eyedrops. This might be resulted from hydrolysis reaction. This reaction occured because chloramphenicol in watery media will be degraded by hydrolysis breakdown in amides circle and forming appropriate amides and dichloracetic acid. Reaction rate is occured under the first order and independent from ionic power of the media [9]. The hydrolysis reaction could be seen in figure 4.

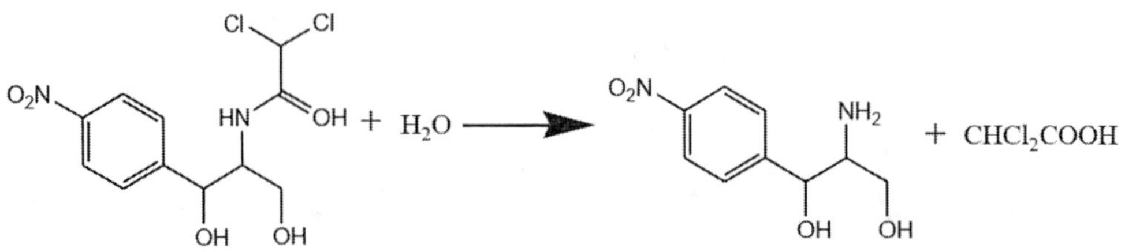

Figure 4. Chloramphenicol hydrolysis reaction

Besides, chloramphenicol in solution also sensitive to photodegradation reaction. Chloramphenicol under direct sunlight slowly changed into yellow color and formed yellowish orange residu. In this photodegradation reaction, chloramphenicol will changed into 4-nitrobenzaldehyde, 4-nitrobenzoic acid, and 4,4'-azoxybenzoic acid through oxidation, reduction, and condensation reaction [9].

Storage in room temperature resulted in relatively lower concentration of chloramphenicol than in direct sunlight exposure and refrigerator temperature. This might be caused by moist which commonly found higher in room temperature. Factors inducing degradation reaction of a drug i.e. temperature, light, moist, and air condition. Esters and amides compounds like chloramphenicol is easily hydrolized in a presence of moisture [10].

Storage in refrigerator temperature resulting the least reduction of chloramphenicol concentration compared to direct sunlight expossure and in room temperature storage. This showed that chloramphenicol in eyedrops will be most stable if stored in refrigerator temperature. The temperature of refrigerator used in this study is 3–7 °C, and included into cold temperature. The lower the temperature, the slower the reaction rate will be. Reaction rate is linear to number of collision per unit of time.

Result of statistic analysis using SPSS showed that there was a correlation between length of storage, light, and temperature to chloramphenicol concentration. Based on normality test using Kolmogorov-smirnov, the data were distributed normally by sig. 2 tailed value was 0.482. The value of $p > 0.05$ showed that data were distributed normally.

Statistical analysis by one way anova resulted that significance value of length of storage, light, and temperature to chloramphenicol concentration were 0.000; 0.925; dan 0.140. Based on these result, significance value of length of storage was < 0.05. Thus, can be conclude that length of storage can affect the reduction of chloramphenicol concentration. While sunlight and temperature were not affecting chloramphenicol concentration significance.

Insignificance effect of light to chloramphenicol concentration might be caused by the package of eyedrops that are white colored and not translucent so the effect of sunlight will be smaller, and the incidence of photodegradation reaction also being lower.

4. Conclusion

Based on chloramphenicol concentration analysis in eyedrops for 30 days storage in 3 various condition which are exposed directly to sunlight, room temperature, and refrigerator temperature using UV Vis Spectrophotometer, could be concluded that length of storage having a significant effect to chloramphenicol concentration reduction, while lights and storage temperature were not having a significant effect to chloramphenicol concentration reduction.

Acknowledgement

The authors acknowledge to Sebelas Maret University which given the facilities for the research and publication.

References

[1] Rahal JJ 1979 *Antimicrob Agents Chemother* **16**(1) 13.

[2] Virani S, Rewri P and Vyas A 2016 *Del J Ophtalmol* **26**(3) 176.
[3] Bajaj S, Singla D and Sakhuja N 2012 *J. Appl. Pharm. Sci.* **2**(3) 129.
[4] Shafaat K, Hussain A, Kumar B, Hasan R, Prabhat P and Yadav VK 2013 *World J. Pharm. Pharm. Sci.* **2**(5) 2499.
[5] Cantrell L, Suchard JR, Wu A and Gerona RR 2012 *Arch Intern Med.* **172**(21) 1685.
[6] Karthikeyan S 2011 *Arch. Phy. Res.* **2**(4) 72.
[7] Chan CC, Lam H, Lee YC and Zhang XM 2004 *Analytical Method Validation and Instrument Performance Verification* (New Jersey: John Wiley & Sons, Inc.)
[8] US Pharmacopeial Convention, Inc. 2007 *United States Pharmacopeia 30* (Rockville, MD: US Pharmacopeial Convention, Inc.)
[9] Ali SL 1978 *J. Chromatogr. A* **154**(1) 99.
[10] Honmane SM, Dange YD, Osmani RAM and Jadge DR 2017 *Asian J. Pharm.* **11**(3) 479.

IOP Conf. Series: Materials Science and Engineering **578** (2019) 012056 doi:10.1088/1757-899X/578/1/012056

Optimatitation of mole ratio for synthesis reaction of Ketoprofen Eugenol Ester

A Yugatama[1], A H Ramelan[2] and E D W Novitasari[1]
[1]Pharmacy Department, Faculty of Mathematics and Natural Sciences, Sebelas Maret University, Indonesia
[2]Physics Department, Faculty of Mathematics and Natural Sciences, Sebelas Maret University, Indonesia

Email: Adiyugatama.apt@gmail.com

Abstract. Ketoprofen is a NSAID which often used, but has been shown to induce gastrointestinal irritation. The side-effect arises because of the presence of carboxylic acid groups. The carboxylic acid groups can be changed by estherification reaction between ketoprofen and eugenol using an acid catalyst. This study was purposed to find out the effect of mole ratio of ketoprofen : eugenol on the synthesis Ketoprofen Eugenol Ester (KEE). The mole ratio of ketoprofen : eugenol which used in the study were 1:3 ; 1:4 ; 1:5 ; 1:6 ; 1:7 ; 1:8. The synthesis was done by heating up ketoprofen, eugenol, and HCl as the catalyst. The results of synthesis were characterized using TLC, FTIR, UV-Vis Spectrophotometer, and HPLC. Then, the data was analyzed statistically using One Way ANOVA (SPSS). The result of sample analysis using TLC showed the Rf value ±0.07. The FTIR analysis indicated that there were peak C=O ester and C-O ester. The highest yield value, absorbance of UV-Vis spectrophotometer and peak area of HPLC were found in the mole ratio 1:6. The One-Way ANOVA statistical analysis indicated that there was a significant difference between the variations of mole ratio which were used.

1. Introduction

Ketoprofen (aryl propionic acid derivative) is a Nonsteroidal Anti-Inflammatory Drug (NSAID) that has been widely used to treat analgesic, antipyretic and anti-inflammatory [1]. Ketoprofen has side effects such as gastrointestinal irritation. Gastrointestinal irritation arises because of the presence of carboxylic acid group in the structure that irritant on the gastrointestinal and also ketoprofen inhibit the cytoprotective COX-1[2]. One way that can be used to reduce gastrointestinal irritation is to change the carboxylic acid group with an estherification reaction [3].

Eugenol is a volatile compound extracted from *Eugenia caryophyllata* Thunb., which has analgesic, antipyretic, anti-inflammatory, anti-ulcer, anti-bacteria and anti-oxidation effect [4]. Eugenol has a hydroxyl group that can be conjugate with ketoprofen. Therefore, based on prodrug principal, ketoprofen and eugenol can be combined into ketoprofen eugenol ester or KEE [2]. Eugenol is selected because it has a synergic effects with ketoprofen and reduced gastrointestinal irritation [5].

KEE can reduce side effect gastrointestinal irritation and increased therapeutic effect of ketoprofen [2]. KEE can be obtain with esterification reaction of ketoprofen and eugenol with acid catalist (HCl). Esterification reaction ketoprofen and eugenol is a reversible reaction, where reaction can be to the right or towards the products and to the left or towards the reactants [6]. Maximum yield of KEE can be obtain by changing the reaction condition must be to the right. To change reaction to the right, the number of reactant can be increased [7], so the appropriate mole ratio of ketoprofen and eugenol is needed. This study was aimed to optimized mole ratio of ketoprofen and eugenol on the synthesis ketoprofen eugenol ester to achieve maximum yield (%) of KEE.

2. Experimental

2.1. Materials

Eugenol (PT. Indesso), Ketoprofen (PT. Kimia Farma), analytical grade solvents like hydrochloric acid, methanol, n-hexane, ethyl acetate, chloroform, petroleum ether, TLC silica gel 60 F_{254} (Merck), sodium bicarbonate, aquadest, analytical balance (Mettler Toledo), micropipette (Thermo), UV-Vis spectrophotometer (Genesys^{TM}), FTIR Spectrometer (Shimadzhu FTIR-8201 PC), HPLC (Waters) with a UV-Vis detector and a C-18 column.

2.2. Methods

2.2.1. Synthesis of the KEE

According to Rifqi [8] a mixture of ketoprofen 2.55 g, eugenol 6.3 mL and hydrochloric acid 1 mL was refluxed for 8 hours at a temperature of 50 ± 5 °C. The mixture was added water and decanted to separated organic layer. The organic layer washed with $NaHCO_3$ solution until the pH becomes neutral, followed by water 15 mL and the mixture separated with liqiud-liquid extraction (LLE). The crude oil distilled out under reduced pressure for purification. The oil was then dissolved in petroleum ether and dried in a oven overnight. Synthesis was carried out on the mole ratio of ketoprofen : eugenol at 1:3 ; 1:4 ; 1:5 ; 1:6 ; 1:7 ; 1:8.

2.2.2. Characterization of KEE

2.2.2.1. Thin Layer Chromatography (TLC)

TLC plates were prepared by silica gel 60 GF_{254} as stationary phase. Synthesis sample 0.1 g was dissolved up to 10 mL with methanol. Using micropipette, about 2 µL sample and standard were loaded into plates. Ketoprofen standard, eugenol standard and sample analysis were developed in TLC plate with mobile phase n-hexane: ethyl acetate: chloroform (2:5:1 v/v) [2]. The chromatograms were observed under UV lamp at 254 nm and were protographed. The Rf value was obtained by using the following formula:

$$Rf = \frac{\text{distance traveled by the center of a spot}}{\text{distance traveled by the solvent front}} \quad (1)$$

2.2.2.2. Fourier-Transform Infrared Spectroscopy (FTIR)

Synthesis sample dropped into KBr plates and stacked with other plates [9], then sample analyzed and FTIR spectra were obtained.

2.2.2.3. UV-Vis Spectrophotometer

A 0,1 g synthesis sample was dissolved up to 10 mL with methanol then dilution was carried out to obtain a 10 ppm concentration solution. Absorbance then was analyzed by UV-Vis spectrophotometer at 261 nm [2].

2.2.2.4. High Performance Liquid Chromatography (HPLC)

Synthesis sample 0.1 g was dissolved up to 10 mL with methanol then dilution was carried out to obtain a 100 ppm concentration. HPLC analysis was carried out with C18 as stationary phase and methanol: water (65:35 v/v) as mobile phase. The measurements used were flow rate 1 mL/minute, UV-Vis detector and sample injection volume was 20 µL.

2.3. Data Analysis
Statistical analysis was carried out for absorbance data UV-Vis spectrophotometer using analysis of variance (ANOVA) test, followed by LSD test for determining level of significance. If P-value < 0.05 were considered statistically significant

3. Results and Discussion
3.1. Synthesis of KEE
The conjugation of ketoprofen and eugenol have many advantages were to synergistic analgesic and anti-inflammatory effects and reduced gastrointestinal irritation [2]. Ketoprofen has carboxylic acid group that can be esterification with hydroxyl group of eugenol with HCl as catalyst, the esterification reaction KEE was reversible (Figure 1). A carboxylic acid can be esterified using excess alcohol that may be reaction shifts to the right or towards the product [6].

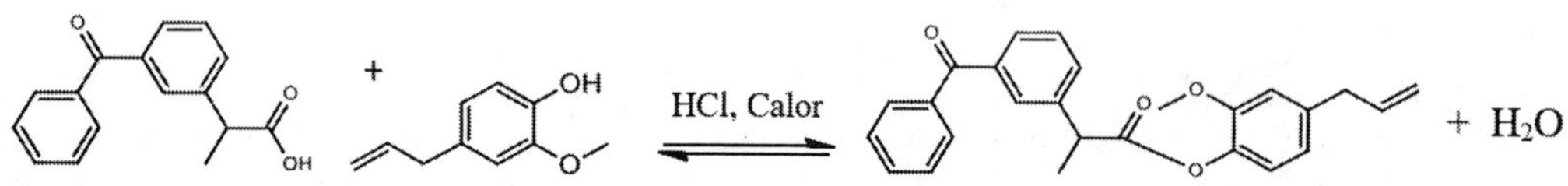

Figure 1. Esterification reaction ketoprofen and eugenol with HCl as catalyst.

Synthesis of KEE carried out several variations of mole ratio of ketoprofen and eugenol to obtain maximum yield (%), the results of the calculation of the average yield increased from KEE 3 to KEE 6 and dropped to KEE 7 and 8 (Table 1). The maximum value of the yield at KEE 6 is 65.60% and minimum value at KEE 3 is 31.85%.

Table 1. Average yield (%) of sample

Sample	Average Yield (%) ± SD
KEE 3 (KTP : EUG = 1:3)	31.85±3.58
KEE 4 (KTP : EUG = 1:4)	38.93±1.58
KEE 5 (KTP : EUG = 1:5)	40.50±8.50
KEE 6 (KTP : EUG = 1:6)	65.60±2.91
KEE 7 (KTP : EUG = 1:7)	53.67±1.92
KEE 8 (KTP : EUG = 1:8)	50.80±2.35

3.2. Characterization of KEE

3.2.1. Thin Layer Chromatography (TLC)
Synthesis sample were analyzed using TLC with a silica gel 60 F_{254} as stationary phase and a n-hexane: ethyl acetate : chloroform (2:5:1 v/v) as mobile phase [2]. The mobile phase use of a solvent mixture aims to obtain optimal separation [10]. Before elution process, chamber was developed with mobile phase.

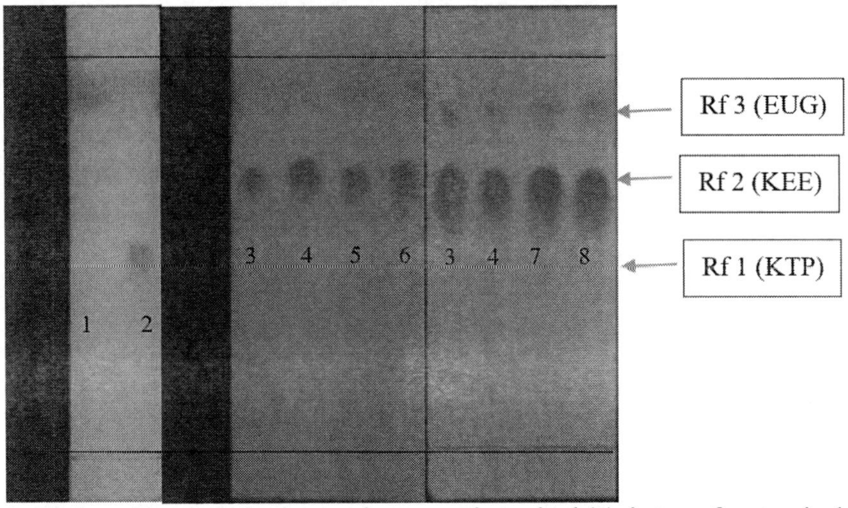

Figure 2. TLC profile of synthesis sample, eugenol standard (1), ketoprofen standard (2) under UV lamp at 254 nm

Table 2. Rf value of TLC sample, standard ketoprofen and standard eugenol

Sample	Rf 1	Rf 2	Rf 3
Eugenol	-	-	0.9
Ketoprofen	0.51	-	-
KEE 3 (KTP : EUG = 1:3)	-	0.70	0.85
KEE 4 (KTP : EUG = 1:4)	-	0.70	0.89
KEE 5 (KTP : EUG = 1:5)	-	0.70	0.89
KEE 6 (KTP : EUG = 1:6)	-	0.71	0.89
KEE 7 (KTP : EUG = 1:7)	-	0.69	0.86
KEE 8 (KTP : EUG = 1:8)	-	0.69	0.85

Visualized TLC spot under UV lamp at 254 nm (Figure 2) with average Rf (2) value 0.70 was identified as KEE (Table 1). According to Dhokchawle et al. [2], Rf value of KEE was 0.71 with n-hexane: ethyl acetate : chloroform (2:5:1 v/v) as mobile phase that same with this study.

3.2.2. Fourier-Transform Infrared Spectroscopy (FTIR)

The synthesized samples were analyzed using FTIR (Fourier-Transform Infrared Spectroscopy) to determine the functional groups of the compound. Different bonds (C-C, C = C, C-O, C = O, O-H) in organic compounds have different vibration frequencies and can be detected by identifying frequencies as absorption in infrared spectra [11].The result of FTIR spectra (Figure 3, Table 2) show that there are ester groups in the synthesized samples which are characterized by peaks at wavenumber of carbonyl C=O ester (1750-1725 cm^{-1}) and C-O ester (1300-1100 cm^{-1}) [12]. The presence of an ester group indicates that the esterification reaction ketoprofen with eugenol was successful.

IOP Conf. Series: Materials Science and Engineering **578** (2019) 012056 doi:10.1088/1757-899X/578/1/012056

Table 3. Result of analysis sample with FTIR

Sample	KEE 3	KEE 4	KEE 5	KEE 6	KEE 7	KEE 8
C-H aromatic	3065	3062	3065	3062	3073	3068
C-H aliphatic	2975	2937	2937	2937	2937	2938
C-H aliphatic	2841	2910	2910	2909	2909	2908
C=O ester	1735	1729	1735	1729	1729	1730
C-O ester	1274	1270	1273	1269	1268	1271
C=C aromatic	1602	1600	1603	1601	1601	1604
C-H aromatic	3065	3062	3065	3062	3073	3068

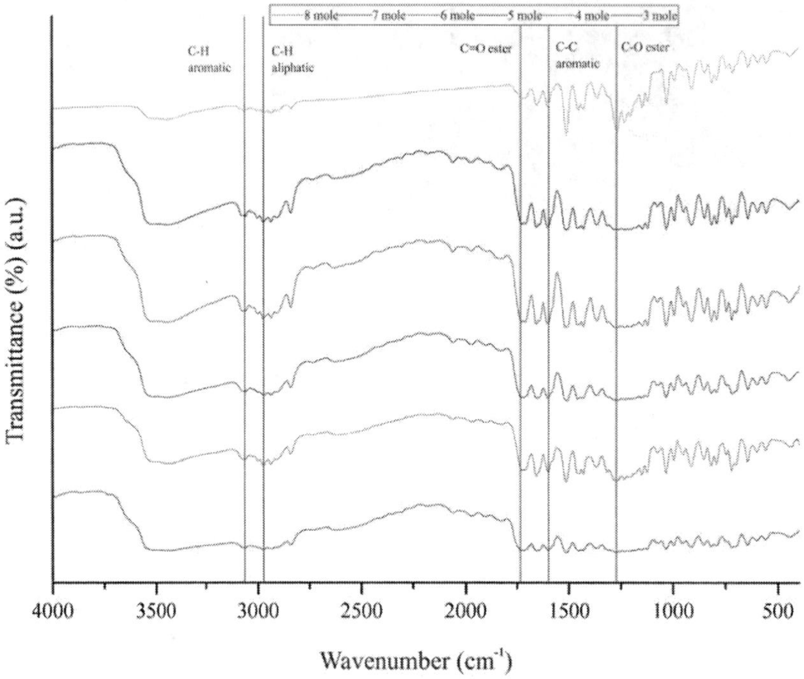

Figure 3. FTIR spectra sample

3.2.3. UV-Vis Spectrophotometer

Synthesis sample was calculated absorbance at the maximum wavelength of KEE at 261 nm [2]. Analysis of samples using UV-Vis spectrophotometer showed that absorbance at 261 nm with KEE 3 to KEE 6 would increase, while in the absorbance of KEE 7 and KEE 8 decreased (Figure 4). Absorbance are directly proportional to the quantity of solute according to Lambert-Beer equation, if the absorbance value of the same compound increases, the concentration of the compound also increases [9]. The maximum absorbance value was 0.308 at KEE 6 which also shows the greatest KEE concentration, while minimum absorbance 0.19 was KEE 8 that mean KEE 8 has minimum concentration of KEE.

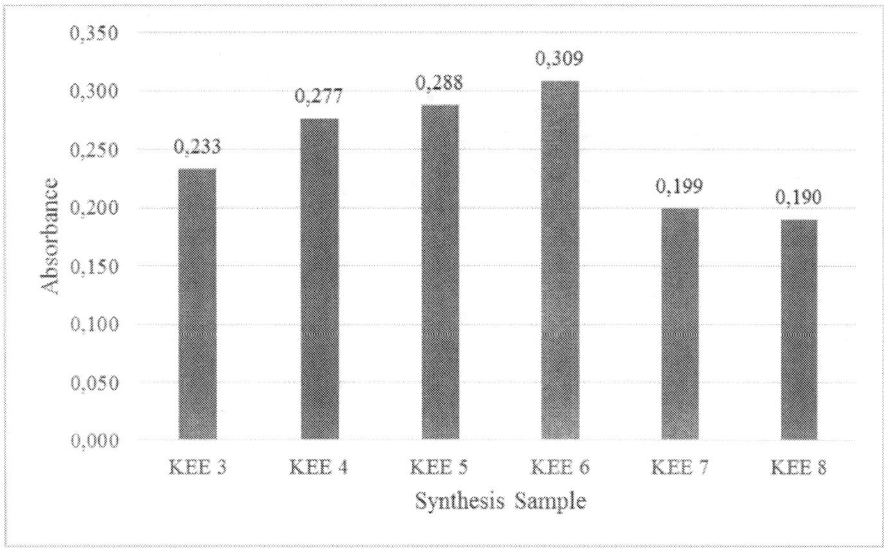

Figure 4. Result of absorbance sample at 261 nm

Absorbance data from different sample were compared by one-way analysis of variance (ANOVA). All result showed that significant difference between sample because have p value <0.05. Statistically it can be said that if mole of reactants changed caused significantly different results of reaction product.

3.2.4. High Performance Liquid Chromatography (HPLC)

Analysis by HPLC to confirm the maximum KEE in the sample. The decrease area HPLC indicated decrease of concentration of compound [10]. According to Figure 5, the highest area at KEE 6 that mean KEE 6 has the highest concentration of KEE. These results are consistent with the results of the analysis using UV-Vis spectrophotometer.

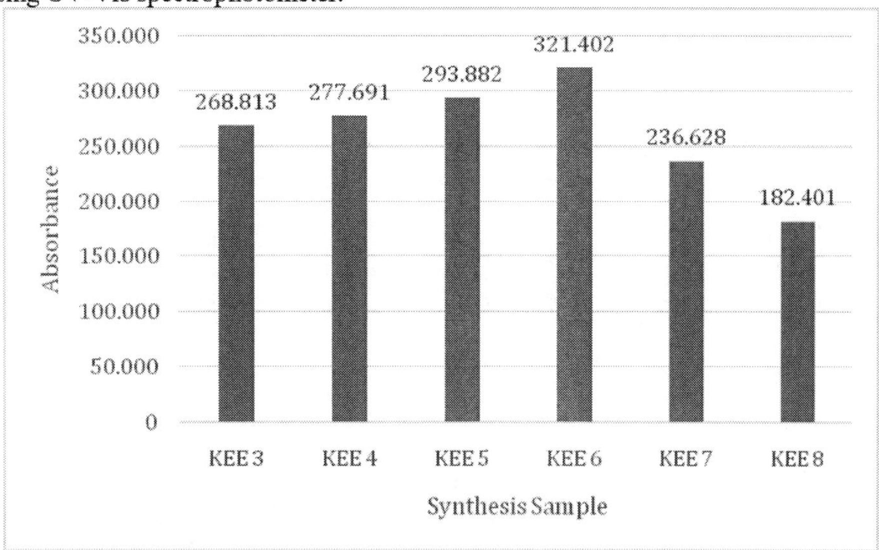

Figure 5. Result of area sample with HPLC

The results of sample analysis at variance mole ratio of ketoprofen : eugenol from 1:3 to 1:6 increased. This is accordance with the Le Chatalier principle that the reaction rate is proportional to the concentration of the reactant, if the reactants are added, the equilibrium of the reaction will shift to the right so that the number of products increases [13]. While at the ratio of ketoprofen : eugenol at 1:7 and 1:8 the value of KEE decreased. It can be said that in the ratio of 1:6 the equilibrium of the reaction has been reached. If the reaction has reached equilibrium, the addition of reactants will not have an impact on the number of reaction results [13]. The decrease can occur due to the hydrolysis reaction of the ester to the initial product. Excessive use of eugenol can produce excess water content that can hydrolyze KEE [14].

4. Conclusion

The optimum of ratio mole of ketoprofen : eugenol in the synthesis KEE was 1:6, with 65.60% of yield, 0.71 as Rf value with n-hexane: ethyl acetate : chloroform (2:5:1 v/v) as mobile phase, peak at 1729 cm^{-1} wavenumber of carbonyl C=O ester (1750-1725 cm^{-1}) and peak at 1268 wavenumber of C-O ester (1300-1100 cm^{-1}) of FTIR spectra, 0.308 of absorbance at 261 nm, and 321,402 of area HPLC.

Acknowledgment

The authors acknowledge to Sebelas Maret University which given the funding for the research.

References

[1] Mc Evoy G K 2002 *AHFS: American Society of Health System, Book 4* (Bethesda: Pharmacist Inc)
[2] Dhokchawle B V, Tauro S J, and Bhandari A B 2016 *Drug Res* **66** 46-50
[3] Li J, Yu Y, Wang Q, Zhang J, Yang Y, Li B, et al. 2012 *MCRE* **21** 995-999
[4] Ji Y B 1999 *Pharmacological Action and Application of Anticancer Traditional Chinese Medicine* (Heilongjiang: Science & Technology Publish Station)
[5] Zhao X L, Chen D W, Gao P, Luo Y F and Li K X 2005 *Pharmazie* **60** 883-887
[6] Fessenden R J and Fessenden J S 1990 *Organic Chemistry, 4th edition* (California: Brooks/Cole)
[7] Zhou Y Y, Li Y, and Yu A N 2016 *Food Sci. Technol* **36(2)** 268-274
[8] Rifqi H D 2017 Synthesis of Ketoprofen Eugenol Esters as Prodrugs Analgetik [Thesis], Surakarta: Sebelas Maret University
[9] Rouessac F and Rouessac A 2007 *Chemical Analysis Modern Instrumentation Methods and Techniques* (Chichester: John Wiley & Sons Ltd)
[10] Kealey D and Haines P J 2002 *Analytical Chemistry* (United Kingdom: BIOS Scientific Publishers Ltd)
[11] Silverstein R M, Webster F X, and Kiemle D J 2005 *Spectrometric Identification of Organic Compounds* (New York: John Wiley & Sons)
[12] Socrates G 2001 *Infrared and Raman Characteristic Group Frequencies* (United Kingdom: John Wiley and Sons Ltd)
[13] Daintith J 2008 *A Dictionary of Chemistry* (Oxford: Oxford University Press)
[14] Hoffman R V 2004 *Organic Chemistry An Intermediate Text* (New Jersey: John Wiley & Sons)

IOP Conf. Series: Materials Science and Engineering **578** (2019) 012057 doi:10.1088/1757-899X/578/1/012057

Highly efficient one pot synthesis of triacetylphloroglucinol: an analogue of acylphloroglucinol natural product

T Kusumaningsih[1], M Firdaus[1], A N Artanti[2], W E Prasetyo[3]
[1] Department of Chemistry, Faculty of Mathematics and Natural Sciences, Sebelas Maret University, Jl. Ir. Sutami No.36A, Surakarta, 57126, Indonesia
[2] Department of Pharmacy, Faculty of Mathematics and Natural Sciences, Sebelas Maret University, Jl. Ir. Sutami No.36A, Surakarta, 57126, Indonesia
[3] Master of Chemistry Program, Graduate School, Sebelas Maret University, Jl. Ir. Sutami No.36A, Surakarta, 57126, Indonesia

E-mail: triana_kusumaningsih@staff.uns.ac.id

Abstract. A green route fast synthesis of triacetylphloroglucinol analogue has been described in detail. This process efficiently achieved by a Friedel-Craft acylation in a one-pot method using phloroglucinol and appropriate acetic anhydride or acyl chloride as acylation agents in the presence of biodegradable catalyst methanesulfonic acid under solvent-free condition. This novel protocol clearly demonstrated green chemistry concept in term of simplicity, good yields and environment-friendly conditions.

1. Introduction

By the virtue of their numerous important applications of phloroglucinol, modification of the phloroglucinol into its derivatives believed to the one important key to obtain better various compounds. Acylphloroglucinol, one of the important class of phloroglucinol derivatives, has been explored in many fields in recent years [1,2]. Both monoacylphloroglucinol and diacylphloroglucinol derivatives have been investigated and shown potential importance as antimicrobial, enzyme inhibiting, and anti-allergy pharmacological agent [3–8]. On the other hand, triacetylphloroglucinols (TAPG), a minor product of secondary metabolite, which is still rarely explored due to its less availability. TAPG is well known of their selective anti-proliferative potential and supported to form some important ligands [8–11]. Based on this reality, some researchers developed various methods to generate TAPG in excellent yield. To only mention recent examples, TAPG has been synthesized using various catalysts such as $AlCl_3$, BF_3-etherate and $ZnCl_2$ employing nitrobenzene as the solvent [2,12]. In term of green chemistry, the mentioned catalysts and solvent can be categorized as toxic materials. In addition, these catalysts cannot be easily recovered and recycled. Therefore, there is an urgent need to attempt greener synthetic way to obtain TAPG.

Green chemistry field derived from the increasing request for more sustainable processes in the chemical industry. This concept designed for chemicals processes that aim to eliminate or minimize the use or generation of harmful materials. The use of toxic organic solvents tends to generate waste and consume more energy towards synthesis [13–15]. Chauthe et al [16], proposed a greener way to

[1] To whom any correspondence should be addressed.

synthesize diacylphloroglucinol (DAPG) using methanesulfonic acid (MSA) catalyst although the DAPG was not the only product a small amount of TAPG was also observed. By adopting green chemistry concepts, we present a new simple green route approach to the preparation of TAPG analogue over biodegradable MSA catalyst. The reaction was designed in one pot step in order to limit the synthetic waste, to spare time, and to simplify practical aspects via the efficiency and environmental sustainability.

2. Experimental

2.1. Materials and Equipment.

All of the reagents were purchased from Merck and Sigma-Aldrich at the greatly commercial quality and directly used without additional purification. The tools used are Infra-red (FT-IR, Prestige 8201 PC, with KBr Pellet), and Nuclear Magnetic Resonance (^{1}H NMR, Agilent 400 MHz). Chemical shifts were reported in parts per million (ppm) relative to the solvent signal for $(CD_3)_2CO$ (^{1}H NMR; 2.09 ppm). Thin-layer chromatography (TLC) was conducted with TLC Silica gel 60G F_{254} 25 Glass plates 20 x 20 cm (E-Merck) and visualized by exposure to UV Light (254 nm) or stained with potassium permanganate.

2.2. *General Method for the Synthesis of TAPG.*

A The synthetic route for the series of TAPG analogues 3 performed previously reported procedures with a slight modification [12]. A flame-dried 10 mL three-necked round bottom flask was charged with phloroglucinol (3 mmol), acetic anhydrides or acyl chlorides (9 mmol) and acetic acid (1.5 mmol) before MSA (9 mmol) was added drop wise at room temperature, stirred 20 min under nitrogen atmosphere and refluxed at 80 °C for 45 min-2 h. The reaction progress was monitored by TLC until the disappearance spot of phloroglucinol. 25 mL of water added to the mixture and allowed to cool at room temperature. The mixture was extracted with ethyl acetate (25 mL). The organic layer was separated and then the aqueous phase was extracted with ethyl acetate (2 × 25 mL) and brine water solution (25 mL), then dried over sodium sulphate anhydrous and concentrated in vacuo to give a crude solid. The resulting crude product was further purified by silica gel (#100–200) column chromatography utilizing eluent the mixture of n-hexane and ethyl acetate (9:1; 8:2; 7:3; and 6:4, respectively) to afford pure product **3**, **5a** and **5b** in 42-72 %.

3. Results and Discussions

In this Herein, the synthetic approach for acetylation of phloroglucinol with acetic anhydride and catalyzed by MSA via Friedel-Craft followed by Fries Rearrangement in acidic condition with short time, high conversions and selectivity, without the use of a solvent is reported. Phloroglucinol (1) constitutes a highly symmetrical and electron rich aromatic system that contains three chemically equivalent active sites that are favourable for electrophilic aromatic substitution reactions. By direct Friedel-Crafts C-acylation, we performed solvent-free acetylation of phloroglucinol via Fries rearrangement of the intermediate O-acetyl derivatives and/or simultaneously both of them as depicted in Figure 1. This synthetic approach is expected as an alternative to substitute the conventional method that still used toxic catalysts, needed longer time, and a large ammount of solvents with relatively high energy intake.

Initially, the reaction investigated at room temperature with acetic anhydride as an acylation agent. Unfortunately, the reaction was incomplete and did not result in the desired product even after 3 h reaction. Carrying out this reaction until 24 h was still not yielding the expected product. TLC analysis showed that the spot of starting material was not completely disappeared even after 24 h reflux. This reaction was controlled by TLC until completion using hexane-ethyl acetate (6:4). In this case, this result proved that solvent-free Friedel-Craft acylation reaction could not be performed at room temperature. Based on previous work [16], the phloroglucinol acylation commonly was done at 80 °C. Thus, in this work, the reaction was then carried out at 80 °C. In order to find optimized condition, varying the amount of acetic acid and MSA was performed in this reaction via a single step for direct acylation of phloroglucinol.

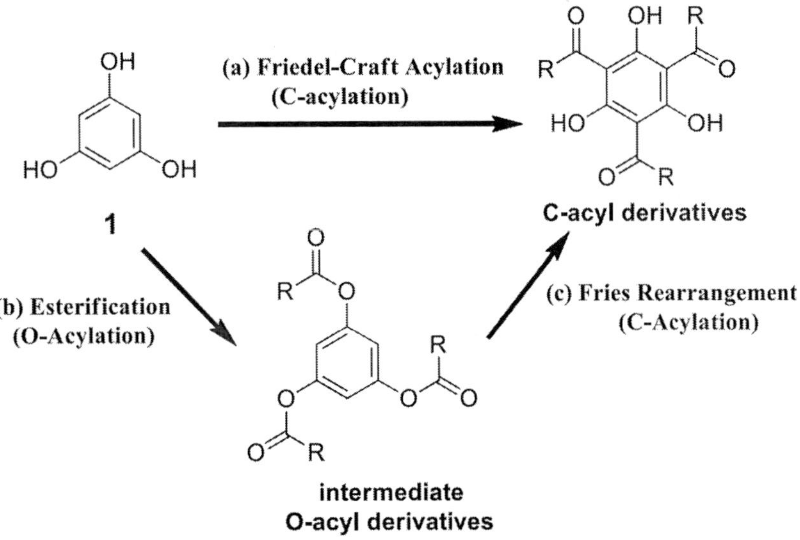

Figure 1. General proposed reaction Friedel-Craft C-acylation of phloroglucinol into triacylphloroglucinol.

The direct acylation of phloroglucinol using 1 mol equivalent of acetic acid and 3 mol of MSA was reported by Chauthe et al [16]. Unfortunately, the TAPG was obtained in lower yield and the reaction tended to give more than 3 products. In the optimum MSA 3 mol, the use of acetic acid less than 0.5 mol equivalent was then investigated and resulted in an incomplete reaction and gave other products such as MAPG and DAPG. In other condition, employing acetic acid more than 1 mol equivalent resulted in TAPG in lower yield followed by MAPG and DAPG. Finally, the use of 0.5 mol equivalent of acetic acid gave the desired product in good yield without MAPG formation although still gave a small amount of DAPG as minor product after 45 minute reaction under a nitrogen atmosphere. By using various solvents such as diethyl ether, chloroform, and diethyl ether also observed in this reaction and did not affect the corresponding target. This is obviously a simple, high yielding and environmentally benign phloroglucinol acylation procedure compared with previous work which uses a toxic catalyst and hazardous solvents. It is clearly differentiating from the other methods which allow the preparation of TAPG with the relatively quick procedure, insignificant waste, and no halogenated constituents.

For the best result, these fascinating initial results prompted us to expand with various TAPG derivatives under raised conditions as outlined in Figure 2.

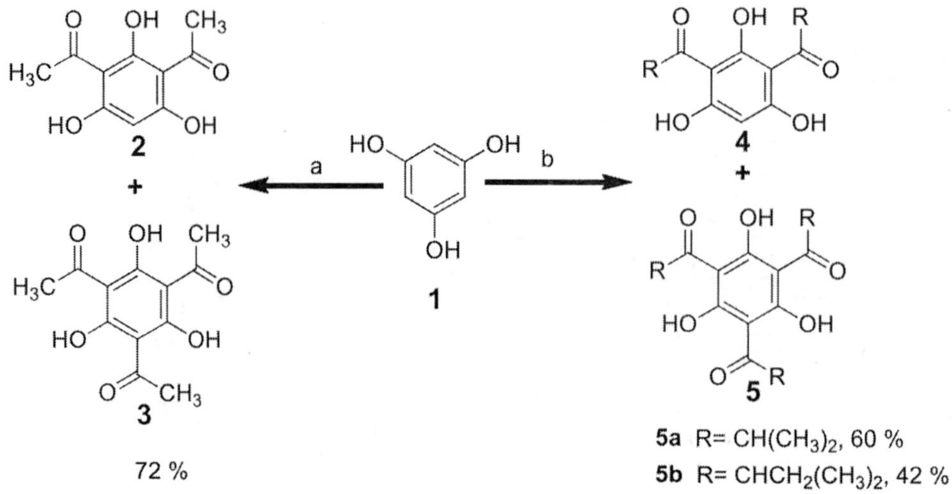

Figure 2. General reaction one-pot synthesis of triacetylphloroglucinol analogue. Reagent and conditions: (**a**) acetic anhydride, CH_3COOH, CH_3SO_2OH, 80 °C, 45 min, (**b**) RCOCl, CH_3COOH, CH_3SO_2OH, 80 °C, 1 h.

Surprisingly, lower yield observed in the presence of bulky acyl group substituents compared to short chain acyl chlorides or acetic anhydride. A stimulating correlation between the size of the acyl chain and the reaction time was observed. By using acetic anhydride and smaller acyl chain, good yield of the products was obtained only in 45 min. These results obviously suggest that the bulk of the acyl chain plays a significant role. A steric hindrance is a possible factor influenced to substitute a benzene ring with three acyl groups. Principally, a reaction with higher atom economy is desirable to one with lower atom economy. This is due to the increasing of atom economy incomes a less waste produced to obtain the desired product. In this work, we found relatively lower value both of the theoretical atom economy (58.34 %) and the experimental atom economy (50.89 %). This indicates that formation of TAPG did not arise easily. However, this method is still gently embedded in green chemistry corridor, because of ability to reduce the reaction time and energy for heating, a small amount of catalyst required and able to eliminate of solvent.

***1, 1', 1''-(2, 4, 6-trihydroxybenzene-1, 3, 5-triyl) triethanone* (3)**
Yield 72%; Colourless needle crystal; IR (KBr, cm^{-1}): ν = 3472.41-3356 (OH), 3000-2800 and 1367 ($C_{sp}{}^{3}$-H), 1623 (C=O), 1418 (C-OH), 1299 (C-O). ^{1}H NMR (400 MHz, $(CD_3)_2CO$) δ 8.11 (s, 1H), 2.71 (s, 1H).
***1, 1', 1''-(2, 4, 6-trihydroxybenzene-1, 3, 5-triyl)tris(2-methylpropan-1-one)* (5a)**
Yield 60 %; Yellow oil; IR (KBr, cm^{-1}): ν = 3483-3387 (OH), 3000-2800, 1419 and 1383 ($C_{sp}{}^{3}$-H), 1619 (C=O), 1418 (C-OH), 1299 (C-O). ^{1}H NMR (400 MHz, $(CD_3)_2CO$) δ 15.30 (*s*, 3H), 3.34-3.26 (*m*, 3H), 1.10-1.13 (*d*, 18H, *J* = 6.5 Hz).
***1, 1', 1''-(2, 4, 6-trihydroxybenzene-1, 3, 5-triyl)tris(3-methylbutan-1-one)* (5b)**
Yield 42%; White solid; IR (KBr, cm^{-1}): ν = 3499-3425 (OH), 3000-2800, 1412 and 1367 ($C_{sp}{}^{3}$-H), 1625 (C=O), 1412 (C-OH), 1299 (C-O). ^{1}H NMR (400 MHz, $(CD_3)_2CO$) δ 8.13 (s, 3H), 3.01 (*d*, 6H, *J* = 6.5 Hz), 2.35-2.18 (*m*, 3H), 0.94-0.97 (*d*, 18H, *J* = 6.5 Hz).

4. Conclusion
In conclusion, a green route fast and convenient method for synthesis TAPG analogue compounds through the direct Friedel-Crafts acylation using a biodegradable MSA catalyst and under solvent-free condition has been developed. This protocol provides a better option because the corresponding reaction afforded in good yield with negligible waste and containing no metallic or halogenated

constituents. The use of a green chemistry concept in this approach because it is believed to be a strategic key to the sustainability of the chemical industry.

References

[1] Singh I P, & Bharate S B 2006 *Nat. Prod. Rep.* **23** 558
[2] Singh I P, Sidana J, Bharate S B & Foley W J 2010 *Nat. Prod. Rep.* **27** 393
[3] Marchand P A, Weller D M & Bonsall R F 2000 1882–1887
[4] Gong L, Tan H, Chen F, Li T, Zhu J, Jian Q, Yuan D, Xu L, Hu W, Jiang Y and Duan X 2016 *Sci. Rep.* **6** 1–9
[5] Tan H, Liu H, Zhao L, Yuan Y, Li B, Jiang Y, Gong L, Qiu S 2017 *Eur. J. Med. Chem.* **125** 492-9
[6] Bharate S B, Khan S I, Yunus N A, Chauthe S K, Jacob M R, Tekwani B L, Khan I A, Singh I P 2007 *Bioorganic Med. Chem.* **15** 87–96
[7] Wu X F, Wang Y D, Yu S S, Jiang N, Ma J, Tan R X, Hu Y C, Qu J 2011 *Tetrahedron* **67** 8155–8159
[8] Spielberg E T, Plass W 2010 *Eur. J. Inorg. Chem.* **20** 3093–3096
[9] Mondal M, Lakshmi P, Krishna R, Sakthivel N 2017 *J. Lumin.* **192** 990–998
[10] Maurya M R, Rana L, Avecilla F 2017 *Polyhedron* **126** 60-71
[11] Feldscher B, Stammler A, Bögge H, Glaser T 2010 *Dalton Trans.* **39** 11675-11685
[12] Choksakulporn S, Punkvang A, Sritana-Anant Y 2015 *J. Mol. Struct.* **1082** 97–102
[13] Anastas P T & Warner J C 2000 *Green Chemistry: Theory and Practice* (Oxford University Press)
[14] Curzons A D, Constable D J, Mortimer D N, Cunningham V L 2001 *Green Chem.* **3** 1–6
[15] Wilkinson M C 2011 *Org. Lett.* **9** 2232-5
[16] Chauthe S K, Bharate S B, Periyasamy G, Khanna A, Bhutani K K, Mishra P D, Singh I P 2012 *Bioorganic Med. Chem. Lett.* **22** 2251–2256

IOP Conf. Series: Materials Science and Engineering **578** (2019) 012058 doi:10.1088/1757-899X/578/1/012058

Optimization SNEDDS (Self-Nano Emulsifying Drug Delivery System) of ZnO that dispersed into Hydrogel Matrix as UV-Protective

W Wulandari[1,*], D E Ermawati[1] and A Yugatama[1]
[1]Department of Pharmacy, Matematics and Natural Science Faculty
Sebelas Maret University, Surakarta, Indonesia

[*]Email : dianekaerma@gmail.com

Abstract. World Health Organization in 2015, estimated the case of non-melanoma cancer increasing due to the depletion of the ozone layer. The case of skin cancer in Indonesia reaches 6-8%, so the skin needs effective protection. ZnO is a sunscreen with SPF [Sun Protecting Factor] value of 50 and sun exposure reduction to the skin, especially UV A [320-400 nm] and UV B [290-320 nm]. ZnO is difficult to dissolve in water and hard to dispersed in the hydrogel matrix, so that it is formulated a nanoemulsion preparation into SNEDDS. SNEDDS consists of olive oil, tween 80 [surfactant], and propylene glycol [cosurfactant]. Optimization of the SNEDDS ZnO with the D-Optimal method was using Design Expert Software producing 16 formulas that tested for physical stability response including transmittance values [%] and pH values. The formula that has the highest response value with the desirability value approaching one was the optimum formula, then analyzed done with PSA. The optimum SNEDDS ZnO was dispersed in the hydrogel matrix of carbopol, thentest included pH, viscosity, dispersion and SPF value. The optimization shows that the optimum composition of olive oil : tween 80 : propylene glycol of 1:9:1. The result of t-test of SNEDDS ZnO optimum formula were not significantly different (*p-value*> 0,05) with *Design Expert* prediction. Optimum formula can load 2 mg ZnO with transmitance value is 92%, SPF value 26, PSA 150 nm, and polydispers size of droplets.

1. Introduction

Indonesia is a tropical country with high intensity of sun exposure. Sunlight contains ultraviolet (UV) that harmful for unprotected skin because it could cause skin cancer, eye damage as cataract and melanoma, skin aging, hyperpigmentation, erythema, and immune system damage [9]. UV light consisted of UV A (320-400 nm), UV B (290-320 nm), and UV C (100-290). *World Health Organization* (WHO) in 2015, estimate non-melanoma cancer cases is increasing to 300,000 cases and melanoma is increasing to 4,500 cases due to depletion of the ozone layer. In Indonesia, skin cancer cases reach 6-8% of all types of cancer [15]. Skin needs an effective protector to reduce the UV light exposure to the skin.

A high SPF sunscreen have an ability to protect skin from UV exposure [7]. SPF or light protection factor have range value of 2-60. Dermatologist recommend the use of sunscreen with the SPF of 15-30 [3]. One of sunscreen compound that contains SPF more than 30 are zinc oxide (ZnO) and titanium oxide (TiO_2). ZnO is a material that could absorb sun spectrum with better light quantum than TiO_2

[8]. *Cosmetics* (2015) explains that ZnO have SPF value of 50 so it could protect skin from UV A and UV B effect. ZnO is difficult to dissolve in the water and have cloudy white colored material, those make it difficult to dispersed in hydrogels. ZnO could also cause skin irritation if the concentration is too large, therefore it is formulated in the form of Nano emulsion with the SNEDDS (*Self-Nanoemulsifying Drug Delivery System*) technique.

SNEDDS is composed of olive oil because it has unsaturated fatty acids consisting of oleic acid (80%), so it has high self-emulsifying ability and a large drug loading capacity. The surfactant used is tween 80, where the greater the surfactant concentration the particle size decreases. In addition, tween 80 is hydrophilic with 15 HLB values and the resulting SNEDDS preparations have transmittance percentages above 80% [9]. Propylene glycol is chosen as co-surfactant because it can reduce the interfacial tension of oil in water and help surfactant work [14].

Optimization of the SNEDDS formula from ZnO in this study was conducted using a D-Optimal mixture design (D-Optimal mixture design). The main advantage of using the D-Optimal method is that it can reduce the number of experimental treatments [2]. In addition, the costs used are lower, because the number of experimental treatments is relatively small [21].

The manufacture of ZnO SNEDDS preparations which are dispersed in the hydrogel matrix is intended to improve the physical properties of turbid hydrogels when added with ZnO. Cosmetic preparations on the market containing ZnO are in the form of creams and emulsions combined with other active substances to provide a synergistic effect. The requirements for hydrogel preparations are semi-solid, clear, stable, translucent, and contain active substances which are dispersed in a colloidal system [1]. Therefore, made in the preparation of ZnO SNEDDS hydrogel to be easily applied to the skin, has a clear and elegant appearance, after dry leaving the film invisibility, easily washed with water, good drug release and good dispersion ability [18].

2. Experimental

2.1. Materials

Mortar, stamfer, sudip, stirring rod, drop pipette, thermometer, porcelain cup, hot plate, flacon, object glass, pH meter (Eutech), analytical scale (OhausPA413), vortex mixer (Maxi Mix II Thermolyne), sonicator (Branson 1510), UV / Vis spectrophotometer (GenesysTM), particle size analyzer (HORIBA SZ100), gel pot, Eppendorf, incubator shaker and glassware (Pyrex).

ZnO, Olive Oil, Tween 80 (Brataco) for surfactants, Propylene Glycol (Brataco), co-surfactant, Carbopol 940 for suspending agents, Glycerin (Brataco), TEA, Nipagin, Aquabidest (IKA), Aquadest, and 70% Ethanol. Materials used by pharmaceutical grade.

2.2 Research Implementation

2.2.1 The making of Carbopol hydrogel matrix. A number of Carbopol were developed with warm water until the dispersion was all assisted by stirring, then added some TEA and stirred until a transparent gel period formed. Added glycerin and nipagin which have been dissolved in propylene glycol and put it until homogeneous. Finally, the addition of residual TEA and aquadest stirred at a constant speed until homogeneous, as in Table 1 [12].

Table 1. Hydrogel Basis Formulation in 20 grams of Hydrogel.

Material	F1	F2	F3	F4
Carbopol	0.10 g	0.08 g	0.07 g	0.06 g
Glycerin	2.00 g	2.00 g	2.00 g	2.00 g
Propylene Glycol	1.00 g	1.00 g	1.00 g	1.00 g
Nipagin	0.01 g	0.01 g	0.01 g	0.01 g
TEA	0.75 g	0.75 g	0.75 g	0.75 g
Aquadest	Ad 20.00 mL	Ad 20.00 mL	Ad 20.00 mL	Ad 20.00 mL

2.2.2 SNEDDS Formula Component Consideration. The design of the SNEDDS formula was carried out using the Design Expert D-Optimal software method. The main components in the manufacture of SNEDDS were oil (Olive oil), surfactants (Tween 80), and co-surfactants (Propylene Glycol). Prior orientation has been made so that the lower and upper limits are set (Table 2).

Table 2. Upper and Lower Limit SNEDDS Making Material Component.

Material	Lower Limit	Upper Limit
Olive Oil (Oil)	9.09 %	11.11 %
Tween 80 (Surfactant)	77.77 %	81.81 %
Propylene Glycol (Co-Surfactant)	9.09 %	11.11 %

2.2.3 Loading Dose Determination. The determination of the ZnO dose (loading dose) was done by dissolving a number of ZnO into the SNEDDS formula based on the Optimal D-method. The ZnO dose series used was 10 mg, 8mg, 5 mg, 3 mg and 2 mg in each gram of the system [11]. The highest ZnO content in cosmetics found on the market was 10%. ZnO homogenized into the SNEDDS formula using aid vortex for 3 minutes, sonication for 5 minutes, and incubated with a water bath at 45 ^{o}C for 15 minutes. ZnO observations that dissolved in SNEDDS were carried out visually (Table 3). Perfectly dissolved ZnO was determined as maximum loading dose every 1,00 grams of SNEDDS [1]. SNEDDS was stored at room temperature during the characterization process [13].

2.2.4 SNEDDS Characteristics (Response Test). SNEDDS characterization results from the formulation was using the D-Optimization method based on the physical response of transmittance values (%) and pH, then inserted into the response variable in the Expert Design software to obtain the optimal formula composition.

Transmittance Test: A total of 1,0 mL of SNEDDS was dissolved up to 50,0 mL with aquabidest at room temperature. The solution was vertexing for 3 minutes until homogeneous. The transmittance percentage was measured using a UV / Vis spectrophotometer at λ650 nm with blank aquabidest [16]. pH test: a number of hydrogens dissolved in 9 parts of aquabidest and measured using a pH meter [13].

2.2.5. Verification and Optimal Formula Determination. The optimal SNEDDS formula was analyzed using S-Optimal Design Expert Method. The three variable components used were oil, surfactant, and co-surfactant evaluated by the transmittance response (%) and the pH value. ANOVA statistical analysis with p-value 0.05 [16]. Then it was characterized and verified the results of optimal formula testing using an analysis of one sample t-test with IBM SPSS Statistics 22 software.

2.2.6. SNEDDS Formula Evaluation and Hydrogel SNEDDS ZnO Physical Characteristics

2.2.6.1. Particle size. The optimal SNEDDS formula was evaluated by measuring particle size distribution using Particle Size Analyzer (PSA) in the Nanotechnology section of the Mathematics and Natural Sciences Faculty, Indonesian Islamic University, Yogyakarta.

2.2.6.2. Organoleptic Test

Organoleptic test was done by observing visually using five senses to describe the shape, color, and smell of the gel [1].

2.2.6.3. Disperse Test. Disperse test was done by weighing 0,5 grams of gel and mixing with a 150,0 grams load for 1 minute [16].

2.2.6.4. Viscosity test. Viscosity test was carried out with gel samples inserted into the glass speaker and placed under the spindle hanger, then the spindle lowered to the immersed limit into the gel

IOP Conf. Series: Materials Science and Engineering **578** (2019) 012058 doi:10.1088/1757-899X/578/1/012058

preparation. Rotor signaled and observed of the red needle on the scale, then read the term shown by the needle [16].

2.2.6.5. Determination of SPF values by In Vitro. A total of 100 mg of SNEDDS ZnO hydrogel was diluted using Aquadest up to 10 mL, then take 1.0 mL and put in 10 mL plus aquadest to the limit. The filtered solution and the absorbance measurements were carried out using a UV/ Vis spectrophotometer in the range of 290-320 nm, every 5 nm interval. The blank used was aquadest. The absorbance results obtained in each dose were recorded and calculated by the SPF value [17].

Table 3. Oil, Surfactant, and Co-Surfactant D-Optimization Formula Design.

Run	Olive oil (%)	Tween 80 (%)	Propylene Glycol (%)
1	10.105	80.805	9.090
2	10.100	78.790	11.110
3	9.090	79.800	11.110
4	9.090	80.805	10.105
5	11.110	79.800	9.090
6	11.110	78.790	10.100
7	9.497	81.005	9.497
8	11.110	77.780	11.110
9	11.110	77.780	11.110
10	10.000	79.495	9.090
11	11.110	79..800	9.090
12	9.090	81.810	9.100
13	11.110	78.790	10.100
14	9.900	80.200	9.900
15	9.090	79.800	11.110
16	9.090	81.810	9.100

2.2.6. SNEDDS Formula Evaluation and Hydrogel SNEDDS ZnO Physical Characteristics

2.2.6.1. Particle size. The optimal SNEDDS formula was evaluated by measuring particle size distribution using Particle Size Analyzer (PSA) in the Nanotechnology section of the Mathematics and Natural Sciences Faculty, Indonesian Islamic University, Yogyakarta.

2.2.6.2. Organoleptic Test

Organoleptic test was done by observing visually using five senses to describe the shape, color, and smell of the gel [1].

2.2.6.3. Disperse Test. Disperse test was done by weighing 0.5 grams of gel and mixing with a 150.0 grams load for 1 minute [16].

2.2.6.4. Viscosity test. Viscosity test was carried out with gel samples inserted into the glass speaker and placed under the spindle hanger, then the spindle lowered to the immersed limit into the gel preparation. Rotor signaled and observed of the red needle on the scale, then read the term shown by the needle [16].

2.2.6.5. Determination of SPF values by In Vitro. A total of 100 mg of SNEDDS ZnO hydrogel was diluted using Aquadest up to 10.0 mL, then take 1.0 mL and put in 10.0 mL plus aquadest to the limit.

The filtered solution and the absorbance measurements were carried out using a UV/ Vis spectrophotometer in the range of 290-320 nm, every 5 nm interval. The blank used was aquadest. The absorbance results obtained in each dose were recorded and calculated by the SPF value [7].

2.3 Statistical Analysis

Optimal formula verification done with one sample t-test analysis using IBM SPSS Statistic 22 software. SNEDDS optimal Formula characteristics result includes transmittance presentation and SPF value analyzed using IBM SPSS Statistic 22 software in trust value of 95% where the result was not significantly different when the result of D-Optimal method prediction and the result of the observation have p-value >0,05 [5].

Optimal formula determination was using Design Expert Version 9 Trial – Stat Ease software. The main component of SNEDDS were oil, surfactant, and co-surfactant variated with D-Optimal method in Design Expert Version 9 Trial – Stat Ease.

3. Results and Discussion

The making of hydrogel matrix was using Carbopol 940 as a gelling agent because it has a clear appearance. The acid properties possessed by Carbopol was neutralized by adding Triethanolamine (TEA) to a concentration of 1%. In addition, in the manufacture of hydrogels the addition of nipagin as a preservative was carried out and the addition of glycerin and propylene glycol as humectants and emollients. Determination of the optimum formula was carried out by pH test and viscosity of hydrogel preparations. The pH test was carried out to see the acidity level of the hydrogel preparation to avoid skin irritation, while the viscosity test was carried out to see the level of viscosity. The results of pH evaluation (Table 4) showed formula 1 and formula 2 was unqualified, while formula 3 and 4 was qualified to the topical preparations in the range of 4.50 to 6.50 [13]. The viscosity test was showing significant differences in each formula because the amount of Carbopol affects the viscosity of hydrogel preparations, so that only formula 4 qualified to the viscosity requirements of gel preparations in the range of 20,00-40,00 dPas [16]. From those two parameter tests, it concluded that formula 4 qualified to the requirements of the hydrogel preparation basis.

Table 4. The Result of pH test and Viscosity of Hydrogel Preparation.

Formula	pH	Viscosity (dPas)
Formula 1	6,60	1,00
Formula 2	6,54	10,00
Formula 3	6,48	18,00
Formula 4	6,48	23,00

Optimization of the optimum formula was done with the D-Optimal method using Design Expert 9 Trial software to determine the upper and lower limit values that were possible to produce SNEDDS formulas according to the desired criteria. Three components in determining the formula composition of SNEDDS were olive oil (A), surfactant tween 80 (B), and co-surfactant propylene glycol (C) determined as independent variables with transmittance values (%) and pH as response variables. The free variables that have been done in the previous orientation have the lower upper and upper limits of 9.09-11.11 (%), 77.77-81.81% (B), and 9.09-11.11% (C) The use of oil and co-surfactant to produce a good SNEDDS formula was <20% and surfactant reaches 60% [3]. The results obtained from D-Optimal were 16 run formulas with different compositions where the total of each component is 100%.

Loading dose determination done with ZnO variation of 10 mg, 8mg, 5 mg, 3 mg and 2 mg in every gram of SNEDDS. The aim of loading dose determination was to know the maximal value of ZnO that could dissolve in the SNEDDS system. The result of ZnO loading dose test that could

IOP Conf. Series: Materials Science and Engineering **578** (2019) 012058 doi:10.1088/1757-899X/578/1/012058

dissolve perfectly if the visual observation turns into clear, homogeny and there no deposits on the SNEDDS system.

The SNEDDS physical characteristics used in optimal formula determination was transmittance value (%) and pH. Transmittance test result from 16 run test formula was about 54,752% to 95,021%. The higher transmittance value signed the smaller particle size so that the solution become clear. SNEDDS Transmittance test result using Design Expert software shown that the data was normally distributed because every result point placed near and followed the diagonal line as predicted by the software so can be continued to the ANOVA test stage. Transmittance test result was modelized in the quadratic form because p value <0,05 of 0,0025 that shown a meaningful difference to the different use of the composition.

Response test using the pH value of hydrogel SNEDDS ZnO preparation was used to determine the pH of the hydrogel SNEDDS ZnO preparation to avoid skin irritation. The pH test results (Figure 1) of 16 run formulas were between 6,816 and 7,678. According to SNI 16-43399-1996 the pH for sunscreen preparation ranged from 4,50 to 8,00. The result of Hydrogel SNEDDS ZnO pH test shown that the data normally distributed because the plot points still on the diagonal line as predicted in the software, so could be continued to the ANOVA test stage. The pH test result modelized as quadratic form with p-value <0,05 of 0.0043 that shown a meaningful difference in the use of composition formulation

Optimal formula for SNEDDS determination was obtained based on the determined limits so that the desirability value approaching to 1 [4] shown the formula that produced was more in line with the optimal formula desired. The optimal formula suggested by Design Expert software was one solution with desirability value of 0,838. Optimal formula verification was done to compare the results of the observation with the results of the Design Expert software response using the one sample t-test analysis (Table 5).

Table 5. Verification Result of SNEDDS Optimal Formula based on D-Optimal Method using Software.

Response	Prediction	CI	Observation ±SE	p-value
Transmittance value (%)	92.59	82.05-103.14	4.66-7.99	0.05
pH	7.37	7.19-7.56	0.082-0.16	0.07

Each response shown low average standard error (SE), where the position near to zero was imaging the average result appeared almost the same to the result in every test. Statistical analysis of transmittance value response (%) and each pH shown un-significance difference (p-value>0.05), so it could be concluded that optimum formula was suggested by verified Design Expert software. The test result obtained was lower than the prediction of transmittance value (%) 92 to pH 7.30.

Optimal formula evaluation was done using particle measurement determination to ensure the formula in the nano-emulsion range of 20-200 nm [6]. The test result shown that SNEDDS optimal formula have the expected nano-emulsion particle size of 150, with polydisperse index (PI) of 0.53. PI value was used to image the nano-emulsion particle homogeneity where the SNEDDS ZnO optimal formula have same particle size distribution in the range of 0.0-1.0.

The physical characteristics test of Hydrogel SNEDDS ZnO preparation was carried out by organoleptic test, dispersion, viscosity, and SPF test. Organoleptic test results of Hydrogel SNEDDS ZnO were transparent bone white, typical tween 80 odor, and semi-liquid. These results were in accordance with the requirements of the gel preparation which it was semi-solid in shape with clear color [1]. The dispersion test of the Hydrogel SNEDDS ZnO preparation showed an increase along with a decrease in viscosity caused by temperature and humidity during storage.

Table 6. Disperse Test and Hydrogel SNEDDS ZnO Preparation Viscosity.

Test Parameter	Week-			
	1	2	3	4
Disperse	5.35	5.60	6.05	6.35
Viscosity	38.00	35.00	32.00	27.00

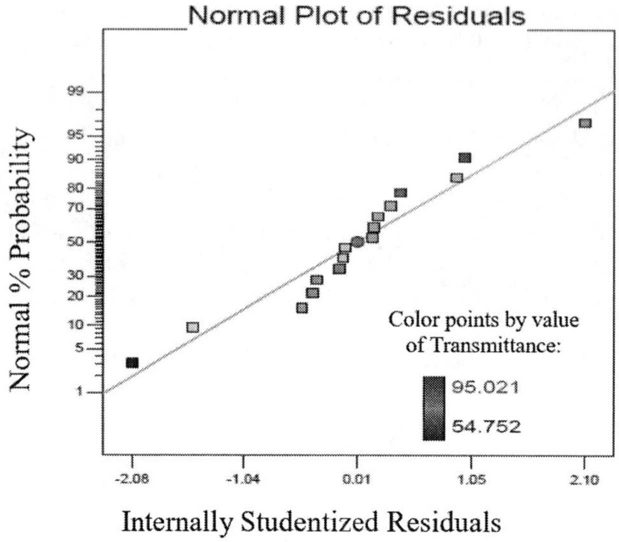

Figure 1. Normality Graphic of Residual Value SNEDDS Transmittance Test. Detecting normal residual values is done by looking at the resulting plot points whether or not they are around the diagonal line.

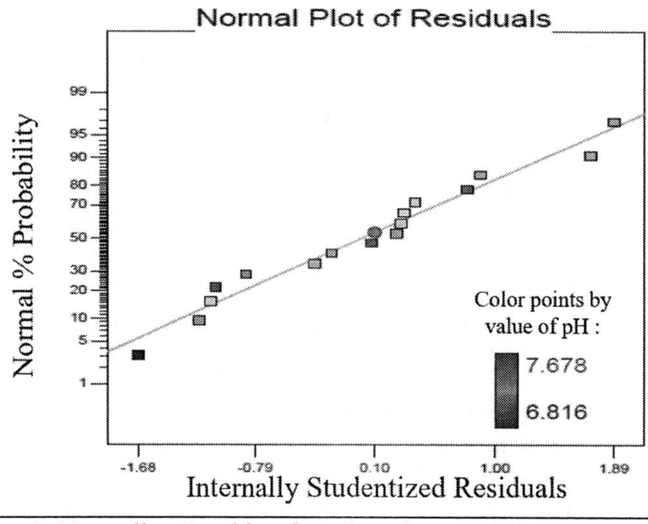

Figure 2. Normality Graphic of Hydrogel SNEDDS ZnO pH Residual Value. Detecting normal residual values is done by looking at the resulting plot points whether or not they are around the diagonal line.

Table 7. SPF Value Test Result Using UV/Vis Spectrophotometer [10]

Wavelength	EE x I	Absorbance
290	0.015	0.430
295	0.0817	0.405
300	0.2874	0.381
305	0.3278	0.357
310	0.1864	0.320
315	0.0837	0.298
320	0.018	0.267
Total	1	2.458
SPF Value	24.58	

SPF testing was carried out in vitro using UV / Vis spectrophotometer and obtained SPF 26 results. These results prove that with small doses of ZnO was still able to provide SPF 15-30 which was able to provide moderate protection to the skin. SPF calculations were performed using the method of Mansur *et al*. (1986), as shown in Table 7.

4. Conclusion

ZnO SNEDDS with the dose of 2 mg/g SNEEDS produce a qualified particle size to the nano-emulsion requirement of 150 nm. After it was disperse hydrogel matrix, it was obtained a qualified preparation of organoleptic test requirement, dispersion, and viscosity with SPF value of 26 (moderate protection).

References

[1] Ansel H C 1969 *Introduction To Pharmaceutical Dosage Forms* English Edition
[2] Borhan F P, Gani S S and Shamsuddin R 2014 *The Scientific World Journal* **2014** 1-8
[3] Cerpnjak K, Zvonar A, Gasperl M and Vrecer F 2013 *Acta Pharm.* **63** 427-445
[4] Dash R N, Habibuddin M, Humaira T and Ramesh D 2015 *Saudi Pharm. J.* **23** 528-540
[5] Gohel M, Purohit A, Patel A and *Int. J. Pharm. Pharm. Sci.* **8** 213-20
[6] Gutierrez J M, González C, Maestro A, Solè I M P C, Pey C M and Nolla J 2008 *Current Opinion in Colloid Inter. Sci.* **13** 245-251
[7] Kaur C D and Saraf S 2010 *Pharmacognosy research* **2** 22
[8] Khan A W, Kotta S, Ansari S H, Sharma R K and Ali J 2015 *Drug Delivery* **22** 552-561
[9] Lolo W A, Sudewi S and Edy H J 2017 *J. Pharm. Sci. Clinical Res.* **2** 1-5
[10] Mansur J S, Breder M N, Mansur M C and Azulay R D 1986 *An. Bras. Dermatol.* **61** 121-4
[11] Pazokifard S, Mirabedini S M, Esfandeh M, and Farrokhpay S 2012 *Advanced Powder Technology 23* 428-436
[12] Rathapon A, Sirivat A, and Vayumhasuwan P 2005 *Pharmaceutical Research* **22** 2134-2140
[13] Savale K S 2015 A *Int. J. Res. Pharm. Nano Sci.* **4** 385-97
[14] Solans C, Izquierdo P, Nolla J, Azemar N, and Garcia-Celma MJ 2005 *Current Opinion In Colloid and Interface Science* **10** 102-110
[15] Tan S S H 2015 The Relationship Between Occupational Sunlight Exposure And Non-Melanoma Skin Cancer *Thesis* **2015**
[16] Winarti L, Martien R, Suwaldi and Hakim L 2016 *Int. J. Pharm. Clinical Res.* **8** 425-32
[17] Zen N I M, Gani S S A, Shamsudin R and Masoumi H R F 2015 *The Scientific World Journal* **2015**
[18] Lachman L and Lieberman H 1994 *Theory and Practice of Industrial Pharmacy* English Edition

Frequency of ABCB1 gene 3435 polymorphism on patients with coronary stent in Surakarta

R Niruri[1], H Sulastomo[2], R Kusumati[2], A N Famila[2]
[1] Department of Pharmacy, Faculty of Math and Science, Universitas Sebelas Maret, Surakarta, Indonesia
[2] Faculty of Medicine, Universitas Sebelas Maret, Surakarta, Indonesia

E-mail: rasmaya@staff.uns.ac.id

Abstract. The use of antithrombotic drugs on patients with coronary stent have reduced the risk of stent thrombosis. ABCB1 Gene 3435TT may limit the drugs availability and response. Frequencies of ABCB1 Gene 3435 Polymorphism varied among different populations. This research aimed to identify frequency of 3435 polymorphism on patients with coronary stent in Surakarta. All the patients (who underwent coronary stent, received antithrombotic therapy, and signed for informed consent forms on the period of May 7th – June 8th 2018 in doctor Moewardi Hospital) were included in this study. The single nucleotide polymorphism (SNP) was detected by PCR and sequencing. As a result, the frequencies of 3435 TT, 3435 CC, and 3435 TC on 30 patients were 50 % (15 subjects), 33 % (10 patients), and 17 % (5 subjects) respectively. In this study, the mutant T allele carriers was found in majority of the patients.

1. Introduction

Antithrombotic drugs, including clopidogrel, have become the standard therapy for protection against ischemic events and prevention of stent thrombosis. However, it was reported that around one third of patients were non-responsive to clopidogrel due to the interindividual genetic variability [1,2]. ATP-binding cassette, subfamily B, member 1 (ABCB1) gene polymorphisms might lead to increased P-glycoprotein (P-gp) expression and diminished clopidogrel absorption [3]. ABCB1 3435 polymorphism involved in Clopidogrel and its active metabolite absorption [1]. Frequencies of ABCB1 Gene 3435 Polymorphism differed among population groups. The mutant genotype frequency of ABCB1 gene 3435TT was observed in 28%, 32%, and 43% of Malays, Chinese, and Indians [4]. In pediatric patients with acute Lymphoblastic Leukemia (ALL) at Cipto Mangunkusumo Hospital Jakarta, the most common genotype was 3435TT (84%) [5]. This research aimed to identify frequency of the 3435 polymorphism on patients with coronary stent in Surakarta.

2. Experimental

The method of this study was approved by health research ethics committee of doctor Moewardi General Hospital – School of Medicine Universitas Sebelas Maret Surakarta Indonesia. All the patients (who underwent coronary stent coronary stent, received antithrombotic therapy, and signed for informed consent forms on the period of May 7th – June 8th 2018 in doctor Moewardi Hospital) were included in this study.

Deoxyribonucleic acid (DNA) was isolated from the subjects' buffy coat with a standard salting out protocol. The primers to identify 3435 polymorphism were: 5'ACT CTT GTT TTC AGC TGC TTG 3' (forward(F)) and 5' AGA GAC TTA ACT TAG GCA GTG ACT 3' (reverse (R)), yielding 206 base *pairs (*bp) polymerase chain reaction(PCR) product [6]. PCR amplification protocol was initiated with pre-denaturation (5 minutes - 94°C), then it was followed by 35 cycles of denaturation (90 seconds - 94 °C), annealing (60 seconds - 56 °C), and extension (60 seconds - 72 °C). The final extension was at 72 °C(10 minutes). The PCR products were determined by electrophoresis in agarose gel(1,5%) that was stained with ethidium bromide(0.5 µg/ml), and then were visualized under UV transilluminator. The rest of PCR products were sequenced. The nucleotide sequences were compared with ABCB1 wild-type (WT) sequence [7].

3. Results and Discussions

Thirty subjects were included in this research (Table 1 and 2). Heterozygous 3435TC had double peaks on sequencing chromatogram. Homozygous 3435TT and 3435CC had a single peak (Table 2). Frequencies of 3435 TT and 3435 TC were 50 % (15 subjects) and 17 % (5 subjects) respectively. The rest 10 patients were wild type (3435CC). The mutant T alleles carriers were observed in the majority of patients.

Table 1. Patients' characteristic (N_{total} = 30)

Characteristic	N (%)
Gender	
Female	9 (30%)
Male	21 (70%)
Age (years)	
48-54	9 (30%)
55-64	12 (40%)
65-74	8 (27%)
>75	1(3%)

ABCB1 3435 C>T is a silent single nucleotide polymorphism, but each allele that makes haplotype can give contribution and alter mRNA level, protein folding, and protein expression. Several studies reported the impact of 3435 polymorphism and its haplotype on MDR-1 structure / function. The mutation could alter the stability of MDR1 mRNA. The use of rare codon could influence the translation kinetics and alter the protein- folding dynamic. Silent polymorphism result ribosome-stalling. Therefore, it can affect drug disposition [8,9,10]

Many cardiovascular drugs, that interact with P-gp, were widely used, including clopidogrel, ticagrelol, digoxin, bepridil, verapamil, aliskiren dabigatran, apixaban, edoxaban, rivaroxaban, warfarin, celiprolol, diltiazem, labetolol, losartan, nadolol, propranolol, talinolol, timolol, atorvastatin, lovastatin, and ambrisentan. Interaction between drugs and P-gp transporter altered drugs absorption and elimination. It was reported that 3435 T allele carrier influenced oral digoxin pharmacokinetic by affecting duodenal P-gp transport of digoxin [3]. Clopidogrel absorption could be reduced by P-gp mediated efflux. P-gp was encoded by ABCB1 gene. The 3435TT polymorphism was significantly associated with lower Area Under Curve (AUC) and Maximum Concentration (C_{max}) of clopidogrel in patients who suffered from coronary artery disease [1]. In subjects who underwent coronary angiography and stent, Clopidogrel AUC was also significantly decreased in 3445 TT compared with CC and TT genotypes [11].

It was reported that blood tended to thrombose after coronary stent placement. Anti-thrombotic drug, including clopidogrel and aspirin, could reduce the risk of stent thrombosis and cardiovascular-cerebrovascular events [12]. However, a meta- analysis study showed that TT homozygous was

IOP Conf. Series: Materials Science and Engineering **578** (2019) 012059 doi:10.1088/1757-899X/578/1/012059

associated with the risk of major adverse cardiovascular events (MACE) in subjects who received clopidogrel 300, compared with CC genotype [13]. In other meta-analysis, 3435 TT might increase the risk of short term (≤ the first 30 days) recurrent ischemic events. On platelet reactivity, ABCB1 3435 variants had a higher impact at baseline, but the impact was decreased during follow up [14]. Therefore, further evaluation on the patients of this study are necessary in order to explore the risk of MACE.

Table 2. ABCB1 gene 3435 polymorphism (N_{total} = 30)

	3435TT	3435CC	3435TC (double peak)
N (%)	15 (50%)	10 (33 %)	5 (17%)
Text sequence / Sequencing chromatogram	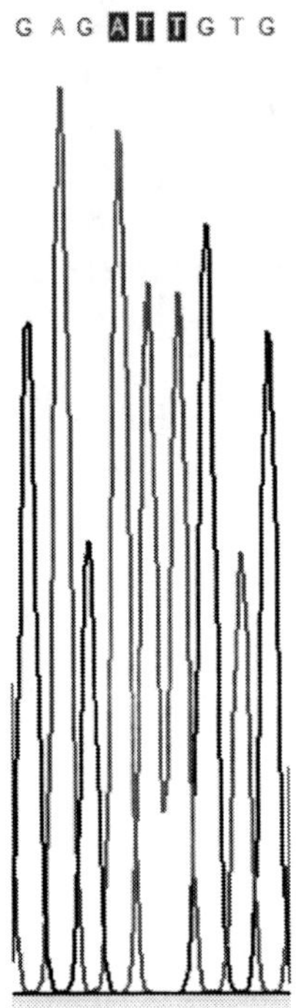	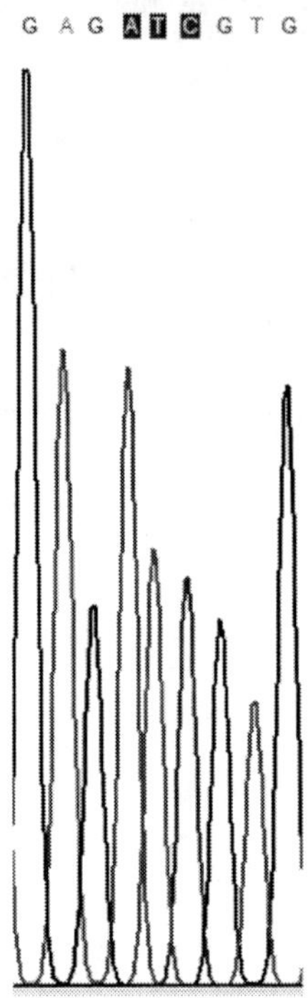	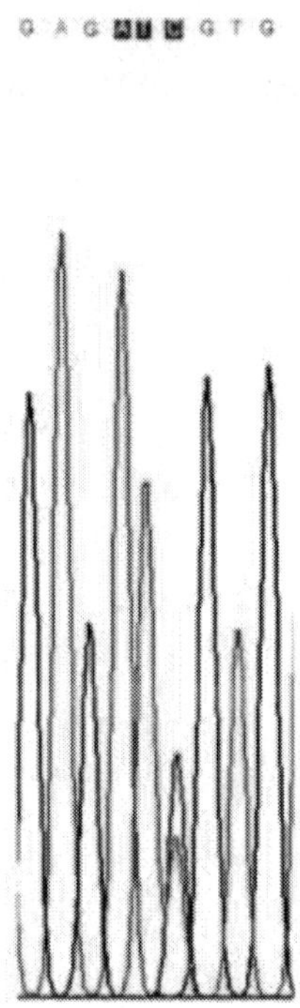

4. Conclusions

Frequencies of 3435TT, 3435CC, and 3435TC were 50 % (15 subjects), 33 % (10 subjects), and 17 % (5 subjects) respectively. The mutant T alleles carriers were found in the majority of patients.

Acknowledgment
This work was supported by the research grant from Institute of Research-Community Service, Universitas Sebelas Maret, Surakarta, Indonesia

References

[1] Taubert D, Von Beckerath N, Grimberg G, Lazar A, Jung N, Goeser T, Kastrati A., Schömig A., and Schömig E. 2006 *Clin Pharmacol Ther* **80(5)** 486-501
[2] Gurbel PA, Bliden KP, Hiatt BL, and O'Connor CM. 2003 *Circulation* **107** 2908-13
[3] Wessler JD, Grip LT, Mendell J, and Giugliano R P 2013 *J Am Coll Cardiol* **61** 2495–502.
[4] Balram C, Sharma A, Sivathasan C, and Lee E J D 2003 *Br J Clin Pharmacol* **56** 78-83
[5] Mutiara R, Agustinus B, Sitompul C B, Malik A, and Gatot D. 2015 *Asian Biomedicine* **9** 625-630

[6] Miladpour B, Shokouhi A M, Shirdel A, Heravi R E, Banihashem A, Esmaeili H, and Khedri A, Behravan J. 2009 *IJBC* **2** 63-67
[7] Chen CJ, Clark D, Ueda K, Pastan I, Gottesman M M and Roninson I B 1990 *J.Biol.Chem.* **265** 506-514
[8] Fung KL and Gottesman M M 2009 *Biochim Biophys Acta* **1794** 860–871
[9] Kimchi-Sarfaty C, Oh JM, Kim IW, Sauna ZE, Calcagno AM, Ambudkar SV, and Gottesman MM. 2007 *Science* **315** 525–528
[10] Kim D.H, Park J.Y., Sohn S.K., Lee N.Y., Baek J.H, Jeon S.B, Kim J.G., Suh J.S., Do Y.R and Lee K.B. 2006 *Int. J. Cancer* **118** 2195–2201
[11] Karazniewicz-Łada M., Danielak D., Rubi B., Burchardt P., Komosa A., Lesiak M. and Glowa F. 2015 *J Clin Pharm Ther* **40** 226-31
[12] Almadah N and Khouzam R N 2017 *J Thorac Dis* **9** 4806-4807
[13] Su J., Xu J., Li X., Zhang H., Hu J., Fang R., et al. 2012 *PLOS ONE* **10** e46366
[14] Luo, M., Li, J., Xu, X., Sun, X., & Sheng, W. 2012 *Thrombosis Research* **129** 754–75

Evaluation of antibacterial activity of sesquiterpene Ar-Turmerone from *Curcuma soloensis* Val. rhizomes

S D Marliyana[1], F R Wibowo[1], M W Wartono[1] and G Munasah[1]
[1]Department of Chemistry, Faculty of Mathematics and Natural Sciences, Sebelas Maret University, Jl. Ir. Sutami 36 A Surakarta, Indonesia

E-mail: msoerya@staff.uns.ac.id

Abstract. *Curcuma soloensis* Val. is a family plant Zingiberaceae and is known as "temu glenyeh". This plant has potential as a medicinal plant but has not been widely reported about its bioactivities. One of them was antibacterial activity. Therefore, the aim of this study was to evaluate the antibacterial activity of ar-turmerone, a sesquiterpenoid which was successfully isolated from *Curcuma soloensis* Val. rizhomes. Antibacterial test was carried out by using disc diffusion method and evaluated against four bacteria namely *Staphylococcus aureus* ATCC 25923, *Escherichia coli* ATCC 25922, *Klebsiella pneumonia* ATCC 13883, and *Pseudomonas aeruginosa* ATCC 27853. The result showed that ar-turmerone was not active against all of these bacteria.

1. Introduction

Infection caused by various drug-resistant bacteria is an increasing problem due to the appearance of and microbial drug resistance propagation and lack of new antimicrobial development [1]. With the increase in resistance of bacteria to antibiotic treatment, it is necessary to find new compounds that have antibacterial activity as alternatives to the antimicrobial therapies [2]. One source of the search for new compounds is from natural products, especially from plants because they have a variety of secondary metabolites with antimicrobial activities [3, 4]. From time to time many medicinal plants exhibit a source of new antimicrobial agents [5].

Curcuma is member of the family Zingiberaceae. It has been reported to have various bioactivities, including as an antioxidant [6], anti-inflammatory [7] antimicrobial [8], antiarthritic, antidepressant, antiatherosclerotic, antiaging [9], antidiabetic [10] and anticancer [11]. One of them is *Curcuma soloensis* Val. which has the potential to be widely used as phytopharmaca. It is a medicinal plant that grows in Java, Indonesia, and the rhizome of *C. soloensis* Val. is called "temu glenyeh". Phytochemically, the bisacuraon, and curcuminoid were compounds that successfully isolated from *C. soloensis* Val. have been reported previously [12]. In addition, we have successfully isolated ar-turmerone, a sesquiterpene derivative.

Given the increasing resistance to the antibiotics, the search for new antimicrobial compounds with variety of chemical structures and novel mechanism of action are urgently needed [13]. Therefore, this study aimed to evaluate the antibacterial activity of ar-turmerone that has possibility antibacterial activity from *C. soloensis* Val. rizhome against four common pathogenic bacteria by disc diffusion method.

2. Experimental

2.1 General procedures

We have previously obtained ar-turmerone from the *C. soloensis* Val. Rhizome. All chemicals were purchased from E. Merck and used directly without any purification. They were Mueller-Hinton Agar (MHA), DMSO, NaCl, ethanol, and paper disc.

2.1.1 Antibacterial testing

The antibacterial test of the compound was determined by the disc difussion method against four bacteria, i.e *Klebsiella pneumonia* ATCC 13883, *Pseudomonas aeruginosa* ATCC 27853, *Staphylococcus aureus* ATCC 25923, and *Escherichia coli* ATCC 25922,. The sample was prepared to be 1000, 500, 250 dan 125 µg/mL in DMSO. The media used was Mueller-Hinton agar. The bacteria were cultured on agar media and incubated for 24 hours at 37 °C. The bacterial cultures were mixed with 9% NaCl solution and turbidity compared to McFarland 0.5 turbidity standard (108 CFU / mL). Afterward, the agar to be poured over the Petri dishes and the bacterial suspension was swab upon the agar. As much as 10 µL of the compound was added to sterile paper discs of 6 mm in diameter, which were put in agar that has been inoculated with bacterial strain. Then the plates were incubated at 37 °C for 24 hours. The test was done in triplicate. The inhibition zone represents as the antibacterial action. Chloramfenicol was used as positive control antibiotic.

3. Results and Discussion

Previously, we have been isolated ar-turmerone, a sesquiterpenoid from *C. soloensis* Val. The structure of ar-turmerone shows in Figure 1.

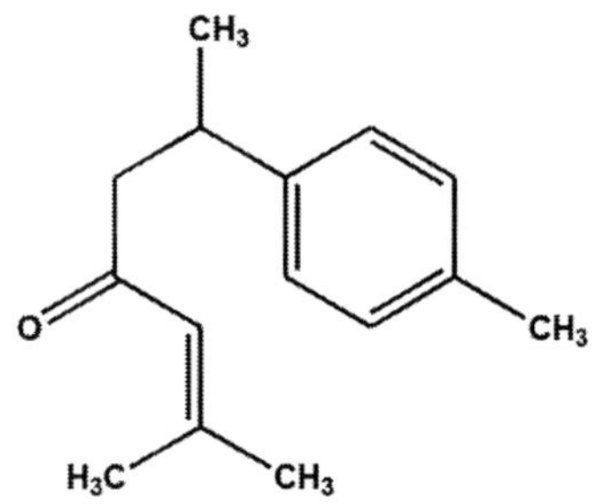

Figure 1. Structure of ar-turmerone

The results of the antibacterial testing of ar-turmerone from *C. soloensis* Val. rizhome against four bacterial species mainly *Klebsiella pneumonia* ATCC 13883, *Pseudomonas aeruginosa* ATCC 27853, *Staphylococcus aureus* ATCC 25923, and *Escherichia coli* ATCC 25922 are presented in Table 1 (inhibition zones).

Table 1. Antibacterial activity of ar-turmerone

Bacteria	**Inhibition zones (mm)**				
	1000	500	250	125	Chloramfenicol
K. pneumonia ATCC 13883	-	-	-	-	33.05
S. aureus ATCC 25923	-	-	-	-	28.10
E. coli ATCC 25922	-	-	-	-	27.05
P. aeruginosa ATCC 27853	-	-	-	-	-

The concentration of compound was in µg/mL.

The data in Table 1 shows that ar-turmerone did not have antibacterial activity against all test bacteria. It is because the activity of the bacteria is influenced by the structure of the compound and the characteristics of the bacteria. Each bacterium has a different level of sensitivity to certain compounds. Besides that, it can also be caused by the nature of synergism, which will be antibacterial if it was in the extract instead of pure compounds [14]. The mechanism of action of sesquiterpenoids cannot be known with certainty. Various mechanisms of action of antibacterial activity of sesquiterpenoids have been proposed. Sesquiterpenoids will first penetrate the cell wall and damage the membrane, disrupting cell activity such as membrane transport, increased permeability, and other metabolic regulatory functions [15].

Chloramphenicol as a positive control inactive to *P. aeruginosa*, maybe the chloramphenicol is resistant to *P. aeruginosa*. Skocibusic' *et al.* stated that gram-negative *P. aeruginosa* is known to have a high level of intrinsic resistance to almost all recognized antimicrobials and antibiotics for the extremely tight combination of outer membrane barriers, even very resistant to synthetic drugs [16].

4. Conclusion

The evaluation of antibacterial activity showed that ar-turmerone, a sesquiterpenoid from *Curcuma soloensis* Val. rizhome was not active against four bacteria, namely *Klebsiella pneumonia* ATCC 13883, *Pseudomonas aeruginosa* ATCC 27853, *Staphylococcus aureus* ATCC 25923, and *Escherichia coli* ATCC 25922.

Acknowledgments

The financial support from a Grant for Fundamental Research with contract number: 543/UN27.21/PP/2018, from the Universitas Sebelas Maret, is gratefully acknowledged.

References

[1] Abreu, AC, McBain, AJ, Simões, M 2012 *Nat Prod Rep.* **29**(9) 1007–1021
[2] Santosh, L, Gaonkar, SL, Vignesh, UN 2017 *Res Chem Intermed.* **43**(11) 6043-6077
[3] Payne, DJ, Gwynn, MN, Holmes, DJ, Pompliano, DL 2007 *Nat Rev Drug Discov.* **6** 29–40
[4] Sgariglia, MA, Soberon, JR, Sampietro, DA, Quiroga, EN, Vattuone, MA 2011 *Food Chem* **126**(2) 395–404
[5] Harit, J, Barapatre, A, Prajapati, M, Aadil, KR, Senapati, S 2013 *Int J Life Sci Biotechnol Pharm Res.* **2**(3) 183–189
[6] Caiqin, L, Weiqing, C, Nan, W, Jianchang, J 2018 *J Nat Sci.* **23**(1) 063-069.
[7] Yuan, T, Zhang, C, Qiua, Ch, Xia, G, Wang, F, Lin, B, Lia, H, Chen, L 2017 *Nat Prod Res.* **32**(16) 1887-1892
[8] Pulido-Moran, MM, Moreno, JF, Ramirez-Tortosa, C, Ramirez-Tortosa, MC 2016 *Molecules.* **21**(3) 264-286
[9] Nelson, KM, Dahlin, JL, Bisson, J, Graham, J, Pauli, GF, Walters, MA 2017 *J Med Chem.* **60** 1620–1637
[10] Kato M, Nishikawa S, Ikehata A, Dochi K, Tani T, Takahashi T 2017 *Mol Nutr Food Res.* **61**(3) 1-20
[11] Naksuriya O, Okonogi S, Schiffelers RM, Hennink, WE 2014 *Biomaterials.* **35** 3365–3383
[12] Vitasari, RA, Wibowo, FR, Marliyana, SD, Wartono, MW 2016 *IOP Conf. Ser.: Mater. Sci. Eng.* ***107*** *012063*
[13] Kamazeri, TSAT, Samah, OA, Taher, M, Susanti, D, Qaralleh, H 2012 *Asian Pac. J. Trop. Med.* **5**(3) 202–209
[14] Zacchino SA, Butassi E, Cordisco E, Svetaz LA 2017 *Phytomedicine.* **37**(15)14-26
[15] Swamy MK, Akhtar MS, Sinniah UR 2016 *Evid Based Complement Alternat Med.* 1-22
[16] Skocibusic' M, Bezic' N, Dunkic V 2006 *Food Chem* **96** 20–28

Antibacterial activity of polyeugenol against *staphylococcus aureus* and *escherichia coli*

D S Handayani[1], Pranoto, D A Saputra, S D Marliyana
Department of Chemistry, Faculty of Mathematics and Natural Sciences, Sebelas Maret University

E-mail: desisuci@staff.uns.ac.id

Abstract. Antibacterial activity of polypugenol compounds has been carried out in the form active ingredients and gels against Staphylococcus aureus (ATCC 25923) and Escherichia coli (ATCC 25922). Antibacterial activity assay of polyeugenol in active form with varied concentration 0.0025 g/mL, 0.005 g/mL, 0.01 g/mL, 0.02 g/mL, and 0.04 g/mL and gelling form varied concentration 0.01 g/mL, 0.02 g/mL, and 0.04 g/mL. Antibacterial activity assay using agar disk diffusion method by measuring the diameter of inhibition.The result showed that antibacterial activity not presence at *Escherichia coli* while *Staphylococcus aureus* has presence of antibacterial activity at concentration 0.01 g/mL, 0.02 g/mL, dan 0.04 g/mL with inhibition diameter values 15.72; 17.28 and 20.51 mm. Polyeugenol in gelling form were tested against *Staphylococcus aureus* and *Escherichia coli* at concentration 0.01g/mL, 0.02g/mL, dan 0.04g/mL. The presence of antibacterial activity obtained for *Escherichia coli* only at concentration 0.04 g/mL with the value of inhibition diameter is 0.53 mm while concentration 0.01g/mL dan 0.02g/mL not presence antibacterial activity. The presence of antibacterial activity for *Staphylococcus aureus* obtained at concentration 0.01 g/mL, 0.02 g/mL and 0.04 g/mL with the value of inhibition diameter are 14.46; 15.68 and 18.81 mm that showed relatively strong antibacterial activity. Statistic tested using *one way* ANOVA in *Staphylococcus aureus* shows significant differences at each concentration from polyeugenol in active form and gelling form with F-count> F-tab.

1. Introduction

Severe bacterial infections are a major cause of morbidity and mortality among newborns and young infants in the developing world [1]. One of the causes of infection due to bacteria is poor hand hygiene. Public awareness of the importance of hand hygiene is still lacking. People are not aware that in their activities, their hands are often contaminated with microbes. One disease that can be caused by not maintaining hand hygiene is diarrhea [2].

Hand hygiene is one of the important aspects to avoid various kinds of diseases caused by bacteria. In general, bacteria such as Staphylococcus aureus (S. Aureus) and Escherichia coli (E. Coli) are found in parts of human skin, where these bacteria can cause dangerous diseases [3]. Lack of facilities such as limited water availabilityand soap in public places can be an obstacle to cleaning hands. One alternative solution that can be used is antiseptic gel because its use is easier and more efficient [4].

Antiseptic gel preparations on the market still use alcohol as an antibacterial ingredient, whereas repeated use can cause irritation to the skin [5]. Alcohol can also dissolve layers of fat and sebum on the skin which serves as a protection against microorganism infections [6]. The alternative is to use natural ingredients as antibacterial which can be derived from essential oils from clove leaves (Syzygium aromaticum). The compound that acts as an antibacterial in essential oils of clove leaves is eugenol. Eugenol has excellent bactericidal activity against various organisms such as Escherichia coli, Staphylococcus aureus, Pseudomonas aeruginosa [7] and Listeria monocytogenes [8].

The value of antibacterial activity, eugenol can be increased by various modifications, one of which is to make derivative compounds from eugenol namely 3-(3,4 dimethoxy phenyl)-1-propanol and to test antibacterial activity against gram-negative E. coli bacteria at various concentrations. The results showed that at a concentration of 40% it was more effective at inhibiting the growth of E. coli bacteria with a diameter inhibition zone of 1.9 cm [9]. However, to make derivative compounds from eugenol requires a long processing time of around 70 hours at room temperature [10] and 48 hours at a constant temperature of 45 ° C [9]. The length of time needed indicates that the use of eugenol to be antibacterial tends to be ineffective and inefficient. Therefore, another modification of eugenol is needed but it can still increase its antibacterial activity value, one of which is polymerization. This is because polymers can be used as antibacterial compounds, one of which is chitosan which has antibacterial activity against gram-positive and gram-negative bacteria [11].

Making polyethanol itself is easier and more efficient and has been carefully examined. Handayani [12] has synthesized polyeugenol with concentrated H_2SO_4 catalyst resulting in a yield of 69.31%, insoluble in water but soluble in acetone, chloroform, and benzene. However, the application of polieugenol as an antibacterial compound has never been done. In fact, these compounds have the potential as antibacterial compounds. Therefore, in this research the synthesis of polieugenol was tested for its antibacterial activity against S. aureus and E. coli.

2. Experimental

Antibacterial activity was carried out by testing bacterial growth inhibition against S. aureus and E. coli using agar disk diffusion method (Kirby bauer) [3]. Stages of research include Tool Sterilization, Media Preparation, Bacterial Rejuvenation and Antibacterial Activity Test. The solid media used is 6.8 grams of MHA (Mueller Hinton Agar) which has been autoclaved and heated on a hot plate by stirring using a magnetic stirrer, the process is carried out in the LAF room [13]. Rejuvenation of pure bacterial cultures is carried out on solid media so that it is tilted by scratching an ose needle containing E. coli and S. aureus with a zigzag pattern while being held near to the flame. Each bacterium is made 2 cultures of bacteria. This stage is carried out in a sterile state in the LAF room and incubated at 37 °C for 24 hours [13].

Before testing the bacterial inhibitory activity against polieugenol, the autoclave MHA media was prepared and placed in the Laminar Air Flow (LAF) room. The method used is the agar diffusion method using paper disks (6 mm diameter) [14]. Disc paper was dripped with 20 µL of polieugenol extract in 6 petri dishes with a variation of 0.4 g; 0.2 g; 0.1 g; 0.05 g; 0.025 g and 0.0125 g respectively in 10 mL of methanol solvent [15] and methanol solvents as negative controls [16]. To make a bacterial suspension, one batch of bacteria was dissolved in 4 mL of 9% sterile NaCl with a color ratio according to the standard McFarland 0.5 [17]. After that 100 µL of suspension and 15 mL of sterile bacterial media were vortexed to be homogeneous. Then the solution was put into a sterile petri dish for ± 15 minutes until the solution solidified. Dry discs that have been extracted are placed on the media so that the petri dish is regular. Each type of bacteria is carried out 3 times. Petri dishes are then wrapped in plastic wrap so they remain sterile and incubated for 24 hours at ± 37 °C. The inhibitory power of bacterial growth can be determined by measuring the diameter of the clear zone around the disc paper. The diameter of the inhibitory zone formed was measured using a calipers to determine the effectiveness of antibacterials [18].

After that the gel preparation process was carried out by dissolving the polieugenol compound with several concentrations of concentrations of 0.4 g, 0.2 g and 0.1 g with 5 mL methanol solvent and 5 mL of distilled water for each variation of concentration. After that, CMC 0.5% is slowly added to each variation of its concentration [19] which has been dissolved in distilled water as a gel base in warm conditions, then stirred using mortar [15]. After becoming a gel preparation, anti-bacterial testing was carried out with 3 replications and compared with its antiseptic power with positive control antiseptic power (hand antiseptic gel preparation with ethanol and triclosan active ingredients).

3. Results and Discussion

The antibacterial activity test of polieugenol was carried out in the form of a preparation gel active ingredient. Antibacterial testing was carried out on E. coli bacteria (ATCC 25922) which are gram negative bacteria [3] and S. aureus (ATCC 25923) which are gram-positive bacteria [20].

The media used in the test is Mueller Hinton Agar (MHA) which is a suitable medium for antimicrobial testing [21]. MHA has been recommended by the FDA and WHO for antimicrobial testing because it contains sulfonamides, trimethoprim and low tetracycline inhibitors so that it can provide good pathogenic growth [22]. Before the testing begins the tools and agar media that will be used must be sterilized first using an autoclave at a temperature of 121°C [23]. Sterilization is carried out so that living microbes and spores can be destroyed so as not to interfere with and contaminate antibacterial testing [24].

The negative control used in the test is methanol, because methanol can dissolve polieugenol and has no antibacterial activity [25]. Natheer [26] said that substances which are used as negative controls are solvents used as compound diluents. The purpose of using negative controls is to compare that the solvent used does not affect the antibacterial test results. The positive control used in the antibacterial activity test is triclosan. The selection of triclosan as a positive control is due to triclosan having a broad spectrum of activity against gram-positive and mostly gram-negative bacteria [27], so it is often used as a mixture that is often found in antiseptic preparations on the market.

Tests are carried out by making several different variations of concentrations, for active ingredients there are 5 variations of concentration namely 0.0025 g/mL, 0.005 g/mL, 0.01 g/mL, 0.02 g/mL and 0.04 g/mL [15], while for sedian gels there are 3 variations of concentration determined from the results of the diameter of the largest inhibition zone in testing active ingredients. Tests took place during the 24-hour incubation period the results of antibacterial activity tests were characterized by the presence of clear areas around the disc that indicated there was no bacterial colony growing in the area, this clear zone called the inhibition area diameter. The results of testing for antibacterial activity and measuring the diameter of inhibitory zones of polieugenol compounds against E. coli and S. aureus can be seen in Table 1.

Table 1. Results of measuring the diameter of the inhibitory zone of the polyeugenol compound against E. coli and S. aureus

Sample	Concentration (g/mL)	Average diameter of inhibition zone (mm ± SD)	
		Escherichia coli ATCC 25922	*Staphylococcus aureus* ATCC 25923
active ingredients	0.0025	0 ± 0	0 ± 0
	0.005	0 ± 0	2.12 ± 0.16
	0.01	0 ± 0	15.72 ± 0.22
	0.02	0 ± 0	17.18 ± 0.14
	0.04	0 ± 0	20.51 ± 0.03
	Trichlosan (positive control)	23.30 ± 0.32	27.39 ± 0.45
	Methanol (negative control)	0 ± 0	0 ± 0
Preparation gel	0.01	0± 0	14.46 ± 0.34
	0.02	0± 0	15.68 ± 0.26
	0.04	0.53 ± 0.92	18.81 ± 0.34
	Trichlosan (positive control)	23.58 ± 0.17	27.76± 0.18
	Methanol (negative control)	0 ± 0	0 ± 0

Based on Table 1, it can be seen that polieugenol compounds in the form of active ingredients do not have antibacterial activity against E. coli bacteria, while in S. aureus bacteria, polieugenol compounds in the form of active ingredients and preparation gels have antibacterial activity. Polieugenol in the form of active ingredients at a concentration of 0.005 g/mL has an average inhibition diameter of 2.12 mm ± 0.16.

Concentrations of 0.01 g/mL and 0.02 g/mL had a mean inhibition diameter of 15.72 mm ± 0.22 and 17.18 mm ± 0.14, while for a concentration of 0.04 g/mL it had an average average inhibitory diameter of 20.51 mm ± 0.03. Based on Rosyita [28] and Ambarwati [29] the classification of antibacterial activity can be seen from the value of the diameter zone <5 mm including weak, 5-9 mm including the medium category, 10-19 mm including the strong category and >20 mm including the very strong category, so that at a concentration of 0.04 g/mL included in the very strong category, the concentration of 0.01 g/mL and 0.02 g/mL in the strong category and 0.005 g/mL in the very weak category.

The concentration of active ingredients which has the greatest inhibitory power was 0.01 g/mL; 0.02 g/ mL and 0.04 g/mL, so this concentration was used in testing the antibacterial activity of polyeugenol compounds in the form of preparatory gels. Preparation of gel preparation from polieugenol compounds using CMC as a gelling agent, because CMC has no antibacterial activity [30] so it will not affect the testing of antibacterial activity. The measurement results showed that the polyeugenol compounds in the dosage form had almost no antibacterial activity against E. coli bacteria, whereas the polieugenol compounds in gel dosage forms had antibacterial activity which was classified as strong at a concentration of 0.01 g / mL; 0.02 g / mL and 0.04 g / mL for against S. aureus.

The results of measurements of inhibitory zone diameters generally showed that polyeugenol compounds in the form of active ingredients and dosage gels only had antibacterial activity against S. Aureus (a gram positive) compared to E. coli (a gram negative). This is due to differences in cell wall structure between gram positive and negative bacteria. Gram negative bacteria have two cell membranes, namely the outer membrane and cytoplasmic membrane, whereas gram-positive bacteria only have cytoplasmic membranes [3].

The difference in composition and structure of cell walls in gram negative and gram positive bacteria can affect the antibacterial activity of a chemical compound [31]. The structure of simpler, single layered gram-positive bacterial cell walls contains 90% peptidoglycan layer [32]with low lipid content (1-4%) facilitating bioactive material into cells. Gram negative bacteria itself has more complex cell walls consisting of three layers, namely the outer layer of lipoprotein, the middle layer of lipopolysaccharide which acts as a barrier to the entry of antibacterial bioactive ingredients, and the inner layer of peptidoglycan with lipid content reaching 11-12% [31].

The dominant peptidoglycan content in the cell wall of S. aureus, which includes gram-positive bacteria, is a water-soluble polymer. This characteristic shows that bacterial cell walls are polar. Polieugenol compounds are also polar so it is easier to penetrate the polar peptidoglycan layer compared to the non-polar lipid layer [28]. E. coli bacteria which are gram negative bacteria contain more lipids, less peptidoglycan and have an outer membrane in the form of a bilayer which serves as a selective defense of compounds that enter or enter cells [33]. This is what causes polieugenol compounds can not provide antibacterial activity against E. coli bacteria.

The antibacterial activity in polymers is related to the ability to absorb bacterial cell walls. A compound can interact better with gram-positive bacteria compared to gram negative because the negative charge on the cell surface of gram-positive bacteria is more than gram negative. The positive charge of polieugenol which is distributed to the surface of the cell wall of gram negative bacteria which in turn will inhibit bacterial activity being tested. This also causes greater antibacterial activity of polieugenol compounds to bacteria than S. aureus with E. coli.

The hydroxyl group (-OH) in this polieugenol compound has an important role for polieugenol compounds in inhibiting bacterial growth [35]. The group damages the cell wall so that it can inhibit the metabolism of bacteria. Based on the hypothesis of Pei [36], that the hydroxyl group (-OH) will interact with proteins present in bacteria so that it can prevent the action of enzymes in bacteria. This causes the growth of bacteria to be blocked and bacteria can die.

The testing of antibacterial activity against S. aureus polieugenol in the form of dosage gel has a smaller inhibition zone value compared to the active ingredient, which indicates that polieugenol in the form of active ingredients is slightly more effective than polieugenol compounds in the form of preparatory gel. This is due to the formulation and use of gelling agents that have not been appropriate in preparing gel preparations. According to Selvia [37] the formulation of adding the right concentration and gelling agent will increase the value of the antibacterial activity of a compound. The selection of gelling agent or gel base used in making gel preparations will affect contact time, speed of release of active substances, and binding capacity to active substances [38]. So that it can affect the spread of power when adding active compounds [39].

Data on the testing of antibacterial activity obtained were then analyzed using statistics. Statistical testing is only done for S. aureus bacteria in the form of active ingredients and preparation gels. The aim is to determine the difference in inhibitory power of each concentration. This statistical test uses the One way Analysis of Variance (ANOVA) test with the results of the results (Table 2) and the following hypothesis.

Ho: There was no significant difference in antibacterial activity between the series of concentrations of polynugenol compounds in the form of active ingredients or in the form of dosage gels against S. aureus.

Ha: There is a significant difference in antibacterial activity between the series of concentrations of polynugenol compounds in the form of active ingredients or in the form of dosage gels against S. aureus.

Table 2. Oneway ANOVA calculation results of variations in the concentration of polyeugenol to S. aureus.

Form of Polyeugenol	F count	F table	Description
Active ingredients	778.57	5.14	Significantly different
Preparation gel	146.15	5.14	Significantly different

Based on the results of variance analysis with a comparison of tablesthe value of F-count> F-table is obtained. These results state that Ho is rejected and Ha is accepted, which means that there is a significant difference in antibacterial activity between the series of concentrations of polynugenol compounds in the form of active ingredients or in the form of gel preparations against S. aureus. This indicates that the antibacterial activity of polieugenol in the form of active ingredients and gel preparation against S. aureus in the concentration series was significantly different. Shown by the value of the diameter of the inhibition zone which increases with increasing concentration, according to the research conducted by Baydar [40].

4. Conclusion

Polyeugenol compounds in the form of active ingredients and gel preparations don't have antibacterial activity against the bacteria Escherichia coli, while Staphylococcus aureus has strong antibacterial activity.

Acknowledgment

The financial support from Indonesian Directorate General of Higher Education through the research project of Penelitian Terapan Unggulan Perguruan Tinggi (PTUPT) 2018 is greatly appreciated.

References

[1] Liu L, Johnson HL, Cousens S, Perin J, Scott S, Lawn JE, Rudan I, Campbell H, Cibulskis R, Li M, Mathers C, Black RE. 2012. *Global, Regional, and National Causes of Child Mortality: An Updated Systematic Analysis for 2010 with Time Trends since 2000.* Lancet 379: 2151-2161.

[2] Ministry of Health of the Republic of Indonesia 2011 Situation of Diarrhea in Indonesia, *Window Bulletin Data and Health Information.* ISSN 2088-270X

[3] Jawetz E, Melnick J L and Adelberg E A 2007 *Medical Microbiology,* Publisher: McGraw-Hill; *First published 1954*, 24th edition. ISBN-13: 978-0 -07-128735- 7

[4] Ningsih W, Firmansyah, and Anggraini S 2016 *Jurnal Ilmiah Farmasi.* **12(2)** 79-85
[5] Dyer D L, Shinder A and Shinder F 2000 *Family Medicine.* **32** 633-638
[6] Jones R D 2000 *Association of Perioperative Registered Nurses.* **71** 584-599
[7] Walsh S E, Maillard J Y, Russell A D, Catrenich C E, Charbonneau D L and Bartolo R G 2003 *J Appl Microbiol.* **94** 240–247
[8] Filgueiras C T and Vanetti M C D 2006 *Braz Arch Biol Technol.* **49** 405–409
[9] Noviansari R, Sudarmin and Siadi K 2013 *Indo. J. Chem Sci.* **2** 114-118
[10] Tahyal C Y, Hanoch J S, Serly J S 2014 *Ind. J. Chem. Res* . **2** 124 - 130
[11] Du W L, Niu S S, Xu Y L, Xu Z R and Fan C L 2009 *Carbohydr Polym.* **75** 385-389
[12] Ngadiwiyana 2005 Jurnal Kimia Sains dan Aplikasinya. **8(2)** 43-47
[13] Islam M, Masumb S, Mahbuba K R and Haque Z 2011 *J. Adv. Sci. Res.* **2** 63-66
[14] Hermawan A 2007 *Scientific Article* (Surabaya: UNAIR Veterinary Faculty)
[15] Sulastrianah, Imran and Fitria E S 2014 *Medula.* **1** 76-84
[16] Ortenz J H 2005 *American Society for Microbiology*, Marie B. Coyle Coord, Ed.
[17] NCCLS 2007 *Performance Standarts for Antimicrobial Susceptibility Testing: Seventeeth informational Supplement.* 26
[18] Roslizawaty, Ramadani N Y and Fakhrurrazi H 2013 *J. Med.* Vet. **7** 91-94
[19] Sugita P, Sjahriza A, and Rachmanita 2007 *Proceedings of the Indonesian Chemical Association National Seminar* (Department of chemistry Institut Pertanian Bogor) p 437-443
[20] Kumala W 2006 *Diagnosis of Clinical Microbiology Laboratory* (Jakarta: Publisher Trisakti University)
[21] Bio Merieux Clinical Diagnostics 2012 *Antimicrobial Resistance*, USA: New York.
[22] Acumedia 2004 *Mueller Hinton Agar* (7101). PI 7101, Rev 02, 11/14/04 (www.neogen.com)
[23] Nayak BK, V Suchitra and Anima Nanda 2015 *Der Pharmacia Lettre.* **7**(4) 51-54
[24] Stefanus L 2006 *Formulation of Sterile Preparations* (Yogyakarta: Publisher Andi Offset)
[25] Ortenz J H 2005 J. *American Society for Microbiology*, Marie B. Coyle Coord, Ed.
[26] Natheer S E, Sekar C, Amutharaj P, Rahman M S A and Khan K K 2012 *Afr J Pharm Pharmacol.* **6** 783-788
[27] Gruendeman B J 2005 *Textbook on Nursing* (Jakarta:ECG Publisher)
[28] Hayati Z, Yulia W, Karmil TF and Azmy A 2012 *Proceedings of the 2nd Annual International Conference Syiah Kuala University.* **2 (1)** 416-420
[29] Panjaitan RS, Madayanti F 2018 *Edu Chemia.* **3(1)** 29-39
[30] Pratiwi S T 2008 *Pharmaceutical Microbiology*. Jakarta: Erlangga Publisher
[31] Xie W, Xu P, Wang W and Liu Q 2002 *Carbohydr Polym.* **50** 35-40
[32] Brooks G F 2005 *Medical Microbiology*, Salemba Medika Publisher: Jakarta.
[33] Feng Q L, Wu J, Chen G Q, Cui F Z, Kim T N, Kim J O 2000 *J Biomed Mater Res.* **52** 662-668
[34] R. C. Goy and O. B. G. Assis 2014 *Braz. J. Chem.Eng.* **31(03)** p 643 - 648
[35] Dorman H J D, and Deans S G 2000 *J Appl Microbiol.* **88** 308-316
[36] Pei R S, Zhou F, Ji B P and Xu J 2009 *J. Food Sci.* **74** 379-383
[37] Selvia R W, Dina M and Sri P F 2015 *Proceedings of the National Conference on Mathematical Science and its Applications* (Faculty of Mathematics and Natural Sciences, Bandung Islamic University) p 351-355.
[38] Taurina W and Rafikasari 2014 *Traditional Medicine Journal.* **19** 70-73
[39] Maulina L and Sugihartini N 2015 *Pharmaciana.* **5** 43–52
[40] Baydar H, Osman S, Gulcan O and Tahsin K 2004 *Food Control.* **15** 169-172

IOP Conf. Series: Materials Science and Engineering **578** (2019) 012062 doi:10.1088/1757-899X/578/1/012062

Purification of Cassiterite Mineral Bangka Indonesia by Pyrometallurgy and Hydrometallurgy Methods

E W Trisnawati[1], L Andriyah[2], S Wahyuningsih[1], L H Lalasari[2], T Arini[2] and F Firdiyono[2]

[1] Chemistry, Faculty of Mathematics and Natural Sciences, Sebelas Maret University, Surakarta, Indonesia.
[2] Metallurgy and Material Research Center, Indonesian Institute of Sciences, South Tangerang, Banten, Indonesia.

E-mail: liaa001@lipi.go.id

Abstract. The cassiterite mineral is producing mineral of tin. It is widely used in the manufacture of catalysts such as FTO (Fluorine Tin Oxide) glass coatings or as a SnO_2 pigment. Tin is usually used in energy and electronics sector. Cassiterite can be found on the islands of Bangka, Belitung, Kundur, Singkep, Karimun and Kampar. In Indonesia, cassiterite mixed with other minerals such as ilmenite, quartz sand, zircon, rutile, pyrite, calcite, lanthanum and monazite. Those mixtures able to give rise of problems if used as raw chemical materials in tin production. The purifying process of cassiterite is done by pyrometallurgy and hydrometallurgy methods. The XRF and XRD analysis of raw cassiterite show that the main components are Sn and O. The decomposition process is done by mixing cassiterite and kalium sulfate (K_2SO_4) with a ratio of 2:3. The leaching process is performed using KOH to dissolve SnO_2 because the compound is soluble in a strong base. The optimum result is obtained at decomposition temperature 800 °C and time of 4 hours. The intermediate product kalium stannate (K_2SnO_3) is not formed because cassiterite has a structure which may cause some difficulties to react with the salt kalium sulfate (K_2SO_4).

1. Introduction

Indonesia is a rich nation with abundant natural resources such are metallic minerals species, one which is cassiterite included in tin ore (raw sand) in Bangka Island [1]. Cassiterites are widely distributed on the islands of Bangka, Belitung, Kundur, Singkep, Karimun and Kampar [2]. Cassiterite has two types of accumulation, primary and secondary type. The primary type formation is hard rock without weathering process from the source rock and the secondary types is weathered rock type, each forming the alluvial deposit and other associated heavy minerals [7]. Cassiterite (SnO_2) is tin oxide mineral with many possible merits one of which is found at catalyst manufacture (tin chemical), for instance in coating of conductive glass FTO (Flourine Tin Oxide) or as SnO_2 pigments [2]. Cassiterite is natural mineral that can be extracted economically [8]. Tin extraction method depends on the type of the ore itself according to the oxidic or sulfide chemical composition [9]. Extraction method of tin from its ore usually utilized by reducing cassiterite with aqueous inorganic acid [11]. Tin is usually utilized in energy and electronic materials sector. The percentage of tin utilization in the world is shown in Figure 1 [3].

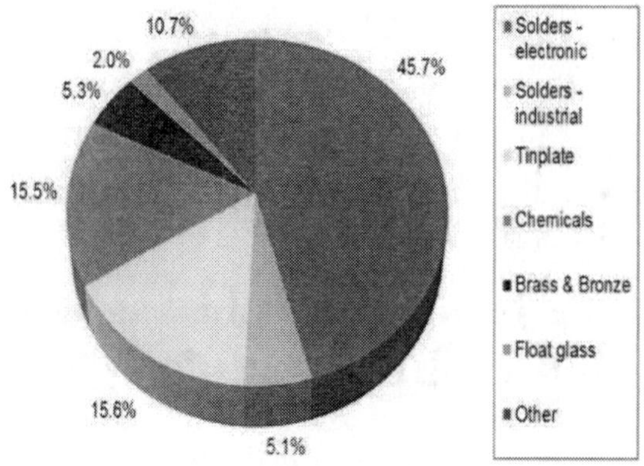

Figure 1. Graph of Tin Utilization in the World

Cassiterite in pure state contains 78,6% of tin, but pure cassiterite is rarely found, it explains the contain varied between 73-75% [10]. Tin metal in nature generally exists as oxide. Its primary sources are cassiterite mineral (SnO_2) which in Indonesia is mostly found as alluvial deposits resulted from weathering, erosion and transportation process for a very long time along with some of its associated impurities such as quartz, ilmenite, monazite, rutile and zircon [1]. Various processes have been developed to obtain new valuable materials from cassiterite including a decomposition process of cassiterite using sodium carbonate. Another process is reported in decomposing cassiterite using sodium carbonate with addition of flowing CO/CO_2 mixture and solvation in NaOH solution [2].

2. Experiemntal

2.1 Research Material

The ingredients are cassiterite powder from Bangka, aquadest, KOH pellets (p.a., Merck), K_2SO_4 powder (p.a., Merck), Whatman filter paper number 41, tissue and weighing paper.

2.2 Research Tools

The tools are X-Ray Diffraction (XRD), X-Ray Fluorescence (XRF), carbolite furnace, 47900 type furnace, Radwag analytical balance wps 4000/c/1, accu plate hotplate stirrer, magnetic stirrer, watch glass, thermometer, mortar, spatula, glass stirring rod, ceramic crucible, graphite crucible, beaker, filter funnel, erlenmeyer flask and volumetric flask.

2.3 Stages of Research

Cassiterite 30 g and K_2SO_4 45 g (2:3) were evenly mixed by stirring. The mixtures were placed into ceramic crucible and decomposed as subsequent treatment. The prepared samples were put into 47900 type furnace. Decomposition process was carried out 9 times using 5 temperature variations ranging from 500-900 °C and 4 time variations (1,2,3,4 hours) of the most optimum temperature. The sample was cooled in furnace to be processed the following day. It was weighed to determine the loss of mass after decomposition process. The results were partly used for leaching process and the rest were analyzed using XRD. Samples 40 g were taken from decomposed results. Leaching process was carried out using KOH solution 2 M for 2 hours at room temperature and stirred using magnetic stirrer proceeded with filtration using two sheets of Whatman filter paper number 41. The filtrate was further proceeded with evaporation while the residue was dried for XRD analysis material. Filtrate was evaporated at temperature 100 °C until it became saturated. The saturated solution was crystallized at temperature of 100-110 °C. The formed crystals were calcined using carbolite furnace at 800 °C for 3 hours and analyzed using XRD. The remaining decomposed samples, leaching residues and the resulting crystals were analyzed using X-Ray Difraction to determine the intermediate formation of potassium stannate and SnO_2. Cassiterite was characterized using XRD and XRF.

3. Results and Discussion

Initially, cassiterite was characterized using XRD and XRF to identify its elements. The XRF analysis results were shown in Table 1, wherein the most abundant element in cassiterite were Sn 57.82% and O 23.39% [2].

Table 1. XRF Analysis Results of Cassiterite

No	Element	Content (%)	No	Element	Content (%)
1.	Sn	57.87	9.	Ca	0.63
2.	O	23.39	10.	Mg	0.62
3.	La	3.43	11.	Al	0.52
4.	Ce	3.37	12.	Si	0.45
5.	Nd	2.18	13.	Cl	0.37
6.	Fe	1.79	14.	Zr	0.29
7.	P	1.68	15.	S	0.11
8.	Ti	1.45	16.	Pb	0.04

Cassiterite was also analyzed using XRD which results were shown in Figure 2 indicating clear and sharp peaks of SnO_2 at 2θ (degree) of (26.64; 33.94; 38.02; 51.84; 54.84; 61.94; 64,82; 66,04)° [2].

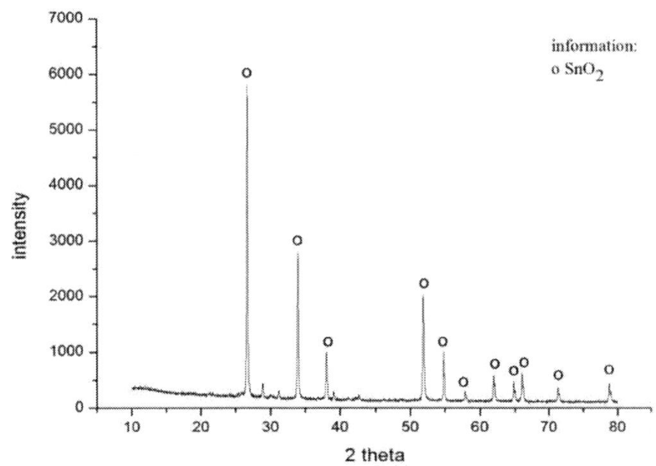

Figure 2. Diffractogram of Cassiterite

The first process is decomposition, decomposition process was aimed to form intermediates of potassium stanate (K_2SnO_3) by means of added heat under its melting point. The occuring reaction in decomposition process is show in equation 1.

$$SnO_{2(s)} + K_2SO_{4(s)} \rightarrow K_2SnO_{3(s)} + SO_{3(g)} \quad (1)$$

Cassiterite and potassium sulfate were mixed with a ratio of 2:3. Temperature variations were carried out to determine decomposition effects on the decreasing mass of decomposed sample. The graph of decomposition temperature effects were shown in Figure 3.

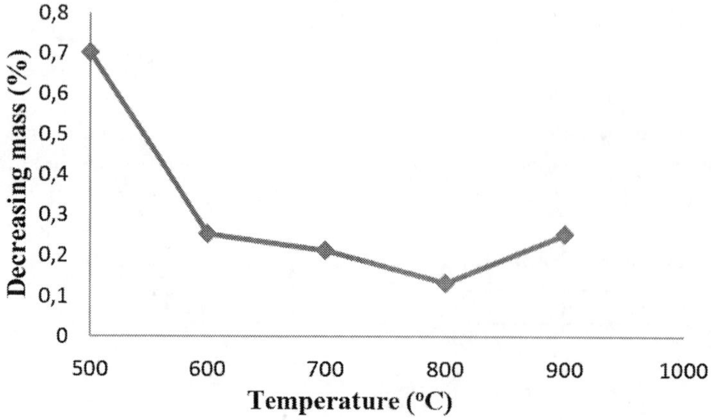

Figure 3. Graph of Decomposition Temperature Effects on Decreasing Mass of Decomposed Sample

The percentage of decreasing mass was smaller along with the increasing temperature, however at temperature 900 °C the percentage was increasing because the probability of particles to evaporate during heating process, optimum yield was obtained at temperature 500 °C for its largest percentage 0.70% and optimum temperature of reaction. The small percentage of decreasing mass due to the occured reaction between SnO_2 and K_2SO_4 was not yet maximum on temperature variations, because SnO_2 was an alluvial sedimentary rock that hardly reacted with salt. Moreover, decomposition process was carried out in atmosphere contained air rather than atmosphere contained CO/CO_2 mixture [4]. The results of decreasing mass from decomposed sample were shown in Table 2 and calculated using equation 2.

$$\text{Decreased mass} = \frac{\text{initial mass} - \text{final mass}}{\text{initial mass}} \times 100\% \quad (2)$$

Table 2. Results of Decreasing Mass from Decomposed Sample

Sample	Initial Mass (g)	Final Mass (g)	Decreased Mass (%)
KS-500-3	75.53	75.00	0.70
KS-600-3	75.00	74.81	0.25
KS-700-3	75.00	74.84	0.21
KS-800-3	75.00	74.9	0.13
KS-900-3	75.00	74.81	0.25

The resulting desomposed sample was characterization using XRD to determine newly formed absorption patterns after decomposition process with temperature variations. The diffractogram of decomposed sample was shown in Figure 4.

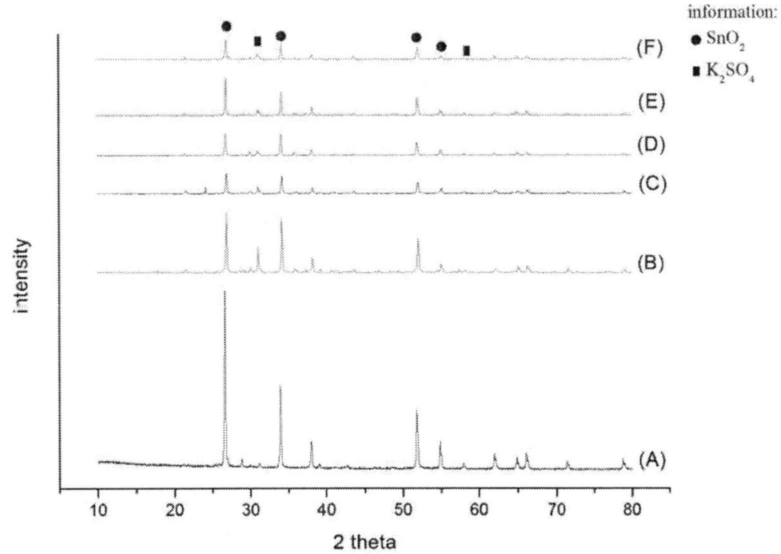

Figure 4. Diffractogram of Decomposed Sample (A) Initial Cassiterite, (B) 500 °C, (C) 600 °C, (D) 700 °C, (E) 800 °C, (F) 900 °C

The diffractogram displayed yield of diffraction patterns from decomposition process using temperature variations of 500 °C - 900 °C, they were matched toward cassiterite's diffraction pattern and devoid of new pattern. Temperature variations solely affected the diffraction intensity of SnO_2 and K_2SO_4. The intensity of SnO_2 diffraction decreased at 2θ (degree) of (26.8; 34.15; 52.02 and 54.9)° and increased at temperature 800 °C. In contrast to SnO_2, the intensity of K_2SO_4 increased at 2θ (degree) of (30.9; 43.55 and 58.1)° then decreased at temperature 800 °C. It could be concluded that after the reaction underwent decomposition process there were none of new phase formed. Temperature 800 °C in this study was the optimum decomposition temperature since at that condition the sample had the highest SnO_2 and least K_2SO_4 content. Another study mentioned the optimum decomposition temperature of SnO_2 and Na_2CO_3 was 875 °C in atmosphere contained CO/CO_2 mixture [5]. Contents of SnO_2 and K_2SO_4 from the decomposed samples were shown in Table 3.

Table 3. Contents of Decomposed Samples

Sample	Content (%)		GOF Value
	SnO_2	K_2SO_4	
KS-500-3	50.5	48.2	3.57
KS-600-3	50.2	49.0	2.19
KS-700-3	49.1	50.4	2.02
KS-800-3	54.0	43.3	2.33
KS-900-3	44.4	50.1	2.15

Effect of decomposition temperature variations on leaching process was identified from the solubility of decomposition sample. The leaching process was carried out using KOH solution to dissolve SnO_2 as it could be dissolved in a strong base. The resulting filtrate was evaporated and crystallized to form pure SnO_2 crystals while the residue was characterized using XRD. Figure 5 showed graph of the effect of decomposition temperatures on solubility of decomposed sample.

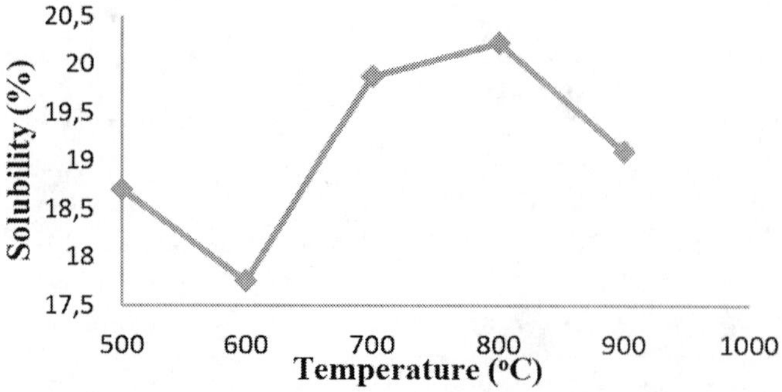

Figure 5. Graph of the Effect of Decomposition Temperatures on Solubility of Decomposed Sample

The leaching optimum temperature was 800 °C, as it had the biggest solubility 20.225%. At temperature 900 °C the solubility decreased possibly because the small amount of SnO_2 was formed during decomposition process. The solubility datas of affected sample by decomposition temperature were shown in Table 4 and calculated using equation 3.

$$\text{Solubility} = \frac{\text{initial mass} - \text{final mass}}{\text{initial mass}} \times 100\% \quad (3)$$

Table 4. Solubility Percentage of Decomposed Samples

Sample	Initial Mass (g)	Final Mass (g)	Solubility (%)
KS-500-3	40	32.52	18.7
KS-600-3	40	32.9	17.75
KS-700-3	40	32.05	19.88
KS-800-3	40	31.91	20.23
KS-900-3	40	32.36	19.1

Solubility percentage of decomposed sample was low, as SnO_2 in the decomposed sample only dissolved partially. The residue from leaching was characterized using XRD to indetifiy its contents of unsolved SnO_2 and find out new diffraction patterns after leaching process using KOH. Diffractogram of residue from leaching process at decomposition temperatures of 500 °C and 800 °C was shown in Figure 6.

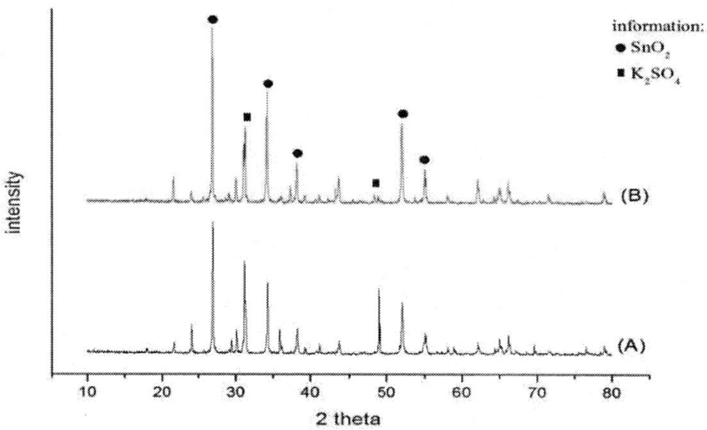

Figure 6. Diffractogram of Residue From Leaching Process (A) KS-500-3 and (B) KS-800-3

Diffractogram displayed that the diffraction pattern of residue had the same pattern and gave no indication of new patterns. Those diffractogram only showed increasing intensity of SnO_2 and decreasing intensity of K_2SO_4 due to temperature rise. The intensity of SnO_2 diffraction pattern increased at 2θ (degree) of (26.8; 34,1; 38.2; 51,9 and 55,2)° while intensity for K_2SO_4 decreased at 2θ (degree) of (31.1 and 48.3)°. Contents of SnO_2 and K_2SO_4 in residue product of leaching process were shown in Table 5.

Table 5. Contents of Residue Product from Leaching Process

Sample	Content (%)		GOF Value
	SnO_2	K_2SO_4	
KS-500-3	34.2	64.2	3.39
KS-800-3	39.8	59.3	2.61

Content of SnO_2 in samples KS-500-3 and KS-800-3 after decomposition process in Table 3 were 50.5% and 54.00% respectively while after underwent the leaching process they decreased to 34.2% and 39.8% as shown in Table 5.

Time variations were applied when decomposition process was reaching temperature 800 °C since it was the optimum decomposition temperature with the highest SnO_2 content and solubility after leaching process. Effects of decomposition time on decreasing mass of decomposed sample were shown in Figure 7.

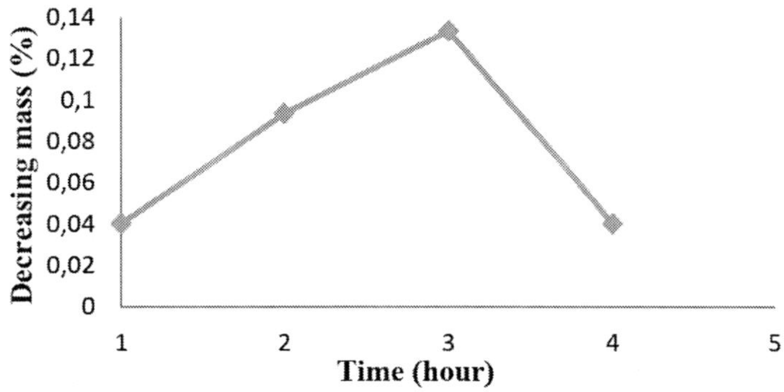

Figure 7. Graph of Effect on Decreasing Mass of Decomposed Sample

The optimum decomposition time was 3 hours with the highest decreasing mass percentage 0.13%. Following the optimum time, percentage on decreasing mass was lessening since its reaction ran suboptimally. The decreasing mass datas were shown in Table 6 and calculated using equation 4.

$$\text{Decreased mass} = \frac{\text{initial mass} - \text{final mass}}{\text{initial mass}} \times 100\% \quad (4)$$

Table 6. Results of Decreasing Mass of Decomposed Sample

Sample	Initial Mass (g)	Final Mass (g)	Decreased Mass (%)
KS-800-1	75	74.97	0.04
KS-800-2	75	74.93	0.09
KS-800-3	75	74.9	0.13
KS-800-4	75	74.97	0.04

A relatively small percentage result was possible because of the non-optimal decompositon reaction between cassiterite and potassium sulfate.

Decomposition time affected solubility in leaching process. The longer decomposition time, the greater solubility of decomposed products. Temperatures at leaching process also affected solubility [6]. The graph of the effects of decomposition time on solubility of decomposed samples was shown in Figure 8.

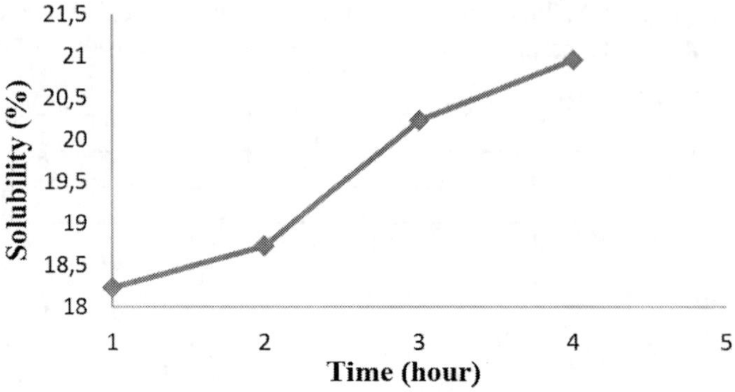

Figure 8. Graph of the Effect of Decomposition Time on Solubility of Decomposed Sample

The optimum time from leaching process was 4 hours with solubility percentage 20.95%. These relatively small solubility result was probably caused by too little time on stirring process, thus the solution had saturated. Moreover, leaching process was carried out at room temperature. Previous research mentioned the optimum temperature of leaching tin (Sn) using NaOH 1 M at temperature 90 °C for 2 hours resulted in percentage 62.4% [6]. Effects of decomposition time on solubility were shown in Table 7 and calculated using equation 5.

$$\text{Solubility} = \frac{\text{initial mass} - \text{final mass}}{\text{initial mass}} \times 100\% \tag{5}$$

Table 7. Effects of Decomposition Time on Solubility Data

Sample	Initial Mass (g)	Final Mass (g)	Solubility (%)
KS-800-1	40	32.71	18.23
KS-800-2	40	32.51	18.73
KS-800-3	40	31.91	20.23
KS-800-4	40	31.62	20.95

The filtrate from leaching process was evaporated to remove H_2O at temperature 100 °C until it became saturated solution. Those saturated solution was crystallized at temperature of 100-110 °C to evaporate the still existing water molecules to obtain crystals. Its form was hydrate and hygroscopic, hence the calcination at temperature 800 °C for 3 hours using calbolite furnace was conducted. White and non-hygroscopic crystals which were produced after calcination process possibly were purer cassiterite mineral crystals.

4. Conclusion

The conclusions from the research results are the optimum temperature of cassiterite purification was obtained at 800 °C because decomposed sample had the highest SnO_2 content and highest solubility 20.95% during leaching process and the optimum time in cassiterite purification process was obtained at 4 hours with potassium stannate intermediate expectedly not to be yielded through temperature variations of 500-900 °C since the process of forming K_2SnO_3 was accompanied by cassiterite's (SnO_2) unability to react with salt (K_2SO_4) because of its form as rock alluvial sediment which structure caused difficulty to react with salt.

Reference

[1] Harjanto S Virdhian S and Afrilinda E 2013 *Proceedings of the 52nd Conference of Metallurgists (Com), Hosting by Materials Science Technology Conference (Ms&T)*. **27** 1-9
[2] Andriyah L Lalasari LH and Manaf A 2017 *IOP Conf. Ser.: Mater. Sci. Eng.* **176** 012037
[3] Pearce J and Wallace T 2015 *Tin Chemical RoadMap 2015* International Tin Report
[4] Liu B Zhang Y Su Z Li G and Jiang T 2016 *Powder Technol.*. **301** 102-109
[5] Yang J Wu Y and Zhang X 2014 *Mineral Processing and Extractive Metallurgy*. **123** 228-233
[6] Ranitović M Kamberović Ž Korać M Jovanović N and Mihjalović A 2016 *Metalurgija*. **55** 153-156
[7] Saisinchai S Boonpramote T and Meechumna P 2016 *Engineering Journal*. **20** 41-49
[8] Su Z Zhang Y Liu B Lu M Li G and Jiang T 2017 *Journal of the Minerals, Metals & Materials Society* **69** 2364-2372
[9] El Deeb A B Morsi I M Atlam A A Omar A A and Fathy W M 2015 *Int. Journal of Sci. Eng. Research* **6** 54-64
[10] Falcon L M 1982 *Journal of the Southern African Institute of Mining and Metallurgy* **82** 112-117
[11] Gerald O Priscilla U Martin O Emmanuel E Chukwuka N and Hwang J-Y 2012 *Journal of Minerals and Mater. Characterization and Eng.* **11** 730-734

IOP Conf. Series: Materials Science and Engineering **578** (2019) 012063 doi:10.1088/1757-899X/578/1/012063

Synthesis of silica from rice husk as cement substitution for high strength concrete

D Dhaneswara, J F Fatriansyah, D P Kusumawardhani[1], S A Khairunnisa
Departement of Metallurgical and Materials Engineering, Faculty of Engineering, Universitas Indonesia, Kampus UI, Depok, Indonesia 16424

E-mail: dessy.prawesti@ui.ac.id

Abstract. This research was conducted to determine the effect of adding the rice husk ash (RHA) from mechano-chemical process as cement replacement for concrete. Mechano-chemical process was carried out using planetary ball mill with the rational speed at 22 rpm for 15 minutes and adding the 1M Hydrochloric Acid. This process aims to reduce the particle size of RHA. RHA will be used as cement replacement material with the composition that will be displayed in this study. concrete was made on 10 x 10 cm cube mold and a compressive strength test was carried out after 28 days curing. Beside RHA, in this study silica fused is used as cement replacement material. Concrete with the addition of RHA from mechano-chemical process has higher compressive strength than concrete with addition of silica fused.

1. Introduction

Indonesia is known as a country that has rice as stable food. The needs of rice as a source of stable food will be increasing along with the population growth. According to data from the Indonesia Ministry of Agriculture, in 2017 Indonesia produced about 81.382.451 tons of paddy. Paddy plants consist of rice grain as the main product and rice husk as by-product that can be separated by milling process. Rice husk from milling process has weight around 20 % of total rice weigth. As a by-product, rice husk are often left or burned directly in environment which can cause air pollution.

Rice husk as by product has around 90-98% silica after complete combustion [1]. Rice husk that has been burned is called rice husk ash (RHA). RHA can play role as silica source because of high value of silica after burning process. RHA can be called as Pozzolanic materials because it has high value of silica [2]. Several studies reported that RHA can be used as material for cement replacement in concrete making process. Silica contained in RHA can increase the mechanical properties of concrete like compressive strength and durability [3,4].

RHA as pozzolanic materials depends on silica value in RHA, crystallization phase of silica, size, and surface area of RHA. Controlled combustion can help to obtain more amorphous phase of silica and larger surface area of RHA. Some studies reported that combustion process below 800^{0}C can produce rice husk ash with mainly content of amorphous silica [5]. RHA with small particle size and large surface area can be obtained by grinding process. Grinding process with planetary ball mill can be used to reduce particle size of RHA

Beside using RHA, silica fuse also can be added in concrete making process. Silica fuse is made from high purity silica, using unique fusion technology, to ensure the highest quality. It has the same

[1] To whom any correspondence should be addressed.

effect of RHA because it has high silica content and amorphous silica. Around 85-99% amorphous silica can be found in silica fused [6].

In this study, we tried to combine the the mechanical and chemical treatment for RHA to investigate its effect on RHA particle size. Further investigation the RHA particle size is used to analyze its effect on mechanical properties of concrete. The variable we used are 22 rpm grinding speed and 1 M hydrochloric acid.

2. Experimental

2.1 Materials

2.1.1 Rice Husk Ask Preparation

The rice husk was obtained from Paddy Mill Industry,Leuwiliang, Bogor and was dried for 8 hours under sun. NaOH, HCl, distlilled water was obtained from C.V. Mitra Bersama and used as it is without any further purification.

2.1.2 Acid solution preparation

The Hydrochloric acid was obtained from CV Mitra Bersama and used without further purification. 1 M hydrochloric acid was obtained by diluting 20.72 ml of 37% HCl in 229,28 ml of water

2.1.3 Cement

The cement we used for this study is ordinary Portland cement type I from PT Indocement. This cement specifications conform to ASTM C 150 (Standard Specification for Portland Cement).

2.1.4 Aggregate

There were 2 types of aggregates used for this study. The fine aggregate is sand from Bangka province. The sand particle size range is 0 to 5 mm sieving grade. The coarse aggregate is gravel from Rumpin with particle size range from 10 to 20 mm.

2.1.5 Silica Fused

Silica fused was obtained from Chem&Size Minerals Limited, China. This silica fused would be used as cement replacement to make a concrete.

2.2 Methods

2.2.1 Grinding Proces

Rice husk was burned at 700°C for 5 hours. RHA from burning process was was left in the furnace for cooling process and removed the next day. Then, RHA was added 1 M hydrochloric acid and ground using planetary ball mill with rotational speed at 22 rpm for 15 minutes.

2.2.2 Sampling

RHA, cement, fine and coarse agregate were placed in correct proportions in the basin. They were mixed manually and slowly add water into the basin. The same procedure was also carried out to make concrete with addition of silica fused. Before the concrete is casted to mould, first moulds have to coated with oil to prevent concrete from sticking to moulds and makes it easier to remove. Then concrete samples were casted into 10 cm x10 cm moulds and compacted with rod and hummer to remove trapped air in the samples. After that samples were left for 24 hours.

In this study, the effect of adding RHA and silica fuse were futher studied. There were three samples in this study, concrete normal without addition of RHA (C-Normal), concrete with RHA that were ground at 22 rpm with addition of HCl (C-Treatment) and silica fuse (C-SF) as a replacement of cement.

Table 1. Composition of concrete

	Cement	Water	Sand	C.A	RHA
C-Normal	483	267	973	1200	0
C - Treatment	464.5	267	973	1200	18.5
C – SF	464.5	267	973	1200	18.5

3 Results and Discussion

3.1 Particle Size Distribution

Table 2. The result of Particle Size measurement

	Particle Size (μm)
RHA non Treatment	51.02
RHA Treatment	23.24
Silica Fused	33.20

Particle size of RHA was measured using Particle Size Analyzer. Based on table 1 and Figure 1, the average particle size of RHA non treatment is 51.02 μm, RHA treatment is 23.24 μm, and silica fused is 33.20 μm. RHA non treatment has largest particle size. The large particle size can give an adverse effect. This adverse effect due to particle failure in filling the void in concrete which can reduce the compressive strength. The improvement of RHA non treatment can be done by mechano-chemical process using planetary ball mill and addition of Hydrochloric acid. Chemical treatment with hydrochloric acid can higher content of silica and lower content of impurities also chemical treatment can affect particle size of RHA because acid can react with the RHA and diminish the intercalation between the particles [7]. This process proved that mechanical and chemical mixing process was sucessful in reducing particle size. In this study, RHA from mechano-chemical has smaller particle size than silica fused. Smaller particle size can improve compressive strength because it can make a excellecent pozzolanic reactivity and ability to fill the void in concrete [8,9].

3.2 Compressive Test

Compressive test were performed on samples after curring process for 28 days. The samples were given constant progessing force. The maximum strength of concrete were shown when the concrete changes its shape due to failure.

Table 3. The result of Compressive Strength Test of Concrete

Concrete	Compressive Strength (MPa)	Compressive Strength Average (MPa)
C-Normal	27.23 24.14 27.77	26.38
C-Treatment	30.4 26.41 28.78	28.53
C-SF	28.24 29.75 27.23	28.41

RHA non treatment has lower properties than RHA treatment and SF like larger particle size and smaller surface area. It would make lower compressive strength due to larger particle size of RHA. So RHA non treatment is not suitable for direct use as a cement replacement in concrete making process. RHA non treatment needs additional process to improve compressive strength of concrete. In this study, mechano chemical was used to improve the properties of RHA. The result was higher compressive strength of C-Treatment. The improvement occured because of reduction of particle size as a mechano chemical process effect. The particle size could be changed into smaller size by increasing the grinding time [10].

Based on table 3, C-Treatment had highest compressive strength than other concrete. C-Treatment had higher compressive strength than C-Normal. The result proved that addition of RHA could improve the compressive strength of concrete. C-Treatment had higher compressive than C-SF. This study has same result with study of Srinivasreddy [11]. This can be due to smaller particle size and bigger surface area which can generate excellent pozzolanic reactivity and fill the void in concrete. C-SF had lower compressive strength, it may be caused by larger particle size that could cause failure to fill the small void in concrete. The higher compressive strength with smaller particle size can prove that RHA as the result of mechano chemical process can be used as silicafused replacement in concrete making process.

4 Conclusion

The addition of RHA can improve the compressice strength of concrete. But RHA can be used directly because it has larger particle size. RHA needs addtion process, mechano-chemical process, to reduce the particle size. Mechano-chemical process is proven to reduce the particle size of RHA from 51.02 μm to 23.24 μm. Based on study, concrete with addition of RHA from mechano-chemical process has highest compressive strength than normal concrete and concrete with addition of silica fused.

References

[1] Todkar B S, Deorukhkar O A, and Deshmukh S M 2016 *IJERD* **12** (3) 2278–67
[2] A Kumar, K Mohanta, D Kumar, and O Parkash 2012 *Int. J. Emerg. Technol. Adv. Eng.* **2** 86–90
[3] Kulkarni M S, Mirgal P G, Bodhale P P, and Tande S N 2014 *JCEET* **1** (1) 26–29
[4] Salavati-Niasari M, Javidi J, and Dadkhah M 2013 *Comb. Chem. High Throughput Screen.* **16** (6) 458–462
[5] Habeeb G A, and Mahmud H Bin 2010 *Mater. Res.* **13** (2) 185–190
[6] Siddique R 2011 *Resour. Conser. Recy.* **55** 923–932
[7] Santana Costa J A, and Paranhos C M 2018 *J. Clean. Prod.* **192** 688–697
[8] Ferraro R M, and Nanni A 2012 *Constr. Build. Mater.* **31** 220–225
[9] Rukzon S, Chindaprasirt P, and Mahachai R 2009 *Int. J. Min. Met. Mater.* **16** (2) 242–247
[10] Bakar B H A 2011 *Mag. Concrete Res.* **63** (5) 313–320
[11] Srinivasreddy A B, Mccarthy T J, and Lume E 2013 *26th Biennial Concrete Institute of Australia's National Conference (Concrete 2013)* 1–10

IOP Conf. Series: Materials Science and Engineering **578** (2019) 012064 doi:10.1088/1757-899X/578/1/012064

Effects of Martempering Heat Treatment on Corrosion Rate in 1.5Ni Laterite Steel

S Herbirowo[1], D Aryanto[2], A L Indharto[3], I Saefuloh[3], B Adjiantoro[1]

[1]Research Center for Metallurgy and Materials – Indonesian Institute of Sciences, 470 building PUSPIPTEK. Tangerang Selatan, Indonesia
[2]Research Center for Physics – Indonesian Institute of Sciences, PUSPIPTEK. Tangerang Selatan, Indonesia
[3]Mechanical Engineering, Universitas Sultan Ageng Tirtayasa, Cilegon

*E-mail : satr006@lipi.go.id

Abstract. Laterite steel is commonly used for applications in the industrial worldwide. The behavior of this characteristics of laterite steel can be repaired in various ways, one of them by special heat treatment. This research aims to determine the value of corrosion rates, hardness, and transformation in microstructure that occur in laterite steel contain 1.564% Ni after heat treatment. Heat treatment carried out is austenizing at 950 °C and then quenching water; oil and normalizing in the air. And also tempering temperature variations 100°C, 200°C, and 300°C. Tests carried out were CMS 100 corrosion test which is 3.5% NaCl solution media, hardness testing with Vickers scale, and metallographic testing using an optical microscope. The results obtained showed that a microstructure transformation in 1.564% Ni laterite steel after HT consisted of martensite, ferrite, pearlite, residual austenite, and bainite. The lowest value of corrosion rate obtained in the sample normalizing 300°C tempering air and the highest in raw material. Whereas the highest hardness value was obtained 459.7 HV in the non-tempered water quenching sample.

1. Introduction

The development of laterite steel is currently being discussed by various research institutions, one of which is the Indonesian Institute of Sciences Center for Metallurgy and Materials Research for its potential to become national steel. According to Barkas in Hasbi et al 2016 study, in general, laterite ore is used as an alloy in the manufacture of stainless steel, superalloys and metal alloys applications to increase the corrosion resistance. With 35-45 % iron content and 0.8-1.5 % nickel, after being processed into steel, laterite nickel ore in the form of Nickel Pig Iron (NPI) will produce steel with a nickel content of 2-3 %[1].

In addition to Fe and Ni content, laterite steel also contains Chromium and Manganese elements, where both elements will have an impact on the properties of steel surface hardness, and corrosion resistance[2]. Low alloys steel is widely used as the main material in construction steel and railway steel because of its long service life, corrosion resistance, easy coating treatment, and economical[3].

Annisa et. al. explained that the optimum hardness value was found in martensitic stainless steel 13Cr which experienced austenization at 1050 °C with tempering temperature 600 °C. The high tempering temperature rise in stainless steel 13Cr does not have a linear negative effect on the wear resistance of the steel. So in this study using low tempering temperature[4]. The purpose of this study was to determine the effect of quench media variations and tempering temperature to characterize hard properties, corrosion rates and metallography of laterite nickel steel.

2. Experimental

Research experiments began with the preparation of plate cast raw materials with cutting and milling based on the standards of hard testing, corrosion and metallography. Then heating austenization at 950 °C with 60 minutes holding time and cooled with 3 different media such as air, water and oil which then heated tempered at a temperature of 100; 200; 300 °C see the effect of the steel characteristics.

Characteristics of mechanical properties using hardness Vickers using loading 5kgf methods, for corrosion testing with Corrosion Measurement System 100 instrument and metallographic testing with grinding, polishing preparation and special etching solutions for Fe alloys with optical microscope analysis tools magnification 500x.

3. Results and Discussions

Chemical composition testing is carried out to determine the content of chemical elements contained in laterite-based alloy cast steel resulting from the casting process with material balance and integration calculations.

Table 1. Chemical composition of the experimental steel

Type	**Chemical Element**								
	C	Si	Mn	P	S	Cr	Ni	Mo	Fe
As-Cast	0.203	0.24	0.3088	0.01	0.0075	0.4520	1.564	0.001	Bal.

From the the chemical composition testing of alloy steel, it can be seen that the main chemical composition contained in the laterite steel samples of As-Cast is (Ni) = 1.564 %, Ni is the main constituent element. From the results of this test the carbon content (C) = 0.203 % with this carbon content can be concluded that this steel is categorized in medium carbon steel and low alloy steel, medium carbon steel that is steel containing carbon elements from 0.10% - 0.30% approached the modified steel standard AISI 4140.[5], [6]

Corrosion measurement testing aims to determine the effect of variations of quench media in laterite steel and variations on temperting temperature.

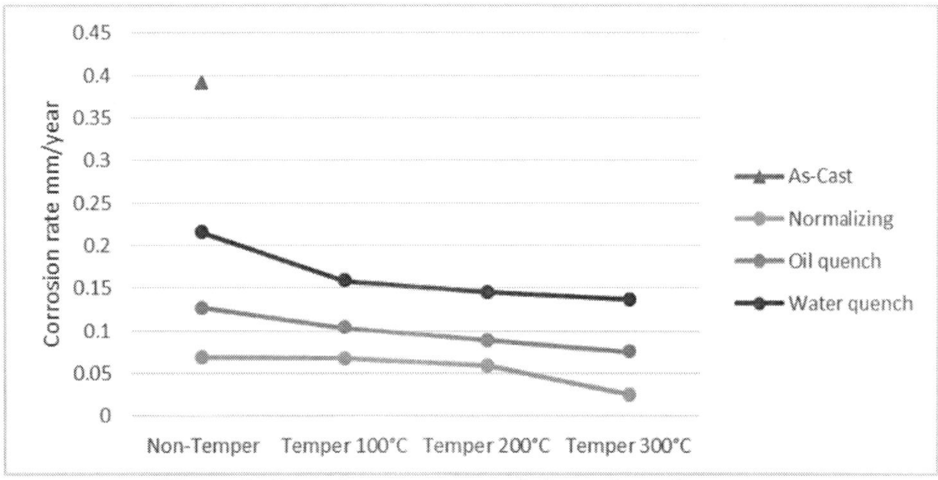

Figure 1. Corrosion rate test result for variety of treatment

From The Figure 1, it can be concluded that the cooling medium affects the corrosion rate. Because with a very rapid decrease in temperature, in the dyeing medium, it also provides a high residual stress on the material of the test material. Therefore, the material dipped in water has a higher corrosion rate than that dipped in oil and allowed to stand in the air. Tempering treatment also affects the corrosion rate, because given tempering treatment, the residual stress after the heat treatment process is reduced.

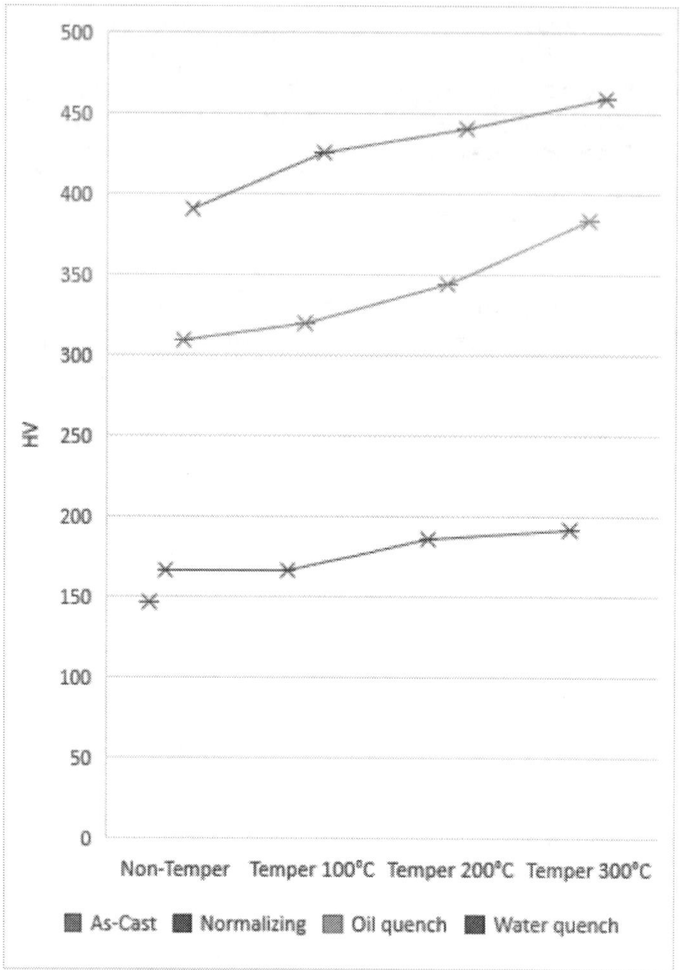

Figure 2. Hardness Vickers test result

In Figure 2 showed that the cooling medium affects the material's hardness. Due to immersion into water, the microstructure of the material is transformed into hard martensite. While the moderate temperature reduction (dipping into the oil), the microstructure of the material turns into a pretty hard bainite. And in the air normalizing grain refinement only occurs without a change of phase, ie the combination of pearlite and ferrite phase which makes the material is soft [7]. Besides the tempering process also affects the hardness of the material. The higher the tempering temperature is given, then the material hardness obtained will be lower.

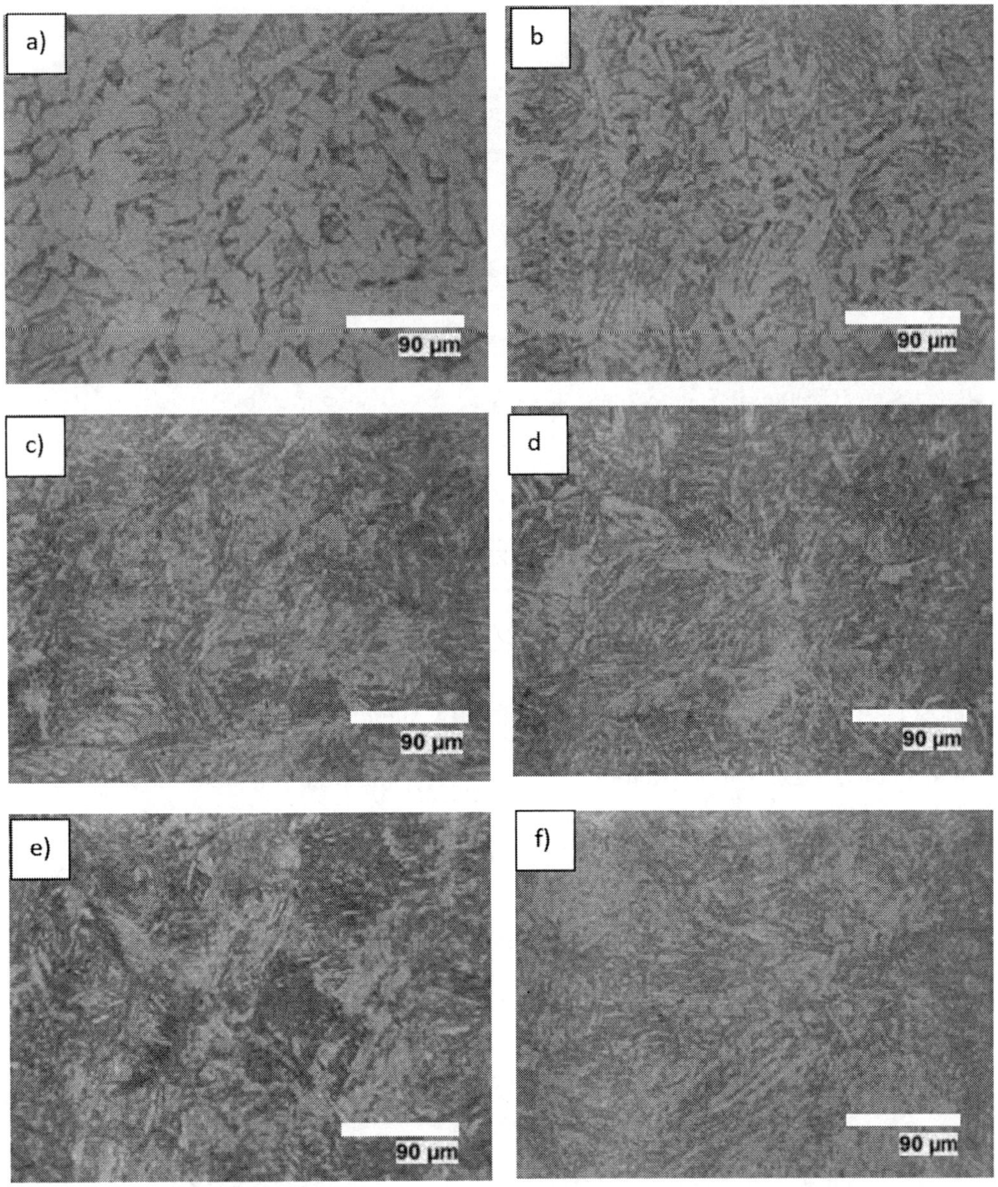

Figure 3. The microstructure of lateritic steel a) normalizing non-temper b) normalizing temper 300°C; c) oil quench non-temper d) oil quench temper 300°C; e) water quench non-temper; f) water quench temper 300°C

Figure 3 show The microstructure formed in the normalizing samples is ferrite and pearlite. The grain size still looks great because in this sample is still in effect casting product with a percentage of pearlite phase 20.03% - 29.56%. The percentage of the pearlite phase affects the hardness of samples, the higher the percentage of perlite the hardness will increase[8].

Grain structure formed has a smaller form factor and dense compared to the grain structure of the as-cast sample. Microstructure formed in the test sample is the oil quenching bainite structure, the bainite percentage of 45.36% - 50.49%. Microstructure formed in the test sample is a water quenching martensitic structure, with martensite percentage of 42.36% - 56.90%. The percentage of bainite and martensite phase affects the hardness value of the sample, the higher the percentage of the phase, the higher the hardness value of the sample [9,10].

4. Conclusion

The higher the tempering temperature, affecting the decrease in residual energy that occurs. So that it affects the value of the corrosion rate. The percentage of test material hardness values tend to increase, and the increase that occurred in the test material is different when compared to the as-cast. In the air normalizing test material an increase of 13.36% - 31.22%. In the oil quenching test material by 110.97% - 161.96%. And on the water quenching test material by 166.53% - 213.36%. Variations of the most influential in this test is the result of quenching medium cooling rate different test objects.

Acknowledgement

The authors would like to acknowledge research facility from Research Center for Metallurgy and Materials, Indonesian Institute of Sciences and this research was supported by Ministry of Research, Technology and Higher Education, Indonesia under Grant No. 13/E/KPT/2018.

References

[1] Herbirowo S, Hakim L and Adjiantoro B 2018 *MATEC web Conf.* **204(05007)** 1.
[2] Jena PK *et al.* 2010 *Int. J. Impact Eng.* **37(3)** 242.
[3] Herbirowo S, Adjiantoro B, Romijarso TB and Widya A 2018 *AIP Conf. Proc.* 020043
[4] Annisa Siti Apriania EM, Anwarb MA and Rusnaldya 2018 *Maj. Metal.* **3(2017)** 115.
[5] Hadi S, Adi S and Syariati A 2018 *Civil Eng. Journal* **4(2)** 294.
[6] Suryo SH, Bayuseno AP, Jamari J and Rahmat MA 2018 *AIP Conf. Proc.* **030022** 1977
[7] Praguna FD, Anwar MS, Mabruri E and Sunardi 2018 *J. Sains Mater. Indones.* **19(3)** 125.
[8] Prifiharni S, Anwar MS and Mabruri E 2016 *Widyariset* **2(1)** 9.
[9] Dwiwandono R, Firmansyah L, Herbirowo S, Hasbi MY and Citrawati F 2017 *Metalurgi* **32(2)** 67.
[10] Fan H, Zhao A, Li Q et al. 2017 *Int J Miner Metall Mater.* **24** 264

IOP Conf. Series: Materials Science and Engineering **578** (2019) 012065 doi:10.1088/1757-899X/578/1/012065

A comparative study between the addition of NH_4Cl and KCl in $NaNO_3$ and NaF-based degassers in aluminum casting

D Dhaneswara[1], J F Fatriansyah[1], A R Basa[1]
[1]Departement of Metallurgical and Materials Engineering, Faculty of Engineering, Universitas Indonesia, Kampus UI, Depok 16424, Indonesia

E-mail: agatha.rona@ui.ac.id

Abstract. The quality of Al-Si casting product is important for its applications, however the molten aluminum has high solubility of H_2 at melting temperature in the casting process, so the formation of gas porosity during aluminum solidification is inevitable. This can cause a major drawback to the mechanical properties. The most common method to remove hydrogen gases in the molten metal is degassing. In this experiment, a conventional degassing method using Sodium Nitrate-Sodium Fluoride based degasser is conducted. Al-Si12 was melted at 720 °C then the degasser was plunged into the furnace bottom and held for three minutes. The results show better mechanical properties by the addition of degasser. This is because the degasser binds H_2 gases which is formed in the melting process. The mechanical properties between casting products that use NaF-$NaNO_3$ degasser and NaF-$NaNO_3$ with addition of NH_4Cl or KCl were measured and compared. The lowest mechanical properties were obtained when using degasser without NH_4Cl or KCl. This is because NH_4Cl or KCl is reactive and helps with the efficiency of degasser.

1. Introduction

Aluminum is the most widely used non-ferrous metal in the world being present in sectors as diverse as transportation, packaging, construction, electricity, and medicine [1]. The use of cast aluminum alloys are increasing in the automotive industry due to their significant advantages of lightweight, attractive appearance, excellent processability and corrosion resistance [2,3]. However, in the industry, aluminum is often used as an alloy because aluminum itself is not particularly strong. One of the alloying element often used is silicon, and the silicon content normally varies from 5 to 12 wt%. Al-Si cast alloys have been used widely to produce automotive components because of the excellent characteristics, one of those is excellent castability. The higher content of silicon in aluminum, the lower is its thermal expansion coefficient. Silicon is also a very hard phase, so its presence in the aluminum alloy will improve the alloy's wear resistance significantly [4].

However, numerous studies shown that cast aluminum alloys are very sensitive to casting defects and the presence of large pore near the specimen's surface will be the dominant cause of fatigue crack initiation [5,12]. Molten aluminum is prone to hydrogen adsorption and oxidation, so gas porosity and oxide inclusions are inevitably found in aluminum castings [2]. Porosity is one of the major defects on aluminum alloy casting and its presence causes a major drawback to the mechanical properties of the casting products. Porosity can either formed by hydrogen precipitation in the liquid solution during solidification, or the inability of the liquid metal to feed through the interdendritic regions [6,7], or other factors like mold-metal reaction, high temperature oxidation, blowholes, or the combinations of these effects [10,11]. Hydrogen is the only significant gas soluble in molten aluminum [6,7,8], so it is important to control the dissolved hydrogen. Once solidification of the alloy progresses, any excess hydrogen, not soluble in the solid and not able to escape the solidified section, will form porosity within the casting, and reducing its properties [4].

Some effective ways to reduce hydrogen porosity is through degassing and fluxing [6]. These methods include the application of nitrogen, argon, or even a mixture of both with chlorine and hexachloroethane (C_2Cl_6) tablets [6,9]. In a smaller industry, degassing is usually conducted by using tablet degasser, and the common tablet degasser is from hexachloroethane (C_2Cl_6) [13] which generates Cl_2 and gaseous $AlCl_3$ [14]. But the use of C_2Cl_6 was prohibited in EU in 1998 because it is also moderately irritating to the skin, and classified as possible human carcinogen [14]. Thus, C_2Cl_6 tablet degassing has been replaced by degassing with dry nitrogen or argon using a lance.

In this experiment, degassing is conducted using tablet degasser, but this degasser is sodium nitrate-sodium fluoride based. In this paper, the tablet degasser was added with ammonium chloride (NH_4Cl) or potassium chloride (KCl). From there, we can determine which one is better at yielding higher mechanical properties from the casting products. The expected result is that by adding NH_4Cl or KCl into the degasser, the mechanical properties of casting products will be improved. This is because the degassing and fluxing can improve the cleanliness of the molten alloy [14].

2. Experimental

2.1 Materials

The degasser used here is mainly a combination of $NaNO_3$ in powder form and NaF with the concentration of 18 wt% and 40 wt% respectively. The other 42 wt% consists of NaCl, NH_4Cl or KCl, Na_2SO_4, and food colorants. In this study some of the tablet degassers contain NH_4Cl or KCl, as shown on Table 1.

Table 1. Chemical composition of $NaNO_3$-NaF based degasser.

Composition	Degasser 1 Weight (kg)	Degasser 2 Weight (kg)	Degasser 3 Weight (kg)
$NaNO_3$	0.8	0.8	0.8
NaF	1.25	1.25	1.25
NaCl	1.6	1.6	1.6
Na_2SO_4	1.2	1.2	1.2
NH_4Cl	0.175	0	0
KCl	0	0	0.175

2.2 Methods

2.2.1. Making of $NaNO_3$ and NaF based degasser. The materials for degasser were then mixed using mixer with the addition of water, and the stirring time is about 30 to 60 minutes so the mixture is well homogenized. Furthermore, the mixed material then processed and formed into a cylindrical tablet shape using a printing machine. Then this degasser tablet was put into the oven at temperature 100 °C for about 60 minutes to evaporate the moisture content so we achieve the tablet degasser with minimum moisture content. The last step if the finishing process of the tablet degasser where it was cut, or polished, and eventually packed.

2.2.2 Casting process of Al-Si12. The chemical composition of Al-Si12 was obtained from conducting Optical Emission Spectrometry (OES) testing and the result was shown in Table 2. The main component is aluminum with majority alloying element is silicon.

Table 2. Chemical composition of the experimental Al-Si12 (wt%).

Al	Si	Fe	Cu	Mn	Mg	Zn	Cr
87.3	12.4	0.0938	0.0015	0.0045	0.0003	0.0024	0.0048

The casting procedure started with melting the Al-Si12 in an electric furnace at 720 °C until the material was melted completely. While the materials were melted, the mold was coated with a mixture of zircon and thinner then preheated at 350 °C for 15 minutes. The tablet degasser was also preheated using a muffle furnace at 100 °C for 1 hour to make sure there was no moisture content left in the degasser. After the aluminum alloy melted completely, the preheated degasser then plunged into the bottom of the furnace using a coated plunger, and held until the tablet degasser totally dissolved and mixed with the molten metal which takes about 3 - 5 minutes. This degassing process was effective in removing the gas inside the molten aluminum which then bring the undesired hydrogen gas up to the surface of the molten metal. Next, this degassed molten metal was poured into a metal mold at pouring temperature 690 °C and also into a porosity testing cup. While the molten metal inside the mold underwent solidification, the one in the porosity testing cup was put into a vacuum condition at a constant pressure of 100mmHg for 315 seconds using Ostek Motorized Porosity Tester. The schematic casting design can be seen in figure 1.

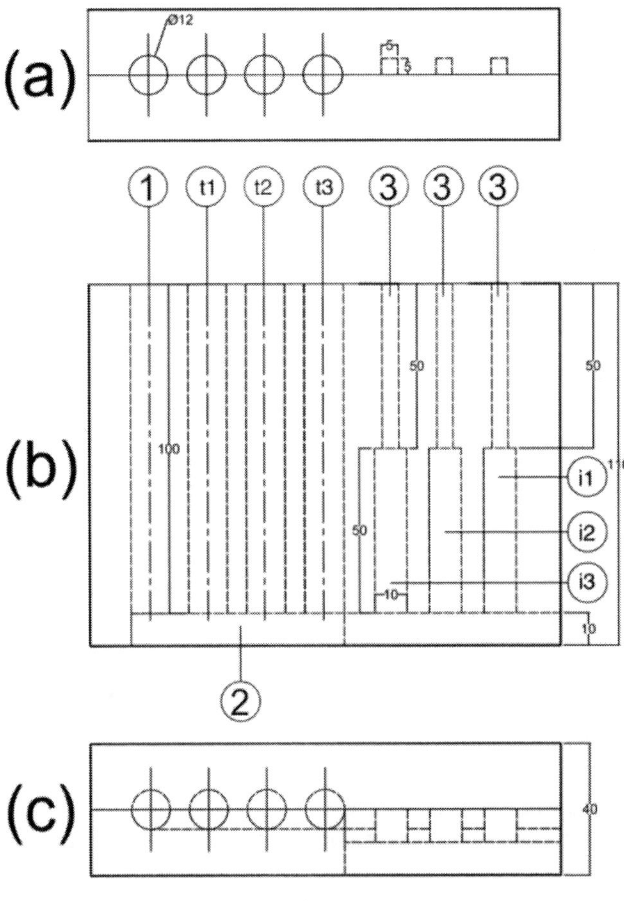

Figure 1. Schematic of casting design. (a) Top View; (b) Front View; (c) Bottom View; (1) Down Sprue; (2) Runner; (3) Gas Tunnel; (t1, t2, t3, t4) Tensile Specimens; (i1, i2, i3) Impact Specimens.

IOP Conf. Series: Materials Science and Engineering **578** (2019) 012065 doi:10.1088/1757-899X/578/1/012065

From this mold we can obtain two types of testing specimen, including the tensile test specimens obtained from t1, t2, and t3, and also impact test specimens obtained from i1, i2, and i3. On the far left side is the down sprue with a diameter of 12 mm, which is where the molten metal was poured. Show as number 2 is the runner, which is where the molten metal flows and will fill in all of the specimen's casting. The other component of the mold is gas tunnel which placed in the top right side of the mold as the place for removing gas from within the mold, here shown as number 3. After the casting process was completed, the specimen then prepared and observed to investigate the mechanical properties of the casting products.

3. Results and Discussion

3.1 Mechanical properties of casting Al-Si12 with degasser

The mechanical properties of the Al-Si12 casting products were characterized by tensile test, impact test, hardness test and microstructure observation. The tensile test was conducted using GOTECH AI-7000 LA 10 machine with JIS Z2241 standard. In order to obtain the impact value, the impact test was conducted using Frank impact testing machine with the charpy method. The hardness test was conducted by using LCB3100 LECO brinell hardness tester with 10 mm diameter of steel ball indenter and the force parameter of 1000 kgf. For the porosity testing sample, they were solidified in a vacuum condition using Ostek Motorized Porosity Tester, then quenched in water to fasten the solidification of the remaining liquid metal. This sample was cut, prepared and then observed to investigate the microstructure, so the amount of gas porosity in the sample can be determined.

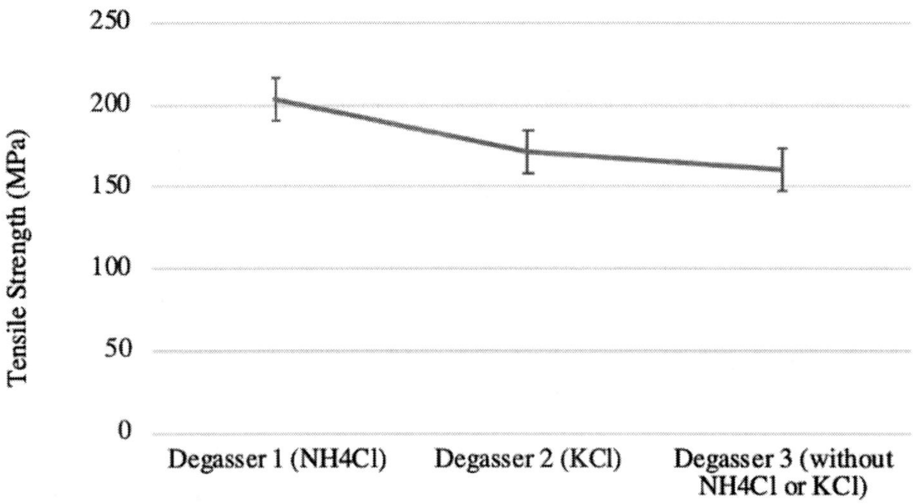

Figure 2. Effect of the addition of NH_4Cl and KCl in $NaNO_3$ and NaF based degasser on tensile strength of Al-Si12.

After conducting the tensile test on each specimen, we obtained the results, as shown in figure 2. The addition of NH_4Cl and KCl surely plays a significant role in improving the tensile strength of Al-Si12 casting products. From figure 2, we can see that the highest tensile strength was obtained when adding NH_4Cl to the tablet degasser. Meanwhile, the lowest tensile strength value was obtained with the basic tablet degasser, without the addition of NH_4Cl or KCl. Reasonably, reducing the hydrogen content will certainly improve the mechanical properties of the resultant aluminum alloys. The effect of degassing on the quality of aluminum alloys include not only controlling the hydrogen content but also the resultant quality of the melt and its cleanliness [14]. Therefore, with the addition of NH_4Cl or

KCl in this degasser helps to achieve a better aluminum casting products quality, and from this tensile result, the addition of NH_4Cl resulted in the best tensile strength, means it works the best on removing the porosities of the Al-Si12 casting products.

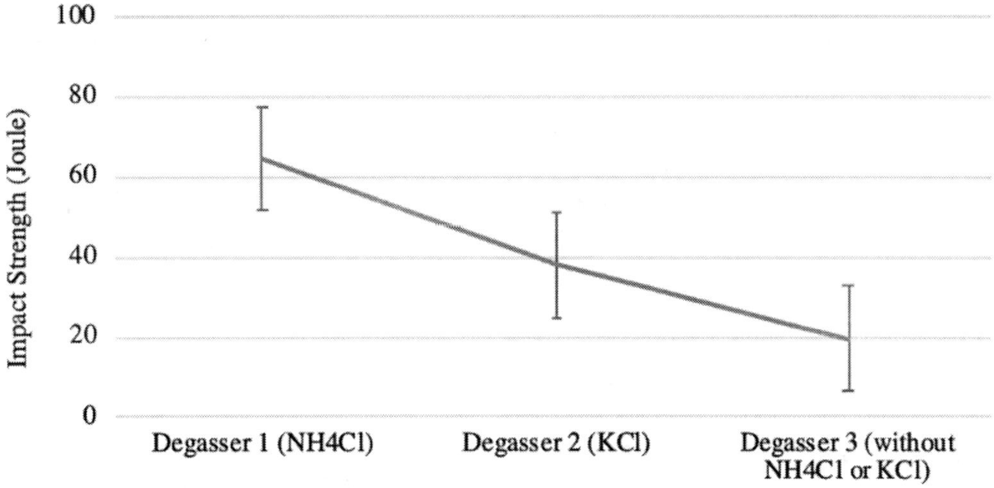

Figure 3. Effect of the addition of NH_4Cl and KCl in $NaNO_3$ and NaF based degasser on impact strength of Al-Si12.

Figure 3 above shows the effect of addition of NH_4Cl or KCl in the degasser on the impact strength. From this result we can also see that by the addition of NH_4Cl the impact value is increased and the lowest value is from the degasser without any addition of NH_4Cl or KCl. The highest value is indeed obtained when using the addition of NH_4Cl because it helps to eliminate gas porosities and other oxide inclusions, resulted in clean molten metal, thus able to withstand any sudden load, and has higher impact value.

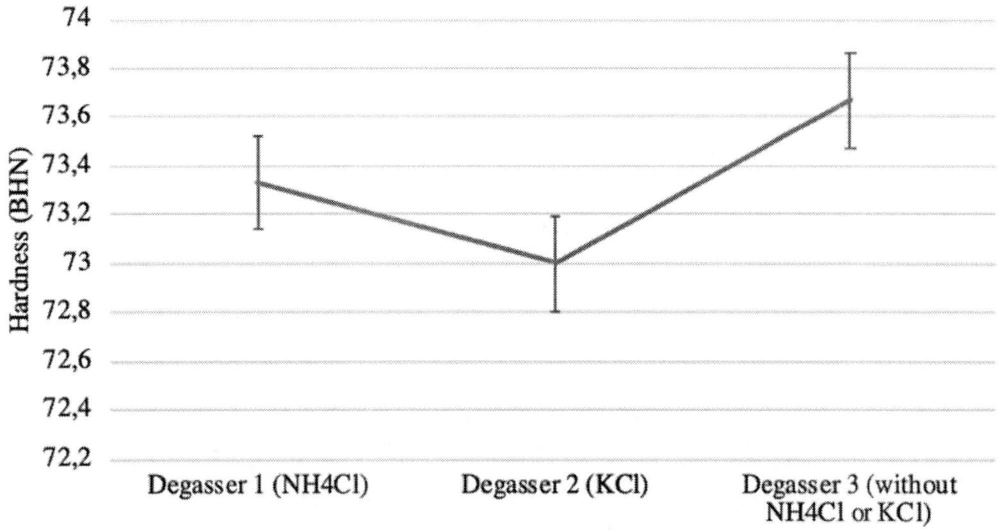

Figure 4. Effect of the addition of NH_4Cl and KCl in $NaNO_3$ and NaF based degasser on hardness of Al-Si12.

The result of hardness test was shown in Figure 4 above. It shows that the highest hardness value was obtained from the degasser without the addition of NH_4Cl or KCl, but still the hardness values are relatively the same, around 73 BHN, and there are no significant changes. The sample produce relatively high hardness values because all of them use degasser. Effect addition of degasser is to reduce the porosities and increasing the hardness [15], and from this result it is also proved.

3.2 Microstructural Observation of Al-Si12

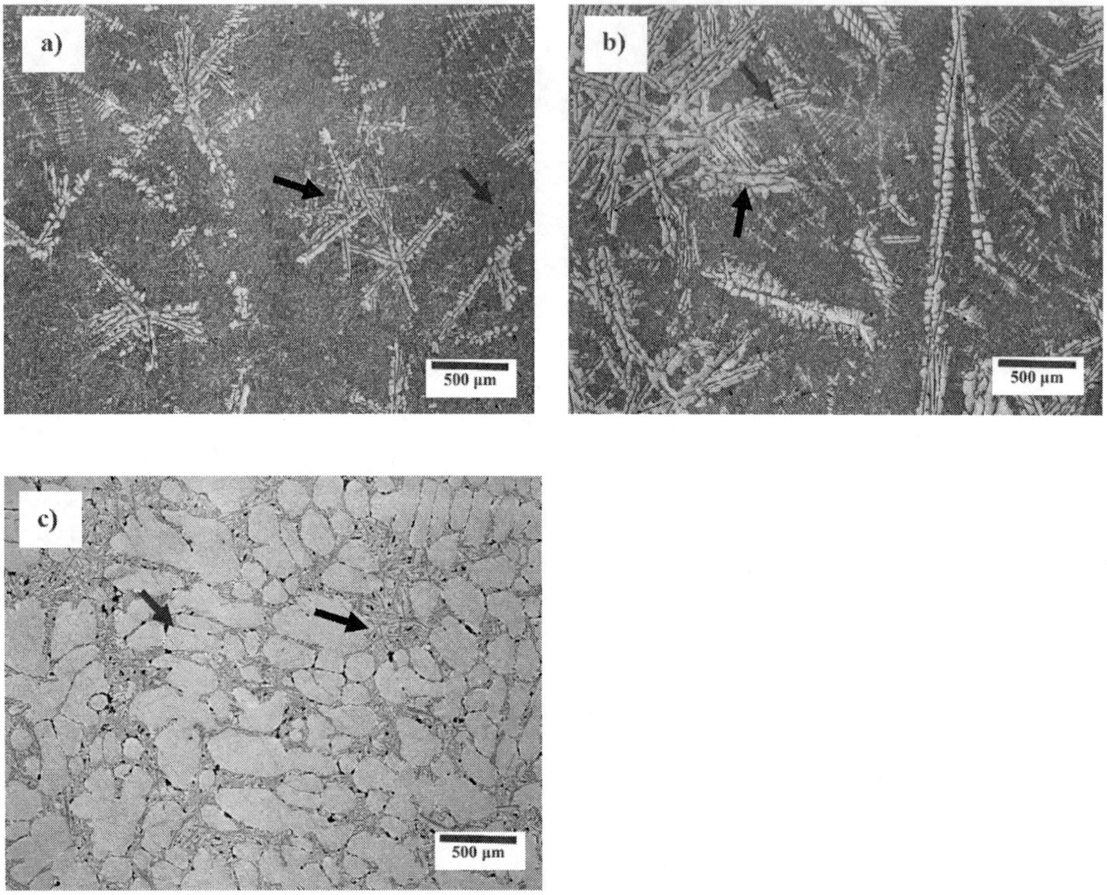

Figure 5. Microstructure of Al-Si12 casting product with variation of degasser (a) with the addition of NH_4Cl; (b) without the addition of NH_4Cl or KCl; (c) with the addition of KCl.

Figure 5 shows the optical micrograph results for all the tablet degasser variations. This microstructure affects the mechanical properties of the alloy. Al-Si solidify by a primary precipitation of dendrites. Dendrites are often drawn having about four secondary arms growing around the primary stem at each junction [4,16]. The illustration of primary aluminum dendrite (α-Al) is shown in figure 5 pointed by the black arrow, and the black dots that are pointed by the red arrows are the porosities that occurred in the casting products. From the microstructure we can see that the number of porosities formed in the casting product is relatively low. It is possible because the tablet degassers are effective to remove gas. By comparing all the microstructures from figure 5, we can see the lowest amount of porosity was obtained by the addition of NH_4Cl to the tablet degasser.

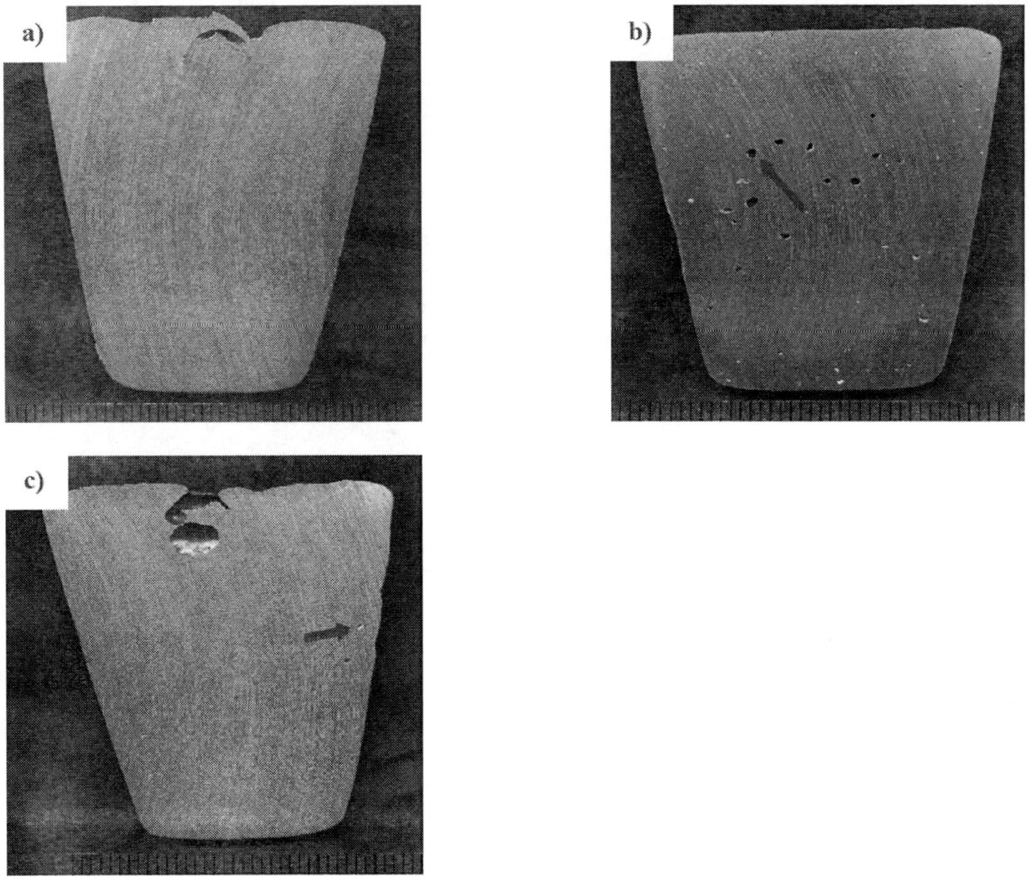

Figure 6. Macrostructure of Al-Si12 casting product with variation of degasser (a) with the addition of NH_4Cl; (b) without the addition of NH_4Cl or KCl; (c) with the addition of KCl

Figure 6 shows the macrostructure appearance of Al-Si12 with all the tablet degasser variations. From the macrostructure, it could be seen that every cast result has low amount of gas porosities, but on figure 6b it shows much more gas porosities than the others, which is true because it resulted from the basic degasser without the addition of NH_4Cl or KCl. Meanwhile the macrostructure on figure 6a shows a very low even almost nonexistent content of porosities, due to the addition of NH_4Cl to the tablet degasser. It could be concluded that the addition of NH_4Cl in the tablet degasser works more effective in removing gas because of the micro and macrostructure, which also correlates to the mechanical properties results.

3.3 Effect of $NaNO_3$ and NaF tablet degassing with addition of NH_4Cl or KCl

Gas porosities and oxides are two common defects in Al-Si which are initiated through poor melt treatment or non-optimized casting route. When the molten aluminum meets with oxygen, it will immediately oxidize forming a skin of oxide. If the oxide skin is removed or cracked, a new oxide layer will immediately form. This oxide film has an amorphous structure and low permeability, therefore creates a protective layer at the surface of the liquid aluminum. As long as the oxides are at the surface of the molten metal, they do not present any significant threat to the quality of casting product.

However, melting, pouring and transferring liquid metal will most certainly entrain newly formed thin oxides into the melt making them possibly harmful [4,17]. The most effective way to solve this is by degassing and fluxing, and the simplest degassing method is by using tablet degasser. In this study, we use $NaNO_3$ and NaF based degassers, and the chemical reaction with hydrogen gas can be inferred in equation (1) and (2).

$$2\ NaNO_3 + H_2 \rightarrow 2\ Na + 2\ HNO_3 \quad (1)$$
$$2\ NaF + H_2 \rightarrow 2\ Na + 2\ HF \quad (2)$$

The tablet degasser when plunged under the metal surface will produce nitrogen gas. The presence of nitrogen gas and bubbles will bind hydrogen gas and carry it to the surface of molten aluminum. The degasser can also remove impurities within the molten aluminum and carry it to the surface as a dross.

The addition of NH_4Cl or KCl in the degassing tablet is also used as a fluxing agent. Solid fluxes are mainly blends of chloride and fluoride salts, which forms a low-temperature (665 ^oC) eutectic. A low melting point is important since it will improve the fluidity of the flux [14]. These salt fluxes are often used to eliminate or reduce the formation of oxides, so we can obtain a clean molten metal which also gives high mechanical properties. The mechanism for cleaning the metal is to wet the oxides with flux so that they will either float or sink, but will not stay suspended in the melt.

4. Conclusion

From this experiment, the use of $NaNO_3$ and NaF based degasser in Al-Si12 casting process produces a relatively low amount of gas porosity. From the mechanical properties test, this low amount of porosity also improves the mechanical properties such as tensile strength, impact strength, and hardness. The addition of NH_4Cl or KCl in the tablet degasser also has better results than the basic $NaNO_3$-NaF degasser. By comparing all the results, we obtain the best mechanical properties from using $NaNO_3$-NaF degasser with the addition of NH_4Cl, due to the least amount of gas porosity based on the microstructure result. It is possible because the addition of NH_4Cl not only remove the hydrogen gas, but also reduce oxidation and other impurities.

Acknowledgements

The authors gratefully acknowledge financial support provided by Hibah PITTA with contract No: 2421/UN2.R3.1/HKP.05.00/2018.

References

[1] Capuzzi S, Timelli G, Capra L and Romano L 2017 *Int. J. Sust. Eng.* 1–9
[2] Wang Q G, Apelian D and Lados D A 2001 *J. Light Met.* 73–84
[3] Cao H, Hao M, Shen C and Liang P 2017 *Vacuum* **146** 278.
[4] Zamani M 2015 *JTH Diss. Series* **21**
[5] Zavodska D, Tillova E, Guagliano M, Kucharikova L and Chalupova M 2017 *Mat. Today. Proceeding* **4**(5) 6001.
[6] Haghayeghi R, Bahai H and Kapranos P 2012 *Mater. Lett.* **82** 230.
[7] Campbell J 2003 *Casting 2nd Edition* (Oxford: Butterworth Heinemann Elsevier)
[8] Han Q and Viswanathan S 2002 *Metal Mater Trans A* **33A** 2067.
[9] Samuel AM and Samuel FH 1992 *J. Mater. Sci* **27** 6533.
[10] Li B, Shen Y and Hu W 2011 *Mater. Desg.* **32** 2570.
[11] Kaufman J G and Rooy EL 2004 *Aluminum alloy castings: properties, processes, and applications. The influence and control of porosity and inclusions in aluminum castings* (Ohio: ASM International)
[12] Gao Y X, Yi J Z, Lee P D and Lindley T C 2004 *Acta Mater.* **52**(19) 5435.
[13] Utigard T A, Friesen K, Roy R R, Lim J, Silny A and Dupuis C 1998 *J. Min. Met. Mater Soc.* **50**(11) 38.
[14] Shih T S and Wen K Y 2005 *Mater. Trans.* **46**(2) 263
[15] Ceschini L, Morri A, Toschi S, Johansson S and Seifeddine S 2015 *Mater. Sci. Eng. A* **648** 340.
[16] Liu Z Y, Xu Q Y, Liu B C 2007 *Int. J. Cast Met. Res.* **20** 109.
[17] Campbell J 2011 *Complete Casting Handbook: Metal Casting Processes, Techniques and Design* (Oxford: Butterworth-Heinemann)

Effect of addition of sodium chloride in sodium nitrate-sodium fluoride-based degasser in aluminum casting

D Dhaneswara[1], J F Fatriansyah[1], M R Firmansyah[1]

[1]Departement of Metallurgical and Materials Engineering, Faculty of Engineering, Universitas Indonesia, Kampus UI, Depok, Indonesia 16424

E-mail : muhammad.reza53@ui.ac.id

Abstract. The effect of addition of Sodium Chloride (NaCl) in increasing degasser efficiency on Al-Si12% casting has been carried out. Al-Si 12% casting product mechanical properties are a very important parameter that indicates the quality of the casting product. However, it is a very common phenomenon that the properties of the casting product was reduced by gas porosity defects, which is formed due to high solubility of H_2 at melt temperature of Al-Si 12%. In this investigation, we used NaF and $NaNO_3$ as degasser with the addition of NaCl in the casting process. The addition of NaCl was variated to see the capability of NaCl as a degasser. The casting process were conducted at melting temperature of 800 °C and pouring temperature of 690 °C. The mechanical properties between casting products that use NaF-$NaNO_3$ degasser with variated ratio of NaCl addition were measured and compared. The result shows that mechanical properties of Al-Si12 cast product in this research will be decreased along with increasing number of NaCl addition ratio in degasser composition.

1. Introduction

Aluminum is one of the most used material in manufacturing industries such as transportation or construction industries because of its light weight properties. Unfortunately, this properties itself is not enough in industrial aspect, because of the nature of pure aluminum that has low strength. Therefore, it has to be alloyed with other elements like silicon to increase its mechanical properties, such as good castability, high strength-to-weight ratio, and good corrosion resistance [1-3]. Silicon usually was added up to 12%, and it will give good castability properties. The higher number of Si added into aluminum, its thermal expansion coefficient will be lowered. Also, Si was a hard phase, so addition of Si can increase aluminum alloy's wear resistance [4].

Porosity is one of the major defects in aluminum alloy casting that is detrimental to mechanical properties of the casting products [4-9]. The presence of porosity inside molten aluminum can be found in microstructure test that is represented by black hole on cast product. It can occur because of some factors, either because of gas precipitation in solidification, inability of the liquid metal to feed through interdendritic regions, or turbulent flow during filling process [5,6,8]. Hydrogen is considered to be the only significant gas dissolved in molten aluminum, so it is very important to prevent the precipitation of hydrogen gas inside molten aluminum to reach high quality of the casting product [7]. By increasing the temperature used in casting process, the number of hydrogen gas that can be trapped in molten aluminum will be higher. Hydrogen is the only gas that does not form compounds with aluminum, so it has high solubility in liquid aluminum, but very low solubility inside solid aluminum, so when casting product is solidified, it will expelled forming gas pores.

To eliminate hydrogen gas inside molten aluminum, several methods has been studied and applicated in industries. Fluxing and degassing are the most common methods that are used in aluminum casting. These methods can lower the number of hydrogen gas by binding the gas and then carried the slag into the surface in the furnace. There has been many studies about degassing methods, such as ultrasonic degassing, re-melting degassing, vacuum degassing, or tablet degassing.

In many researches, the most common tablet degasser that were used are tablet degassing by hexachloroethane (C_2Cl_6). C_2Cl_6 will decompose to form $AlCl_3$ gas bubbles that will collect hydrogen gas, but because of its tendency to produce noxious odor that will become an environmental problem, its production was prohibited by EU in 1998 [10]. Since the prohibition of C_2Cl_6 tablet degasser, degassing method has been replaced by another method such as purging, lance degassing, and many else.

In this paper, composition of tablet degasser used is changed into sodium nitrate-sodium fluoride-based degasser with the addition of NaCl. NaCl addition in tablet degasser composition was variated, with expectation that it can replace sodium nitrate-sodium fluoride as degasser's main composition, because NaCl has a lower price, so industries can set a lower cost to produce casting product but with same quality.

2. Experimental

2.1. *Making of Degasser*

The tablet degasser used is containing 18%wt $NaNO_3$ in powder forms, 40%wt NaF, and the other 42%wt consists of NaCl, NH_4Cl, Na_2SO_4, and dye. These ingredients were placed in container and then water is poured. Mixer machine was used to mix all of them together. The stirring process then conducted for approximately 30 minutes to help homogenized the mixture. After the mixing process is finished, the mixture then processed and formed into tablet shape using printing machine. The last process in this step is to heat the tablet degasser inside oven at 100 °C for 1 hour to evaporate moistures inside, mostly water.

For this research, the ratio number of NaCl that was added is variated into 4 different weight ratio, it is 0.2%*wt*, 0.4%*wt*, 0.8%*wt*, and 1.5%.*wt*. The purpose want to be achieved in this research is to know the effect of increasing ratio number of NaCl addition in degassing process.

Table 1. Chemical Composition of Tablet Degasser in Casting Process.

	Degasser 1	**Degasser 2**	**Degasser 3**	**Degasser 4**
Composition	Weight (Kg)	Weight (Kg)	Weight (Kg)	Weight (Kg)
$NaNO_3$	0.8	0.8	0.8	0.8
NaF	1.25	1.25	1.25	1.25
NaCl	0.4	0.8	1.6	3.2
Na_2SO_4	1.2	1.2	1.2	1.2
NH_4Cl	0.175	0.175	0.175	0.175

2.2 *Characterization of Al-Si 12*

Table 2. Chemical composition of Al-Si12 (%).

Al	Si	Fe	Cu	Mn	Mg	Zn	Cr
86.7	12.7	0.123	0.12	0.0186	0.0011	0.0428	0.0123

The chemical composition of material that were used was obtained by OES test and was shown in table 2 above. From the table, it can be seen that the main composition of the material are aluminum and silicon.

The casting process starts with melting the material in electric furnace at temperature 800 °C. While waiting for the material to completely melt, the mold and tablet degasser were being prepared. The mold is coated with zircon and then preheated up to 350 °C in muffle furnace, while the tablet degasser is preheated at 100 °C in an oven. After the material melted completely, tablet degasser was inserted into the furnace and pressed by plunger until it dissolved completely inside the molten aluminum. The inserting time of degasser that is used are 3 minutes. Degassing process is needed to remove hydrogen gas that is trapped inside, also it was conducted to remove slag layer on the surface of moltenaluminum. After the degasser and molten aluminum completely mixed, it was poured into mold and porosity test cup at pouring temperature of 690 °C. Poured metal inside mold then undergo solidification step, while poured metal inside porosity cup was tested in vacuum condition with constant pressure of 100 mmHg for 315 seconds. The schematic design of the mold can be seen in figure 1 below.

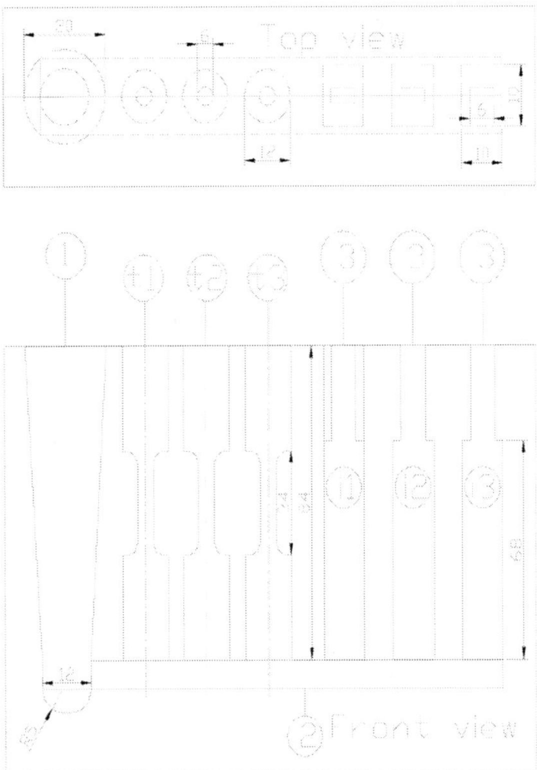

Figure 1. Schematic design of mold. (a) Top View; (b) Front View ; (1) Down Sprue ; (2) Runner ; (3) Gas Tunnel ; (t1,t2,t3) Tensile Specimens ; (i1,i2,i3) Impact Specimens.

The specimens for mechanical properties testing were obtained by using metal mold with design on figure 1. The tensile specimens can be obtained from t1, t2, and t3, while impact specimens can be obtained from i1, i2, and i3. The down sprue is located at the left side of the mold with diameter of 12 mm, and molten aluminum will flow into runner that is located at the bottom of the mold. The molten aluminum then will rise into tensile and impact area, and heat gas will be released from the mold via gas tunnel. After it solidified, the casting process was considered done, and the samples will be observed to investigate the mechanical properties of the casting products.

To characterize the mechanical properties of Al-Si12% as cast, 4 testing methods were chosen. It is tensile test, hardness test, impact test, and microstructure test. In order to get tensile test result, tensile

IOP Conf. Series: Materials Science and Engineering **578** (2019) 012066 doi:10.1088/1757-899X/578/1/012066

test was conducted by using GOTECH AI-7000 LA machine with JIS Z2241 standard. For hardness test, LCB 3100 LECO brinnel machine was used with 10 mm diameter of steel ball indentor and 1000 kgf force parameter, using ASTM E10 standard. Impact test was conducted by using Frank impact testing machine with charpy method, using ASTM E23 standard. Microstructure of Al-Si12% cast sample was obtained from SEM test, and it shows the phase that appears from casting process and also shows the number of porosity formed inside the molten aluminum.

3. Results and Discussion

3.1 Microstructure Observation and Phase Identification of Al-Si12

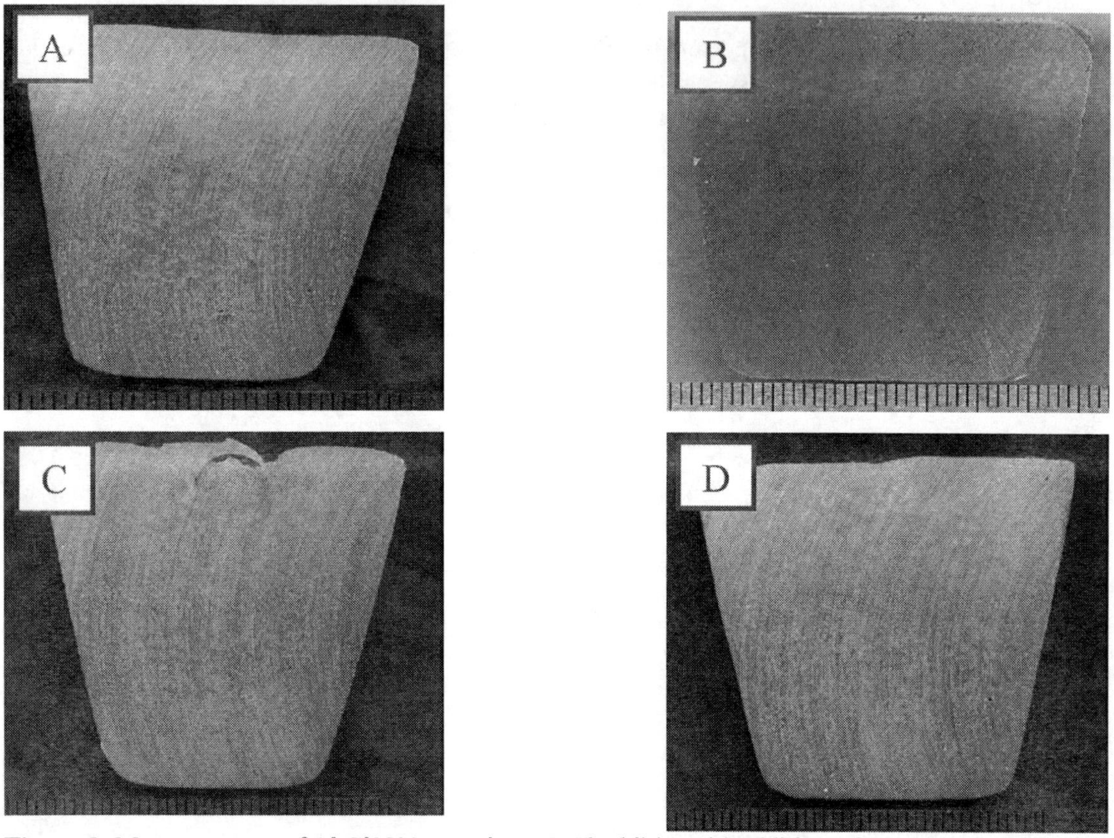

Figure 2. Macrostructure of Al-Si12% at various NaCl addition: (a) addition of 0,2%*wt* NaCl; (b) addition of 0,4%*wt* NaCl ; (c) addition of 0,8%*wt* NaCl ; (d) addition of 1,5%*wt* NaCl.

From figure 2 above, it can be seen that the number of porosities that formed in as-cast product with addition of NaCl in tablet degasser is very low. By comparing all of the macrostructure pictures from all of different NaCl addition in tablet degasser, it could be concluded that all of the results shows that by adding NaCl on tablet degasser, no matter how much the addition ratio is, it all have the same effect on reducing gas porosity that can be formed on as-cast products above.

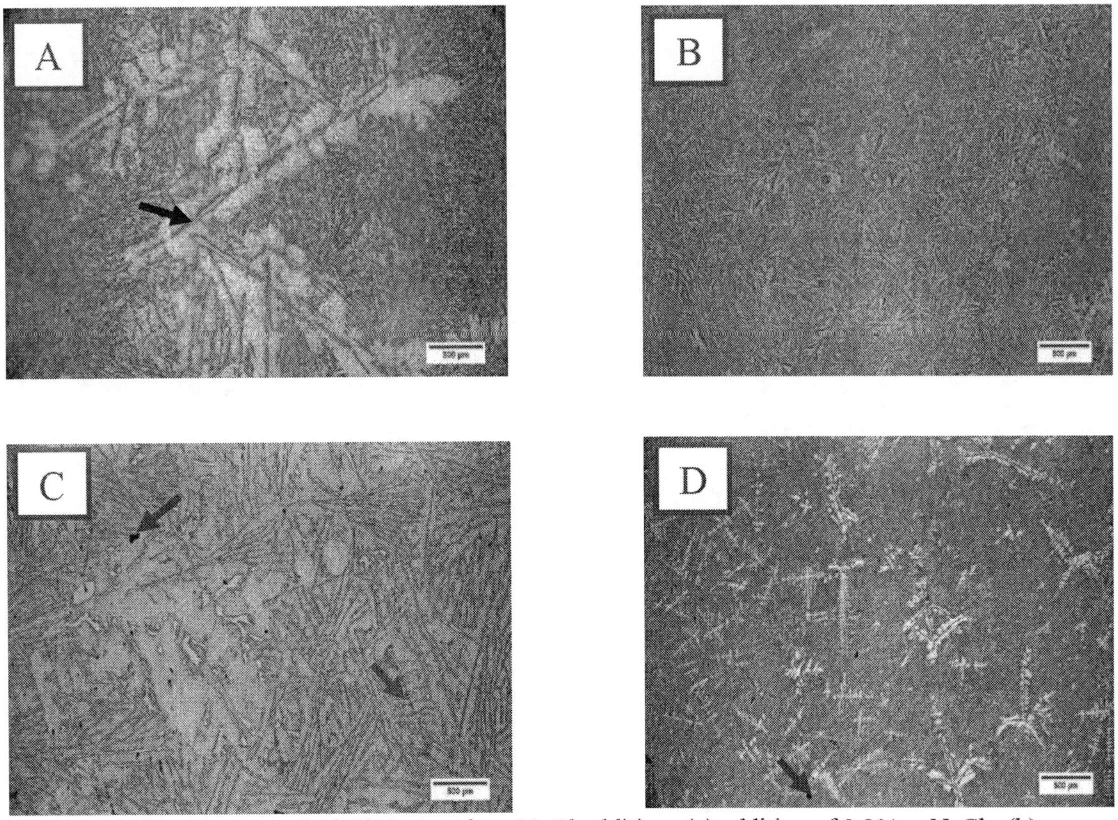

Figure 3. Microstructure of Al-Si12 at various NaCl addition: (a) addition of 0,2%*wt* NaCl ; (b) addition of 0,4%*wt* NaCl ; (c) addition of 0,8%*wt* NaCl ; (d) addition of 1,5%*wt* NaCl.

Al-Si alloys microstructure are affected by microstructural characteristics such as dendritic cell size and grain size. Al-Si will solidifies by a primary precipitation of aluminium dendrites (α-Al) [4]. Based on microstructure result on figure 3, α-Al area are pointed by black arrow, and the porosity that occurs in Al-Si12 casting products can be seen as black dots pointed by red arrow on picture C and D. It can be seen that most porosity are formed at casting product with higher NaCl addition, it is at addition of 0,8%*wt* NaCl. At microstructure D, porosity are formed too, but the number of porosity that appears is lower than microstructure C. With low addition of NaCl inside tablet degasser composition, the number of porosity inside casting products are decreased, especially at addition of 0,2%*wt* NaCl, where the number of porosity that appeared are very low, as a matter of fact it cannot be seen on the microstructure above.

3.2 Mechanical Properties of Al-Si12

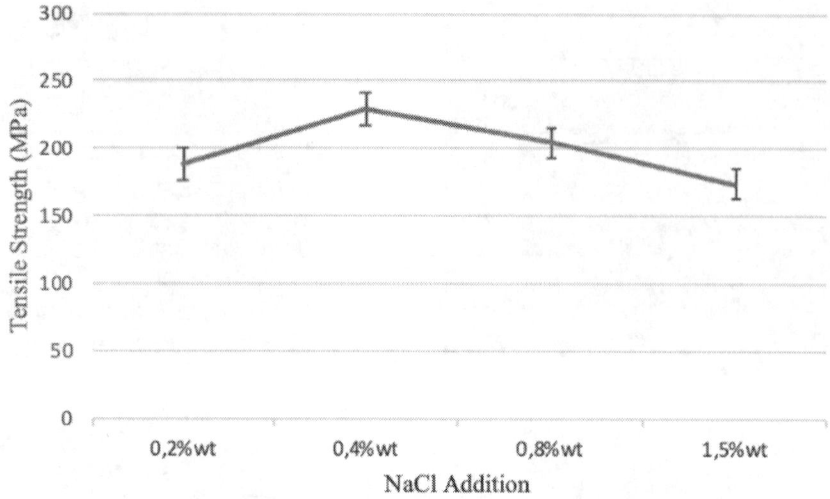

Figure 4. Effect of NaCl addition ratio on tensile strength of Al-Si12.

From figure 4 above, it can be seen that by adding more NaCl on tablet degasser composition, the tensile strength of Al-Si12 will increase at first from 189,14 MPa into 229,83 MPa, but after that the graphic will be showing a reducing number of tensile strength until the lowest tensile strength which is 174,67 MPa that belongs to addition of 3,2 kg NaCl into tablet degasser composition.

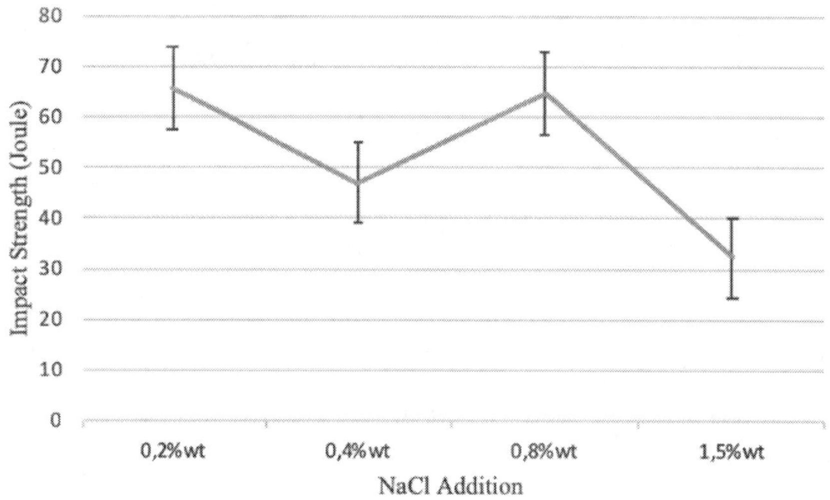

Figure 5. Effect of NaCl addition ratio on impact strength of Al-Si12.

Figure 5 shows the impact strength of each casting product with different NaCl ratio inside tablet degasser composition. From the result above, fluctuative graphic appeared. When increasing the number of NaCl inside tablet degasser, impact strength will decrease into 47 Joule, but after that it will increase again into 64,67 Joule. Highest addition of NaCl shows the lowest impact strength on Al-Si12, which is 32,3 Joule.

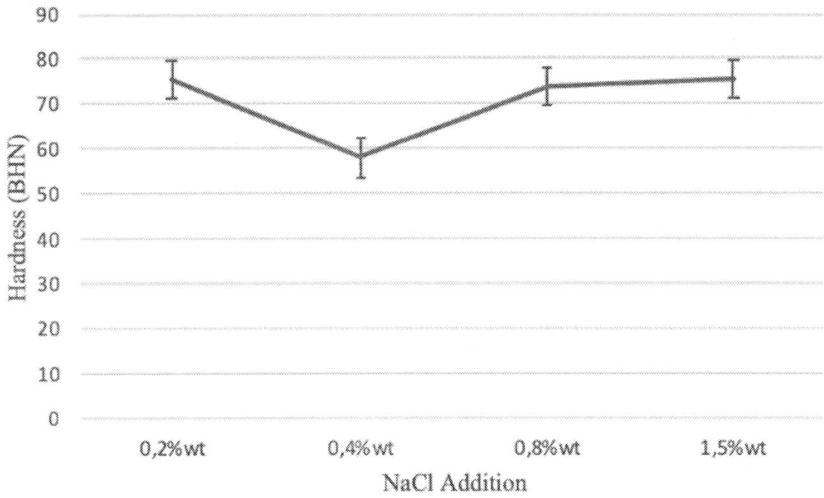

Figure 6. Effect of NaCl addition ratio on hardness of Al-Si12.

Figure 6 above represents the hardness of Al-Si12 casting product. From the result, it can be seen that all of the sample has a slightly similar hardness strength, except for sample with addition of 0,4%*wt* NaCl. It shows that 0,4%*wt* NaCl addition is not compatible for degassing method when the hardness of each casting product is compared.

3.3 Effect of Sodium Nitrate, Sodium Fluoride, and Sodium Chloride on Degassing

To remove dissolved hydrogen inside molten metal, tablet degassing is the simplest method that can be used in casting process. By holding the degasser with plunger inside molten aluminium, there will be a reaction between degassing particle and hydrogen gas that will cause binding process of hydrogen happened. The most common tablet degasser used is C_2Cl_6 tablet degasser, but because it can cause an environmental damage, it is prohibited to be used again.

In this research, NaF and $NaNO_3$ were added as main composition of the tablet degasser to substitute the usage of C_2Cl_6. From equation (i) and (ii) below, it can be seen that each NaF and $NaNO_3$ addition have a role in reducing the number of hydrogen gas trapped inside molten aluminium.

$$2\ NaF + H_2 \rightarrow 2\ Na + 2\ HF \qquad (1)$$

$$2\ NaNO_3 + H_2 \rightarrow 2\ Na + 2\ HNO_3 \qquad (2)$$

When adding tablet degasser into molten metal, nitrogen gases will be produced, and it will bound hydrogen gas, thus it will form gas bubbles and will float into the molten surface. It will be considered as a slag in casting and it needed to be removed immediately from the surface of molten aluminium.

The addition of NaCl have a purpose to see whether NaCl can be a substitute for NaF and $NaNO_3$, because it has lower price and it was categorized as a salt flux just like NaF and $NaNO_3$, so it will react with hydrogen gas with the same principle as NaF and $NaNO_3$, as shown in equation 3 below.

$$2\ NaCl + H_2 \rightarrow 2\ Na + 2\ HCl \qquad (3)$$

NaCl will forms low-temperature eutectic (665°C), which is needed to improve the fluidity. Higher fluidity will cause the diameter of flux particle from tablet degasser to decrease, so it will increase the filtration efficiency of degasser particle inside molten metal [11,12].

4. Conclusion

Based on microstructure test and mechanical properties test that has been done in this research, it can be concluded that NaCl addition in sodium nitrate-sodium fluoride based degasser didn't have any significant effect on reducing porosity in Al-Si12 casting product. On microstructure test, by adding more ratio of NaCl addition in tablet degasser composition, the number of porosity that appears are increasing, and when comparing mechanical properties of Al-Si12 casting product at different NaCl addition, majorities of the result shows that the smallest addition of NaCl actually have the highest mechanical properties.

Acknowledgement

The authors gratefully acknowledge financial support provided by Hibah PITTA with contract No: 2421/UN2.R3.1/HKP.05.00/2018.

References

[1] Santosh MV, Suresh KR and Kiran AS 2017 *Mat. Today: Proc.* **4**(10) 10987.
[2] Fan KL, He GQ, Liu XS, Liu B, She M, Yuan YL and Lu Q 2013 *Mater. Sci. Eng. A.* **586** 78.
[3] Cao H, Hao M, Shen C and Liang P 2017 *Vacuum.* **146** 278.
[4] Zamani M 2015 *JTH Diss. Series.* **21**.
[5] Zhao L, Pan Y, Liao H and Wang Q 2012 *Mater. Lett.* **66**(1) 328.
[6] Haghayeghi R, Bahai H and Kapranos P 2012 *Mater. Lett.* **82** 230.
[7] Ren Y, Ma W, Wei K, Yu W, Dai Y and Morita K 2014 *Vacuum.* **109** 82.
[8] Cao H, Hao M, Shen C and Liang P 2017 *Vacuum.* **146** 278.
[9] Puga H, Barbosa J, Tuan NQ and Silva F 2014 *Trans. of Nonfer. Met. Soc. China (English Edition).* **24**(11) 3459.
[10] Utigard TA, Friesen K, Roy RR, Lim J, Silny A and Dupuis C 1998 *Jom.* **50**(11) 38.
[11] Hosouli S and Tamizifar M 2016 Effects of fluxing operations on the properties and microstructure of LM2 alloy *5th Int. Conf. on Mater. Eng. Met.*
[12] Majidi O, Shabestari SG and Aboutalebi MR 2007 *J Mater Process Technol.* **182** 450.

Magnesium removal from brine water with low lithium grade using limestone, rembang, indonesia

L H Lalasari[1,a], F R Fatahillah[2], D R G Rahmat[2], E Tarmizi[2], A R Rhamdani[1], E Sulistiyono[1], L Andriyah[1], T Arini[1], N C Natasha[1], F Firdiyono[1]

[1]Research Center of Metallurgy and Materials, Indonesian Institute of Sciences, Kawasan PUSPIPTEK Gedung 470, Tangerang Selatan, 15314, Indonesia

[2]Department of Chemical Engineering, Institut Teknologi Indonesia, Jl Raya Puspiptek Serpong-Tangsel, Indonesia

E-mail: [a]lati003@lipi.go.id ; [a]ifa_sari@yahoo.com

Abstract. Experiments have been carried out to remove magnesium ions from brine water using limestone, Rembang, Indonesia. The aim of the study was to produce brine water concentrates that were rich in lithium and did not contain magnesium elements. Brine water used has the following chemical composition: 74.67 ppm Li; 877.891 ppm Na; 1549.81 ppm K; 147.23 ppm Mg; 38.49 ppm Ca and others. The initial stages were 200 g of natural lime calcined at 900 °C for 3 hours using a furnace as a precipitation agent. It is then added to 1000 ml of brine water with a variation of 0.336 g, 1 g, 10 g, 20 g, 30 g, 40 g, 50 g by stirring for 3 hours at atmospheric pressure. The results showed that the magnesium removal from brine water began to be seen in the addition of roasted limestone of 1 g with the dominant phase as $Mg_{0.03}Ca_{0.97}CO_3$ in the precipitated residue. On the addition of 10 g and 20 g of roasted limestone into brine water, the percentage of magnesium removal was almost maximum of 98.8% and 99.8% with the precipitated residues as $Mg(OH)_2$ phases. This experiment was successful to remove magnesium from brine water so that the lithium concentration of brine water increased to 104.32 ppm Li and 105.86 ppm Li with the addition of roasted limestone of 10 g and 20 g, respectively. These results indicate that the use of roasted limestone to eliminate magnesium from brine water with low lithium grade is recommended.

1. Introduction

Lithium consumption in the world continues to increase every year along with the increasing use of lithium in various applications. This condition can be seen from the data that shows the lithium consumption level in 2016 is projected to be around 37,800 tons. Where this requirement increased from 33,300 tons in 2015. The following are distribution of marketing and lithium utilization in 2017 in a variety of applications including batteries (39%), ceramics and glass (30%), grease lubricants (8%), flux powder in casting and polymer production (5%), air treatment (3%), and other uses (10%) [1]. It is seen that the consumption of lithium for batteries has increased significantly along with the increasing need for electronic devices [2]. Based on market opportunities, lithium is the first rank in the metal needed (± 594x of total zinc requirements) and then followed by titanium (130x), coal (109x), bauxite (102x), uranium (91x), iron ore (41x), nickel (39x), copper (37x), and gold (18x) [3]. Currently, the availability of lithium in the world is not comparable to the consumption of lithium

[1] To whom any correspondence should be addressed
[2] To whom any correspondence should be addressed

which is expected to increase continuously until 2025. Lithium resources in the world that have been identified in the United States, Argentina, Bolivia, Australia, Chile, China, and others.

Lithium reserves are shown as follows: The United States has a reserve of 6.9 million tons, Argentina and Bolivia are around 9 million tons, Australia is 2 million tons, Chile is more than 7.5 tons and China is about 2 million tons, Congo, Russia, and Serbia each has about 1 million tons, Brazil and Mexico each have around 200,000 tons, and Austria and Zimbabwe each have more than 100,000 tons [4]. On the other hand, the mapping of lithium resources in Indonesia is not yet known with certainty, so there needs to be an exploration to indicate the lithium deposits in Indonesia. In recent years, the LIPI Center for Metallurgy and Materials Research has tried to identify sources of lithium deposits based on Indonesia's geological structure which is rich in mineral resources. In decades, commercial lithium production still relies on sources of mineral ores such as spodumene, petalite, and lepidolite. In fact, extracting lithium from these sources is significantly more expensive than extracting lithium from brine containing lithium. The cost of extracting lithium from the rock is estimated to be twice that compared to producing from brine water. Though it is known that the potential of brine water as a raw material to produce lithium is very large [5].

As we know that besides lithium, there are other associated components of brine water such as magnesium, calcium, sodium, potassium, sulfate, boron and others. These components must be minimized in order to obtain a lithium-rich brine water concentrate that can be used to produce lithium carbonate. If this component is not removed, it will cause contaminants in the production process of lithium chemical compounds. Of the many constituent components contained in brine water, magnesium is one of the biggest challenges in processing brine water to produce high purity lithium carbonate. Magnesium removal is the initial step before extracting lithium from brine water. The success in eliminating magnesium from brine water with cheap and environmentally friendly technology will offset the production costs of lithium chemical compounds such as lithium carbonate from brine water.

Stephan Bogner (2017) [6] reported that MGX technology had successfully eliminated high magnesium levels (76,000 ppm) of lithium brine samples from a US site to undetectable levels (<1 ppm). This MGX technology uses a combination of nano-filtration and nano-flotation systems where it is sufficiently flexible and effective in providing continuously increasing solutions for lithium demand, Lalasari et.al. (2017) [7] was able to precipitate magnesium from brine water by chemical precipitation techniques using NH_4OH chemical reagents. According to Kang & Yoon (2015) [8], magnesium can be removed by chemical precipitation using calcium hydroxide ($Ca(OH)_2$). Calcium hydroxide reacts with magnesium to form magnesium hydroxide ($Mg(OH)_2$) as a precipitate.

This study tried to remove magnesium from brine water by utilizing Indonesian mineral resources, namely limestone from Rembang. Limestone is pre-prepared and mixed into brine water. The expected goal is magnesium from brine water precipitated to form $Mg(OH)_2$ so that brine water is rich in lithium and free of magnesium.

2. Experimental

The materials used in the study were limestone from Rembang, Indonesia and Brine Water from Bogor, Indonesia. The research stages included the preparation of raw materials, magnesium removal processes, filtration, drying, and characterization as shown in Figure 1. The preparation stage of the raw material was 200 g of natural limestone to be burned at 900 °C for 3 hours using a muffle furnace to produce CaO compounds. The roasted limestone product was then characterized using X-Ray Diffraction (XRD) to investigate the formation of CaO as shown in Figure 2 and X-Ray fluorescence (XRF) to identify its chemical composition as shown in Table 1. In the preparation of the brine sample was carried out as follows: natural brine water was evaporated to around 40 % volume using a distillation apparatus equipped with a reflux condenser. The aim was to increase the lithium level from the brine water sample. The brine water samples from the evaporation were then characterized using inductively coupled plasma - optical emission spectrometry (ICP-OES) to determine its chemical composition as shown in Table 2.

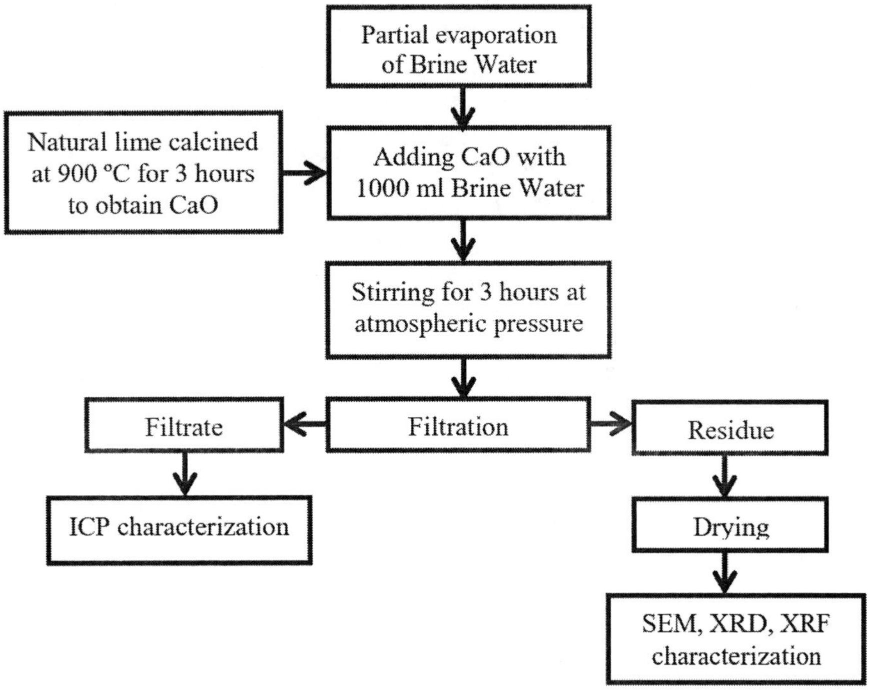

Figure 1. The flow chart of research

Table 1. The chemical composition of roasted limestone (XRF analysis)

Ca	Mg	K	Si	Others
99.33%	0.34%	0.04 %	0,24 %	0.05%

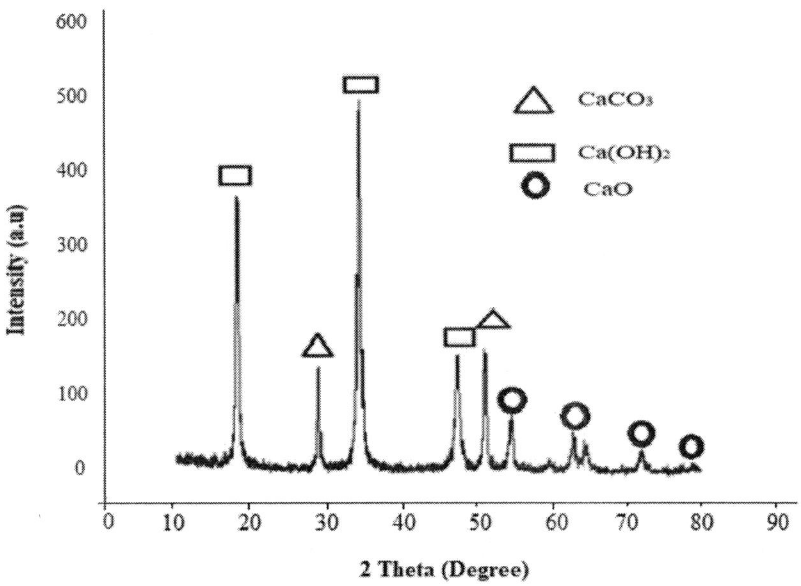

Figure 2. The XRD pattern of roasted limestone, Rembang

Table 2. The chemical composition of brine water samples (ICP-OES analysis)

Li	Na	K	Ca	Mg
74.67 ppm	877.891 ppm	1549.81 ppm	38.49 ppm	147.23 ppm

The second stage was magnesium removal from the brine water samples. The procedure was as follows: The roasted limestone (CaO) varied 0.336 g, 1 g, 2 g. 10 g, 20 g, 30 g, 40 g, 50 g was put into an Erlenmeyer containing 1000 ml samples of brine water. Then the stirring process is carried out for 3 hours at 1 atmospheric pressure. In this experiment, the use of 0.336 g roasted limestone is the minimum amount obtained through stoichiometric calculations. The third stage was the filtration and drying of samples. Filtration aims to separate lithium-rich brine filtrate and magnesium-rich residues. The residue was then dried in a muffle furnace at 110 °C for 2 hours. The last stage was characterization where the filtrate was analyzed using ICP-OES, while the residue was analyzed by Scanning Electron Microscopy-Energy Dispersive X-Ray Spectroscopy (SEM-EDS), XRD and XRF.

3. Results and Discussion

Based on Table 1, the composition of roasted limestone from Rembang was dominated by Ca compounds with levels of 70.22% with the most impurities being Mg (0.34%) and Si (0.24%). This impurity can affect the purity of the product resulted during the removal process of Mg. On the other hand, the XRD result in Figure 2 shows that the roasted limestone of Rembang was dominated by $Ca(OH)_2$, $CaCO_3$, and CaO compounds. The diffraction peaks which showed $Ca(OH)_2$ compounds were at the 2Θ position of 18.228°; 36,939°; 47,547° with ICDD card no. 4-0733. In this experiment, $Ca(OH)_2$ and CaO compounds act as precipitate agents in the process of separating Mg elements from brine. Supposedly, in the roasting process at 900 °C, limestone (CaCO3) decomposes completely into CaO by releasing CO_2 gas. But in fact, the XRD characterization results show that there is still $CaCO_3$ which did not decompose completely into CaO. It also showed the existence of the $Ca(OH)_2$ phase in Figure 2. This $Ca(OH)_2$ compound was formed due to the hydration process of CaO with air containing H_2O.

In Table 2, it can be seen that the main components of brine water, Bogor after evaporation of 40 % volume are Na (877.891 ppm), K (1549.81 ppm), Mg (147.23 ppm), while the minor element is Li (74.67 ppm) and Ca (38.49 ppm). The Ca level in Brine shown in Table 2 was relatively low because, during partial evaporation, element Ca was partly precipitated to form $CaCO_3$. This result was consistent with the previous research conducted by Lalasari et al (2017) [7]. To be able to produce lithium-rich brine concentrate, it was necessary to remove the main element from brine, especially IIA class in a periodic system such as Mg. According to Stephan Bogner (2017) [6], the removal of Mg from lithium brine with traditional solar evaporation was less effective and not economical because Mg and Li have almost similar ionic properties so they were difficult to separate from each other. Magnesium separation from lithium brine needed to use other technologies. On this occasion will be explained in detail the results of research related to the Mg removal from brine water, Bogor using natural limestone, Rembang. Initial characterization will be carried out first on residual samples (precipitated products) using XRD, XRF and SEM-EDS analysis. Furthermore, characterization of brine filtrate samples using ICP-OES analysis to find out how much magnesium can be removed in lithium brine. The following was the amount of residue obtained after the addition of variously roasted limestone.

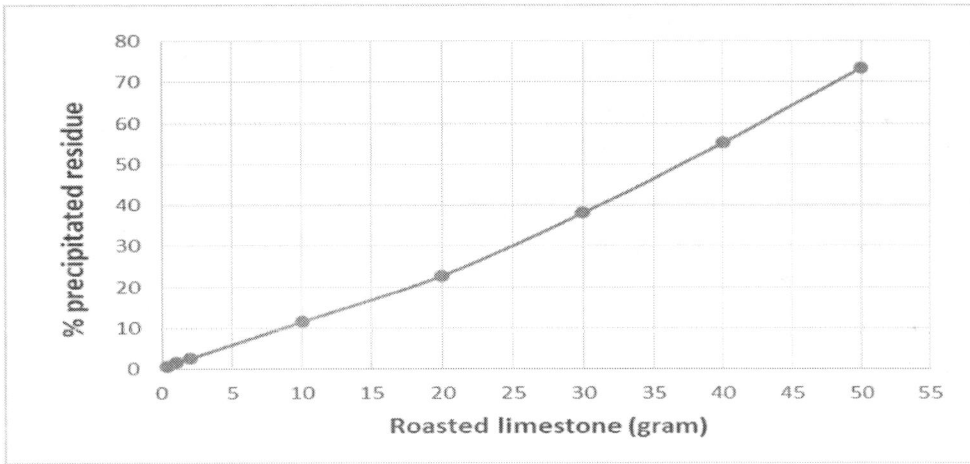

Figure 3. The percentage of precipitated residues in variously roasted limestone

In Figure 3, it can be seen that the more roasted limestone was added to the brine water, the more obtained precipitate increases. This indicated that many associated components of brine water settled together with precipitate agents. Supposedly, the roasted limestone precipitated Mg ions into $MgCO_3$ or $Mg(OH)_2$ instead precipitating other compounds from brine water. Therefore it was necessary to optimize the use of precipitate agents such as a roasted limestone.

3.1 Structure, morphology and chemical composition of residues (precipitated products)

The following were the results of the characterization of the residue sample produced by chemical precipitation techniques using the addition of roasted limestone varied 0.336 gr, 1 gr, 10 gr, 20 gr, 30 gr, 40 gr, 50 gr into 1000 ml brine water. To find out the structure or phase formed from the residual sample, XRD analysis was carried out as shown in Figure 3.

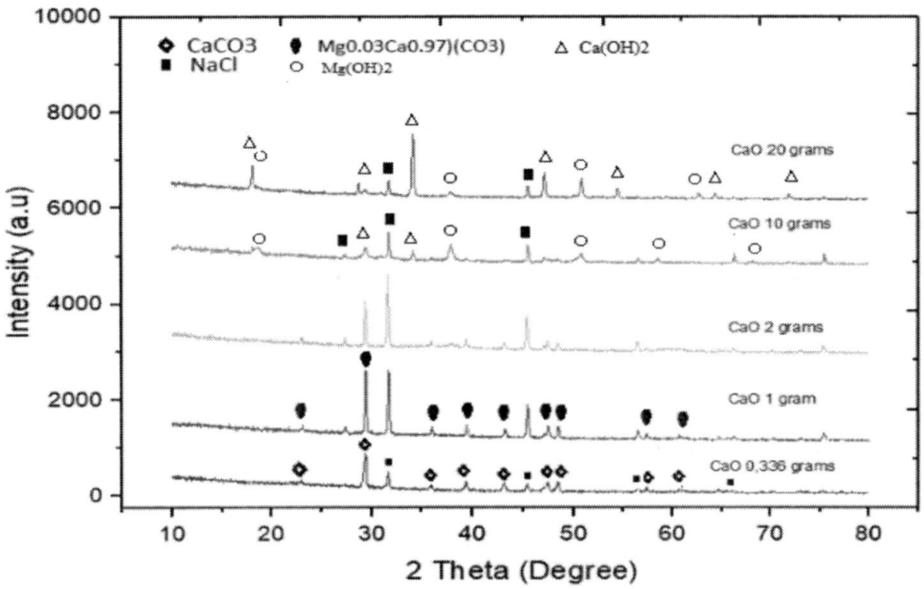

Figure 4. The XRD pattern of residue (precipitated product) in variously roasted limestone (CaO) mass

IOP Conf. Series: Materials Science and Engineering **578** (2019) 012067 doi:10.1088/1757-899X/578/1/012067

Figure 4 clearly looked the phase changes shown in the precipitated residues after the roasted limestone addition (compared to Figure 2). At the addition of 0.336 g roasted limestone, the peaks that appeared on the residue have $CaCO_3$ and NaCl phases. The phase of $MgCO_3$ or $Mg(OH)_2$ had not been apparent in a residue. This indicated that the roasted limestone containing CaO and $Ca(OH)_2$ with an amount of 0.336 gr (according to stoichiometry) had not been able to remove some of the magnesium in brine water. In addition to 1 g of roasted limestone in the brine water, the reaction shifted to form $Mg_{0.03}Ca_{0.97}CO_3$ phase. This shows that adding 1 g of limestone can remove magnesium in brine water even though it was not perfect. If the amount of limestone burned was increased by about 10 g and above, then magnesium has undergone precipitation into the $Mg(OH)_2$ phase with excess $Ca(OH)_2$ phase in the residues. Precipitation reactions that may occur are as follows:

$$CaO_{(s)} + H_2O_{(l)} \rightarrow Ca(OH)_{2\,(aq)} \quad (1)$$

$$Ca(OH)_{2\,(aq)} + Mg^{2+}_{(aq)} \rightarrow Mg(OH)_{2\,(s)} + Ca^{2+}_{(aq)} \quad (2)$$

To better explain and know the chemical composition of the precipitated residue, it was necessary to do XRF analysis as shown in Table 3.

Table 3. The chemical composition of residue (precipitated products) (XRF analysis)

Element	Wt. %				
	Roasted limestone (raw material)	Roasted limestone addition			
		1 g	2 g	10 g	20 g
Na	-	34.88	41.09	10.1	15.3
Mg	0.34	21.13	28.94	27.24	11.06
Si	0.24	10.71	4.62	2.52	2.36
S	-	0.01	0.01	0.04	0
Cl	-	20.39	16.56	18.88	10.45
K	0.04	0.76	0.45	0.91	0.38
Ca	99.33	11.42	7.92	38.65	59.75

Based on Table 3, it can be seen that the addition of roasted limestone can remove magnesium in brine water. This was indicated by the percentage of Mg element in the residue at the addition of 1 g of roasted limestone which was increased to 21.13% from 0.34% Mg from the initial roasted limestone and the main element of the roasted limestone was Ca that became reduced from 99.33% to 11, 42%. The removal of Mg from brine water was seen to be great at the addition of roasted limestone of 2 g which was shown in percentages of Mg from the residue to increase to 28.94% and the element Ca reduced to 7.92%. This condition was satisfactory because of the possibility that $Ca(OH)_2$ from the roasted limestone reacts with Mg from brine water to form precipitate $Mg_{0.03}Ca_{0.97}CO_3$ or $Mg(OH)_2$. The $Mg(OH)_2$ phase was not apparent in the addition of 1 g and 2 g of limestone because of the possibility of having peaks with low relatively intensity, this was like the research conducted by Meshram (2014) [9]. Table 3 also showed that if the roasted limestone mass was increased to 10 g, the residue contains a high Ca element of about 38.65% and the Mg element of 27.24%. This result was in accordance with the XRD characterization shown in Figure 4 where the diffraction peaks were found on the addition of 10 g and 20 g of roasted limestone containing the dominant compound with $Ca(OH)_2$ phase, other than the expected Mg $(OH)_2$ phase. This showed that adding 10 g and 20 g of roasted limestone to brine water was excess.

Figure 5. The morphology of precipitated residues by addition roasted limestone (a) 0.336 g, (b) 1 g, (c) 10 g, (d) 20 g

Figure 5 showed that precipitated residues in various precipitation agents (roasted limestone) had varying morphology such as cubic, nearly round, and irregular at 1000x magnification. The cubic shape was indicated as NaCl. The round shape was indicated as $CaCO_3$. The shape of the needle and discs were indicated as $Mg(OH)_2$ [10]. In Figure 5a, the morphology of the particles was dominated by a cubic, irregular round shape, and agglomeration. This showed that precipitated particles at the roasted limestone addition of 0.336 g were more dominant in containing NaCl and $CaCO_3$. The results of this analysis were in accordance with the XRD characterization shown in Figure 4. The addition of 1 g of roasted limestone (Figure 5b) showed that particle morphology was predominantly resembling sponges with a fairly heterogeneous form including round, thin needles, cubic and irregular shapes that tend to agglomerate. This explained that the particles in this precipitated residue contained several compounds including $MgCO_3$, $CaCO_3$, $Mg(OH)_2$, NaCl. Whereas, in addition to 10 g and 20 g roasted limestone (Figure 5c and Figure 5d), the morphology of the particles became large grains. The large particles were formed because mutually agglomerating particles form large chunks with a thin needle above the chunk. The morphology shown in this residue was dominated by $Ca(OH)_2$ and $Mg(OH)_2$.

3.2 The chemical composition of brine filtrate samples

To investigate the effect of adding roasted limestone to the removal of magnesium and other associated elements in brine water, the filtrate sample was characterized using ICP-OES as shown in Table 4.

Table 4. The chemical composition of brine filtrate samples in various roasted limestone addition

Elements	**Brine Water concentrate(ppm)**	**Filtrate concentration (ppm) in various addition roasted limestone (ppm)**			
		10 g	**20 g**	**30 g**	**40 g**
Li	74.67	104.32	105.858	93.2973	88.9843
Na	877.891	737.309	771.395	815.548	902.229
K	1549.81	1987.07	2025.75	1827.78	1777.99
Mg	147.23	1.7412	0.278966	0.04782	0,035921
Ca	38.49	32.8911	33.074	33.6691	33.8250

Table 4 showed that the addition of roasted limestone to brine water can remove magnesium from brine water. The removal of magnesium was quite successful at the addition of roasted limestone 10 g where the magnesium content of brine water was reduced from 147.23 ppm to 1.7412 ppm. The same conditions occurred with the addition of 20 g, 30 g, and 40 g of roasted limestone wherein the brine filtrate contained magnesium at 0.279 ppm, 0.048 ppm, and 0.036 ppm respectively. This showed that the magnesium removal in the addition of 10 g of roasted limestone and above was very significant. The elimination of magnesium elements from brine water also caused an increase in the levels of other elements such as lithium and potassium from brine water. This was indeed expected to produce lithium-rich brine concentrates. From the previous characterization, it was explained that magnesium was detected in the residue with the addition of 1 g of roasted limestone. This indicated that the removal of magnesium from brine water also occurred with the addition of 1 g roasted limestone and above. Therefore, it was necessary to further characterize brine filtrate in the next study. It aimed to produce the optimum conditions for adding roasted limestone to brine water so that obtained lithium-rich brine concentrate and free of magnesium.

4. Conclusion

Limestone burned at 900 °C for 3 hours was used as a precipitation agent containing several compounds including CaO, $Ca(OH)_2$ and $CaCO_3$ phases. The magnesium removal from brine water was detected at the roasted limestone addition of 1 g which shown from XRD analysis on the precipitated residue indicating the dominant phase as $Mg_{0.03}Ca_{0.97}CO_3$. While the addition of 10 g and 20 g of roasted limestone, the dominant phase was $Mg(OH)_2$ and $Ca(OH)_2$. The apparent morphology of precipitated residues was fairly heterogeneous forms including round, thin needles, cubic and irregular shapes that tend to agglomerate. The result of XRF analysis also showed that the magnesium content in the residue increased to 21.13%, 28.94% and 27.24% in roasted limestone addition to 1 g, 2 g, and 10 g compared to the magnesium content of the initial burnt limestone was 0.34%. The results of ICP-OES analysis on brine filtrate samples also showed that the roasted limestone as a precipitation agent was quite successful in removing magnesium from brine water up to 0.2-0.03 ppm and able to increase lithium levels from brine water so that this method is suitable to be applied to produce lithium-rich brine.

Acknowledgments

This research was carried out at hydrometallurgy laboratorium and pyrometallurgy laboratorium in Research Center for Metallurgy and Material, Indonesian Institute of Sciences. The authors are grateful for the main financial supports from INSINAS budget 2018.

References

[1] H Hao, Z Liu, F Zhao, Y Geng, and J Sarkis 2017 *Resour. policy* **51**100–106.
[2] William H. Werkheiser 2018 *Mineral Commodity Summaries 2018* (U.S. Geological Survey).
[3] M Hocking, J Kan, C Terry, and D Begleiter 2016 *Industri Litium 101: Welcome to the Lithium-ion Age Global* (Deutsche Bank Markets Research) 1–179.
[4] B W Jaskula 2017 *Lithium in Mineral Commodity Summaries 2017* (U.S. Geological Survey).
[5] L A Munk, S A Hynek, D C Bradley, D Boutt, K Labay, and H Jochens 2016 *Reviews in Economy Geology* **18** 339–365.
[6] S Bogner 2017 MGX *Minerals Solves the Magnesium Problem of the Lithium Industry Rockstone Research, Report 26* (Canada: MGX Minerals Inc.)
[7] L H Lalasari, M K Widowati, N C Natasha, E Sulistiyono, and A B Prasteryo 2017 *IOP Conf. Ser. Mater. Sci. Eng.* **176** 012040
[8] D J Kang, M H Yoon, and J W An 2015 *Method for Producing High Purity Lithium Carbonate* (Patent No: US 9,169,125 B2, Date of Patent: Oct. 27, 2015).
[9] P Meshram, B Pandey, and T Mankhand 2014 *Hydrometallurgy* **150** 192–208
[10] P Wang, C Li, H Gong, H Wang, and J Liu 2011 *Ceram. Int.* **37** 3365–3370

The pyrometallurgical recovery of zinc from electric arc furnace dust (EAFD) with active carbon

G S T Sinaga[1], A S Wismogroho[1], A Fitroturokhmah[1], R Kusumaningrum[1], W B Widayatno[1],G Hadiko[2], M I Amal[3]
[1] Research Center for Physics, Indonesian Institute of Sciences, Serpong 15314, Tangerang Selatan, Banten, Indonesia
[2] Center for Technology for Mineral Resources, Agency for The Assessment and Application of Technology, Setu 15314 Tangerang Selatan, Banten, Indonesia
[3] Research Center for Metallurgy and Materials, Indonesian Institute of Sciences, Serpong 15314, Tangerang Selatan, Banten, Indonesia

E-mail: muha137@lipi.go.id

Abstract. The reduction of minor impurities (e.g., Ferrite, Magnesium, Lead, Chromium) from Zinc rich-EAFD with active carbon addition was investigated at temperature between 900-1300 °C. The result was powder with higher zinc content which then analysed by X-ray fluorescence spectroscopy (XRF), X-ray diffraction (XRD) and scanning electron microscope (SEM) equipped with X-Ray energy dispersive spectrometry (EDS). The thermogravimetry (TG) and differential thermal analysis (DTA) of EAFD and carbon mixture was conducted previously to grasp the temperature of thermic reaction of mixture. Recovery of 90wt% zinc from EAFD available was achieved up to 99.2wt% after the pyrometallurgy process at temperature 900 °C in atmosphere pressure while the extension of the time corresponds to the amount of recovered zinc dross.

1. Introduction

The prodigious expansion of steel manufacture grows rapidly because the advancement of electric arc furnace (EAF) technology. EAF technology has contributed to more than 25% of the world's total crude steel production [1]. In typical EAF operation, approximately 10-20 kg dust, called EAFD is produced per ton of steel [2]. Dust formation essentially associated to the thermic process of the charged material that includes scrap materials, lime, dolomitic lime and carbon which then will determine on the quality of steel product [3]. When the scrap is melted in the EAF, most of the zinc from the steel scrap ends up in the dust and fume due to its low solubility in molten steel and slag, beside the zinc vapour pressure is higher than iron vapour pressure at steel production temperature. Iron and other slag particles in an oxidizing atmosphere are always found as impurity in the released zinc particles. Subsequently, the forming compounds like ZnO and ZnFe2O4[4] is ensued on top of the furnace and flows through the ventilation system. Thereafter, all evaporated material are collected in the dusting system originating the EAFD [5,12].

Generally, EAFD contains two major elements zinc and iron [6,7], nevertheless EAFD is always mixed with other minor impurities which are categorized as harmful heavy metal, such as Pb, Cr and Cd [8]. Hence, EAFD is classified hazardous waste that difficult to store and transport [9]. EAFD requeres further treatment to be released in the environment, because it contains hazardous, leachable elements [10]. The accumulation of EAFD in landfilling area is not only causing harmful effect to plan and animal as well as human environment. However, EAFD has several valuable elements such as zinc oxide and iron, thereof many EAFD recovery has been studied and developed.

There are two method of metal recovering that has been widely used for purifying EAFD, the first is the leaching method [13]. This method involves chemical reaction between leachable element in EAFD with acid solution. Although this method is less energy required, the process results another waste [14]. The second is treating the EAFD with pyrometalurgy process [4-10]. The pyrometallurgy process is considered as the best method of recovering the zinc oxide from EAFD though it requires a lot of thermal energy because this method does not result any more waste material. The puspose of this research paper is to present the results of laboratory scale pyrometallurgy process established for recovering zinc oxide from EAFD. Information regarding physical and chemical properties of recovered zinc oxide were studied by XRD, XRF and SEM-EDX.

2. Experimental

2.1. Materials

EAFD from steelmaking industry, was in form of lump with dark blue colour. The concentration of elements in the EAFD is shown in Table 1. The content of zinc metal appertains to 90 wt% and the remaining were considered as impurities. The reductant material chosen for recovering the zinc was active carbon which was synthesized in laboratory of research centre for physics Indonesian institute of sciences.

Table 1. The EAFD chemical composition (wt%)

Zn	Fe	Mg	Al	Ca	S	Cl	K	Ba	Other
90	4.09	2.53	0.49	0.19	0.1	0.09	0.08	0.07	<0.05

Other: Si, P, Ti, Cr, Sr, Sn

2.2. Methods

The mixture of EAFD and active carbon were prepared using ball milling method. The compositions of the EAFD and active carbon were varied 7:1 and 3:1. Each 100 gram of mixture was dissolved with 200 ml demineralized water and milled in planetary ball mill for 40 minutes. The result was slurry mixture which then filtered to reduce the water content. The less water slurry then dried in oven at 100 °C for 24 hours. The drying process was aimed to get the lowest possible moisture, since hydrogen in the furnace off-gas could have a detrimental effect on the efficiency of zinc condensation [15]. The dried mixture was then heated in tube furnace at 1200 °C in atmospheric pressure for 2.5 hours. The temperature and heating duration was chosen based on the preliminary experiment to achieve both zinc oxide and residual material. The result and residual was then analysed by XRF, XRD and SEM-EDX.

The experimental system consisted of an alumina crucible inside an alumina tube were set into electric furnace 3.3 kW as shown in figure 1. The thermocouple was set close to sample to measure the real time temperature of the sample. One tip of the tube was connected to a low pressure blower while the other side was set longer in order to trap the zinc vapour crystalized on the lower temperature zone on the inside of the tube wall.

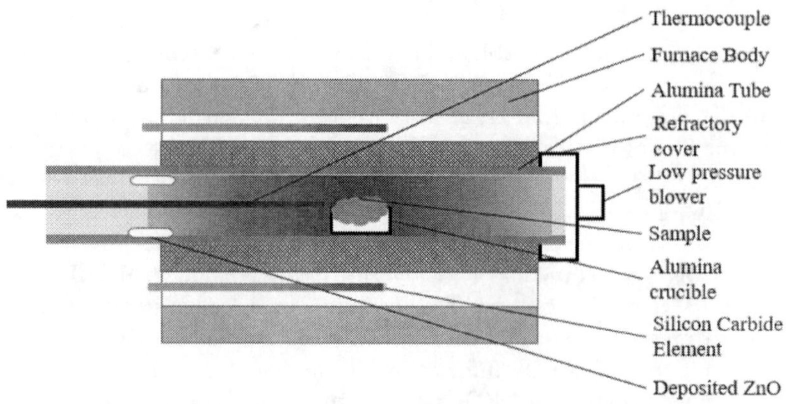

Figure 1. Tube furnace prepared with low pressure blower.

2.3. Thermodynamic Calculation

The reaction temperature involved in pyrometallurgical processing of EAFD are summarized by FactSage™ software version 7.0 in Table 2 [16]. The data shows that zinc oxide can be reduced into metallic forms at temperatures below 1000 °C, it assists recycling EAFD via pyrometallurgical process. After reduction, zinc, lead and cadmium will evaporate because of their high vapor pressures at the EAF operating temperature. The metal vapor can then be condensed, separated and recovered using a condenser system. Conversely, iron and chromium will be retained in the residue after thermal reduction. The result of thermodynamic analysis is shown in Fig 1.

Table 2. The reaction temperature in pyrometallurgical processing.

No	Reaction	Reaction temperature (°C)
1	$ZnO_{(s)} + C_{(s)} = Zn_{(g)} + CO_{(g)}$	955.88
2	$ZnFe2O4_{(s)} + C_{(s)} = Zn_{(g)} + Fe_{(s)} + CO_{(g)}$	791.35
3	$PbO_{(s)} + C_{(s)} = Pb_{(l)} + CO_{(g)}$	302.10
4	$CuO_{(s)} + C_{(s)} = Cu_{(l)} + CO_{(g)}$	15.18
5	$MnO_{(s)}+C_{(s)} = Mn_{(s)}+CO_{(g)}$	1426.59
6	$3Fe_2O_{3(s)}+C_{(s)} = 2Fe_3O_{4(s)}+CO_{(s)}$	319.78
7	$Fe_3O_{4(s)}+C_{(s)} = 3FeO_{(s)}+CO_{(g)}$	696.39
8	$FeO_{(s)}+C_{(s)} = Fe_{(s)}+CO_{(g)}$	726.41
9	$ZnO_{(s)}+Fe = Zn_{(g)}+FeO_{(s)}$	1198.88
10	$ZnO_{(s)}+CO_{(g)} = Zn_{(g)}+CO_{2(g)}$	1317.90
11	$ZnFe_2O_{4(s)} +CO_{(g)} = Zn_{(g)}+Fe_{(s)}+CO_{2(g)}$	1330.96
12	$MnO_{2(s)}+2C_{(s)} = Mn_{(s)}+CO_{(g)}$	555.14
13	$Fe_3O_{4(s)}+CO_{(g)} = 3FeO_{(s)}+CO_{2(g)}$	637.10
14	$FeO_{(s)}+CO_{(g)} = Fe_{(s)}+CO_{2(g)}$	570.96
15	$CrO_{3(s)}+3C_{(s)} = Cr_{(s)}+3CO_{(g)}$	192.42
16	$CrO3_{(s)}+3CO_{(g)} = Cr_{(s)}+3CO_{2(g)}$	31151.28
17	$FeCr_2O_{4(s)}+C_{(s)} = Fe + Cr_2O_3+CO_{(g)}$	1054.47
18	$CdO_{(s)}+C_{(s)} = Cd_{(g)}+CO_{(g)}$	598.94
19	$CdO_{(s)}+CO_{(g)} = Cd_{(g)}+CO_{2(g)}$	434.80
20	$(ZnO)+[C] = Zn_{(g)}+ CO_{(g)}$	864.92
21	$(PbO)+[C] = Pb_{(g)}+ CO_{(g)}$	765.25
22	$(FeO)+ [C] = [Fe] +CO_{(g)}$	692.56
23	$ZnFe_2O_{4(s)}+CaCl_{2(l)} = CaFe_2O_{4(s)}+ZnCl_{2(g)}$	1332.15
24	$Fe_2O_3+3CaCl_{2(l)} = 2FeCl_{3(g)}+3CaO_{(s)}$	2434.90
25	$ZnO_{(s)}+CaCl_{2(l)} = ZnCl_{2(g)}+CaO_{(s)}$	1468.48

IOP Conf. Series: Materials Science and Engineering **578** (2019) 012068 doi:10.1088/1757-899X/578/1/012068

3. Results and Discussion

Laboratory scale of pyrometallurgy devices for separating zinc from EAFD was successfully constructed. Recovery of 90wt% zinc from EAFD available was achieved up to 99.2wt% after the pyrometallurgy process at temperature 1200 °C in atmosphere pressure. The product was a condensed white powder adhere on the inside surface of alumina tube. The product was analysed by XRF, XRD and SEM. The result of XRF analysis are presented in Table 3.

Table 3. Chemical composition of product.

Metal Oxide	Content (ppm)
ZnO	992000
Al_2O_3	1440
Fe_2O_3	979
SiO_2	1550
MgO	2020

Characteristics of the products was analysed using XRD, the result are shown in figure 2. The result confirm that the product is Zincite and Hydrozincite. Zincite is the mineral form of ZnO while the Hydrozincite is carbonite mineral consisting $Zn_5(CO_3)_2(OH)_6$.

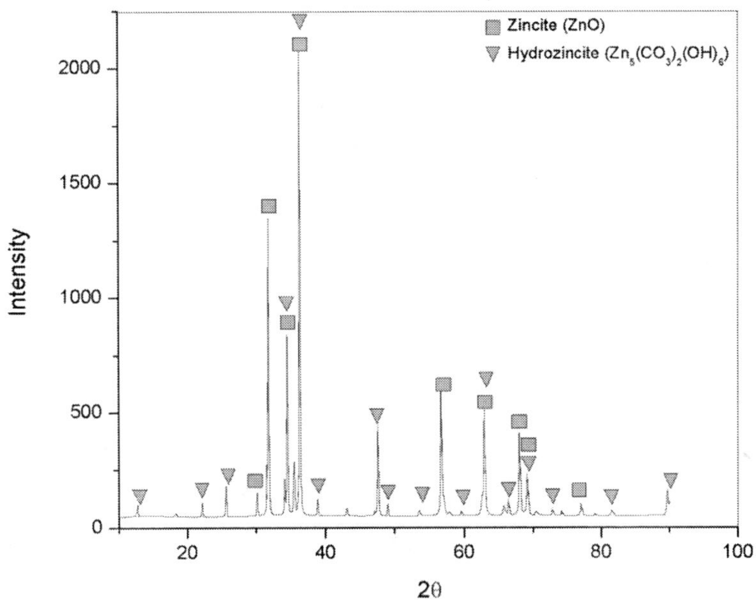

Figure 2. Diffractogram of the product

SEM was also carried out to study the morphology characteristic of the product. The BSE image of the product is shown in figure 3. As shown in figure 3a, the particle seems agglomerate become a bigger particle. The bigger particle consists of random shaped smaller particle where the smaller particles are dominated by rod shape particle.

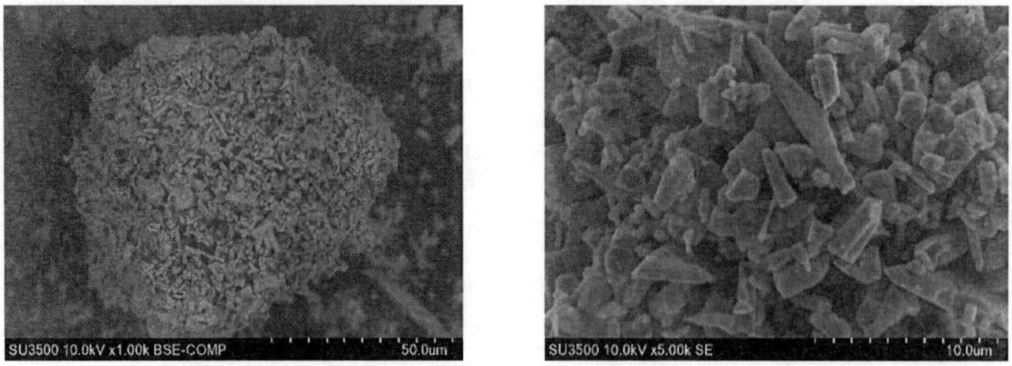

(a) (b)

Figure 3. SEM image of the product (a) 1k (b) 5k magnification

4. Conclusion

EAFD with 90% zinc content from steel industry have been given treatment in pyrometalurgi process with addition active carbon at 1200 °C. The result was solids with 99.2% zinc content in chemical form ZnO and $Zn_5(CO_3)_2(OH)_6$. The increment of heating period is responsible to the amount of the achieved product.

Acknowledgment

This work was supported by Insinas Kemitraan contract number 058/P/RPL-LIPI/INSINAS-1/III/2018, "Development of Aluminium Metal Powder for Pigments and Hebel Additive". The authors are gratefully acknowledged the Laboratory of High Resistant Materials, Research Center for Physics-LIPI for providing research facilities.

References

[1] Steel Statistical Yearbook of World Steel Association, 2016 http://www.worldsteel.org/steel-by topic/statistics/steel-statistical-yearbook-.html
[2] Dutra A J B, Paiva, P R P and Tavares L M 2006 *J. Min. Eng.* **19** 478
[3] Pretorius E, Oltmann H and Jones J 2010 *EAF fundamentals* (New York: LWB Refractories)
[4] Machadoa J G M S, Brehm F A, Moraes C A M, Santos C A D, Vilela A C F and Cunha J B M D C 2006 *J. Haz. Mat.* **B136** 953-960
[5] Leclerc N, Meux E and Lecuire J M 2003 *J. Hydrometallurgy* **70** 175-183
[6] Sobrinho P J N, Espinosa D C R and Tenorio J A S 2003 *J. Ironmak. Steelmak.* **30**(1) 11-17
[7] Orhan G 2005 *J. Hydrometallurgy* **78** 236-245
[8] Salihoglu G and Pinarli V 2008 *J. Haz. Mat.* **153** 1110-1116
[9] Lis T, Nowacki K, Zelichowska M and Kania H 2015 *Metalurgija* **54** 283-285
[10] Guézennec A G, Huber J C, Patisson F, Sessieq P, Birat J P, Ablitzer D 2005 *Powder Tech.* **157** 2–11.
[11] Compendium of Chemical Terminology 2009 *International Union of Pure and Applied Chemistry* 1824
[12] Richardson G G, Brimacombe J K, Toop G W 1985 *Metall Mater.Trans* **B 16**(3) 513-527.
[13] Sebag M G, Korzenowski C, Bernardes A M, Vilela A C 2009 *J. Haz. Mat.* **166 (2-3)** 670–675.
[14] Oustadakis P, Tsakiridis P E, Katsiapi A, Agatzini-Leonardou S 2010 *J. Haz. Mat.* **179** (1-3) 1-7
[15] Abdel-latif M A 2002 *Min. Eng.* **15**(11) 945–952.
[16] Xiaolong L, Zhiwei P, Jiaxing Y, Zhizhong L, Jiann-Yang H, Yuanbo Z, Guanghui L, Tao J 2017 *J. of Cleaner Production* **149** 1079-1100

Recycled Polypropylene/Kenaf Fiber/Halloysite (rPP/rNR/KF/Hall) smart composites: The influence of bio-compatibilizer addition on Tensile, Morphological and Heat Combustion Properties

N S Suharty[1], H Ismail[2], K Diharjo[3], D S Handayani[1], A Sholihati[1] and A I Nisa'[1]

[1] Chemistry Department, Faculty of Mathematics and Natural Sciences, Sebelas Maret University, Jl. Ir Sutami 36A Surakarta, Indonesia

[2] School of Materials and Mineral Resources Engineering, Engineering Campus, University Sains Malaysia, 14300 Nibong Tebal, Pulau Pinang, Malaysia

[3] Faculty of Engineering, Sebelas Maret University, Surakarta, Indonesia

E-mail: nsuharty@yahoo.com

Abstract. Had been successfully synthesized rPP/DVB/PP-AA/rNR/KF//Hall+ZB smart composite in solution reactive processes with various recycled Natural Rubber (rNR) bio-compatibilizer concentration. The addition of rNR bio-compatibilizer improved the total capacity loading (Hall and KF) filler to reach the optimum concentration. The addition of 3% rNR is capable to increase capacity of total loading filler KF and Hall up to 46% as the optimum condition (KH4). At optimum conditions it can increase the Tensile Strength (TS) and Elongation (Elo) up to 16% and 17% respectively, and decrease the Young Modulus (YM) to 1% compared to that the composites without rNR (KH1). The rNR plasticizer serves as a bio-compatibilizer that can increase the loading filler capacity and well interaction, finally, it can increase the TS properties. The increase of total loading filler KF and Hall to 46% (KH4) can also improve the inflammability of the composites. The Heat combustion (Hg) of KH1 without rNR was 42.3 kJ/g and at optimum condition the Hg of KH4 increases 4.8%, compared to that of KH1. The time to Ignition (TTI) increases by 6.9%, Burning Rate (BR) decreases by 3.7% and the Heat Release (HR) increases by 2.1% compared to that of the composites without rNR (KH1).

1. Introduction

World car production is increasing every year but on the other hand metal as raw material is depleting. Therefore, some automotive products replace metal materials with composite materials that are strong, lightweight, and combustible [1]. Composites are very attractive materials because, they combine different materials properties in ways not found naturally [2]. The use of 35% lightweight composite in automotive can save around 20% fuel usage [3]. Composites are very suitable to be applied in automotive components such as upholstery, door panels, interiors and car bodies [4]. One of the well-known natural reinforcement is kenaf fiber (KF), which has a high cellulose content thus has high tensile strength, ductile and not brittle [5]. In addition, any clay types can also be used for composite reinforcement. Lee et al. made PP/WF/MMt composites with PP-g-MA coupling compounds internally resulting in composites with TS, MY and high melting temperatures [6]. A combination at least two polymeric materials produces composites that have superior functions and properties than their constituent materials. Fiber material provides a temporary strength of polymer matrix, providing dimensional stability and transfer of shear stress

IOP Conf. Series: Materials Science and Engineering **578** (2019) 012069 doi:10.1088/1757-899X/578/1/012069

between fibers [7]. The result is a composite of a strong functional material that can be easily adjusted for a variety of uses [8]. To improve the mechanical properties and flame resistance of a composite can be done in addition together of fiber with clay such as halloysites (Hall) [9]. Ismail and Suryadiansyah [7] made PP/recycled natural rubber (rNR) composites reported to have high elasticity and mechanical properties compared to PP. Elastic composite engineering is one of them with the addition of a plastic such as rubber. Xie et al. made PP/styrene-butadiene-rubber (SBR)/SGF composites produce composites with high elasticity and mechanical properties [10]. Balakrishnan et al. made composites PP/epoxydized natural rubber (ENR)/OMMt [11] and other researcher Wang et al. made PP/EPDM/SGF composites to produce composites with high elasticity and mechanical properties [7]. Riyajan et al. [12] made composites PS/recycle rubber glove (rRG)/sugarcane leaves produced by composites with high elasticity and mechanical properties compared to starting materials. Based on this description, in this study studied the effect of the bio-compatibilizer rNR addition in various concentrations in a composite with Hall and natural fibers fillers: that have strong, lightweight and elastic properties compared to the starting material without plastic rNR.

2. Experimental

2.1 Materials

Recycled polypropylene (rPP) was collected from plastic recycling center in Surakarta Indonesia. The rPP prepared by cleaning up the waste cup and then chopped approximately in 5 mm x 2 mm. KF was obtained from the Kenaf Fiber Processing Center in Lamongan Indonesia. The KF was then overnight alkalized with 4% NaOH, and washed with distilled water and then dried in an oven at 60°C before chopped into 5 mm in sized. The recycle natural Rubber (rNR) was collected from waste natural rubber glove in Surakarta, Indonesia. The halloysite (Hall) clay was supplied by Applied Minerals Inc., USA. The Hall prepared by sieving with a 200 mesh and calcinated at 800°C for an hour. Preparation of rNR, after natural rubber glove cleaned up and dried then was decrosslinked. The crosslinked was done by using two roll mill and then crushed and sieved with a 40 mesh. All chemicals used in this study were pro-analysis from Aldrich; while benzoyl peroxide (BPO) and zinc borate (ZB) purchased from E.Merck.

2.2 Composites Manufacturing Process

Formulations of composites as follow: rPP/20%KF/0%rNR/20%Hall+ZB (KH1), rPP/23%KF/ 0%rNR/23%Hall+ZB (KH2), rPP/23%KF/1%rNR/23%Hall+ZB (KH3), rPP/23%KF/3%rNR /23%Hall+ZB (KH4), rPP/23%KF/5%rNR/23%Hall+ZB (KH5), and rPP/23%KF/7%rNR/ 3%Hall+ZB (KH6), composites synthesis are given Table 1. Composites synthesis is reactively done by solution process in xylene, equipped with a condenser, mechanical stirrer and N_2 gas. All starting materials: rPP, DVB, PP-AA, Hall, and KF blended in boiled xylene and mixed perfectly for 1 hour. Acrylate acid or AA is a multifunctional coupling compound that functions as a coupling agent for bonding of non-polar compounds (PP, rNR) with polar compounds (KF, Hall). In order for AA to interact more with non-polar compounds, AA coupling compounds are first converted to PP-AA. The mixture product was then evaporated to release the xylene to form masterbatches (MB's) composites. The MB's was hot pressed for 20 minutes at 180 °C to produce the specimen for physical and mechanical test [13].

Table 1. Formulation of rPP/DVB/PP-AA/rNR/KF/Hall+ZB composites

Code	Ingredients (perhundred, phr)					
	rPP	PP-AA	rNR	KF	Hall	ZB
KH1	65	15	0	20	20	5
KH2	65	15	0	23	23	5
KH3	65	15	1	23	23	5
KH4	65	15	3	23	23	5
KH5	65	15	5	23	23	5
KH6	65	15	7	23	23	5

2.3 Testing of materials
Characterization functional groups of raw materials as well as new material produced were performed using Shimadzu FTIR-8201 PC. Mechanical testing of TS, %Elo and Young Modulus (YM) was carried out using universal testing machine (UTM) by following ASTM D638. The testing of flame resilience of composites in accordance with ASTM D635 to determine the value of time to ignition (TTI), burning rate (BR) and %Heat Release (%HR). Bomb Calorimeter model 1341 Parr was used to measure the heat combustion (Hg).

2.4. Morphology Analyzing
The fracture surface of specimens after TS testing was characterized using SEM-EDX JEOL JSM-6360 LA to analysis morphology, especially on the cross section area of the fracture surface.

3. Results and Discussion

3.1. FTIR Analysis Functional Groups of Composites
The result of FTIR characterization is as follow. (a) In the rPP was obtained absorption at 2953-2839 cm^{-1} was >CH- and (>CH_2) groups and typical methyne group (>CH-) of PP uptake at 1723 cm^{-1}. (b) Broad absorption at 3433 cm^{-1} represents a hydrogen bonded of –S-H group of rubber, and at 1630 cm^{-1} of butadiene rubber (-HC=CH-). (c) Broad absorption at 3458 cm^{-1} represents a hydrogen bonded of hydroxyl groups of cellulose KF, Hall and –S-H group of rubber. Absorption at 2953-2839 cm^{-1} was >CH- and (>CH_2) groups. Absorption peaks at: 2723 (>CH- typical PP), 1629 cm^{-1} of butadiene rubber (-HC=CH-), and 1373 cm^{-1} methyl (-CH3) [14,15,16,17]. Absorption at 1730 cm^{-1} represents absorption of carbonyl groups of ester which were produced between hydroxyl cellulose of KF and carboxylic groups of AA. Absorption at 896-1046 cm^{-1} shown the Al-O and Si-O groups, representing interaction of Hall and PP in smart composites. It can be seen, that had been occurred the interaction among the starting material to form the new rPP/DVB/PP-AA/rNR/KF/Hall+ZB composites. This is supported by the previous work [18,19,20]. This interaction is expected to result in the mechanical properties and heat combustion better than the starting one.

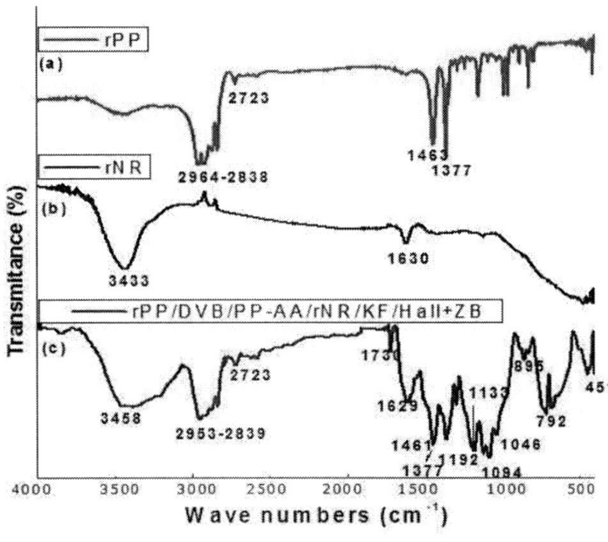

Figure 1. FTIR spectra of (a) rPP, (b) rNR, (c) rPP/DVB/PP-AA/rNR/KF/Hall+ZB

3.2. Mechanical properties of composites
Testing of mechanical properties was carried out with ASTM D638. To prove the mechanical properties of various (KH1-KH6) composites ASTM D635 is used for testing composites. The result is presented by tensile strength (TS), % Elongation (Elo) and Young Modulus (YM) are presented in Figure 2-7. The effect of adding rNR bio-compatibilizer on mechanical properties

based on the results of testing with ASTM D638 is as follows: (a) Tensile strength of rPP/KF20%/Hall20%+ZB (KH1) composite formulation without addition of rNR plasticizer was 53.9 MPa, an increase of 76% compared to that of raw material rPP (30.6 Mpa). (b) Meanwhile, the addition of rNR plasticizer achieved the optimum condition of composites rPP/3%rRG/KF23%/Hall23%+ZB (KH4). (c) In the composite formulations of KH2, KH3 and KH4 there was a variation in the increase in rNR concentration as follows: 0%, 1% and up to 3% there was a significant increase in TS, 2%, 5% and 16% respectively compared to KH1 without rNR. (d) However, the composite formulations of KH5 (5% rNR) and KH6 (7% rNR) the TS decreased to 3% and 4% respectively. The increasing of mechanical properties was caused by both KF and Hall act as a reinforcing material [21,22,23].

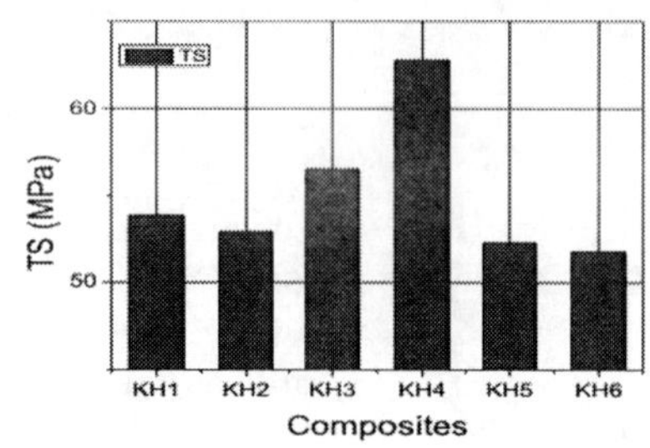

Figure 2. TS values of TPE PP/rNR/KF/Hall+ZB with various rNR concentration.

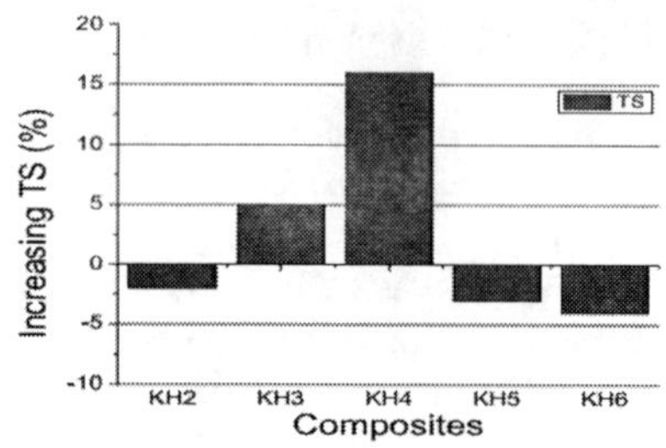

Figure 3. TS percentages comparison to that of their KH1 composite.

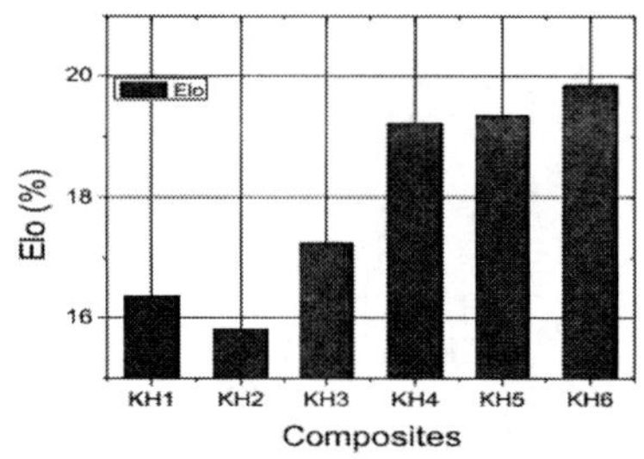

Figure 4. Elo values of TPE PP/rNR/KF/Hall+ZB with various rNR concentration.

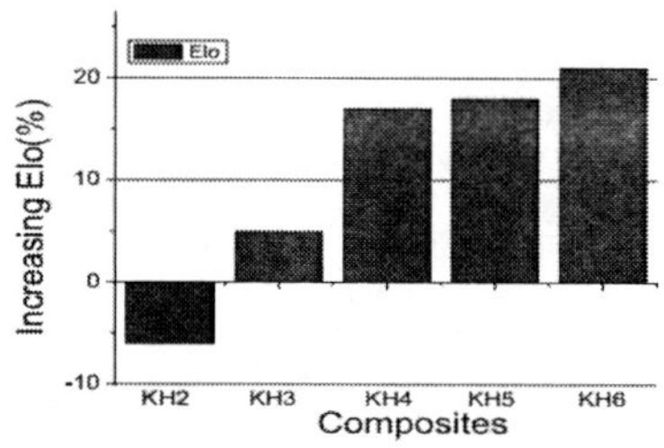

Figure 5. Elo percentages comparison to that of their KH1 composite.

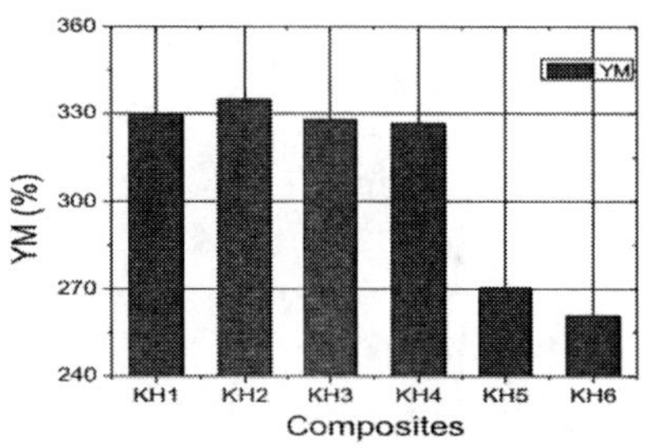

Figure 6. YM values of TPE PP/rNR/KF/Hall+ZB with various rNR concentration.

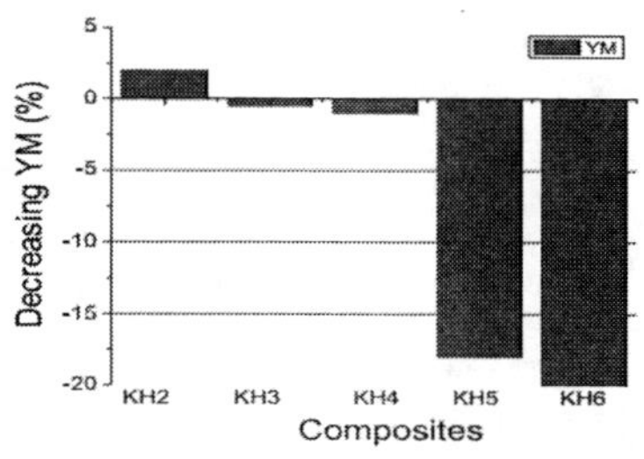

Figure 7. YM percentages comparison to that of their KH1 composite.

The addition of rubber plasticizer improves softness which results in the ability to increase the total amount of filler receipts by 15%. The increase in fillers has a consequence on the improvement of mechanical properties: TS up to 16%, Elo up to 17% and a decrease in stiffness down to 1% compared to that of KH1 composites without rNR plasticizer. Increasing the amount of KF filler in a mixture of PP matrices will increase the additive interaction between KF and PP which results in increased tensile. The existence of the rNR will increase the elongation which means the % Elo increases, this will reduce the stiffness (YM).

3.3. Morphologies Analysis

The morphological study carried out after TS testing and then characterized using SEM image especially on the cross section area of the fracture surface as can be seen at Fig. 8.

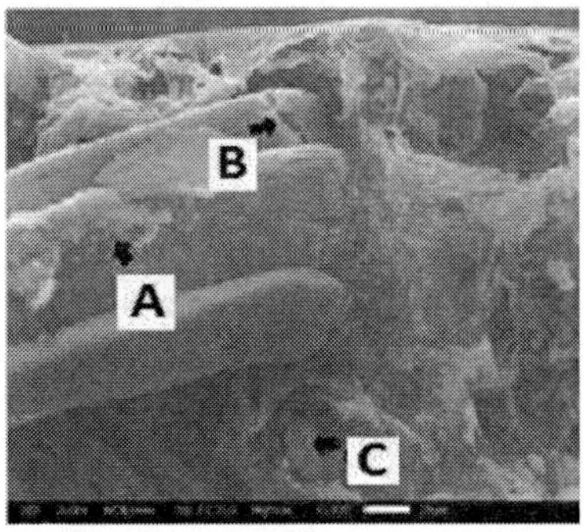

Figure 8. SEM of the cross section area of the fracture surface PP/3%rNR/23%KF/23%Hall+ZB (KH4) composite after Tensile Strength (TS) testing

SEM observations showed that the PP/3% rNR mixture matrix was able to bind the KF reinforcement well so that the TS of PP/3% rNR/23%KF/23%Hall+ZB (KH4) is 63.8 MPa. This KH4 composite strength was higher compared to that of PP/20%KF/20%Hall+ZB (KH1) without rNR plasticizer (53.9 MPa). Thus, the addition of 3% rNR can increase the total capacity loading of fillers 15% and larger reinforcements with better binding strength which reflected the increase of TS to 16% compared to that of KH1 without rNR plasticizer. (1) This is indicated by the presence of PP/3% rNR matrix which is firmly bound to the surface of kenaf fiber (can be seen in the SEM image on points A) and (2) the presence of kenaf fiber fracture near the end of the PP/3% rNR matrix bond (can be seen in the SEM image on points B). (3) In addition, the mixture matrix of PP/3% NR is also able to attach strongly to Hall (can be seen in the SEM image on points C); (4) so that the presence of Hall can function properly as an reinforcing filler which is simultaneously a flame retardant. The addition of rNR bio-plasticizer influences the softness which results in the ability to increase the total amount of filler receipts by 15%. The increase in fillers has a consequence on the improvement of mechanical properties: such as the TS up to 16%, the Elo up to 17% and a decrease in stiffness which was down to 1% compared to that of KH1 composites without rNR plasticizer. Increasing the amount of KF filler in a mixture of PP matrices will increase the additive interaction between KF and PP which results in increased tensile. The existence of the rNR will increase the elongation which means the % Elo increases, this will reduce the stiffness.

3.4. Heat Combustion, TTI, BR, and %HR of rPP/DVB/PP-AA/rNR/KF/Hall+ZB

To understand the heat combustion values of rPP/DVB/PP-AA/rNR/KF/Hall+ZB (KH1, KH2, KH3, KH4) composites, its carried out heat combustion testing and also to determine the influence of rNR bio-compatibilizer. The results of heat combustion (Hg) composites (KH1-KH6) using Bomb Calorimeter are presented in Fig. 9 and 10. The Hg of rPP raw material was 45.67 kJ/g. The Hg of rPP/KF20%/Hall20%+ZB (KH1) composites at optimum condition without rNR plasticizer were 42.3 kJ/g, it decreases 7.4% compare to that of rPP. The effect of adding rNR bio-compatibilizer on heat combustion (Hg) is as follows: the Hg of rPP/rNR0%/KF23%/Hall23%+ZB (KH2) composite formulation with addition of 0%rNR plasticizer was 41.3 kJ/g, a decrease 2.4% compared to that of KH1 without rNR. However, in the composite formulations of KH3, KH4. KH5 and KH6 there was containing a variation in the rNR concentration as follows: 1%, 3%, 5%, and 7% there was a significantly increase of the Hg, 2.4%, 4.8%, 9.5%, and 14.3%, respectively, compared to KH1 without rNR.

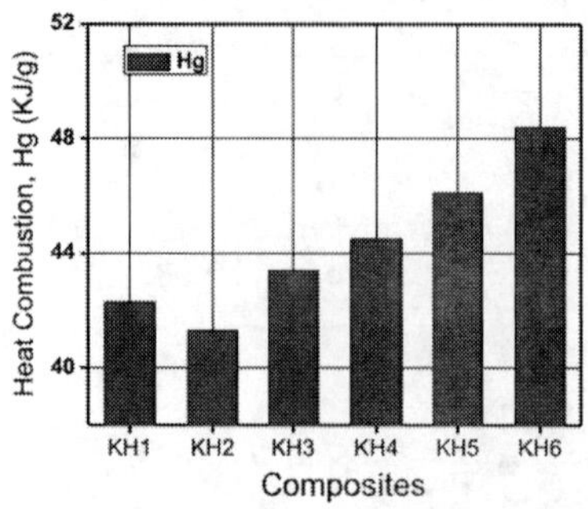

Figure 9. Hg values of TPE PP/rNR/KF/Hall+ZB with various rRG concentration.

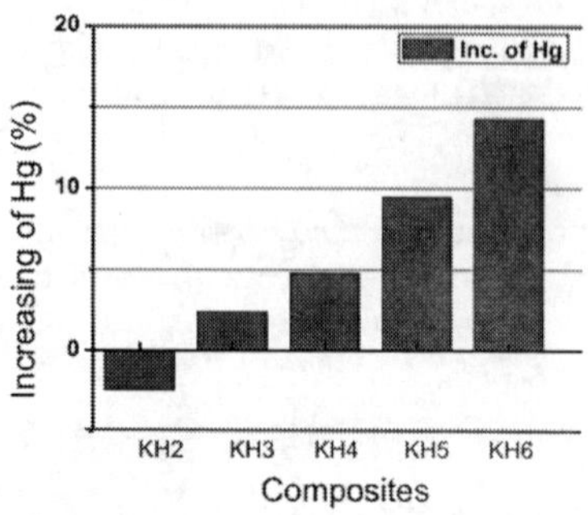

Figure 10. Hg percentages comparison to that of their KH1 composite.

To prove the inflammability of various (KH1-KH6) smart composites it performed resistance combustion testing. The method horizontal burning of ASTM D635 is used for testing composites. The result is presented by the time to ignition (TTI), the burning rate (BR) and % Heat Release (%HR), which can be seen in Figure 11-16.

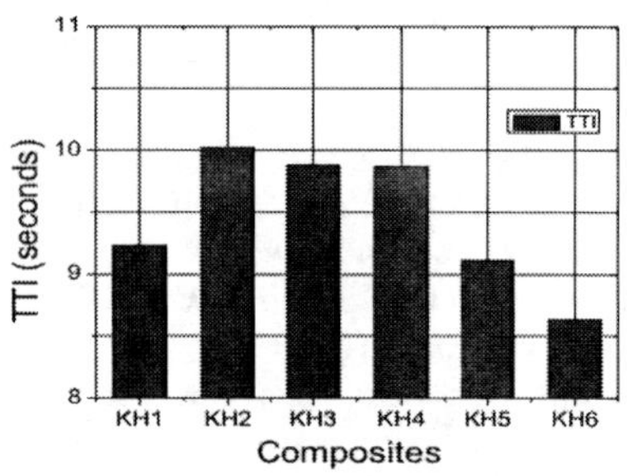

Figure 11. TTI values of TPE PP/rNR/KF/Hall+ZB with various rRG concentration.

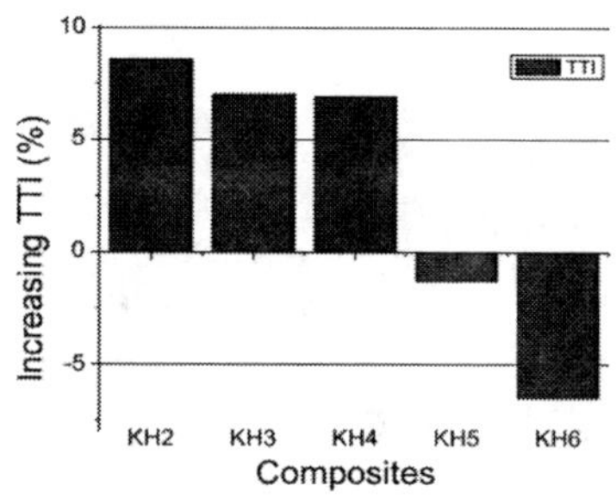

Figure 12. TTI percentages comparison to that of their KH1 composite.

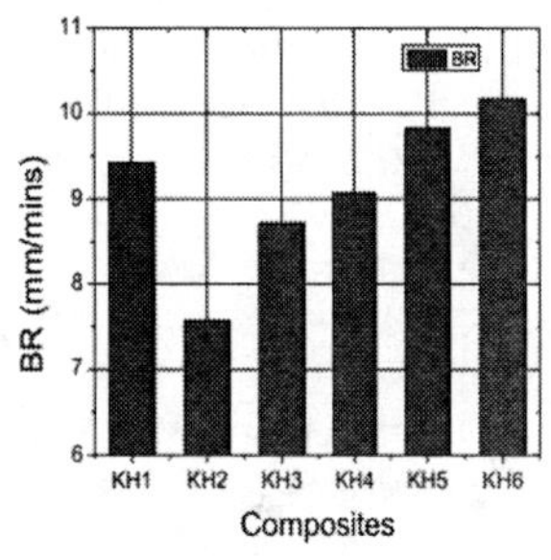

Figure 13. BR values of TPE PP/rNR/KF/Hall +ZB with various rRG concentration.

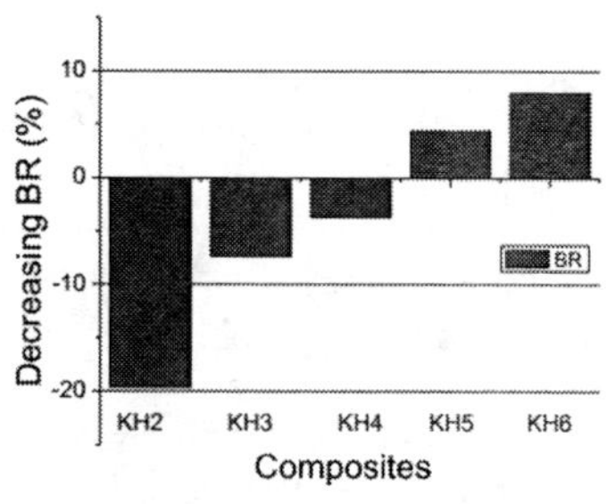

Figure 14. BR percentages comparison to that of their KH1 composite.

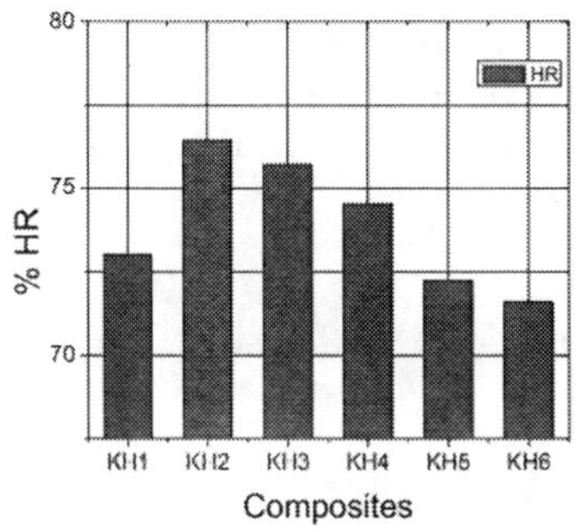

Figure 15. % HR values of TPE PP/rNR/KF/ Hall+ZB with various rRG concentration.

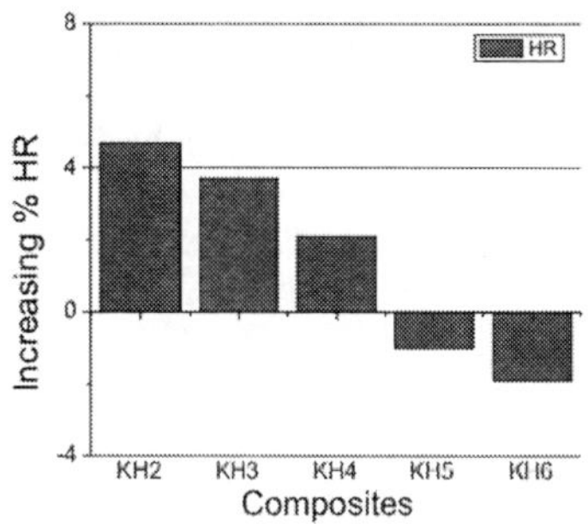

Figure 16. HR percentages comparison to that of their KH1 composite.

The addition of rNR as bio-plasticizer influences the softness which results in the ability to increase the total amount of total filler receipts by 15%. The increase in fillers has a consequence on the improvement of heat combustion. In the presence of 23% KF and 23% Hall+ZB (KH4) it increases Hg to 4.8%, this because KF and Hall+ZB can inhibit oxygen in the burning process. Cellulose fibers of KF during the combustion process formed charcoal as a barrier that prevents the entry of oxygen at the surface. The Hall undergo endothermic reaction which can absorb heat energy during the combustion process takes place, while ZB can release water vapor to cool the system [24,25,26,]. The synergy of these three components causing the flame resistance of the composite rPP/KF/rNR/Hall+ZB (KH4) increased significantly. The TTI, BR and %HR of the rPP/KF20%/Hall20%+ZB (KH1) composites without rNR plasticizer was 9.23 second, 9.42 mm/mins and the %HR is 73. In general, the presence of rNR in composites will accelerate combustion. The higher the rNR concentration the fuel resistance decreases, because of the nature of combustible rubber. However, the presence of KF and Hall is able to withstand these flammability. Meanwhile, the addition of 3% rNR plasticizer achieved the optimum condition of composites rPP/rRG/KF23%/Hall23%+ZB (KH4). The presence of 5% ZB as fire retardant additive can also inhibited TTI and also slow down the BR. The ZB compound if burned is also formed B_2O_3 which is moist and form a stable charcoal. Water that is released by ZB during combustion can cause charcoal to form a foam that inhibits combustion, this is supporting by some previous work [2,22,27].

4. Conclusion

The compositions of synthesis rPP/DVB/PP-AA/rNRKF/Hall+ZB by using 3% (w/w) rNR plasticizers to reach the optimum condition can increase total input capacity (KF+Hall) is 15% (w/w. The increased 15% of total loading filler capacity can increase the TS and % Elongation to 16% and 17% respectively and decrease young modulus (YM) to 1% compared to that of the optimum condition without rNR plasticizers (KH1). It can also improve the Heat Combustion 4.8% compared to that of KH1. The fuel resistance of composites also better as follow: increasing TTI by 6.9%, decreasing BR to 3.7%, and %HR increasing to 2.1% compared to that of the optimum conditions without rNR plasticizers (KH1).

Acknowledgment

The author would like to thank for financial support in this project through International Research Collaboration and Scientific Publication Grants to Directorate General of Higher Education Indonesia.

References

[1] Gordon R B, Bertram M, and Graedel T E 2006 *Proc Natl Acad Sci.* **103** 1209-1214
[2] Suharty N S, Almanar I P, Sudirman, Dihardjo K, and Astasari N 2012 *Procedia Chemistry* **4** 282-287.

[3] Bandivadekar A, Bodek K, Cheah L, Evans C, Groode T, Heywood J, Kasseris E, Kromer M, Weiss M 2008 *On the road in 2035: reducing transportation's petroleum consumption and GHG emissions* (Cambridge, MA: Massachusetts Institute of Technology) Report no. LFEE 2008-05 RP.
[4] Lyu M and Choi T G 2015 *Int J Precis Eng Man.* **16** 213-220.
[5] Mwaikambo LY 2006 *J. Food Sci. Technol.* **7** 120-133.
[6] Lee S Y, Kang I A, Doh G H, Kim W J, Yoon H G, and Wu Q 2008 *Express Polym Lett.* **2** 78-87
[7] Ismail H and Suryadiansyah 2002 *Polym. Test.* **21** 389-395
[8] Wang W, Longxiang T, and Baojun Q 2003 *Eur. Polym. J.* **39** 2129–2134
[9] Suharty N S, Ismail H, Diharjo K, Handayani D S, and Firdaus, M 2016 *Procedia Chemistry* **19** 253-258.
[10] Xie H Q, Zhang S, and Xie D 2005 *J Appl Polym Sci.* **96** 1414-1420
[11] Balakrishnan H, Attaran S A, Imran M, Hassan A, and Wahit M U 2012 *J. Thermoplast. Compos. Mater.* **27** 233-250
[12] Riyajan S A, Intharit I, and Tangboriboonrat P 2012 *Ind Crops Prod* **36** 376–382.
[13] Suharty N S, Wirjosentono B, Maulidan F, Handayani D S, Sholikhah J, and Maharani Y A 2008 *J. Phys. Sci.* **19** 105-115.
[14] Suharty N S, Hanafi I, Fajar R W, Kuncoro D, Desi S H, Maulidan F, and Lupi L 2015 *App. Mech. Mater.* **736** 13-18.
[15] Mustafa I 2010 *Natural* **10** 7-13.
[16] Benmesli S and Riahi F 2014 *Polym. Test.* **36** 54-61.
[17] Suharty N S, Ismail H, Diharjo K, Nizam M, and Firdaus M 2014 *Adv. Mater. Res.* **950** 18-23
[18] Yuan P, Peter D S, Zongwen L, Malcolm E R G, James M H, Sarah J A, and Camerom J K 2008 *J. Phys. Chem. C.* **112** 15742-15751.
[19] Larena A, Urreaga J M, and de la Orden M U 1991 *J. Colloid Interface Sci.* **142** 111-115.
[20] Petersen H N, Kusano Y, Bronsted P, and Almdal K 2013 *Risoe International Symposium on Materials Science Proceedings.* **34** 333-340.
[21] Pasbakhsh P, Ismail H, Ahmad Fauzi M N, and Abu Bakar A 2009 *Polym. Test.* **28** 548-559.
[22] Sain M, Balatinecz J, and Law S 2000 *J. Appl. Polym. Sci.* **77** 260.
[23] Al-Hassany Z, Genovese A, and Shanks R A 2010 *J. Polym. Lett.* **4** 70-93.
[24] Suharty N S, Ismail H, Wibowo F R, Handayani D S, Firdaus M, and Lathifah L 2015 *Adv. Mater. Res.* **1105** 56-61.
[25] Al-Hassany Z, Genovese A, and Shanks R A 2010 *J. Polym. Lett.* **4** 70-93.
[26] Delhom C D, Ghoorahoo W, and Pang S S 2010 *Composites: Part B.* **41** 475-481.
[27] Suharty N S, Ismail H, Diharjo K, Nizam M, and Firdaus M 2014 *Adv. Mater. Res.* **950** 18-23.

IOP Conf. Series: Materials Science and Engineering **578** (2019) 012070 doi:10.1088/1757-899X/578/1/012070

Compatibility and Thermal Properties of Maleic Anhydride-grafted-Polystyrene Containing Microcrystal cellulose (Avicel) as Raw Material for Soil Binder

B Wirjosentono[1*], Tamrin[1], A H Siregar[1], Sukatik[2], D A Nasution[1]
[1)] Department of Chemistry, University of Sumatera Utara, Medan 20155, Indonesia
[2)] Department of Civil Engineering, Padang State Polytechnics, Padang, Indonesia

*E-mail: basuki@usu.ac.id

Abstract. Polymeric resins have been utilized widely as soil or dust binders in building and road construction industries. In this work maleic anhydride (MA) was grafted into polystyrene (PS) resins in a bench-scale xylene reflux reactor in the presence of benzoyl peroxide (BPO) to improve their adhesion properties, whereas the grafting efficiency was characterized using FTIR spectroscopy. Further processings were then carried out with addition of varied loading of microcrystal cellulose (Avicel), 1-5%, to improve their mechanical and thermal properties. It was found that addition of Avicel into the MA-g-PS resins up to 4% improved tensile strength and Young's modulus. The presence of BPO may initiate chain scission of the MA-g-PS resins, which decreased the melting temperature as well as heat of combustion. Morphological analysis of surface of the resin specimens using electron microscopy (SEM) indicated rougher surface when BPO and Avicel were added in the resins. It was expected that binding of MA-g-PS/Avicel resins onto surface of sand and soil aggregates will be higher.

1. Introduction

Utilisations of polymeric resin in building and road constructions is not limited as structural and engineering materials, but also as resin emulsion (water and oil-based latexes) for polymer concretes, aggregate binders (soil stabilizers), and paint binders (water-borne binders), [1]. Especially as water-based resin emulsion (WRE), it has been utilized in strengthening foundation in road constructions in the USA, [2]. And generally, the WRE used as soil stabilisers are polyacrylates and polyacrylamides with solid polymer contents up to 55%, [3]. When compared to oil-based resin emulsion (ORE), the WRE is not only more economical, but also more environment friendly and healthy since it reduces the use of (sometimes a toxic and carcinogenic) organic solvents, [4].

On the other hand, polystyrene as commodity (cheap and abundantly produced) polymer resins (fresh as well as recycled), recently has only been used as packagings and general purposes consumer's goods, instead of more valued-added engineering materials [5]. Whereas when used in aggressive and higher temperature outdoor environments (under UV-radiation in sunlight), the resin easily undergoes degradations and deteriorations, which decreases its physical and mechanical properties [6],[7]. After modification, stabilization, and emulsification processes (using additives comonomers, antioxidants, and surfactants), the polymer resin can be produced as water-borne resin latex emulsion (WRE) for paint binders, and soil stabilisers [8], [9], [10]. In this work, the commodity polystyrene resins (fresh and recycled) have been processed to produce raw material for water-based resin emulsion, with higher mechanical and thermal properties for application as raw materials for soil and dust stabilizers in building and road constructions. The technique involves processing of polystyrene resin in a bench scale modification xylene reflux reactor, with addition of maleic anhydride (MA) as modifier, microcrystal cellulose (Avicel) as reinforcement, and in the presence of benzoyl peroxide (BPO) as initiator.

2. Experimental

The work start with selections of comonomers (maleic anhydride, styrene, and divinylbenzene) used as compatibilizers, and reactive processing radical generator (benzoyl peroxide, dicumyl peroxide) as initiators. The modification processes were carried out in a bench-scale reflux reactor intensive stirring for (1-6 hours), using (toluene, xylene) solvent under nitrogen atmosphere, [11]. Further processings were then carried out with addition of varied loading of microcrystal cellulose (Avicel), 1-5%, to improve their mechanical and thermal properties [12]. The reaction mixtures were then evaporated in vacuum evaporator, dried in vacuum oven at 80 °C to constant weight. The dried resins were then compression moulded at 170 °C and cut to produce film specimens (thickness: 0.1-0.2 mm) for mechanical testing according to ASTM D638. Efficiency of the modification reactions were investigated using infrared spectroscopy (FTIR) analysis of the specimens before and after exhaustive Soxhlet extraction in dichloromethane. Compatibility, morphology and thermal properties of the modified-polystyrene resins were studied using Scanning Electron Microscopy (SEM), and Differential Scanning Calorimeter (DSC) analysis.

3. Results and Discussion

3.1 Grafting efficiency of maleic anhydride (MA) in polystyrene (PS) matrix

Grafting efficiency of maleic anhydride (MA) after reactive modification in reflux of polystyrene (PS) solution in xylene was investigated using FTIR spectroscopy of film specimens before and after exhaustive Soxhlet extraction in dichloromethane for 4 hours. FTIR spectra of the film (PS-g-MA) before and after extraction were shown in Figure 1 (a and b).

It was shown that FTIR spectra of both specimens (before and after extraction) indicated the existence of absorption peaks at wave number 1604 cm^{-1} (stretching), and at 1373 cm^{-1} (bending) of carbonyl group of maleic anhydride (>C=O). And when intensities of the absorption peaks were compared to that of respective corresponding absorption peaks of hydrocarbon (C-H) group at 3464 cm^{-1}, the ratio did not change significantly. This revealed that the maleic anhydride compatibilizer has been unextractable and efficiently grafted (physico-chemically) into polystyrene resin. In addition both FTIR spectra did not show any absorption peak in the wave number area of 1700-1800 cm^{-1}, which is particularly representing the existence of carbonyl group of carboxylic acid [-(C=O)-OH]. And this is indicating that during the modification reactions the maleic anhydride directly bounded onto the polystyrene backbone without side reaction to produce maleic acid. However, there is no absorption peak data shown which indicating how the MA grafted chemically onto PS macromolecular backbone. And it is inevitable that BPO radical formed at the beginning of the reaction may initiate various possible side reactions, such as: homopolymerisation to produce poly/olygomaleic anhydride [-$(MA)_n$-], and this may further grafted and even crosslinked onto PS backbone.

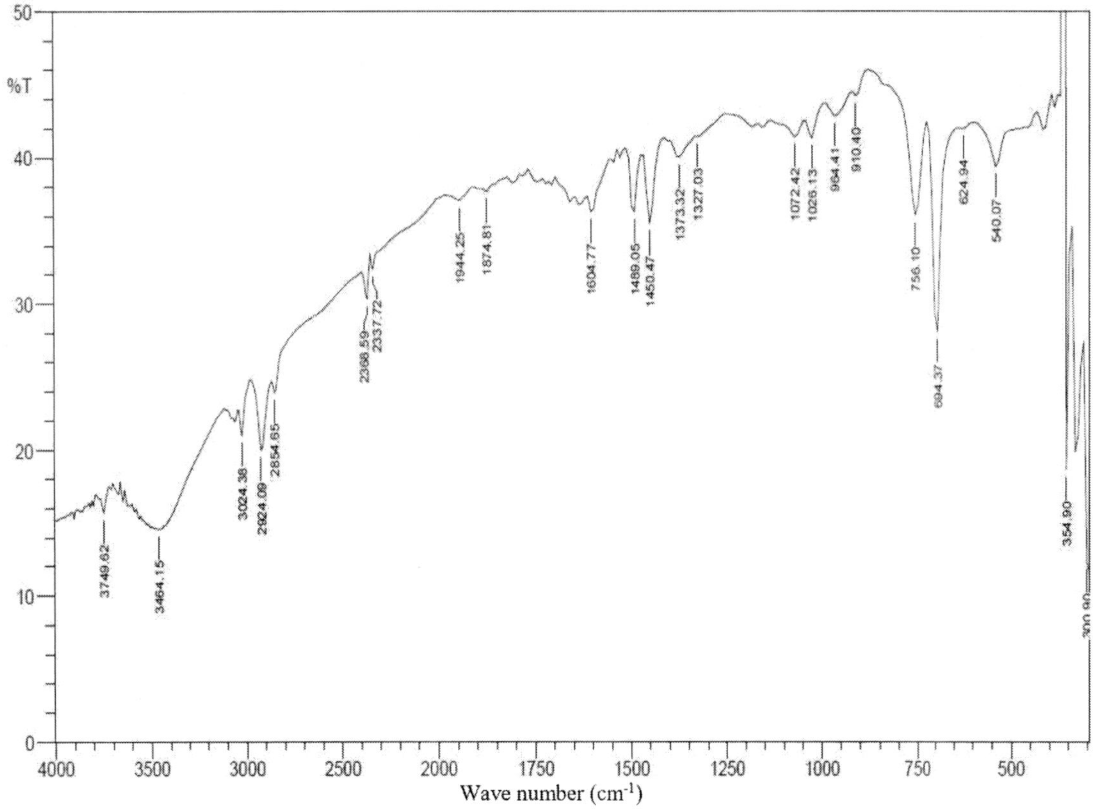

Figure 1a. FTIR spectrum of PS-g-MA film specimen

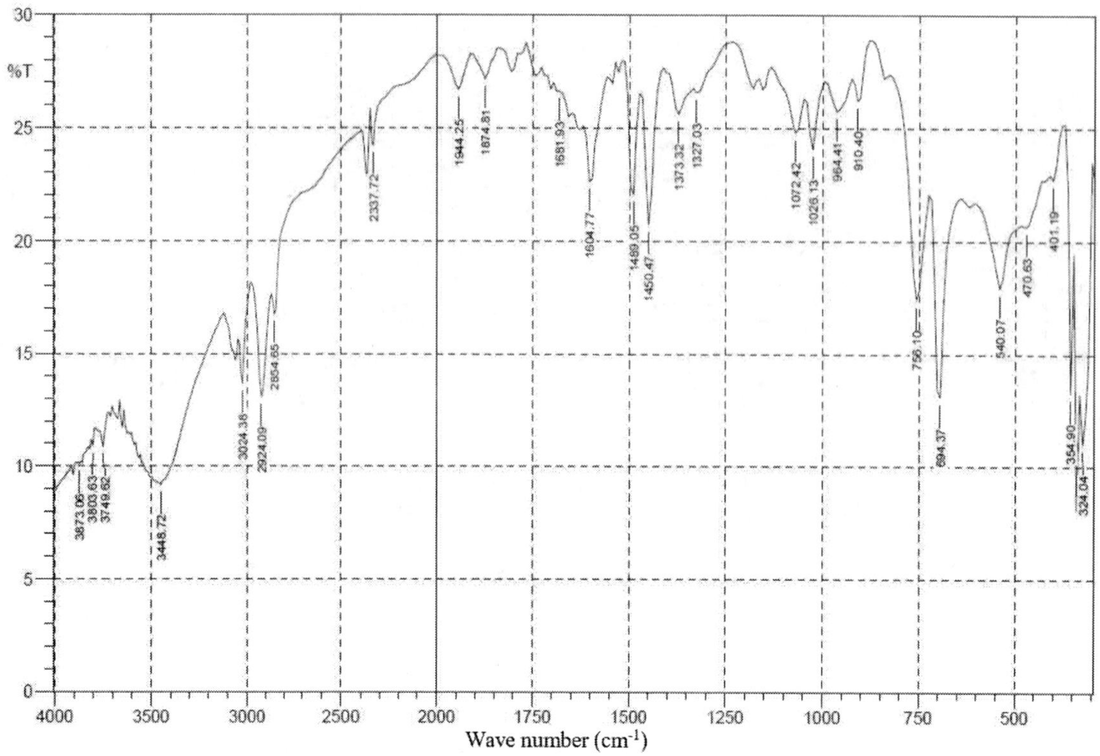

Figure 1b. FTIR spectrum of PS-g-MA film specimen after exhaustive Soxhlet extraction in dichloromethane for 4 hours

3.2 Mechanical properties and compatibility of MA-grafted-polystyrene containing Avicel (PS-g-MA/BPO/Avicel) resins

Mechanical properties of maleic anhydride-grafted-polystyrene resins containing Avicel (PS-g-MA/Avicel) with various loading (1-5%), as shown in Table 1, were tested according to ASTM D638 using Universal Tensile Testing RTF, Load Cell 5000 N, rate 5 mm/minute.

Table 1. Mechanical properties testing of (PS-g-MA/BPO/Avicel) samples using Universal Tensile Testing RTF, Load Cell 5000 N, testing rate 5 mm/minute.

No	Samples	Avicel (%)	Tensile Strength (MPa)	Elongation (%)	Modulus (MPa)
1	PS/MA/BPO	0	7.72	2.59	170.70
2	PS/MA/BPO/Avicel1	1	13.71	10.76	189.77
3	PS/MA/BPO/Avicel2	2	14.03	9.16	190.43
4	PS/MA/BPO/Avicel3	3	13.58	7.77	210.77
5	PS/MA/BPO/Avicel4	4	16.34	6.15	211.64
6	PS/MA/BPO/Avicel5	5	10.14	5.59	194.05

It was shown from Table 1 and Figure 2a-2f, that in general addition of Avicel (1-5%) into the PS-g-MA resins increased their mechanical properties. Especially using addition of 4% Avicel the tensile strength and modulus increased from 7.72 to 16.334 MPa, and from 170.70 to 211.64 MPa, respectively. This is indicating that the PS-g-MA resin has interacted well on the surface of the Avicel particles, although when the Avicel loading up to 5% the mechanical properties decreased slightly, which doe to agglomeration of the Avicel particles.

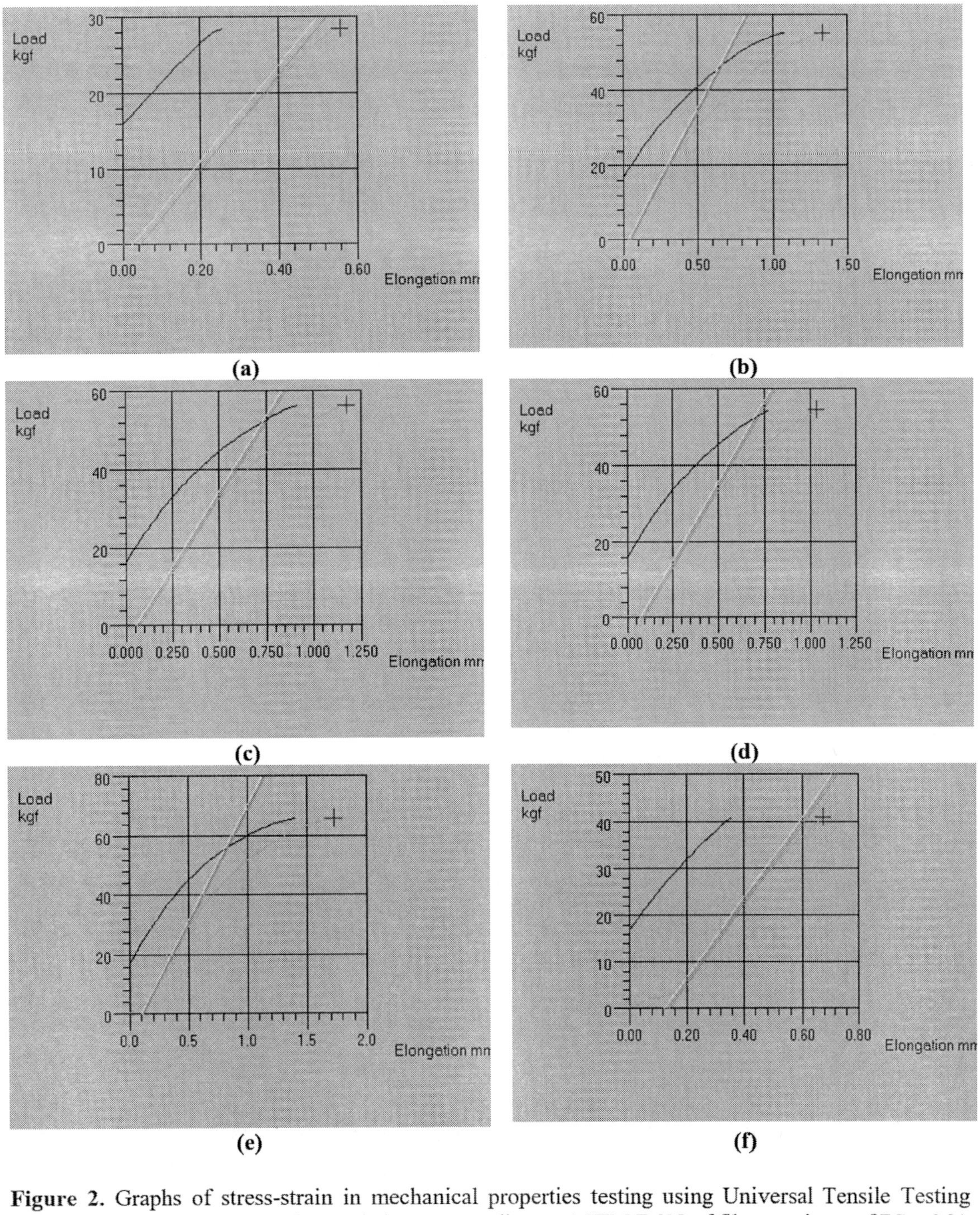

Figure 2. Graphs of stress-strain in mechanical properties testing using Universal Tensile Testing RTF, Load Cell 5000 N, speed 5 mm/minute, according to ASTM D638 of film specimen of PS-g-MA without Avicel (a) and with varied loadings of Avicel: 1% (b), 2% (c), 3% (d), 4% (e), and 5% (f)

IOP Conf. Series: Materials Science and Engineering **578** (2019) 012070 doi:10.1088/1757-899X/578/1/012070

3.3 Thermal properties of PS-g-MA/BPO/Avicel using Differential Scanning Calorimeter (DSC)

Figures 3a to 3c and Table 2 showed thermographs and the data of endothermic and exothermic processes of PS/MA, PS-g-MA/BPO, and PS-g-MA/BPO/Avicel samples, measured using: DSC-60; C30935300214SA, temperature range 30°-500 °C, temperature rate 15°/minute, in Nitrogen gas atmosphere, flow rate 30 mL/minute, Cell Alumina, and sample weight 5 mg. They showed endothermic as well as exothermic peaks, which represented melting and decomposition processes of the heated resins. It was shown that addition of BPO and Avicel decrease endothermic peak temperature slightly (from 164.45° to 159.00 °C), followed by slight decrease on the heat capacity (-63.38 to -72.25 J/g), which is related to chain scission of the resins macromolecules. The exothermic process also showed decrease on temperature peaks when the BPO and Avicel were added (from 424.78° to 407.81 °C), and followed by substantial increase on heat capacity (from 280.38 J/g to 903.10 J/g). It seemed that, especially with addition of BPO efficiency of the decomposition process is higher.

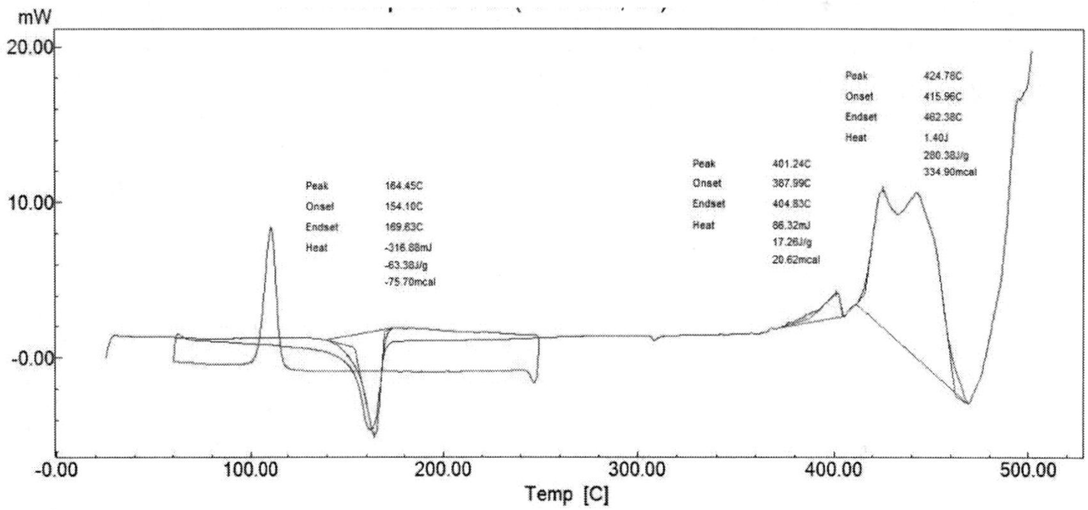

Figure 3a. Thermograph of PS/MA sample, measured using: DSC-60; C30935300214SA, temperature range 30°-500 °C, temperature rate 15 °/minute, in Nitrogen gas atmosphere, flow rate 30 mL/minute, Cell Alumina, and sample weight 5 mg.

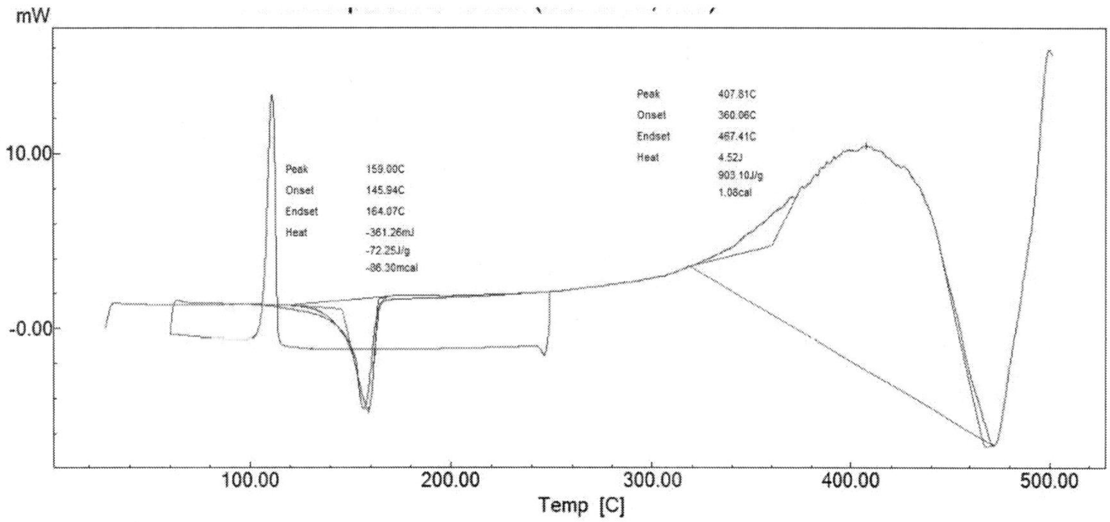

Figure 3b. Thermograph of PS-g-MA/BPO sample, measured using: DSC-60; C30935300214SA, temperature range 30°-500 °C, temperature rate 15 °/minute, in Nitrogen gas atmosphere, flow rate 30 mL/minute, Cell Alumina, and sample weight 5 mg.

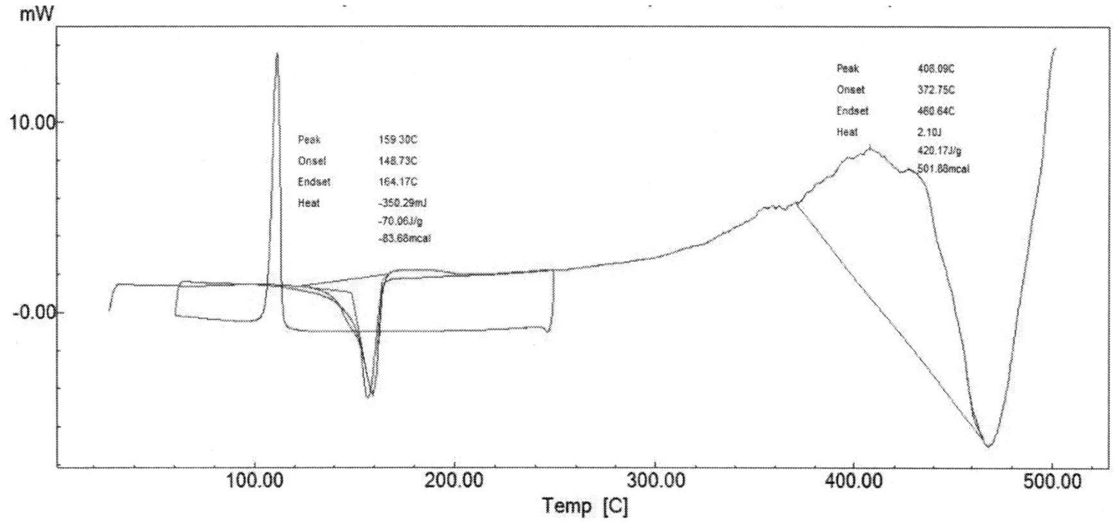

Figure 3c. Thermograph of PS-g-MA/BPO/Avicel (4%) sample, measured using: DSC-60; C30935300214SA, temperature range 30°-500 °C, temperature rate 15 °/minute, in Nitrogen gas atmosphere, flow rate 30 mL/minute, Cell Alumina, and sample weight 5 mg.

IOP Conf. Series: Materials Science and Engineering **578** (2019) 012070 doi:10.1088/1757-899X/578/1/012070

Table 2. Thermograph data of (a) PS/MA, (b) PS-g-MA/BPO, and (c) PS-g-MA/BPO/Avicel (4%) samples, measured using: DSC-60; C30935300214SA, temperature range 30°-500 °C, temperature rate 15°/minute, in Nitrogen gas atmosphere, flow rate 30 mL/minute, Cell Alumina, and sample weight 5 mg.

No	Samples	Endothermic Process (Melting)				Exothermic Process (Decomposition)			
		Peak (°C)	Start (°C)	End (°C)	Calory (J/g)	Peak (°C)	Start (°C)	End (°C)	Calory (J/g)
1	PS/MA	164.45	154.10	169.63	- 63.38	424.78	415.96	462.38	280.38
2	PS-g-MA/BPO	159.00	145.94	164.07	- 72.25	407.81	360.06	467.41	903.10
3	PS-g-MA/BPO/ Avicel(4%)	159.30	148.73	164.17	- 70.06	408.09	372.75	460.64	420.17

3.4 Morphology characteristic of PS-g-MA/BPO/Avicel samples

Microscopic surface images of film specimens were shown in Fugure 4(a-c), i.e. that of: (a) (PS/MA) without BPO, and with addition of BPO, (b) (PS-g-MA/BPO), and with Avicel 4% (c) (PS-g-MA/BPO/Avicel). It was clear tht without BPO the blending process of PS/MA did not resulted in homogenous and physic-chemical interaction blends, which showed porous surface with diameter 10-30 μm (Figure 4a). When BPO was added in the reactive blending process the PS-g-MA/BPO experienced physic-chemical interaction, in which the MA comonomer grafted onto the PS backbone initiated by BPO and formed more homogenous blend (pore diameter $<$ 10 μm, Figure 4b). The grafted MA has also shown to function as compatibilizer to finely distribute the Avicel particle as shown in Figure 4c.

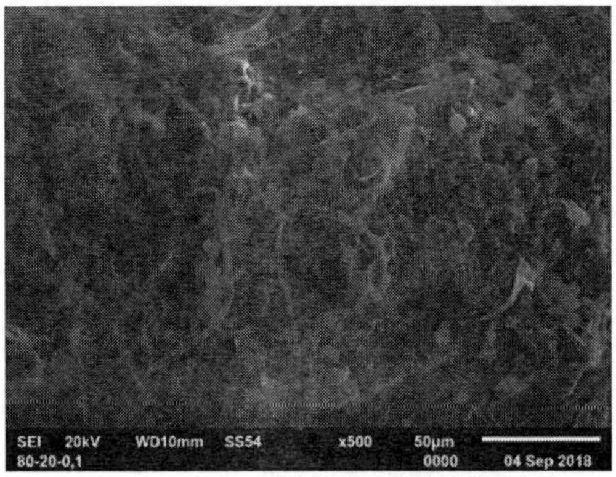

(a)

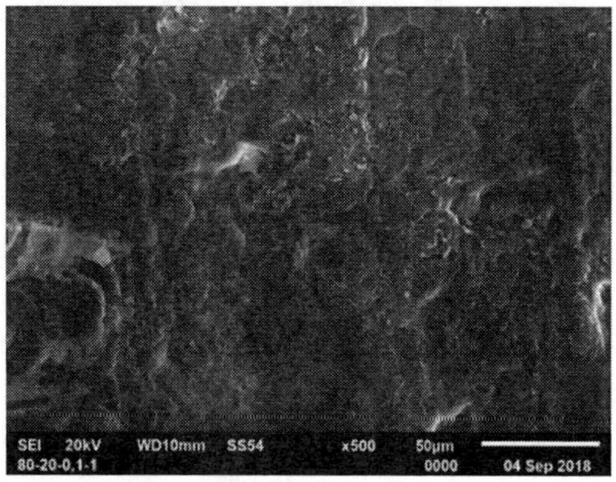

(b)

(c)

Figures 4. Micrographs using electron microscope (SEM) JEOL *series* JSM-6510LA (magnification 500 times) on film specimen surface of: (a) PS/MA without BPO, (b) PS-g-MA/BPO, and (c) PS-g-MA/BPO/Avicel (4%).

4. Conclusion

It was found that addition of Avicel into the MA-g-PS resins up to 4% improved tensile strength and Young's modulus. The presence of BPO may initiate chain scission of the MA-g-PS resins, which decreased the melting temperature as well as heat of combustion. Morphological analysis of surface of the resin specimens using electron microscopy (SEM) indicated rougher surface when BPO and Avicel were added in the resins. It was expected that binding of MA-g-PS/Avicel resins onto surface of sand and soil aggregates will be higher.

IOP Conf. Series: Materials Science and Engineering **578** (2019) 012070 doi:10.1088/1757-899X/578/1/012070

Acknowledgements
The authors would like to thanks to Universitas Sumatera Utara and General Directorate of Higher Education, Ministry of Research and Higher Education of Indonesia for granting the research grant on research scheme of "*Penelitian Terapan Unggulan Perguruan Tinggi (PTUPT)*" 2018.

References
[1] Fauzia A, Wan-Abdul-Rahman WMN and Jauhari Z 2013 *Procedia Eng.* **53** 42-47.
[2] Howard J, Hanley M, Eyad A, Srinath M, Iyengar R, Rodriguez AK and Bazi HS 2014 Co-Polymer Soil Subgrade Binders *USA Patent*, US20140169879 A1.
[3] Sharifah-Zaliha SZ, Kamarudin H, Mustafa Al Bakri AM, Binhussain M and Siti-Salwa MS, 2013 *Aust. J. Basic Appl. Sci.* **7**(5) 258-265.
[4] Ilies NM, Circu AP, Nagy AC, Ciubotaru VC and Kisfaludi-Bak Z 2017 *Procedia Eng.* **181** 444-451.
[5] Al-Malaika S, Scott G and Wirjosentono B 1993 *Polym. Degrad. Stab.* **40**(2) 233-238.
[6] Arnaud R, Dabin P, Lemaire J, Al-Malaika S, Chohan S, Coker M, Scott G, Fauve A and Maaroufi A 1994 *Polym. Degrad. Stab.* **46**(2) 211-224.
[7] Yehyea WA, Abdul-Rahman N, Ariffin A, Abd. Hamid SB, Alhadi AA, Kadir FA and Yaeghoobi M 2015 *Eur. J. Med. Chem.* **101** 295-312.
[8] Muhpidah, Hambali E, Suryani A, Kartika IA 2017 *IOP Conf. Ser.: Earth Environ. Sci.* **65** 012033.
[9] Wirjosentono B, Tamrin, Siregar A H, Nasution T I, Dalimunthe K Z, Nasution D A 2018 *J. Phys. Conf. Ser* **1116** 042043.
[10] Poljanšek I, Fabjan E, Burja K, Kukanja D 2013 *Prog. Org. Coat.* **76**(12) 1798-1804.
[11] Wirjosentono B, Guritno P, Ismail H, 2004, *Int J Polym Mater Po* **53**(4) 295-306.
[12] Sumaiyah, Wirjosentono B and Karsono 2016 *Int. J. PharmTech. Res.* **9(7)** 130-139.

Thermal behaviour properties and corrosion resistance of carbon/polyurethane film

O Kurniawan[1], I H Ramadhan[2] and B Soegijono[2]
[1.] Graduate Materials Science Program Department of Physics, Universitas Indonesia, Depok, Indonesia
[2.] Department of Physics, Universitas Indonesia, Depok, Indonesia

Email: naufal@ui.ac.id

Abstract. Carbon/polyurethane film composite were prepared by adding carbon with different content in polyurethane as a matrix. The materials were characterized by means of thermogravimetric analysis (TGA), differential scanning calorimetry (DSC), Fourier transform infrared spectroscopy (FT-IR), Optical microscope and salt spray test. TGA and DSC results showed decomposition temperature shifted to a lower point and the FT-IR spectra showed chemical reaction of carbon/polyurethane matrix. The corrosion resistance overall decrease with the increasing carbon content.

1. Introduction

Efforts to use Carbon as a filler added to the polymer have been carried out to determine and obtain the desired material properties. Adding carbon to polyurethane is possibe to obtaining material such as heat resistance, flame retardance and proper of mechanical properties [1-4].

To obtain polymer materials with performance related to heat resistance and corrosion resistance, it is necessary to understand the decomposition of polyurethane materials. Thermal decomposition of polyurethane is recognized as a heterogeneous complex [5]. Thermal decomposition of polyurethane (heat energy feature absorbed into its degradation) and corrosion resistance are important occurrence of fundamental and industrial practice. To obtain high-performance polymeric materials, understanding the degradation process allow the determination of optimal conditions for designing polyurethanes. Specific temperatures or temperature limits where the material can be used without losing its properties are often called thermal stability.

Research of Carbon as a filler of polymer (as a base of composite) have been conducted [6-8]. Carbon works well as a filler in a polyurethane matrix that has the results of thermal resistance and good electrical conductivity [9]. Research of composite polyurethane filled by carbon was conducted on wear resistance of polyrethane/carbon with pigment concentration and NCO/OH ratio as parameters that affect the potential of wear resistance and the result found that the most effective factor in wear resistance is type polyols due to CO's potentiality to form hydrogen bonds with hard segments of polyurethane and also that the weight of carbon pigments has an average effect on wear resistance [10].

Research to improve advanced properties such as mechanics, permeability and surface area in polyurethane / carbon composites has been frequently and widely investigated. The addition of carbon nanotubes or graphite in an effort to improve thermal mechanical properties can also increase the mechanical strength of about 20% by adding up to 0.05% carbon nanotube and also increase resistance to fire [1-11]

This paper describes the results of the characterization of carbon-polyurethane nanocomposite. Polyurethane used is a commercial product with the addition of aluminum. The addition of this aluminum makes it different from other polyurethane. Observations on thermal behavior and corrosion resistance were carried out based on different carbon content and it was estimated that there was a change in thermal behavior and corrosion resistance due to changes in carbon content

2. Experimental

Polyurethane used is a commercial product where there is an additional aluminum which makes it different from other polyurethane. The Polyurethane composition consists of diisocyanate (MDI), MDI based Polyisocuanate, 4.4' – Diphenylmethane, Aluminum, Aromatic Solvent Blend and Isobutyl Acetate. Figure 1 show the urethane groups –NH–(C=O)–O– link the molecular units.

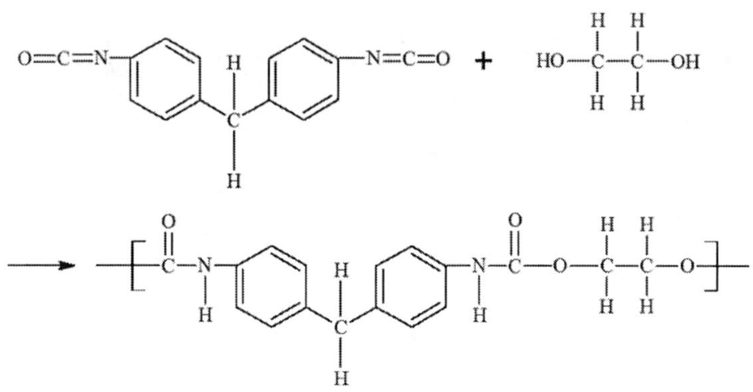

Figure 1. The urethane groups –NH–(C=O)–O– link the molecular units

Carbon used as fill material is a carbon allotrope, which is graphite. Micro-carbon comes from pure carbon graphite which is then made in powder size with a diameter of 30 Mesh (0.23 inch or 595 microns) by mashing using mortar. The composite were prepared by stirring polyurethane solution with various content of Carbon, 1%, 3%, 5%wt. The mixing of fillers begins by preparing each ingredient according to the percentage fraction of mass and mixtures needed.

The method used to mix fillers into polyurethane as a matrix by the mixing method for obtaining carbon/polyurethane. The melt mixing method was chosen because the process is simple by mixing the fillers into the liquid polyurethane matrix so that a dispersed particle composite is formed. The mixing process was carried out using a magnetic stirrer at Cimet SP131635 Barnstead Thermolyne Magnetic Stirrer (Apogent Technologies Inc., US) with a stirring speed of 500 rpm for 120 minutes.

Furthermore after stirring process, sample was printed on the aluminum plate (diameter of plate 23 mm) using gun spray which exposed by High Volume Low Pressure (HVLP) method. HVLP method is one method of making composite-based open mold, namely the printing process carried out in an open space, does not need to be in a vacuum chamber. The solution of the polyurethane matrix and its filler put into a spray gun and then sprayed evenly onto the mold. The spray gun used has a strong air pressure of 15 Psi. During the spraying process, it is assisted by a small spatula to flatten the mixture of ingredients on the plate so that no air is trapped in the matrix and filler solution. After the sample thickness is in accordance with the desired amount of 0.35 mm, the sample is dried at room temperature for 24 hours. After drying, the layer is removed from the mold and prepared for testing.

Thermogravimetric analysis (TGA) and Differential scanning analysis is carried out to obtain information of mass loss during the heating period which is needed to see process of decomposition. We expected that thermogravimetric analyzer can obtain information of polyurethane degradation through forming various products and a number of decomposition processes. TGA describes information on loss of assumption during heating as a function of temperature of the sample. DSC is in the same process to identify the thermal stability of the test sample when temperature increases. The pattern of data displayed after the DSC test can represent the state of the test material such as decomposition, oxidation, dehydration or crystallization. These conditions are some thermal resistance analysis parameters of each sample. TGA and DSC test used 6000 Simultaneous Thermal Analyzer (STA) (Perkin-Elmer Inc., USA) with a test temperature range of 30 - 800 °C at heating rate 10^0C/minute.

To identify of organic material, Fourier Transform Infrared Spectroscopy (FTIR) analytical technique is applied. The FTIR technique measures the absorption of wavelengths of infrared light emitted by the material being tested. This infrared absorption band shows the structure and components of certain molecules. FTIR testing was carried out using Spectrum Two FTIR Spectrometer L160000A (Perkin Elmer Inc., USA). This non-destructive test uses specimens measuring 2 cm x 2 cm starting from radiation wavelengths of 300 - 4000 cm^{-1}.

Corrosion resistance testing is carried out by the salt spray method according to ASTM B117 standards. For the salt spray test a sample measuring 2 cm x 2 cm was prepared. This method was chosen because it is the right and efficient method for carbon/polyurethane samples. In this method the sample is sprayed with 0.5 M NaCl solution with the spraying period every 3 hours in a total 72 hour test time.

Optical Microscopy was carried out using Zeiss Axio Lab.A1 (Carl Zeiss Co., German). For the Optical Microscopy test a sample measuring 2 cm x 2 cm was prepared which was observed using 100 times magnification. This test was carried out to observe the morphology of the composite layer of composite samples at the initial state of the sample before salt spray testing and sample conditions after the salt spray test was carried out.

3. Results and Discussion

3.1. Thermogravimetric (TGA) analysis

Thermogravimetric analysis is carried out to obtain information of mass loss during the heating period which is needed to see process of decomposition. We expected that thermogravimetric analyzer can obtain information of polyurethane degradation through forming various artifact and a several processes of decomposition. Thermogravimetric analysis describes information on loss of mass during heating as a function of temperature. Figure 2 display TGA curve as a function of temperature

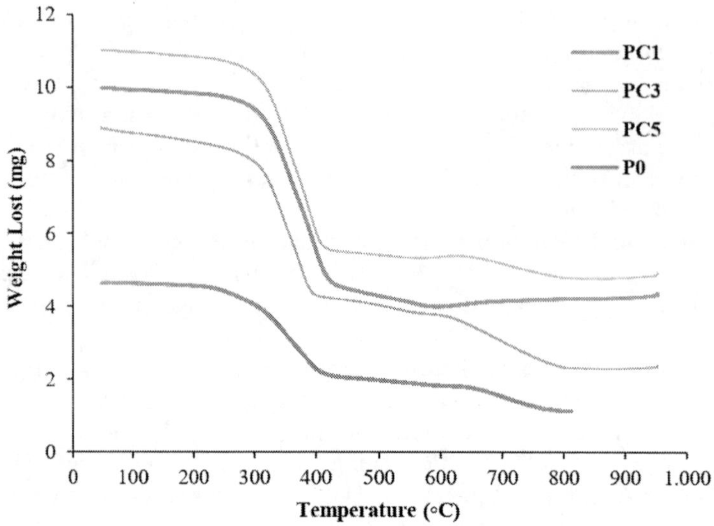

Figure 2. TGA curve of carbon/polyurethane composite sample with various content of carbon and pure polyurethane

Figure 2 show degragation of sample in two step, the first step is round 270-300 °C (onset temperature) then the second step is round 390-430 °C (endset temperature). The first stage is the main stage where there is a sharp decline, while in the second stage the composite with different content carbon , the smallest loss of mass occurs on PC5 following PC1 and PC3.

3.2. Differential Scanning Calorimetry (DSC) analysis

Figure 3 display heat consumed or loss versus temperature of the composite samples. The graph obtained a poor value in the PC3 and the trend on the graph has not looked good. DSC test results are always related to how the bonds in the material chain are. The crosslinking or cross-linking in the chain will affect the temperature response. There is a melting temperature shift and crystallization temperature caused by crosslinking which results in reduced chain flexibility and inhibits the movement of polymer chains, so that thermal transitions occur at high temperatures [12].

Samples PC1, PC3 dan PC5 show that heat flow suddenly decrease on around 300°C, however P1 is more excellently in thermal stability

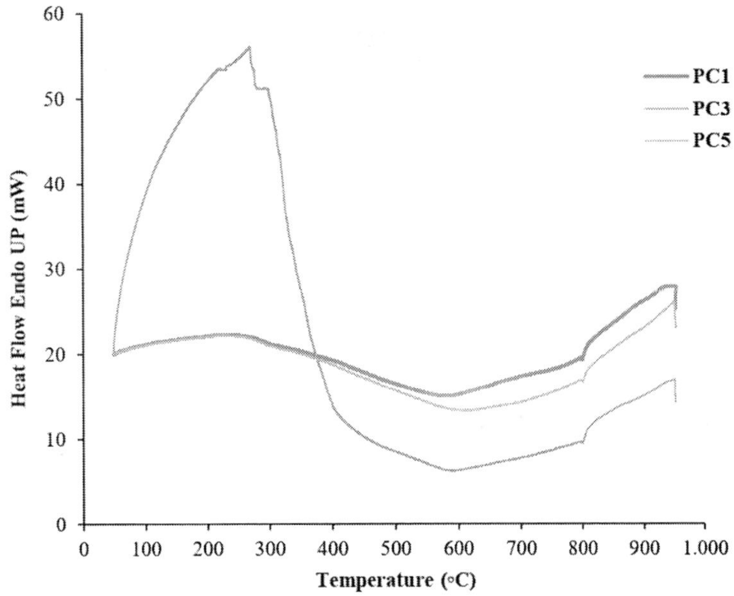

Figure 3. DSC curve of Carbon/Polyurethane composite sample with various content of Carbon

The study of the behavior of thermal degradation that has been carried out by several researchers shows that at high temperatures it can give fingerprints of the material that do not have to be only with the characteristics of the original material, the kind of fillers but also with the processing and final quality of end products [13]. Due to the complexity and degradation of polyurethane has provided motivation for researchers to investigate it endlessly.

3.3. Fourier Transform Infrared (FTIR) Analysis

Figure 4 show FTIR pattern of the carbon/polyurethane composite samples. The transmittance as a function of wave number of. Pure Polyurethane show the excellence transmittance for wave number 500-4000 cm^{-1} than PC1 and PC5.

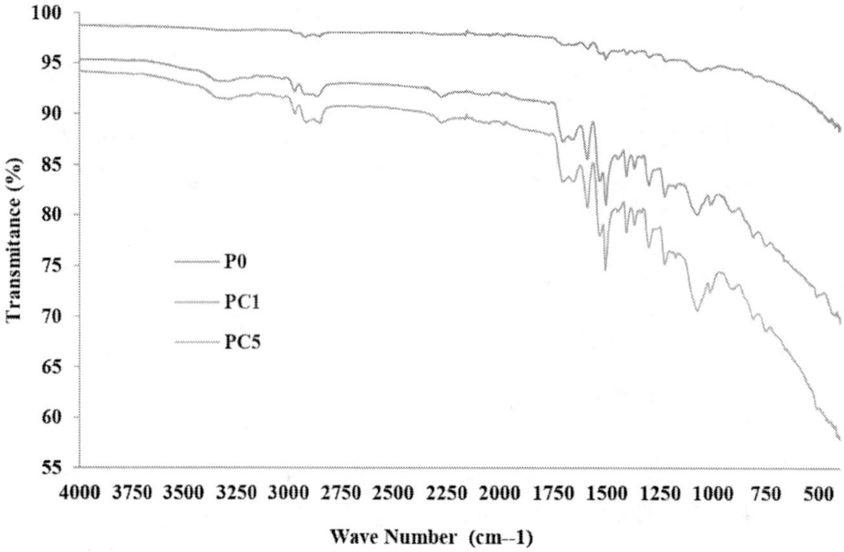

Figure 4. FTIR curve of the carbon/polyurethane composite samples Carbon/Polyurethane with various content of Carbon

Hydrogen bond in Figure 4 can be seen from the N-H and O-H bonds around 3200 - 3400 cm^{-1} and 1600 - 1700 cm^{-1}. At other peaks that indicate the existence of carbon bonds are shown in Figure 4 the N-H region of the urethane chain in the PC1 is deeper and larger than the other PC sample variations, it can be interpreted that the PC1 bond is stronger.

3.4. Salt spray and Morphology analysis

Long-term corrosion protection is very important and is needed in many industries. Specifically when facing with steel structures due to they are permanently and for the long time exposed to water in the form of sheet pile walls, water gates, barrages, dams and breakwater structures. A huge quality elastic polyurethane coating system based on raw materials is required. Composite of Polyurethane with various carbon contents were subjected to salt spray. Figure 5 and 6 show surface morphology of the carbon/polyurethane composite samples before and after salt spray test to find corrosion resistance properties.

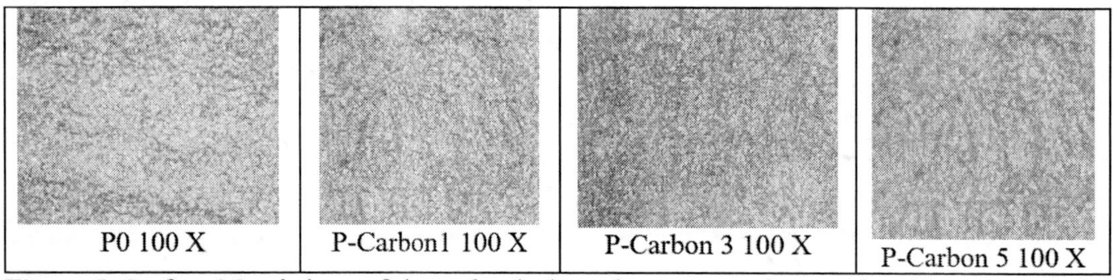

Figure 5. Surface Morphology of the carbon/polyurethane composite sample with various Carbon content before corrosion

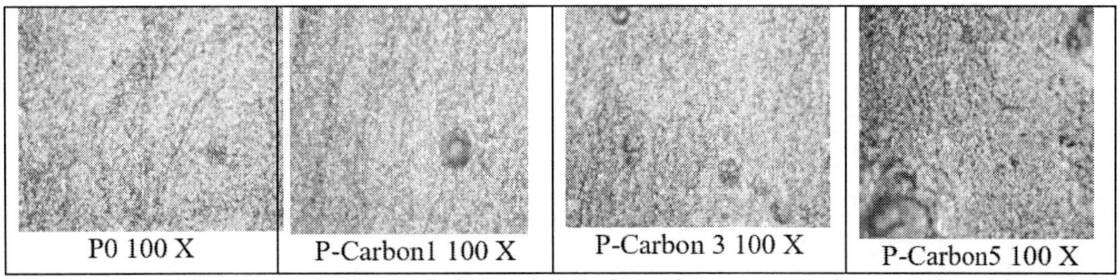

Figure 6. Surface Morphology of the carbon/polyurethane composite sample with various Carbon content before corrosion

Carbon / polyurethane composite samples P0, PC1 and PC3 can be seen that have a smoother of surface morphology than PC5 before the salt spray test. After the corrosion test, the surface morphology seemed to change. It seems that samples PC5 and PC3 are more rough than PC1 and P0. It seem that P1 is more resistance than the others.

4. Conclusion

Carbon/Polyurethane composite with various Carbon content have been prepared. The study obtain that PC1 have more a better thermal stability and good corrosion resistance than PC3 and PC5. It seem that addition of carbon was decreased the properties of pure polyurethane and it is probably was not good to use as a coating materials.

Acknowledgment

The author acknowledge with gratitude the financial support from the Ministry of Research, Technology and Higher Education of Indonesia (RISTEK-DIKTI), under grant *Penelitian Dasar Unggulan Perguruan Tinggi (PDUPT)* No. 366/UN2.R3.1/HKP05.00/2018).

References

[1] Ciecierska E, Jurczyk-Kowalska M, Bazarnik P, Gloc M, Kulesza M, Kowalski M and Lewandowska M 2016 *J. Composite Structures*. **140** 67-76
[2] Goh P S, Ismail A F and Ng B C 2014 *J. Composites Part A: Applied Science and Manufacturing* **56**, 103-126
[3] Fujigaya T and Nakashima N 2015 *J. Science and Technology of Advanced Materials*. **16** 024802
[4] Mittal G, Dhand V, Rhee KY, Park SJ and Lee WR 2015 *Journal of Industrial and Engineering Chemistry* **21** 11-25
[5] Song L, Hu Y, Tang Y, Zhang R, Chen Z and Fan W 2005 *J. Polymer Degradation and Stability* **87**(1) 111-116
[6] Zhang P, Xiao X and Ma Z W 2016 *J. Applied Energy* **165** 472-510
[7] Nomura T, Tabuchi K, Zhu C, Sheng N, Wang S and Akiyama T 2015 *J. Applied Energy* **154** 678-685
[8] Karaipekli A, Biçer A, Sarı A and Tyagi V V 2017 *J. Energy Conversion and Management* **134** 373-381
[9] Njuguna J and Pielichowski K 2004 *Journal of Materials Science* **39**(13) 4081-4094
[10] Mirhosseini SS, Razavi RS, Taheran M and Barekat M 2016 *J. Progress in Organic Coatings* **97** 37-43
[11] Goclon J, Panczyk T and Winkler K 2018 *J. Applied Surface Science* **433** 213-221
[12] Sattar R, Kausar A and Siddiq M 2015 *Chinese Journal of Polymer Science* **33**(9) 1313-1324
[13] Prime RB, Burns JM, Karmin ML, Moy CH and Tu HB 1988 *J. Coatings Technology* **60**(761) 55-60.

IOP Conf. Series: Materials Science and Engineering **578** (2019) 012072 doi:10.1088/1757-899X/578/1/012072

One-pot solvothermal synthesis and characterization of UiO-66/HKUST-1 composites

R Ediati[1*], L L Zulfa[1], K A Nugroho[1], A Mukminin[2], D O Sulistiono[1], M Nadjib[1] and Y Kusumawati[1]

[1] Department of Chemistry, Institut Teknologi Sepuluh Nopember, Surabaya, Indonesia

[2] Sekolah Tinggi Teknologi Minyak dan Gas Bumi, Balikpapan, Indonesia

E-mail: rediati@chem.its.ac.id

Abstract. Metal-organic frameworks (MOFs) are promising crystalline materials for various application due to their controllable porous structure and large specific surface area. Their properties can be further improved by several means, such as making a binary MOF comprising of two different MOFs. In this research, UiO-66/HKUST-1 composites have been synthesized by solvothermal method, in which the UiO-66 was prepared at 120 °C for 24 h in N,N-dimethylformamide (DMF) prior to preparation of HKUST-1 at 100 °C for 10 h in solvent mixture of DMF, ethanol and water. Amounts of Zr^{2+} for the synthesis of UiO-66 added during the synthesis process were varied at 5, 10 and 20% toward the amount of Cu^{2+} for synthesis of HKUST-1. Diffractogram patterns of the obtained composites showed dominant characteristic peaks, similar to that reported for HKUST-1. Furthermore, FTIR spectra of all solids showed absorption bands at the same wavenumbers as reported for HKUST-1. Transmission Electron Microscopy (TEM) images showed the spread of Zr metal evenly on the pore of composites.

1. Introduction

Metal Organic Framework (MOF) is a group of porous materials that are unique and superior compared to other porous materials. The presence of metals and organic ligands makes MOF become an efficient adsorbent in which the size and volume of the pore of the MOF can be adjusted with the material to be adsorbed [1]. MOF has been widely used as catalysts, adsorbent, drug delivery, and material templates [2]. MOF has several sub-groups including IRMOF, ZIF, HKUST-1, and UiO-66.

One type of MOF that is often studied is HKUST-1 or commonly known as Cu-BTC. MOF type HKUST-1 was first synthesized by Chui et al., by use of cupric nitrate trihydrate and trimesic acid ligands. HKUST-1 has a characteristic surface area of 692.2 m^2/g, pore size and pore volume of 0.9 nm and 0.33 cm^3/g, respectively, with thermal stability of 240 °C [3]. Besides having advantages, the HKUST-1 also has disadvantages, namely the low thermal stability and micro pore characteristics that limit the application performance [4]. The low thermal stability of MOF is caused by the bond between metals and their organic ligands [5].

Several methods have been carried out to improve the thermal stability of MOF including the addition of metals or other compounds [6]. Several previous studies have introduced MOF in its composite form. Furtado et al. has introduced the synthesis of HKUST-1 with addition of MCM-41 support [7]. Kondo et al. have successfully reported the preparation of HKUST-1/SBA-15 composite and its application as ethanol adsorption [8]. Liu et al. reported that performance of $H_6P_2W_{18}O_{62}$@Cu-

IOP Conf. Series: Materials Science and Engineering **578** (2019) 012072 doi:10.1088/1757-899X/578/1/012072

BTC composite increased in dye adsorption applications [9]. However, some of these materials are expensive if applied on a large scale. Recently, there is an innovation to increase the thermal stability of the MOF by synthesis of binary MOF. UiO-66 is one type of MOF which has high thermal stability up to 540 °C [10]. Besides, some interesting properties are shown by UiO-66, namely a large surface area and high porosity and stable in water media. The composting of the two types of MOF is expected to be able to cover the shortcomings of one of the MOF properties so as to improve its performance in certain applications.

2. Experimental

2.1. Materials
All chemicals used in this study, namely copper(II)nitrate trihydrate [$Cu(NO_3)_2.3H_2O$, 99%], zirconium tetrachloride ($ZrCl_4$, 99.0%), benzene-1,3,5-tricarboxylic acid (H_3BTC, 98%), and benzene-1,4-dicarboxylic acid (H_2BDC, 98.9%), were obtained from Sigma Aldrich. Ethanol (EtOH, 99.8%), methanol (MeOH, 99.8%), chloroform ($CHCl_3$, 99.9%) and N,N- Dimethylformamide (DMF, 99.88%) were purchased from Merck and aqua demineralized (aqua DM).

2.2. Synthesis HKUST-1
HKUST-1 was synthesized using the procedure reported by Chowdury et al. [11]. First, 2.077 g of $Cu(NO_3)_2.3H_2O$ was dissolved in 15 mL of aqua DM, while 1 g of H_3BTC was dissolved in 30 mL solvent mixture of ethanol and DMF (v:v = 1:1). The reaction mixture was then stirred with magnetic stirrer for 30 min, followed by heatting process in oven at 100 °C for 10 h. The reaction mixture was allowed to cool at room temperature in two days. The solid was then separated from solvent by decantation. The solid was washed by adding 40 mL of DMF, soaked overnight, and the solvent was removed by decantation. The washing process was repeated by adding 60 mL of methanol and placed in oil bath at range temperature of about 40 °C. The resulting powder was dried in oven at 55 °C for 4 hours.

2.3. Synthesis UiO-66
UiO-66 was synthesized using a procedure reported by Rahmawati and Ediati [12], in which 0.3496 g of $ZrCl_4$ and 0.2492 g H_2BDC were dissolved in 30 mL DMF, then solution was stirred with magnetic stirrer for 30 min in a vial bottle. The reaction mixture was then placed in oven at 120 °C for 24 h. The reaction mixture was allowed to cool at room temperature overnight. The solid was then separated by decantation, washed with 40 mL of DMF and 30 mL chloroform, respectively. The obtained powder was dried in oven at 60 °C for 4 hours.

2.4. Synthesis UiO-66/HKUST-1 Composites
The method for synthesis of binery MOFs of UiO-66/HKUST-1 is the same as for synthesis of HKUST-1 and UiO-66. UiO-66 was first synthesis by solvothermal method at 120 °C for 24 h in DMF solvent, so that a mixture of solid in solvent was obtained. The reaction mixture was then added to solution for HKUST-1 synthesis. The reaction mixture was then treated as for HKUST-1 synthesis, i.e. heated in oven at 100 °C for 10 h. In this study, the amount of UiO-66 added was varied at 5, 10 and 20 (w/w %) in order to determine the effect of UiO-66 addition on HKUST-1 crystal formation. The obtained solids were labelled as U/H(X) where X was the amount of UiO-66 added.

2.5. Characterization
The crystallinity of the obtained solid was examined using X-ray diffractometer with Cu Kα radiation (λ = 1.5406 Å), and angle range of 3 - 50°. Determination of functional groups was carried out using FTIR instrument in the wavenumber of 400 - 4000 cm^{-1}. TEM images were obtained using a JEOL JEM1400 transmission electron microscope at an accelerating voltage of 200 kV. The thermal gravimetric analysis was recorded in the temperature range of 30 – 900 °C under air flow with heating rate of 10°/min.

IOP Conf. Series: Materials Science and Engineering **578** (2019) 012072 doi:10.1088/1757-899X/578/1/012072

3. Results and Discussion

3.1. X-ray diffraction (XRD) analysis

The XRD patterns of the obtained solids are shown in Figure 1. The formation of HKUST-1 structure was confirmed by the appearance of characteristic peaks at 2θ of 6.7, 9.5, 11.6, 13.8 and 17.0°, respectively, as reported by Lin et al. [2]. Likewise, the UiO-66 showed characteristic peaks at 2θ of 7.3° with the highest intensity and at 2θ of 8.5° with lower intensity, which are similar to that reported by Gokpinar et al. [13] for UiO-66. The XRD patterns of the composite with 5 %w/w UiO-66 was dominated by the characteristic peaks of HKUST-1. When the amount of UiO-66 was increased, however, the characteristic peak of UiO-66 was seen at 2θ of 8.5°. These results indicated that the UiO-66 structure in the composites remained unchanged, even though the existence of UiO-66 in the composites decreased the intensity of characteristic peaks of HKUST-1.

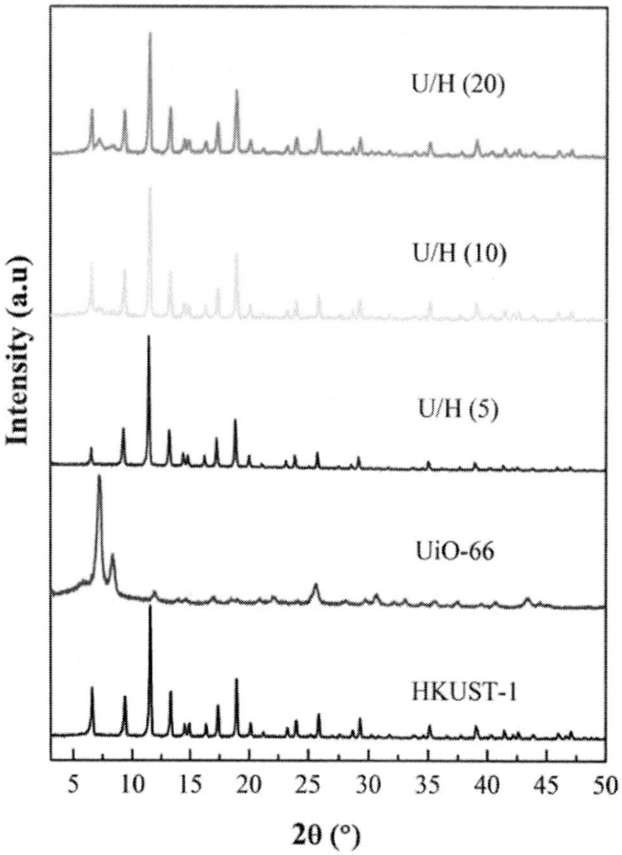

Figure 1. XRD pattern of HKUST-1, UiO-66 and Composites.

3.2. The Fourier transform infrared (FT-IR)

The FTIR spectra of all solids are shown in Figure 2. As can be seen, the HKUST-1, UiO-66 and all UiO-66/HKUST-1 composites show absorption bands at wavenumber similar to that of binary MOF of UiO-66 and HKUST-1, as reported by Azhar et al. [14]. Absorption bands appeared in the wavenumber ranging of 1450 - 1650 cm^{-1} indicated the presence of C = C groups of aromatic groups. The presence of C-O stretching carboxylic groups was shown by absorption bands at around 1300-1400 cm^{-1}. The absorption bands at around 3200 - 3600 cm^{-1} indicated the presence of O-H groups. The finger print absorption bands of about 700 cm^{-1}, showed the presence of bonds between Cu-O and

IOP Conf. Series: Materials Science and Engineering **578** (2019) 012072 doi:10.1088/1757-899X/578/1/012072

Zr-O in all the obtained solids [15][16]. In general, there has been no significant difference between FTIR spectra of the HKUST-1, UiO-66 and binary MOFs.

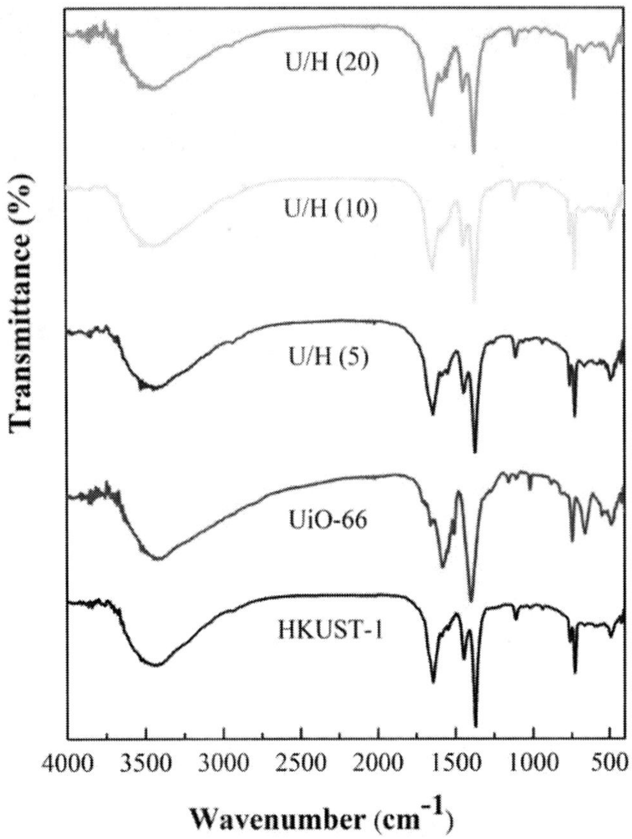

Figure 2. FT-IR spectra of HKUST-1, UiO-66 and Composites.

3.3. Transmission Electron Microscope (TEM)

TEM analysis was carried out to determine the pore morphological characteristics of a material. In this study TEM analysis was performed on composite samples U/H(20) and HKUST-1 as a comparison. As is shown in Figure 3, the pure HKUST-1 exhibited more uniform pore morphology when compared to composite U/H(20). Furthermore, the composite U/H(20) showed the spread of black dots which might indicate the spread of Zr atoms in HKUST-1 pores.

IOP Conf. Series: Materials Science and Engineering **578** (2019) 012072 doi:10.1088/1757-899X/578/1/012072

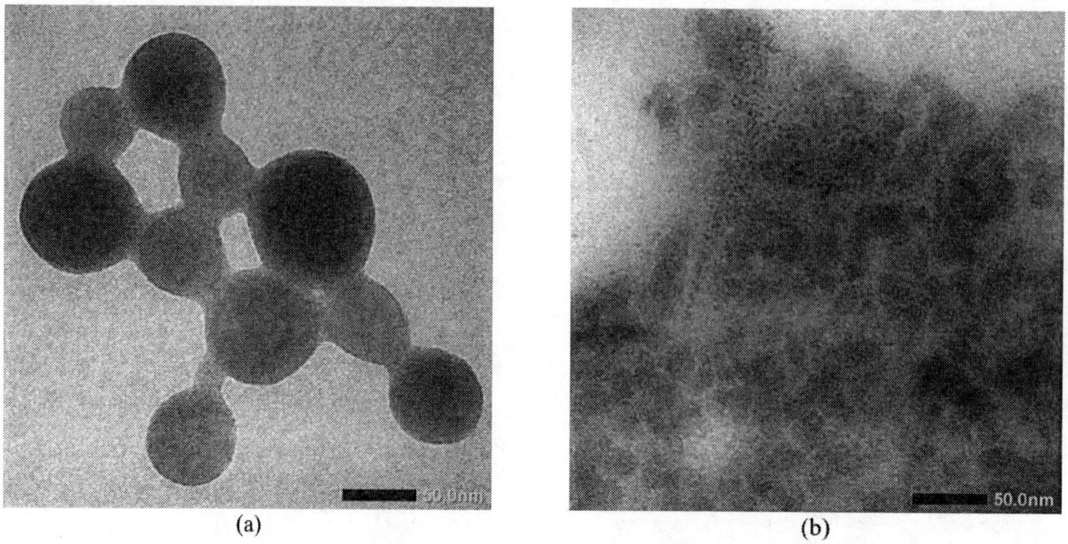

Figure 3. TEM image of pure HKUST-1 (a) and composite U/H(20) (b).

3.4. Thermal Gravimetric Analysis

The results of thermal gravimetric analysis of the obtained HKUST-1, UiO-66 and U/H(20) composite are shown in Figure 4. It can be seen that the thermal stability of the HKUST-1 reached a temperature of 292 °C with two stages of mass reduction [17]. A high thermal stability of 510 °C was exhibited by UiO-66. The thermal stability of the U/H(20) composite, however, was lower than those of HKUST-1 and UiO-66. The decrease in the value of thermal stability of the composite U/H(20) may be as a result of crystal growth disturbance of the HKUST-1 due to the presence of UiO-66 in the reaction mixture of composite synthesis.

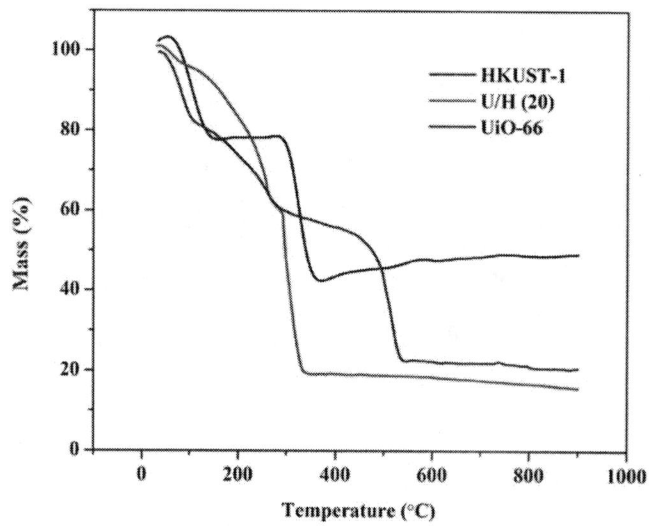

Figure 4. The thermal gravimetric analysis of materials synthesized

4. Conclusion

In this study, the UiO-66/HKUST-1 composite was successfully synthesized by the solvothermal method with a one-step synthesis. The results of XRD analysis showed that the addition of UiO-66 during the synthesis of HKUST-1 did not change the crystal structure of either HKUST-1 or UiO-66, significantly. The results of FTIR analysis showed that the composites exhibited absorption bands at same wavenumber to those of HKUST-1 or UiO-66. The thermal stability of the U/H(20) composite, however, was lower when compared to those of HKUST-1 and UiO-66.

Acknowledgement

We gratefully acknowledge funding from Institut Teknologi Sepuluh Nopember (ITS) Surabaya through EPI-UNET Research Grants 2018, Material Chemistry and Energy Laboratory, Department of Chemistry, as well as The Energy Laboratory of ITS for facilities supporting this work.

References

[1] Tari N E, Tadjarodi A, Tamnanloo J and Fatemi S 2016 *Microporous Mesoporous Mater* **231** 154–62
[2] Lin S, Song Z, Che G, Ren A, Li P, Liu C and Zhang J 2014 *Microporous Mesoporous Mater* **193** 27-34
[3] Chui S S Y, Lo S M F, Charmant J P H, Orpen a G and Williams I D 1999 *Science* **283** 1148–1150
[4] Qin L, Zhou Y, Li D, Zhang L, Zhao Z, Zuhra Z and Mu C 2016 *Ind. Eng. Chem. Res.* **55** 7249-7258
[5] Zhao Q, Yuan W, Liang J and Li J 2013 *Int. J. Hydrogen Energy* **38** 13104–13109
[6] Bosch M, Zhang M and Zhou H-C 2014 *Advances in Chemistry* 1-8 (http://dx.doi.org/10.1155/2014/182327)
[7] Furtado A M B, Liu J, Wang Y and LeVan M D 2011 *J. Mater. Chem.* **21** 6698-6706
[8] Kondo A, Takanashi S and Maeda K 2012 *J. Colloid Interface Sci.* **384** 110-115
[9] Liu X, Luo J, Zhu Y, Yang Y and Yang S 2015 *J. Alloys Compd.* **648** 968-993
[10] Cavka J H, Jakobsen S, Olsbye U, Guillou N, Lamberti C, Bordiga S and Lillerud K P 2008 *J. Am. Chem. Soc.* **130** 13850–13851
[11] Chowdhury P, Bikkina C, Meister D, Dreisbach F and Gumma S 2009 *Microporous Mesoporous Mater* **117** 406–413
[12] Ediati R, Rahmawati I D and Prasetyoko D 2014 *Iptek Journal of Proceeding Series* **1** 42–46 (DOI: http://dx.doi.org/10.12962/j23546026.y2014i1.417)
[13] Gökpinar S, Diment T and Janiak C 2017 *Dalt. Trans.* **46** 9895–9900
[14] Azhar M R, Abid H R, Sun H, Periasamy V, Tadé M O and Wang S 2017 *J. Colloid Interface Sci.* **500** 88-95
[15] Senthil Kumar R, Senthil Kumar S and Kulandainathan M A 2013 *Microporous Mesoporous Mater* **168** 57–64
[16] Zhu X, Gu J, Wang Y, Li B, Li Y, Zhao W and Shi J 2014 *Chem. Commun.* **50** 8779–8782
[17] Schlichte K, Kratzke T and Kaskel S 2004 *Microporous Mesoporous Mater* **73** 81–88

Fabrication of Graphene Oxide/Calcium Carbonate/Chitosan Nanocomposite Film with Enhanced Mechanical Properties

M Handayani[1,*], E Sulistiyono[1], F Rokhmanto[1], N Darsono[1], P L Fransisca[1], A Erryani[1] and J T Wardono[1]
[1]Research Center of Metallurgy and Materials, Indonesian Institute of Sciences

[*]E-mail: murni.handayani@lipi.go.id

Abstract. Nanocomposites are an attractive field of materials providing novel performance owing to their remarkable properties at low filler loading. Currently, they are being increasingly used by industry whilst displacing the use of conventional filler materials. A lot of nanoscale fillers are capable to enhance mechanical and thermal properties of nanocomposites. Graphene, a single layer of carbon atoms in a hexagonal lattice, has recently attracted much attention due to their novel electronic and mechanical properties. Graphene and graphene oxide show very high mechanical properties with well biocompatibility, and they have potential application as biomaterials. Therefore, 2D graphene or graphene oxide sheets are expected to offer promising nanoscale filler for the next generation of nanocomposite materials. Here, Graphene Oxide/Calcium Carbonate/Chitosan (GO/$CaCO_3$/CS) Nanocomposite Film has been fabricated by the blending aqueous solution. The graphene oxide used in this study was synthesized by chemical way from pure graphite using modified hummer's technique. The graphene oxide resulted from hummer's method was characterized by IR, XRD and SEM-EDX. The measurement result of the mechanical properties of the nanocomposite film showed that with adding GO content in the composite enhanced the tensile strength and peaks at the content of GO 5 % due to good dispersion of the graphene oxide as nano-scale fillers.

1. Introduction

Nanocomposites are an attractive field of materials providing novel performance owing to their remarkable properties at low filler loading. Recently, they are being increasingly used by industry whilst displacing the use of conventional filler materials. A lot of nanoscale fillers are able to enhance mechanical and thermal properties of nanocomposites [1].

Graphene is one atomic layer sheet of carbon atoms with sp^2 hibridized two dimensional honeycomb structure, has recently attracted much attention due to their fascinating physical, chemical, optics and mechanical properties. It's a single atomic layer of the graphite with each carbon atom bonded with strong covalent bonds [2]. The potential applications of this unique nanostructure hold novel promises for various industrial and professional fields such as nanoelectronics, sensor applications, composite materials, energy storage, and hydrogen storage [3]. Besides that, graphene and graphene oxide exhibit very high mechanical properties with well biocompatibility, and very promising as functional biomaterials [4]. 2D graphene or graphene oxide sheets are expected to offer very potential nanoscale reinforcements for the next generation of nanocomposite materials [5]. Nevertheless, the neat graphene does not have functional groups which restrics the dispersibility in the matrix and limits the applications of graphene as nanoscale filler in nanocomposite materials.

Graphene oxide is more appropriate than the neat graphene owing to the higher contents of oxygen functional groups such as hydroxyl and epoxides [6].

Chitosan is a natural polymer found in the exoskeleton of crustaceans, produced by de-acetylation of chitin, composed mainly of b-(1,4)-linked 2-deoxy-2-amino-D-glucopyranose units. Chitosan is one of the most abundant natural biopolymer which is an antibacterial, biocompatible, non- toxic, edible biodegradable polymer with good mechanical properties and it is widely used for food packaging, biosensor, water treatment, biomedicine, drug delivery and as a matrix of biomaterial for bone tissue engineering [1,5].

Calcium carbonate, a natural mineral with great biocompatibility, is one of the most common and inexpensive inorganic filler that has been widely used in industry, technology, and many other fields. Along with technological developments, the use of calcium carbonate, especially with particle size in nanoscale, is not only for fillers but also for biocompatible materials with has advantage that is easily adaptable in human body [5,7,8].

In this work, we reported the facile preparation of graphene oxide and used as nano-fillers material of graphene oxide/calcium carbonate/chitosan nanocomposites by blending and casting method. The effect of the loading of graphene oxide contents to the mechanical properties of the nanocomposites was investigated.

2. Experimental

2.1. Materials

Graphene oxide (GO) was synthezied by modified hummer's method using graphite powder which was purchased from Merck. The other chemical such as $KMnO_4$, $NaNO_3$, 98% H_2SO_4, NaOH, Hac and 30 % H_2O_2 are analytical grade chemical. Chitosan (CH) was purchased from Merck. Calcium carbonate ($CaCO_3$) nanoparticles with everage of size around 100 nm was obtained from natural limestone which was synthesized using utrasonic method reported in the previous study [7, 8].

2.2. Experiment

2.2.1. Preparation of Graphene Oxide

Graphene oxide was synthesized by well known modified Hummer's method. 1 g natural graphite powder was added into a 250 mL beaker, then 0.5 g $NaNO_3$ and 23 mL H_2SO_4 were added into it under stirring in an ice-bath. 5 gr KMnO4 was added slowly into the beaker with stirring and the temperature was controlled under 20 °C. After maintaining the mixture stirred below 20 ^oC for 2 hours, increase the temperature to 35-40 °C and the reaction was kept for around 14 hours. Next step, add hydrogen peroxide and distilled water into the brown paste mixture and stir for 30 mins. Then, add some amount of distilled water into the mixture to neutralize the pH and dilute the mixture. After that, 1 M HCl was added and followed by adding distilated water to neutralized the solution. The result was filtered and dried in oven at 80 °C. To exfoliate the graphite oxide become graphene oxide, we applied the ultrasonication for around 30 minutes and then centrifugation.

2.2.2. Preparation of Graphene Oxide/Calcium Carbonate/Chitosan

Desired concentration of oxidized graphene oxide powder from natural graphite obtained from the above synthesis was dispersed in destillated water and sonicated for around 30 minutes. After that, 0.4 g powder of nano calcium carbonate precipitates was added and followed with adding by 100 ml of ultrapure water. The solution was then sonicated for 90 minutes then stirred for one night at room temperature. After that, 1 ml acetic acid glassial (HAc) was added to the solution and 1.5 g chitosan were added into the suspension under stiring for 60 minutes. The mixture solution was cast on the glass plate and it was kept at room temperature and then dried at 50 ^oC. The dry nanocomposite films were then immersed in 0.1 NaOH and followed by washing the film with water and dried at room temperature for 24 hours. Finally the nanocomposites films were dried in oven at 50 °C. We prepared 4 samples nanocomposites films as follows : Chitosan film without adding the reinforacement of graphene oxide and calcium carbonate (CH), calcium carbonate/chitosan films with content of graphene oxide is 1 % (GO1/CAL/CH), graphene oxide/calcium carbonate/chitosan with content of graphene oxide is 5 % (GO5/CAL/CH) and graphene oxide/calcium carbonate/chitosan with content of graphene oxide is 10 % (GO10/CAL/CH).

2.2.3. Characterization

The caractherization of the crystal structure of graphene oxide was done by X-Ray Diffraction (XRD) The morphology and elemental analysis of graphene oxide were performed by SEM-EDX JEOL with Type JSM-639OA. The mechanical properties of nanocomposite films such as tensile properties, young's modulus and break elongation were charactherized by using Tinius Olsen, 300 SL, Super L-60.

3. Results and Discussion

3.1. Analysis of graphene oxide using Fourier Transform Infrared (FT-IR)

The results of the analysis using *FT-IR* for graphene oxide is depicted in Figure 1. FT-IR technique is used in this analysis to investigate the functional group which can confirm the bond interaction of graphene oxide. A strong and broad band at 3425.58 cm^{-1} is indicated the O-H strecthing vibration. The peak at in the range 1630 cm^{-1} to 1650 cm^{-1} show the C=C bond. Peak at 1620 cm^{-1} is attributed to C=O stretching vibrations, and the peak at 1057 cm^{-1} is attributed to C-O-C bond which is appropriate with the previous report [9].

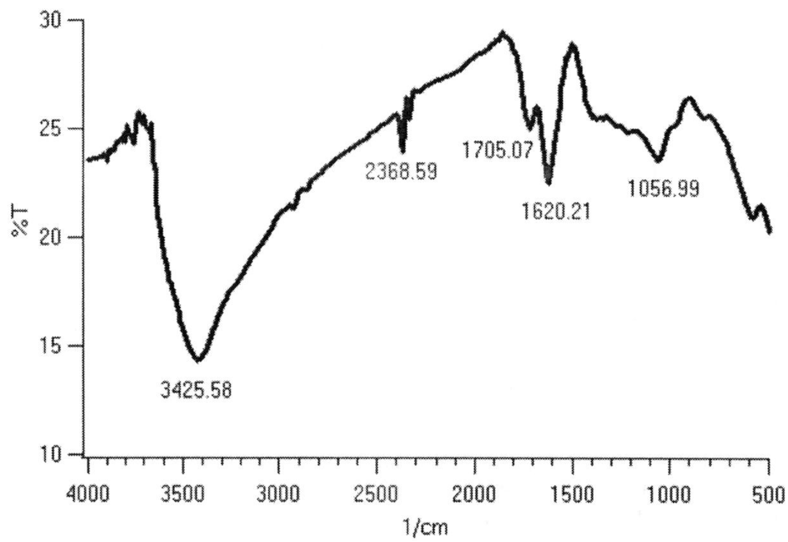

Figure 1. FT-IR spectra of graphene oxide.

3.2. Analysis of graphene oxide by X-Ray Diffraction (XRD)

The X-ray Diffraction Analysis / XRD is the most common used technique to characterize crystalline matrial. The average spacing's between layer or row of atoms can be measured by this technique. Besides that, XRD can be applied to determine the orientation of a single crystal. The analysis of graphene oxide syntesized by hummer's method is demonstrated in Figure 2.

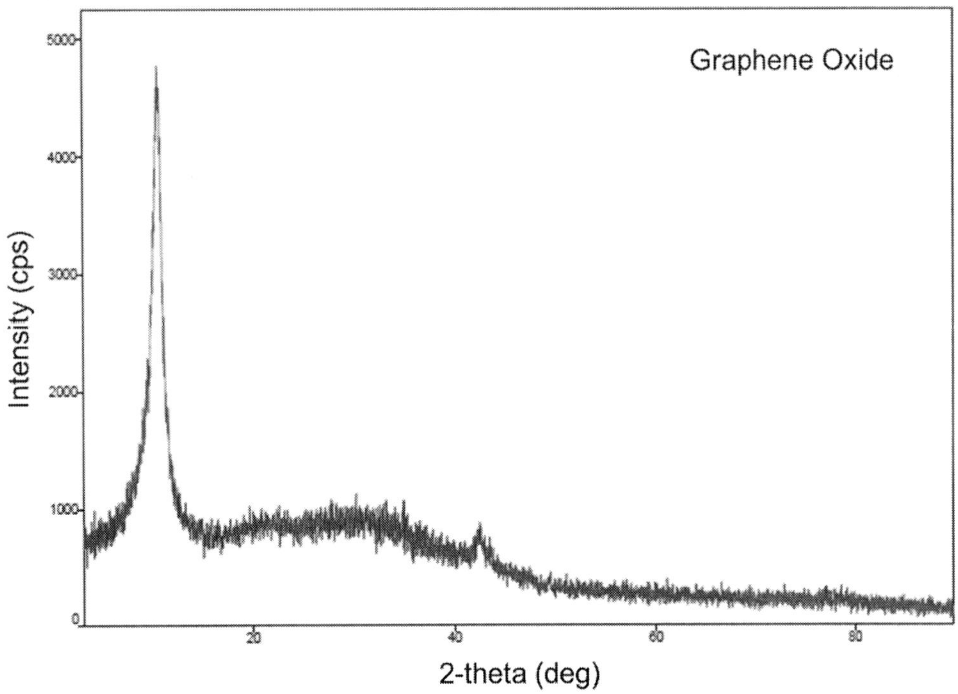

Figure 2. XRD Analysis result of graphene oxide.

It is known that the diffraction peak of pure graphite is at 2θ around 26 ° due to the interlayer distance of 0.335 nm [9,10]. Here we confirm that graphene oxide demonstrate a typical diffraction peak at 2θ around 9.9 ° with d-spacing = 0.89 nm which is indicated to the typical peak of GO. The disappearance of the peak at 26 ° proves that the compound result is completely oxidized after treatment of the oxidation and exfiolation of graphite resulted the formation of oxygen-containing functional groups.

3.3. Analisys of graphene oxide by scanning electron microscopy-Energy-dispersive X-ray spectroscopy (EDX)

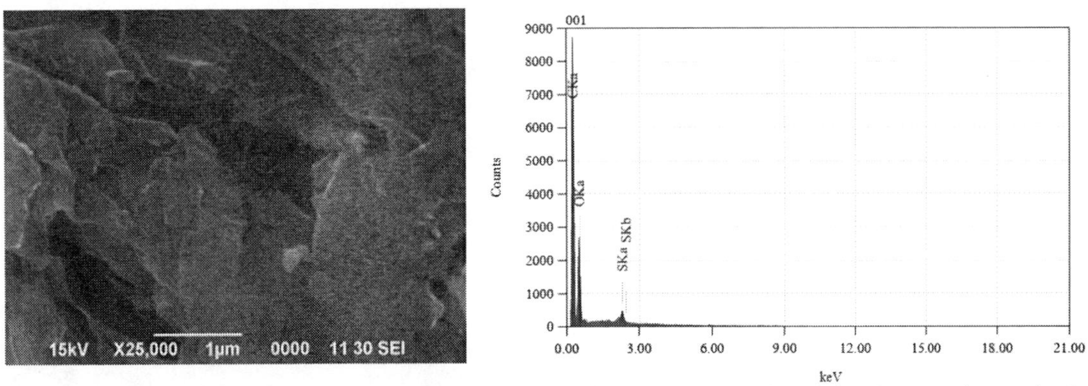

Figure 3. (Left) SEM analysis result of the graphene oxide produced by hummer's method, (Right): EDX analysis result of the graphene oxide.

The result of SEM analysis for graphene oxide is shown in figure 3. The magnification of the image was taken at 25,000 x. The result indicates that the typical surface morphology of the graphene oxide have well observed and shoe a porous interlinked three dimensional graphene layer. The elemental analysis result obtained from EDX analysis shows that the majority of the content of the compound is carbon (with 62.93% of mass) and following with oxygen (with 35.44% of mass) which proves the existence of graphene oxide. The existence of sulfur element eventhough it is a very small content (1.63 % of mass) indicated that the neutralization of the sample is not performed completely.

3.4. Mechanical properties of nanocomposites films

We investigate the mechanical properties of the nanocomposites films to observe the effect of adding the different of the content of graphene oxide in the nanocomposites. We compared also to the adding the chitosan matrix with calcium carbonate only without adding the graphene oxide. The results of the mechanical properties of the nanocompoites such as tensile strength, young's modulus, and elongation at break are shown in Table 1.

Table 1. Mechanical properties of the nanocomposites films.

Sample Code	Ultimate Force	Tensile Strength	Young's Modulus	Elongation
	(N)	(Mpa)	(Gpa)	(%)
CH	19.8	16.5	4.52	4.3
GO1/CAL/CH	85.2	71	3.45	9.8
GO5/CAL/CH	102	85.2	5.68	14.8
GO10/CAL/CH	88.5	73.8	6.21	4.19

The tensile strength of chitosan without adding the nanoscle filler give the result of 16.5 Mpa and increase to 71 MPa with adding the fillers nanoparticles. The result demonstrate that the nanoscale filler can increase the tensile strength significantly compared to the pure chitosan. The calcium carbonate supports the graphene oxide filler to the chitosan matrix so that increases the tensile strength significantly. Note that we use the same weight of the calcium carbonate and the results show that the increasing of the tensile strength of the nanocomposites is mainly due to the content of GO addition on the nanocomposites films. The detail graphic of the mechanical properties of the nanocomposites films was shown in Figure 4.

Increasing the percentage of GO on the nanocomposite make the tensile strength higher with the maximum tensile strength is 85.2 Mpa with the loading of the GO content of 5%. This means that increasing GO content enhances the tensile strength due to the good dispersibility of the nanofiller into matrix and resulting the higher reinforcement effect. However, when the content of GO is increased further to 10%, the tensile strength is decreasing to 73.8 % as a result of the aggglomeration of the GO particles at higher contents and give rise the dispersion the fillers in the matrix is not homogenous.

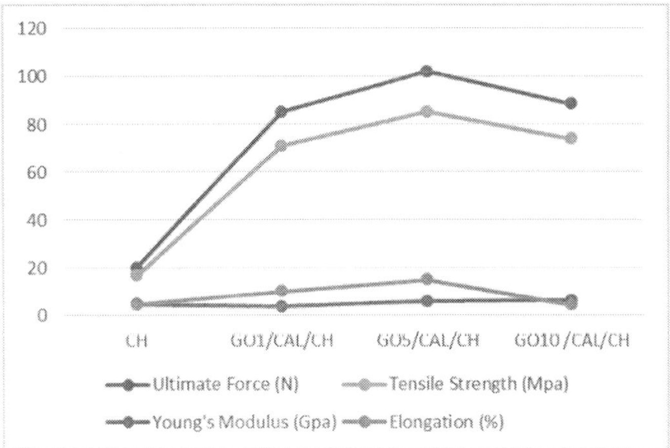

Figure 4. Measurement results of the mechanical properties of nanocomposite films.

The addition of the nanoscle fillers on the nanocomposite film causes the young modulus increases gradually with the maximum 6.21 Gpa by adding of 10% GO in the nanocomposites films. Besides, the elongation at the break point is increasing gradually until the maximum at the addition of 5% GO content and then drastically reduces after the addition of GO content 10 %. The result can be attributed to the interaction between higher content of GO with the matrix which restrict the movement of the chitosan chain.

4. Conclusion

The graphene oxide was successfully synthesized by hummer's method and the result was confirmed by FT-IR, XRD and SEM-EDX. The obtained graphene oxide was used as nanoscale reinforcement materials in the nanocomposites films. The mechanical properties of the nanocomposites were affected by the addition of the graphene oxide as reinforcement. Based on the result of tensile strength measurement, the tensile strength of the nanocomposite increased significantly and reached the maximum with value is 85.2 Mpa with the addition of 5 % GO due to good dispersion of graphene oxide in the nanocomposites. The young modulus increases gradually with the maximum 6.21 Gpa by adding of 10% GO in the nanocomposites films. Besides, the elongation is increasing gradually until the maximum at the addition of 5% GO content and then drastically reduces after the addition of GO content 10%.

References

[1] Sanaz A, Samira S, Nor A I, Wan Md Z W Y, Mohammad Z A R, Susan A and Asma F 2012 *Int. J. Mol. Sci.* **13** 4508-22

[2] Sohail M, Saleem M, Sana U, Noor S, Ayesha A, Majid K and Arif M 2017 *Modern Electron. Mater.* **3** 110-16

[3] Satish B, Bikshalu K, Venkateswara R K, Niveditha R, Barray and Shilpa C C 2018 *Int. J. Current Sci., Eng. Tech.* **1** 1-8

[4] Ping-Ping Z, Hua-Feng F, Zhi-Zhen X, Ling-Fan Z, Yu-Long Z, Wei X and Wen-Qing Z 2013 *Chem. Central J.* **7** 1-11

[5] Donglin H, Lifeng Y, Wuhang C and Wan L 2011 *Carbohydr. Poly.* **83** 653-58

[6] Paullchamy B, Arthi G and Lignesh BD 2015 *J. Nanomed. Nanotechnol.* **6** 1-4

[7] Handayani M, Sulistiyono E, Firdiyono F and Fajariani E N 2018 *IOP. Conf. Ser.: Mater. Sci. Eng.* **333** 01243

[8] Sulistiyono E, Handayani M, Firdiyono F and Fajariani E N 2018 *IOP. Conf. Ser.: Mater. Sci. Eng.* **333** 012042
[9] Mohammad A K, Fatemeh B and Mehdi R 2017 *J. Mater. Sci.: Mater. Electron.* **28** 1844-51
[10] Chujiang C, Nannan S, Zhigang S and Xiaohu Z 2017 *J. Exp. Nanosci.* **12** 1 247-62

Alginate-chitosan hydrogel as controlled release of NPK macronutrient

A Suratman[1], A D Oktaviani[1], N H Aprilita[1] and A H Wibowo[2]
[1] Department of Chemistry, Faculty of Mathematics and Natural Sciences, Universitas Gadjah Mada, Yogyakarta
[2] Department of Chemistry, Faculty of Mathematics and Natural Sciences, Universitas Sebelas Maret, Surakarta

E-mail: adhitasari@ugm.ac.id

Abstract. Alginate-chitosan hydrogel as controlled release of Nitrogen, Phosphorus, and Potassium (NPK) macronutrient has been studied. Variations of alginate and chitosan mass, N; P and K concentration, pH solution, swelling ratio and mass ratio of alginate and chitosan were optimized. Hydrogel was prepared by mixing alginate, chitosan and $CaCl_2$ as a cross linker in the water media. Hydrogel product was characterized by Fourier Transform Infrared (FTIR). N; P; and K release were measured by UV-Vis spectrophotometer and Atomic Absorption Spectrophotometer (AAS). The result showed that addition of chitosan in the hydrogel caused a decrease of swelling ratio hydrogel and a decrease of N; P and K release. The release of N; P and K was affected by N; P and K concentration and pH solution. The higher N; P and K concentrations and lower pH solution caused an increase of release rate of N; P and K. The mechanism of N; P and K release from the alginate-chitosan hydrogel followed the Korsmeyer-Peppas model. The ratio of alginate and chitosan in the hydrogel also affected to the the release rate of N; P and K.

1. Introduction

Plants are basicly composed of 14 essential macronutrients as nitrogen, phosphorus, and potassium [6]. Generally, plants are capable to adapt at low pH in uptaking NH_4^+ and at high pH solution in uptaking NO_3. Inorganic phosphorus as nutrient for plant in soil is found in the form of $H_2PO_4^-$, HPO_4^{2-} and potassium cation[5]. Nutrients were taken by plants through step by step absorption. In this case, the utilisation of conventional fertilizers may not effective as the rest nutrients which are not absorbed may loose due to leaching, run-off, gaseous emission, and fixation by soil and the following effects, soil degradation, reducing the quality of water and environmental degradation may occur[2]. Therefore, the use of nutrient-control release may help to solve the problems. The control release makes nutrients are slowly released to the media following by nutrients absorption by plants without any waste due to leaching[3]. Here, a media that is to keep and to release nutrient in nutrient uptaking of plants is needed. In this study, bio-copolymer of alginate is studied as the media that acts for nutrient control release. Alginate is a polysaccharide consisting of 1-4 linked α-D-mannuronat and β-L-guluronat where in the presence of divalent cations will form a crosslinked cation-alginate gel and form

[1] Corresponding Author

distinctive eggbox structure[13][10]. The nutrition leaching during preparation of hydrogel can be solved by synthesizing bi-polymeric alginate beads with other natural polysaccharides[8]. Due to owing negative-charge surface, alginate can interact with chitosan which has positive charge to form polyelectrolyte complex[1]. Here, the aim of the study is to form hydrogel from alginate and chitosan with the presence of $CaCl_2$ crosslinker as controlled release of nutrient.

2. Experimental

2.1. Materials

The material used in the study are Na-Alginate, Chitosan, Calcium Cloride ($CaCl_2$), Ammonium nitrate (NH_4NO_3), Diammonium Hydrogen Phosphate $(NH_4)_2HPO_4$, Potassium Nitrate (KNO_3) from Merck (all are analytical grade). Ammonium molibdat ($(NH_4)_6Mo_7O_{24}.H_2O$), ethanol 96%, ammonium vanadate (NH_4VO_3), Nitric Acid (HNO_3), Hydrochloric Acid (HCl), para dimetilamino benzaldehide (p-DAB) were purchased from Mallinckrodt.

2.2. Instrumental

Fourier Transform Infrared/FTIR (Shimadzu FT-IR-821 PC) was used to characterize the hydrogel product. Spectroscopy UV-Visible (Thermo Scientific Evolution 201), and Atomic Absorption Spectroscopy/AAS (Parkin Elmer 3110) was used to measure the release performance.

2.3. Preparation of alginate-chitosan hydrogel

The fertilizer solution was prepared by dissolving ammonium nitrate, diammonium phosphate, and potassium nitrate in 100 mL deionized water[4]. The alginate-chitosan hydrogel was prepared by dissolving alginate 2% (b/v) and chitosan into fertilizer solution with mass ratio of alginate: chitosan (10:0); (8:2); (7:3); (6:4). Acetic acid (2%) was added into solution under constant stirring. The mixture was injected into syringe and then dropped into $CaCl_2$ 0.5 M solution to form viscous gel. Gel formed was rinsed with distilled water and dried in the room temperature.

2.4. Swelling ratio of alginate-chitosan hydrogel

Alginate-chitosan hydrogel was weighed firstly in the dry form. The hydrogel was then immersed into deionization water and various pH solution for 100 minutes before filtration and weighing. The swelling ratio of hydrogel was calculated using equation:

$$SR = \frac{Wx - Wd}{Wd} \times 100\% \qquad (1)$$

Where *Wx* referred to weigh of swollen hydrogel and *Wd* is dried hydrogel.

2.5. Optimisation of N; P and K release in various condition

The hydrogels made with various mass ratio and various N; P; K concentration were immersed into 20 mL deionized water. The solution (3 mL) was sampled and measured for N; P; and K concentration every 24h until for 12 days. The mother solutions were added with 3 mL deionized water after each measurement done. The content of N; P; and K was determined with UV-Vis Spectrophotometer and AAS. N; P; and K content was also determined at pH 3; 5; and 6.8.

2.6. Modelling and release kinetics

The release mechanism of N; P; and K was investigated using a semi-empirical model, known as Korsmeyer-Peppas which correlates nutrient release from polymer as following equation:

$$\frac{M_t}{M_\infty} = kt^n \qquad (2)$$

Where M_t and M_∞ represent the amount of released nutrient at a time t and equilibrium condition, k is constant characteristic of fertilizer-polymer system, and n is the diffusion exponent characteristic of the release mechanism (Quasi-Fickian diffusion n = 0.5; non-Fickian or anomalus transport n = 0.5 – 1.0 and Case II transport n = 1.0[7])

3. Results and Discussion

3.1. Characterization of alginate-chitosan hydrogel by FTIR

The IR spectra of alginate (figure 1) showed a peak at 3448 cm^{-1} indicated -OH group from carboxyl and peaks at 1620 and1420 cm^{-1} showed asymmetric and symmetric vibration of -COOH groups and strong peak at 1126 cm^{-1} indicated the presence of C-O vibration from carboxylic acid. Chitosan was identified with N-H peak at 1651 cm^{-1} and at 3448 cm^{-1} (overlapping of O-H and NH_2 vibration), at 1080 cm^{-1} (C-N stretching vibration) and at 1309 – 1419 cm^{-1} (deformation from C-OH vibration). The complex formation of alginate-chitosan as polyelectrolyte was indicated by the decrease of absorption intensity at 3425 cm^{-1} (-OH and NH_2 vibration). It was caused by protonation of -COOH group into COO- and NH_2 into NH_3^+ which affects to ionic interaction between the carbonyl group of alginate and the amino group of chitosan. The peak shift at 1651 cm^{-1} to 1635 cm^{-1} and 1420 cm^{-1} to 1381 cm^{-1} indicated that complexation between alginate and chitosan occured.

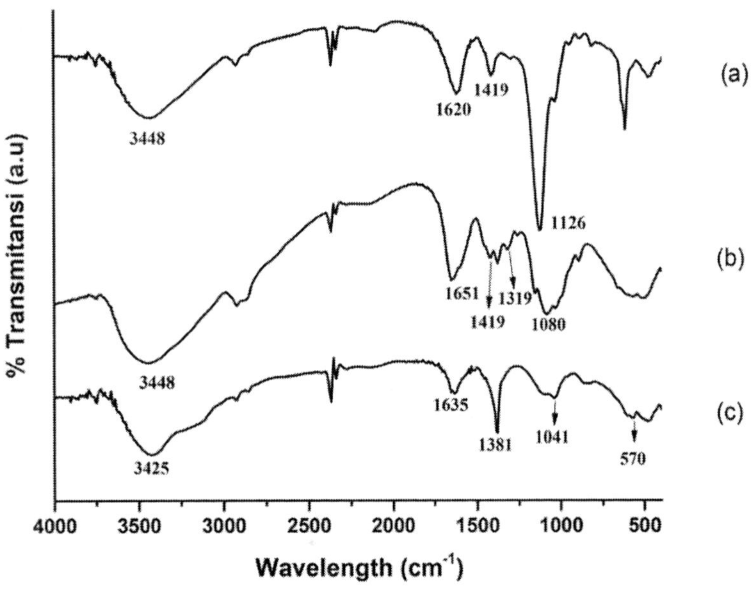

Figure 1. FTIR Spectra of (a) chitosan, (b) alginate, (c) alginate-chitosan

3.2 Swelling ratios of alginate-chitosan hydrogel

The swelling ratio of alginate-chitosan hydrogel in deionized water is shown in figure 2. The swelling ratio of hydrogel decreased with the increase of chitosan in the hydrogel. This decrease of swelling ratio is well known, due to hydrophobic of chitosan more than alginate. The lowest swelling ratio was achieved in the alginate-chitosan hydrogel with 8:2 mass ratio. Here, the large number of alginate with negative groups that were bonded by cross linker caused the increase of hydrogel stability in deionized water.

IOP Conf. Series: Materials Science and Engineering **578** (2019) 012074 doi:10.1088/1757-899X/578/1/012074

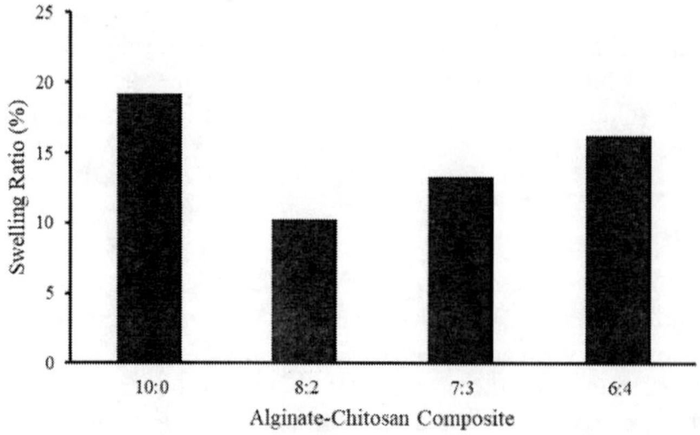

Figure 2. Swelling ratio of alginate-chitosan hydrogel with different composition

The swelling ratio of alginate-chitosan hydrogel with 8:2 mass ratio in various pH solutions is shown in figure 3. From the figure, it was conclused that the swelling ratio decreased with the increase of pH solutions. It was because of unreacted NH_2 group that was protonated to NH_3^+ in acidic solution and leaded to dissociation of the hydrogen bonding of the amino groups[9]. As consequence, the entrance of solvent into hydrogel and the release of nitrogen, phosphorus and potassium was higher. The increase of pH solution caused weakness of the protonation of unreacted NH_2 group which also caused lower swelling ratios of hydrogel.

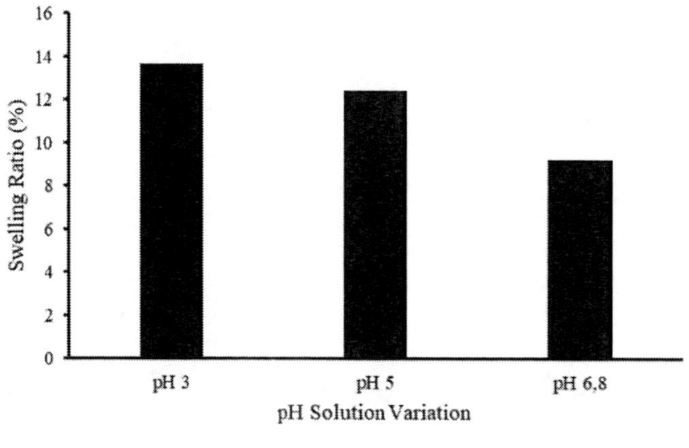

Figure 3. Swelling ratio of 8:2 alginate-chitosan hydrogel at pH solution variations

3.3 Optimisation of N; P and K release in various condition.

The N; P; and K release from the hydrogel in the media of deionized water at variations of alginate-chitosan mass ratios are shown in figure 4. From the data, the release of N; P; and K decreased with the increase of chitosan content in hydrogel. The alginate-chitosan hydrogel at 8:2 mass ratio showed the lowest release of N; P; and K. The release of N; P; and K out of hydrogel occured after absorption of water by hydrogel and then N; P; and K was released by diffusion out of hydrogel. The diffusion of N; P; K is more easily at higher swelling ratio than lower swelling ratio. On the other hand, the hydrogel with the higher chitosan content caused the increase of ionic interaction between amine groups of chitosan and carboxylic groups of alginate and decrease of interaction between alginate and Ca^{2+}[12]. This phenomenon contributed to the decrease of hydrogel swelling.

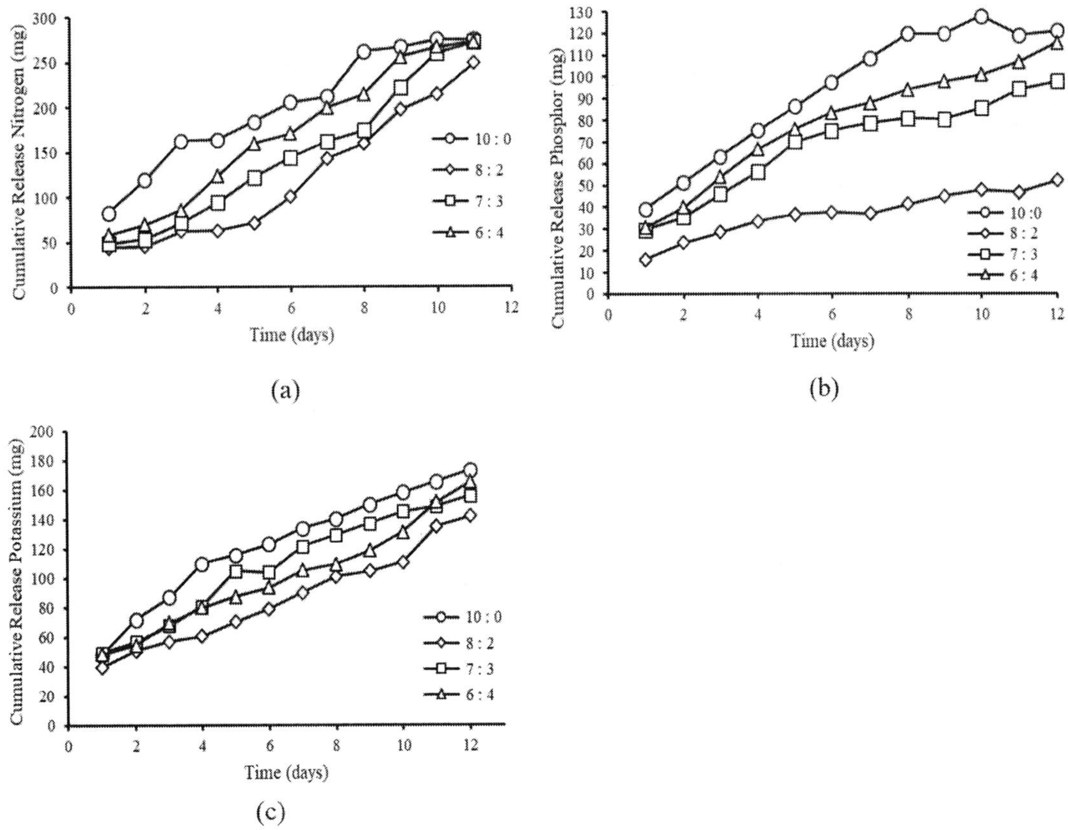

Figure 4. The release of (a) N; (b) P; and (c) K with the variations of alginate-chitosan mass ratios

The release of N; P; and K with the variations of N; P; and K concentration are shown in figure 5. It was shown that the released amount of N; P; and K was increased with higher content of N; P; and K concentration in hydrogel.

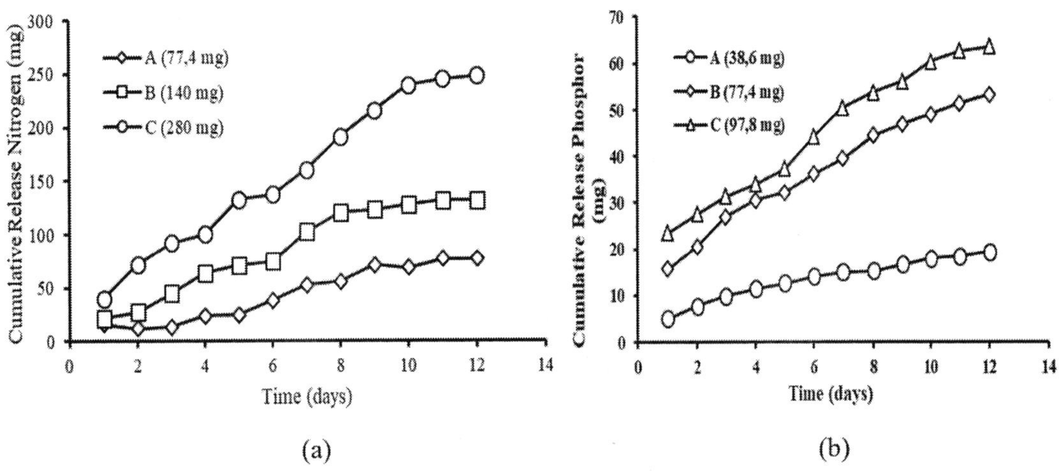

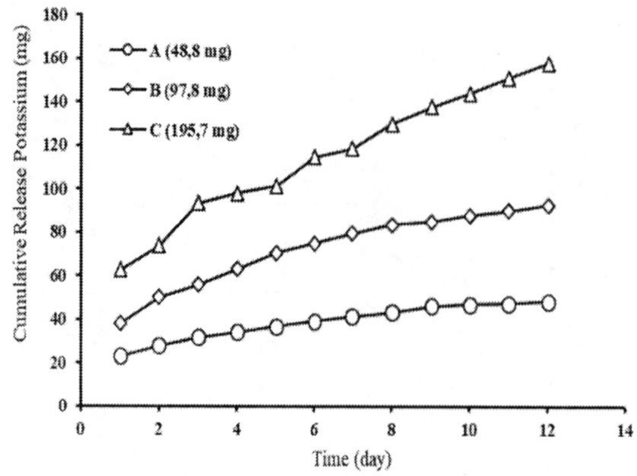

(c)

Figure 5. The release of (a)N; (b)P; and (c) K with the variations of N; P; and K concentrations in the hydrogel

The release of N; P; and K macronutrient at the variation of pH is shown in figure 6. It was showed that the release of N; P; and K at lower pH was higher than at higher pH. It was due to the unreacted NH_2 group was protonated in acidic solution, that also caused polymer chains exert repulsive forces and dissociation of secondary interaction. As consequence, the hydrogel could absorb more water and increase the swelling ratio of hydrogel.

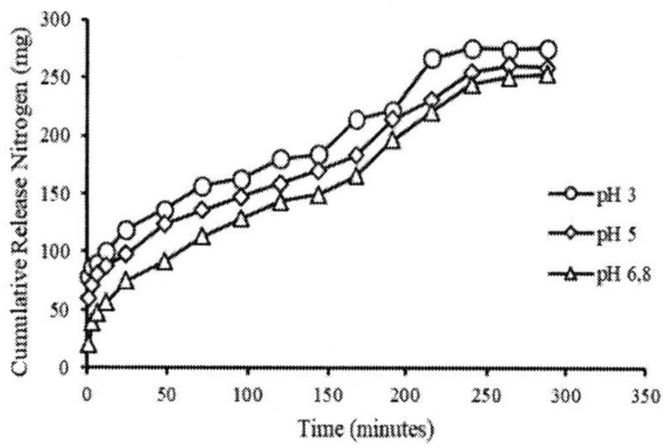

(a)

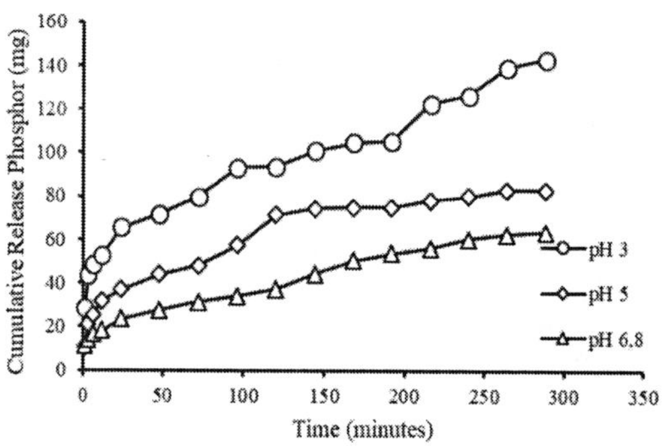

(b)

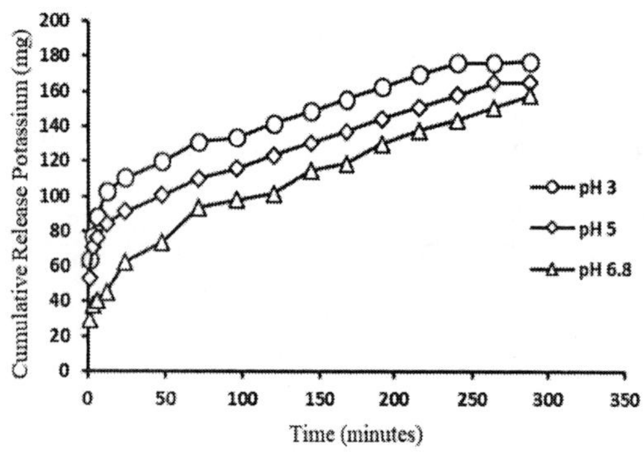

(c)

Figure 6. The release of (a) N; (b) P; and (c) K with pH solution variation

The release mechanism of N; P; and K macronutrient that was investigated using semi empirical equation known as Korsmeyer-Peppas was correlated with the release exponent, *n*, correlation coefficient (R^2) and release factor (*k*) of each hydrogel. The values were obtained by potting ln (M_t/M_∞) vs ln *t*. The values of *n* for released of N; P; and K with mass variation are listed in table 1 and table 2

Table 1. Release exponent (*n*), correlation coefficient (R^2) and release factor (*k*) of N; P; K release by alginate:chitosan mass variation in deionized water.

Nutrient	Alginate: chitosan	Corelation coefficient R^2	Korsmeyer-Peppas	
			n	k (mg/day)
N	10:0	0.99	0.50	0.25
	8:2	0.85	0.78	0.10
	7:3	0.94	0.78	0.13
	6:4	0.95	0.77	0.17
P	10:0	0.97	0.50	0.25
	8:2	0.98	0.44	0.11
	7:3	0.98	0.54	0.19
	6:4	0.97	0.50	0.18
K	10:0	0.99	0.50	0.25
	8:2	0.93	0.51	0.17
	7:3	0.96	0.51	0.22
	6:4	0.95	0.50	0.21

Table 2. Release exponent (*n*), correlation coefficient (R^2) and release factor (*k*) of N; P; K release by N; P; K concentration variation in deionized water.

Nutrient	Concentration	Corelation coefficient R^2	Korsmeyer-Peppas	
			n	k (mg/hari)
N	A	0.82	0.85	0.11
	B	0.97	0.82	0.13
	C	0.98	0.74	0.14
P	A	0.99	0.50	0.035
	B	0.97	0.58	0.098
	C	0.95	0.42	0.13
K	A	0.99	0.30	0.11
	B	0.99	0.36	0.19
	C	0.98	0.37	0.30

The mechanism release of nitrogen, phosphorus, and potassium showed that *n* value was 0.5 – 1.0, indicated that the release of N; P; and K was a Non-Fickian diffusion. This mechanism indicated that N; P; and K was diffused by controlled swelling and controlled diffusion. The different *n* value was seen in the phosphorus release at 8:2 mass variation. It was affected by the different chemical structure of the hydrogel. The *n* value for the release of nitrogen and phosphorus by variation of concentration of N; P; and K was 0.5 -1.0 indicated that the release of nitrogen and phosphorus was Non-Fickian diffusion. *n* value for potassium release was smaller than 0.5, indicated that the release of N; P; and K from hydrogel was a quasi Fickian. This mechanism indicated that N; P; and K diffused partially through swollen hydrogels and water filled pores of the hydrogel.

4. Conclusion

The alginate-chitosan hydrogel as control release of N; P and K showed as good media for nutrient release. The swelling ratio of hydrogel decreased by the content increase of chitosan in the hydrogel. The release of N; P; and K decreased with the content increase of chitosan. The increase of N; P; K concentration in the hydrogel affected the release increase of N; P; and K. The lower pH solution affected to the higher swelling ratio of hydrogel and therefore, the release of N; P; and K also increased. From the mechanism study, it was found that the mass variation generally exhibited the small value of release exponent that is only 0.5-1.0, indicated a non-Fickian diffusion mechanism. N; P; and K concentration was following Quasi-Fickian and Non-Fickian diffusion. Release of N; P and K could be controlled by ratio of the alginate-chitosan content in the hydrogel.

References

[1] Abdou E S, Osheba A and Sorour M 2012 *Int. J. Appl. Sci. Technol* **7** 158-169
[2] Balgiar V C, Fageria N K and He Z L 2001 *J Commun Soil Sci Plant Anal.* **32** 7-8
[3] Byung-Su K, Young-Sang C and Hyun-Ku H 1996 *Ind. Eng. Chem. Res.* **35** 250-257.
[4] Jamnongkan T and Kaewpirom 2010 *Sci. J. UBU* **1** 43-50
[5] Maathius F 2009 *Cur Opin Plant Biol.* **12** 250-258
[6] Marschner H 1995 *Mineral Nutrition in Higher Plants* (London Academic Press) p 681-861
[7] Singh B 2007 *Int J Pharm* **334** 1
[8] Singh B, Sharma D and Gupta A 2009 *J Hazard Mater.* **161** 208-216
[9] Shu X Z, Zhu K J and Song W 2001 *Int J Pharm.* **212** 19
[10] Smidsrod O and Skjak-Braek G 1990 *TIBTECH* **8** 71-78
[11] Sonia T A and Sharma C P 2011 *Adv Polym Sci.* **243** 23-54
[12] Tahtat D, Mahlousa M, Benamera S, Khodjaa A N, Oussedik-Oumehdib H and Laraba-Djebarib F 2013 *Int J Bio Macro.* **53** 160-168
[13] Uzaşçı S, Tezcan F and Erim F B 2013 *Int J Environ Sci Technol.* **11** 1861–1868

The Role of Magnesium(II) Salt Precursor in Controlling the Structure of Coordination Polymers based on Magnesium(II) and benzene 1,3,5-tricarboxylic acid

W W Lestari, R A Tedra, T H Novita, and Sakinah
Research Group of Porous Materials for Sustainability, Department of Chemistry, Sebelas Maret University, Jl. Ir. Sutami 36A Kentingan, Jebres, Surakarta, Indonesia 57126

E-mail: witri@mipa.uns.ac.id

Abstract. Synthesis of coordination polymers based on magnesium(II) and benzene 1,3,5-tricarboxylic acid (H_3BTC) with two different types of magnesium(II) salts precursors, namely $MgCl_2{\cdot}6H_2O$ and $Mg(OAc)_2{\cdot}4H_2O$ in ethanol under solvothermal condition was successfully carried out. The synthesized materials were characterized by several methods such as X-Ray Diffraction and FTIR spectroscopy, thermogravimetry and scanning electron microscopy (SEM). The synthesis results showed that the crystallization of the materials was highly influenced by the type of metal ion precursor Mg(II), in this case the anion. XRD analysis revealed that the use of $MgCl_2{\cdot}6H_2O$ salt tend to produce a coordination polymer $[Mg_3(BTC)_2(H_2O)_{12}]$ which called also as *National Institute of Chemistry Slovenia-4* (NICS-4, CCDC 768993) with a 1-dimensional structure, while the use of $Mg(OAc)_2{\cdot}4H_2O$ salt produced 3-dimensional coordination polymer $[Mg_3(BTC)_2]$ or *well known as National Institute of Chemistry Slovenia-6* (NICS-6, CCDC 768995). NICS-4 and NICS-6 have thermal stability up to 530 °C and 542 °C based on the thermogravimetric analysis. A significant shift from absorption peak at 1721 to 1622 dan 1628 cm^{-1} correspond to the deprotonation of trimesic acid and coordinated to the Mg^{2+} cation. The coordination of Mg^{2+} with H_2O molecules in NICS-4 is indicated by the presence of broad absorption peak at 3472 cm^{-1} which was not found in NICS-6. After exposed to the air, the structural dimension of NICS-6 could change due to the coordination with water molecules. SEM image exhibited that NICS-4 has flat-needle-like morphology with an average particle size of 12.42 ± 0.78μm, while NICS-6 has a hollow spherical morphology with an average particle size of 63.21 ± 0.12 μm.

1. Introduction

Coordination bonding between metal ions and bidentate or polydentate ligands tends to produce coordination polymer structures in the form of one, two, or three dimensions [1]. What distinguishes between coordination polymers with porous coordination polymers or well known as metal-organic framework (MOFs) itself is the presence of permanent porosity in MOFs after activation and removal of solvent molecules [2]. Design of MOFs and coordination polymers can be achieved by varying metal ions and organic ligands according to the desired application. Therefore, this type of material has a rich variety of topologies and functions [3] in various fields such as in the catalysis[4], gas storage and separation [5] [6], and slow release of drugs [7].

Synthesis of MOFs and coordination polymers is able to be pursued by several methods including microwave assisted synthesis [8], sonochemistry [9], electrochemistry [10], mechanochemistry [11], and solvothermal or hydrothermal [12]. Synthesis under solvothermal condition has several advantages of producing high crystallinity material, high surface area and porosity material, reproducible and easy method for screening new MOFs structure [13]. Some reaction parameters that influence during the synthesis include metal salt precursors [14], reaction time [15], ligand types [16], stoichiometry [15], temperature [17], and solvents [18]. In the synthesis, temperature is an external stimulus that will influence the topology and structure of the material [19]. The greater the reaction temperature, the higher the structural dimension is produced because the tendency of bond formation in the M-O-M metal oxide cluster is stronger [20]. In addition, the use of solvents also plays an important role in crystallization both in the kinetics and thermodynamics of the reaction during the coordination process [21]. Mazaj et al, [18] reported the dimensions of the hybrid material in the form of a coordination polymer based on Mg-BTC (BTC: benzene 1,3,5-tricarboxylate) influenced by the type of solvent. The presence of water molecules as solvent in the synthesis will reduce the dimensions of the produced Mg-BTC. Biemmi et al. [22] reported that zinc acetate salt and zinc nitrate precursors lead to produce MOF-5 structure, however the use of zinc(II) sulfate will produce unknown structure. In line with these researches, herein we study the effect of different types of magnesium(II) salt precursors, namely $MgCl_2{\cdot}6H_2O$ and $Mg(OAc)_2{\cdot}4H_2O$ with the H_3BTC as organic linker under solvothermal method in ethanol that have never been previously reported.

2. Experimental

2.1. Materials

All chemicals used are commercial materials with analytical grade quality and are used without further purification. Magnesium acetate tetra hydrate [$Mg(OAc)$ $2{\cdot}4H_2O$] (99%), magnesium chloride hexahydrate [$MgCl_2{\cdot}6H_2O$] (≥98%), and benzene 1,3,5-tricarboxylic acid (H_3BTC, 95%) were purchased from Sigma Aldrich, Germany. Ethanol (96%) purchased from Merck EMSURE.

2.2. General Characterization Methods

The obtained materials were analyzed for its phase purity using X-ray diffraction (XRD) type Pan Analytical Expert with Kα1 1.5406 Å Cu detector. The diffractogram of NICS-4 and NICS-6 were further refined by the Le Bail method. The shift of functional group of the resulting materials were characterized using Fourier transform infrared (FTIR) Shimadzu IR Prestige-21 at a wave number of 4000-400 cm^{-1} with KBr pellets. The thermal stability of the materials were measured using Thermal gravimetric analysis (TGA) type of STA Linseis PT-1600 at a temperature of 0-600 °C with a heating speed of 10 °C / minute. The particle size and morphology were observed with the FEI Inspect-S50 SEM.

2.3 Synthesis of [$Mg_3(BTC)_2(H_2O)_{12}$] (NICS-4) and [$Mg_3(BTC)_2$] (NICS-6)

$MgCl_2{\cdot}6H_2O$ (0.609g, 3 mmol) and H_3BTC (0.630g, 3 mmol) were dissolved in 5 mL and 12.5 mL ethanol, respectively. The solution was then mixed in a teflon line volume of 25 mL and stirred for 15 minutes then reacted under solvothermal condition in an Memmert UN-30-1060 E708 oven at 190 °C for 24 hours. The synthesis procedure for NICS-6 was carried out in the same way except that the metal ion precursors were replaced using magnesium acetate ($Mg(OAc)_2{\cdot}4H_2O$). The yield of synthesis of NICS-4 and NICS-6 is 40.4 and 84.6%, respectively.

3. Results and Discussion

The X-Ray diffractogram in Figure 1 shows that the coordination polymer based on Mg (II) -BTC with $MgCl_2{\cdot}6H_2O$ as metal salts precursor produces a material [$Mg_2(BTC)_2$ $(H_2O)_{12}$] (NICS-4) which has a one-dimensional structure according to CCDC 768993. This is evidenced by the appearance of characteristic peaks at 2θ 18.58° (1 1 1); 18.87° (3 1 -1) and 28.60° (5 1 0) which represent the coordination of trimesate anion and H_2O as ligands with magnesium(II) as the central metal ion forming one-dimensional structure and its chains of trimming anions and isolated octahedral with magnesium centers. There are two types of Magnesium here, which are monodentate fashion (μ_1) and the other bidentate fashion (μ_2) forms [$Mg_2(BTC)_2(H_2O)_{12}$] [18]. While the use of $Mg(OAc)_2{\cdot}4H_2O$ precursor produces [$Mg_3(BTC)_2$] (NICS-6) according to CCDC 768995 with a typical peak found at 2θ 25.46° (1 2 0); 27.22° (2 1 2); and 29.43° (2 2 2) (Figure 1) has a three dimensional structure in which magnesium(II) metal ions coordinate with 6 trimesate ligands. Its structure consists of a linear Mg-centered octahedra face-sharing trimers connected through trimestate anion forming a 3-D framework structure. In [$Mg_3(BTC)_2$], magnesium(II) occupies two crystallographic sites (Mg1 and Mg2). The Mg1 atom coordinates with three oxygen atoms with monodentate fashion (μ_1) from the carbonyl group and three oxygen atoms with bidentate fashion (μ_2) with a longer Mg-O bond. Mg2 atoms coordinate with six oxygen atoms with bidentate fashion (μ_2) [18]. More clearly the description of the structure of NICS-4 and NICS-6 is shown in Figure 2. The results of the refinement explain that NICS-4 and NICS-6 have been successfully synthesized with a residual phase and Rwp (residual weight phase) of 5.578%; 8.483% and 4.259%; 7.532% respectively, which shows high compatibility with standard pattern as shown in Figure 3; where the standard used in NICS-4 is CCDC 768993 and ligand H_3BTC JCPDS 45-1880. Whereas the used standard for NICS-6 is CCDC 768995 and H_3BTC JCPDS 45-1880.

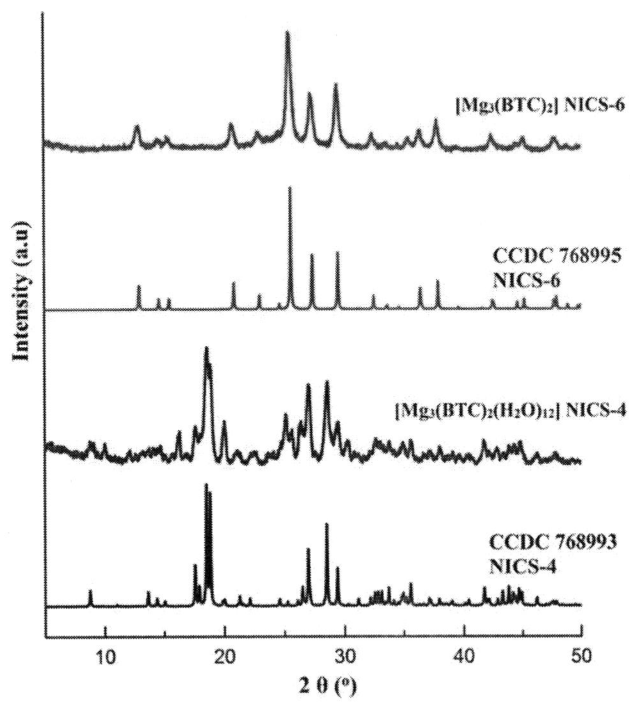

Figure 1. X-Ray Diffractogram of synthesized NICS-4 and NICS-6 compared to standard pattern.

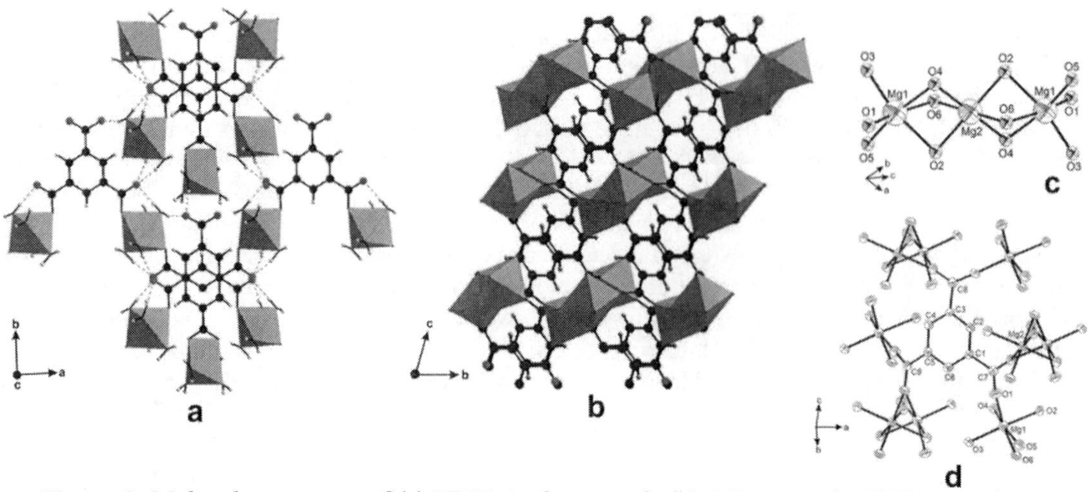

Figure 2. Molecular structure of (a) NICS-4 a long a axis (b) 3-D network of NICS-6 a long a axis (c) Linear Mg3 face-sharing building unit in NICS-6 structure and (d) connectivity of BTC ligand to Mg(II).

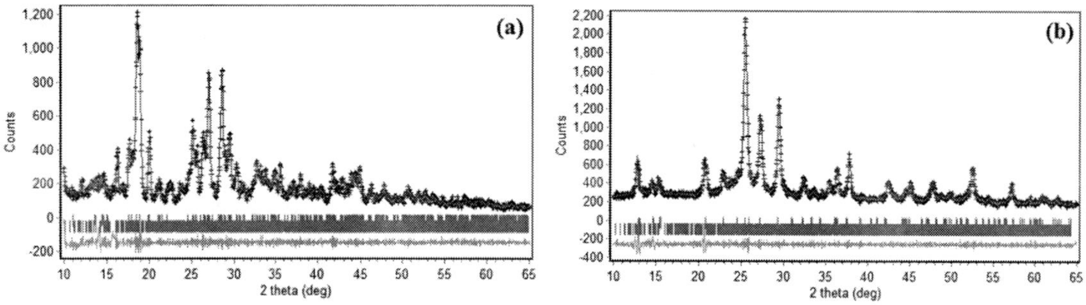

Figure 3. Refinement result of diffractogram (a) NICS-4 dan (b) NICS-6 compared to standard pattern

The successful synthesis of $[Mg_2(BTC)_2(H_2O)_{12}]$ (NICS-4) and $[Mg_2(BTC)_2]$ NICS-6 was further confirmed by FTIR spectra as shown in Figure 4. Absorption peak of C=O groups in carbonyl at 1722 cm^{-1} shifts significantly to 1622 cm^{-1} (NICS-4) and 1628 cm^{-1} (NICS-6) which correspond to the deprotonation of proton (H^+) from carboxylate to form carboxylate moiety (COO^-) which then bind to Mg^{2+}. The coordination of Mg-OH_2 (from water molecules) in NICS-4 can be observed by the presence of absorption at 3472 cm^{-1} and 712 cm^{-1} which is not found in NICS-6.

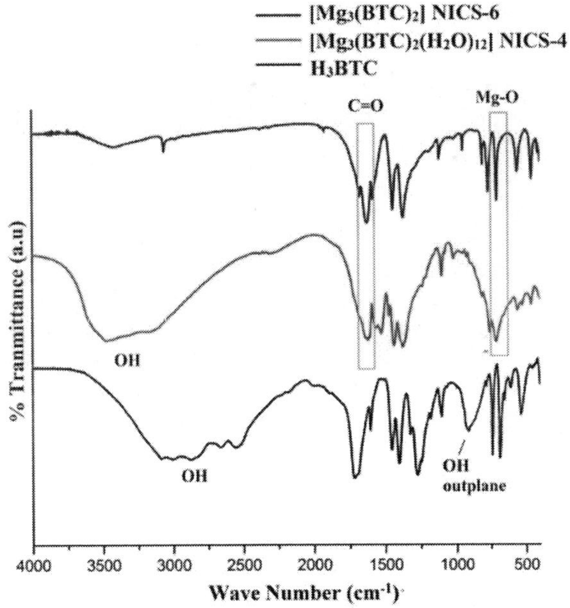

Figure 4. FTIR spectra of NICS-4 and NICS-6

To determine water molecule content as hydrated or coordinated to the metal centre of Mg^{2+} a thermogravimetric analysis was performed. The TG profiles of NICS-4 and NICS-6 are shown in Figure 5. At 222.6 °C there was the first decrease in NICS-4 by 31.6% indictated the loss of twelve water molecules coordinated on NICS-4. Whereas in the mass reduction diagram on NICS-6, the material has been exposed to moisture and resulting in a ligand exchange between BTC and H_2O in the air so H_2O is coordinated with Mg^{2+} metal [23]. The results showed that NICS-4 and NICS-6 have thermal stability up to 530.6 °C and 542.7 °C, respectively, where at this temperature the structure of the coordinated BTC ligand began to decompose and initiate to the total decomposition of the referred polymers. In the TGA analysis, the material measurement process has not been completed as a whole, leaving residues in NICS-4 and NICS-6 of ±26.95% and ±55.19% respectively. Mazaj et al reported that the magnesium oxide residues in NICS-4 and NICS-6 were ± 39.3% and ± 54.5% [18].

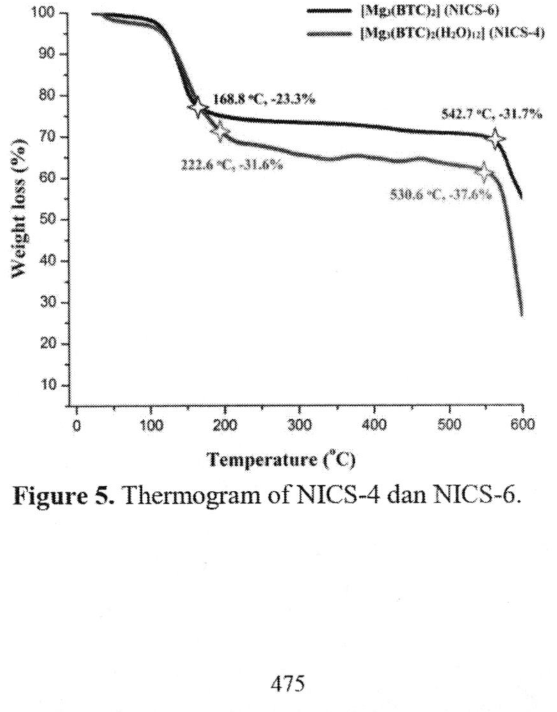

Figure 5. Thermogram of NICS-4 dan NICS-6.

The SEM image (Figure 6) shows that $[Mg_2(BTC)_2(H_2O)_{12}]$ (NICS-4) has a flat-needle-like morphology that forms like a flower with a particle size of 12.42 ± 0.78 µm,while $[Mg_2(BTC)_2]$ (NICS-6) has a hollow spherical morphology with particle size of 63.21 ± 0.12 µm.

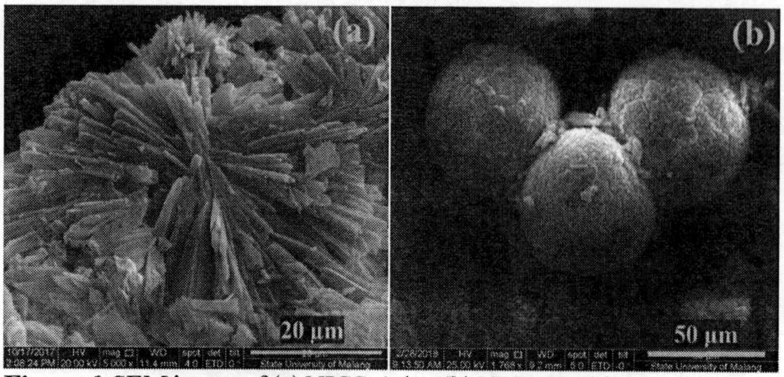

Figure 6. SEM image of (a) NICS-4 dan (b) NICS-6 with magnification 5000x and 2500x respectively.

The presence of precursors with different anions in this case acetate and ion chloride ions tend to give a different tendency of reaction which is possible because the solubility of these ions in ethanol also has different strengths. Acetate ions tend to orient Mg(II) ions more coordinated with trimesic acid, while Cl^- ions which are weak ligands tend to exchange easily with water so that Mg(II) ions are more coordinated with water and partly coordinated with trimesate ligand to form the different geometry and dimensions of the polymer. In addition, chloride ions also have higher solubility in water than acetate.

4. Conclusions

The use of different types of magnesium(II) metal salt precursors in the synthesis of coordination polymer based on ligand benzene 1,3,5-tricarboxylic acid will produce different polymer dimensions. $Mg(OAc)_2 \cdot 4H_2O$ reacted with H_3BTC lead to produce $[Mg_2(BTC)_2]$ (NICS-6), meanwhile the utilization of $MgCl_2 \cdot 6H_2O$ tend to produce $[Mg_2(BTC)_2(H_2O)_{12}]$ (NICS-4). Nonetheless, this 3-dimensional polymer is quite sensitive to air moisture which changes its dimensions, therefore in the future it is interesting to study in more detail the changes in this structure as a function of time.

Acknowledgement

We gratefully acknowledge to Ministry of Research, Technology and High Education of Republic Indonesia (RISTEK DIKTI) in the scheme International Research Collaboration (PKLN) project number 474/UN27.21/PP/2018 for funding.

References

[1] Biradha K, Ramanan A and Vittal J J 2009 *Cryst. Growth. Des.* **9(7)** 2969
[2] Betten S R and Champness 2016 *Phil. Trans.* **375** 20160032
[3] Gangu K K, Maddila S, Mukkamala S B and Jonnalagadda S B 2016 *Inorg. Chim. Acta.* **446** 61
[4] Chen B, Xiang S and Qian G 2010 *Accounts Chem. Res.* **43** 1115
[5] Phan A, Doonan C J, Uribe-Romo F J, Knobler C, O'keeffe M and Yaghi O M 2010 *Accounts Chem. Res.* **43** 58
[6] Ferey G 2008 *Chem. Soc. Rev.* **37** 191
[7] Allendorf M D, Bauer C A, Bhakta R K and Houk R J 2009 *Chem. Soc. Rev.* **38** 1330
[8] Chen Y, Ni D, Yang X, Liu C, Yin J and Cai K 2018 *J. Electacta* 278 114
[9] Lestari W W, Arvinawati M, Martien R, and Kusumaningsih T 2018 *Mat. Chem. Phys.* **204** 141
[10] Xuan W, Ramachandran R, Zhao C and Wang F 2018 *J. Solid State Electrochem.* 1
[11] Huang Y, Lo W, Kuo Y, Chen W, Lin C and Shieh F 2017 *Chem. Commun.* **53** 5818

[12] Barton H F, Davis A K, Lee D T and Parsons G N 2018 *J. Visualized Experiment* **13** 136
[13] Wang Q M, Shen D, Bülow M, Lau M L, Deng S, Fitch F R, Lemcoff N O and Semanscin J 2002 *Micropor Mesopor Mater.* **55**(2) 217
[14] Sayed F N and Polshettiwar V 2015 *Sci. Rep.* **5** (0973) 1
[15] McKinstry C, Cussen M J, Fletcher A J, Patwardhan S V and Sefcik J 2013 *Cryst.Growth Des.* **13**(12) 5481
[16] Sánchez-Sánchez M, Getachew N, Díaz K, Díaz-García M, Chebude Y and Díaz I 2015 *Green Chem.* **17**(3) 1500
[17] Gurunatha K L, Uemura K and Maji T K 2008 *Inorg. Chem.* **47** 6578
[18] Mazaj M, Celic T B, Mali G, Rangus M, Kaucic V and Logar N Z 2013 *Cryst. Growth Des.* **13**(8) 3825
[19] Forster P M, Burbank A R, Livage C, Frey G and Cheetham A K 2004 *Chem. Commun.* 368
[20] Stock N and Bein T 2005 *J. Mater.Chem.* **15**(13) 1384
[21] Birsa-Celic T, Rangus M, Lazar K, Kaucic V and Logar Z 2012 *Angewandte Chem. Int Ed* **51** 12490
[22] Biemmi E, Christian S, Stock N and Bein T 2009 *Micropor Mesopor Mater* **117** 111
[23] Mali G, Trebosc J, Martineau C and Mazaj M 2015 *J. Phys. Chem. A* **119**(14) 7831

The Flexural Strength of Post-Fire Concrete with the use of Waste Banner Fibers

A Rahmawati[1,*], A Nurhidayati[1], Roemintoyo[1], and M R Prakoso[1]
[1]Building Engineering Education, Faculty of Teacher Training and Education Universitas Sebelas Maret

[*]E-mail: anisrahmawati@staff.uns.ac.id

Abstract. The use of nylon fiber from the waste banner in this study was expected to be an alternative effort to reduce the negative impact of waste banner while improving the property of concrete that was a flexural strength as an effect of the character of nylon fiber that has high tensile strength. The purpose of this study was to determine the maximum value of the flexural strength of post-fire concrete resulting from the addition of waste banner fiber. 64 of beam samples in 150 mm of width, 150 mm of thickness, and 600 mm of length used to test the flexural strength of concrete. The variation of the waste banner fiber's lengths was 30 mm, 60 mm and 90 mm while the combustion temperature used were 200 °C, 300 °C and 400 °C. The results of the study showed that 60 mm fiber's length gave the highest flexural strength results compared to other fiber lengths. The higher the combustion temperature produces the decreasing of concrete's flexural strength.

1. Introduction

Fiber Reinforced Concrete (FRC) has been successfully used in various civil engineering applications such as slabs [1] pre-cast poducts [2], lightweight concrete [3], structures in seismic regions [4], and many other applications. The main purpose of using FRP in concrete is to increase the energy absorption capacity and toughness of the material, as well as to increase the tensile and flexural strength of concrete [5]. Fibers of various shapes and sizes [6] produced from steel [7,8], polypropylene [9], glass [10], plastic [11] and natural materials [12] have been commonly used in FRC. With the increasing awareness of environmental conservation, the use of fibers made from waste material come to gain a lot of attention from the researcher as well as concrete industries. In this research, waste banner chosen as row material for fiber.

The banner sheet is consists of nylon fiber which covered by polyvinyl and used to called as Polyvinyl chloride (PVC) banner. PVC banner broadly used as momentary advertising media or an event's backdrop. PVC is one of the plastic categories which generally made of thermoplastic resins [13]. The wastes of plastics have been becoming a serious problem in the world. Scientists estimate that during 1950 untill 2015, approximately 6300 Mt plastic wastes were produced around the world, in which mostly (79%) was fiil up the landfills or disposed in the environment [14]. The use of waste banner as fiber in FRC could be an alternative way to decrease the cumulation of plastic pollutant in a landfill as well as in the ocean.

The most severe hazards that may occur during the service life of a structure is fire. Many recent studies had reported the collapse of a structure due to fire. During 2017 there were 2055 building fire cases In Jakarta, Indonesia [15]. Most of the fire incidents caused the structure to collapse. Even if a structure survives from a fire incident, the next question is whether or not the structure is safe to be used, and whether or not it needs repair. When the fire occurred in a structure made from fiber reinforced concrete, the uncertainties as to the remained strength would be higher because of the lack of knowledge on the performance of FRC reinforcing bars at high temperatures [16]. Since Polyvinil Chloride (PVC), banner's raw materials, is a material with a very poor heat stability with maximum operating temperature around 140 °F (60 °C) when heat distortion begins to occur [17], so the assessment of the remaining strength of FRC with the waste banner after the fire is crucial to observe on the further use of the affected buildings and structures. The current study investigates the flexural strength of post-fire FRC with the use of waste banner as fiber.

2. Experimental

This study conducted by doing an experimental laboratory investigation of speciments prepared with various variations of fiber content for the flexural test.

2.1. Materials

Concrete main ingredients were fine aggregate, coarse aggregate, and Cement. The natural fine aggregate was used and taken from Merapi mountain's quarry, which selected by sieving on sieves size and selected the fine gradation as shown in figure 1, while the physical properties of fine aggregate are listed in Table 1. This property confirmed with the standard from S-04-1989-F. The coarse aggregates were also taken from Merapi mountain's quarry which crushed with the maximum size of 20 mm, and the physical properties of coarse aggregate are presented in Table 2. The Portland Cement Composite (PCC) manufactured by HOLCIM was used. It's conformed to Indonesian specification of SNI 7064-2004 [18].

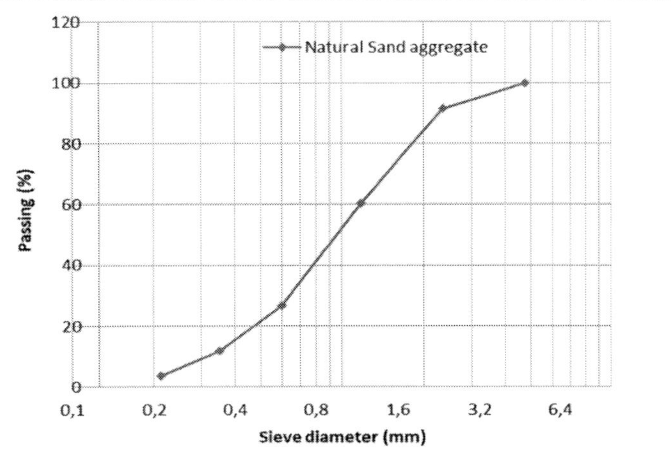

Figure 1. The gradation of fine aggregates.

Table 1. The Physical Properties of Fine Aggregates.

Characteristic	Point	Standard from SNI	No. of SNI
Water content	1.18%	1-3%	SNI 03-1970-1990
Absorbtion	1.09%	<2%	SNI 1970:2008 [19]
Specific gravity	2.57	2.5-2.7	SNI 1970:2008
Fineness modulus	3.06	2.5-3.8	SNI S-04-1989-F

Table 2. The Physical Properties of Coarse Aggregate.

Property	Point	Standard from SNI	No. of SNI
Specific gravity	3.38	2.5 – 2.7	SNI 1969:2008
Absorbtion	1.07	>3%	SNI 1969:2008
Abration	9.32	<50%	SNI 2417:2008
Fineness modulus	3.38	2.5-3.8	SNI 03-1968-1990

Waste banner used as fiber materials in a rectangular shape with a dimension of 2 mm in width and three variations lengths that were 30 mm, 60 mm, and 90 mm as shown in Figures 2.

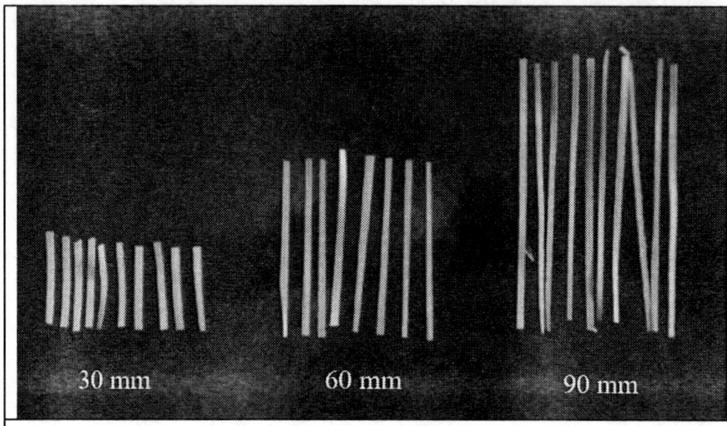

Figure 2. Waste Banner Fiber.

2.2. Test specimens preparation
The dimension of specimens for the flexural test according to SNI 4154-2014 [20] was 150mm x 150mm for the cross section and 600mm for the length. Mix design concrete conducted according to SNI 7656:2012 [21]. Table 3 summarizes the result of mix design concrete.

Table 3. Mix design result.

Materials	Amount	Unit
Cement	197.101	Kg/ m^3
Fine Aggregate	874,883	Kg/ m^3
Coarse Aggregate	988.393	Kg/ m^3
Water	204.523	Kg/ m^3

Three variations of waste banner length, that were 30 mm, 60 mm, and 90 mm, added in the concrete mixture as much as 0.2% of concrete weight. The specimens prepared to be fired in three variation temperatures, that were 200 °C, 300 °C and 400 °C for one ours. Normal concrete with no fiber added nor firing also prepared as a control variable. Each variation consists of four specimens.

2.3. Castings of test specimens
Castings of beam concrete specimens were prepared to determine the flexural strength of waste banner FRC. The proses of concrete making start with mixing cement and natural sands together in the dry

state for about 3 minutes to obtain a uniform mixture. The next step was adding the water to the blend slowly and mixed into a homogeneous mixture Then the coarse aggregates and waste banner fiber added, blend together till a uniform mixture was obtained. The mixing proses were performed using a 350-litre concrete mixer. The fresh concrete mixture then poured into the moulds and compacting using vibrator table.

After casting, the specimens were covered with wet burlap and left in the casting room at a temperature of (20±5) °C for a period of 24 h. The specimens were then de moulded and submerged in water for 26 days, and then placed in room temperature for a day before the specimens were burned in 200 °C, 300 °C, and 400 °C.

2.4. Testing
Four specimens of normal concrete and 60 specimens of waste banner FRC were tested to determine the residual flexural strength after exposure to high temperatures. All tests were performed at room temperature after the specimens had cooled down from the defined temperatures for each specimen. The flexural test using a simple beam with center-point loading was performed according to standard SNI 4154-2014 [20]. All concrete specimens were tested using Universal Testing Machine (UTM) with max capacity 50 Ton.

3. Results and Discussion
Table 4 summarizes the results of the residual flexural strength for all variations of waste banner length. The flexural strength data came from the average of four samples tested specimens for each type of concrete. The retained ratio in table 4 was calculated with respect to the flexural strength of concrete without combustion. The residual flexural strength of FCR with 60 mm length of waste banner fiber after combustion in 200 °C was higher than the other residual flexural strength of all variation of FRC which were observed in this study, that reached 3.58 N/mm^2. This point was approximately 76% and 7.8% greater than the retained strength of FCR with 30 mm and 90 mm length of the waste banner fiber, respectively. The retained ratio of all variation was above 50%. This result was in line with previous research which obtained that the strength of the concrete will be decreased to below 30% at 500 °C [16]. Hamdi&Kuba [22] from their research concludes that the mechanic degradations of normal concrete post-fired in 100 °C, 300 °C, and 600 °C were about 4%, 52%, and 65%, respectively. This degradation was caused by the decreasing in attachment between aggregates and cement paste, which is characterized by cracks and fragility of concrete or in the other word, concrete which is easily cracking.

Figure 3 shows the trend of the flexural ratio of all concrete variations in this study respect to the normal concrete as a variable control. From the figure noted that the higher the combustion temperature will reduce the flexural strength of the concrete, with the residual strength was around and above 50%.

Table 4. Flexural strength test result.

Waste banner length (mm)	Combustion Temperature (°C)	Flexural Strength (N/mm^2)	Retained ratio
0	Without Combustion	4.72	1.00
	200	3.48	0.74
	300	2.73	0.58
	400	2.65	0.56
30	Without Combustion	2.98	1.00
	200	2.03	0.68
	300	2.27	0.76
	400	2.32	0.78
60	Without Combustion	4.39	1.00
	200	3.58	0.82
	300	2.80	0.64
	400	2.71	0.62
90	Without Combustion	4.18	1.00
	200	3.32	0.79
	300	2.54	0.61
	400	2.32	0.56

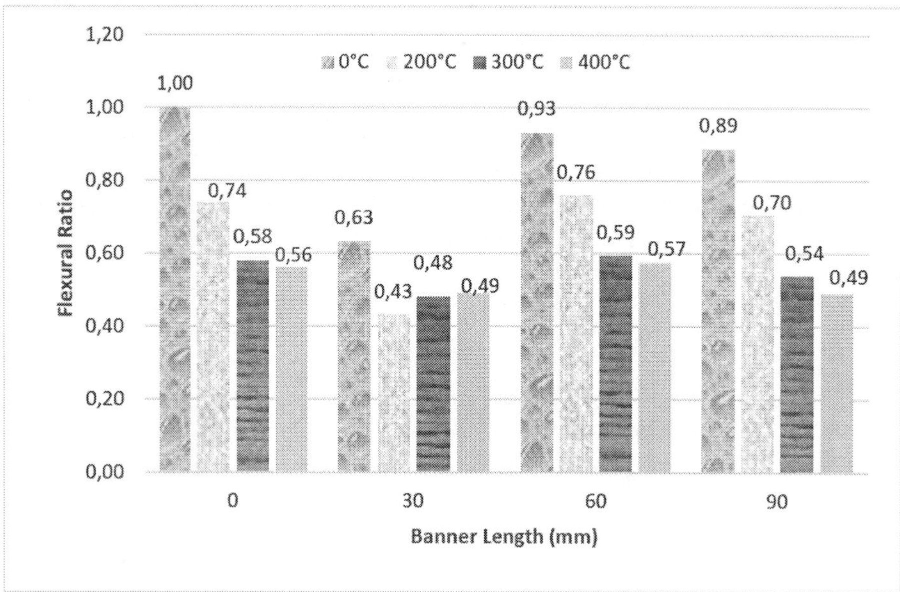

Figure 3. The ratio of flexural strength.

Fiber used as substitution material in the concrete mixture will shrink and lose ductility drastically if it is exposed to temperatures above 200 °C which causes the fiber material can no longer withstand the given flexural load. The resistance of FRC to high temperatures is lower than that of normal concrete. The decrease of the flexural strength of the beam in post-fire concrete is due to the water content that will evaporate when the concrete is heated. Tjokrodimulyo [23] said that the loose of strength occurs also due to gradual changes in chemical composition in the cement paste.

4. Conclusion

An experimental study has been carried out to investigate the flexural strength of post-fire FRC with the use of waste banner as fiber. 60 specimens prepared with three variations of waste banner length, that were 30 mm, 60 mm, and 90 mm, then fired in three variation temperatures, that were 200 °C, 300 °C and 400 °C. According to the experimental result, the variation in fiber length that produces the highest flexural strength of all variations used in this study was at 60 mm fiber length with a combustion temperature of 200 °C that was 3.58 Mpa.

References

[1] Wlodarczyk M and Jedrzejewski I 2016 *Procedia Eng* **153** 866–873
[2] Budi T 2015 *Procedia Eng.* **125** 1121–8
[3] Ali A, Iqbal S, Holschemacher K and Bier T A 2016 *Period. Polytech. Civ. Eng.* **60** 97–102
[4] Cai G and Degée H 2017 *Arch. Civ. Mech. Eng.* **17** 96–120
[5] Nemati K M 2015 *Fiber Reinforced Concrete* (Washington: University of washington)
[6] Meddah A, Belagraa L and Beddar M 2015 *Procedia Eng.* **108** 185–92
[7] Ali A, Iqbal S, Holschemacher K and Bier T A 2017 *Period. Polytech. Civ. Eng.* **61** 128–34
[8] Domski J, Katzer J, Zakrzewski M and Ponikiewski T 2017 *J. Clean. Prod.* **158** 18–28
[9] Mohammadhosseini H, Awal A S M A and Mohd J B 2017 *Constr. Build. Mater.* **143** 147–57
[10] Mastali M and Dalvand A 2016 *Compos. Part B Eng.* **92** 360–76
[11] Bhogayata A C and Arora N K 2017 *Constr. Build. Mater.* **146** 455–63
[12] Kumar S, B V T and Dutta R K 2017 *Period. Polytech-Civ.* **61**(3) 554–563,
[13] Hahladakis J N and Aljabri H M S J 2019 *Sci Total Environ.* **653** 294–299
[14] Shayene A, Bomfim C De, Milanez M, Dias Á, Jacobus H, Voorwald C, Cristina K, Carvalho C De, Magalhães D, Oliveira D, Odila M and Cioffi H 2019 *Waste Manag.* **83** 123–30
[15] Service J F and R 2017 *Fire Statistics Based on Causes* Media 113 (Jakarta)
[16] Hajiloo H and Green M F 2018 *Compos. Struct.* **201** 398–413
[17] Joyce M A and Joyce M D 2004 *Residential Construction Academy : Plumbing* (Clifton Park, NY: Thomson/Delmar Learning)
[18] Indonesian National Standardization Agency 2004 *SNI 7064:2014. Portland Cement Composite*
[19] Indonesian National Standardization Agency 2008 *SNI 1970:2008; Test methode for the specific gravity and water absorption of fine aggregate*
[20] Agency I N S 2014 *SNI 4154-2014. Test method for flexural strength of concrete using a simple beam with center-point loading.* (Jakarta: BSNI)
[21] Agency Ii N S 2012 *SNI 7656:2012 The procedure of selecting proportion for normal, heavyweight, and mass concrete* (Jakarta: BSNI)
[22] Hamdi F and Kuba M S S 2017 The Mechanical Degradation of Post-Fire Concrete (Degradasi Mekanik Beton Normal Pasca Bakar) *Simposium Nasional Teknologi Terapan (SNTT)* **5** 469–474
[23] Tjokrodimuljo 2004 *Teknologi Bahan Konstruksi* (Yogyakarta: Teknik Lingkungan Fakultas Teknik Universitas Gadjah Mada)

IOP Conf. Series: Materials Science and Engineering **578** (2019) 012077 doi:10.1088/1757-899X/578/1/012077

Effect of citric acid on setting-time and compressive strength of concrete

J Purnomo[1], S Sumarni[2], I N Saputro[2]
[1] Department of Civil Engineerng Education, Faculty of Teacher Training and Education, Sebelas Maret University
[2] Department of Civil Engineerng Education, Faculty of Teacher Training and Education, Sebelas Maret University
Sebelas Maret University, Jl. Ir. Sutami 36A Surakarta 57126 Indonesia

E-mail: srisumarni@staff.uns.ac.id

Abstract. Citric acid is an organic acid naturally present in fruits, one of the ingredients that can be used as a retarder in a concrete mixture. As retarder serves to slow the timing of concrete hardening. In the implication the use of this material affects the time of hardening and compressive strength, therefore need a better control to get the optimal value. The test sample used was 36 with variation of addition of citric acid 0%, 0.15%, 0.3%, 0.45% for the weight of cement. The results obtained the greater the addition of citric acid the longer the hardening of concrete and the greater workability of concrete. The maximum value of compressive strength of concrete is in the addition of 0.15% citric acid with an increase in compressive strength of 82.2% of the nominal concrete.

1. Introduction

Chemical Retarder Admixture brand is an additional chemical that serves to slow down the initial cement seting time so that the mixture will still be workable for a longer time. The way a retarder works is by wrapping the cement grains with OH^- thus slowing the initial reaction of the hydration. The formation of Ca salts in water reduces Ca concentration and slows crystallization during the hydration phase [1].

The additives that can be used as retarders include sugar, sucrose, sodium gluconate, glucose, citric acid and tartaric acid [1]. The addition of sugar to the concrete can extend the optimum initial seting time of the concrete. In the variation of 0% to 0.45%, the optimum length of the initial seting time continues to increase [1]. The addition of sugar levels can longer the time of hardening of the cement to the level of 0.15%, after which the hardening time again decreases when more sugar is given [2]. Portland / calcium aluminate blended cement (PC / CAC) was combined with citric acid or lactic acid as additives to the antioxidant effects of PC / CAC as a potential fast hardening and low cost repair material for concrete [3]. Mortar specimens with the carboxylic acid additives of either 0.5%, 1% or 3% by weight, prepared with a binder: sand: water ratio (by weight) of 1: 3: 0.5 experiments analyzed revealed Portlandite crystals were the major hydrate phase in PC / CAC with lactic and citric acids.

Lactic acid below 2% wt. improved both compressive and flexural strength at early ages due to improved crystallinity of the calcium hydroxide crystals. Combined with its inherent rapid setting

[1] Corresponding author: marnis_ri@yahoo.com

time, PC / CAC blended cements have a potential to be developed into a suitable repair material for concrete.

The Citric acid effectiveness depends strongly on the amount added to the interaction with other raw materials in particular, in particular calcium aluminate cement [4]. Citric acid partially inhibits the hydration binder from hemihydrate to dihydrate, as observed by X-ray diffraction analysis [5].

Sugar content can be used as a retarder has been researched by researchers. Citric acid is an organic acid that is naturally found in fruits such as oranges, pineapple, and pear[6], [7], [8], [9]. Citric acid is first extracted and crystallized from citrus fruits, so extracted citric acid from these fruits is known as natural citric acid. Citric Acid has the chemical formula $C_6H_8O_7$ or CH_2 (COOH) • COH (COOH) • CH_2 (COOH) [10].

Citric acid is an organic acid that is widely used in the food, beverage, pharmaceutical, cosmetic, agricultural and chemical industries. Citric acid can be produced through the fermentation process of citric acid-producing microorganisms. Aspergilus niger is a microorganism that can be used in the production of citric acid. The use of citric acid as a retarder is an environmentally friendly material.

Environmentally friendly alternative materials that could substitute cement both partially or totally have been developed using waste material such as fly ash [11]. According to the Building Constructive Materials and Civil Engineering Guidelines Number 22 / SE / M / 2015 concerning guidelines for the use of chemical admixture in concrete, the added ingredients are divided into several categories according to their chemical composition, including: 1) lighnosulphonic acids and other salts for example Na, Ca, or NH_4, 2) Hydro-carboxylic acids and their salts, 3) Carbohydrates - carbohydrates and sugars, 4) Inorganic salts based on flourates, phosphates, oxides, borax, and magnesium salts [12].

The purpose of this study was to determine the optimal value of adding citric acid to concrete with cement bonding time, sloughness and concrete compressive strength.

2. Theoretical Underpinnings

2.1. Cement Setting Time (Vicat Test)

The time of cement bond is the time it takes for cement to react with water. The time of cement bonding is divided into two, namely the initial seting time and the final seting time. The definition of initial seting time in accordance with Indonesian Standart is the time required by cement paste to change its properties from liquid to solid conditions. While the final seting time of cement is the time when the penetration of the vicat needle is not visually visible [13]. The initial seting time can be seen when the vicat needle gets 25 millimeters of penetration while the final seting time is when the vicat needle cannot penetrate [14].

2.2. Slump Test

Slump test is an empirical test used to determine consistency / stiffness from a mixture of fresh concrete to determine the level of workability. Stiffness in a concrete mixture shows how much water is used. For this reason the slump test shows whether the concrete mixture is less, more, or enough water. In a mixture of concrete, the water content is very considered because it determines the level of workability or not. Concrete mixtures that are too liquid will cause the concrete quality to be low and long dry time. While the concrete mixture that is too dry causes the mixture to be uneven and difficult to print [15].

2.3. Compressive Strength of Concrete

The compressive strength of concrete will increase with increasing age of concrete. The strength of the concrete will increase quickly up to 28 days, but after that the increase will be law. The compressive strength of concrete in certain cases will continue to increase for several years upfront. Usually the compressive strength of the concrete plan is calculated at 28 days.

Concrete compressive strength identifies the quality of a structure. The higher the level of the desired structural strength, the higher the quality of the concrete produced. Nawy [24] argues that the parameters that most influence the strength of concrete are: Quality of cement, proportion of cement to mixture, strength and cleanliness of aggregates, interaction or adhesion between cement paste and

IOP Conf. Series: Materials Science and Engineering **578** (2019) 012077 doi:10.1088/1757-899X/578/1/012077

aggregate, sufficient mixing of materials concrete forming, correct placement, solution, concrete compaction, and concrete treatment.

3. Experimental

3.1. Materials
The materials used in this research were cement, sand, gravel, and citric acid. The cement used had to meet prescribed requirements in the specification of group A construction material. The cement used was Gresik cement. The sand used was taken from River Kaliworo. It met the prescribed requirements in the specification of group A construction material. [16]. The coarse aggregate used was taken from PT Pancadarma Puspawira, it met the prescribed requirements in Indonesian Standart [17], [18]. Retarder or added material used in this research is Elephant brand citric acid produced by PT Budi Starch & Sweetener Tbk.

3.2. Experimental Procedure
3.2.1. Examination of Material
The sand examination was done through the following steps: the examination of water level, specific gravity, unit weight, organic content, mud content, and sand gradation which had to meet the prescribed requirements of Indonesia standard. The gravel examination was done through the following steps: the examination of specific gravity, abration, and gravel gradation which had to meet the prescribed requirements of Indonesia standard.

3.2.2. Mix design concrete
At the calculation stage of the concrete mix planning with or design of a mixture of concrete adding citric acid as a retarder with variations in addition of 0%, 0.15%, 0.30%, 0.45% of the weight of cement The concrete mix plan calculation is done to determine the need for gravel, sand, cement and water. The mixed method used in this study is to use the Indonesian standart mix design method [19].

3.2.3. Setting Time Test
Testing step of the initial cement setting time using vicat test apparature. This test is carried out using American standard. [20] The following is Vicat Apperature or Vicat Tool which is used to know the time of binding of cement:

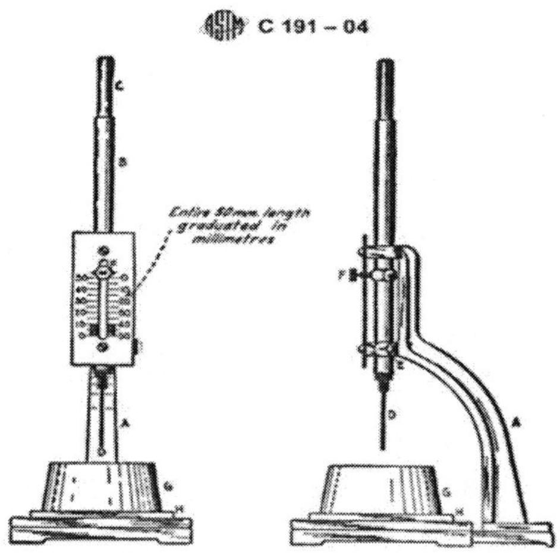

Figure 1. Vicat Tool (1 mm diameter needle)

3.2.4. Workability Test
Testing is done using Abrams cone and steel rod as crusher, testing is conducted to know the value of slump concrete (workability) based on Indonesian Standart [6].

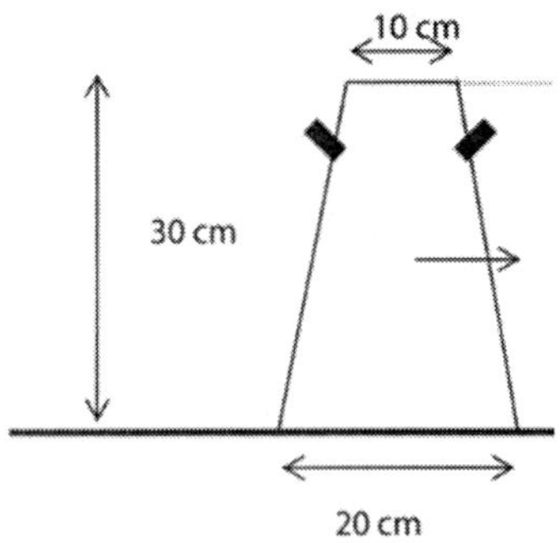

Figures 2. Abram Cone

3.2.5.The Compressive Strength Test of Concrete
Concerning the compressive strength test of concrete. Compressive strength of concrete bricks is the amount of load per area unit which cause the specimen to break if it is loaded with certain amount of compressive force produced by compressing machine. The compressive strength test of concrete was conducted 28 days after the production of the specimens, pursuant to the prescribed requirements of Standar Nasional Indonesia by using concrete compressive test apparatus.

The compressive strength was calculated by using equation as follows:

$$\sigma = P/A$$

where σ= compressive strength, P= load, A= area of specimen

4. Results and Discussion
The result of sand material test is shown in Table 1. Based on the examination result, the sand already met the requirements to be used as construction material. One of the requirements of sand, which is good to be used as construction material is having a specific gravity of 2.4-2.9 [6].

Table 1. Properties of Sand

Properties	Values
Water Content	6.67%
Specific Gravity	2.5
Fineness Modulus	3.69
Zone	2
Absorption	1.8 %

The result of gravel material test is shown in Table 2. Based on the examination result, the gravel already met the requirements to be used as construction material. One of the requirements of gravel, which is good to be used as construction material is having a specific gravity of 2.4-2.9 [10].

Table 2. Properties of Gravel

Properties	Values
Abration	5.206 %
Specific Gravity	2.53
Fineness Modulus	4.905
Zone	2
Absorption	1.493%

The cement testing carried out in the study included normal consistency testing and initial cement binding test. Normal consistency testing is done to determine the water content needed to react to each unit of cement needs. [21] The method of obtaining a normal consistency value is to do a minimum of 5 trials and with the same results, so that in this study a normal consistency value of 27.33% was obtained. The initial setting time test is conducted to determine the time needed for cement to react with water to be plastic. The data of the results of initial setting time test is shown in Figure 3.

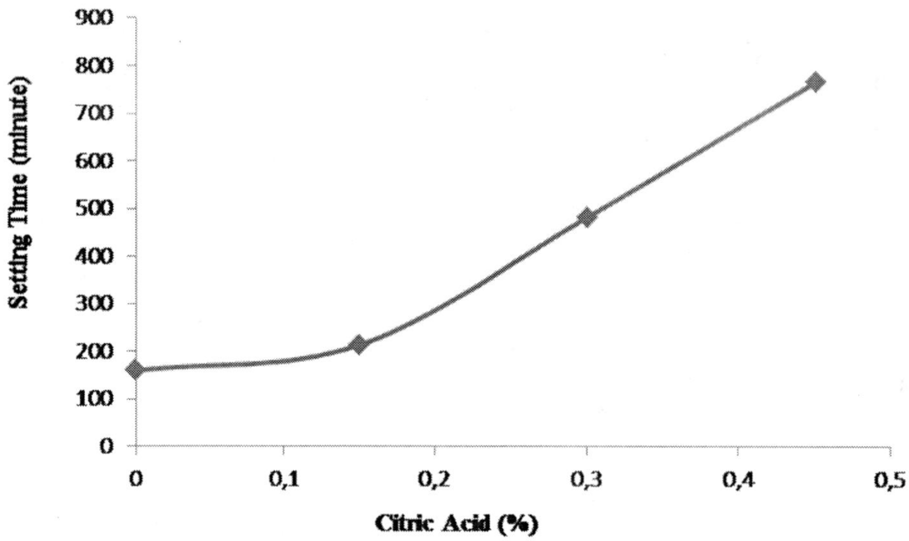

Figure 3. Result of Setting Time

Based on the results of data analysis in testing the first hypothesis it was concluded that there was an effect of adding citric acid as a retarder with a percentage of the weight of cement 0.00%; 0.15%; 0.30%; 0.45% of the initial setting time of cement. This conclusion is in accordance with the results of the research that has been conducted. It can be defined that with more and more levels of addition of citric acid as a retarder in cement, the length of time the initial setting of cement is produced. Slowing the setting time on cement occurs because the citric acid reacts with water and then the water reacts with cement so that a different hydration process occurs according to the variation of addition or level given. In simple terms the reaction that the retarder will wrap the cement grain with OH- so that it causes a slowdown in the initial reaction of the hydration of cement. Another factor that influences the time of cement bonding or setting time other than the ambient temperature is the water content or FAS (cement water factor) used to react a number of cement [22].

The hydration process of cement in fresh concrete requires approximately 25% of the weight of cement [23]. To determine the amount of water or water content needed in the setting time test (vicat test), normal consistency testing is carried out. In this study using ASTM C-187 standard so that the required water content is 82 milliliters or 27.33% of the weight of cement. Whereas for the need for citric acid as an added ingredient is to calculate the percentage percentage of cement content multiplied by the weight of the cement to be tested at time (according to ASTM).

Examination of fresh concrete slits is done by testing the slump using the Abrams cone. Data of the slump test is shown in Figure 4.

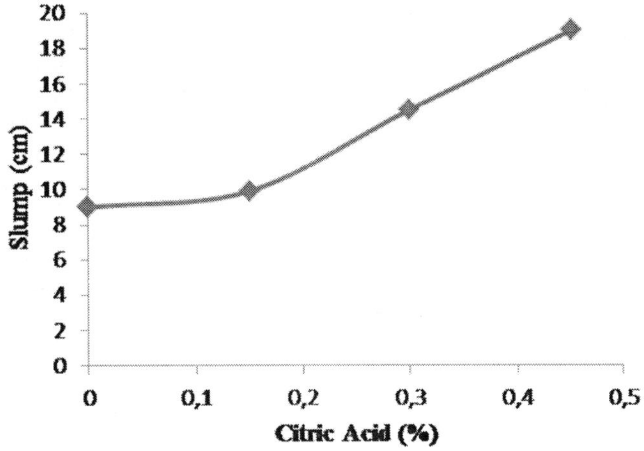

Figure 4. Slump Test

Based on the results showed that there was an increase in concrete slope which was directly proportional to the addition of citric acid. Workabbility can be interpreted as concrete thickness. The thicker the fresh concrete, the more difficult the workability level. [24] The elements that influence workability are the amount of mixing water, cement content, gradation of sand and gravel mixtures, coarse aggregate granular forms, and compaction methods or methods. The difference in slump values that occur in this study can be caused by the type of added material and the percentage of added material given. This is consistent with the results of research the addition of certain levels of sugarcane drops resulted in an increase in slump value. [23] On the addition of sugarcane drops with an addition of 0%, 0.2%, 0.8%, respectively, the slump values increased by 7.5 cm, 10.8 cm and 13.4 cm.

The result of compression strenght test of concrete is shown in Figure 5.

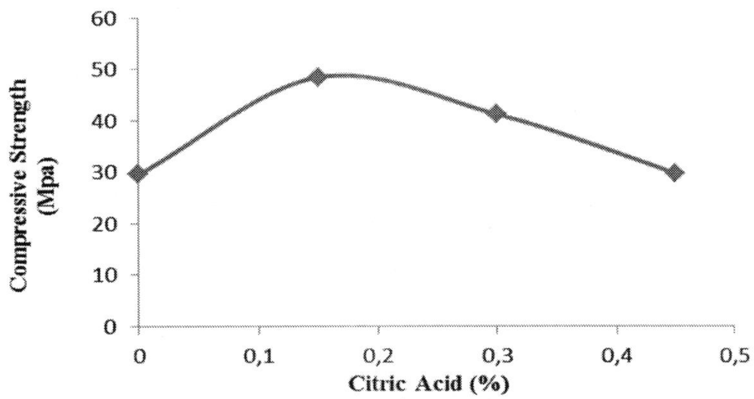

Figure 5. Variations of brick versus compressive strength

From the result of experiment it is apparent that it can be seen that the addition of certain levels of citric acid has an effect on the concrete compressive strength at 28 days. Concrete with variations in the addition of 0% citric acid produced an average compressive strength of 26.59 Mpa as well as variations in the addition of citric acid 0.15%, 0.30% and 0.45% respectively resulting in an average compressive strength of 48, 46 MPa, 41.20 MPa, and 29.70 MPa. Based on the graph, it can be seen that the compressive strength value after the variation of 0.15% has decreased, it happened because of some influencing factors. According to Nurlita Pertiwi the factors that influence the strength or quality of concrete are the cement water factor or the ratio of water and cement weight, aggregate type and gradation, cement quality and concrete curing. In addition, as one of the other factors that influence the compressive strength of concrete is added material (admixture) and the level or percentage of material use added. Improper addition of retardering admixture can result in the nature of the concrete to be ugly or not as expected. Based on the graph figure 3-5 test results the average compressive strength of concrete aged 28 days produced different values. This happens because the percentage of addition of citric acid as a different retarder. The results of the 28 days age concrete compressive strength test can be proven by looking at the results of the setting time testing and slump testing because they are interrelated. In principle, the longer the initial setting time of cement occurs, the higher the slump value. However, a high slump value is not necessarily followed by a high compressive strength value. This happened because the percentage of retarder admixture was added improperly so that in the reaction process cement was unable to hydrate and bind aggregate well because it was wrapped by too many retarders. This inability results in aggregates not being distributed well and also resulting in a squeeze or separation between aggregates of fresh concrete dough. The result of the failed reaction process is that after dry the concrete has a low compressive strength.

5. Conclusion

The results obtained the greater the addition of citrit acid the longer the hardening of concrete and the greater concrete (workability) of concrete. The maximum value of compressive strength of concrete is in the addition of 0.15% citrit acid with an increase in compressive strength of 82.2% of the nominal concrete.

References

[1] Desmi A 2017 *Journal Terrace* **4** (2)
[2] Puryanto, Absor M, and Subrianto A 2014 *Pilar Journal of Civil Engineering* **10** (2) 118
[3] Kastiukas G, Zhou X, Castro-Gomes J, Huang S and Saafi M 2015 *Constr. Build. Mater.* **101** 389-395
[4] Salomão R and Pandolfelli V C 2011*Ceram. Int.* **37** (6) 1839-1842
[5] Lanzón M and García-Ruiz P A 2012 *Constr. Build. Mater.* **28** (1) 506-511
[6] Abalaka E A 2011 *AU J. T.* **14** (3) 226
[7] Okafor F O 2008 *NIJOTECH* **27** (1) 1-8
[8] Eko R M and Riskowski G 2001 *Agricultural Engineering International: the CIGR Journal of Scientific Research and Development* Manuscript BC **99** 001 (3)
[9] Pertiwi H 2011 *The Effect of Sugar-Based Additional Materials to Pressure Strength and Concrete Elasticity Concrete (the Influence of Sugar-Based Admixture on Concrete's Compressive Strength and Elasticity Modulus)* Doctoral Dissertation (Surakarta: Sebelas Maret University)
[10] Khairul A et al 2010 *Isolation and Selection of Citric Acid Producing Fungi, Microbial Practicum and Potential Report* (Bogor: Bogor Agricultural University)
[11] Sasmitaloka K S 2017 *J. Proc. Integ.* **6** (3)
[12] Kristiawan S A 2011 *Toward Environmentally Friendly Construction Material With Class C Fly Ash-Based Geopolymer*
[13] Indonesian product standardization SNI 03-6827-2002
[14] American standart ASTM C191-92
[15] Indonesian product standardization SNI 03-1972-2008
[16] Indonesian product standardization SNI S-04-1989-F
[17] Indonesian product standardization SNI 1969-2008
[18] Indonesian product standardization SNI 2417-2008

IOP Conf. Series: Materials Science and Engineering **578** (2019) 012077 doi:10.1088/1757-899X/578/1/012077

[19] Indonesian product standardization SNI 03-3449-2002
[20] American standart ASTM C-191-92
[21] American standart ASTM C187
[22] Nugraha P and Antoni 2007 *Concrete technology: from material, manufacture, to high-performance concrete* (Yogyakarta: Andi Publisher)
[23] Santoso A 2012 Utilization of Sugarcane Drops as an Alternative to Substitute Retarders *INERSIA* **2** VIII, Desember 2012. Retrieved from; https://journal.uny.ac.id/index.php/inersia/article/viewFile/4002/3458
[24] Mulyono T 2004 *Concrete Technology* (Yogyakarta: Andi Publisher)

IOP Conf. Series: Materials Science and Engineering **578** (2019) 012078 doi:10.1088/1757-899X/578/1/012078

The Compressive Strength of Post-Fire Concrete Prepared with Bamboo Fiber

A Nurhidayati[1], A Rahmawati[2], F D Isnantyo[3], C F R Putri[4]
Building Engineering Education, Faculty of Teacher Training and Education Universitas Sebelas Maret.

E-mail: aryanti_n@staff.uns.ac.id

Abstract. The purpose of this research was to identify the compressive strength of the concrete post-fire with the addition of bamboo fiber length variation. This research used quantitative experimental method. The independent variables were bamboo fiber length variation (2 cm, 3 cm and 4 cm by 1.5% proportion of cement weight, and the combustion temperature variation (200 °C, 300 °C and 400 °C). The dependent variable was compressive strength of concrete. The result showed that the addition of bamboo fiber length variation and combustion temperature variation producing the maximum unit of concrete compressive strength was found in bamboo fiber with length of 2 cm and burned at 200 °C. The value decreased with the increase of combustion temperature. The length of 4 cm bamboo fiber produced the highest compressive strength with the maximum compressive strength 20.93 MPa.

1. Introduction

The need for high-quality materials for construction is increasing every day. This is due to the increasing demands for development growth. These have made the architectural engineers look for different material that can be used as an alternate material in construction [1]. Fiber Reinforced Concrete (FRC) is one of the technology developments of concrete by adding fiber in the concrete mix. This type of concrete is one of the solutions for improving concrete quality [2]. The interest of using fiber in terms of composite strengthening has raised over the last few years [3]. The structural characteristics of concrete can be improved by the addition of natural or artificial fibers [4]. One of the natural fibers [5] which can be used as a material in a concrete mix is bamboo fiber [6]. Bamboo is low-cost material, fast-growing [7], and widely distributed in many zones [8].

One of the failures on the structure of concrete construction is when there is a fire on the construction. In the burned concrete structure, concrete's strength is influenced by temperature change, heating level, and period [9]. The changes in the mechanical properties of concrete with high temperature are also related to the level of heating [10]. Concrete can also undergo significant physicochemical changes and loss of durability, through the appearing of cracks, spalling and a gradual deterioration of the hardened cement paste [11]. The decrease in concrete resistance and the expansion generated in reinforced

[1] Building Engineering Education Department of Universitas Sebelas Maret.
[2] Building Engineering Education Department of Universitas Sebelas Maret.
[3] Building Engineering Education Department of Universitas Sebelas Maret
[4] Building Engineering Education Department of Universitas Sebelas Maret.

concrete structures by direct exposure to fire at 400 °C maximum temperature serves as the basis for the present research [12].

The fire causes structures in a building to collapse. Even though the structure is intact, the strength remaining from this building is necessary to investigate so that it can be seen whether the building needs to be renovated or otherwise. When the fire is in structures made of fiber-reinforced concrete, uncertainty in the remaining strength would be higher due to lack of knowledge about the performance of FRC steel reinforcement at high temperatures [13]. In this study, bamboo fiber addition in a concrete mix was done by the existing optimum percentage and the concrete was combusted at the specific temperatures. It aimed to identify the strength produced by bamboo fibrous concrete after the combustion process.

2. Experimental

The research method used is quantitative experiments, by conducting laboratory investigations of specimens prepared with various fiber lengths for concrete compressive strength test.

2.1 Material

The main constituent of concrete is fine aggregate, coarse aggregate and cement. Natural fine aggregates were taken from the Mount Merapi mine, which had been tested by gradation tests and the results are shown in Figure 1. Meanwhile, the physical properties of fine aggregates are listed in Table 1. The test results are in accordance with *SNI* (Indonesian National Standards) S 04-1989-F criteria. Coarse aggregates were also taken from the destroyed Merapi volcano mine with a maximum size of 20 mm. The physical properties of coarse aggregates are presented in Table 2. Portland Cement Composites (PCC) used in this study is produced by Semen Gresik and in accordance with Indonesian specifications of SNI 7064- 2004 [14].

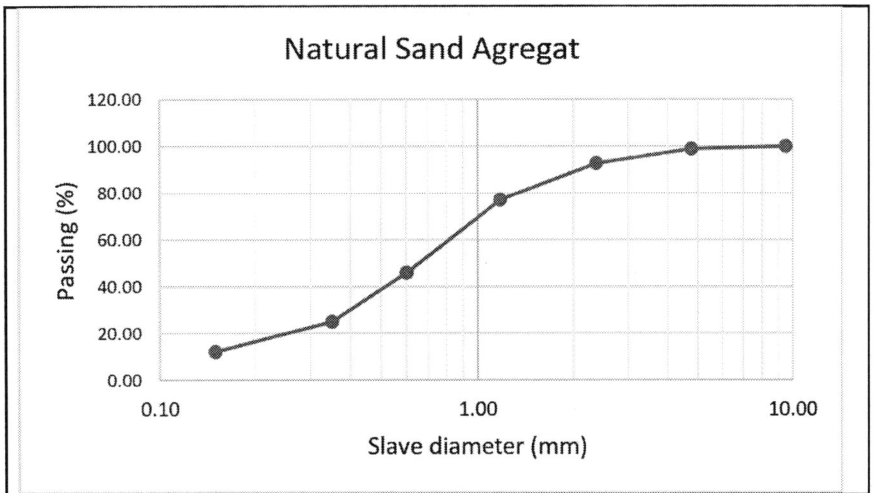

Figure 1. The gradation of fine aggregates.

Table 1. The Physical Properties of Fine Aggregates

Characteristic	Point	Standard from SNI	No. of SNI
Water content	0.2%	1-3%	SNI 03-1970-1990
Absorbtion	1.09%	<2%	SNI 1970:2008 [15]
Specific gravity	2.54	2.5-2.7	SNI 1970:2008[15]
Fineness modulus	2.48	2.5-3.8	SNI 03-1968-1990

Table 2. The Physical Properties of Coarse Aggregate

Property	Point	SNI Criteria	No. of SNI
Specific gravity	2.62	2.5 – 2.7	SNI 1969:2008
Absorption	0.94	>3%	SNI 1969:2008
Abrasion	46.47	<50%	SNI 2417:2008
Fineness modulus	3.58	2.5-3.8	SNI 03-1968-1990

The bamboo fiber used has a diameter of 0.2 mm with three variations in length of 20 mm, 30 mm and 40 mm as shown in Figure 2.

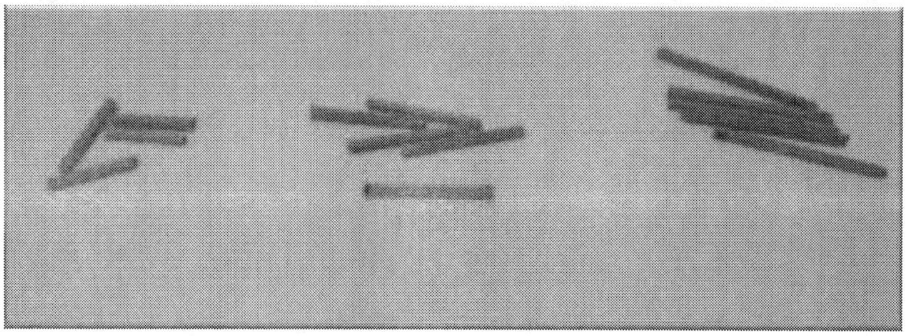

Figure 2. Bamboo Fiber

2.2 Test specimens preparation

Specimen dimensions for compressive strength test according to SNI 1974: 2011 [16] are cylindrical in shape with a diameter of 150 mm and a height of 300 mm. The design of the concrete mix was in accordance with SNI 7656: 2012 [17]. Table 3 summarizes the results of the concrete mix design.

Table 3. Mix design result

Material	**Amount**	**Unit**
Cement	297.101	Kg/ m^3
Fine Aggregate	868	Kg/ m^3
Coarse Aggregate	984.639	Kg/ m^3
Water	273.588	Kg/ m^3

Three length variations of bamboo fiber, 20 mm, 30 mm, and 40 mm, were added to the concrete mixture as much as 1.5% of the weight of cement. Specimens were prepared to be burned in three temperature variations namely 200 °C, 300 °C, and 400 °C. Normal concrete without additional fiber and combustion is also prepared as a control variable. Each variation consisted of four specimens.

2.3. Casting of test specimens

Concrete cylinder specimens were prepared to determine the compressive strength of FRC. The process of mixing the ingredients began by inserting half the sand into a cement mixer, inserting half the cement and waiting for them to mix. Then half of the bamboo fiber was put in consistently bit by bit and waited until evenly distributed. Half the cricket was put into the mixture. The next step was putting the remaining of sand, cement, bamboo fiber, and gravel in the same way. After everything was mixed well, then the water was poured bit by bit. The mixing process was performed so that bamboo fiber could be mixed consistently in the concrete mixture. Fresh concrete mixture was poured into the mold and compacted using a vibrator table. Based on SNI 2493-2011, maintenance of concrete fiber test objects was performed by opening the mold after 24 hours ± 8 hours after printing. Concrete would be tested after the concrete reached 28 days. Soaking in concrete was done as a concrete treatment. Specimens were soaked in water for 26 days then placed at room temperature for a day before the specimens were burned in 200 °C, 300 °C, and 400 °C.

2.4. Testing

Four normal concrete specimens and 60 FRC bamboo fiber specimens were used to determine compressive strength at high temperature exposure. All tests were executed at room temperature after the specimen was cooled from the temperature specified for each specimen. Testing of concrete compressive strength was carried out in accordance with SNI 1974: 2011 testing procedures [17]

3. Results and Discussion

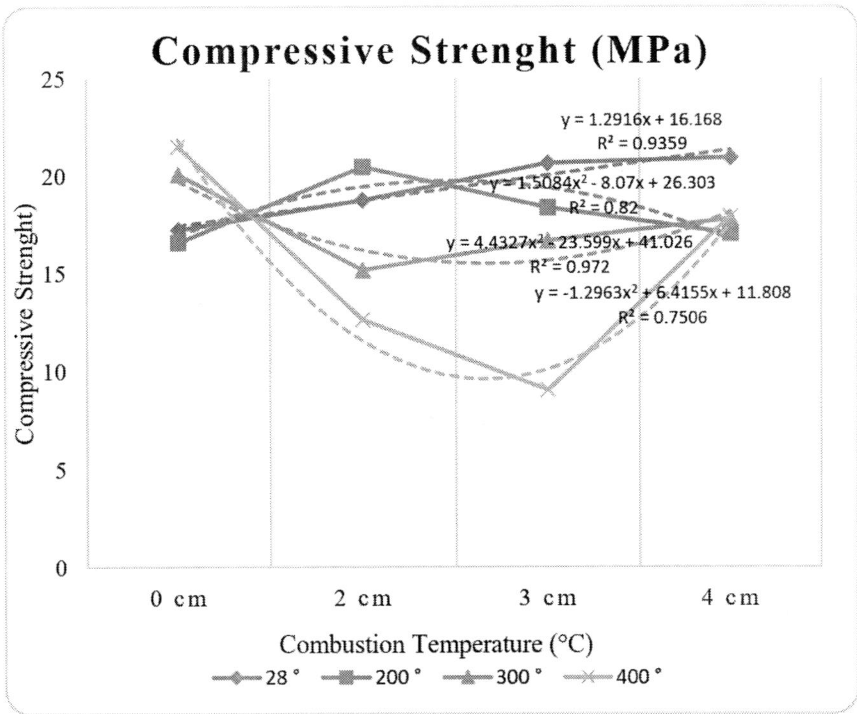

Figure 3. Compressive strength in length fiber variations and combustion temperature

Based on Figure 3, it could be seen that the addition of bamboo fiber increased the compressive strength of concrete. It was directly proportional to the increase in fiber length. The highest increase in compressive strength was 21.31% in the addition of 4 cm bamboo fiber length.

The compressive strength of concrete with increasing length of bamboo fiber had increased. Nevertheless, the combustion process at a temperature of 200 °C, 300 °C, and 400 °C made the strength of the concrete change. At combustion temperature of 200 °C, the value of concrete compressive strength could reach the maximum value with an increase of 23.30% in the addition of 2 cm bamboo fiber. The process was supported by an opinion which states that if the cement paste is heated in room temperature to about 200 °C, its strength appears to increase slightly.

Combustion of concrete at a temperature of 300 °C and 400 °C did not meet the maximum value on the compressive strength of concrete. The greatest decrease in compressive strength was 24.41% at the addition of bamboo fiber length of 2 cm and temperature of 300 °C. At the temperature of 400 °C, the greatest decrease in strength occurred at 57.89% at the addition of 3 cm fiber length.

In short, based on the research results and discussion concerning bamboo fiber length toward the value of post-combustion concrete compressive strength, it can be concluded that (1) the variation of bamboo fiber length by a diameter of 0.2 cm and a length of 2 cm, 3 cm, and 4 cm resulted a decrease of the compressive strength at 300 °C and 400 °C. At a temperature of 300 °C, the reduction percentage was 24.41%, 16.90%, and 11.27%. At a temperature of 400 ° C, the reduction percentage was 41.23%, 57.89%, and 16.67%. Besides, (2) the variation of the length of the bamboo fiber and the variation in combustion temperature producing maximum value of concrete compressive strength occurred on the bamboo fiber by a diameter of 0.2 cm and a length of 2 cm combusted at 200 °C. The compressive strength produced was 20.46 MPa.

4. Conclusion

Based on the research results and discussion concerning bamboo fiber length toward the value of post-combustion concrete compressive strength, it can be concluded that:

- The variation of bamboo fiber length by a diameter of 0.2 cm and a length of 2 cm, 3 cm, and 4 cm resulted in a decrease of the compressive strength at 300 °C and 400°C. At a temperature of 300 °C, the reduction percentage was 24.41%, 16.90%, and 11.27%. At a temperature of 400 °C, the reduction percentage was 41.23%, 57.89%, and 16.67%.
- The variation of the length of the bamboo fiber and the variation in combustion temperature producing maximum value of concrete compressive strength occurred on the bamboo fiber by a diameter of 0.2 cm and a length of 2 cm combusted at 200 °C. The compressive strength produced was 20.46 MPa.

References

[1] Krishna N K, Prasanth M, Gowtham R, Karthic S, and Mini K M 2018 *Mater Today Proc.* **5** 23816–23823

[2] ACI Committee 544 1982 *State of the Art Report on Fiber Reinforced Concrete- Report: ACI 544 1R-82* (Farmington Hills: American Concrete Institute)

[3] Tharunkumar K, Vignesh T P, Vignesh D, and Suriya kumar M 2017 *Int. J. Latest Eng. Res. Appl.* **2**(3) 98-102

[4] Khan M and Ali M 2018 *Constr Build Mater* **182** 703-715\

[5] Sen T, Rai R N, and Paul A 2016 *Damage and Degradability Study of Pretreated Natural Fiber-Reinforced Polymers Composites and Its Comparative Analysis with Artificial Fiber-Reinforced Polymers Composites* in book: Reference Module in Materials Science and Materials Engineering (Publisher: Elsevier)

[6] Buson R F, Melo L F L, Oliveira M N, Rangel G A V P, and Deus E P *2018 Science and Technology of Materials* **30**(2) 67-73

[7] Terai M, and Minami K 2011 *Procedia Eng.* **10** 2967-2972

[8] Ahmad S, Raza A, and Gupta H 2014 *Mechanical Properties of Bamboo Fibre Reinforced Concrete* (Dubai: ICRSET)

[9] Xiao J Z, Li J, and Huang Z F 2008 *Aci Struc J.* **105**(5) 531-540

[10] Rao C B K and Kumar R 2015 *World Acad Sci Eng Technol* **9**(3) 283-287

[11] Dias A R D O, Amancio F A, Rafael M F D C, and Cabral A E B 2018 *Procedia Structural Integrity* **11** 84-90
[12] Serrano R, Cobo A, Prieto M I, and González M N 2016 *Constr. Build. Mater.* **122** 302-309
[13] Hajiloo H and Green M F 2018 *Compos. Struct.* **201** 398-413
[14] Agency I N S 2004 SNI 7064 2014. *Portland Cement Composite* (Jakarta: BSNI)
[15] Agency I N S 2008 SNI 1970 2008 *Test method for the specific gravity and water absorption of fine aggregate* (Jakarta: BSNI)
[16] Agency I N S 2011 1974 2011. *Standard Test Method for Compressive Strength of Cylindrical Concrete Specimens* (Jakarta: BSNI)
[17] Agency Ii N S 2012 SNI 7656 2012 *The procedure of selecting proportion for normal, heavyweight, and mass concrete* (Jakarta: BSNI)

IOP Conf. Series: Materials Science and Engineering **578** (2019) 012079 doi:10.1088/1757-899X/578/1/012079

Design and properties of asphalt concrete with various plastic waste in dry processed

RA Pamungkas[1], **C Bagaskara**[1], **BA Prabanto**[1], **AN Rahmat**[1], **WB Pambayun**[1], **YS Anggoro**[1], **A Setyawan**[1,2]

[1] Civil Engineering Department Engineering Faculty
Sebelas Maret University, Jl. Ir. Sutami 36 A Solo 57126 Indonesia

[2] Roadmate Research Group, Civil Engineering Department
Sebelas Maret University, Jl. Ir. Sutami 36 A Solo 57126 Indonesia

E-mail: arysetyawan@staff.uns.ac.id

Abstract. Plastic is a waste that cannot be decomposed easily, plastic need about a ten to hundreds years to be decomposed perfectly depends on the plastic type. According to CNN, Indonesia is a second largest plastic waste contributor after China with more than 1 million in a minute, 50 percent are just single use then throw away and 5 percent are the truly recycle plastic waste. Along with the current development, there are a lot of innovations to the all things, one of them is substitute the aggregate with plastic, in this thing, plastic will substitute the aggregate pass the ¾ aggregate filters and strained the number 4 aggregate filter substitute with LDPE, PP and PET Plastic each. This research is aimed for evaluating and comparing the effect of adding or substitute the aggregate with LDPE, PP and PET plastics type for designing asphalt concrete. There are four test that related to strength and durability of Asphalt Cocrete: Unconfined Compressive Strength (UCS), Indirect Tensile Strength (ITS) test, Cantabro test, and Permeability test, each of them has a value and each of test will be comparing the specimen test whose aggregate is substitute on LDPE, PP and PET. The result of this research show that in the UCS test PP miture speciment test can hold the vertical pressure 253.41% stronger than PET and 399.12% stronger than LDPE. In ITS test PP mix object 56.29% stronger than PET and 396.90% stronger than LDPE. In the Cantabro test PP mix has durability 1.33% better than PET and 1.60% better than LDPE and in the Permeability test PP has a better classification to be the Asphalt Concrete Wearing Course with Practically Impervious classification.

1. Introduction

Plastic is a waste that is very difficult to decompose, plastic requires decades or even hundreds of years to decompose completely depending on the type of plastic. Indonesia is the second largest contributor to plastic waste in the world after China with a total of 187.20 million tons. More than one million plastic bags are used every minute, and 50 per cent of the plastic bags are used only once and immediately discarded, only five per cent are completely recycled. Asphalt and aggregates are the main material that must exist in a pavement, asphalt as an aggregate adhesion media that compiles into a flexible pavement that is comfortable to use. Quality improvement needs to be carried out in terms of resilience and stability, reducing asphalt material or adding additional materials or substitute materials on aggregates can also be an option to improve the quality of this pavement. Along with the development of time, innovation in all things also exist, one of them is innovation on highway hardness. This innovation can be a bridge for quality improvement in pavement. Innovation of substitute materials on pavement can be in the form of plastic, fuel oil or resin. In this study polymer or plastic materials were used. Plastics are one type of commercial product or waste that is widely available in Indonesia. There are also many types of plastic types, namely PETE or PET (Polyethylene terephthalate) [1,2,3], HDPE (High Density Polyethylene), PVC (Polyvinyl chloride), LDPE (Low Density Polyethylene) [4,5], PP (Polypropylene), PS (Polystyrene). In this study, PP seeds and PET plastic seeds will be used from the plastic processing plant CV. Flamboyant Plastic which is located in Mojosongo. Polyethylene is a soft, transparent and flexible film has good impact strength and tear strength. PET has a low absorption of water vapour, as well as absorption of water. PP contains versatile ingredients because it is hard and strong but still has flexibility. PP also provides strong chemical resistance[6]. This platinum PET and PP type is expected to be a substitute for 100% aggregate and make bitumen more resistant to water. This study aims to evaluate and compare the effect of addition or substitute of aggregate with PETE or PET type and PP type plastic on permeability test [7], Unconfined Compressive Strength test using UCS tool, tensile strength test using ITS (Indirect Tensile Strength) , and the Cantabro test to evaluate the resistance to disintegration on asphalt mixture[8].

2. Experimental

The research method used was an experimental method conducted at the Sebelas Maret University Civil Engineering Study Highway Laboratory. The type of data used are primary and secondary data. The tests were performed to the samples that have been produced at optimum bitumen content. Three types of asphalt concrete modified with plastic.

- PP Mixtures with 4.5% bitumen content and 100% replacement of fine aggregates.
- PET Mixture with 5% bitumen content and 100% replacement of fine aggregates
- On plastic waste LDPE plastic only as added by 6% of bitumen content on the pavement mixture was taken from the application of asphalt plastic on the UNS dormitory road carried out by the Center for Research and Development of Roads and Bridges (Pusjatan), Research and Development Agency (Balitbang) and Ministry of Public Works and Public Housing (PUPR) .

The samples required for this experiment are presented in Table 1.

Table 1. The number of speciments required for the experiments

No	Testing	AC with PP	AC with PET	AC with LDPE
1	Indirect Tensile Test	3	3	3
2	UCS	3	3	3
3	Cantabro	3	3	3
4	Permeability	3	3	3
	Total	12	12	12

2.1. Indirect Tensile Test

This test uses the Marshall loading principle with 12.5 mm wide concave loading strip. The loaded cylindrical specimen is then connected in parallel along the diameter plane vertically. This results in a tensile stress perpendicular to the direction of loading and along the vertical plane of diameter, which automatically causes the test object to fail / be damaged along the vertical diameter.

2.2. Unconfined Compressive Strength

The Compressive strength is the ability of the pavement layer to hold the load vertically, expressed in kg or lb. The amount of vehicle load that is channeled through the wheels of the vehicle is the compressive load received by the pavement, while the loading takes place in various temperature variations due to changes in weather and time. This temperature change will affect the viscosity of asphalt as a binder so that it also affects the compressive strength of the pavement.

2.3. Cantabrian Test
Test cantabro aims to determine the resistance of the pavement mixture to wear using a Los Angeles machine. The specimens which are allowed to stand for 48 hours at room temperature and at least 6 (six) hours before temperature testing must be kept at room temperature. Before the specimen was inserted into the Los Angeles machine drum, it was first weighed to get the weight before being abrasive. Then the specimen was inserted into the Los Angeles machine drum one by one without a steel ball. Los Angeles machines run with speeds between 30-33 rpm as many as 300 rounds. Then the test object is removed and weighed to get the weight after abrasion.

2.4. Permeability test
Permeability is the ability of porous media to drain liquid. Any material that has an empty space between them is called porous, and if the empty space is interconnected, the material will have permeability properties. The size of permeability is denoted as K (cm^2) and the permeability coefficient is denoted ask (cm/second).

3. Results and Discussion
This research has the purpose to find out and analyze the results of the UCS, ITS, Cantabro and Permeability tests on a mixture of Asphalt Concrete with the replacement of aggregate through the 1/2 filter and suspended filter no. 4 was replaced with LDPE, PP and PET plastic waste respectively. Of the 4 tests each Asphalt Concrete mixture with the replacement of aggregate with plastic waste compared for each test to find out which plastic waste is the best and can be applied

3.1. Indirect Tensile Test
This test aims to determine the indirect tensile strength of the results of the emphasis on the speciments with kg units using ITS test equipment, after that, it is calculated to find out the strength of ITS with the KPa unit. This test is carried out at room temperature of 25 ^{0}C with a speed of 51mm per minute. Table 2 presents the calculations of ITS on asphalt concrete modified plastic mixed specimens. The properties were then illustrated in Figure 1.

Table 2. The comparison of ITS properties on plastic modified asphalt concrete mixtures

Type of mixtures	ITS (KPa)	Strain (micro strain)	Static Modulus Elasticity (KPa)
LDPE	1,204.21	0.0023	525,409.297
PET	3,828.70	0.01073	468,543.70
PP	5,983.74	0.00315	2,428.40

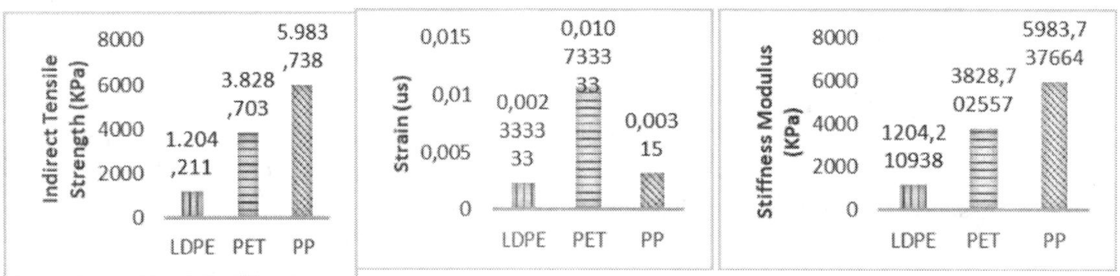

Figure1. The ITS Properties of plastic modified asphalt concrete

In the comparison of the tensile strength of the mixture type PP has the greatest tensile strength than the other types of plastic mixture with a large tensile strength of 5983,738 Kpa. In the Mixed Type PET has the highest strain with 0.0107 inversely proportional to the voltage which is relatively smaller than the other types of mixture. From the graph the comparison of PP modulus of elasticity has the greatest modulus of elasticity because it has the largest voltage and a smaller strain which is 5983.7376. On general, the type of PP mixture has the highest tensile strength due to the nature of PP plastic which changes shape to become soft when the heat conditions make the aggregate attachment become more and make the test object denser when after compacting and causes tensile strength to be greater.

3.2. Unconfined Compressive Strength

The UCS test was carried out to determine the vertical compressive strength of a deep pavement supporting the weight. For comparison of the UCS values for each mixture will be presented in Table 3 and illustrated in Figure 2.

Table 3. The UCS comparison for three types of Asphalt concrete

	LDPE	PET	PP
UCS (KPa)	2372.24	3350.26	11840.37

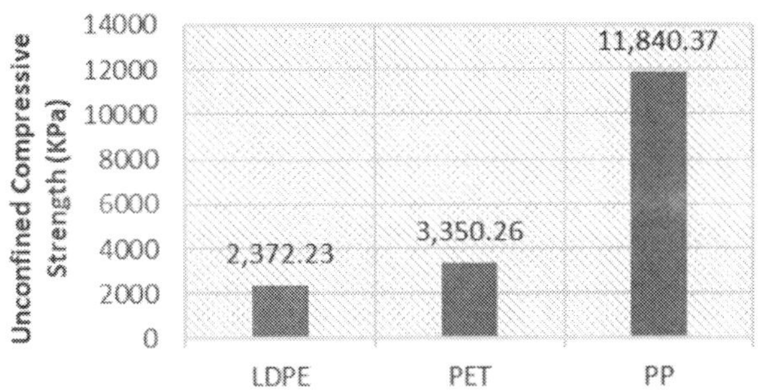

Figure 2. The UCS comparison for three types of asphalt concrete

The results in the graph show that the type of mixture using PP has the highest vertical compressive strength compared to other types of mixtures with a result of 11840.37. With these results PP is superior in vertical compressive strength rather than LDPE and PET mixture types. The type of PP mixture has the highest compressive strength due to the nature of PP plastic that changes shape to become soft during hot conditions and in conditions it will be composted to make the PP type mixture becomes more dense with less air space, after compacting.

3.3. Cantabrian Test

Cantabile testing is carried out to determine wear or weight loss on specimens expressed in grains. Comparisons of 4 types of mixtures will be presented in Table 4 and Figure 3.

Table 4. The Cantabro Loss comparison for three types of asphalt concrete

Type of mixture	LDPE	PET	PP
Cantabro Loss (%)	4.89	1.60	1.33

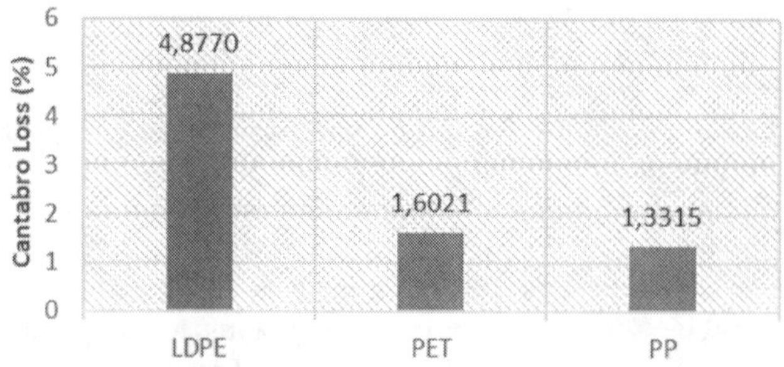

Figure 3. The Cantabro Loss comparison for three types of asphalt concrete

It can be seen in the graph that LDPE has a high level of weight loss and vice versa the loss of assets of mixed PET and PP specimens has a loss of 1.6% and 1.3%. From the graph PP has a better weight loss rate than other types of mixtures. With more plastic content in PET and PP, the aggregate attachment has more value and causes the cantabro loss value to be smaller, and the nature of PP itself which is easier to soft can better cover and embed the aggregates.

3.4. Permeability

Permeability testing is carried out to determine the ability of water to pass through the pavement in this case in the form of a test object, a comparison of the results of 4 types of mixtures will be presented in Table 5 and Figure 4.

Table 5. The coefficient permeability for three types of asphalt concrete

Type of mixture	LDPE	PET	PP
Coefficient Permeability	1.7E-04	1.4E-04	2.9E-05

Seen on the PET graph has higher permeability results and is easier to pass through with the results of the permeability coefficient of 0.000141 with the classification of poor drainage. PET has high permeability results because heated PET plastic will tend to be like fibers that make pavement or in this case the test specimen has more cracks to pass through and vice versa PP has properties when heated it will easily soften making when compacting aggregates mixed and attached and make the test object have fewer slit paths.

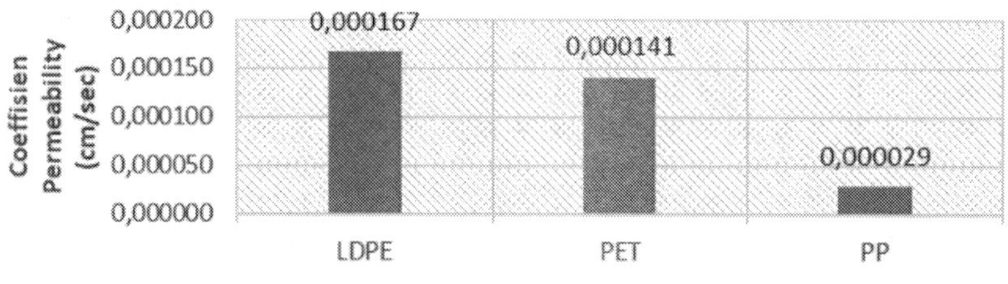

Figure 4. The coefficient permeability for three types of asphalt concretes

4. Conclusion

In the UCS test the PP Mixed Type has the power to withstand a vertical compressive load of 11,840.37 KPa, PET 3,350.26 KPa and LDPE 2,372.24. In the ITS test, the PP Mixed Type has a tensile strength of 5,983.74 KPa, with PET 3,828.70 KPa, with LDPE 1,204.21 KPa. In the Cantabro test the mixture type PP has lost weight 1.33%, PET 1.60% and for LDPE 4.88%. In the Permeability test PP has 0.000029 for PET 0.000141 and for LDPE 0.000167.

In the UCS Type PP Mixed test has the highest compressive strength with a strength ratio of 253.42% stronger than PET and 399.12% stronger than the LDPE Mixed Type. In the ITS test the Mixed Type has the strongest tensile strength with a strength ratio of 56.27% stronger than the PET Mixed Type and with a strength ratio of 396.90% stronger than the LDPE Mixed Type. In the Cantabro Loss Test Type of PP Mix have the least weight loss with a PET ratio which has a Cantabro Loss value of 1.60% and LDPE with a Cantabro Loss value of 4.88%. And in Permeability Test PP has a smaller Permeability result than the mixture type LDPE and PET with the classification Practically Impervious. 386.72% smaller than PET and 475.63% smaller from LDPE.

5. References

[1] Rahman W M N W A and Wahab A F A 2013 *Procedia Eng.* **53** 124-128
[2] Hassani A, Ganjidoust H, and Maghanaki A A 2005 *Waste Manag. Res.* **23** (4) 322-327
[3] Gürü M, Çubuk M K, Arslan D, Farzanian S A, and Bilici I 2014 *J. Hazard. Mater.* **279** 302-310; ASTM 2012 *Standard Test Method for Indirect Tensile (IDT) Strength of Asphalt Mixture D 6931*
[4] Gawande A, Zamare G, Renge V C, Tayde S, and Bharsakale G 2012 *JERS* **3** (2) 01-05
[5] Tapkın S 2008 *Build. Environ.* **43** (6) 1065-1071
[6] Zoorob S E and Suparma L B 2000 *Cement Concrete Comp.* **22** (4) 233-242
[7] Tarefder R A, & Ahmad M 2014 *J. Mater. Civil Eng.* **27** (5) 04014172; Doyle J D, and Howard I L 2011 Evaluation of the cantabro durability test for dense graded asphalt. In *Geo-Frontiers 2011: Advances in Geotechnical Engineering* 4563-4572
[8] Cox B C, Smith B T, Howard I L, and James R S 2017 *J. Mater. Civil Eng.* **29** (10) 04017174

The Physical and Mechanical Properties of Ethylene Vinyl Acetate Modified Binder

A Setyawan[1], M D Sistra[1], D Sarwono[1], Djumari[1], Zulfadly[2]
[1]Roadmate Research Group, Civil Engineering Department
Sebelas Maret University, Jl. Ir. Sutami 36 A Solo 57126 Indonesia
[2]The Ministry of Public Work Republic of Indonesia

E-mail: arysetyawan@staff.uns.ac.id

Abstract. During its service life, road pavement can easily crack, perforated or deformed. These damages happen as a result of low quality asphalt which indicated by its penetration index and stiffness value. 60/70 penetration asphalt modified by EVA polymer is expected to increase penetration index and stiffness value. Some experiments in the laboratory were carried out based on specified requirements and SNI-2011 asphalt testing procedures. The solid *EVA* was mixed with the liquid asphalt pen 60/70 at a medium to high temperature until it reaches 250 °C. During the mixing, this mixture was stirred for approximately ±15 minutes until homogeneous. These experiments consist of ductility test, penetration test, softening point test, flash point test, fire point test, specific gravity test, lose on heating test and asphalt stickiness to aggregate test. From the experiment, it can be concluded that the addition of 3.5% EVA in the asphalt pen 60/70 is generally acceptable based on the ductility value of SNI-2011 qualification. The addition of 3.5% give the ductility value of 130 cm, flash point of 300 °C, fire point of 310 °C, softening point of 54 °C, specific gravity of 1.020 gr/cc, asphalt stickiness to aggregate of 100% covered, and lose on heating of 0.037%. Nevertheless the experiment results in penetration value of 37.27 which doesn't conform the qualification of asphalt pen 60/70. This penetration value resulting in penetration index value of -0.9. This shows that the asphalt resist to the temperature changing. Theoretical predictions of asphalt characteristics using BANDS 2.0 showed that the addition of EVA can increase the penetration index and the stiffness of the bitumen.

1.Introduction

The highway is a land transportation infrastructure that is very important to facilitate community activities in economic movements and other social activities. In general, one of the highway pavement binders that are often used is an asphalt mixture because asphalt is a viscoelastic material. However, during its service life the asphalt in the mixture was easily damaged due to weather changes, overloading and heavy traffic volume. With the consideration of the resistance to load and weather it can determine the pavement design. If the road design uses flexible pavement design, the planned quality and thickness must be good so that the pavement is not easily damaged due to the load of vehicles passing through it. A method is needed to improve the ability of asphalt in the mixture to maintain or improve the properties of asphalt by modifying asphalt with added ingredients of elastomeric polymer. The type of elastomeric polymer used in this study is Ethylene Vinyl Acetate (EVA) [1]. Asphalt modification with EVA is expected to increase the stiffness of the asphalt, because of its good stiffness in holding the load and can reduce the penetration value so that the asphalt becomes harder and can increase the softening point value so as to produce a good Penetration Index value so that the asphalt is more resistant to change weather [2,3].

2. Experimental

The method used in this study is the experimental method, namely research in the Highway Laboratory of the Civil Engineering Study Program of the Faculty of Engineering, University of Sebelas Maret, Surakarta based on SNI-2011 terms and conditions. The results of this study will illustrate the effect of adding EVA polymer on 60/70 penetration bitumen. The resulting value can be used as a basis for the analysis of EVA modified asphalt characteristics so that a conclusion can be drawn.

2.1 Sample Production

Before the manufacture of specimens, asphalt mixing was carried out this was added with EVA with variations in the addition of 0% to 4% at 0.5% intervals of asphalt weight. Making test specimens is done after obtaining one of the best mixture methods or mixtures which do not clump from the three methods carried out, and then the modified EVA asphalt is poured in the medium of the test object for each test. The specimens to be made in this study can be clarified using Table 1.

Table 1. The Number of Sample Unit

Percentage of EVA (%)	Number of Sample (Unit)							The number of sample (Unit)
	Penetration	Softening Point	Flash Point/ Burning Point	Ductility	Specific Gravity	Lost Weight	Afinity	
0	3	4	2	3	3	3	2	20
0.5	3	4	2	3	3	3	2	20
1	3	4	2	3	3	3	2	20
1.5	3	4	2	3	3	3	2	20
2	3	4	2	3	3	3	2	20
2.5	3	4	2	3	3	3	2	20
3	3	4	2	3	3	3	2	20
3.5	3	4	2	3	3	3	2	20
4	3	4	2	3	3	3	2	20
The total number of sample								180

2.2 Sample Testing

The Testing of the specimens is to examine the properties of 60/70 penetration asphalt which has been modified with EVA. The test results used are the average value of the test results of the test object. If there is a deviation from the test results then it must be retested. The tests carried out are Ductility (SNI 2432: 2011), Penetration (SNI 2456: 2011), Softening Point (SNI 2434: 2011), Flash Point and Burn Point (SNI 2433: 2011), Specific Weight (SNI 2441: 2011), Asphalt attachment on aggregates (SNI 2439: 2011), and weight loss (SNI 2440: 2011). Based on the results of penetration testing and asphalt softening point, it can predict modified stiffness modulus and penetration index with BANDS 2.0 computer program.

3. Results and Discussion

3.1 The Physical properties of EVA modified bitumen

The mixing method used in this study is a hot-cold method. Mixing is done using medium heat until it reaches a temperature of 250° C and is stirred until the mixture does not clot. The way to find out the homogeneous mixture or not is done visually by looking at the mixture of asphalt and EVA mixed

IOP Conf. Series: Materials Science and Engineering **578** (2019) 012080 doi:10.1088/1757-899X/578/1/012080

perfectly and there is no clumping of EVA in the asphalt mixture. The results of testing the basic characteristics of EVA modified asphalt can be seen in Table 2.

Table 2. The Test Results of Asphalt 50 Pen Modified with EVA at 5% Interval

The Test	Unit	The Condition of 60/70 pen	Percentage of EVA (%)								
			0%	0.5%	1%	1.5%	2%	2.5%	3%	3.5%	4%
Penetration (25°C, 100gr, 5 sec)	0.01 mm	50 - 70	62.13	44.20	41.60	39.67	38.53	37.67	37.47	37.27	36.53
Softening Point (*ring and ball*)	°C	Min. 50	51.0	51.5	51.9	51.9	52.0	52.1	53.0	54.0	56.5
Flash Point(*Cleveland open cup*)	°C	Min. 200	295	315	325	307.5	305	320	315	290	290
Burning Point(*Cleveland open cup*)	°C	Min. 200	310	322.5	332.5	316	312.5	325	325	300	300
Ductility 25°C	Cm	Min. 100	150	150	150	150	150	150	150	130.0	68.0
Specific Gravity	1	Min. 1	1.049	1.046	1.029	1.028	1.024	1.022	1.021	1.020	1.011
Weight Loss	%	≤ 0,8	0.005	0.01	0.013	0.017	0.020	0.023	0.033	0.037	0.037
Afinity	%	≥ 97	100	100	100	100	100	100	100	100	100

3.2 The Prediction of the Mechanical Properties.

The results of the research on EVA modified asphalt in the laboratory were obtained Penetration values and Softening Points which were used as input data in the BANDS 2.0 Program. Then the assumption data are vehicle speed with a value of 0.02 (± 30 km / h), asphalt temperature of 20-60 °C (condition of road surface temperature in Indonesia) and the temperature at Penetration testing is 25 °C. The prediction results of penetration index and bitumen stiffness using the BANDS 2.0 program can be seen in Table 3.

Table 3. Predictions of the Penetration Index and Stiffness Bitumen with the BANDS 2.0 Program.

Prediction of Asphalt Modification EVA with BANDS 2.0 Program			*Penetration Index*	*Bitumen Stiffness (MPa)*				
Kadar *EVA*	Penetration	Softening Point		20°C	30°C	40°C	50°C	60°C
0%	62.13	51.0	-0.4	21.6	3.97	0.664	0.146	0.0344
0.5%	44.20	51.5	-1.1	45.6	7.44	0.974	0.173	0.038
1%	41.60	51.9	-1.1	50.9	8.34	1.07	0.185	0.0404
1.5%	39.67	51.9	-1.2	56.9	9.12	1.13	0.189	0.0404
2%	38.53	52.0	-1.3	60.4	9.63	1.17	0.193	0.0411
2.5%	37.67	52.1	-1.3	63.2	10	1.21	0.197	0.0417
3%	37.47	53.0	-1.1	59	10.1	1.32	0.221	0.048
3.5%	37.27	54.0	-0.9	53.9	10.2	1.44	0.252	0.0552
4%	36.53	56.5	-0.4	47.3	10.5	1.76	0.331	0.0774

Penetration index is obtained from the properties of asphalt, namely the softening point and penetration values that describe the sensitivity level of asphalt to changes in temperature [4]. Based on The Shell Bitumen Handbook that a good penetration index value is between -3 to +7. If the

IOP Conf. Series: Materials Science and Engineering **578** (2019) 012080 doi:10.1088/1757-899X/578/1/012080

penetration index value is smaller than -3 then it indicates that the material is easily affected by the surrounding temperature, if the penetration index value is greater than +7 then the material is not affected by the surrounding temperature [5]. The prediction results of the Penetration Index with the BANDS 2.0 program can be seen in Figure 1.

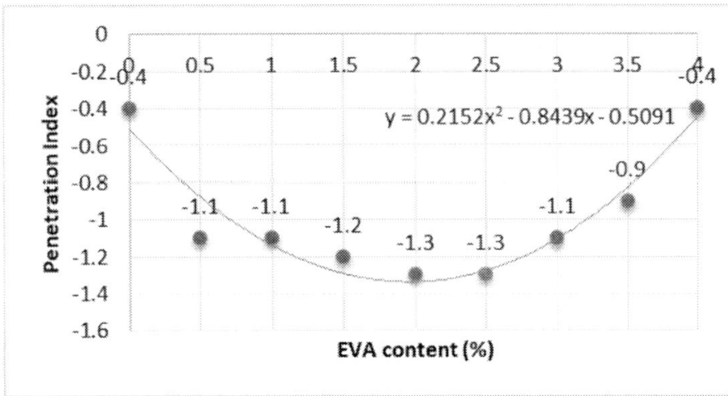

Figure 1. Relationship Graph between EVA Levels and Penetration Index

The prediction results of Stiffness Bitumen with the BANDS 2.0 program are strongly influenced by the asphalt surface temperature in the field. This can be seen in the results of the prediction of Bitumen stiffness in Table 7. The value of Stiffness Bitumen decreases with increasing temperature. This phenomenon is due to the nature of asphalt which is viscoelastic where asphalt will become liquid when heated and become hard at cold temperatures, so that at high temperatures the asphalt will have a small stiffness and at low temperatures the asphalt has a large stiffness [6].

The reading of the relationship between EVA levels and Stiffness Bitumen at each temperature has increased results in line with the value of EVA levels. The greater the EVA level, the bitumen stiffness increases. For more details the results of the prediction of Stiffness Bitumen can be seen in Figure 2.

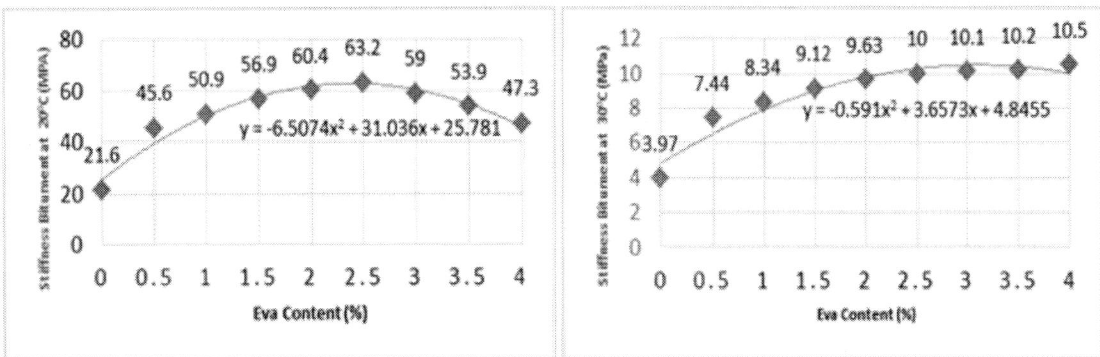

Figure 2. Relationship Graph between EVA Levels on Bitumen Stiffness at 20 °C and 30 °C

The reading of the graph and the quadratic function on the 20 °C Stiffness Bitumen shows the maximum value is when the addition of EVA is 2.5% with the Stiffness Bitumen value of 63.2 MPa. The reading of the graph and the 30 °C Stiffness Bitumen quadratic function shows the maximum value is when the addition of EVA is 3.094% with a value of Stiffness Bitumen of 10.5 MPa.

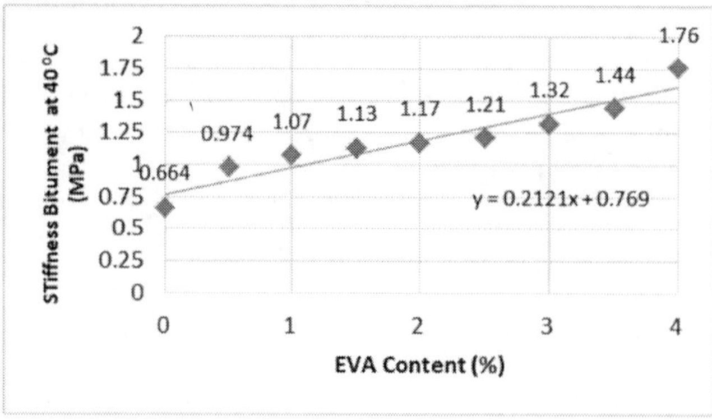

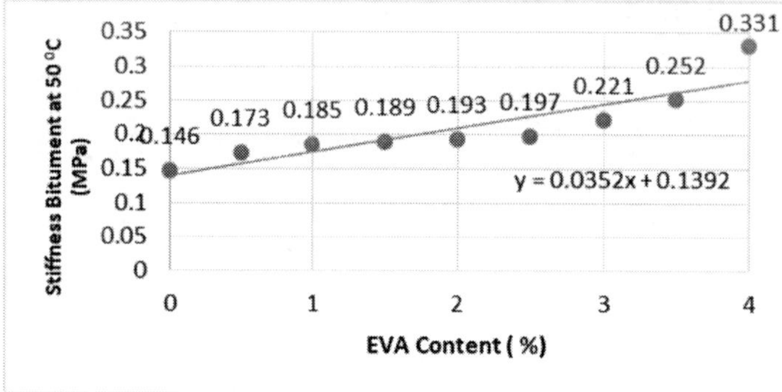

Figure 3. Relationship Graph between EVA Levels on Bitumen Stiffness At Temperatures of 40 °C and 50 °C

The results of predictions of bitumen stiffness 40 °C and 50 °C show that the value of bitumen stiffness decreases from the result of prediction of 30 °C stiffness bitumen or in other words there is a decrease in bitumen stiffness value with increasing asphalt surface temperature. The addition of EVA is positively correlated with bitumen stiffness value; the greater the percentage of EVA added to the asphalt, the greater the bitumen stiffness value. This shows that with the addition of EVA, it can make asphalt more resistant to resist loads at temperatures of 40 °C and 50 °C.

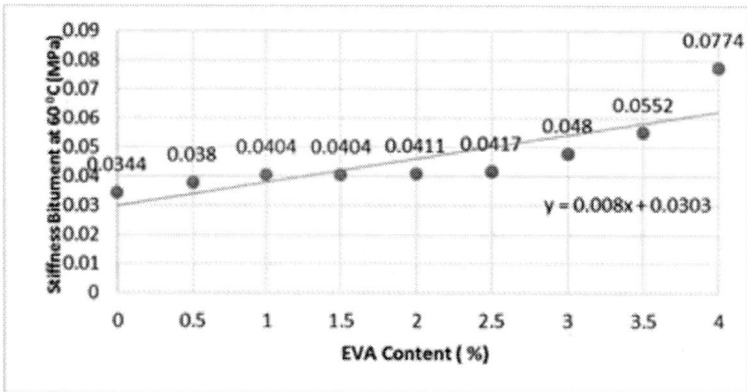

Figure 4. Graph of Relationship between EVA Levels and Bitumen Stiffness at 60 °C Temperature

Asphalt modification of EVA can increase asphalt stiffness at a temperature of 60 °C, although not significant can be seen from asphalt with a level of 0% EVA of 0.0344 MPa to 4% EVA of 0.0774 MPa experiencing an increasing trend of asphalt stiffness so that EVA modified bitumen is used for materials road pavement in Indonesia with an average road pavement temperature of around 60 °C.

IOP Conf. Series: Materials Science and Engineering **578** (2019) 012080 doi:10.1088/1757-899X/578/1/012080

4. Conclusion

The best mixing temperature is 250 °C with a hot-cold mixing method that can produce a homogeneous mixture or not clump between asphalt and EVA. Asphalt characteristics with the addition of EVA have better bitumen properties compared to bitumen without using EVA with ductility values, softening points, flash point, burn point, aggregate attachment on asphalt and weight loss testing all meet 60/bit penetration requirements 70.

Asphalt penetration values do not meet the requirements for 60/70 penetration asphalt. Percentage of optimum EVA levels that still meet the requirements of 60/70 asphalt penetration is at a percentage of 3.5% with ductility values reaching a length of 130.0 cm, flash point 300°C, burn point 310 °C, softening point at 54°C, density of material 1.020 gr/cc, asphalt attachment to aggregate is 100% covered, and loses weight 0.037%.

Penetration value of 3.5% does not meet the requirements of 60/70 penetration asphalt, which are 37.27 with a penetration index of -0.9 so that stronger asphalt against hot weather changes. The theoretical prediction of asphalt characteristics with the BANDS 2.0 program shows that the addition of EVA 3.5% on asphalt can increase the penetration index value and can increase the bitumen stiffness value.

References

[1] Garcıa-Morales M, Partal P, Navarro FJ, Martínez-Boza F, Gallegos C, González N and Muñoz ME 2004 *Fuel* **83(1)** 31.
[2] Kurniawan R, Setyawan A and Djumari D 2016 *Matriks Teknik Sipil*, **4(2)**.
[3] Panda M and Mazumdar M 1999 *J. mater. civil eng.* **11(2)** 131.
[4] Gonzalez O, Munoz ME, Santamarıa A, Garcıa-Morales M, Navarro FJ and Partal P 2004 *Eur. Polym. J.* **40(10)** 2365.
[5] Read J and Whiteoak D 2003 *The shell bitumen handbook*. Thomas Telford
[6] Sengoz B and Isikyakar G 2008 *Construction and Building Materials*, **22(9)** 1897.

The Attack of Alkali Silica Reaction on The Strength of Sea Water Resistance Concrete Pavement

A Sumarsono[1], NM Labib[2], Suryoto[1], A Setyawan[1], Zulfadli[3]

[1]Roadmate Research Group, Civil Engineering Department, Sebelas Maret University
[2]Civil Engineering Department, Sebelas Maret University Jalan Ir.Sutami No.36A Surakarta 57126.
[3]Staff at Indonesian Ministry of Public Work

E-mail: arysetyawan@staff.uns.ac.id

Abstract. Aggregate is the main material in concrete slab for rigid pavement. Quality of the aggregate should be eligible for the rigid pavement. Roads with rigid pavement structure can be damaged by non-technical factors including the effect of tidal flood. That damage due to the attack of alkali silica reaction in concrete. So, it required a research to select the type of aggregate that is more resistant to alkali silica reaction that will produce concrete more resistant to sea water. The method used is this research is experimental one, some sample in alkali silica reaction test using mortar bar of (25x25x285) mm dimensions. For the testing compressive strength and flexural strength of concrete using a cube-shaped test specimens of 150mm x 150 mm and a beam shape (150x160x600)mm. For testing performed after the age of concrete 28 days for normal conditions and after a concrete immersed fresh water 28 days, then soaked in seawater 13 days, and 26 days for testing the effect of sea water on the concrete strength. The test was performed using aggregates from PT. Armada Hada Graha, Magelang and PT. Panca Dharma Puspawira , Surakarta. The conclusion at this experiment is aggregate basalt (Surakarta) is more resistant to alkali silica reaction that caused by sea water than the aggregate andesite (Magelang).

1. Introduction

Today the use of concrete as one of the construction options in road pavement is commonly called rigid pavement. In concrete mixes, the aggregate proportion is 70% - 75% of the total volume of the concrete plate. Aggregate quality and quality must meet the requirements for road pavement. If the aggregate quality does not meet the specified requirements, then the level of durability and durability of construction becomes low. In addition, there are non-technical problems including the effects of rob floods. Rob floods often occur in the northern coastal areas, one of them is Semarang City or other areas located on the coast of the North Coast. One of the causes of floods is caused by a decrease in land surface so that the sea level becomes higher [1]. Rob floods caused some damage to existing infrastructure, especially on roads using rigid pavement. This damage can be caused by an alkali silica reaction.

Alkali silica reaction is a reaction between silica content (SiO_2) in aggregate and alkali in cement. Silica-containing aggregates can be reactive or non-reactive to alkaline elements in cement. This reaction causes expansion in concrete. This expansion can cause cracking, porous and spalling surfaces. Alkali silica reactions usually occur in near-shore construction that experience direct contact with sea water or those affected by rob flooding. Seawater has sodium content which is an alkaline element which can cause alkali silica reactions in concrete

Because the alkali silica reaction can occur at any time, the selection of the aggregate to be used needs to be selected first so that it will not cause a decrease in performance and damage to rigid pavement. Including aggregate use from the Asphalt Mixing Plant (AMP) in Magelang Regency and Surakarta City. The aggregates from AMP PT. Panca Dharma Puspawira in Surakarta City is basalt and aggregate rock originating from AMP PT. Hada Graha Fleet in Magelang Regency is andesite rock.

Alkali silica reaction is a chemical reaction in concrete and mortar between hydroxyl ions (OH-) from alkali (Na^+ and K^+) from cement, and certain silica-containing rocks and minerals found in some aggregates. This reaction and the gel resulting from the alkali silica reaction under certain conditions can cause abnormal expansion and cracks in the concrete [2]. Identifying the vulnerability of aggregates to alkali-silica reactions before using them in concrete is one of the most efficient attempts to prevent damage caused by alkali-silica reactions. [3]

The presence of sea water can accelerate the damage to concrete, because sea water contains a lot of salt solution, around 78% is sodium chloride (NaCl) and 15% is magnesium sulfate ($MgSO_4$). The presence of salts in seawater can also reduce the strength of concrete up to 20% and decrease the strength and durability of the construction that was built [4].

This study aims to determine the effect of aggregate types namely basalt and andesite rocks which are aggregates of the Asphalt Mixing Plant (AMP) at PT. Fleet of Hada Graha, Magelang and PT. Panca Dharma Puspawira, Surakarta for the alkali silica reaction that occurs and determine the effect of sea water immersion on alkali silica reaction in terms of compressive strength and concrete flexural strength. So that it will be known which mixtures use which aggregates are more reactive and which have higher durability to the alkali silica reaction.

2. Experimental

For testing the reaction of alkali silica aggregate to be carried out refers to ASTM C 1260: Standard Test Method for Potential Alkali Reactivity of Aggregates (Accelerated Mortar-Bar Method). The testing phase of the Aggregate Alkali Silica Reaction [5] is as follows:

- Printing mortar bars with sizes (25x25x285) mm with standard mix design ASTM C 1260
- After 24 hours the test object is removed from the mold
- then soaked in water with room temperature for 24 hours.
- The test object is then soaked in water with a temperature of 80 °C for 24 hours, after which it is recorded how much the size of the specimen.
- Transfer and soak the specimen into 1N NaOH solution at 80 °C
- Subsequent measurements were carried out on days 1, 4, 7, 11, and 14.
- After 14 days of immersion with a 1N solution of 80 °C NaOH, we can know the criteria of the aggregate clarification if the specimen increases in length:
  - <0.10% including non-reactive aggregates.
  - 0.10% to 0.20% including potentially reactive aggregates.
  - 0.20% including reactive aggregates.
- To determine the effect of seawater on the alkali silica reaction that occurs then the manufacture of test specimens and test steps as above is done, but in step 5 immersion is changed with the use of sea water at a temperature of 80 °C.

For compressive strength testing carried out refers to SNI 03-1974-1990 with a test object in the form of cubes measuring 15 cm x 15 cm x 15 cm and testing the concrete flexural strength carried out referring to SNI 03-4154-1996 with a test object shaped beam size 15 cm x 16 cm x 60cm with curing process using 3 types of immersion, including: a) 28 days fresh water bath, b) fresh water 28 days + 13 days soak using sea water, c) 28 days fresh water soak + 26 days soaking using sea water.

IOP Conf. Series: Materials Science and Engineering **578** (2019) 012081 doi:10.1088/1757-899X/578/1/012081

3. Results and Discussion

3.1. The Alkali Silica Reaction

Testing of alkali silica reactions to aggregates from PT. Fleet of Hada Graha, Magelang and PT. Panca Dharma Puspawira, Surakrata to find out the reactivity of both types of aggregates and compare which aggregates are more resistant to alkali silica reactions. Tests are carried out using the ASTM C 1260 (Accelerated Mortar Bar Method) method. The test results are presented at Table 1.

Table 1. Increase in length after immersion in NaOH 80° solution

Type of Rock	No	Length Expansion (mm)				
		D+1	D+4	D+7	D+11	D+14
Basalt (SKA)	1	0.02	0.05	0.11	0.15	0.18
	2	0.02	0.08	0.11	0.16	0.17
	3	0	0.03	0.08	0.15	0.17
Andesit (MGL)	4	0.24	0.28	0.34	0.36	0.41
	5	0.15	0.22	0.29	0.31	0.33
	6	0.02	0.08	0.19	0.21	0.23

From the results above it can be seen how the average length of increase of each specimen after the 14th day immersion with 80 °C sea water, and from the results of the study can be calculated how much the percentage of the increase in the second length of the aggregate. The results are presented at Table 2.

Table 2. Aggregate Reactivity Tests on Alkali Silica Reactions

Aggregate Type	Initial Length (mm)	Average Expansion D+14 (mm)	% Expansion	Reaction
Basalt (SKA)	286.50	0.173	0.060	Non Reactive
Andesit (MGL)	285.83	0.323	0.113	Potential Reactive

The alkali silica reaction is then tested with ASTM C 1260 method, but the solution used to soak the specimens using seawater, the purpose of replacing the solution is to find out how the impact of sea water on the alkali silica reaction on the test object. And the results of this test are presented in Table 3.

Table 3. Increase in length after soaking with 80° Seawater

Aggregate type	No	Length Expansion (mm)				
		D+1	D+4	D+7	D+11	D+14
Basalt (SKA)	1	0.04	0.09	0.12	0.15	0.15
	2	0.05	0.13	0.21	0.23	0.24
	3	0.04	0.06	0.07	0.08	0.09
Andesit (MGL)	4	0.01	0.05	0.11	0.12	0.15
	5	0.12	0.14	0.16	0.18	0.19
	6	0.07	0.12	0.18	0.2	0.22

From the results of the experiments of alkali silica reaction using sea water, the results are obtained as presented in Figure 1.

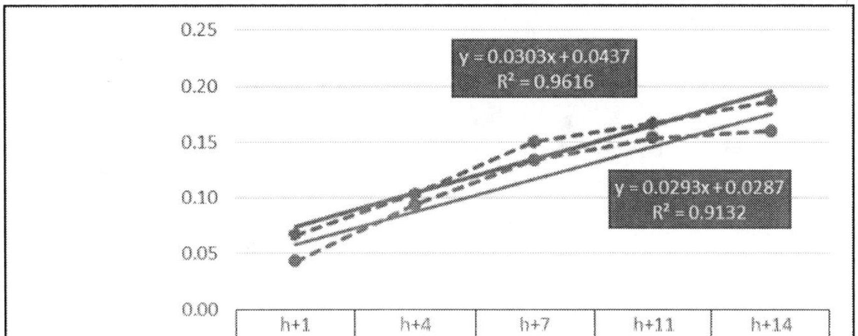

	h+1	h+4	h+7	h+11	h+14
Basalt (SKA)	0.04	0.09	0.13	0.15	0.16
andesit (MGL)	0.07	0.10	0.15	0.17	0.19

Figure 1. The Length Expansion Immersed in Sea Water

3.2. Concrete Compressive and Flexural Strength Test

Treatment with this condition is intended to find out how the characteristics of the test specimens in the seawater environment, and to determine the comparison of the durability of the specimens consisting of different aggregates.

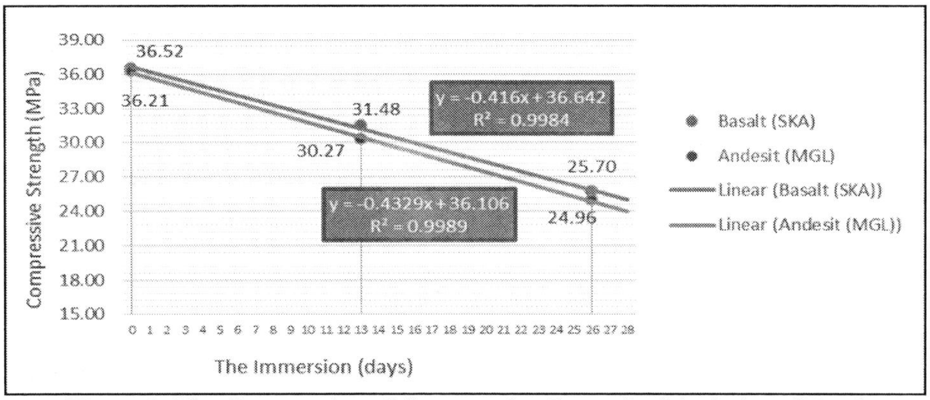

Figure 2. The Comparation of Compressive Strength of Andesit and Basal

Table 4. Concrete Strength Test Results

No	Immersion Type	Aggregate Type	Code	Strength at 28 days(MPa)	Average Flexural Strength(MPa)	Condition
1	Fresh Water 28 days	Basalt (SKA)	A	36.89	36.52	Fullfil
2			B	38.89		
3			C	33.78		
4		Andesit (MGL)	A	35.71	36.21	Fullfil
5			B	34.80		
6			C	38.13		
7	Fress Water 28 days + Sea Water 13 days	Basalt (SKA)	A	30.76	31.48	Fullfil
8			B	31.33		
9			C	32.36		
10		Andesit (MGL)	A	28.44	30.27	Fullfil
11			B	26.18		
12			C	36.18		
13	Fress Water 28 days + Sea Water 26 days	Basalt (SKA)	A	24.67	25.70	Fullfil
14			B	28.44		
15			C	24.00		
16		Andesit (MGL)	A	27.02	24.96	Fullfil
17			B	22.89		

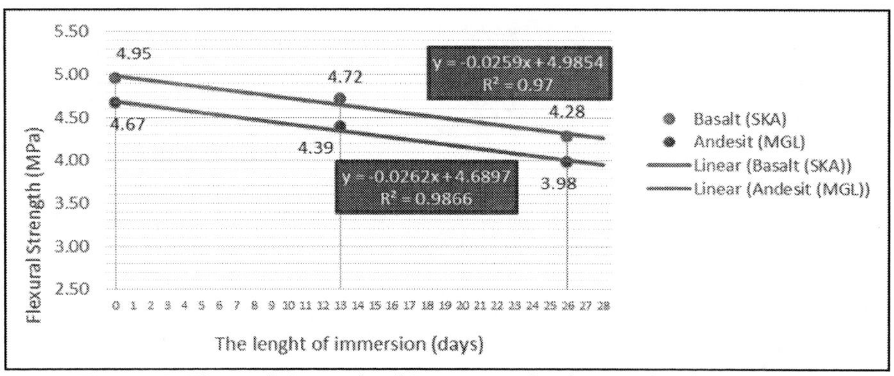

Figure 3. Flexural Strength of Concrete after immersion in sea water

Table 5. The Regression Lina and R^2

No	Test	Aggregate Type	Regression Line	R^2
1	Alkali Silica Reaction	Basalt (SKA)	y = 0.0293x + 0.0287	0.9132
		Andesit (MGL)	y = 0.0303x + 0.0437	0.9616
2	Compressive Strength	Basalt (SKA)	y = -0.416x + 36.642	0.9984
		Andesit (MGL)	y = -0.4329x + 36.106	0.9989
3	Flexural Strength	Basalt (SKA)	y = -0.0259x + 4.9854	0.9700
		Andesit (MGL)	y = -0.0262x + 4.6897	0.9866

It can be seen in Table 5, the relationship between the line equation gradient on the alkali silica reaction test and the compressive strength testing and concrete flexural strength. In testing the alkali silica reaction, Basalt aggregate (Surakarata) shows a gradient or a ratio of length increase which is smaller than the Andesite aggregate (Magelang) and in testing the strength of the aggregate concrete Basalt (Surakarta) decreases the strength of concrete which is smaller than the Andesite aggregate (Magelang). From this relationship it can be concluded that the greater the length increase ratio in testing the alkali silica reaction will result in a greater decrease in compressive strength and flexural strength. The similar results as the research conducted by Ahmet et al [6].

4. Conclusion

Based on ASTM C 1260 Standard Test Method for Potential Alkali Reactivity of Aggregates (Accelerated Mortar-Bar Method), aggregates originating from PT. Panca Dharma, Surakarta (basalt) is not a reactive aggregate against attacks of alkali silica reactions, while aggregates from PT. Armada, Magelang (andesite) is potentially reactive to attacks of alkali silica reactions. That is, the aggregate from PT. Panca Dharma Puspawira, Surakarta is more resistant to attacks by alkaline silica reactions than aggregates from PT. Armada Hada Graha, Magelang.

Results of testing the compressive strength and flexural strength of concrete with variations in sea water immersion time, shows that concrete made from Basalt (Surakarta) aggregates is superior when used as material for rigid pavement compared to Andesite aggregates (Magelang), including for rigid pavement Basalt (Surakarta) aggregate seawater is more resistant and better in performance than the Andesite aggregate (Magelang).

Concrete strength performance made of Basalt (Surakarta) aggregates and Andesite aggregates (Magelang) can meet the value of compressive strength and concrete flexural strength required by SNI Pd T-14-2003 as road repair materials that are resistant to seawater.

References

[1] Abdullahi M 2012 *Int. J. civil structural eng.***2**(3) 791.
[2] Thomas M, Fournier B, Folliard K, Ideker J and Shehata M 2006 *Cement Concrete Res.* **36**(10) 1842.
[3] Swamy RN and Al-Asali MM 1988 *Mater. J.* **85**(5) 367.
[4] Ramachandran VS 1996 *Concrete admixtures handbook: properties, science and technology.* (William Andrew).
[5] Astm C 2007 1260 *Section* **4** 676.
[6] Ahmed T, Burley E, Rigden S and Abu-Tair AI 2003 *Constr. Build. Mater.* **17**(2) 123.

IOP Conf. Series: Materials Science and Engineering **578** (2019) 012082 doi:10.1088/1757-899X/578/1/012082

Performance Comparison of Modified AODV-ETX with AODV and AODV-ETX Routing Protocol in an MANET

A Purnomo[1], Widyawan [2], W Najib[2], R Hartono[1] and Hartatik[1]

[1] Department of Informatics Engineering, Faculty of Mathematics and Natural Sciences, Sebelas Maret University, Surakarta, Indonesia

[2] Department of Electrical Engineering And Information Technology, Faculty of Engineering, Gadjah Mada University, Yogyakarta , Indonesia

E-mail: agus.purnomo@staff.uns.ac.id

Abstract. Expected Transmission Count (ETX) is one of the many metrics developed in the ad hoc routing protocol. Most research, studied on modified AODV-ETX by varying the value of MY ROUTE TIMEOUT (MRT) and ACTIVE ROUTE TIMEOUT (ART) values in the RREQ and RREP packet headers. Those studies resulted the optimum value of MRT and ART was 80 seconds. These results has not shown optimal performance of AODV and AODV-ETX. This paper discusses performance comparison of modified AODV-ETX with AODV and AODV-ETX on MANET. The result is that the throughput on the modified AODV-ETX has increases 5.72 %, PDR increases 5.80 %, 4.31 % overhead reduction, 10.19 % increase on end to end delay, and 3.09 % increase on route discovery process.

1.Introduction

Mobile ad hoc network (MANET) consists of independent nodes. A node can communicate with each other without the presence of a network infrastructure. A node can function as a transmitter and receiver as well as a router [1]. MANET was developed to reduce cable infrastructure and support for mobile communications. The problems is when one of node moves outside the transmission range of another nodes, the link can be broken so that it will affect the quality of video stream, video surveillance, and streaming of data. The Internet Engineering Task Force (IETF) has standardized two types of routing protocols on ad hoc networks that is reactive types, e.g. AODV, DSR, TORA, and proactive types, e.g. DSDV, WRP, GSR, CGSR [2]. Both types of protocols each have their strengths and weaknesses.

This paper focussed on ad hoc On Demand Distant Vector (AODV) protocol. AODV protocol is a routing protocol in ad hoc networks which uses hop count as routing metric. The AODV protocol has more minimal overhead compared to the type of proactive protocol [3] and has better energy efficiency compared to the type of proactive protocol and other reactive protocols [4]. The metric that is widely used in ad-hoc networks is the hop count, because it is the most convenient and simple.

The metric hop count selects the path that has the smallest number of hops even though other paths with a larger number of hops offer greater throughput [5]. One of the approaches to overcome this problem is to replace the hop count metric with Expected Transmission Count (ETX). ETX metric is a new route metric for finding high-throughput paths in multi-hop wireless networks. The ETX of a path is the expected total number of packet transmissions (including retransmission) required to successfully deliver a packet along that path [6]. The path with the minimum ETX value is the path with the largest throughput value. AODV-ETX is a modified AODV protocol with ETX metric [7].

The AODV protocol has MY ROUTE TIMEOUT (MRT) and ACTIVE ROUTE TIMEOUT (ART). MRT is the life time of the reverse lane and ART is the life time of the forward lane. If the reverse and forward paths stored in the routing table are not used during the MRT and ART times, the reverse and forward paths will be removed from the routing table and the route discovery process will be repeated if a route is needed again. If the reverse path and forward path are stored in the routing table, then MRT and ART will be updated. The time set on MRT and ART will affect the route discovery process. Another paper wrote the results of tunning the value of MRT and ART on the AODV-ETX protocol in MANET. The results obtained by the MRT and ART 80s results in a stable AODV-ETX Protocol performance [8]. We have not found another paper that provides information on improving the performance of the AODV protocol by modifying the MRT and ART 80s values. The solution proposed in this paper is to compare the performance of the AODV-ETX protocol that has been modified, called the AODV MRT ART 80s with the AODV and AODV-ETX protocols.

2. Experimental

Performance testing simulations are carried out using the Network simulator (NS2 version 2.35). Other supporting devices include: Gedit, setdest, NS2 Scenarios Generator 2 (NSG2), Cbrgen, NS2 Trace Analyzer and AWK script. MANET is simulated by making a topology consisting of 25 nodes that move randomly at a speed of 1.38 m/sec in an area of 1500 m x 300 m and a simulation time of 300 seconds shown in Table 1. To get a different pattern of movement, it is tested with a different pause time. Different pattern of movement obtained by testing with a different pause time.

Table 1. List of system parameters in simulation

Parameter	Value
Channel type	Wireless channel
TX Power	0.28 W
TX dan RX Gain	1 dB
Posisi antena dari tanah	1.5 m
Propagation model	shadowing
MAC layer	IEEE 802.11b
Routing protocol	AODV, AODV-ETX, AODV MRT ART 80s
Traffic type	UDP
Application type	CBR
Number of trensmiter and receiver node	1
Packet size	1000 byte
Packet interval	0.005 s
Simulation time	0-300 s dan 300 s
Number of nodes	25
Transmission range	250 m
Movement models	Random Way Point
Node Speed	1.38 m/s
Pausetime	0 s, 1 s, 5 s, 10 s, 20 s, 30 s
Probe interval	1 s
Probe window	10 s
Area	1500 m x 300 m
Location	Outdoor

A small pause time reflects a constant node movement. Contrarily, a large pause time reflects the movement of nodes that tend to be idle. By changing the pause time value, different node movement patterns could be obtained even though they use the same speed shown in Figure 1.

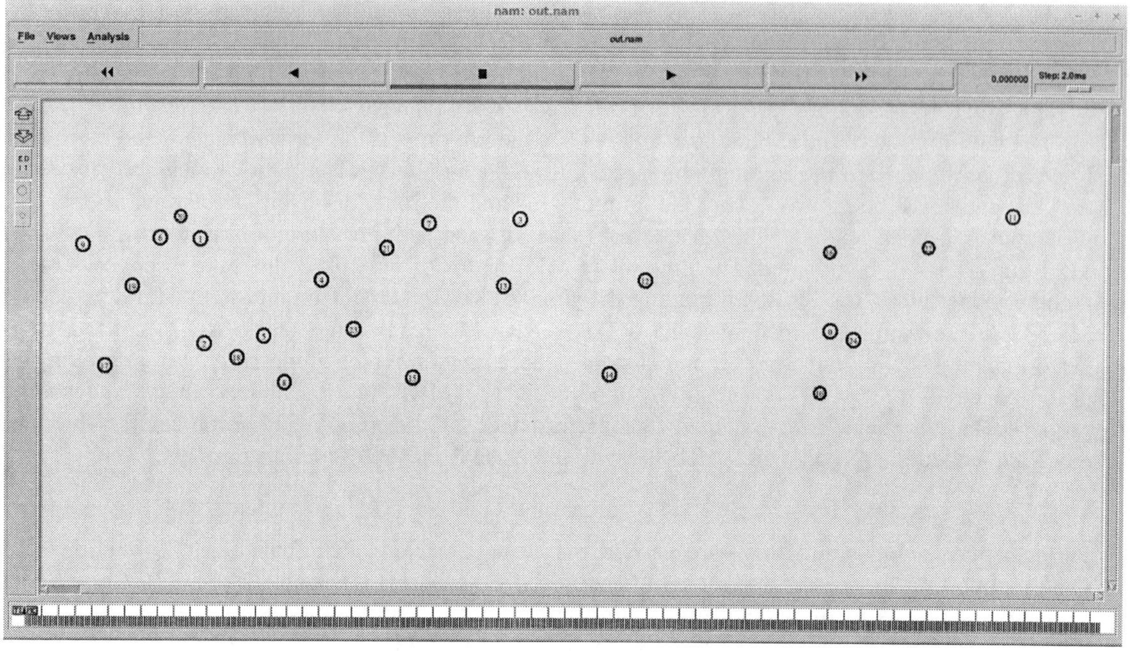

Figure 1. MANET in simulation

To determine the performance of all protocols, the throughput, end to end delay and routing overhead, packet delivery ratio (PDR) and total route discovery processes are measured and compared.

3. Results and Discussion

From the tests that have been done, a comparison of the performance of all protocols had been concluded. In this paper, the AODV modification protocol is named AODVETX MRT = 80s ART80s.

Different pause time values is not part of this research. its is just to make different simulation conditions.

3.1 Throughput

This is the total data packet received by the recipient from the sender, calculated in kbps [9]. In Figure 2, if the performance of AODV-ETX MRT = ART 80s = 80s compared to AODV-ETX performance, the average throughput in AODV-ETX MRT = 80s ART = 80s increases = 5.72 %. Compared with AODV, the average throughput increases = 15.84 %.

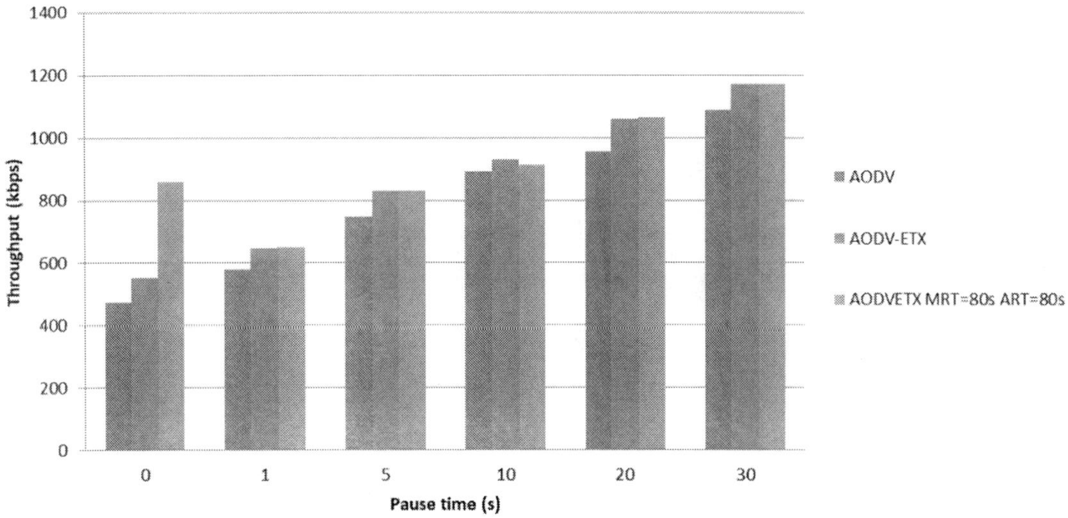

Figure 2. Comparison of throughput values

3.2 end to end delay

End to end delay is the time needed for the package to be sent until the package is received and recognize the packet sender [9]. In Figure 3, the average end to end delay performance of AODV-ETX MRT = ART 80s = 80s compared with the AODV-ETX has increased = 10.19 %. Compared with AODV, the average end to end delay is increased 30.65 %.

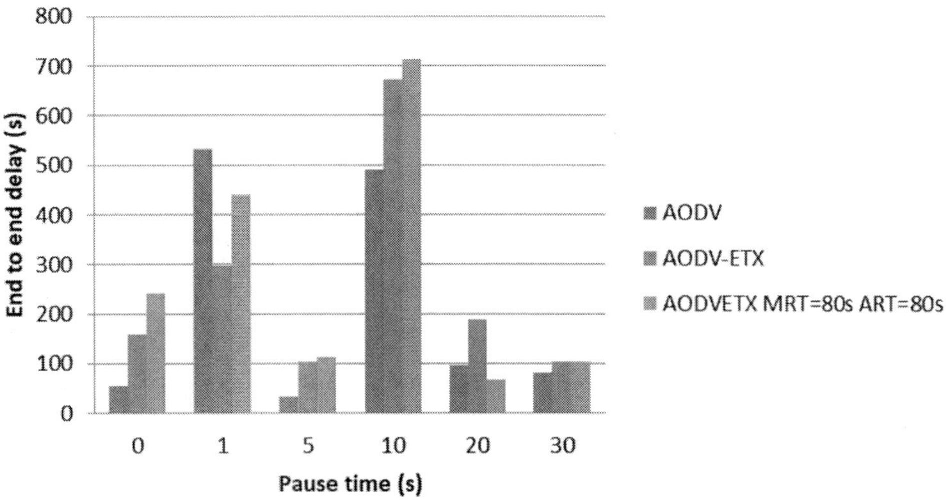

Figure 3. Comparison of end to end delay values

3.3 packet delivery ratio (PDR)

PDR is the ratio between the packet that was successfully received by the destination and the packet sent by the source [10]. In Figure 4, the average PDR performance of AODV-ETX MRT = ART 80s = 80s compared to the AODV-ETX has increased = 5.80 %. Compared with AODV, the average PDR has increased by 15.87 %.

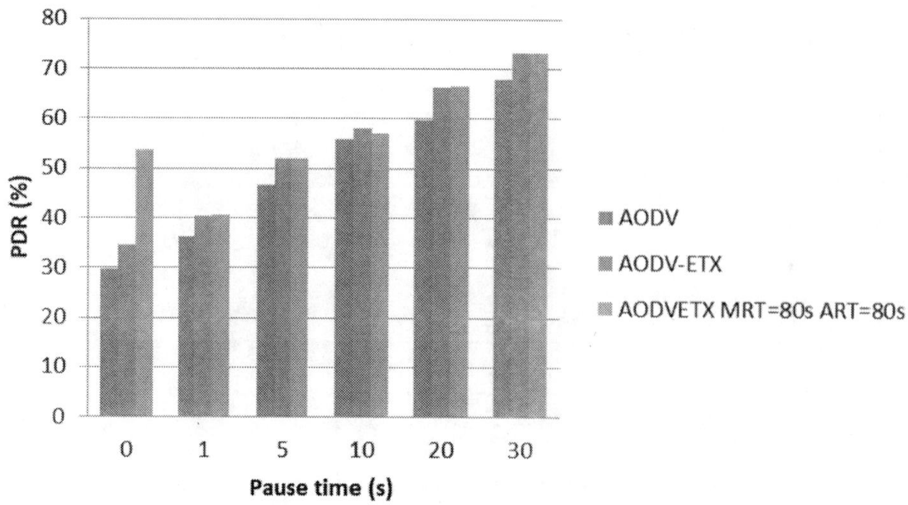

Figure 4. Comparison of PDR values

3.4 Routing overhead

Routing overhead is the total routing package proportional to the number of packets received by the recipient. This routing overhead can be obtained by calculating the number of routing packets divided by the number of packets received by the receiver [9]. In Figure 5, the average routing overhead performance of AODV-ETX MRT = 80s ART = 80s compared AODV-ETX has decreased = 4.31 %. Compared with AODV, the average routing overhead is increased by 28.34 %.

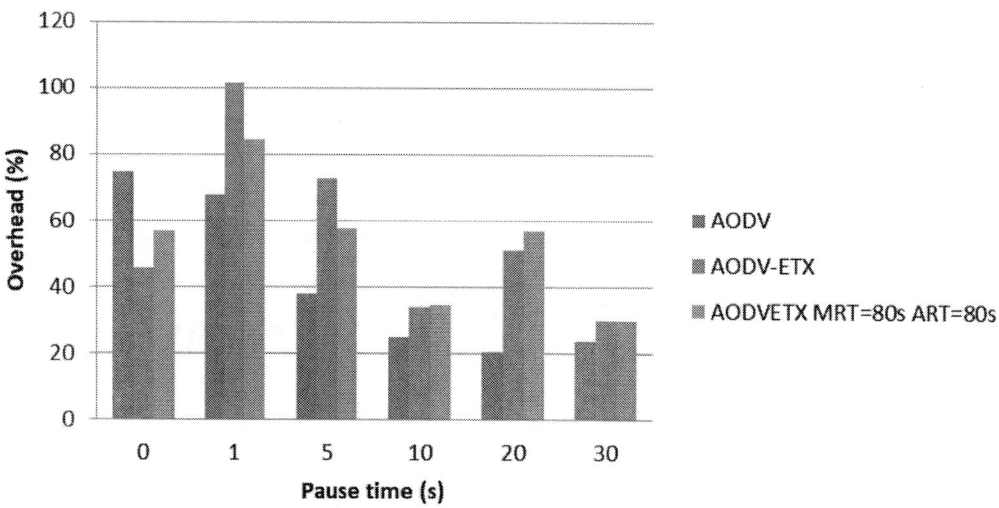

Figure 5. Comparison of overhead values

3.5 Total process route discovery

Total process route discovery is how often routing protocols do route discovery processes. In Figure 6, If more of route processes discovered then the routes are often disconnected. The average route discovery process performance of AODV-ETX MRT = 80s ART = 80s compared to the AODV-ETX, has increases = 3.09 %. Compared with AODV, the average route discovery process on MRT = 80 seconds ART = 80 seconds decreases = 8.94 %.

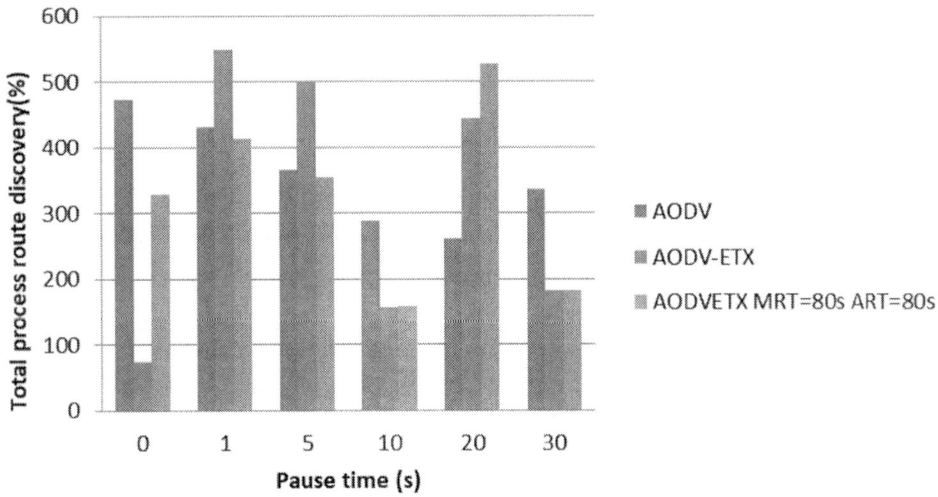

Figure 6. Comparison of route discovery processes

4. Conclusion

The conclusion of this study is that the modification of the AODV-ETX protocol by applying the MRT and ART values of 80s increases throughput by 5.72 %, increasing PDR by 5.80 %, decreasing routing overhead by 4.31 %, but has an increase in end to end delay of 10.19 % and a route discovery increase of 3.09 % compared to the AODV protocol.

References

[1] Roy R 2010 *Handbook of mobile ad hoc networks for mobility models* (New York: springer)

[2] Raut, Surendra H, Ambulgekar H P 2013 Proactive and Reactive Routing Protocols in Multihop Mobile Ad hoc Network *IJARCSSE* **3** 152–7

[3] Richard C, Perkins C E, and Westphal C 2015 *Second Annual IEEE Communications Society Conf. on Sensor and Ad-Hoc Communications and Networks* (California:IEEE)

[4] Gouda B S, Dass A K, and Narayana 2013 K.L *IEEE Autom. Comput. Commun. Control Compress. Sens. Imac4s* (Kottayam: IEEE) 306 – 312.

[5] Dubey V and Dubey N 2014 Performance evaluation of AODV and AODVETX *Proc. - 2014 6th Int. Conf. Comput. Intell. Commun. Networks, CICN 2014* 482–5

[6] Liu S, Wu M, Chen C, Lv B and Li S 2013 A High-Throughput Routing Metric for Multi-Hop Ad hoc Networks based on Real Time Testbed *2013 IEEE International Conference of IEEE Region 10 (TENCON 2013)* (Xi'an, China: IEEE) pp 4–7

[7] Hatti S, Kamakshi M.B 2013 *Springer* **131** 817-826

[8] A Purnomo, Widyaqwan, W Najib, R Hartono and Hartatik 2018 *IOP Conf. Ser.: Mater. Sci. Eng.* **333** 012107

[9] Rohal P, Dahiya R, and Dahiya P 2013 *IJARET* **1** 54-58

[10] Lee S J and Gerla M. 2000 *WCNC* (Chicago: IEEE) 1311-1316

IOP Conf. Series: Materials Science and Engineering **578** (2019) 012083 doi:10.1088/1757-899X/578/1/012083

Automatic Electric Outlet Using Internet of Things and Web Service

R Hartono [1], A Purnomo [1], M A Safiie[1] and M R Erliyanto [1]
[1] Departement of Informatics Engineering , Faculty of Mathematics and Natural Sciences , Sebelas Maret University, Surakarta, Indonesia

E-mail: rudi.hartono@staff.uns.ac.id

Abstract. Internet of Things ("IoT") is a term that refers to the ability of objects to be connected to the internet, so the objects can send or receive data through it. This research proposed using IoT to develope a smart home system can controll devices on it. By using this smart home systems, users can controled and monitored devices remotely. IoT could be applied by using a minimum system of devices such as Atmega, Raspberry Pi, Arduino Uno, etc. This research used ESP8266 minimum systems as a gateway device between devices / sensors and server, server used to record data and information. The data communication is using wireless transmission media with a frequency of 2.4 GHz. The results is obtained a socket device can be controlled from the internet which has 4 ports and each them can be controlled separately. This devices is equipped with PIR sensor to detect human attandance in order to control automatically.

1. Introduction

The use of energy more effectif and effisien in a building or a house are an important subect in Internet of Things (IoT) research[1],[2]. According to Indonesian Ministry of Energy and Mineral Resources (2017), national electricity consumption continues to show an increase as electricity access increases or electrification and changes in people's lifestyles. The data shows that Indonesia's electricity consumption in 2017 reached 1,012 Kilowatts per Hour (KWH) / capita, up 5.9% from the previous year. In this year, the government is targeting public electricity consumption to increase to 1,129 kwh/capita. The data shows, the electric usages is continues to increase every year. In order to get less energy consumtion, we could create an automated devices in the building[3],[4]. Therefore, to obtain users can monitored and controlled devices remotely need gateway used to communicate between devices to the internet[5],[6],[7]. To develope an IoT, we can use a minimum system such as Atmega, Rasberry Pi, Arduino Uno, etc and sensors used to interact directly to devices[8]. All of devices in the house, could connected to the internet such as AC, TV, lamp, house security, etc, therefore all of them can controlled remotely using smartphone. This research is developt an Internet of Things (IoT) tools based on ESP-8266 using PIR sensors, on/off control in electric outlet and gateway. Communication between the gateway and internet using wireless transmission with frequency of 2,4 GHz. The on/off status that appear in the dashboard are saved on the database server. User canuse android or web to control remotely.

IOP Conf. Series: Materials Science and Engineering **578** (2019) 012083 doi:10.1088/1757-899X/578/1/012083

2. Experimental

Generally, Automatic Electric Outlet System are using an Internet of Things and Web Services, so it can control an online outlet from anywhere and anytime through internet connection. System will retrieve the data from the API Server which is connected to the database server, its function is to read the status on the database whether the port is on or off. The system could send and receive the message on realtime because it is directly connected to the database server. Figure 1 is describing the general architecture of the system.

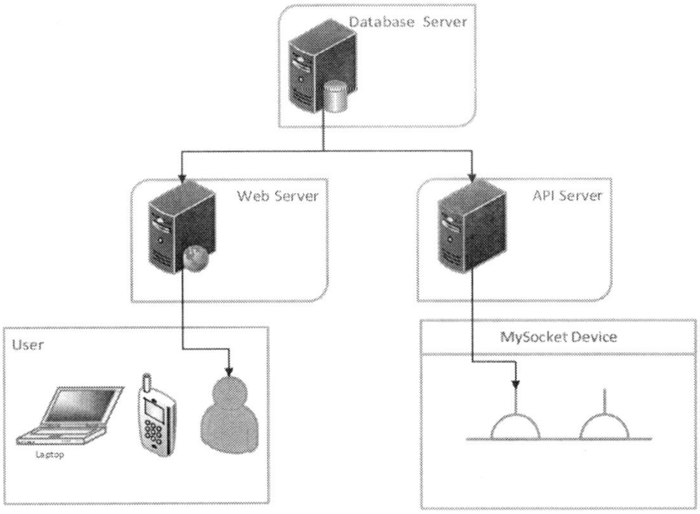

Figure 1. General architecture of Automatic Electric Outlet System.

2.1 System architecture

2.1.1 Hardware Prototype

The hardware are consist of WeMos ESP-8266 as a microcontroller that can control the actuaor, an adapter as a power supply, relays module as an actuator, LED as an indicator, PIR module as a sensor, transistors and resistors. All of them are described as MySocket Device in figure 1. PIR module is a *Passive Infra Red* sensor, it can sense an appearance of human in several distance and send the information to microcontroller. Relays are an actuator used to turn on or off an AC circuit through microcontroller. ESP-8266 is a wireless microcontroller, used to read the sensor message and instruct the actuator to apply its command.

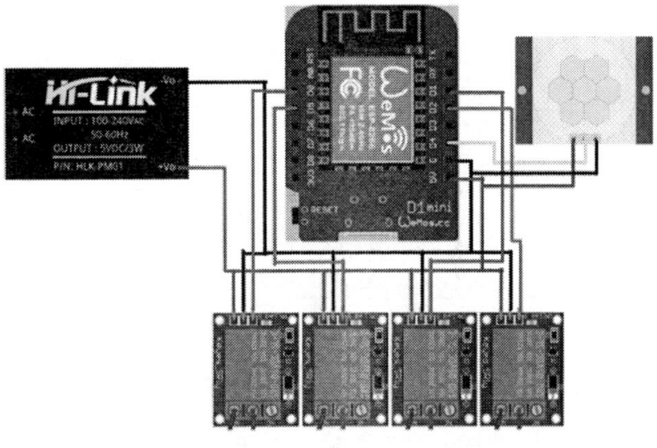

Figure 2. MySocket Device hardware design

2.1.2 Flowchart Control Device

Flowchart when controlling the device on an account such as turning on or off the socket in Figure 3.

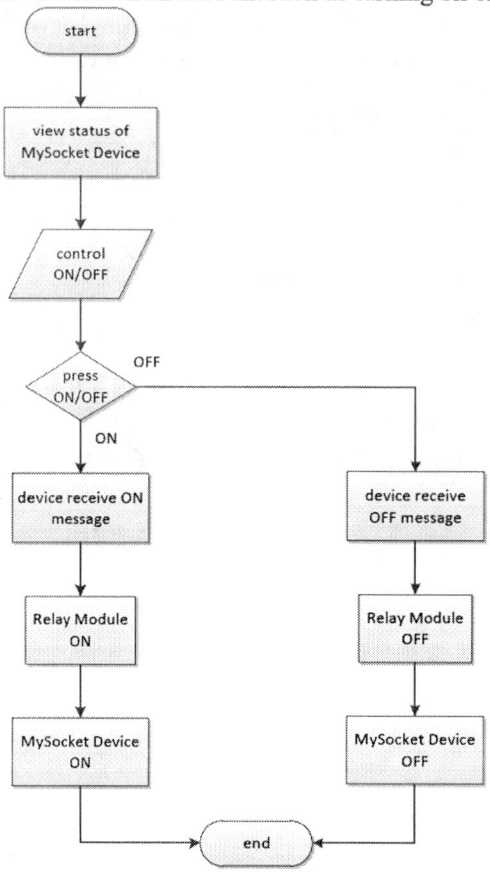

Figure 3. Basic flowchart of the system

3. Results and Discussion

To test the system that has been made, a test is performed on this system. In this test there are 2 parts, namely hardware testing and software testing. ESP8266 microcontroller device is the brain of connected hardware. For the test results can be seen in Table 1.

Table 1. ESP8266 Testing

No	Case / Form tested	Test Scenario	Expected results	Test result
1	Turn on ESP8266	Connect the appliance plug to the socket	ESP8266 will live and emit wifi	**Succeed**
2	Configure ESP8266 Connection	Connect wifi from ESP8366 then enter wifi information for the main connection	ESP8266 is connected to the main wifi for internet connection	**Succeed**
3	ESP8266 can be connected to the server	The device can receive messages from the server	The device runs a server command	**Succeed**
4	ESP8266 will reset the configuration if it does not detect possible connections	If the configured wifi cannot be found or die	If the device cannot be connected to wifi, or the main wifi turns off. Then the tool will reboot and can be reconfigured	**Succeed**

3.1. Hardware testing

Hardware is the main component of this system because without the hardware, this system will not work properly. The hardware that will be tested is, ESP8266 connection, Relay, PIR Sensor, and the electric current at the socket.

3.2 Relay

This hardware is a command executor from the ESP-8266 microcontroller. The command from ESP8266 will be run by this Relay so that it can cut off the electricity to the socket. Relay testing results in Table 2.

Table 2. Relay control

No	Tested UI Form	Test Scenario	Expected Result	Test Result
1	Turn on Port 1	Press the on button in the application on port 1	The relay will turn on and the power supply can be connected to the port 1 socket	Succeed
2	Turn on Port 2	Press the on button in the application on port 2	The relay will turn on and the power supply can be connected to the port 2 socket	Succeed
3	Turn on Port 3	Press the on button in the application on port 3	The relay will turn on and the power supply can be connected to the port 3 socket	Succeed
4	Turn on Port 4	Press the on button in the application on port 4	The relay will turn on and the power supply can be connected to the port 4 socket	Succeed
5	Turn off Port 1	Press the off button in the application on port 1	The relay will turn off and disconnect the power on the port 1 socket	Succeed
6	Turn off Port 2	Press the off button in the application on port 2	The relay will turn off and disconnect the power on the port 2 socket	Succeed
7	Turn off Port 3	Press the off button in the application on port 3	The relay will turn off and disconnect the power on the port 3 socket	Succeed
8	Turn off Port 4	Press the off button in the application on port 4	The relay will turn off and disconnect the power on the port 4 socket	Succeed

3.3 PIR Sensor

This device is a feature that makes motion detection or motion sensor. If a motion is detected, the relay on the specified port will turn on. The results of PIR Sensor testing can be seen in Table 3.

Table 3. Testing of PIR Sensors

No	Tested UI Form	Test Scenario	Expected Result	Test Result
1	Detect Motion for port 1	Motion is detected and turns on the relay on port 1	When motion is detected, the relay on port 1 will turn on	Succeed
2	Detect Motion for port 2	Motion is detected and turns on the relay on port 2	When motion is detected, the relay on port 2 will turn on	Succeed
3	Detect Motion for port 3	Motion is detected and turns on the relay on port 3	When motion is detected, the relay on port 3 will turn on	Succeed
4	Detect Motion for port 4	Motion is detected and turns on the relay on port 4	When motion is detected, the relay on port 4 will turn on	Succeed

IOP Conf. Series: Materials Science and Engineering **578** (2019) 012083 doi:10.1088/1757-899X/578/1/012083

4. Conclusion

Research has been carried out and resulted in the conclusion that in making Automatic Electric Outlet devices using ESP8266-based Webservice that implements Internet of Things has been successfully carried out and implemented. The main function is to make the user can control the socket can be controlled (turned off and turned on) from wherever and whenever. So that users can be flexible to regulate electrical equipment connected to this device and can save electricity consumption.

References

[1] J Pan, R Jain, S Paul, T Vu, A Saifullah, and M Sha 2015 *IEEE Internet Things J.* **2**(6), 527–537

[2] C. Bekara 2014 *Procedia Comput. Sci.* **34** 532–537,.

[3] V S Gunge and P S Yalagi 2016 *Int. J. Comput. Appl. Natl. Semin. Recent Trends Data Min.* Rtdm 975–8887

[4] V Marinakis and H Doukas 2018 *Sensors (Switzerland)* **18**(2)

[5] V K. Sagar N and K S M 2015 *Int. Res. J. Eng. Technol.* 2395–56,.

[6] K Sarath, C Kotcherlakota, V Bharadwaj and R Chinta 2016 *J. Eng. Res. Appl.* **6**(5) 77–80

[7] D Kunal, D Tushar, U Pooja, Z Vaibhav and V Lodha 2016 *Int. J. Adv. Res. Comput. Commun. Eng.* **5**(2) 576–578

[8] R Piyare 2013 *Int. J. Internet Things* **2**(1) 5–11

IOP Conf. Series: Materials Science and Engineering **578** (2019) 012084 doi:10.1088/1757-899X/578/1/012084

The development of Student Attandance System using RFID and Internet of Things (IoT) technology

M A Safi'ie[1], R Hartono[1] and G Pratama[1]
[1] Infomatics Engineering, Sebelas Maret University

Email: safiie99@staff.uns.ac.id

Abstract. The attandance of student in the classroom is reflecting their academic performance, the Department of Informatics Engineering in the Universitas Sebelas Maret used a manual way to record student attendance. The manual way is using papers to record student attendance and lecture activities, it may suffer to human error, because attandee can manipulate their presence and the departments can't monitor them. Radio Frequency Identification (RFID) can be applied to identify various things, because it has unique information codes saved in a special chip called tags. RFID can be applied to developed an attendance system prototype. The term of Internet of Things (IoT) can be described as an embedded system with internet data exchange, by using this technology we can monitored and controlled the devices, also we can integrate the database. The aim of this research is to measure the data speed of the prototype of RFID attendance system, the prototype has function to stored, process, and record the student attendance in the Department of Informatics Engineering.

1. Introduction

Recording the presence of students is one of the important factors to show their achivement in learning process. In addition, in-depth and detailed information about the students presence can determine achievement, productivity or progress of the institution in general[1]. Generally, the institution records the student's presence using a conventional recording tool by paper. Some of students often manipulate their attendance data, if a continuous monitoring is not done properly. Moreover the tradional way take's time consuming and prone of error [1],[2],[3]. Therefore to avoid this problem, we can use Information and Communication Technology (ICT) to create autonomous systems to record students attandance with their presence in the course subjects[4].

In recent years, Student Attandance have been developed by using a various things such as magnetic stripe, RFID, barcode scanner, QR scanner, biometric and Computer Vision. Pireva on his research, proposed using RFID to create a management systems for student attandace [5]. The paper shows that the systems has presented a good solution web application using RFID technology for attending student classes. However the paper is mentioned that the systems need improvement which can send annaucement via bluetooth thus the students will know immediately their attandance has been recorded successfully. Meanwhile Kishore, etc proposed their research using Computer Vision technolgies to develop an IoT based student's attandance systems [2]. According to their paper, they used a face recognition on Raberry Pi to represent the student's attandance. Sutar, etc in the research

IOP Conf. Series: Materials Science and Engineering **578** (2019) 012084 doi:10.1088/1757-899X/578/1/012084

built a smart attance system using RFID and IoT, but in their woks they used SMS Gateway to report the student attandace [6].

This paper proposed using a RFID to record the studentss attandance and IoT as an infrastructure to handle the microcontroller and the management systems. Student's use the RFID to tags on the reader which consist of Arduino Uno and Ethernet Shield, the information sent to the management system and recorded to the database. The system checked the information and calculate them to it's course schedule. Furthermore in this research we propose to build a feature with automatic email send to each students to check their availability in the class.

2. Experimental

2.1. System Architecture

The term of Internet of Things or usually abbraviated IoT was proposed in 1999 by Kevin Ashton, which firtsly describe a RFID identify the number of objects to a reader through wireless then send the information via internet as the base of it's communication network [7]. Atabekov also stated that nowdays IoT became a new network paradigm which allows various physical entities in the world connect each other. In the other word we can conclude that IoT consist of specific purpose computer added with internet as their communication network.

This research built an IoT based on Arduino Uno and Wiznet Ethernet Shield, in addition Mifare RFC522 is used to read RFID tags and then sent the information to the management system via internet communication. The Management system's is custommaly built with a several functions such as student management, lecturer management, course managment, schedule management, course log data, sent recap of presence via email, etc. The general architecture of the systems is describe in figure 1 bellow:

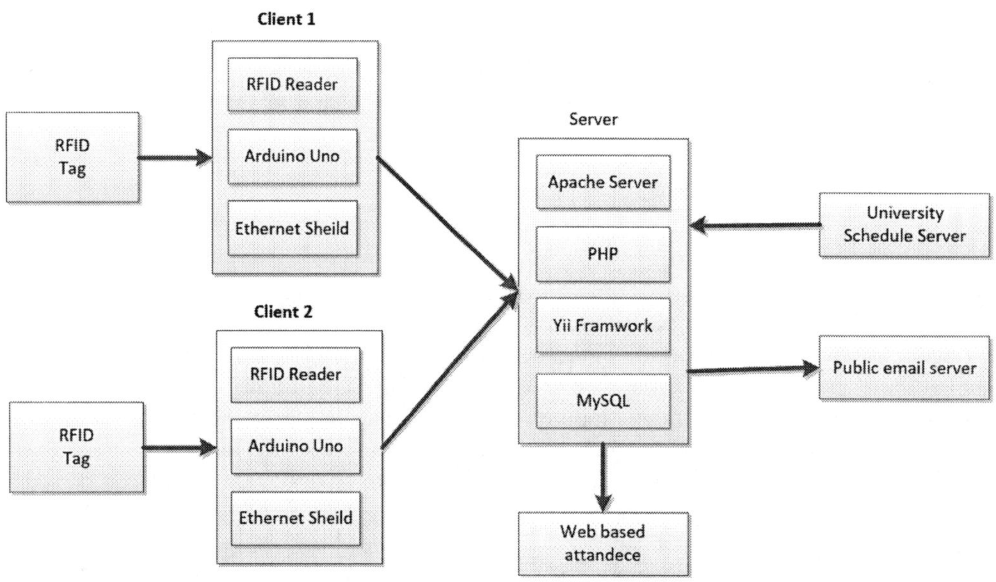

Figure 1. General architecture of the system

The development of management systems is using XAMPP as the bundled software that consist of an Apache Web Server to distribute the web file, PHP for the language programming, MySQL as a database engine and phpmyadmin as a database management systems. Here we use Yii Framework for the language framework that much more stabil than other framework. Based on figure 1, in breaf we can explanations about the process, RFID tags is for representing student's attandance read by the RFID, and the server had an independent job to get the University course schedule and public email server, meanwhile the web based attandance system used to display the student attandance and course log data.

2.2. Implementation

The hardware system is shown in figure 2, it is consist of Arduino Uno as the mainboard which we put the programming code in, it is mixed with the Wiznet Ethernet Shield to communicate and sent the information through internet communication network. Meanwhile the Mifare RFC522 is used to read the student's tags and by using the ethernet shield we can use it's library that contains of Client and Server Class to sent and receive the information to the host server.

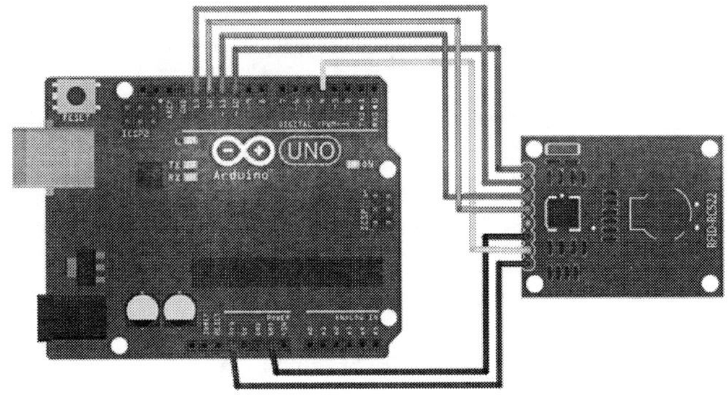

Figure 2. Schematic of the students attandance systems hardware

The software management system contains of several functions that designed and operated to handle backend process such as user management, lecturer management, course log data management, synchronizing course schedule, sent the presence recapitulation to email, etc. Generally the operation is show in figure 3. First the server check te course schedule if true than activate the course log data so the lecturer and the students can tags their id. The systems would record the data to the database, and later it would analyze and sent the result to their email.

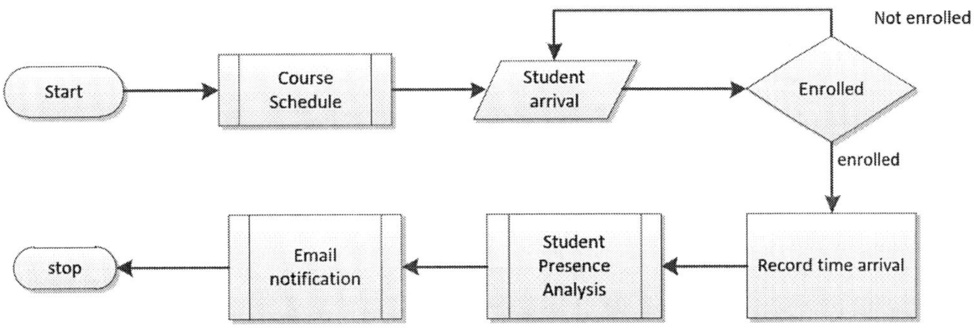

Figure 3. Flowchart of the operational system

3. Results and Discussions

The testing scenario, there are 2 testing and that is a reading test for the hardware and an average speed test of whole system. In the reading test we used to determine the distance that covered by the hardware, in the reading test we get the result as shown in figure 4.

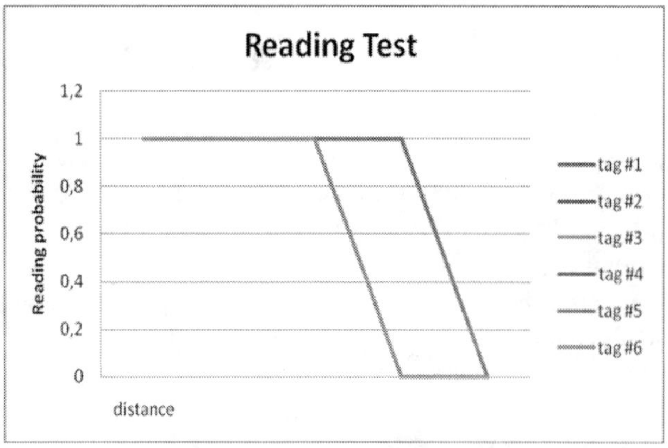

Figure 4. Hardware reading test

Figure 4 is discribing distance(cm) in x-axis and reading probability in y-axis. This testing used 6 different RFID tags, tapped to the harware and measured the probability. The picture ilustrate that the tags can optimally read at 3-4 cm, above of it the hardware can not read.

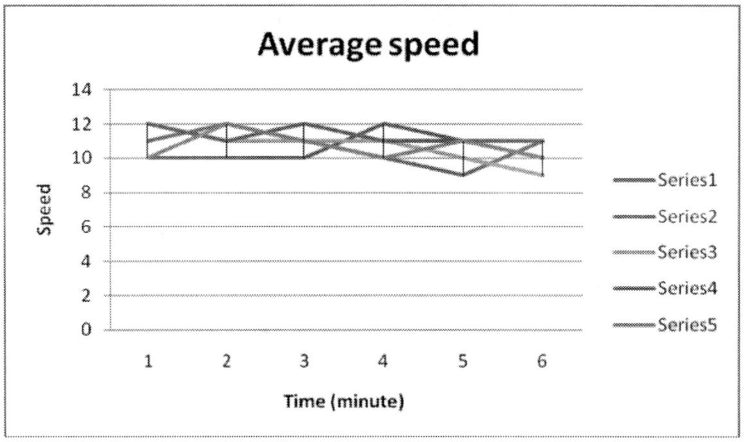

Figure 5. Average speed test of the whole system

Figure 5 is describing the average speed test of whole system, first the tags is read by hardware and then the information is send to the management system. Secondly, the managemement system is record into the database and show them to user, so user can validate their presence in here. The test used 6 different tags and tapped to the hardware, furthermore it shown to the user. The above figure is show time (minute) in x-axis and speed of reading tags in y-axis. It ilustrate the system can read the tags aproximately 9-12 times in 1 minute. The inteval of the system may have long interval, because it is set to avoid flooding on tags read.

4. Conclusion

The test shows that the harware can read optimally at 3-4 cm when the tags tapped to it, and it shows the management system can read the tags within 9-12 times per minute. On the other hand, the system can handle to log the attandee, and it can manage the course log. The system can recap the presence of attandee and send them to their email privately.

IOP Conf. Series: Materials Science and Engineering **578** (2019) 012084 doi:10.1088/1757-899X/578/1/012084

References

[1] H D, N Salih, A Al, B Al-Sadawi, and H Alsharqi 2018 *Int. J. Adv. Comput. Sci. Appl.* **9**(1) 266–274

[2] K. Kishore, O N. Lobo, and P. Saraiwala 2018 *International Journal of Innovations & Advancement in Computer Science* (IJIACS) **7**(4), 506–511.

[3] T. Victor & T. Alexandru 2016 *Romanian Economic Business Review*, **10**(1), 253-263

[4] A Dey, K Stuart, and M E Tolentino 2018 *Proc* .**2018** 505–510

[5] K R Pireva, J Siqeca, and S Berisha 2013 *RFID: Management system for students' attendance* 15(1). IFAC

[6] S W M Sutar, M Patil 2016 *Int. J. Adv. Res. Comput. Eng. Technol.* **5**(4) 2278–1323

[7] A Atabekov 2016 *Proc. 31st Annu. ACM Symp. Appl. Comput. SAC '16* 746–747

IOP Conf. Series: Materials Science and Engineering **578** (2019) 012085 doi:10.1088/1757-899X/578/1/012085

Smart Energy Meter based on Arduino and Internet of Things

N M Yoeseph, M A Safi'ie, and F A Purnomo
Diploma 3 Teknik Informatika Sebelas Maret University, Jl. Ir. Sutami 36A Kentingan Jebres Surakarta 57126, Indonesia

Email : nanang.my@staff.uns.ac.id

Abstract. Energy meter is a measuring instrument to calculate the amound of electric energy consumed by an electrically powered device. The energy metering which measure line voltage, current, and calculating active power can be used to increase usage efficiency of electricity. This paper presets a system which provide real time energy meter reading using ACS712 as current sensor, ZMPT101B as voltage sensor, arduino and an IoT server. The energy meter can display consumed energy in kWh and electricity bills to be paid. system's user interface is in Android. the error in current measurement is less than 2 % and error for peak to peak voltage measurement is 4 %. A notification for the amount of electricity energy usage or electricity cost can be generated for user to be displayed.

1. Introduction

The power meters are one of important aspect in the smart grid concept [1]. They are the devices which measure electricity and also connecting different equipment over the electric grids. The concept that introduce the automation of measurements of electronic meters was called AMR (Automated Meter Reading). AMR is a concept that enables devices to be accessed remotely and collect the electronic data generated by the meter at consumer units. The data then transmitted from the meter to the electric company using radio frequency, telephone, power line or satellite communication [1].

A smart energy meter (SEM) is electronic device having energy meter chip for electric energy consumed measurement, protocols for data communication, security purposes, interface for data display and other function[2]. The difference of smart meter from traditional energy meter devices is by its communication ability. A smart meter should be able to record active power consumption and also other information of an electronic device, such as: voltage and current phasors, reactive power, maximum power demand, frequency, power factor and other information, in real time [1].

The Internet of Things (IoT) is defined by ITU and IERC as a dynamic global network infrastructure with self-configuring capabilities based on standard and interoperable communication protocols where physical and virtual "things" have identities, physical attributes and virtual personalities, use intelligent interfaces and are seamlessly integrated into the information network [3]. Following the utilization of IoT, the smart meters needs to be developed with the intention of using device and communication technology to attain consumer satisfaction. This paper present a design of a smart energy meter, which provide accurate measurement of power consumption and display it to the consumer.

2. Experimental

There are two basic types of electronic meters, electro-mechanical and electronic. The most common type of electricity meters are the electro-mechanical induction watt-hour meter. It works by counting the revolutions of a non-magnetic metal disc which is made to rotate at a speed proportional to the power passing through the meter. The number of the disk revolutions is proportional to the device's energy usage[2]. The electrical power is the ratio of electric energy per unit time. In other words, it is the amount of power used or absorbed by a device within a given time. The giving unit by the International System of Units is the Watt [4]. The real power calculated from the instantaneous power signal. The instantaneous power signal is generated by a direct multiplication of the current and voltage signals [2]. Electronic power is calculate using equation (1)

$$P = V * I \tag{1}$$

where: V is the voltage (V); I is the current (A); P is the power (W)

The system's design for the smart meter using arduino and Internet of Things (IoT) is shown on Figure 1. In our system, the voltage was measured using ZMPT101B, the current was measured by ACS712, and the estimated power is calculated using the effective value of both voltage and current signal. Arduino is used to read the sensors readings and send the data via ethernet shield to the server. The server will receive the data, calculate the estimated power and display it for user. User can access the data using mobile devices.

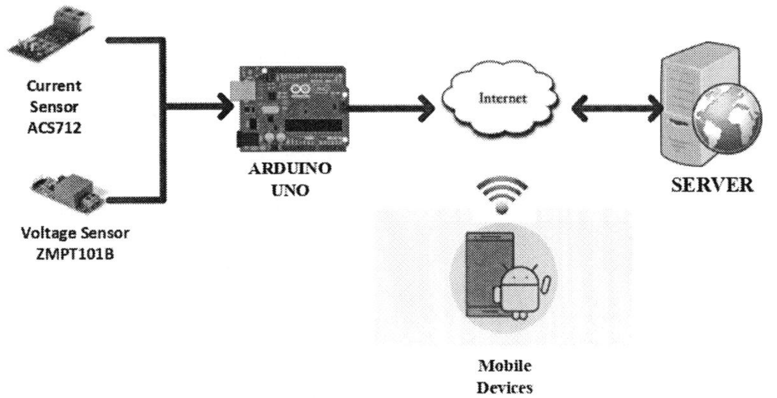

Figure 1. The Design System of The Smart Meter.

ZMPT101B is a voltage sensor made from the ZMPT101B voltage transformer. It has high accuracy, good consistency for voltage and power measurement and it can measure up to 250V AC [5], [6]. ACS712 is a Hall effect-based linear current sensor Integrated Circuit (IC) with 2.1 kVRMS isolation and a low resitance current conductor. This sensor has a maximum 185 mV/A output sensitivity with extremely stable output offset voltage [7]. Both voltage and current units are isolated, very cheap, and easy to use [6]. Arduino is an open-source platform used for building electronics projects. Arduino consists of both a physical programmable circuit board and a sofware or IDE (Integrated Development Environment), used to write and upload computer code to the physical board.

Figure 2. ZMPT101B Voltage Sensor Module.

Figure 3. ACS712 Hall Effect Current Sensor.

IOP Conf. Series: Materials Science and Engineering **578** (2019) 012085 doi:10.1088/1757-899X/578/1/012085

After implementing the system design, sample of voltage and current were taken by connecting the sensors to electric source. Afterward, the value from the sensor is compare with data obtained from a multimeter (Heles UX866TR). A bank of 7W, 15W, 18W, 42W, and 100 W light bulb is connected as load.

3. Results and Discussion

A hardware prototype had implemented according to the design. The hardware prototype for the Smart Energy Meter is shown on Figure 4.

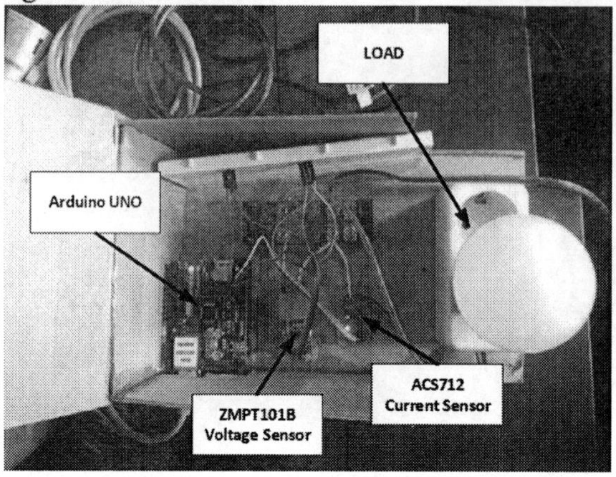

Figure 4. Prototype of Implemented Smart Energy Meter.

The voltage measured by the ZMPT101B voltage sensor and the multimeter were read and recorded. The values and are represented as a graph in Figure 5. The accuracy of ZMPT101B voltage sensor's reading is 77 %.

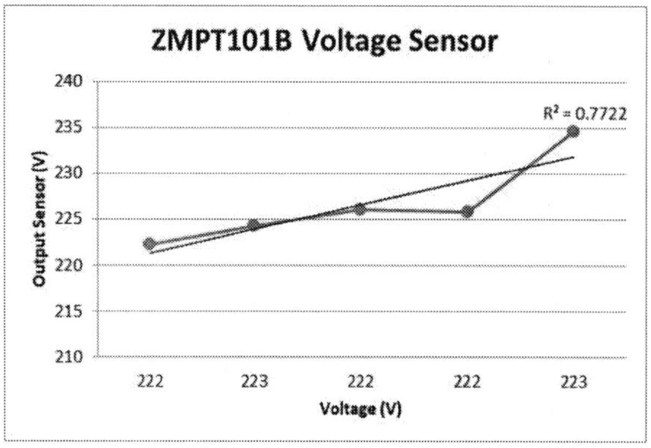

Figure 5. Result of Voltage Reading by the ZMPT010B Voltage Sensor.

The current measured by the ACS712 Current sensor and multimeter were read and recorded. The values and percentage errors are represented as a graph in Figure 6. The accuracy of ACS712 current sensor's reading is 86 %.

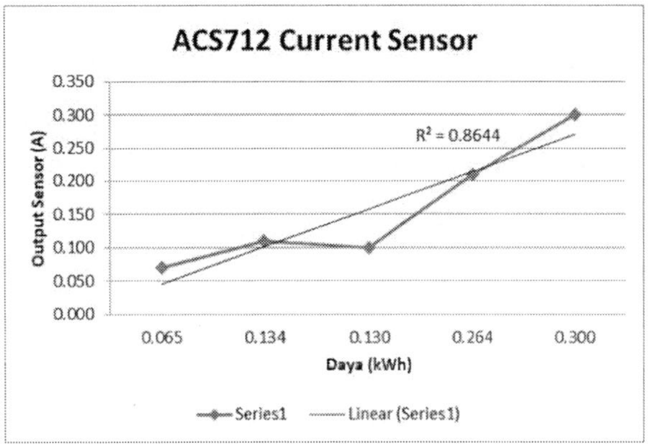

Figure 6. Result of ACS712 Current Sensor's reading.

User interface is design so that user can monitor usage of electricity by the equipment. The user interface also shows the cost spent for the electricity. Smart energy meter's user interface is shown on Figure 7.

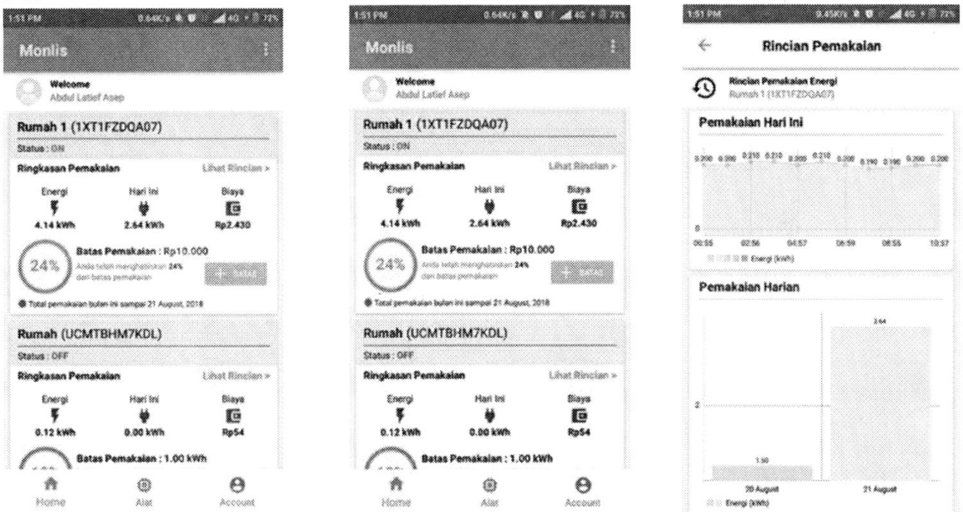

Figure 7. User Interface of Smart Energy Meter.

4. Conclusion

As conclusion, a smart energy meter using arduino and IoT has been designed and tested. The meter is design to facilitate electronic consumer to monitoring its electrical utilization. The meter consist of ZMPT101B as voltage sensor, ACS712 as current sensor and Arduino UNO with ethernet shield. The data from sensor's reading will be sent to server. Server will calculate the electrical power used by the equipment. User can monitor all of this data using mobile devices. The sensor testing shows 77 % accuracy for voltage sensor and 86 % accuracy for current sensor. The future work is to improve the sensor accuracy that the energy reading become more precise.

References

[1] Garcia F D, Marafao F P, Souza W A De and Silva L C P Da 2017 Power Metering: History and Future Trends *IEEE Green Technol. Conf.* 26–33
[2] Rahman M, Islam M O and Salakin S 2015 Arduino and GSM Based Smart Energy Meter for Advanced Metering and Billing System 21–3
[3] Vermesan O and Friess P 2014 *Internet of Things – From Research and Innovation to Market*

Deployment (Denmark: River Publishers)
[4] M M S, Borre J D and Santos J C 2014 Design and Construction of a Power Meter to Optimize Usage of the Electric Power *International Congress of Engineering Mechatronics and Automation (CIIMA)* pp 1–5
[5] Abubakar I, Khalid S N, Mustafa M W, Shareef H and Mustapha M 2017 Calibration of ZMPT101B voltage sensor module using polynomial regression for accurate load monitoring *ARPN J. Eng. Appl. Sci.* **12** 1076–84
[6] Mnati M J, Van den Bossche A and Chisab R F 2017 A smart voltage and current monitoring system for three phase inverters using an android smartphone application *Sensors* **17**
[7] Allegro MicroSystems L 2012 *ACS712 - Integrated, Fully Linear, Hall Effect-based Sensor, Current*

IOP Conf. Series: Materials Science and Engineering **578** (2019) 012086 doi:10.1088/1757-899X/578/1/012086

Study of Weight in Motion Sensor for Railroad Crossing Warning System Using Artificial Neural Network

N M Yoeseph[1], A Marzuki[2] , M Yunianto[2], and F A Purnomo[1]
[1] Diploma 3 Teknik Informatika Sebelas Maret University, Jl. Ir. Sutami 36A Kentingan Jebres Surakarta 57126, Indonesia
[2] Physics Department of Post Graduate Program Sebelas Maret University, Jl. Ir. Sutami 36A Kentingan Jebres Surakarta 57126, Indonesia

E-mail: nanang.my@staff.uns.ac.id

Abstract. IThe risk of accidents in the railway, especially on unmanned railroad crossing, are still high. A train arrival on railroad crossing warning system is indispensable. The warning system will detect train arrival on the railroad crossing and issue a warning to the crossers. A sensor based on optical fiber Weight in Motion (WIM) is proposed. This paper presents a design of a railroad crossing warning system based on WIM sensor. The sensor placed under the railroad tracks and detects vibrations footprints transmitted along those rails. Artificial Neural Network (ANN) is used to process the signal pattern of the WIM system output. By using ANN, the vibration footprint pattern was classified and used as on-sensor train detection. As the train detected, the arrival time of the train on the raiload crossing was estimated. The result is a warning system which can detect train arrival on the railroad crossing and issue a warning signal.

1. Introduction

Accident in railroad crossing are considered more dangerous than other transportation accidents in term of death rate dan casualties. Thus risk of accidents are increase, especially on unmanned railroad crossing. Therefore, a warning system in railroad crossing is needed to detect train arrival and issue warnings to the crossers.

In a railroad crossing warning system, train detection are main components. Information from detector device are used by the warning system to initiate warning for crossers. In automatic level crossing control systems, the warning are interrelated to train signalling system, but triggered by the train movements, especially on wheel detection technology for synergystic effect [1]. Wheel detector function by detecting the travelling direction and presence of the train wheels at the strike in point [2]. Wheel detection can be used on detection point, tract location, direction of traffic and train speed [1].

Fiber optic sensors have the advantage of high sensiitivity, anto-electromagnetic interference, high temperature and corrosion resistance compared with traditional sensors [3] . Weight-in-Motion (WIM) system measure the vehicle weight by detects the the shift light intensity output of optical fiber when the vehicles passing through on fiber sensor [4].

Pattern recognition is perhaps the most common use for neural networks. the neural networks the attempts to determine if the input data matches a pattern that the neural network has memorized [5]. This paper proposed a design of warning system which uses a sensor based on optical fiber Weight-in -Motion (WIM) and Artificial Neural Network (ANN).

IOP Conf. Series: Materials Science and Engineering **578** (2019) 012086 doi:10.1088/1757-899X/578/1/012086

2. Experimental

There are many scenarios and designs for train detection on railroad crossing. Such as, using sensor network [6] [7] [8], RFID [9] [10], IR [11][12][13][14], Ultrasonic [15], Accelerometer sensor [16], Android and GSM [17] [18].

The vibrations generated by the railway wheel will be detected by the WIM sensor placed under the railway bearing. Data from WIM sensors will be sent via Arduino UNO. The signal obtained from the WIM sensor will be processed first so it can be displayed in graphical form. The design of the railroad crossing warning system is shown on Figure 1.

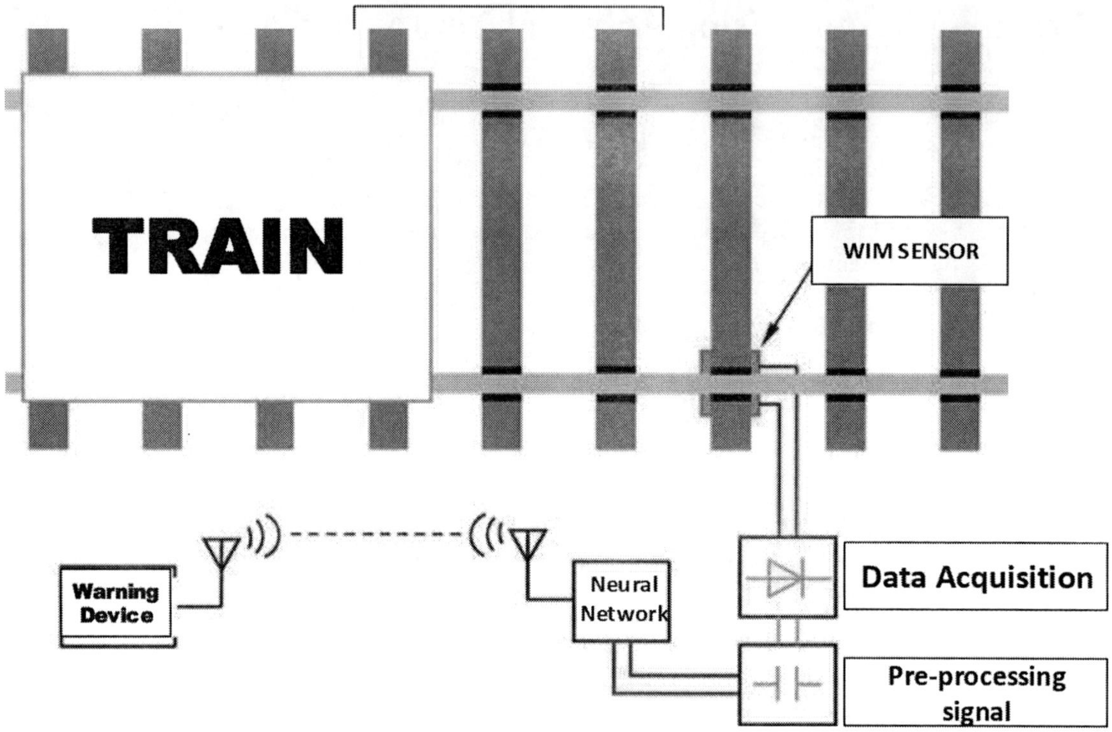

Figure 1. Design of Railroad Crossing Warning System.

The graph data then processed by Neural Network (NN). Neural Network will determine whether the signal being processed is indicating a train or not. If Neural Network detects that the signal is a train arrival signal, the Neural Network will calculate the time it takes for the train to arrive at the intersection. Based on the time calculation, the warning device will issue a warning signal for the arrival of the train.

The sensor will continue to read vibrations from the railway to the rail and send a signal to the Neural Network. If Neural Network detects the signal generated by the WIM sensor indicating that the train is running away from the WIM sensor, then the warning device will be turned off. The train detection processes is shown by Figure 2.

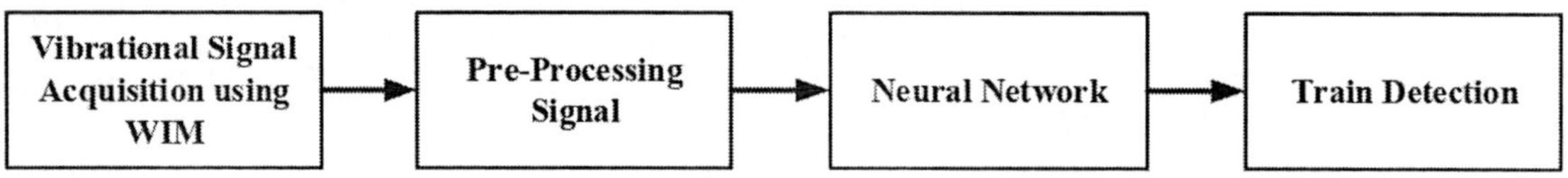

Figure 2. Train detection processes.

IOP Conf. Series: Materials Science and Engineering **578** (2019) 012086 doi:10.1088/1757-899X/578/1/012086

3. Results and Discussion

The Neural Network consists of 256 nodes in the input layer, 64 neurons in the hidden layer, and 2 neurons in the output layer. The activation function of each neuron is a binary sigmoid, momentum constant is 0.6, the learning rate is 0.5, and the target error is 0.001. The number of learning and testing data of vibration patterns for train and non-train are shown in Table 1. The identification result of WIM pattern by NN in the testing phase is shown in Table 2. It shows that the NN can distinguish between the train and the non-train with the rate of 100 %.

Table 1. The number of learning and testing data for NN

Condition	Data Learning	Data Testing
Train	5	3
Not Train	10	3
Total	**15**	**6**

Table 2. Identification Result

Target	Testing	MSE	
		0.01	0.001
Train detected	1	Train	Train
	2	Train	Train
	3	Train	Train
	4	Train	Train
	5	Train	Train
Train Not Detected	1	Not Train	Not Train
	2	Not Train	Not Train
	3	Not Train	Not Train
	4	Not Train	Not Train
	5	Not Train	Not Train
Percentage		**100**	**100**

4. Conclusion

This study has been focused to study the feasibility of using Weight In Motion (WIM) sensor to detect the arrival of train in a rail road crossing. The result will be used to design a warning system for a train arrival on railroad crossing. A Neural Network is used to determine whether the signal being processed is indicating a train or not. The result of the experiment shows that the Neural Network can recognize the signal from WIM sensor and determine whether the signal indicating a train or not with success rate 100 %.

5. References

[1] Grundnig G and Pucher C 2014 *Various applications of wheel detection systems with inductive wheel sensors* SIGNAL+DRAHT **106** 45–48 (Austria: Frauscher Sensor Technology GmbH)

[2] Pavithradevi V, Roopini D, Vaseela R, Aleem A and V S S 2014 *Int J Innov Res Sci Eng Technol.* **3** 136–42

[3] Jian X and Bin M 2010 Investigation of Discrete Wavelet Transform for Signal De-noising in Weight-in-Motion System *2010 2nd International Conference on Future Computer and Communication* pp 769–72 (Publisher IEEE)

[4] Syukron A A, Marzuki A and Setyawan A 2017 *J. Phys. Conf. Ser.* **909** 012033

[5] Heaton J 2005 *Programming Neural Networks in Java* 1-6

[6] Anil M D, Sangeetha S, Divya B, Niranjana B and Shruthi K S 2014 *Int. J. Adv. Res. Comput. Commun. Eng.* **3** 3–5

[7] Berlin E and Van Laerhoven K 2013 *Proc.IEEE Int. Conf. Distrib. Comput. Sens. Syst. (DCoSS) 2013* 80–87 (Publisher IEEE)
[8] Madhok S, Chauhan S and Jindal U 2015 *Gyancity J. Eng. Technol.* **1** 12–7
[9] Balog M, Husár J, Knap L and Šoltysová Z 2015 *Int. Sci. J. about Technol.* **1** 9–12
[10] Ilie-Zudor E, Kemény Z, Egri P and Monostori L 2006 The RFID Technology and Its Current Applications *The Modern Information Technology in the Innovation Processes of the Industrial Enterprises-MITIP* pp 29–36
[11] Krishnamurthi K, Bobby M, Vidya V and Baby E 2015 *Int. J. Adv. Res. Comput. Eng. Technol.* **4** 539–43
[12] Siddh S, Baig M S A and Reddy N 2015 *Int. J. Innov. Technol.* **03** 355–7
[13] Kottalil A M, S. A, M. A M, J. A L and Babu A 2013 *Int. J. Adv. Res. Electr. Electron. Instrum. Eng.* 7619–22
[14] Theja K A, Kumaresan M and Kumar K S 2015 *Int. J. Innov. Res. Comput. Commun. Eng.* **3** 11382–91
[15] Vidyasagar K, Babu P S and Ramprasad R 2014 *Int. J. Comput. Apllication* **107** 3–6
[16] Ardiansyah H, Rivai M and Nurabdi L P E 2018 *AIP Conf. Proc.* **1977** 040029
[17] Mapari N, Imran S M, Chaudhari S and Ashraf N 2016 *Int. J. Adv. Electron. Comput. Sci.* **3** 63–7
[18] Prasad R J C, Reddy P A and Praneeta Y S 2016 *Int. J. Sci. Eng. Technol. Res.* **5** 740–5

Design and evaluation on the immersive Virtual Reality system in learning archeology

F A Purnomo, E H Pratisto, F S Bahtiar, B K Riasti, L Puspitasari, N Ardhiana
Informatics Engineering Department, Sebelas Maret University,
Jl. Ir. Sutami 36A Kentingan Jebres Surakarta 57126, INDONESIA

E-mail: fendi_aji@mipa.uns.ac.id

Abstract. This research aims to design and evaluate the three-dimensional Virtual Reality game at showroom 1 and 2 of the Sangiran Museum. This application is designed for the media campaign and increase the appeal of the public to educational tourism contained in Sragen regency, the location of the museum. The game genre is quiz with scoring and it allows the user to guess the fossil name that is displayed on the screen. Twenty visitors were participated to evaluate the application after they finished exploring both showrooms. Questionnaire is used to measure the use of the application. The result shows good acceptance in term of usability. It is expected that visitors can add educational value to their knowledge on the post-visit to the museum.

1. Introduction

Sangiran Museum has been designated as one of the World Heritage by UNESCO on December 6, 1996. Sangiran Museum is an educational place and is often visited by visitors both inside and outside the city. Sangiran Museum, especially in Krikilan, keeps the largest collection of ancient objects compared to the Ancient Museum in the vicinity. However, based on the observation of many of these visitors, they forgot and did not even know what the names of fossils were in the showroom. This is because most of the visitors just look at the fossils and or take pictures near the fossils. From these problems, the value of education from the museum was not right on target.

Based on research by [1], stated that the technology-based interactive entertainment experience is offered by the museum to keep visitors interested and entertained. Several applications have been developed that help visitors when visiting a museum. Like a visitor assistant application called iMuse developed with the mobile UHF RFID Reader [2]. Even to find out the knowledge of visitors [3] has proven through virtual museum games regarding fossil age simulation, that the user's knowledge has improved after the pretest and posttest were performed. Serious Games applications have also been developed [4][5][6,7] and analyzed [8] through the creation of accurate 3D models of cultural heritage presented in fully immersive virtual and interactive environments. This research will make and evaluate the results of the making of Serious Games (SG) regarding the introduction of cultural heritage (CH), namely the introduction of the names of ancient fossils at the Sangiran Museum. Regards this case, Virtual Museum can also be used for preparation of actual museum visits, or after a visit for the purpose of debriefing in the formal learning setting framework [1].

IOP Conf. Series: Materials Science and Engineering **578** (2019) 012087 doi:10.1088/1757-899X/578/1/012087

2. Experimental

There are several stages in making Serious Games Introduction to fossils in users. The application is made using Unity Games 3D, modeling 3D fossils made using Blender. Name The fossil object was obtained at the Sangiran Museum in showroom 1 and showroom 2, each of which took 10 fossil objects so that there were a total of 20 fossil object introductions. Tests were carried out on various devices and questionnaire forms regarding the usefulness of the Serious Games application that was created. The research stages are presented in Figure 1 and modeling 3D fossils in Figure 2.

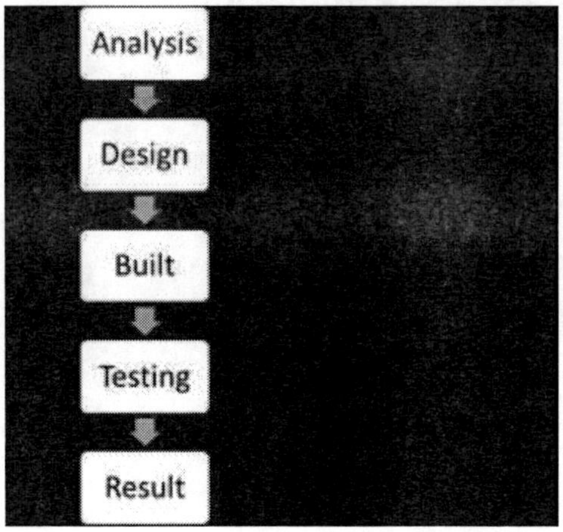

Figure 1. The making of Serious Games is the introduction of fossil objects

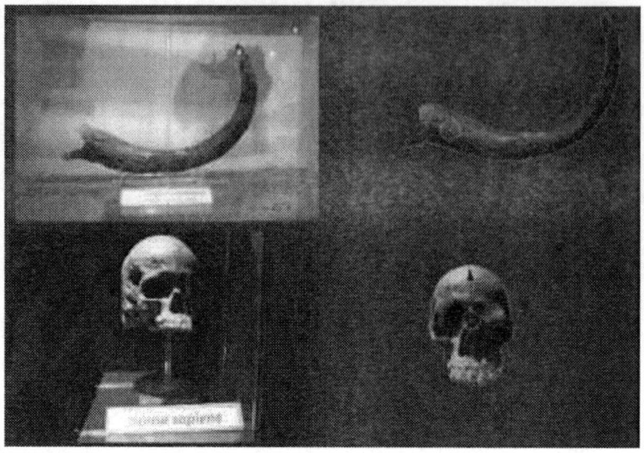

Figure 2. 3D The fossil reconstruction model (right) is almost equal to the original (left)

Interactive buttons on games are performed by the cursor head. So to choose navigation is done by moving the player's head and the help of the gyroscop sensor embedded in the device. So the selection of navigation buttons is done by positioning the cursor for 2 seconds. Games play in Serious Games fossil recognition is presented in the order in Figure 3.

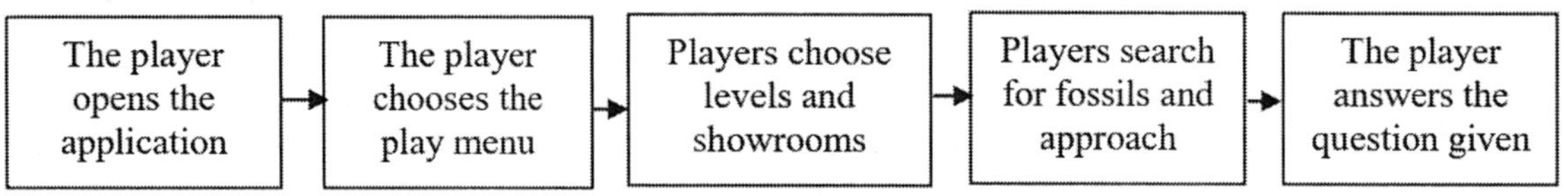

Figure 3. Games play on the Serious Games for fossil recognition

3. Results and Discussion

The making of Serious Games is shown in Figure 4, starting with the spalsh screen, the main menu, the guide menu, the fossil view menu, the play menu, the results if the answers are right and wrong and the game summary results.

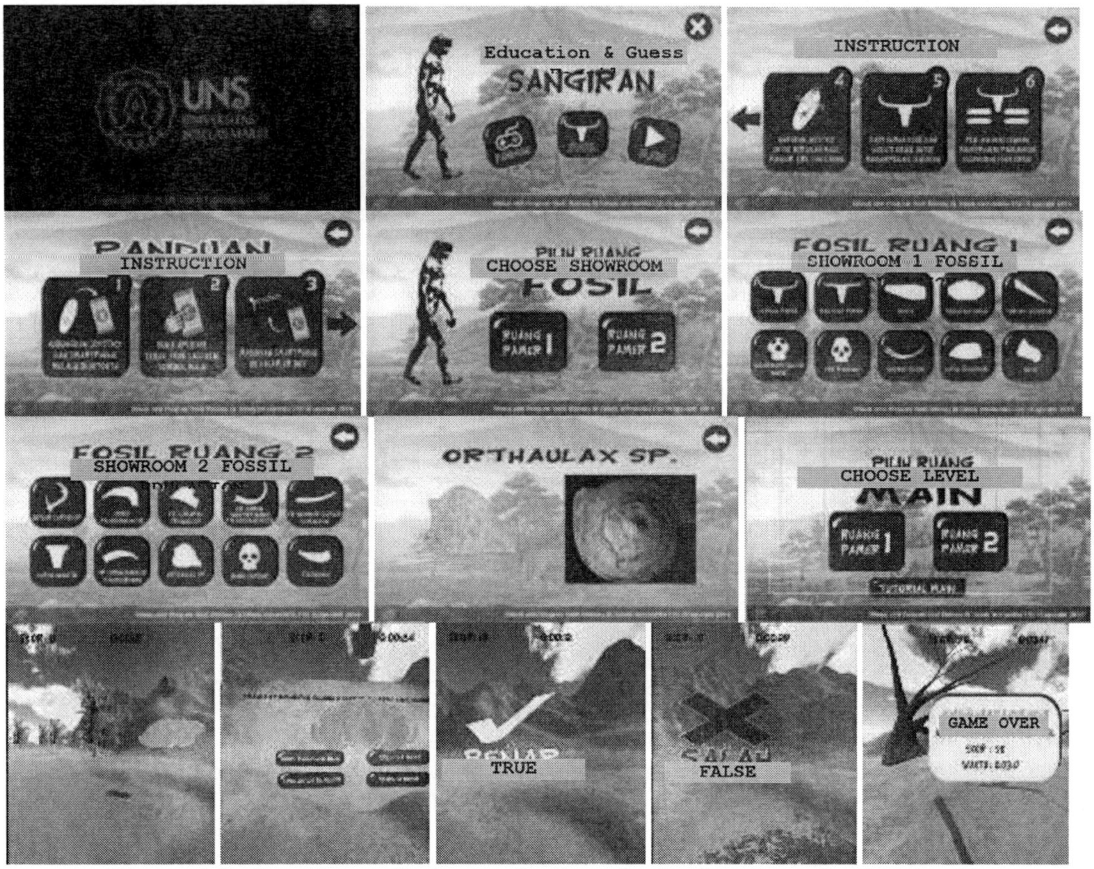

Figure 4. Serious Games display results about fossil education at the Sangiran Museum

The concept of Serious Game, Virtual Reality Technology in playing the Fossil Guess Sangiran Museum is made based on Android, which is a visual game application with interactive objects in the form of fossils in showrooms 1 and 2 which are presented in an environment of desert and forest. Besides that, there are also some information related to the explanation of existing fossils. The application is run with the help of VR Box and Bluethooth joystick to explore. This application complements the results of the manufacture [6,7] in presenting fossil learning in the Sangiran museum. The feature added in this study was that visitors were tested to recognize the name of the fossil in the exhibition hall of the Sangiran cluster in Krikilan.

Testing of 4 devices is done with different processor and GPU specifications but the same RAM memory is 3 GB. The results show the larger the processor clock, the application installation process and game games can run smoothly, which is 40% faster. The test summary is shown in Table 1.

Serious Games Testing about fossil education on users is carried out on 20 respondents on aspects of use, aspects of exploration experience, aspects of information delivery and aspects of goodness of fossil 3D models. Assessments made through a Likert scale strongly agree, agree, simply agree, disagree and disagree. The results of the questionnaire stated that the application was easy to use (strongly agree and agree 17.5% & 55%), players found new experience in recognizing ancient objects at the Sangiran Museum (strongly agree and agree worth 20% & 57.5%), information on introduction of ancient objects/fossils done well (strongly agree and agree worth 15% & 55%) and the results of the reconstruction of 3D fossil models are done to resemble the original form (strongly agree and agree worth 21.6% & 43.3%)

Table 1. Test results on android devices

No	Devices	Can play	Duration of opening the application (Seconds)	Gyro sensor	Information
1	Samsung J5 Pro (Prosessor Octa-Core 1.6 GHz, Mali T-830MP2 GPU, 3 GB RAM)	Yes	3	Yes	The application can run smoothly
2	Asus Zenfone Max Pro (Qualcomm Snapdragon 660 1.80 Hz, Adreno 512 GPU, 3 GB RAM)	Yes	3	Yes	The application can run smoothly
3	Xiaomi Redmi 3 Prime (Qualcomm MSM8937 Snapdragon 430, Adreno 505 GPU, 3 GB RAM)	Yes	5	Yes	The application can run smoothly
4	Asus Zenfone Selfie (Qualcomm MSM8939 Snapdragon 615, Adreno 405 GPU, 3 GB RAM)	Yes	8	Yes	The application can run but is slow

The application tested in Table 1 is an android application with a file size of 307 MB.

4. Conclusion

Research results in the making of Serious Games for fossil recognition at Sangiran Museum have been successfully created. The evaluation results state that devices that can run applications properly have a 1.2 GHz processor clock with 3 GB RAM. The results after playing the game, visitors are interested in studying fossils or ancient objects at the Sangiran Museum through this Serious Games which are presented with Virtual Reality technology. Serious Games about fossil education can be used as an education for users who will visit or after visiting Sangiran Museum.

Reference

[1] Pivec M 2016 Virtual Museum: Playful Visitor Experience in the Real and Virtual World *2016 8th International Conference on Games and Virtual Worlds for Serious Applications (VS-GAMES)* (Barcelona: IEEE) 1–4.
[2] Fevgas A, Fraggogiannis N, Tsompanopoulou P and Bozanis P 2014 The iMuse Virtual Museum : towards a cultural education platform IISA 2014 *The 5th International Conference on Information, Intelligence, Systems and Applications* (Chania: IEEE) 171–5.
[3] Nurhasanah Y, Rohman A S and Prihatmanto A S 2012 The Design and Implementation Determining Age of Fossils Game Simulation at Virtual Museum of Indonesia (A Case Study at a Museum of Geology) *2012 International Conference on System Engineering and Technology (ICSET)* (Bandung: IEEE) 1–6.
[4] Anastasovitis E, Ververidis D, Nikolopoulos S and Kompatsiaris I 2017 Digiart: Building new 3D cultural heritage worlds*3D True Vision v2: Research and Applications in future 3D Media (3DTV-Con2017)* (Copenhagen Denmark: IEEE) 3–6.
[5] Cirulis A, Tommaso L, Paolis D and Tutberidze M 2015 *Proced. Comput. Sci.* **77** 199-206.
[6] Purnomo F A, Santoso I, Hartanto R 2018 *IOP Conf. Ser. Mater. Sci. Eng.* **333** 012103.
[7] Purnomo F A, Pratisto E H and Putra A R M 2018 *AIP Conf. Proc.* **1977** 030014.
[8] Paliokas I and Sylaiou S 2016 The Use of Serious Games in Museum Visits and Exhibitions : A Systematic Mapping Study *2016 8th International Conference on Games and Virtual Worlds for Serious Applications (VS-GAMES)* (Barcelona: IEEE) 1–8 .

IOP Conf. Series: Materials Science and Engineering **578** (2019) 012088 doi:10.1088/1757-899X/578/1/012088

A Study Of Characteristic On The Samop Core Cooling System Using Fluent CFD

D W Lestari[1], Suharyana[1], Riyatun[1], and D Hidayanti[2]
[1]Physics Department, Faculty of Mathematics and Natural Science, Sebelas Maret University, Jl. Ir Sutami 36A Kentingan, Jebres, Surakarta 57126, Indonesia
[2]P2STPIBN, Nuclear Energy Regulatory Agency (BAPETEN), Jl. Gajah Mada No.8 Jakarta 10120 Indonesia

E-mail: suharyana61@staff.uns.ac.id

Abstract. Safety study of the cooling system of the SAMOP facility has been done. The thermal-hydraulics is required to evaluate the heat removal system in order to exhibit that sufficient cooling capacity to prevent fuel solution overheating. In this study, the *inlet* velocities have been varied namely 0 to 0.522 cm/s and 2.090 to 4.179 cm/s. The modelling and numerical simulation had been carried out using the *Computational Fluid Dynamic* (CFD) employing FLUENT 6.3.26 *software*. The results of water velocities were 0 – 1.045 cm/s or natural convection. The simulated maximum water temperature was (39.39 ± 0.07) °C for the *inlet* velocities of 0 cm/s and (32.34 ± 0.85) °C for 0.104 cm/s, respectively. The maximum water temperature was located in the channel between the central tube wall and the inner annular tube wall. The average water temperature was 27.76 °C for the *inlet* velocities of water from 0.209 to 4.179 cm/s. It could be concluded that the SAMOP core cooling system using natural convection was capable and optimal to prevent the fuel overheating even the value of water *inlet* velocity of 0.209 cm/s.

1. Introduction

Technisium-99 metastable (^{99m}Tc) is one of the most important isotopes for medical diagnostics which is obtained from the decay of ^{99}Mo. ^{99m}Tc decays to ^{99g}Tc with a half-life of 6 hours by emitting a 140 keV gamma rays [1]. According to The Nuclear Energy Agency (NEA) from France, the consumption of ^{99m}Tc isotope all the world is approximately 10 kCi every week [2]. It is an urgent need for constant supply of ^{99m}Tc by increase production of ^{99}Mo. The world research reactor facility as a ^{99}Mo producer is *Aqueous Homogeneous Reactors* (AHR). AHR has a potentially advantageous method of ^{99}Mo production because all of the ^{99}Mo can be extracted from the fuel solution without shutdown a reactor [3].

Indonesia is encouraged to carry out a safety study of the Subcritical Assembly for ^{99}Mo Production (SAMOP), one type of AHR. One of the safety aspects that must be reviewed is thermal-hydraulics. It is required to evaluate the heat removal system in order to exhibit that sufficient cooling capacity to prevent fuel solution overheating. The safety limit of the reactor operation refers to the safety regulations in operating the TRIGA Mark II reactor by General Atomic. The regulation states that to limit the ^{16}N radiation level on the surface of the tank, it is not permitted the fuel boiled [4].

According to the previous results of the study, it has been reported that the SAMOP at PSTA-BATAN has been designed having a power range from 563.2 to 860.8 W. It should be noted that the power of 563.2 W is equivalent to heat of 398.39 °C [5]. In other result SAMOP reactor core cooling

IOP Conf. Series: Materials Science and Engineering **578** (2019) 012088 doi:10.1088/1757-899X/578/1/012088

system using the design of forced convection is able to provide sufficient cooling capacity to prevent fuel solution overheating [6]. In order for the SAMOP not damaged, the core should be cooled.

2. Numerical Method

This research was carried out by making the SAMOP geometry using GAMBIT 2.3.6 software and analyzing thermal-hydraulics parameters using Computational Fluids Dynamic (CFD) method based on FLUENT 6.3.26 software. CFD techniques are very reliable and have been used for thermal-hydraulics nuclear reactor systems [7,8]. The geometry design has been made according to the SAMOP geometry developed by PSTA-BATAN (Figure 1). The core consists of an aqueous uranyl nitrate. The fuel is marked with orange color and that of water with white color at figure 1. The uranium concentration was 300 g U/L and the core volume was 22,382 cm^3. The uranyl nitrate volume was 226 cm^3. The water flow rate was set up 0 to 0.522 cm/s and 2.090 to 4.179 cm/s.

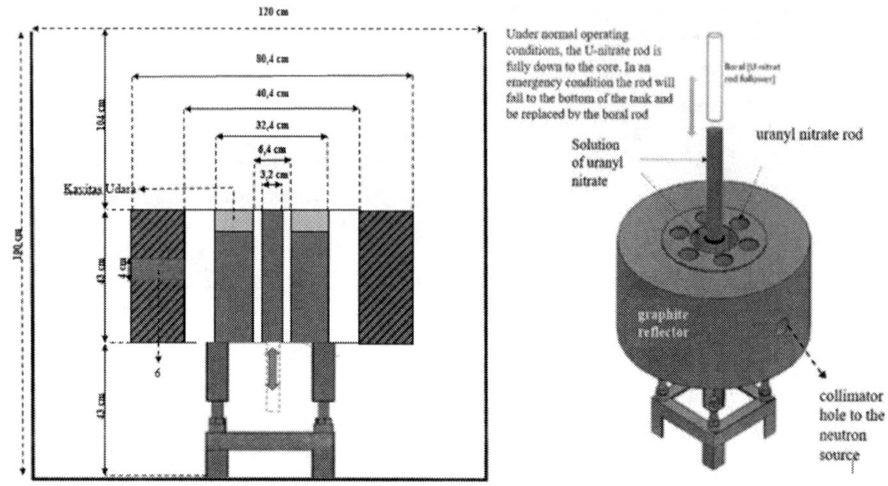

Figure 1. The core design scheme of the SAMOP facility [5].

The main reactor core parameters are written in Table 1.

Table 1. The SAMOP reactor core parameters [5].

Parameter	Value
Fuel solution	$UO_2(NO_3)_2$
Uranyl nitrate tube diameter	3.7 cm
Reflector	Graphite
Reactor vessel	Stainless steel
Thermal power (kW)	1

In this simulation, it has been assumed that the fluid flow was laminar, steady and incompressible conditions. The characteristic of the cooling system was obtained by calculating the Richardson number (Ri), [9]

$$Ri = \frac{Gr}{Re^2} \quad (1)$$

$$Gr = \left\{\frac{\beta\, g\, d^3}{v^2}\right\}(Twall - Tbulk) \quad (2)$$

$$Re = \frac{\rho\, v\, d}{\mu} \quad (3)$$

β is liquid expansion coefficient (/°K), g is gravitational acceleration (m/s^2), d is gravitational acceleration (m/s^2), d is pipe diameter (m), v is kinematic viscosity (m^2/s), T is temperature (T), ρ is water density (kg/m^3), υ is fluid flow rate (m/s), μ is fluid viscosity (kg/m s), Gr is Grashof number and Re is Reynolds number.

3. Result and Discussion

The SAMOP geometry in the XY field has been successfully modeled using the GAMBIT software. The results are shown in Figure 2.a and the geometric mesh in Figure 2.b where the water is marked with yellow area. The geometry on FLUENT is shown in Figure 3.

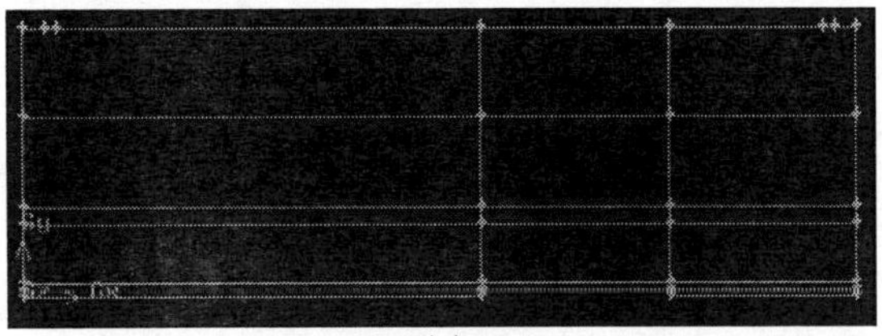

(a)

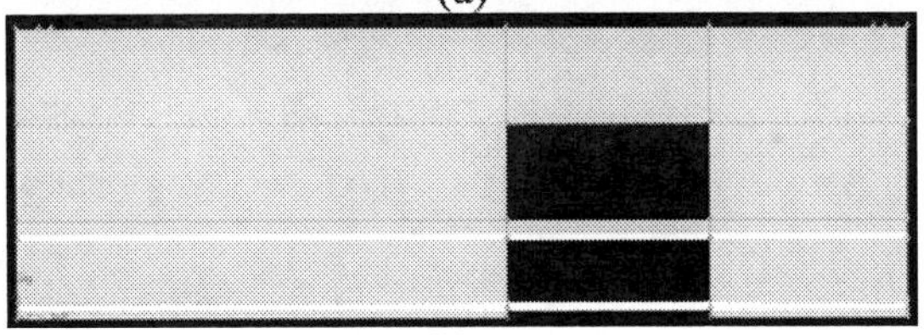

(b)

Figure 2. The SAMOP Geometry [10]

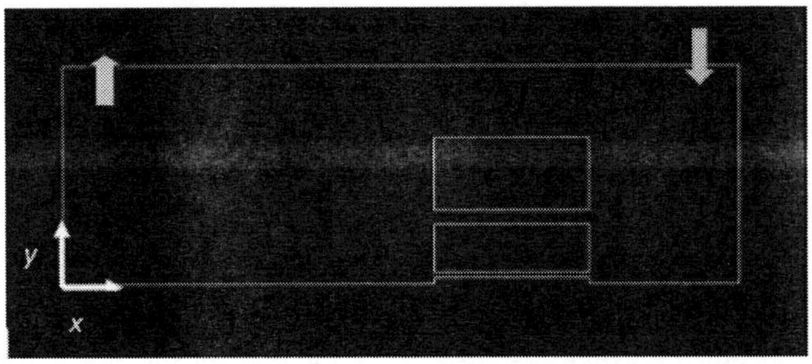

Figure 3. SAMOP geometry on FLUENT [10]

Figure 3 explain that the water as the cooling enters from the inlet flows to the left along the negative x-axis and exits through the outlet. Figure 3 is equivalent to Figure 4.

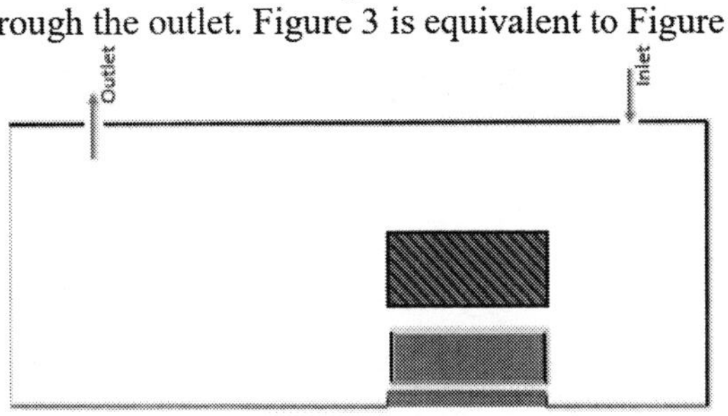

Figure 4. SAMOP Geometry [10]

In Figure 4 it has been shown the parts of the geometry being analyzed. The inlet and outlet can be seen in Figures 3 and 4. In Figure 4, the orange color is fuel of uranyl nitrate solution, a graphite reflector with black striped, and water with white color. The color of the line as a symbol for the parts of the geometry that has been made, namely:

- Red : central tube wall

IOP Conf. Series: Materials Science and Engineering **578** (2019) 012088 doi:10.1088/1757-899X/578/1/012088

- Green inner annular tube wall
- Yellow : outer annular tube wall
- Dark blue : rotary axis
- Light blue : surface of the tank
- Black : SAMOP tank

The results of the maximum and minimum temperature water are presented in Table 2. The maximum water temperature was located in the channel between the central tube wall and the inner annular tube wall. From the Table 2, it can be seen that the operation of the reactor is safe because the maximum water temperature on the core is only (39.39 ± 0.07) °C. Even when the water velocity is 0, the reactor safety limit is still fulfilled. This means that the temperature of the fuel is below the permissible temperature limit under normal operating conditions.

Table 2. The maximum and minimum temperature of several variations

Water velocities (cm/s)	Maximum Temperature (°C)	Minimum Temperature (°C)
0	39.39 ± 0.07	34.24 ± 0.07
0.104	32.34 ± 0.85	27.00 ± 0.85
0.209	29.78 ± 0.28	26.99 ± 0.28
0.313	29.76 ± 0.19	26.99 ± 0.19
0.418	29.76 ± 0.19	26.99 ± 0.19
0.522	29.76 ± 0.19	26.99 ± 0.19
2.090	29.68 ± 0.17	26.99 ± 0.17
3.135	29.62 ± 0.16	26.99 ± 0.16
4.179	29.56 ± 0.15	26.99 ± 0.15

The simulation results of the SAMOP characteristic of the cooling system are written in Table 3. It can be seen that the obtained Ri numbers are greater than 1 for the velocities of water from 0.104 - 1.045 cm/s so it can be concluded as natural convection. Combined convection must be used when the water inlet velocity is 2.090 cm/s while the force starts at 3.135 cm/s and above. From the results it can be seen the design of SAMOP cooling by natural convection has been able to transfer of heat due to fission reactions.

Table 3. The characteristics of cooling system in SAMOP for several variations

Water velocities (cm/s)	Richardson Number	Natural/Combined/ Forced
0.104	68 ± 6	Natural
0.209	127 ± 6	Natural
0.313	88 ± 3	Natural
0.418	56 ± 2	Natural
0.522	568 ± 16	Natural
2.090	1.07 ± 0.03	Combined
3.135	0.44 ± 0.01	Forced
4.179	0.24 ± 0.01	Forced

4. Conclusion

The calculation results show that the SAMOP heat removal systems. designed natural convection is capable and optimal to prevent the fuel overheating. The maximum water temperature was (39.39 ± 0.07) °C for the *inlet* velocities of 0 cm/s and (32.34 ± 0.85) °C for 0.104 cm/s, respectively. The maximum water temperature was located in the channel between the central tube wall and the inner annular tube wall.

5. Reference

[1] Avagyan R, Avetisyan A, Kerobyan I and Dallakyan R 2014 *Nuclear Med. Bio.* 705.
[2] IAEA 2015 *Technical Reports* 478.
[3] Daniel MP, et al. 2015 *J.Thermodyn.* **15(11)**.
[4] Supriyanto AA 2016 *Jurnal Simetris,* **7(2)** 559-566.
[5] Khakim A 2018 *Seminar Keselamatan Nuklir,* **PB14** 458-463.
[6] Syarip and Nugroho THS 2017 *Jurnal Pengembangan energi Nuklir*, **19(1)** 25.
[7] Rouch HA, et al. 2014 *Ann. Nucl. Energy* **64** 449.
[8] Sudjatmi KA 2013 *Jurnal Teknologi Reaktor Nuklir* **15(1)** 51.
[9] Mochizuki H 2009 *Thermal-Hydroulics in Nuclear Reactor*
[9] Syarip and Sarjono Y 2006 *Seminar Keselamatan Nuklir.* 1412-3258
[10] Lestari DW, Suharyana, Riyatun and Hidayanti D 2018 *Prosiding Seminar Keselamatan Nuklir,* **PB16** 469-475.

IOP Conf. Series: Materials Science and Engineering **578** (2019) 012089 doi:10.1088/1757-899X/578/1/012089

Analytical solution of the Bohr-Mottelson equation in the presence of minimal length for yukawa potential using hypergeometric method.

S N Fatimah*[1], A Suparmi[1], C Cari[1] and I L Elviyanti[1]**

[1]Physics Department, Graduate Program, Sebelas Maret University, Surakarta 57126 Indonesia

E-mail: * fatmalida@gmail.com, **soeparmi@staff.uns.ac.id

Abstract. The relativistic energy and wave function of the Bohr-Mottelson equation in the presence of minimal length are investigated by the hypergeometric method. The Bohr-Mottelson equation in the presence of minimal length is influenced by Yukawa potential. The energy relativistic is calculated by using Matlab software, and the wave function is expressed in the Hypergeometric term. The existence of minimal length caused increased of the energy relativistic of Bohr-Mottelson equation.

1. Introduction

The behavior of nucleus is described using the Bohr-Mottelson equation [1-6]. The Bohr-Mottelson equation is studied using potentials such as Infinite Square well potential [1,2], Eckart potential [3], Kratzer potential [4] and Davidson potential [5]. The method is used to solve the Bohr-Mottelson equation such as Supersymmetric Quantum Mechanics (SUSYQM) [5], Nikiforov-Uvarov [3] and Asymptotic Iteration Method (AIM) [6].

In mechanics quantum commutation relations between position and momentum operators is explained by Heisenberg Uncertainty Principle. At Heisenberg Uncertainty Principle is influenced by gravity quantum, Heisenberg Uncertainty Principle with additional a small constant as follow [1,2]

$$[X,P] \geq i\hbar\left(1+\alpha(\Delta P)^2\right) \tag{1}$$

Here α is parameter minimal length that has a very small positive value, the X and P are position and momentum, respectively. The equation (1) is General Uncertainty Principle or minimal length [1,2].

This paper is studied the Bohr-Mottelson equation for rigid deformed nucleus in the presence of minimal length for Yukawa potential. To solve the Bohr-Mottelson equation in the presence of minimal length is used Hypergeometric method. By using Hypergeometric method, the relativistic energy and wave function is obtained. The organized this paper as follows, the Bohr-Mottelson Equation in the presence of minimal length and Hypergeometric method is described in section 2. The result and discussion in section 3 and the last section is conclusion.

2. Experimental

2.1. The Bohr-Mottelson equation in precence of minimal length

The General Uncertainty Principle (GUP) modify of Heisenberg Uncertainty Principle at influenced by gravity quantum. At energy much lower than the Planck mass, the equation (1) is reduced to Heisenberg Uncertainty Principle. By using the deformed canonical commutation relation in equation (1) becomes [2],

$$\hat{X}_i = \hat{x}_i \tag{2}$$

$$\hat{P}_i = \left(1+\alpha \hat{p}^2\right)\hat{p}_i \tag{3}$$

By applying $p^2 = -\hbar^2\Delta$ in equation (3), so we get

$$P^2 = -\frac{\hbar^2}{2B_m}(1-2\alpha\Delta)\Delta \tag{4}$$

where Δ is Laplacian operator. The Laplacian operator for nucleus is given

$$\Delta = \frac{1}{\sqrt{g}}\sum_{i,j}\frac{\partial}{\partial q_i}\sqrt{g}\,g_{ij}^{-1}\frac{\partial}{\partial q_j} \tag{5}$$

with g is determinant and g_{ij}^{-1} is inverse of the matrix g_{ij} that has has three degrees of freedom $: q_1 = \phi, q_2 = \theta, q_3 = \beta$, so we obtain Laplacian operator for nucleus

$$\Delta = \left[\frac{1}{\beta^2}\frac{\partial}{\partial\beta}\beta^2\frac{\partial}{\partial\beta} + \frac{1}{3\beta^2}\left[\frac{1}{\sin\theta}\frac{\partial}{\partial\theta}\sin\theta\frac{\partial}{\partial\theta} + \frac{1}{\sin^2\theta}\frac{\partial^2}{\partial\varphi^2}\right]\right] \tag{6}$$

The Hamiltonian equation which is expressed

$$H = T + V(\beta) = \frac{P^2}{2B_m} + V(\beta) \tag{7}$$

where $V(\beta)$ is potential energy in β function and B_m is a mass parameter. Then, equation (4) is inserted in equation (7), we get

$$\left[-\frac{\hbar^2}{2B_m}\Delta + \frac{\alpha\hbar^4}{B_m}\Delta^2 + V(\beta,\theta,\phi) - E\right]\chi(\beta,\theta,\phi) = 0 \tag{8}$$

The equation (8) is Bohr-Mottelson equation in the presence of minimal length. In the case of Bohr-Mottelson equation without the minimal length is setting $\alpha_{ML} = 0$ [12] for equation (8), so obtain square term is given as [2],

$$\Delta^2 = \frac{4B_m^{\ 2}}{\hbar^4}\left(V(\beta) - E^0\right)^2 \tag{9}$$

Equations (6) and (9) are inserted in equation (8) and multiplied by $-\frac{2B_m}{\hbar^2}$ and $\hbar = 1$ (natural unit), we obtain

$$\left[\left[\begin{array}{l}\frac{1}{\beta^2}\frac{\partial}{\partial\beta}\beta^2\frac{\partial}{\partial\beta}+\\ \frac{1}{3\beta^2}\left(\begin{array}{l}\frac{1}{\sin\theta}\frac{\partial}{\partial\theta}\sin\theta\frac{\partial}{\partial\theta}\\ +\frac{1}{\sin^2\theta}\frac{\partial^2}{\partial\varphi^2}\end{array}\right)\end{array}\right]+2B_m\left(\begin{array}{l}4\alpha_{ML}B_m\left(V(\beta)-E^0\right)^2\\ +\left(V(\beta)-E\right)\end{array}\right)\right]\chi(\beta,\theta,\varphi)=0 \tag{10}$$

By using the separation variable method $\chi(\beta,\theta,\varphi)=R(\beta)\Theta(\theta)\Phi(\varphi)$ in equation (10), we have,

$$-\left(\frac{1}{\Phi(\varphi)}\frac{1}{\sin^2\theta}\frac{\partial^2\Phi(\varphi)}{\partial\varphi^2}+\frac{1}{\Theta(\theta)}\frac{1}{\sin\theta}\frac{\partial}{\partial\theta}\sin\theta\frac{\partial\Theta(\theta)}{\partial\theta}\right)=\lambda \tag{11}$$

Equation (11) is Euler angles part of Bohr-Mottelson equation in presence of minimal length. The $\beta-$ part of Bohr-Mottelson equation in presence of minimal length is given,

$$\left\{\begin{array}{l}\frac{1}{\beta^2}\frac{\partial}{\partial\beta}\beta^2\frac{\partial R(\beta)}{\partial\beta}+2B_m\left(E-V(\beta)\right)R(\beta)\\ -8B_m{}^2\alpha_{ML}\left(E_0^2-2E_0V(\beta)+V^2(\beta)\right)R(\beta)\end{array}\right\}=\frac{\lambda}{3\beta^2}R(\beta) \tag{12}$$

By applying $R(\beta)=Z(\beta)/\beta$ and $\lambda=L(L+1)$ in equation (12) so we have,

$$\left\{\begin{array}{l}\frac{d^2F(\beta)}{d\beta^2}-\frac{L(L+1)}{3\beta^2}Z(\beta)\\ +2B_m\left(E-V(\beta)\right)Z(\beta)\\ -8B_m{}^2\alpha_{ML}\left(E^2-2EV(\beta)+V^2(\beta)\right)Z(\beta)\end{array}\right\}=0 \tag{13}$$

The equation (13) is Bohr-Mottelson equation for a $\beta-$part in presence of minimal length effect for rigid deformed nucleus case.

2.2. Hypergeometric Method

The second-order differential equation of hypergeometric function as follow [7,8],

$$m(1-m)\frac{d^2\Psi}{dm^2}+\left(c-(a+b+1)m\right)\frac{d\Psi}{dm}-ab\Psi=0 \tag{14}$$

The energy eigenvalue is obtained from the condition in equation (14) , [7,8]

$$a=-n \text{ or } b=-n \tag{15}$$

where n=0,1,2,3.... Equation (15) can be finite series of polynomials of rank n by equation (14) . The solution of a wave function is given as

$$\Psi(m)={}_2F_1(a,b,c,\mathrm{m})=\sum_{k=0}\frac{(a)_k(b)_k}{(c)_k n!}m^k=1+\frac{ab}{c}m+\frac{a(a+1)b(b+1)}{c(c+1)}\frac{m^2}{2!}+\ldots \tag{16}$$

By applying the suitable variable change in equation and reduced to standard hypergeometri equation, we get energy eigenvalue and wave function [7,8].

3. Results and Discussion

The Yukawa potential is used to describe nucleon-nucleon interactions in meson theory, the general Yukawa potential as follow [9,10],

$$V(\beta)=\frac{-V_0 e^{-\eta\beta}}{\beta} \tag{17}$$

where V_0 is potential depth and ϖ is range of nucleon force. The function of $1/\beta$ is approximated by

$$\frac{1}{\beta}=\frac{2\eta e^{-\eta\beta}}{1-e^{-2\eta\beta}} \tag{18}$$

$$\frac{1}{\beta^2}=\frac{2\eta^2}{\sinh^2\eta\beta} \tag{19}$$

Then the Bohr-Mottelson in equation (13) is inserted by equations (18) and (19), we get

$$\frac{d^2 Z(\beta)}{d\beta^2}-\left[\begin{array}{l}\left(\frac{2\eta^2 L(L+1)}{3}+2\alpha B_m^{\ 2}V_o^{\ 2}\right)\frac{Z(\beta)}{\sinh^2\eta\beta}\\ -\left(B_mV_o-8\alpha B_m^{\ 2}E^0V_o+4\alpha B_m^{\ 2}V_o^{\ 2}\right)\coth\eta\beta\\ -\left(2B_mE-8\alpha B_m^{\ 2}E^{02}+8\alpha B_m^{\ 2}V_oE^0-4\alpha B_m^{\ 2}V_o^{\ 2}-B_mV_o\right)\end{array}\right]Z(\beta)=0 \tag{20}$$

Equation (20) is the Bohr-Mottelson equation in presence of minimal length for Yukawa potential. The equation (20) is reduced becomes

$$\frac{d^2Z(\beta)}{d\beta^2}-\left[\frac{v(v-1)}{\sinh^2\eta\beta}-2q\coth\eta\beta+k^2\right]Z(\beta)=0 \tag{21}$$

Equation (21) is differential equation with,

$$v(v-1)=\left(\frac{2\eta^2L(L+1)}{3}+2\alpha B_m^{\ 2}V_o^{\ 2}\right) \tag{22}$$

$$2q=-\left(B_mV_o-8\alpha B_m^{\ 2}E^0V_o+4\alpha B_m^{\ 2}V_o^{\ 2}\right) \tag{23}$$

$$-k^2=\left(2B_mE-8\alpha B_m^{\ 2}E^{02}+8\alpha B_m^{\ 2}V_oE^0-4\alpha B_m^{\ 2}V_o^{\ 2}-B_mV_o\right) \tag{24}$$

By reducing equation (21) to hypergeometric differential equation type by using $\coth\eta\beta=1-2y$, we get

$$y(1-y)\frac{d^2Z(\beta)}{dy^2}+(1-2y)\frac{dZ(\beta)}{dy}+\left[v'(v'-1)-\frac{4\alpha_H^2}{4y}-\frac{4\beta_H^2}{4(1-y)}\right]Z(\beta)=0 \tag{25}$$

with,

$$\frac{-2q'+k'^2}{\eta^2}=4\alpha_H^2 \tag{26}$$

$$\frac{-2q'+k'^2}{\eta^2}=4\beta_H^2 \tag{27}$$

$$v'\left(v'-1\right)=\frac{v\left(v-1\right)}{\eta^2} \tag{28}$$

by applying the new wave function is given,

$$Z(\beta)=y^{\alpha_H}(1-y)^{\beta_H}f(y) \tag{29}$$

in equation (25), we obtain

$$\begin{bmatrix} y(1-y)\frac{f^2(y)}{dy^2}+\left[(2\alpha_H+1)-(2\alpha_H+2\beta_H+2)y\right]\frac{f(y)}{dy} \\ +\left[v'\left(v'-1\right)-(\alpha_H+\beta_H)(\alpha_H+\beta_H+1)\right]f(y) \end{bmatrix}=0 \tag{30}$$

The equation (30) is hypergeometric differential equation that obtain parameter hypergeometric, yield

$$a=\alpha_H+\beta_H-\left(v'-1\right),\ b=\alpha_H+\beta_H+v',\ c=2\alpha_H+1 \tag{31}$$

By inserting equations (22)-(24),(26)-(28), and (31), we obtain

$$E=\frac{\eta^2}{2B_m}\left[\left(\sqrt{\left(\frac{2L(L+1)}{3}+\frac{2\alpha B_m{}^2V_o^2}{\eta^2}+\frac{1}{4}\right)}-\frac{1}{2}-n\right)^2-\left(\frac{\left(\frac{-\left(B_mV_o-8\alpha B_m{}^2E^0V_o+4\alpha B_m{}^2V_o^2\right)}{\eta^2}\right)^2}{\left(\sqrt{\left(\frac{2L(L+1)}{3}+\frac{2\alpha B_m{}^2V_o^2}{\eta^2}+\frac{1}{4}\right)}-\frac{1}{2}-n\right)^2}\right)\right]+4\alpha B_mE^{02}-4\alpha B_mV_oE^0+2\alpha B_mV_o^2+V_o \tag{32}$$

The equation (32) is relativistic energy equation of Bohr-Mottelson in the presence of minimal length for Yukawa potential. To get relativistic energy, we calculated numerically equation by using Matlab software. The result is shown in Table 1.

Table 1. The relativistic energy the presence of minimal length for ${}^{132}_{54}Xe$, $\eta=0.5$,n=0, and $V=0.005$

l	E (eV)					
	$\alpha=0$	$\alpha=0.002$	$\alpha=0.004$	$\alpha=0.006$	$\alpha=0.008$	$\alpha=0.01$
0	0	-139.216	-35.6483	-16.2225	-9.34032	-6.11683
2	0.000852	0.000854	0.000855	0.000857	0.000858	0.00086
4	0.010209	0.010306	0.010402	0.010497	0.010593	0.010688
6	0.022921	0.023398	0.023874	0.024349	0.024822	0.025295
8	0.040441	0.041999	0.043554	0.045107	0.046658	0.048208
10	0.062944	0.066839	0.07073	0.074619	0.078504	0.082386

Table 1 show that energy relativistic without minimal length a lower that energy relativistic with minimal length. The relativistic energy is increased when the minimal length parameter and angular momentum quantum number (l) are increased.

Then, to obtain the wave function use equations (16), (29), and $\coth\eta\beta=(1-2z)$, we get

$$Z(\beta)=\frac{(1-\coth\eta\beta)^{\alpha_H}}{2}\frac{(1+\coth\eta\beta)^{\beta_H}}{2}\ {}_2F_1(a,b,c,y) \tag{33}$$

And by applying equation (31) in equation (33), we obtain wave function in Table 2.

Table 2. The wave function of the Bohr-Mottelson equation in the presence of minimal length

n	$Z(\beta)$
0	$Z(\beta)=\frac{(1-\coth\eta\beta)^{\alpha_H}}{2}\frac{(1+\coth\eta\beta)^{\beta_H}}{2}$
1	$Z(\beta)=\frac{(1-\coth\eta\beta)^{\alpha_H}}{2}\frac{(1+\coth\eta\beta)^{\beta_H}}{2}\left[1+\frac{(-n)(\alpha_H+\beta_H+v)\left(\frac{1-\coth\eta\beta}{2}\right)}{(2\alpha_H+1)}\right]$

Table 2 is wave function of Bohr-Mottelson equation in presence of minimal length for n=0 and n=1, respectively. The value of wave function depends on the value of parameter hypergeometric.

4. Conclusion

The Bohr-Mottelson equation is investigated in the presence of minimal length effect for Yukawa potential by using Hypergeometric method. From the investigation of the Bohr-Mottelson equation is obtained relativistic energy and wave function in minimal length effect. From the result can be conclude that relativistic energy of the Bohr-Mottelson equation in the presence of minimal length for Yukawa potential increase duo to increase minimal length parameter.

Acknowledgement

This research was partly supported by Ministry of Research, Technology and Higher Education with contract no. 089/SP2H/LT/DRPM/2018.

References

[1] M. Alimohammadi and H. Hassanabadi 2017 *Nucl. Phys A* **957** 439-449.
[2] M. Chabab, A.l Batoul, A. Lahbas, and M. Oulne 2016 *Phys. Lett. B* **758** 212-218.
[3] L. Naderia and H. Hassanabadi 2016 *Eur. Phys. J. Plus* **131**(5) 133.
[4] M. Alimohammadi, H Hassanabadi and S Zare 2017 *Nucl. Phys A* **960** 78-89.
[5] D. Bonatsos, P. E. Georgoudis, D. Lenis, N. Minkov, and C. Quesne 2011 *Phys. Rev. C* **83** 044321.
[6] M . Chabab, A. Lahbas, and M. Oulne 2015 *Eur. Phys. J. Plus* **51**(131).
[7] Suparmi 2011 *Mekanika Kuantum II* (Surakarta: FMIPA UNS ISBN 978-602-99344-2-7).
[8] I L Elviyanti , A Suparmi and C Cari 2017 *J. Phys.: Conf. Ser.* **909** 012003.
[9] H. Yukawa 1937 *Proc. Phys-Math. Soc. Japan* **19** 1084–1093.
[10] J P Edwards, U Gerber, C Schubert, MA Trejo, and A Weber 2017 *Prog. Theor. Exp. Phys* **2017**(8) 083A01.

Alternative solution of the bohr-mottelson equation in minimal length effect for cotangent potential using the hypergeometric method

A Suparmi*[1], S N Fatimah[1], C Cari[1], and I L Elviyanti[1]

[1]Physics Department, Graduate Program, Sebelas Maret University, Surakarta 57126 Indonesia

E-mail: soeparmi@staff.uns.ac.id

Abstract. The rigid deformed nucleus of minimal length effect is investigated using the Bohr-Mottelson equation that influenced by Cotangent potential. The Bohr-Mottelson equation of minimal length effect is solved by Hypergeometric method to obtain the energy spectrum and wave function. The energy spectrum is calculated by using Matlab software. The wave function is expressed in the Hypergeometric term. The result shown that, the energy spectrum is increased caused existence of minimal length effect.

1. Introduction

The Bohr-Mottelson equation is used to explain the behavior of nucleus such as the rigid deformed nucleus [1-3]. The rigid deformed nucleus occur nucleus rotation motion at low excitation energy [1-3]. The Bohr-Mottelson for various potential has been solved using Super Symmetric Quantum (SUSY QM) for Davidson potential [4], Nikiforov-Uvarov for Eckart potential [5] and Asymptotic Iteration Method (AIM) for Hulthen and Ring Shape potential [1].

The concept of minimal length is connected by commutation relations between position and momentum operators in Heisenberg Uncertainty Principle. The minimal length occur at Heisenberg Uncertainty Principle is influenced by gravity quantum which is called General Uncertainty Principle (GUP) [1-3,6-7]. The General Uncertainty Principle modify Heisenberg Uncertainty Principle with additional a small constant [3].

In this paper, we solved Bohr-Mottelson equation in minimal length effect for Cotangent potential using Hypergeometric method. By using Hypergeometric method, we get energy spectrum and wave function. The Cotangent potential is used to explain nucleon excitation [8-9]. This paper include of 4 sections, there are Bohr-Mottelson Equation in minimal lengt effect and hypergeometric method in section 2, the result and discussion in section 3 and the last section is conclusion.

2. Experimental

2.1. The Bohr-Mottelson equation in minimal length effect

The general canonical commutation between position and momentum is expressed [6-7]

$$\left[X,P\right]\geq i\hbar \tag{1}$$

where X is a position, P is a corresponding momentum. Then, the general canonical commutation between position and momentum is influenced by quantum gravity, it is becomes [1-3,6-7],

$$\left[X,P\right]\geq i\hbar\left(\quad P)^{2}\right) \tag{2}$$

The equation (2) is called General Uncertainty Principle, where α was a minimal length parameter that has very small positive values. The uncertainty relation is caused by commutation relation. The equation (2) can be reduced becomes [1-3,6-7]

$$\hat{X}_i=\hat{x}_i \tag{3}$$

$$\hat{P}_i=\left(1+\alpha\hat{p}^2\right)\hat{p}_i \tag{4}$$

Then, equation (4) can be written [2,3]

$$P^2=-\frac{\hbar}{2B_m}(1-2\alpha\Delta)\Delta \tag{5}$$

where Δ is Laplacian operator for nucleus that has three degrees of freedom: $q_1=\phi, q_2=\theta, q_3=\beta$, the Laplacian operator as follow [2,3]

$$\Delta=\frac{1}{\sqrt{g}}\sum_{i,j}\frac{\partial}{\partial q_i}\sqrt{g}g_{ij}^{-1}\frac{\partial}{\partial q_j} \tag{6}$$

with g and g_{ij}^{-1} are determinant and inverse of the matrix g_{ij}, respectively. We get Laplacian operator, is given as [2,3]

$$\Delta=\left[\frac{1}{\beta^2}\frac{\partial}{\partial\beta}\beta^2\frac{\partial}{\partial\beta}+\frac{1}{3\beta^2}\left[\frac{1}{\sin\theta}\frac{\partial}{\partial\theta}\sin\theta\frac{\partial}{\partial\theta}+\frac{1}{\sin^2\theta}\frac{\partial^2}{\partial\varphi^2}\right]\right] \tag{7}$$

Then equation (5) is inserted in Hamiltonian equation which is expressed,

$$H=T+V(\beta)=\frac{P^2}{2B_m}+V(\beta) \tag{8}$$

where P is momentum operator, $V(\beta)$ is potential energy in β function and B_m is a mass parameter. We obtain,

$$\left[-\frac{\hbar}{2B_m}\Delta+\frac{\hbar}{B_m}\Delta+V(\beta,\theta,\phi)-E\right]\Psi(\beta,\theta,\phi)=0 \tag{9}$$

The equation (9) is Bohr-Mottelson equation in minimal length effect. In the case of Bohr-Mottelson equation without the minimal length effect with $\alpha_{ML}=0$ [12] for equation (9), so yields square term is given as [2],

$$\Delta^2=\frac{4B_m^{\ 2}}{\hbar}\left(V(\beta)-E^0\right)^2 \tag{10}$$

Equations (7) and (10) are inserted in equation (9) and multiplied by $-\frac{2B_m}{\hbar}$ and $\hbar$ (natural unit), is yields

$$\left[\left[\begin{matrix}\frac{1}{\beta^2}\frac{\partial}{\partial\beta}\beta^2\frac{\partial}{\partial\beta}+\\ \frac{1}{3\beta^2}\left(\begin{matrix}\frac{1}{\sin\theta}\frac{\partial}{\partial\theta}\sin\theta\frac{\partial}{\partial\theta}\\ +\frac{1}{\sin^2\theta}\frac{\partial^2}{\partial\varphi^2}\end{matrix}\right)\end{matrix}\right]+2B_m\left(\begin{matrix}4\alpha_{ML}B_m\left(V(\beta)-E^0\right)^2\\ +(V(\beta)-E)\end{matrix}\right)\right]\Psi(\beta,\theta,\varphi)=0 \tag{11}$$

By setting $\Psi(\beta,\theta,\varphi)=R(\beta)\Theta(\theta)\Phi(\varphi)$ which is the separation variable method that used to solve equation (11), we have Euler angles part of Bohr-Mottelson Hamiltonian with minimal length,

$$-\left(\frac{1}{\Phi(\varphi)}\frac{1}{\sin^2\theta}\frac{\partial^2\Phi(\varphi)}{\partial\varphi^2}+\frac{1}{\Theta(\theta)}\frac{1}{\sin\theta}\frac{\partial}{\partial\theta}\sin\theta\frac{\partial\Theta(\theta)}{\partial\theta}\right)=\lambda \tag{12}$$

and $\beta-$ part of Bohr-Mottelson Hamiltonian with minimal length,

$$\left\{\begin{matrix}\frac{1}{\beta^2}\frac{\partial}{\partial\beta}\beta^2\frac{\partial R(\beta)}{\partial\beta}+2B_m\left(E-V(\beta)\right)R(\beta)\\ -8B_m{}^2\alpha_{ML}\left(E_0^2-2E_0V(\beta)+V^2(\beta)\right)R(\beta)\end{matrix}\right\}=\frac{\lambda}{3\beta^2}R(\beta) \tag{13}$$

By applying $R(\beta)=F(\beta)/\beta$ and $\lambda=L(L+1)$ in equation (13) so we have,

$$\left\{\begin{matrix}\frac{d^2F(\beta)}{d\beta^2}-\frac{L(L+1)}{3\beta^2}F(\beta)\\ +2B_m\left(E-V(\beta)\right)F(\beta)\\ -8B_m{}^2\alpha_{ML}\left(E^2-2EV(\beta)+V^2(\beta)\right)F(\beta)\end{matrix}\right\}=0 \tag{14}$$

The Bohr-Mottelson equation for a $\beta-$part in a minimal length effect for rigid deformed nucleus case is expressed by equation (14).

2.2. *Hypergeometric method*

The second-order differential equation of hypergeometric function as follow [10,11],

$$y(1-y)\frac{d^2\Phi}{dy^2}+\left(c-(a+b+1)y\right)\frac{d\Phi}{dy}-ab\Phi=0 \tag{15}$$

The energy eigenvalue is obtained from the condition in equation (15) , [10,11]

$$a=-n \text{ or } b=-n \tag{16}$$

where n=0,1,2,3.... Equation (16) can be finite series of polynomials of rank n by equation (15) . The solution of a wave function is given as

$$\Phi(y)={}_2F_1(a,b,c,y)=\sum_{m=0}\frac{(a)_m(b)_m}{(c)_m n!}y^m=1+\frac{ab}{c}y+\frac{a(a+1)b(b+1)}{c(c+1)}\frac{y^2}{2!}+\ldots \tag{17}$$

By applying the suitable variable change in equation and reduced to standard hypergeometri equation, we get energy eigenvalue and wave function [10,11].

3. Results and Discussion

The Bohr-Mottelson in equation (14) is inserted by Cotangent potential as follow [8-10],

$$V(\beta)=V_o\cot\delta\beta+V_1 \tag{18}$$

where V_o and V_1 are potential constants, and δ is a range of potential. The visualization of Cotangent potential is given figure 1.

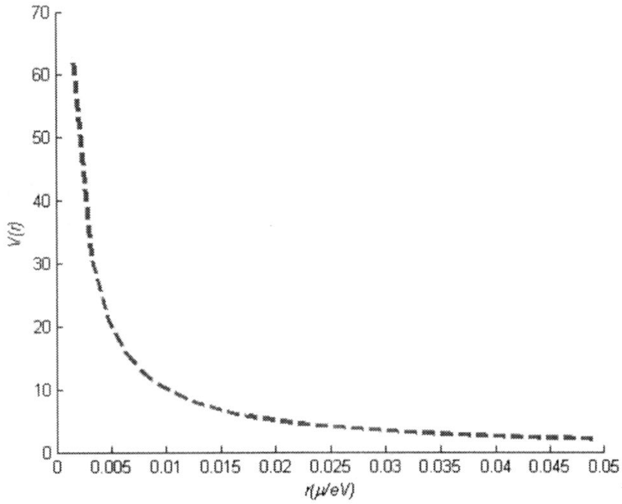

Figure 1. The visualization of cotangent potential

Figure 1 is a visualization of cotangent potential in r function that approximately from 0 *fm* until 14 *fm*. We can see, The Cotangent potential comes to infinity value in a very small value of r, while for the higher value of r, the cotangent potential incline to be constant. Then, The Bohr-Mottelson in equation (14) becomes,

$$\frac{d^2F(\beta)}{d\beta^2}-\left[\begin{array}{l}\left(\frac{\delta^2L(L+1)}{3}+8B_m{}^2\alpha V_o{}^2\right)\frac{1}{\sin^2\delta\beta}\\ -\left(16B_m{}^2\alpha V_oE_0-16B_m{}^2\alpha V_2-2B_mV_o\right)\cot\delta\beta\\ -\left(\begin{array}{l}16B_m{}^2\alpha V_1E_0+8B_m{}^2\alpha V_o{}^2\\ -8B_m{}^2\alpha V_1^2-8B_m{}^2\alpha E_0{}^2-2B_mV_1+2B_mE\end{array}\right)\end{array}\right]F(\beta)=0 \quad (19)$$

by setting,

$$v(v+1)=\left(\frac{\delta^2L(L+1)}{3}+8B_m{}^2\alpha V_o{}^2\right) \quad (20)$$

$$2q=\left(16B_m{}^2\alpha V_oE_0-16B_m{}^2\alpha V_2-2B_mV_o\right) \quad (21)$$

$$-\kappa^2=\left(\begin{array}{l}16B_m{}^2\alpha V_1E_0+8B_m{}^2\alpha V_o{}^2\\ -8B_m{}^2\alpha V_1^2-8B_m{}^2\alpha E_0{}^2-2B_mV_1+2B_mE\end{array}\right) \quad (22)$$

Equation (19) is reduced becomes

$$\frac{d^2F(\beta)}{d\beta^2}-\left[\frac{v(v+1)}{\sin^2\delta\beta}-2q\cot\delta\beta+\kappa^2\right]F(\beta)=0 \quad (23)$$

Equation (23) is differential equation has reduced to the hypergeometric differential equation type by using $\cot(\delta\beta)=i(1-2z)$, we get

$$z(1-z)\frac{d^2F(\beta)}{dz^2}+(1-2z)\frac{dF(\beta)}{dz}+\left[v'(v'+1)-\frac{4\alpha_H^2}{4z}-\frac{4\beta_H^2}{4(1-z)}\right]F(\beta)=0 \quad (24)$$

with,

$$\frac{2qi-\kappa^2}{\delta^2}=4\alpha_H^2,\quad \frac{-2qi-\kappa^2}{\delta^2}=4\beta_H^2,\quad v'\left(v'+1\right)=\frac{v\left(v+1\right)}{\delta^2} \tag{25}$$

by applying the new wavefunction as follow,

$$F\left(\beta\right)=z^{\alpha_H}\left(1-z\right)^{\beta_H}g\left(z\right) \tag{26}$$

in equation (24) we obtain

$$\left[\begin{array}{l} z(1-z)\dfrac{g^2\left(z\right)}{dz^2}+\left[\left(2\alpha_H+1\right)-\left(2\alpha_H+2\beta_H+2\right)z\right]\dfrac{g\left(z\right)}{dz} \\ +\left[v'\left(v'+1\right)-\left(\alpha_H+\beta_H\right)\left(\alpha_H+\beta_H+1\right)\right]g\left(z\right) \end{array}\right]=0 \tag{27}$$

The equation (27) is hypergeometric differential equation that obtain parametric hypergeometric as follow

$$a=\alpha_H+\beta_H-v',\ b=\alpha_H+\beta_H+v'+1,\ c=2\alpha_H+1 \tag{28}$$

By inserting equations (20)-(22), (25), and (28), we get

$$E=\frac{\delta^2}{2B_m}\left[\begin{array}{l}\left(\sqrt{\left(\dfrac{L\left(L+1\right)}{3}+\dfrac{8B_m^{\ 2}\alpha V_o^{\ 2}}{\delta^2}+\dfrac{1}{4}\right)}+\dfrac{1}{2}-n\right)^2 \\ -\left(\dfrac{\left(\dfrac{\left(16B_m^{\ 2}\alpha V_o E_0-16B_m^{\ 2}\alpha V_2-2B_m V_o\right)}{\delta^2}\right)^2}{\left(\sqrt{\left(\dfrac{L\left(L+1\right)}{3}+\dfrac{8B_m^{\ 2}\alpha V_o^{\ 2}}{\delta^2}+\dfrac{1}{4}\right)}+\dfrac{1}{2}-n\right)^2}\right)\end{array}\right]+\left[\begin{array}{l}4B_m\alpha V_1^2+4B_m\alpha E_0^{\ 2} \\ +V_1-8B_m\alpha V_1 E_0-4B_m\alpha V_o^{\ 2}\end{array}\right] \tag{29}$$

The equation (29) is energy spectrum equation of Bohr-Mottelson in minimal length effect for Cotangent potential. To get energy spectrum, we calculated numerically equation by using Matlab software. The result is shown in Table 1.

Table 1. The energy spectrum of Bohr-Mottelson in minimal length effect for ${}^{152}_{62}Sm$, $\delta=0.1$, $V_0=0.07$ and $V_1=0.1$

n	l	$E\ (eV)$		
		$\alpha=0.001$	$\alpha=0.005$	$\alpha=0.01$
0	0	-6.50609	-2.87751	-2.78404
	2	-6.01505	-2.82617	-2.75838
	4	-5.11246	-2.7128	-2.70016
1	0	-10.0042	-3.51843	-3.21467
	2	-9.09471	-3.45048	-3.18338
	4	-7.48489	-3.30104	-3.11250
2	0	-17.1663	-4.38043	-3.74380
	2	-15.1886	-4.28786	-3.70510
	4	-11.8874	-4.08523	-3.61758

The table 1 shows that the energy spectrum is increase duo to increase of minimal length parameter and angular momentum quantum number (l). And the other hand, energy spectrum decreased by increase of quantum number (n).

Then, to obtain the wave function use equations (17), (26), and $\cot(\delta\beta) = i(1-2z)$, we get

$$F(\beta) = \frac{\left(1+i\cot(\delta\beta)\right)^{\alpha_H}}{2} \frac{\left(1-i\cot(\delta\beta)\right)^{\beta_H}}{2} \, {}_2F_1(a,b,c,z) \tag{30}$$

And by applying equation (28) in equation (30), we obtain

$$F(\beta) = \frac{\left(1+i\cot(\delta\beta)\right)^{\alpha_H}}{2} \frac{\left(1-i\cot(\delta\beta)\right)^{\beta_H}}{2} \tag{31}$$

$$F(\beta) = \frac{\left(1+i\cot(\delta\beta)\right)^{\alpha_H}}{2} \frac{\left(1-i\cot(\delta\beta)\right)^{\beta_H}}{2} \left[1 + \frac{(-n)(\alpha_H+\beta_H+\nu+1)\left(\frac{1+i\cot(\delta\beta)}{2}\right)}{(2\alpha_H+1)}\right] \tag{32}$$

The equations (31) and (32) are wave function of Bohr-Mottelson equation in minimal effect for n=0 and n=1, respectively. The value of wave function depends on the value of the hypergeometric parameter.

4. Conclusion

The Bohr-Mottelson equation is investigated in minimal length effect for Cotangent potential by using Hypergeometric method. The Hypergeometric method is using to obtain energy spectrum and wave function of Bohr-Mottelson equation in minimal length effect. The result is shown the energy spectrum increase duo to increase minimal length parameter.

Acknowledgement

This research was partly supported by Ministry of Research, Technology and Higher Education with contract no. 089/SP2H/LT/DRPM/2018.

References

[1] M. Chabab, A. Lahbas, and M. Oulne 2015 *Eur. Phys. J. Plus* **51**(131)
[2] M. Alimohammadi and H. Hassanabadi 2017 *Nucl. Phys A* **957** 439-449
[3] M. Chabab, A. l Batoul, A. Lahbas, and M. Oulne 2016 *Phys. Lett. B* **758** 212-218
[4] D. Bonatsos, P. E. Georgoudis, D. Lenis, N. Minkov, and C. Quesne 2011 *Phys. Rev. C* **83** 044321
[5] L. Naderia and H. Hassanabadi 2016 *Eur. Phys. J. Plus* **131**(5) 133
[6] S Hossenfelder 2004 *Mod. Phys. Lett. A* **19**(37) 2727–2744
[7] L.J. Garay 1994 *Int. J. Mod. Phys. A* **10**(2) 145-165
[8] Cari, Suparmi, Deta A U and Werdiningsih I S 2013 *Makara J. Sci.* **17**(3) 93-102
[9] A Suparmi, C Cari, and Isnaini Lilis Elviyanti 2017 *J. Phys.: Conf. Ser.* **909** 012003
[10] Suparmi 2011 *Mekanika Kuantum II* (Surakarta: FMIPA UNS ISBN 978-602-99344-2-7)
[11] I L Elviyanti , A Suparmi and C Cari 2017 *J. Phys.: Conf. Ser.* **909** 012003

IOP Conf. Series: Materials Science and Engineering **578** (2019) 012091 doi:10.1088/1757-899X/578/1/012091

Bound state energies of a diatomic quantum dot molecule in a magnetic field

M Ma'arif[1]*, A Suparmi[1], C Cari[1]
[1]Physics Department of Graduate Program, Sebelas Maret University, Surakarta, Indonesia

E-mail: miftahulmaarif15@student.uns.ac.id

Abstract. We investigate energy levels of a diatomic quantum dot by using an analytical method. We use Schrodinger equation for spherical quantum dot molecule in the presence of a static magnetic field and a confining potential which is Morse potential. We obtain the ground state and the excited states eigen solutions which are compared with the experimental data.

1. Introduction

Nanotechnology is one of the leader areas of advance materials research to involve nanoscale manipulation of atoms and molecules [1]. Nowadays, one of the major groups of nanoparticles that being developed for various industrial and also biomedical applications are Quantum Dots (QDs) [2-4]. The idea of QDs was firstly proposed by Alexey Ekimov in 1981 and Brus in 1985 [5]. QDs are a new generation of quasi three-dimensional semiconductor in a nanoscale whose structure are such that the hole and the electron are confined within it [6].

Recent researches which different methods were used to study the electronic [7], optic [8], and thermodynamic properties of QDs [9]. Theoretically, many authors studied the Hamiltonian of Quantum Dot-system, then it is solved to get physical properties by using analytical or numerical method. Some of them, Hassanabadi and Rajabi in 2009 had studied the energy levels of a spherical quantum dot in a confining potential which is solved analytically to obtain the ground state and the excited wave function as well as the corresponding eigenvalues for it [10]. Then, Hassanabadi and Rahimov in 2011 had an alternative method for spectrum of a three-electron-quantum dot composed parabolic confinement plus Coulomb terms [11]. In our study, we have calculated the bound state energy of the electron in a diatomic quantum dot within static magnetic field by using Schrodinger equation and Morse potential which is representing the diatomic molecule.

2. Hamiltonian of a diatomic quantum dot in a static magnetic field

The Hamiltonian for an electron confined in a GaAs QD by a Morse potential in a static magnetic field can be written as

$$H=\frac{1}{2m_e^*}\left(\vec{p}+\frac{}{c}A\right)+V(z) \tag{1}$$

Where m_e^* is the electron effective mass, $\vec{p}$ and $\vec{A}$ denote respectively the momentum operator and the vector potential and is given by $\vec{B} = \nabla \times A$ which is a magnetic field applied perpendicular to the growth direction [$\vec{A} = (0, -Bz, 0)$], and $V(z)$ is Morse potential which can be written as follow

$$V(z) = D_e\left(e^{-2az} - 2e^{-az}\right) \tag{2}$$

Where D_e is the well depth of the confining potential and a is a parameter to control the width of the potential. By substituting equation (2) into equation (1) and applying all these particular cases, so the Hamiltonian is

$$H = \frac{1}{2m_e^*}\left(-\hbar \qquad \frac{e\hbar}{c}\frac{}{\partial y} + \frac{e^2B^2z^2}{c^2}\right) + D_e\left(e^{-2az} - 2e^{-az}\right) \tag{3}$$

In this work, we eliminate the x and y variables because we pointed just in transitions between state confined in the QD. So we can use the total wave functions as follows

$$\psi(x, y, z) = e^{ik_x x} e^{ik_y y} \varphi(z) \tag{4}$$

So we can get the Schrodinger equation in the growth plane that can be written as bellow

$$-\frac{\hbar}{2m_e^*}\frac{}{dz^2} + \left[\frac{\hbar}{m_e^* c}z + \frac{e^2B^2}{2m_e^* c^2}z^2 + D_e\left(e^{-2az} - 2e^{-az}\right)\right]\varphi = \left[E - \frac{\hbar \qquad)}{2m_e^*}\right]\varphi \tag{5}$$

We expand the Morse potential in equation (5) about the origin as bellow

$$V(z) = 2D_e\sum_{l=0}^{\infty} c_l z^l \tag{6}$$

Where $c_l = (-1)^l\left(2^{l-1} - 1\right)a^l / l!$ and we take $l = 0$, 1, and 2 because the magnetic field affected just until the second order of polynomials, so we get the new formulation can be written as

$$-\frac{\hbar}{2m_e^*}\frac{}{dz^2} + \left[\frac{\hbar}{m_e^* c}z + \left(\frac{e^2B^2}{2m_e^* c^2} + D_e a^2\right)z^2\right]\varphi = \left[E + D_e - \frac{\hbar \qquad)}{2m_e^*}\right]\varphi \tag{7}$$

Equation (7) is Schrodinger equation with harmonic plus linear potential with

$$\omega_c = \sqrt{\frac{e^2B^2}{m_e^{*2}c^2} + \frac{D_e a^2}{m_e^*}} \tag{8}$$

$$\beta = \frac{\hbar}{m_e^* c} \tag{9}$$

$$E' = E + D_e - \frac{\hbar}{2m_e^*}\left(\kappa_x + \kappa_y^2\right) \tag{10}$$

Which ω_c is cyclotron frequency and β is damping force. Then, we substitute equations (8) and (9) into equation (7) we have

$$-\frac{\hbar}{2m_e^*}\frac{}{dz^2} + \left(\frac{1}{2}m_e^*\omega_c^2 z^2 + \beta z\right)\varphi = E'\varphi \tag{11}$$

Supersymmetric quantum mechanics has been applied to solve Schrodinger equation in equation (11). We have just hypothesized the super potential as written by

$$\phi(a_0, z) = \alpha z + \frac{\beta}{2\alpha} \quad (12)$$

By applying the super partner potential in SUSY QM Hamiltonian, so we can obtain the following equations

$$\alpha = \omega_c \sqrt{\frac{m_e^*}{2}} \quad (13)$$

$$E_0' = \frac{\hbar}{2} - \frac{{}^{\prime 2}}{2m_e^* \omega_c^2} \quad (14)$$

From equations (14) and (10) we get the bound state energy for the nodes $l = 1$ and $l = 2$ is defined by

$$\Delta E = \frac{\hbar}{2} + \frac{{}^{\prime 2}}{2m_e^* \omega_c^2}$$

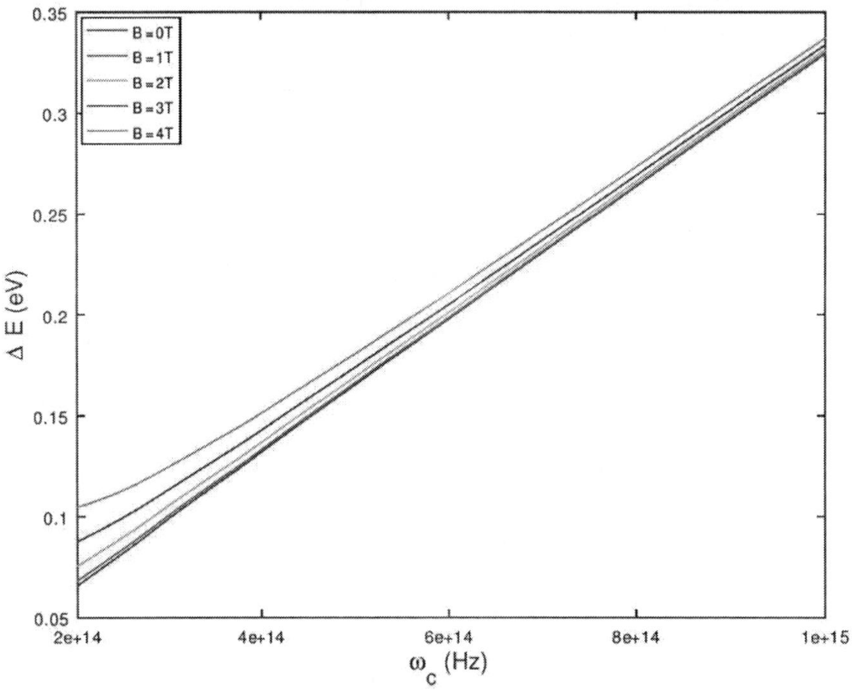

Figure 1. Bound state energies ΔE versus ω_c for various cases of magnetic field values.

We show the bound state energy versus cyclotron frequency in the interval $2 \times 10^{14} \leq \omega_c \leq 1 \times 10^{15}$ for the case $0 \leq B \leq 4$. We have chosen the value of the effective mass of an electron in GaAs semiconductor is $0.05 m_e$ [12]. This result is in good agreement with the previous experimental [13] or theoretical data [10]. The bound state energy has a linear relationship between cyclotron frequency and also the presence of static magnetic field has made it higher than before.

3. Conclusion

The ground state and the first excited state energies of a diatomic quantum dot molecule have been obtained by using Schrodinger equation with Morse potential that has been used to represent the diatomic molecule. The results show that the increase in the values of static magnetic field lead to increase in the bound state energy of the electron, respectively. Our results are also in good agreement with the previous data.

Acknowledgement
This research was partly supported by Ministry of Research, Technology, and Higher Education with contract no. 089/SP2H/LT/DRPM/2018.

References
[1] Mnyusiwalla A, Daar AS and Singer PA 2003 *Nanotechnology* **14**(3) R9
[2] Gorman J, Hasko DG and Williams DA 2005 *Phys. Rev. Lett.* **95**(9) 090502
[3] Lim YT, Kim S, Nakayama A, Stott NE, Bawendi MG and Frangioni JV 2003 *Molecular imaging* **2**(1) 15353500200302163
[4] William WY, Chang E, Drezek R and Colvin VL 2006 *Biochem Biophys Res Commun.* **348**(3) 781-786
[5] Ekimov AI, Hache F, Schanne-Klein MC, Ricard D, Flytzanis C, Kudryavtsev IA, Yazeva TV, Rodina AV and Efros AL 1993 *JOSA B* **10**(1) 100-107
[6] Tavernier MB, Anisimovas E, Peeters FM, Szafran B, Adamowski J and Bednarek S 2003 *Phys. Rev. B* **68**(20) 205305
[7] Elzerman JM, Hanson R, Greidanus JS, Van Beveren LW, De Franceschi S, Vandersypen LMK, Tarucha S and Kouwenhoven LP 2003 *Phys. Rev. B* **67**(16) 161308
[8] Woggon U 1997 *Optical properties of semiconductor quantum dots* Springer.
[9] Wang Y and Lieberman M 2004 *IEEE Transactions on Nanotechnology* **3**(3) 368-376.
[10] Hassanabadi H and Rajabi AA 2009 *Phys. Lett. A* **373**(6) 679-681
[11] Hassanabadi H, Hamzavi M, Zarrinkamar S and Rajabi AA 2010 *Few-Body Syst* **48**(1) 53-58
[12] Cardona M 1961 *Phys. Rev.* **121**(3) 752
[13] Glasberg S, Finkelstein G, Shtrikman H and Bar-Joseph I 1999 *Phys. Rev. B* **59**(16) R10425

IOP Conf. Series: Materials Science and Engineering **578** (2019) 012092 doi:10.1088/1757-899X/578/1/012092

Spatially q-deformed radial momentum of d-dimensional Klein-Gordon equation for Harmonic Oscillator plus Inverse Quadratic potential investigated using hypergeometric method

C Cari[1], D A Dianawati[1] and A Suparmi*[1]

[1]Physics Departement, Graduate Program, Sebelas Maret University
Jl. Ir. Sutami 36A Kentingan Jebres Surakarta 57126, INDONESIA

*E-mail: soeparmi@staff.uns.ac.id

Abstract. Spatially quadratic deformation radial momentum of D-dimensional Klein-Gordon equation with three-dimensional Harmonic Oscillator and Inverse Quadratic potential was investigated using Hypergeometric Method. By using appropriate variable transformation, D-Dimensional Klein-Gordon equation for q deformed momentum was reduced into one order dimensional Schrodinger like equation with Pöschl-Teller like potential. The relativistic energy and wave function equations were obtained from the solution of Schrodinger like equation. Then, the relativistic energy and wave function were calculated and visualized using Matlab R2013 software, respectively. The q deformed parameter had the effect to the relativistic energy and wave function. The increase of q deformed parameter causes the increase of energy spectra and the decrease of amplitude.

1. Introduction

In the 20th century, several scientists debated about the concept of classical physics that refers to Newton's law and the concept only explain some macroscopic materials behavior. Some experiments showed that Newton's physics could not explain some microscopic materials behavior. Therefore, a new theory was founded which was called the theory of quantum mechanics. It could explain about microscopic objects behavior with interaction of energy on the scale of atoms and subatomic particles [1,2].

In quantum mechanics, the relativistic effect will be considered when a particle is moving in a strong potential field [3,4]. The Klein-Gordon equation is used to describe a zero-spin particle [3,4,5], pions and kaons, in large potentials [6]. In relativistic equation consists of four vector potential $V(r)$ and space-time scalar potential $S(r)$ [3]. By using non-trivial nonrelativistic limit $S(r) = V(r)$, we obtain the potential function $2V(r)$, not $V(r)$. Besides that, by using nonrelativistic limit, we obtain the potential interaction $V(r)$, not $2V(r)$ [7].

Hypergeometric method was one of the methods which was used to solve the Klein-Gordon equation. This method was obtained from Hypergeometric function by substituting new variable to get the second-order differential equation [2,5].

Some potentials can be used to obtain the exact solution of Klein-Gordon equation, especially shape invariant potentials, such as Kratzer, Pöschl-Teller, and Manning Rosen potentials [8,9]. Indeed, the

Harmonic Oscillator plus Inverse Quadratic potential can be used to solve the Klein-Gordon equation too [11]. In this research, we investigate and calculate energy spectrum and wave function for D-dimensional Klein-Gordon equation with spatially q-deformed of radial momentum for Harmonic Oscillator plus Inverse Quadratic potential using Matlab R2013 Software.

There are several sections of this paper such as the D-dimensional Klein-Gordon equation, the Hypergeometric method, result and discussion, and conclusion.

2. Experimental

2.1. D-dimensional klein-gordon equation

The motion of particle with mass M and relativistic energy E on scalar $S(r)$ and vector $V(r)$ potentials in natural unit ($\hbar = 1, c = 1$) is given as [7,12,13]:

$$\left\{\nabla_D^{\ 2}+\left(V(r)-E\right)^2-\left(S(r)+M\right)^2\right\}\psi_{\ell\ \ \ell}^{\ell\ \ \ell}\cdots{}_{,}=0 \tag{1}$$

where ∇_D^2 is Laplacian operator which is expressed by [13]

$$\nabla_D^2=\frac{1}{r^{D-1}}\frac{\partial}{\partial r}\left(r^{D-1}\frac{\partial}{\partial r}\right)-\frac{L_{D-1}^2}{r^2} \tag{2}$$

The angular momentum L_{D-1}^2 in equation (2) can be written as

$$L_{D-1}^2=-\left(\sum_{j=1}^{D-2}\frac{1}{\sin^2\theta_{j+1}\sin^2\theta_{j+2}\ldots\sin^2\theta_{D-1}}\times\left\{\frac{1}{\sin^{j-1}\theta_j}\left(\frac{\partial}{\partial\theta_j}\sin^{j-1}\theta_j\frac{\partial}{\partial\theta_j}\right)\right\}+\left\{\frac{1}{\sin^{D-2}\theta_{D-1}}\left(\frac{\partial}{\partial\theta_{D-1}}\sin^{D-2}\theta_{D-1}\frac{\partial}{\partial\theta_{D-1}}\right)\right\}\right) \tag{3}$$

The D-dimensional Klein-Gordon equation for $S(r) = V(r)$ can be written as [7,12]

$$\left\{\nabla_D^{\ 2}+\left(E^2-M^2-(E+M)V(r)\right)\right\}\psi_{\ell\ \ \ell}^{\ell\ \ \ell}\cdots{}_{,}=0 \tag{4}$$

The total of wave function equation is [3,7,12]

$$\psi_{\ell_1\ldots\ell_{D-2}}^{(\ell)}=R_l(r)Y_{\ell_1\ldots\ell_{D-2}}^{(\ell)}(\theta_1,\theta_2,\ldots,\theta_{D-1})\quad\text{with}\quad R_l(r)=\frac{F}{r^{\frac{(D-1)}{2}}} \tag{5}$$

By inserting equations (2) and (5) into equation (4), we obtain

$$\frac{\partial^2F}{\partial r^2}-\frac{\left(\frac{D-1}{2}\right)\left(\frac{D-3}{2}\right)}{r^2}F+\left(E^2-M^2-(E+M)\mathrm{V(r)}\right)F=\frac{\lambda_{D-1}}{r^2}F \tag{6}$$

Equation (6) is the radial part of wave function equation with central potential $V(r)$.

2.2. Spatially q deformed of quadratic radial momentum

In this section, we will discuss about the deformation in quantum mechanics. The q deformation of momentum can be expressed by [16]

$$\frac{\partial^2}{\partial r^2}=D_r^{q^2} \tag{7}$$

The q-deformed of quantum mechanics equation was [14]

$$D_r^q=\left(1+qr^2\right)\frac{d}{dr} \tag{8}$$

The q deformed of quadratic radial momentum D_r^q can be rewritten by

$$D_r^{q^2}=(1+qr^2)^2\frac{d^2}{dr^2}+2qr(1+qr^2)\frac{d}{dr} \tag{9}$$

By using new variable defined as

$$y = \sqrt{q}r \rightarrow dy = \sqrt{q}dr \rightarrow dr = \frac{dy}{\sqrt{q}} \tag{10}$$

and by inserting equation (10) into equation (9), we have

$$D_r^{q^2} = (1+y^2)^2 q \frac{d^2}{dy^2} + 2qy(1+y^2)\frac{d}{dy} = \frac{\partial^2}{\partial r^2} \tag{11}$$

Equation (11) was the general form of deformation momentum of radial part.

2.3. Harmonic Oscillator plus Inverse Quadratic potential

Harmonic Oscillator plus Inverse Quadratic can be expressed by [4]

$$V(r) = kr^2 + \frac{g}{r^2} \tag{12}$$

with k is constant of arbitrary and g is power of inverse quadratic potential.

By changing the r variable in equation (10) into equation (12) and by substituting equation (12) into equation (6), we get

$$(1+y^2)^2\frac{d^2F}{dy^2} + 2y(1+y^2)\frac{dF}{dy} - \left(\frac{\left(\frac{D-1}{2}\right)\left(\frac{D-3}{2}\right) + \lambda_{D-1}}{y^2}\right)F + \left(\left(\frac{E^2-M^2}{q}\right) - (E+M)\left(k\frac{y^2}{q^2} + \frac{g}{y^2}\right)\right)F = 0 \tag{13}$$

Equation (13) can be solved by Hypergeometric method to gain energy and wave function equations.

2.4. Hypergeometric methods

The second order differential equation of Hypergeometric function can be written in form [5,15]

$$z(1-z)\frac{\partial^2\phi}{\partial z^2} + (c-(a+b+1)z)\frac{\partial\phi}{\partial z} - ab\phi = 0 \tag{14}$$

from equation (14), we have singular points at $z = 0$ which was written in series form [8]

$$\phi = z^s \sum\nolimits_{n=0}^{n} a_n z^n \tag{15}$$

By inserting equation (15) into equation (14), we have the recursion formula was expressed by

$$a_n = \frac{(a)(a+1)(b)(b+1)...(a+(n-1))(b+(n-1))}{(c)(1+c)...((n-1)+c)n!} \tag{16}$$

and the solution form of Hypergeometric differential equation from equation (14) was given as [8,15]

$${}_2F_1(a,b,c;z) = \Phi_1(z) = \sum\nolimits_{n=0} \frac{(a)_n(b)_n}{n!(c)_n} z^n \tag{17}$$

If $a = -n$ or $b = -n$, the solution of equation (17) to get the energy eigen value of the system in polynomial rank n.

3. Results and Discussions

The solution for equation (13) can be solved by using variable approximation with

$$y = \tan x \rightarrow dy = \sec^2 x dx \tag{18}$$

By inserting equation (18) into equation (13), we have

$$\frac{d^2F}{dx^2} - \left(\left(\left(\frac{D-1}{2}\right)\left(\frac{D-3}{2}\right) + \lambda_{D-1} + (E+M)g\right)\cot^2 x + \frac{(E+M)k}{q^2}\tan^2 x - \left(\frac{E^2-M^2}{q}\right)\right)F = 0 \tag{19}$$

or

$$\frac{d^2F}{dx^2}+\left(\left(\frac{D-1}{2}\right)\left(\frac{D-3}{2}\right)+\lambda_{D-1}+(E+M)g+(E+M)\frac{k}{q^2}+\left(\frac{E^2-M^2}{q}\right)\right)F$$
$$-\left(\left(\frac{\left(\frac{D-1}{2}\right)\left(\frac{D-3}{2}\right)+(E+M)g+\lambda_{D-1}}{\sin^2 x}\right)+\left(\frac{(E+M)\frac{k}{q^2}}{\cos^2 x}\right)\right)F=0 \tag{20}$$

by setting

$$A_s=\left(\left(\frac{D-1}{2}\right)\left(\frac{D-3}{2}\right)+(E+M)g+\lambda_{D-1}\right) \tag{21}$$

$$B_s=(E+M)\frac{k}{q^2} \tag{22}$$

$$E_s'=\left(\left(\frac{D-1}{2}\right)\left(\frac{D-3}{2}\right)+\lambda_{D-1}+(E+M)g+(E+M)\frac{k}{q^2}+\left(\frac{E^2-M^2}{q}\right)\right) \tag{23}$$

Equation (20) can be rewritten by

$$\frac{d^2}{dx^2}F-\frac{A_s}{\sin^2 x}F-\frac{B_s}{\cos^2 x}F=-E_s'F \tag{24}$$

or

$$\frac{d^2}{dx^2}F-\frac{\eta(\eta-1)}{\sin^2 x}F-\frac{\mu(\mu-1)}{\cos^2 x}F=-E_s'F \tag{25}$$

By using the variable approximation, we get

$$A_s=\eta(\eta-1)\rightarrow\eta=\frac{1\pm\sqrt{1+4A_s}}{2} \quad \text{and} \quad \mu(\mu-1)=B_s\rightarrow\mu=\frac{1\pm\sqrt{1+4B_s}}{2} \tag{26}$$

The form of equation (25) was Schrodinger like equation with Pöschl-Teller potential. Therefore, the form of solution equation (26) resemble with solution of Schrodinger like equation with Pöschl-Teller potential by Hypergeometric method.

We set the variable approximation in equation (24) [8,15]

$$\cos^2 x=z\rightarrow dz=-2\sin x\cos x dx\rightarrow\frac{dz}{dx}=-2\sqrt{z(1-z)} \tag{27}$$

By inserting equation (27) into equation (24), the form of equation (24) becomes

$$z(1-z)\frac{d^2F}{dz^2}+(1-2z)\frac{dF}{dz}+\left(\frac{E_s'}{4}-\frac{A_s}{4(z-1)}-\frac{B_s}{4z}\right)F=0 \tag{28}$$

From equation (28), we gain the general solution of wave function equation

$$F=z^{\alpha}(1-z)^{\beta}f(z) \tag{29}$$

By substituting equation (29) into equation (28), we obtain the second order of Hypergeometric function equation which was

$$z(1-z)f''+\left(\left(2\alpha+\frac{1}{2}\right)-(2\alpha+2\beta+1)z\right)f'+\left(\frac{E_s'}{4}-(\alpha+\beta)^2\right)f=0 \tag{30}$$

Equation (30) equivalent with equation (14). Consequently, we have $c=2\alpha+\frac{1}{2}$ and $a+b=2\alpha+2\beta$ and

$$\alpha=\frac{\mu}{2} \quad \text{and} \quad \beta=\frac{\eta}{2} \tag{31}$$

From equation (30), we can obtain the relativistic energy equation

$$E_n' = (\mu + \eta + 2n)^2 \tag{32}$$

By setting $E_n' = E_s'$ and by substituting equation (23) into equation (32), we obtain the relativistic energy equation

$$\left(E^2 - M^2\right) = q\left(\frac{\sqrt{1+4B_s}}{2} + \frac{\sqrt{1+4A_s}}{2} + 2n\right)^2 - q\left(\frac{D-1}{2}\right)\left(\frac{D-3}{2}\right) - q\lambda_{D-1} - q(E+M)g - (E+M)\frac{k}{q} \tag{33}$$

with A_s and B_s in equations (21) and (22), respectively.

The relativistic energy in equation (33) can be calculated by Matlab R2013 software with several variation parameters that the result of relativistic was showed in Table 1.

Moreover, solution of wave function equation can be gained by substitution equation (27) into equation (29) that can be expressed by

$$F = \left(\cos^2 x\right)^{\alpha} \left(\sin^2 x\right)^{\beta} {}_2R_1(a,b,c;z) \tag{34}$$

with α and β in equation (31), and based on the explanation above that $c = 2\alpha + \frac{1}{2}$ and $a + b = 2\alpha + 2\beta$, equation (34) can be written in form

$$F = \left(\cos^2 x\right)^{\frac{\left(1+\sqrt{1+4B_s}\right)}{4}} \left(\sin^2 x\right)^{\frac{\left(1+\sqrt{1+4A_s}\right)}{4}} C'(-1)^n \left(2\alpha + \frac{1}{2}\right) {}_2R_1(-n, 2\alpha + 2\beta + 1, 2\alpha + \frac{1}{2}; \cos^2 x) \tag{35}$$

or

$$F_n(x) = \left(\cos^2 x\right)^{\frac{\left(1+\sqrt{1+4\left((E+M)\frac{k}{q^2}\right)}\right)}{4}} \left(\sin^2 x\right)^{\frac{\left(1+\sqrt{1-4\left(\left(\frac{D-1}{2}\right)\left(\frac{D-3}{2}\right)+(E+M)g+\lambda_{D-1}\right)}\right)}{4}}$$
$$C'(-1)^n \left(2\alpha + \frac{1}{2}\right) {}_2R_1(-n, 2\alpha + 2\beta + 1, 2\alpha + \frac{1}{2}; \cos^2 x) \tag{36}$$

Equation (36) was the wave function equation for D-dimensional Klein-Gordon equation with spatially q deformed of quadratic radial momentum which can be expressed with quantum number variation that the result of wave function equation was showed in Table 2. After that, the wave function of radial part equation can be visualized by Matlab R2013 software and the result of visualization shows in Figure 1 and Figure 2.

IOP Conf. Series: Materials Science and Engineering **578** (2019) 012092 doi:10.1088/1757-899X/578/1/012092

Table 1. The Relativistic Energy for D-Dimensional Klein-Gordon Equation with Spatially q-Deformed of Quadratic Radial Momentum in Natural Unit $g = 5, k = 5, M = 0.5$

D	q	n	l	En	D	q	n	l	En	D	q	n	l	En	D	q	n	l	En
	0.01			-0.499971168		0.01			-0.499964827		0.01			-0.499959965		0.01			-0.499956137
	0.04			-0.499536735		0.04			-0.499435396		0.04			-0.499368076		0.04			-0.499296456
	0.07			-0.498573384		0.07			-0.498264398		0.07			-0.498054961		0.07			-0.497839721
3	0.1	1	0	-0.497068669	4	0.1	1	0	-0.496442892	5	0.1	1	0	-0.495958588	6	0.1	1	0	-0.495579007
	0.4			-0.487683977		0.4			-0.439255589		0.4			-0.432415071		0.4			-0.426815068
	0.7			-0.471016256		0.7			-0.297536024		0.7			-0.28147968		0.7			-0.267228517
	1			-0.456134621		1			-0.260742085		1			-0.033221057		1			-0.010785701
	0.01		0	-0.499901117		0.01		0	-0.499884828		0.01		0	-0.499870616		0.01		0	-0.499858166
			1	-0.499870616				1	-0.499858166				1	-0.499847202				1	-0.499837491
	0.04		0	-0.498408918		0.04		0	-0.498149714		0.04		0	-0.497922615		0.04		0	-0.49772351
			1	-0.497922615				1	-0.497723510				1	-0.497548132				1	-0.497392802
	0.07		0	-0.495084952		0.07		0	-0.494302652		0.07		0	-0.493610763		0.07		0	-0.493002727
			1	-0.493610763				1	-0.493002726				1	-0.492466691				1	-0.491991777
3	0.1	2	0	-0.489854215	4	0.1	2	0	-0.488294308	5	0.1	2	0	-0.486895064	6	0.1	2	0	-0.485660665
			1	-0.486895064				1	-0.485660665				1	-0.484570768				1	-0.483604448
	0.4		0	-0.424850746		0.4		0	-0.289756944		0.4		0	-0.27290057		0.4		0	-0.256513525
			1	-0.272900570				1	-0.256513525				1	-0.241326335				1	-0.227479737
	0.7		0	-0.330125824		0.7		0	0.223567354		0.7		0	0.254687619		0.7		0	0.288826719
			1	0.254687619				1	0.288826719				1	0.323110119				1	0.356134655
	1		0	-0.24584695		1		0	0.866863603		1		0	1.083235851		1		0	1.127603641
			1	0.595194825				1	1.127603641				1	1.175444165				1	1.224238971
			0	-0.499851138				0	-0.499764712				0	-0.499740707				0	-0.499718762
	0.01		1	-0.49974071		0.01		1	-0.499718762		0.01		1	-0.499698673		0.01		1	-0.499680244
			2	-0.499698673				2	-0.499680244				2	-0.499663300				2	-0.499647683
			0	-0.497622508				0	-0.496217109				0	-0.495834479				0	-0.495483896
	0.04		1	-0.495834478		0.04		1	-0.495483895		0.04		1	-0.495162685		0.04		1	-0.494867937
			2	-0.495162685				2	-0.494867937				2	-0.494596903				2	-0.494347095
			0	-0.489564937				0	-0.488333274				0	-0.487174376				0	-0.486106896
	0.07		1	-0.487174376		0.07		1	-0.486106896		0.07		1	-0.485126765		0.07		1	-0.484226456
			2	-0.485126765				2	-0.484226456				2	-0.483398134				2	-0.482634454
			0	-0.485425386				0	-0.475973911				0	-0.47365138				0	-0.471494699
3	0.1	3	1	-0.473651380	4	0.1	3	1	-0.471494699	5	0.1	3	1	-0.469507743	6	0.1	3	1	-0.467679444
			2	-0.469507743				2	-0.467679444				2	-0.465995681				2	-0.464442394
			0	-0.318560124				0	-0.054929713				0	-0.031606151				0	-0.006470972
	0.4		1	-0.031606151		0.4		1	-0.006470972		0.4		1	0.018639996		0.4		1	0.042897042
			2	0.018639996				2	0.042897043				2	0.065955502				2	0.087701430
			0	-0.096790519				0	1.003874291				0	1.040484579				0	1.084758103
	0.7		1	1.040484578		0.7		1	1.084758103		0.7		1	1.133176679		0.7		1	1.183277382
			2	1.133176679				2	1.183277382				2	1.233476400				2	1.282808807
			0	0.098867212				0	2.372029852				0	2.566924238				0	2.619146378
	1		1	2.153142822		1		1	2.619146377		1		1	2.679551264		1		1	2.745316912
			2	2.679551264				2	2.745316912				2	2.814208802				2	2.884559555

Table 1. shows that the energy values were affected by D-dimension, quantum number, orbital number and deformation parameters. The increase of quantum number, q deformation and orbital number cause the increase of energy values. The energy values were positive and negative that those indicates free particle and bound energy of a particle in atom's shell, respectively. Meanwhile, the q deformation shows about the energy gap in atom's shell.

Table 2. The Wave Function for D-Dimensional Klein-Gordon Equation with Spatially q-Deformed of Quadratic Radial Momentum

n	$F_n(x)$
0	$F_0(x)=\left(\cos^2 x\right)^{\frac{\left(1+\sqrt{1+4\left((E+M)\frac{k}{q^2}\right)}\right)}{4}}\left(\sin^2 x\right)^{\frac{\left(1+\sqrt{1-4\left(\left(\frac{D-1}{2}\right)\left(\frac{D-3}{2}\right)+(E+M)g+\lambda_{D-1}\right)}\right)}{4}}C'$
1	$F_n(x)=-\left(\cos^2 x\right)^{\frac{\left(1+\sqrt{1+4\left((E+M)\frac{k}{q^2}\right)}\right)}{4}}\left(\sin^2 x\right)^{\frac{\left(1+\sqrt{1-4\left(\left(\frac{D-1}{2}\right)\left(\frac{D-3}{2}\right)+(E+M)g+\lambda_{D-1}\right)}\right)}{4}}C'\left(2\alpha+\frac{1}{2}\right)\left(1-\frac{(2\alpha+2\beta+2)}{\left(2\alpha+\frac{1}{2}\right)}\left(\cos^2 x\right)\right)$
2	$F_n(x)=\left(\cos^2 x\right)^{\frac{\left(1+\sqrt{1+4\left((E+M)\frac{k}{q^2}\right)}\right)}{4}}\left(\sin^2 x\right)^{\frac{\left(1+\sqrt{1-4\left(\left(\frac{D-1}{2}\right)\left(\frac{D-3}{2}\right)+(E+M)g+\lambda_{D-1}\right)}\right)}{4}}$ $C'\left(2\alpha+\frac{1}{2}\right)\left(2\alpha+\frac{3}{2}\right)\left(1-\frac{2(2\alpha+2\beta+3)}{\left(2\alpha+\frac{1}{2}\right)}\left(\cos^2 x\right)+\frac{(2\alpha+2\beta+3)(2\alpha+2\beta+4)}{\left(2\alpha+\frac{1}{2}\right)\left(2\alpha+\frac{3}{2}\right)}\left(\cos^2 x\right)^2\right)$

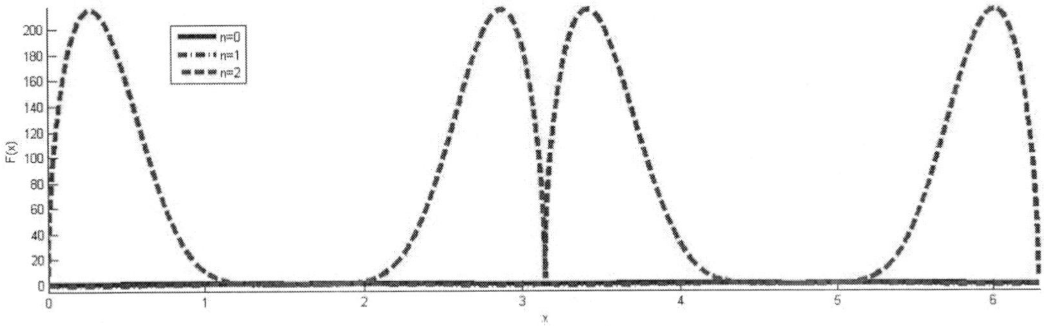

(a)

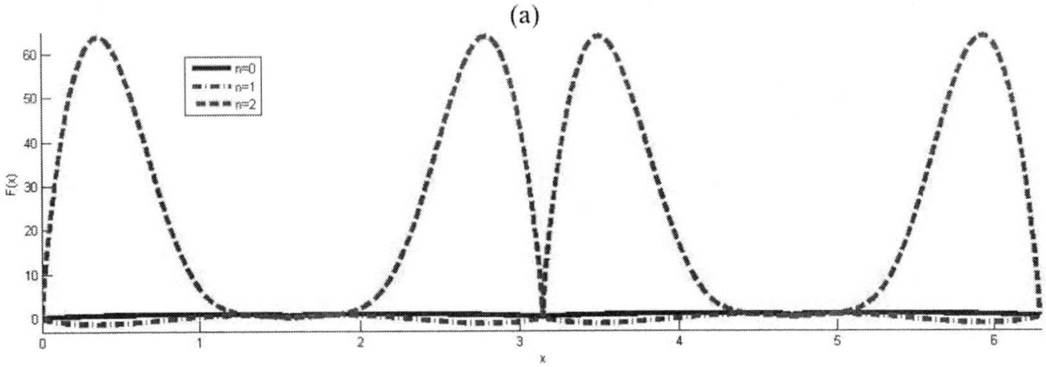

(b)

Figure 1. The wave function unnormalize for Oscillator Harmonic plus Inverse Quadratic Potential in $D = 3$ with (a) $q = 0.01$ and (b) $q = 1$

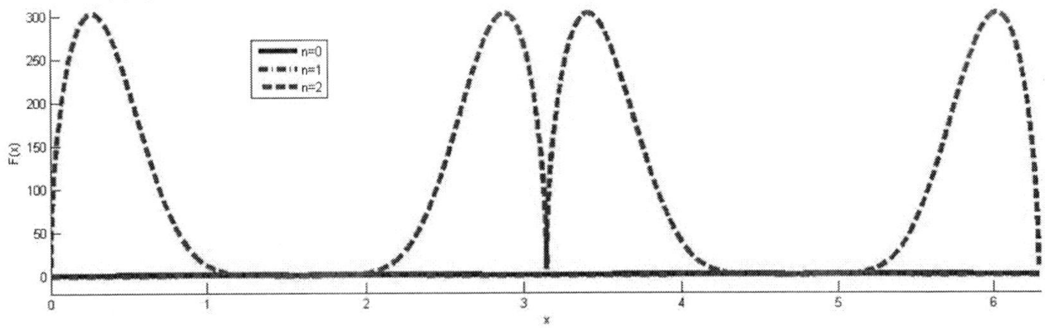

(a)

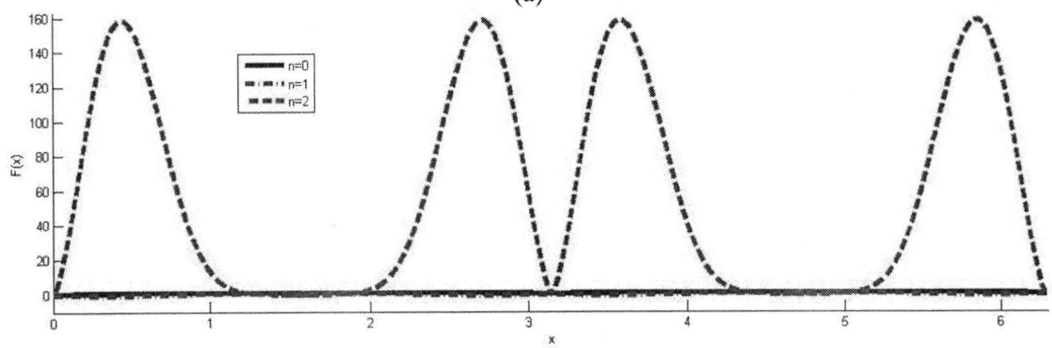

(b)

Figure 2. The wave function unnormalize for Oscillator Harmonic plus Inverse Quadratic Potential in $D = 6$ with (a) $q = 0.01$ and (b) $q = 1$

The radial part of wave function in Figure 1 shows that vibration of particle which was influenced by Oscillator Harmonic plus Inverse Quadratic potential. In Figure 1 and Figure 2 show about the model of wave function with amplitude as radius of particle with Oscillator Harmonic plus Inverse Quadratic potential. As a result, a particle in strong potential field has the chance of finding particle is greater when the radius of a particle is closer to core that shows in the increase of quantum number causes the increase of amplitude. Then, the increase of q deformation quadratic parameters causes the decrease of amplitude from wave function. For instance, $n = 2$ in Figure 1 with (a) $q = 0.01$ has maximum amplitude $\pm 217\ fm$ and (b) $q = 1$ has maximum amplitude $\pm 65\ fm$. Besides that, in Figure 2 with (a) $q = 0.01$ has maximum amplitude $\pm 310\ fm$ and (b) $q = 1$ has maximum amplitude $\pm 160\ fm$. Meanwhile, the increase of D-dimension and quantum number cause the increase of amplitude from wave function. For instance, for $q = 0.01$ in Figure 1 and Figure 2 have maximum amplitude around $0.85\ (n = 0), 1{,}3\ (n = 1), 217\ (n = 2)$ and around $0.9\ (n = 0), 1{,}4\ (n = 1), 310\ (n = 2)$, respectively.

4. Conclusions

The D-dimensional Klein-Gordon equation with spatially q-deformed of quadratic radial momentum for Harmonic Oscillator plus Inverse Quadratic potential can be solved by Hypergeometric Method. The energy values were positive and negative that those indicates free particle and bound energy of a particle in atom's shell, respectively. The increase of D-dimension, quantum number, orbital number and deformation parameters causes the increase of energy values. Then, the increase of q deformation quadratic parameters causes the decrease of amplitude from wave function, while the increase of quantum number and D-dimension parameters cause the increase of amplitude from wave function.

Acknowledgement

This research was partly supported by Hibah Pascasarjana UNS 543/UN27.21/PP/2018.

References

[1] D.A.Dianawati, A.Suparmi, C.Cari 2017 *J. Phys. Conf. Ser.* **795**
[2] A.Suparmi 2013 *Mekanika Kuantum I*: FMIPA UNS ISBN 978-602-99344-1-0
[3] A.D.Antia & A.B.Adeolu 2018 *World Scientific News An International Scientific Journal* **101** 89-107
[4] B.I.Ita, H.P.Obong, C.O.Ehi-Eromosele, A.Edobor-Osoh, A.I.Ikeuba 2014 *AIP Conf. Ser.* 162-166
[5] I.L.Elviyanti, A.Suparmi, C.Cari, D.A. Nugraha, B.N.Pratiwi 2017 *J. Phys. Conf. Ser.* **909**
[6] M.R.Shojaei & A.A.Rajabi 2011 *Int. J. Phys. Sci.* **6** 33 7441-7446
[7] D.A.Nugraha, A.Suparmi, C.Cari 2017 *J. Phys. Conf. Ser.* **1** 1 42-48
[8] C.Cari 2013 *Mekanika Kuantum*: UNS Press ISBN 978-979-498-830-5
[9] Nurhayati, Suparmi, V.I.Variani, Cari, Wahyudi 2012 *Proceding Pertemuan Ilmiah XXVI HFI Jateng & DIY* 229-232
[10] E.A.Bangudu 2003 *Arabian J. Sci. Eng.* **28** 2A
[11] Shi-Hai Dong & M.Lozada-Cassou 2005 *American J. Appl. Sci.* **2** 1 376-382
[12] D.A.Nugraha, A.Suparmi, C.Cari 2017 *J. Phys. Conf. Ser.* **820** 1
[13] Dong, Shi-Hai 2011 *Wave Equations in Higher Dimensions*: New York Spinger Dordecht Heidelberg London
[14] H.Sobhani, W/S.Chung, H.Hassanabadi 2017 *Indian J. Phys.* **03** 65
[15] A.Suparmi 2011 *Mekanika Kuantum II*: FMIPA UNS ISBN 978-602-99344-2-7
[16] H.Hassanabadi, W.S.Chung, S.Zare, S.B.Bhardwaj 2017 Hindawi *Adv. High Energy Phys.* ID 1730834

IOP Conf. Series: Materials Science and Engineering **578** (2019) 012093 doi:10.1088/1757-899X/578/1/012093

Solution of time independent cosmic string for harmonic oscillator plus poschl teller non-central potentials using supersymmetry quantum mechanics methods

D Anggraini[1], A Suparmi[1] and C Cari[1]
[1]Physics Department of Graduate Program, Sebelas Maret University,
Jl. Ir. Sutami 36A Kentingan, Jebres, Surakarta 57126, Indonesia

E-mail : soeparmi@staff.uns.ac.id

Abstract. Time independent cosmic string equation for Harmonic Oscillator and Poschl Teller non-central potential have been solved using supersymmetry quantum mechanics method. The time independent cosmic string with non-central potential was reduced into three one dimensional Schrodinger equations by using variable separation method. The energy levels and radial wave function were analyzed using supersymmetry quantum mechanics method from radial part of Schrodinger like equation. The angular wave function was obtained from angular part of Schrodinger equation. The effect of the non-central potential caused energy levels increasing. Energy spectra for $\lambda = 0$ and $\kappa = 0$ have smaller value than energy levels for $\lambda \neq 0$ and $\kappa \neq 0$. The increased value of parameter λ and κ caused increased in energy levels. However, the decreased of cosmic string parameter α, caused the increased of energy levels.

1. Introduction

String theory is theory of everything which is written in space and time coordinates. This is a theory that combine the theory of quantum mechanics with the theory of relativity, the theory can explain about everything that happen in the universe [1]. Cosmic string theory was studied as a vacuum solution of Kerr general relativity which has been reduced to 1+2 dimensions. Then the three dimensional solution was expanded into four dimensional space time [2]. Cosmic string is linear topology defect which is one of the most important predictions [3,4]. Cosmic string as a stable one dimensional topology defect, whose the length are infinite and straight has an angular parameter α [5].

A lot of studies on the solution of cosmic string as topological defects on the nonrelativistic and relativistic quantum mechanical systems have been investigated [6,7]. Solution of cosmic string can be obtained by reduce cosmic string equation into Schrodinger equation, Dirac equation and Klein Gordon equation. Solution of cosmic string equation was reduced into Klein Gordon equation have been solved by Mederois and Melo [8]. Cosmic string equation was reduced to Dirac equation also have been solved by Bakke [9]. Solution of a two-dimensional Duffin Petiau Kemmer oscillator in the presence of a coulomb potential in the cosmic string have been solved by Boumali and Mesai [10]. In other research, solution of cosmic string equation for quantum particle by reduce that equation into Schrodinger equation have been solved by Hasanabadi et al [11]. Nonrelativistic system be able to analyzed using quantum mechanics. Behavior of quantum particle of non-central potential in cosmic string space time have been analyzed by Afshardoost and Hassanabadi [12], they obtained solution of cosmic string for Poschl Teller Double Ring Shaped Coulomb potential. Lots of potentials like Woods-Saxon, Double Ring Shape Oscillator and other non-central potentials [7,13] have been solved in cosmic string. These potential system were solved using some methods like Supersymmetry Quantum Mechanics method, Confluent Hypergeometric method, Ansatz method. That methods can be used to obtain energy spectrum and wave function of the particle [5,11,14]. Ikot et. al, had been solved solution of cosmic string equation coulomb plus improved ring shaped potentials using Nikiforov Uvarov method [15].

In this paper, we find solution of time independent cosmic string equation using Supersymmetry Quantum Mechanics method for Harmonic Oscillator and Poschl Teller non-central potential. The Harmonic Oscillator and Poschl Teller non-central potential can be written:

$$V(r,\theta)=\frac{1}{2}M\omega^2r^2+\frac{\hbar^2}{2Mr^2}\left(\frac{\kappa(\kappa-1)}{\sin^2\theta}+\frac{\lambda(\lambda-1)}{\cos^2\theta}\right) \tag{1}$$

Where $0\le r\le\infty$, $\lambda>\kappa-1$, and $0\le\theta\le\pi$, r is displacement, M is the reduced mass and ω is the angular frequency [16,17,18]. This study refers to research is conducted by Hasanabadi et.al and Afshardoost and Hassanabadi [11,12]. This paper is organized as follow. In section 2, we begin with experimental which including time independent cosmic string and supersymmetri quantum mechanics method. In section 3, result and discussion are presented. In section 4, we summaize our result.

2. Experimental

The metric of cosmic string space time for spherical coordinates [15] is defined as:

$$ds^2=-dt^2+dr^2+r^2d\theta^2+\alpha^2r^2\sin^2\theta\, d\varphi^2 \tag{2}$$

Where r, θ and ϕ were in the range $0\le r\le\infty$, $0\le\theta\le\pi$ and $0\le\phi\le2\pi$. With parameter $\alpha=1-4G\mu$ in $[0,1]$ interval and μ is density of cosmic string. When α approaches 1, cosmic string space time is reduced to minkowski space time. To solve space component, we can ignore the time component and then the metric become time independent cosmic string. From equation (2) we use separation variable. Matric of the time independent cosmic string is:

$$g_{\mu\nu}=\begin{bmatrix}1 & 0 & 0\\ 0 & r^2 & 0\\ 0 & 0 & \alpha^2r^2\sin^2\theta\end{bmatrix} \tag{3}$$

From this matric, we obtain Laplacian equation:

$$\nabla^2_{LB}=\frac{1}{\sqrt{g}}\frac{\partial}{\partial x^\mu}\left(\sqrt{g}g^{\mu\nu}\frac{\partial}{\partial x^\nu}\right) \tag{4}$$

$$\nabla^2_{LB}=\frac{1}{r^2}\frac{\partial}{\partial r}\left(r^2\frac{\partial}{\partial r}\right)+\frac{1}{r^2\sin^2\theta}\frac{\partial}{\partial\theta}\left(\sin\theta\frac{\partial}{\partial\theta}\right)+\frac{1}{\alpha^2r^2\sin^2\theta}\frac{\partial^2}{\partial^2\varphi} \tag{5}$$

where ∇^2_{LB} is Laplace-Beltrami operator, μ and $\nu=1,2,3$ and $g=\det g^{\mu\nu}$.

Supersymmetry quantum mechanics system as a system that consists of a super charge operator Q_i which is commute with supersymmetry Hamiltonian H_{SS} [19] is:

$$[Q_i,H_{SS}]=0 \text{ with } i=1,2,\dots,N \tag{6}$$

where N is the number of super power that meets the anti-commutation relationship

$$\{Q_i,Q_j\}=\delta_{ij}H_{SS} \tag{7}$$

For the simplest supersymmetry system, N = 2 with super charge operators Q_1 and Q_2. Super charge operators Q_1 and Q_2 are defined as:

$$Q_1=\frac{1}{\sqrt{2}}\left(-\sigma_1\frac{p}{(2M)^{1/2}}+\sigma_2\phi(x)\right) \text{ and } Q_2=\frac{1}{\sqrt{2}}\left(-\sigma_1\phi(x)+\sigma_2\frac{p}{(2M)^{1/2}}\right) \tag{8}$$

where σ_1 and σ_2 are Pauli spin matrices, $p=-i\hbar\frac{d}{dx}$ is a linear momentum operator. By using equation (7) and equation (8), components of supersymmetry Hamiltonian are showed as:

$$H_{SS}=\begin{pmatrix} -\frac{\hbar^2}{2M}\frac{d^2}{dx^2}+\frac{\hbar}{(2M)^{1/2}}\frac{d\phi}{dx}+\phi^2 & 0 \\ 0 & -\frac{\hbar^2}{2M}\frac{d^2}{dx^2}-\frac{\hbar}{(2M)^{1/2}}\frac{d\phi}{dx}+\phi^2 \end{pmatrix}=\begin{pmatrix} H_- & 0 \\ 0 & H_+ \end{pmatrix} \quad (9)$$

We shall write $H_{SS}=H_\pm$

$$H_-=-\frac{\hbar^2}{2M}\frac{d^2}{dx^2}+V_-(x) \quad \text{with} \quad V_-(x)=\phi^2-\frac{\hbar}{(2M)^{1/2}}\frac{d\phi}{dx} \quad (10)$$

$$H_+=-\frac{\hbar^2}{2M}\frac{d^2}{dx^2}+V_+(x) \quad \text{with} \quad V_+(x)=\phi^2+\frac{\hbar}{(2M)^{1/2}}\frac{d\phi}{dx} \quad (11)$$

where $V_-(x)$ and $V_+(x)$ are supersymmetry partner for supersymmetry Hamiltonian $H_\pm$. Supersymmetry Hamiltonian is factorized as

$$H_-=A^+A^- \quad \text{and} \quad H_+=A^-A^+ \quad (12)$$

$$A^-=\frac{\hbar}{(2M)^{1/2}}\frac{d}{dx}+\phi \quad (13)$$

$$A^+=-\frac{\hbar}{(2M)^{1/2}}\frac{d}{dx}+\phi \quad (14)$$

where A^+ is raising operator and $A^-$is lowering operator. Supertpartner potential is said to be shape invariant if these potential are following this conditions:

$$V_+(x:a_j)=V_-(x:a_{j+1})+R(a_{j+1}) \quad (15)$$

$$V_+(x:a_j)=\phi^2(x:a_j)+\frac{\hbar}{(2M)^{1/2}}\phi'(x:a_j) \quad (16)$$

$$V_-(x:a_j)=\phi^2(x:a_j)-\frac{\hbar}{(2M)^{1/2}}\phi'(x:a_j) \quad (17)$$

where $j=0,1,2,...$ and parameter a determined sequentially and R is constant which don't depend on x. Energy level for shape invariant can be find by using:

$$H^{(S)}=-\frac{\hbar}{(2M)^{1/2}}\frac{d}{dx}+V_-(x:a_S)+\sum_{k=1}^{S}R(a_k) \quad (18)$$

if that equation is operated with ground state wave function, than ground state energy level will be obtained from H which is written as:

$$E_n^-=\sum_{k=1}^{n}R(a_k) \quad (19)$$

Finally, total energy level is obtained:

$$E_n=\varepsilon_0+E_n^- \quad (20)$$

where ε_0 is a ground state energy of the system. Based on characteristic of lowering operator, ground state operator wave function is obtained as:

$$A^-\cdot\psi_0^{(-)}=0 \quad (21)$$

The n state wave function can be obtained by multiplying the raising operator with ground state wave function. The n state wave function, following form:

$$\psi_n^{(-)}(x:a_0)=A^+(x:a_0)\cdot\psi_{n-1}^{(-)}(x:a_1) \quad (22)$$

That explanation is a simple procedure to construct the hierarchy of Hamiltonian. In the next section, the solution of time independent cosmic string for harmonic oscillator and poschl teller non-central potential will be investigated.

IOP Conf. Series: Materials Science and Engineering **578** (2019) 012093 doi:10.1088/1757-899X/578/1/012093

3. Results and discussions

In this research, we solve the time independent cosmic string equation use spherical coordinates in the part of space using supersymmetry quantum mechanics method. The time independent cosmic string equation is reduced to Schrodinger Equation. Laplace-Beltrami operator equation (5) and Harmonic Oscillator plus Poschl Teller non-central potential (1) are substituted to the three dimensional Schrodinger like equation, and then we obtain the formulation of Schrodinger equation:

$$\left\{\begin{matrix} -\frac{\hbar^2}{2M}\left[\frac{1}{r^2}\frac{\partial}{\partial r}\left(r^2\frac{\partial}{\partial r}\right)+\frac{1}{r^2\sin^2\theta}\frac{\partial}{\partial\theta}\left(\sin\theta\frac{\partial}{\partial\theta}\right)+\frac{1}{\alpha^2 r^2\sin^2\theta}\frac{\partial^2}{\partial^2\varphi}\right] \\ +\left[\frac{1}{2}M\omega^2 r^2+\frac{\hbar^2}{2Mr^2}\left(\frac{\kappa(\kappa-1)}{\sin^2\theta}+\frac{\lambda(\lambda-1)}{\cos^2\theta}\right)\right] \end{matrix}\right\}\psi(r,\theta,\varphi)=E\psi(r,\theta,\varphi) \tag{23}$$

Separation variable is applied to the Schrodinger equation (23). We obtain three part of equation from separation variable, there are radial, angular and azimuth.

$$\frac{1}{R(r)}\frac{\partial}{\partial r}r^2\frac{\partial R(r)}{\partial r}-\frac{2M}{\hbar^2}\frac{1}{2}M\omega^2 r^4+\frac{2M}{\hbar^2}E-l(l+1)=0 \tag{24}$$

$$\frac{1}{\Theta(\theta)\sin\theta}\frac{\partial}{\partial\theta}\sin\theta\frac{\partial\Theta(\theta)}{\partial\theta}-\left[\frac{\kappa(\kappa-1)}{\sin^2\theta}+\frac{\lambda(\lambda-1)}{\cos^2\theta}\right]+l(l+1)-\frac{m^2}{\alpha^2\sin^2\theta}=0 \tag{25}$$

$$\frac{1}{\Phi(\varphi)}\frac{1}{\alpha^2\sin^2\theta}\frac{\partial^2\Phi(\varphi)}{\partial\varphi^2}+\frac{m^2}{\alpha^2\sin^2\theta}=0 \tag{26}$$

Angular part must be solved, to obtain value of constants l. We assume that $\Theta(\theta)=\frac{H(\theta)}{(\sin\theta)^{1/2}}$ for equation (25), and we obtain:

$$-\frac{\hbar^2}{2M}\frac{d^2H(\theta)}{d\theta^2}+\frac{\hbar^2}{2M}\left[\frac{\kappa(\kappa-1)\alpha^2-\frac{1}{4}\alpha^2+m^2}{\sin^2\theta}+\frac{\lambda(\lambda-1)}{\cos^2\theta}\right]H(\theta)=\frac{\hbar^2}{2M}\left[l(l+1)+\frac{1}{4}\right]H(\theta) \tag{27}$$

Equation (27) is the one dimensional Schrodinger equation which have effective potential. Formulation of the effective potential is:

$$V_{eff}=\frac{\hbar^2}{2M}\left[\frac{\kappa(\kappa-1)\alpha^2-\frac{1}{4}\alpha^2+m^2}{\sin^2\theta}+\frac{\lambda(\lambda-1)}{\cos^2\theta}\right] \tag{28}$$

where:

$$E'=\frac{\hbar^2}{2M}\left[l(l+1)+\frac{1}{4}\right] \tag{29}$$

To obtain the energy levels and the wave function of angular part we must set a superpotential. Formulation of superpotential:

$$\phi=A\tan\theta+B\tan\theta \tag{30}$$

Superpotential is a formulation which like an effective potential. From equation (30) we obtain formulation of ϕ^2 and ϕ'. Then, we obtain parameters of superpotential:

$$A=\frac{\hbar}{(2M)^{1/2}}\lambda \quad\text{and}\quad B=-\frac{\hbar}{(2M)^{1/2}}\kappa' \tag{31}$$

where, the value of κ' and ε_0:

$$\kappa'=\frac{1}{2}+\left[\kappa(\kappa-1)+\frac{m^2}{\alpha^2}\right]^{1/2} \tag{32}$$

$$\varepsilon_0=\frac{\hbar^2}{2M}\lambda^2+\frac{\hbar^2}{2M}\kappa'^2+\frac{\hbar^2}{2M}2\lambda\kappa' \tag{33}$$

To obtain angular wave function we need raising and lowering operators. Then, we can obtain raising and lowering operators, by using equation (13) and (14):

$$A^{+}=-\frac{\hbar}{(2M)^{1/2}}\frac{d}{d\theta}+\frac{\hbar}{(2M)^{1/2}}\lambda\tan\theta-\frac{\hbar}{(2M)^{1/2}}\kappa^{'}\cot\theta \tag{34}$$

$$A^{-}=\frac{\hbar}{(2M)^{1/2}}\frac{d}{d\theta}+\frac{\hbar}{(2M)^{1/2}}\lambda\tan\theta-\frac{\hbar}{(2M)^{1/2}}\kappa^{'}\cot\theta \tag{35}$$

If the lowering operator is multiplied by the ground state wave function, the result well be zero like equation (21). Then we obtain ground state wave function:

$$H_{0}^{(-)}(\theta)=C\cos\theta^{\lambda}\sin\theta^{\left\{\frac{1}{2}+\left[\kappa(\kappa-1)+\frac{m^{2}}{\alpha^{2}}\right]^{1/2}\right\}} \tag{36}$$

Where C was constant. Equation (36) is non-normalized angular ground state wave function. The first state wave function for angular part is obtained by multiplying the raising operator with ground state wave function like equation (22). The first state wave function for angular part is obtained (37).

$$H_{1}^{(-)}(\theta)=\left\{\begin{array}{l}\frac{\hbar}{(2M)^{1/2}}C\left\{(2\lambda+1)\sin\theta^{2}-\left\{2\left\{\frac{1}{2}+\left[\kappa(\kappa-1)+\frac{m^{2}}{\alpha^{2}}\right]^{1/2}\right\}+1\right\}\cos\theta^{2}\right\}\\ \times\left\{\cos\theta^{\lambda}\sin\theta^{\left\{\frac{1}{2}+\left[\kappa(\kappa-1)+\frac{m^{2}}{\alpha^{2}}\right]^{1/2}\right\}}\right\}\end{array}\right\} \tag{37}$$

To obtain solution of energy levels, we use of superpartner potential equation (16) and equation (17). Superpartner potential for angular part are:

$$V_{-}(\theta:\kappa^{'}{}_{0}:\lambda_{0})=\left[\frac{\hbar^{2}}{2M}\lambda(\lambda-1)\sec^{2}\theta+\frac{\hbar^{2}}{2M}\kappa^{'}(\kappa^{'}-1)\csc^{2}\theta-\frac{\hbar^{2}}{2M}(\lambda+\kappa^{'})^{2}\right] \tag{38}$$

$$V_{+}(\theta:\kappa^{'}{}_{0}:\lambda_{0})=\left[\frac{\hbar^{2}}{2M}\lambda(\lambda+1)\sec^{2}\theta+\frac{\hbar^{2}}{2M}\kappa^{'}(\kappa^{'}+1)\csc^{2}\theta-\frac{\hbar^{2}}{2M}(\lambda+\kappa^{'})^{2}\right] \tag{39}$$

Energy levels for angular part based on equation (20), following form:

$$E_{n,\theta}=\frac{\hbar^{2}}{2M}\left[(\lambda+n_{\theta})+(\kappa^{'}+n_{\theta})\right]^{2} \tag{40}$$

By using equation (30) energy levels following form:

$$E_{n,\theta}=\frac{\hbar^{2}}{2M}\left\{(\lambda+n_{\theta})+\left\{\frac{1}{2}+\left[\kappa(\kappa-1)+\frac{m^{2}}{\alpha^{2}}\right]^{1/2}+n_{\theta}\right\}\right\}^{2} \tag{41}$$

By substitution equation (41) to equation (29), we obtain value of constant l:

$$l=(\lambda+n_{\theta})+\left\{\left[\kappa(\kappa-1)+\frac{m^{2}}{\alpha^{2}}\right]^{1/2}+n_{\theta}\right\} \tag{42}$$

To solve radial part, we assume that $R(x)=\frac{X(r)}{r}$ and we obtain:

$$-\frac{\hbar^{2}}{2M}\frac{d^{2}X(r)}{dr^{2}}+\left[\frac{1}{2}M\omega^{2}r^{2}+\frac{\hbar^{2}}{2M}\frac{l(l+1)}{r^{2}}\right]X(r)=EX(r) \tag{43}$$

Equation (43) is one dimension Schrodinger equation which have an effective potential, following formulation:

$$V_{eff}=\frac{1}{2}M\omega^{2}r^{2}+\frac{\hbar^{2}}{2M}\frac{l(l+1)}{r^{2}} \tag{44}$$

We must write a superpotential. From superpotential formulation we also obtain ϕ^2 and ϕ'. which that formulation will be the same formulation with effective potential,

$$\phi = Cr + \frac{D}{r} \tag{45}$$

and then, we obtain parameters of superpotential:

$$C = \left(\frac{1}{2}M\right)^{1/2}\omega \quad \text{and} \quad D = -\frac{\hbar}{(2M)^{1/2}}(l+1) \tag{46}$$

$$\varepsilon_0 = \hbar\omega\left(l + \frac{3}{2}\right) \tag{47}$$

Then, to obtained radial wave function we also need raising and lowering operators. We can write raising and lowering operators for radial part base on equation (13) and (14), following form:

$$A^+ = -\frac{\hbar}{(2M)^{1/2}}\frac{d}{dx} + \left(\frac{1}{2}M\right)^{1/2}\omega r - \frac{\hbar}{(2M)^{1/2}}\frac{(l+1)}{r} \tag{48}$$

$$A^- = \frac{\hbar}{(2M)^{1/2}}\frac{d}{dx} + \left(\frac{1}{2}M\right)^{1/2}\omega r - \frac{\hbar}{(2M)^{1/2}}\frac{(l+1)}{r} \tag{49}$$

By using equation (21) and (22). Ground state wave function for radial part and first state wave function are obtained, following:

$$R_0^{(-)} = Cr^{(l+1)}\exp\left[-\frac{1}{2}\frac{M\omega r^2}{\hbar}\right] \tag{50}$$

$$R_1^{(-)} = \left\{\left\{-\frac{\hbar}{(2M)^{1/2}}C\left[(2l+3)r^{(l+1)} - \frac{M\omega}{\hbar}r^{(l+3)}\right]\right\} + \left(\frac{1}{2}M\right)^{1/2}C\omega r^{(l+3)}\right\}\left[\exp\left[-\frac{1}{2}\frac{M\omega r^2}{\hbar}\right]\right] \tag{51}$$

By using equation (16) and (17), Superpartner potential for radial part can be written:

$$V_-(r:l_0) = \frac{1}{2}M\omega^2 r^2 + \frac{\hbar^2}{2Mr^2}l(l+1) - \hbar\omega\left(l + \frac{3}{2}\right) \tag{52}$$

$$V_+(r:l_0) = \frac{1}{2}M\omega^2 r^2 + \frac{\hbar^2}{2Mr^2}[(l+1)(l+2)] - \hbar\omega\left(l + \frac{3}{2}\right) \tag{53}$$

and by using superpartner potential for radial part, energy level is obtained:

$$E_{n_r} = \hbar\omega\left(l + \frac{3}{2} + 2n_r\right) \tag{54}$$

by using equation (54) and equation (42) energy levels of time independent cosmic string for Harmonic Oscillator and Poschl Teller non-central potential.

$$E_{n_r,n_\theta} = \hbar\omega\left\{(\lambda + n_\theta) + \left\{\left[\kappa(\kappa-1) + \frac{m^2}{\alpha^2}\right]^{1/2} + n_\theta\right\} + \frac{3}{2} + 2n_r\right\} \tag{55}$$

Wave function of angular and radial part can be visualized. Ground state angular wave function correspond to the equation (36). Figure 1, is three dimensional ground state angular wave function in spherical coordinates for several parameter variations. Based on Figure 1, is show that probability of particle distribution based on spherical coordinates which depended by θ. These wave function are result of visualization by assuming $0 \le \theta \le \pi$. Based on Figure 1, variation of parameters α, λ and κ, cause variation of wave amplitude in the ground state wave function for angular part. In the Figure 1(a) and Figure 1(b), value of parameter α affect to the wave characteristic. If value of parameter α decrease, amplitude of wave function also decrease. So, for the smaller value of parameter α, probability of particle distribution decrease.

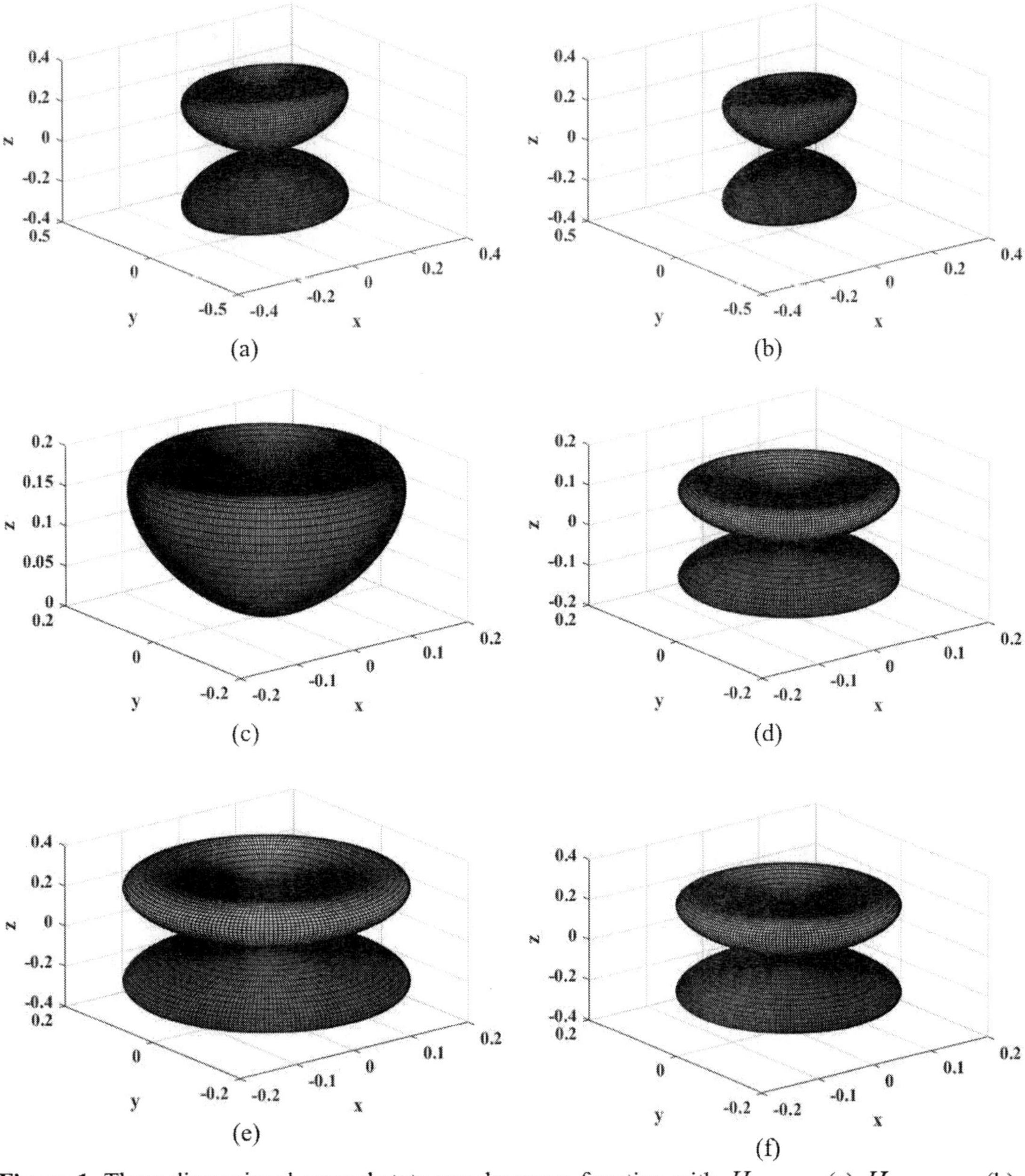

Figure 1. Three dimensional ground state angular wave function with $H_{(n,\lambda,\kappa,\alpha)}$ (a) $H_{(0,1,1.5,0.8)}$, (b) $H_{(0,1,1.5,0.5)}$, (c) $H_{(0,2,1.5,0.9)}$, (d) $H_{(0,3,1.5,0.9)}$, (e) $H_{(0,1,2.5,0.9)}$, (f) $H_{(0,1,3.5,0.9)}$

Value of parameter λ and κ also affect to the amplitude of ground state wave function. In the Figure 1(c) and Figure 1(d), if value of parameter λ increase, amplitude of wave function decrease. But there are different waveform, if value of parameter λ is even number, and odd number. In the Figure 1(e) and Figure 1(f), if value of parameter κ increase then amplitude of wave function decrease. For increasing value of parameter κ, probability of particle distribution decrease.

IOP Conf. Series: Materials Science and Engineering **578** (2019) 012093 doi:10.1088/1757-899X/578/1/012093

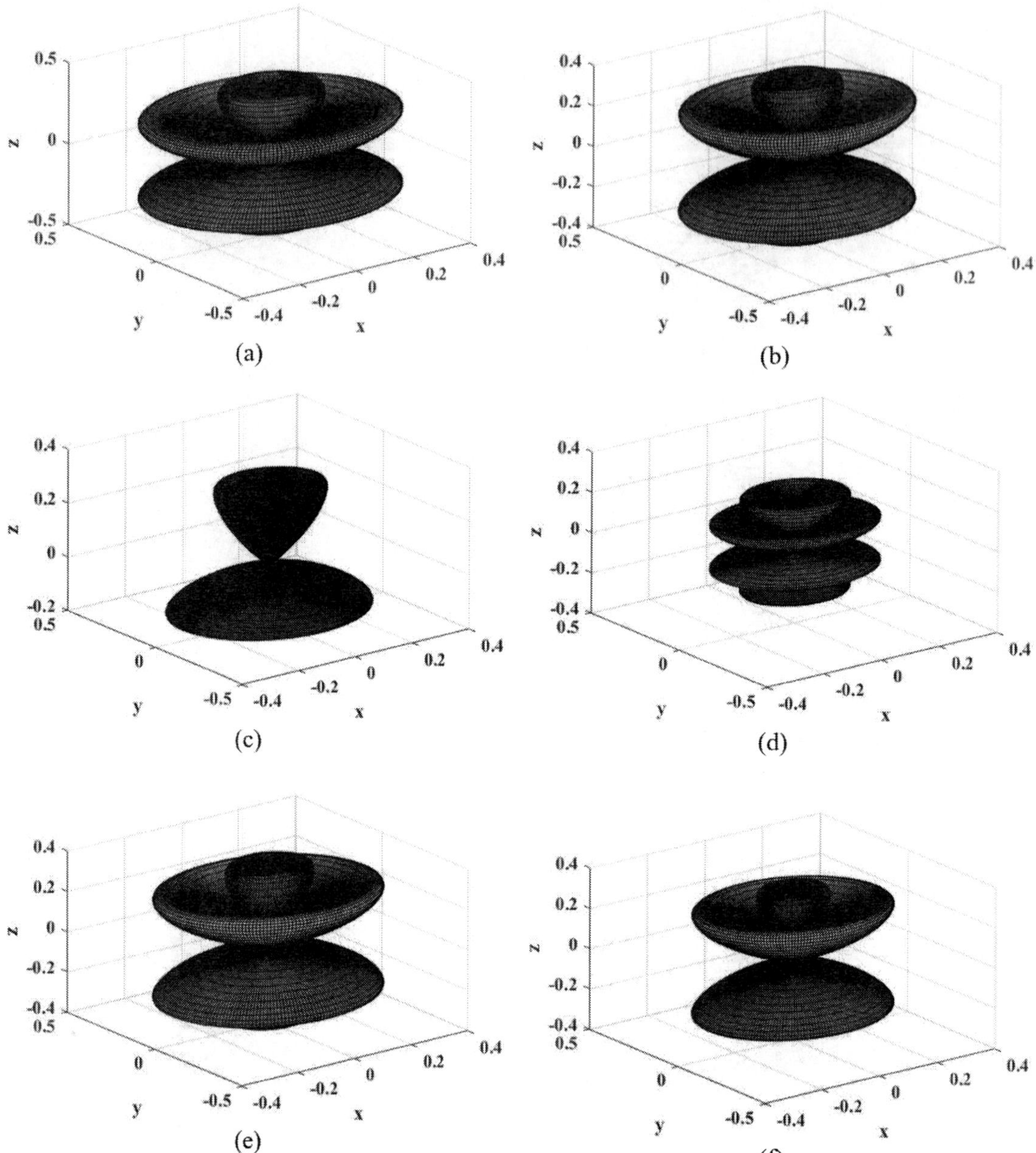

Figure 2. Three dimensional first state angular wave function with $H_{(n,\lambda,\kappa,\alpha)}$ (a) $H_{(1,1,1.5,0.8)}$, (b) $H_{(1,1,1.5,0.5)}$, (c) $H_{(1,2,1.5,0.9)}$, (d) $H_{(1,3,1.5,0.9)}$, (e) $H_{(1,1,2.5,0.9)}$, (f) $H_{(1,1,3.5,0.9)}$

First state angular wave function correspond to the equation (37). Figure 2, is three dimensional first state angular wave function in spherical coordinates for several parameter variations. Based on Figure 2, show that variation of parameters α, λ and κ, obtain variation of wave amplitude in the first state wave function for angular part. In the Figure 2(a) and Figure 2(b), if value of parameter α decrease, amplitude of wave functions also decrease. For the smaller value of parameter α, probability of particle distribution decrease. Value of parameter λ and κ also affect to the amplitude of ground state wave function. In the Figure 2(c) and Figure 2(d), if value of parameter λ increase, amplitude of wave function decrease. For increasing value of parameter λ, probability of particle distribution decrease. In the Figure 2(e) and Figure 2(f), if value of parameter κ increase then amplitude of wave function decrease. For increasing value of parameter κ, probability of particle distribution decrease.

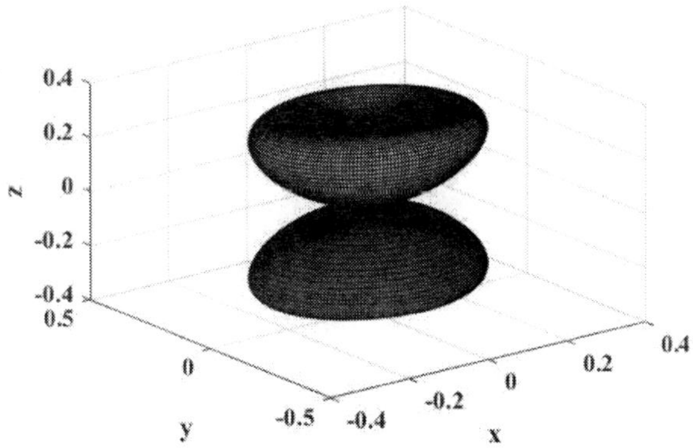

(a)

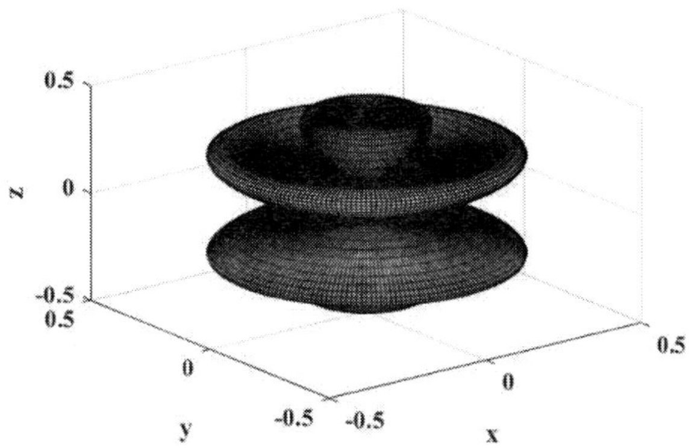

(b)

Figure 3. Three dimensional angular wave function with $H_{(n,\lambda,\kappa,\alpha)}$ (a) $H_{(0,1,1.5,0.9)}$, (b) $H_{(1,1,1.5,0.9)}$,

Based on Figure 3, the difference between ground state wave function and first state wave function is the largest number of wave, that can be concluded that increasing value of n_θ caused the number of wave also increase. Beside that, angular wave function also describe the probability of particles in the spherical coordinate system. The amount of deviation on the x, y axis indicated that probability value of a particle being located. There is a difference between amplitude of ground state wave function and amplitude of first state wave function, so there is a possibility of particles moving.

Ground state wave function in Figure 4 correspond to the equation (50). Based on Figure 4, show that variation of parameters α, λ and κ, obtain variation of wave amplitude in the ground state wave function for radial part. Waveform is damped along of the r axis and probability of particle distribution decrease when it is far from the center. In the Figure 4(a), if value of parameter α decrease, amplitude of wave function increase and the amplitude shift smaller in the r axis. For the smaller value of parameter α, probability of particle distribution increase. In the Figure 4(b), if value of parameter λ increase, amplitude of wave function increase and the amplitude shift further in the r axis. For increasing value of parameter λ, probability of particle distribution also increase. In the Figure 4(c), if value of parameter κ increase then amplitude of wave function increase. For increasing value of parameter κ, the amplitude shift further in the r axis, and probability of particle distribution increase. The biggest probability of distribution in ground state wave function is λ variation.

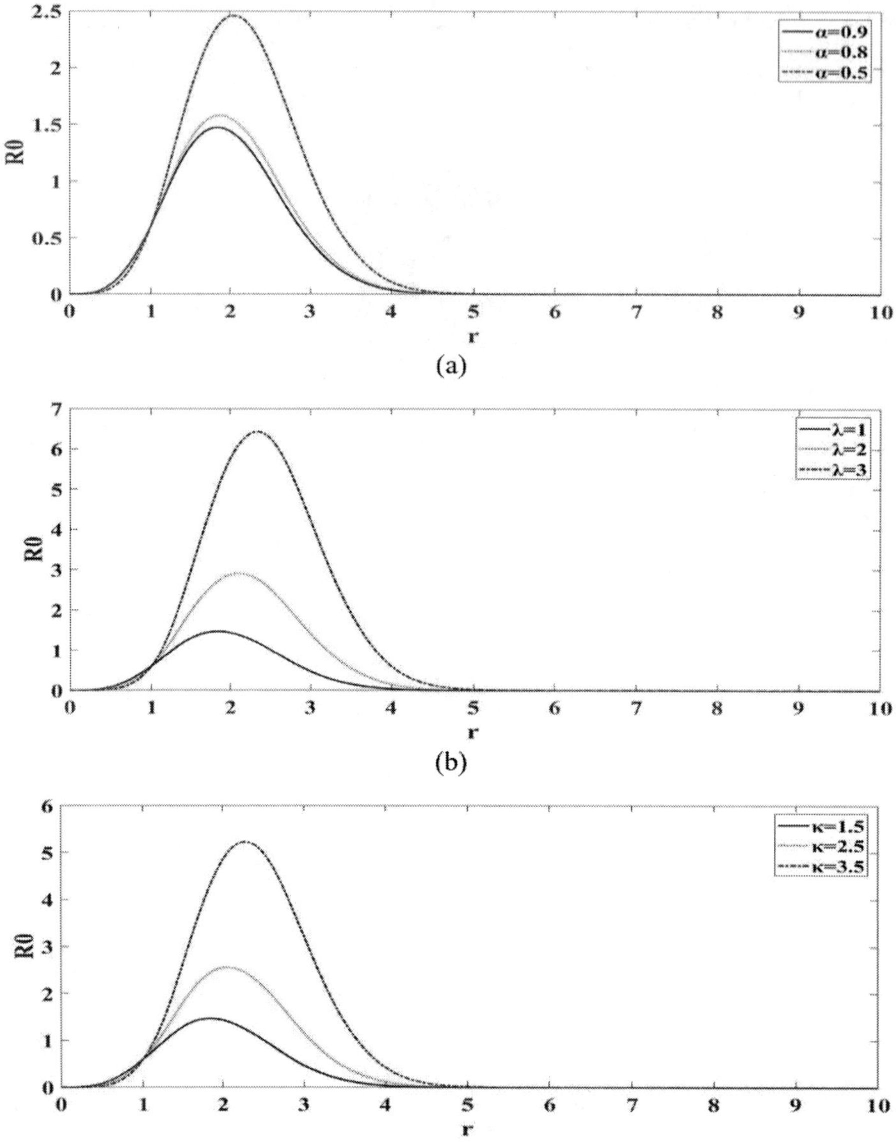

Figure 4. Ground state wave function for radial part with (a) α variation, (b) λ variation and (c) κ variation

First state wave function in Figure 5 correspond to the equation (51). Based on Figure 5, show that variation of parameters α, λ and κ, obtain variation of wave amplitude in the first state wave function for radial part. Waveform is damped along of the r axis and probability of particle distribution decrease when it is far from the center. In the Figure 5(a), if value of parameter α decrease, amplitude of wave function increase but value of amplitude is negative and the amplitude shift smaller in the r axis. For the smaller value of parameter α, probability of particle distribution increase. In the Figure 3(b), if value of parameter λ increase, amplitude of wave function increase but value of amplitude is negative and the amplitude shift further in the r axis. In the Figure 5(c), if value of parameter κ

increase then amplitude of wave function increase but value of amplitude is negative and the amplitude shift further in the r axis. For increasing value of parameters λ and κ, probability of particle distribution increase. The biggest probability of distribution in first state wave function is λ variation. The greater the value of n_r then the greater the wave form is damped. So, the probability of particle distribution is smaller and there is a difference between amplitude of ground state wave function and amplitude of first state wave function. There is a possibility of particles moving

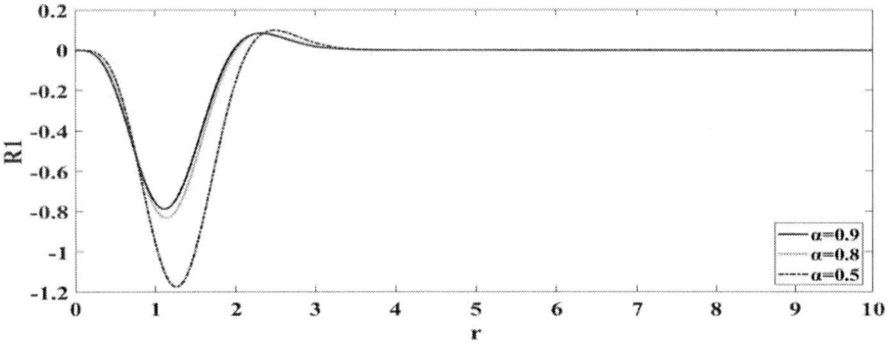

(a)

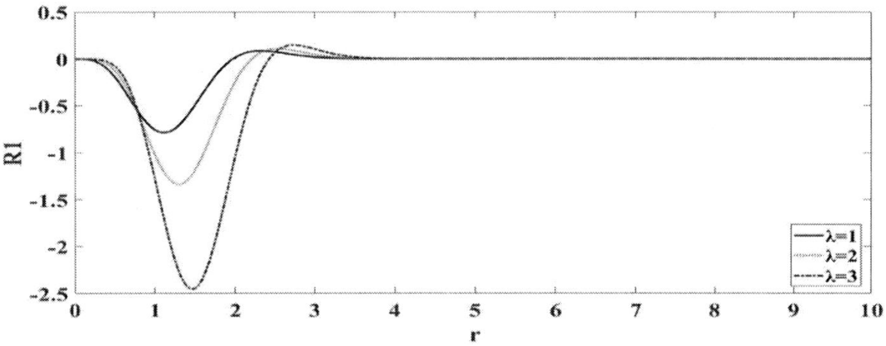

(b)

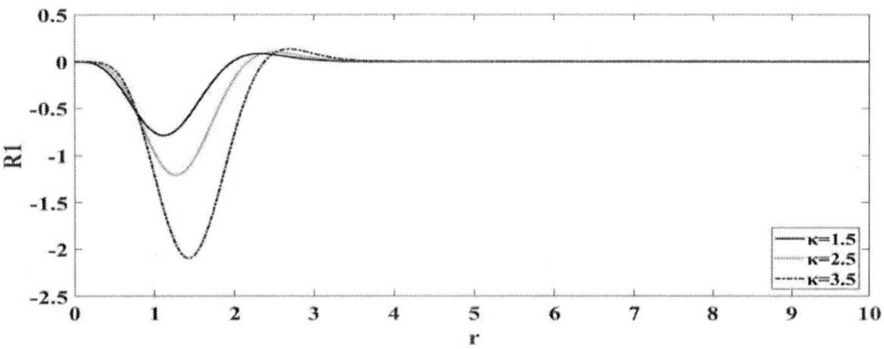

(c)

Figure 5. First state wave function for radial part with (a) α variation, (b) λ variation and (c) κ variation

Based on equation (55), energy levels can be obtained. Based on Table 1, show energy levels time independent for harmonic oscillator plus poschl teller non-central potential. The value of parameter λ and κ have a significant effect. Energy levels for $\lambda = 0$ and $\kappa = 0$ is smaller than energy levels for $\lambda \neq 0$ and $\kappa \neq 0$. Parameter $\lambda \neq 0$ and $\kappa \neq 0$ caused increasing energy levels.

Table 1. Energy levels time independent for harmonic oscillator and poschl teller non-central potential

n_r	n_θ	λ	κ	α	m	l	Energy
0	0	0	0	0.9	1	2.1111	4.6111
1	0	0	0	0.9	1	2.1111	6.6111
2	0	0	0	0.9	1	2.1111	8.6111
3	0	0	0	0.9	1	2.1111	10.6111
0	1	1	1.5	0.9	1	2.4087	5.9087
1	1	1	1.5	0.9	1	2.4087	7.9087
2	1	1	1.5	0.9	1	2.4087	9.9087
3	1	1	1.5	0.9	1	2.4087	11.9087

Table 2. Energy levels time independent for harmonic oscillator and poschl teller non-central potential (κ , λ , α and n_θ variation)

n_r	n_θ	λ	κ	α	m	l	Energy
1	1	1	1.5	0.9	1	2.4087	7.9087
1	1	1	2.5	0.9	1	3.2326	8.7326
1	1	1	3.5	0.9	1	4.1598	9.6598
1	1	2	1.5	0.9	1	2.4087	8.9087
1	1	3	1.5	0.9	1	2.4087	9.9087
1	1	1	1.5	0.8	1	2.5207	8.0207
1	1	1	1.5	0.5	1	3.1794	8.6794
1	2	1	1.5	0.9	1	2.4087	9.9087
1	3	1	1.5	0.9	1	2.4087	11.9087

Based on Table 2, show that the value of parameter κ have a significant effect to the energy levels. The increasing value of parameter κ caused increasing in energy levels. The value of parameter λ also have a significant effect to the energy levels. The increasing in energy levels also given by λ variation. Increasing of the value of parameter λ caused increasing in energy levels. But deferent result is only given by variation of parameter α . The smaller value of parameter α caused increasing in energy levels. Energy levels also associate with n. From variation of quantum number n_r and n_θ , both give the same effect. These number give the same increasing in energy levels. The biggest increasing in energy levels is caused by increasing number of n_r and n_θ .

4. Conclusion

We investigate energy levels and wave function of time independent cosmic string equation for Harmonic Oscillator plus Poschl Teller non-central potential. Equation have been solved using supersymmetry quantum mechanics method. By using variable separation method, the time independent cosmic string with non-central potential be modified into one dimensional Schrodinger equation. Energy levels and wave function of corresponding the Schrodinger equation have been solved using supersymmetry quantum mechanics method. We use superpartner potential to solve the energy levels and use raising and lowering operators to obtain wave function. The energy levels and radial wave function are from radial part of Schrodinger like equation. The angular wave function are obtained from angular part of Schrodinger equation. Energy levels for $\lambda = 0$ and $\kappa = 0$ have smaller value than energy levels by using parameter $\lambda \neq 0$ and $\kappa \neq 0$. Parameters κ , λ , α and n_θ have the significant effect to the energy levels. The effect of the non-central potential caused energy level increase for variation of parameter κ , λ , α and n_θ. Increasing value of parameters λ or κ or n_θ cause increasing in energy levels. But, the decreasing of parameter α cause increase in energy levels.

Acknowledgement
This research was partly supported by Hibah Pascasarjana UNS 543/UN27.21/PP/2018.

References

[1] Kibble T W 1976 *J. Phys. A: Math. Gen.* **9** 1387-98
[2] Jensen B & Soleng H H 1992 *Phys. Rev. D.* **45** 2528-33
[3] Peter P 2001 *New Astron Rev.* **45** 277-281
[4] Bouchet F R, Peter P, Riazuelo A & Sakellariadou M 2001 *Phys. Rev. D.* **65** 1-4
[5] Muniz C R, Bezerra V B & Cunha M S 2014 *Ann. Phys.* **350** 105-111
[6] Salazar-Ramírez M, Ojeda-Guillén D & Mota R D 2016 *Ann. Phys.* **372** 283-296
[7] Jusufi K & Jakimovski D 2012 *Physica Macedonica.* **61** 85-89
[8] Medeiros E F & de Mello E B 2012 *Eur. Phys. J. C.* **72** 2051
[9] Bakke K 2013 *Gen. Relativ. Gravit.* **45** 1847-1859
[10] Boumali A & Messai N 2017 *Can. J. Phys.* **95** 999-1004
[11] Hassanabadi H, Afshardoost A, & Zarrinkamar S 2015 *Ann. Phys.* **356** 346-351
[12] Afshardoost A, & Hassanabadi H 2015 *Can. J. Phys.* **94** 1-4
[13] Chen C Y, You Y, Wang X H, & Dong S H 2013 *Phys. Lett. A.* **377** 1521-25
[14] Santos L C N & Barros Jr C C 2018 *Eur. Phys. J. C.* **78** 13
[15] Ikot A N, Abbey T M, Chukwuocha E O & Onyeaju M C 2016 *Can. J. Phys.* **94** 1-11
[16] Suparmi A, Cari C, Handika J, Yanuarief C & Marini H 2012 *J. Appl. Phys.* **2** 43-51
[17] Suparmi A, Cari C & Handika J 2013 *J. Phys. Conf. Ser.* **423** 1-11
[18] Ita B I, Ikeuba A I & Ikot A N 2014 *Commun. Theor. Phys.* **61** 149-152
[19] Suparmi A 2011 *Mekanika Kuantum II* :FMIPA UNS ISBN 978-602-99344-2-7

IOP Conf. Series: Materials Science and Engineering **578** (2019) 012094 doi:10.1088/1757-899X/578/1/012094

Determination Of First Criticality In TRIGA 2000 Reactor Modificated Using Plate Type Fuel With MCNP6

I S Hardiyanti[1], Riyatun[1], Suharyana[1] and A Khakim[2]
[1]Physics Department, Faculty Mathematics and Natural Sciences, Sebelas Maret University, Jl. Ir. Sutami 36A Kentingan, Jebres, Surakarta 57126, Indonesia
[2]P2STPIBN, Nuclear Energy Regulatory Agency (BAPETEN), Jl. Gajah Mada No.8 Jakarta 10120 Indonesia

E-mail: iza@student.uns.ac.id

Abstract. This research offers a fuel conversion from cylinder type to plate type for solving the operation problems as a solution when the TRIGA reactor fuel production run out. The research has been implemented utilizing MCNP6 software to model and simulate the core of the TRIGA 2000 reactor plate type fuel. The geometry of the simulated reactor consists of a reflector, a moderator and a reactor tank. The fuel was U_3Si_2Al with 19.75% enrichment and the density was 2.96 gU/cm^3, the reflector was graphite and the control rod was AgInCd. The first criticality value with 15 fuel assemblies and the 4 control rods position at 11.97cm from the top surface of the reactor core was (1.00000 ± 0.00016) in agreement with the reference value for initial configuration of TRIGA reactor core was (1.00 ± 0.01).

1. Introduction

Reactors TRIGA (*Training, Research, Isotope Production by General Atomic*) are still used today by several countries to support their nuclear science and technology research. Indonesia has two TRIGA reactors, namely the TRIGA 2000 Bandung and the Kartini TRIGA MARK II. Their power are 2 MW and 250 kW, respectively. The numbers of the TRIGA reactors around the world at 2015 are 69 reactors, there are 38 reactors are operated whereas the rest, 31 reactors, are not operated or under decommissioning conditions [1]. *General Atomic* as the producer TRIGA reactors, plans not to supply UZrH fuel with cylindrical cladding [2]. This is not good for the sustainability of TRIGA operations in Indonesia. In other words, the TRIGA reactor fuel element should be replaced with another type.

Another reactor operated by BATAN is the RSG-GAS (Reaktor Serba Guna- GA Siwabessy). This reactor equiped plate type fuel produced by PT. INUKI for 25 years without problems. The plate type fuel element is U_3Si_2Al with an enrichment of 19.75%. On this basis, it was proposed to replace the cylinder type TRIGA 2000 into the plate type fuel. The use of domestic-made fuels if applied can increase independence in Indonesia's nuclear science [3].

Regarding the level of save and safety of the reactor, before the reactor is built all the components related to the reactor must be carefully tested. The testing or modelling carried out included neutronic, thermal-hydraulics and mechanical aspects.

Neutronic calculations can be performed using two methods, namely deterministic and statistical method. Monte Carlo is an example of the statistical method. The process simulated in Monte Carlo is

suharyana61@staff.uns.ac.id

a trace of neutron particles beginning when neutrons are born until they disappear. The simulation is done by selecting neutrons randomly by taking into account every possible interaction with the material [4].

The reactor is always designed with a certain safety criteria. It is intended that the safety of the reactor is guaranteed, so that the reactor can be operated according to its benefits and the impact or risk posed as small as possible. One of the safety parameters of reactor design that can be studied from the neutronic aspect is the first criticality. In this work, we report our study on the first criticality in the TRIGA 2000 reactor using plate type fuel with MCNP6 software.

2. Experimental

The parameters used to model the TRIGA 2000 core are presented in Table 1. The reactor core model is based on the TRIGA 2000 reactor core geometry. The modeled TRIGA 2000 reactor geometry is the core, graphite reflector, light water moderator and reactor tank. The core geometry and fuel input arrangement are shown in Figure 1 and 2, respectively.

Table 1. Parameter of fuel element and control element [5].

Parameter	Value
Fuel element dimension and control element	77.1 x 81 x 600 mm
Thickness of the fuel plate	1.3 mm
Width of the cooling channel	2.55 mm
Total of U_3Si_2Al plate per fuel assemblies	21
Total of U_3Si_2Al plate per control assemblies	15
Thickness of the fuel cladding	0.38 mm
Fuel material	U_3Si_2Al
Density	2.96 gU/cm^3
Neutron absorber	AgInCd
Material absorbent cladding	SS-321
Thickness absorbent cladding	0.85 mm

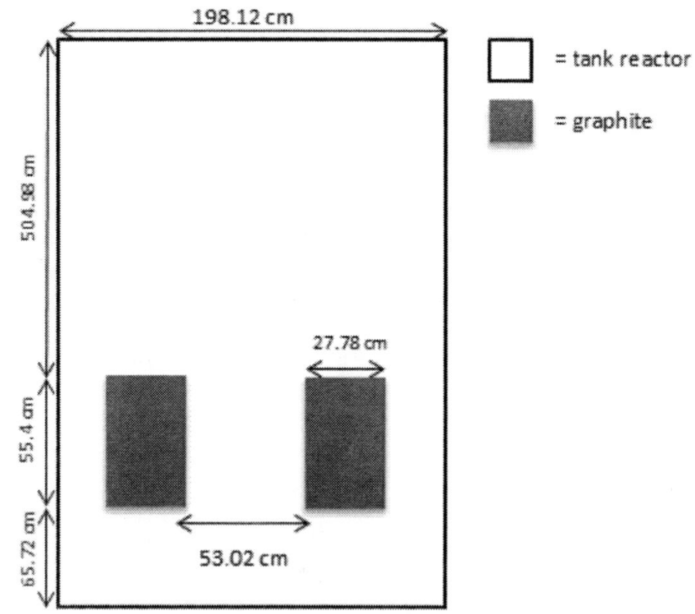

Figure 1. The side view of TRIGA 2000 reactor

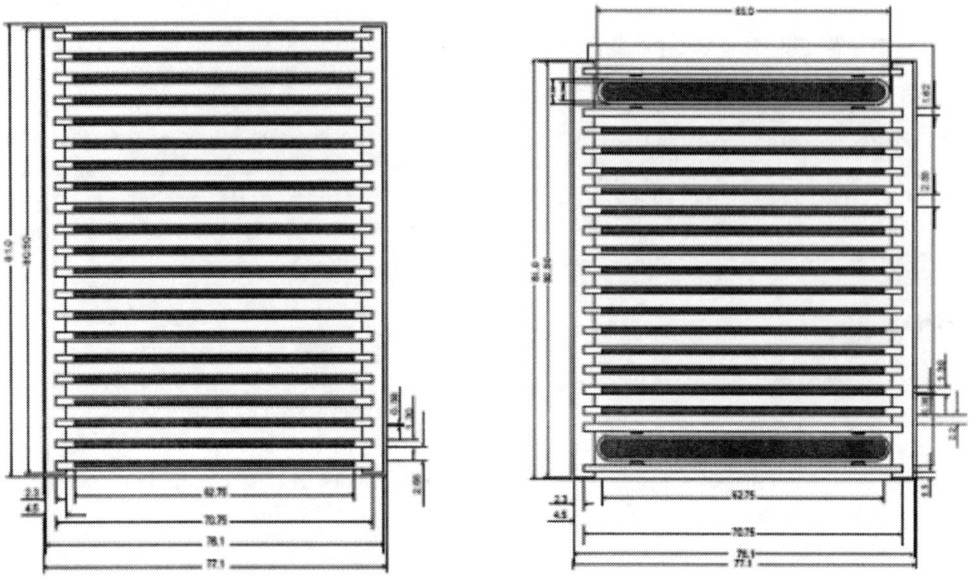

(a) (b)

Figure 2. (a) Fuel element design (b) control element design [5].

In this research, the number of simulated neutrons for each cycle is 100,000 with an initial criticality value is 1. The number of cycles passed is 50 (assuming that the 51th iteration has been convergent and deserves to be simulated statistically in subsequent iterations) while the number of iterations is 200. The position of the initial neutron source to trigger a chain reaction on each fuel assembly is 21 pieces while for each control rod assembly is 15 pieces.

3. Results and Discussion

Previously, we published our research with the maximum number of assemblies namely 20 fuel assemblies. The resulting value of the k_{eff} was (1.07100 ± 0.00017) [6]. The value is considerable higher than of the reference criticality. Accordingly, we carried out on the amount of fuel 14 and 15 assemblies in order to reduce the k_{eff} values. The results are written in Table 2.

Table 2. Calculation of the k_{eff} values

Number of assemblies	k_{eff}	Standard deviation
14	0.99022	0.00017
15	1.01377	0.00016

From these results it can be seen that the critical core configuration can be achieved with a minimum amount of 15 assemblies. However, the criticality value produced is still quite high so the next step is to change the position of the compensation control rods which were originally fully up to partially down. The reactor core geometry for the determination of the first criticality with 15 fuels is displayed in the visual editor MCNP6 as shown in Figure 3, while the results of criticality testing with changes in the position of the control rod are shown in Table 3.

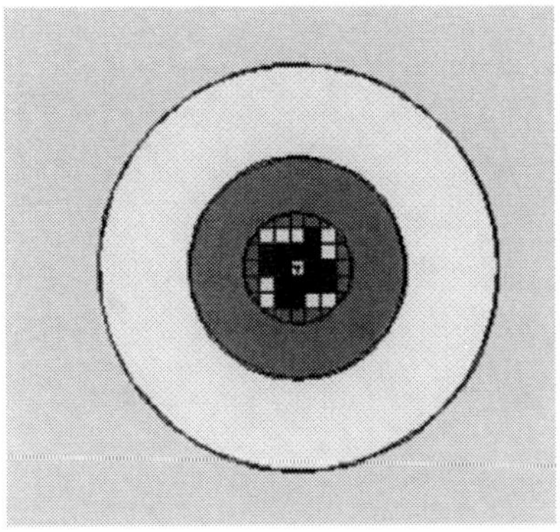

Figure 3. Reactor core geometry with 15 fuel elements from XY's viewing position

Table 3. The position of the compensation control rods on the reactor core

Position partially down (cm)	k_{eff}	Standard deviation
11.35	1.00089	0.00016
11.45	1.00042	0.00016
11.65	1.00032	0.00017
11.85	1.00031	0.00019
11.97	1.00000	0.00016
12.01	0.99958	0.00017
12.05	0.99948	0.00018
12.25	0.99887	0.00017
13.25	0.99744	0.00017

The simulation results show that with 15 fuels and the condition of the compensation control rod fully down at 11.97 cm from the top surface of the core, the first criticality is (1.00000 ± 0.00016). That means, chain fission reactions on the core can be maintained. Also it does mean the reactor are in a safe condition from the point view of neutronic [7, 8, 9, 10].

4. Conclusion

The first critical value in the TRIGA 2000 reactor type plate was obtained with 15 fuel assemblies with the compensation control rod position at 11.97 cm from the top core surface. The calculation value k_{eff} is (1.00000 ± 0.00016) so that the calculation results have met the required standards. In subsequent studies the design of the reactor model will be added to be supported, thermalizing column and thermal column to get a more accurate criticality value.

References

[1] IAEA 2015 *Safety Standard Series No.SSR 3-5: Safety of Research* Reactor (Vienna, Austria: International Atomic Energy Agency).
[2] Hampel G 2010 *The Importance of TRIGA Reactors* Behalf of the European TRIGA community Johannes Guttenberg : Universitat Mainz D-55099
[3] Setiyanto and Surbakti T 2016 *Conference Proceedings PTKRN Batan* 3
[4] MCNP Team 2010 *MCNP - A General Monte Carlo N- Particle Transport Code* Version 5 (United States: University of California, Los Alamos National Laboratory)
[5] Purwadi 2017 *Conference Proceedings PRSG BATAN* **21** 87
[6] Hardiyanti I S, Riyatun, Suharyana and Khakim A 2019 *J. Phys. Conf. Ser.* **1153** 012104
[7] Salam M A, Soner M A M, Sarder M A, Haque A, Uddin M M, Sarker M M and Islam S M A 2014 *Ann. Nucl. Energy* **68** 257
[8] Wu Z, Williams R E, Rowe M, Newton T H and O'Kelly 2017 *Nucl Technol* **199** 67.

[9] Manowogbor V C, Odoi H C and Abrefah R G 2018 *Research & Reviews: Journal of Pure and Applied Physics* **6** 7
[10] Ravnik M, Zagar T and Persic A 1999 *Nucl Technol* **128** 35

IOP Conf. Series: Materials Science and Engineering **578** (2019) 012095 doi:10.1088/1757-899X/578/1/012095

Solution of Schrodinger equation in the presence of minimal length for cotangent hyperbolic potential using hypergeometric method

F Hidayat[1], A Suparmi[2], and C Cari[2]
[1]Physics Department, Graduate Program, Sebelas Maret University, Indonesia
[2]Physics Department, Faculty of Mathematics and Fundamental Science, Sebelas Maret University, Indonesia

E-mail: fajarhidayat2296@gmail.com

Abstract. This study was proposed to get the solution of the Schrodinger equation with cotangent hyperbolic potential in the presence of minimal length by using the hypergeometric method. By getting the numerical results with the help of Octave software, we can analyze nonrelativistic energy. The visualization of wave function was obtained using Matlab software. The result is defined as the larger of minimal length, the width of potential V_1, and the angular momentum was included, the energy is increasing as well. The larger of the width of potential V0 affected the decrease of energy in this system.

1. Introduction

The Schrödinger equation was obtained using the wave function of a freely moving particle. Since then, the Schrodinger equation has been the basic principles of physics, especially in quantum mechanics [1]. Furthermore, the uncertainty principles, whose initiated by Heisenberg, achieved more attention as long as our incapability to find out precisely, the position or momentum of the particle. Until, the topic of minimal length came up to the surface.

The minimal length recently has been an interesting issue in quantum theory, string theory, or in generalized uncertainty principle [2-4]. Various methods has been solved to analyzing the Schrodinger equation which included the minimal length formalism, such as Nikiforov-Uvarov technique [5], Supersymmetry (SUSY) method [6], and Hypergeometry method [7]. The hypergeometric method based on hypergeometric function. The solutions takes the equation into the separation variables, which makes the stationary version of the Schrodinger equation [8].

In this paper, we study the solution of Schrodinger equation in the presence of minimal length for cotangent hyperbolic potential using hypergeometric method. The potential can be written as.

$$\frac{-V_0}{2}(\coth\gamma r - 1) + V_1 \tag{1}$$

The potentials takes place as positive and negative constant numbers, while r expressed the inter-nuclear distance which relate with constant number, γ.

Before we can conclude the equation by using Hypergeometric method, there was two part from spherical coordinate we used to described the wave function, radial and polar parts. As the potential in this reseach, is not three-dimensional potential. Thus, we have only used the radial part.

In order to reach the form of non-relativistic energy. Firstly, we substituted the minimal effect formalism to Schrodinger equation. Thereafter, the hypergeometric method is applied. Thus, the non-relativistiv energy is obtained as well as the wave functions. The numerical result and the figure of wave functions is obtained by including the formula to Octave and Matlab software. Lastly, we can assumed the conclusions of this research well.

2. Experiments

2.1. Schrodinger equation with minimal length applied

The concept of minimal length was a result of Generalized Uncertainty Principle (GUP) approached. If we take Heisenberg algebra which obtained from position and momentum operator, followed by commutative relationship ($\boldsymbol{\alpha > 0}$), then we cut the series until the second order, we can obtained [9]

$$[\mathrm{x}, \mathrm{p}] = \mathrm{ih}[1 + \alpha \mathrm{p}^2] \tag{2}$$

Where α is minimal length parameter we used in this paper. From these equation (2), with some adjustment, and in addition. If we interpreted with string theory, p_i as momentum operator for low energy, and P_i as momentum operator for high energy, momentum can be defined as [10]

$$\hat{P}_i = (1 + \alpha \hat{p}^2)\hat{p}_i \tag{3}$$

By multiplying equation (3), since minimal length parameter varies from interval $0 < \alpha < 1$. Thus, for α approaching zero value, we had

$$P^2 = p^2 + 2\alpha p^4 \tag{4}$$

Where,

$$p = -i\hbar \nabla \tag{5}$$

If we put equation (5) into equation (4), we can obtained [11]

$$P^2 = -\hbar^2 \nabla^2 + 2\alpha \hbar^4 \nabla^4 \tag{6}$$

For Schrodinger equation, we had

$$E\Psi = \frac{P^2}{2m}\Psi + V(r)\Psi \tag{7}$$

By substituting the minimal length equation (6) into equation (7), we obtained

$$E\Psi = -\frac{\hbar^2 \nabla^2}{2m}\Psi + \frac{2\alpha \hbar^4 \nabla^4}{2m}\Psi + V(r)\Psi \tag{8}$$

If the minimal length, are set to zero, ($\alpha = 0$), equation (8) can be written as

$$E_0 \Psi = -\frac{\hbar^2}{2m}\nabla_0^2 \Psi + V(r)\Psi \tag{9}$$

where

$$\nabla_0^2 = -\frac{2m}{\hbar^2}(E^0 - V(r))\Psi \tag{10}$$

Afterwards by substituting these equation (10) to equation (8), and with some adjustment, we got

$$E'\Psi = -\frac{\hbar^2}{2m}\nabla^2 \Psi + 4\alpha m V^2 \Psi + V(1 - 8\alpha m E_0)\Psi \tag{11}$$

E' is expressed as

$$E'\Psi = E\Psi - 4\alpha m (E_0)^2 \Psi \tag{12}$$

2.2. *Cotangent hyperbolic potential*

Cotangent hyperbolic potential we used, can written as

$$V(r) = \frac{-V_0}{2}(\coth\gamma r - 1) + V_1 \tag{13}$$

By putting these potential equation (13) into equation (11), we obtained

$$-\frac{\hbar^2}{2m}\nabla^2\Psi + \left[V_0{}^2\alpha m\coth^2\gamma r - \left(2V_0{}^2\alpha m + 4V_0V_1\alpha m - 4E_0V_0\alpha m + \frac{V_0}{2}\right)\coth\gamma r + 4\alpha m\left(\frac{V_0{}^2}{4} + V_0V_1 + V_1{}^2 - E_0V_0 - 2E_0V_1\right) + \frac{V_0}{2} + V_1\right]\Psi = E'\Psi \tag{14}$$

To simplify equation (14), we assumed

$$E''\Psi = \left[E' - 4\alpha m\left(\frac{V_0{}^2}{4} + V_0V_1 + V_1{}^2 - E_0V_0 - 2E_0V_1\right) - \frac{V_0}{2} - V_1\right]\Psi \tag{15}$$

For the potential,

$$V'\Psi = \left[V_0{}^2\alpha m\coth^2\gamma r - \left(2V_0{}^2\alpha m + 4V_0V_1\alpha m - 4E_0V_0\alpha m + \frac{V_0}{2}\right)\coth\gamma r\right]\Psi \tag{16}$$

Therefore, equation (14) can be written as simple as

$$E''\Psi = -\frac{\hbar^2}{2m}\nabla^2\Psi + V'\Psi \tag{17}$$

From equation (17), Laplacian operator, ∇^2, can be described as

$$\nabla^2 = \frac{1}{r^2}\frac{\partial}{\partial r}\left(r^2\frac{\partial}{\partial r}\right) + \frac{1}{r^2\sin\theta}\frac{\partial}{\partial\theta}\left(\sin\theta\frac{\partial}{\partial\theta}\right) + \frac{1}{r^2\sin^2\theta}\frac{\partial^2}{\partial\phi^2} \tag{18}$$

Since the potensial we used only consist r variable. Thus, we only take radial form. The angular part take changes into

$$\frac{1}{r^2\sin\theta}\frac{\partial}{\partial\theta}\left(\sin\theta\frac{\partial}{\partial\theta}\right) + \frac{1}{r^2\sin^2\theta}\frac{\partial^2}{\partial\phi^2} = l(l+1) \tag{19}$$

As a result with modification, equation (17) can be rewritten as

$$E''\Psi = -\frac{\hbar^2}{2m}\frac{\partial^2}{\partial r^2}\Psi + \frac{\hbar^2}{2m}\frac{l(l+1)}{r^2}\Psi + (V_0{}^2\alpha m\coth^2\gamma r - \left(2V_0{}^2\alpha m + 4V_0V_1\alpha m - 4E_0V_0\alpha m + \frac{V_0}{2}\right)\coth\gamma r)\Psi \tag{20}$$

Furthermore, approximation was made, with

$$\frac{1}{r^2} \approx \frac{\gamma^2}{\sinh^2\gamma r} \tag{21}$$

Thus, equation (21) turned into

$$(E'' - V_0{}^2\alpha m)\Psi = -\frac{\hbar^2}{2m}\frac{\partial^2}{\partial r^2}\Psi + \frac{\hbar^2}{2m}\frac{l(l+1)\gamma^2 + V_0{}^2\alpha m}{\sinh^2\gamma r}\Psi - \left(\left(2V_0{}^2\alpha m + 4V_0V_1\alpha m - 4E_0V_0\alpha m + \frac{V_0}{2}\right)\coth\gamma r\right)\Psi \tag{22}$$

From equation (22), we set

$$\nu(\nu-1) = l(l+1)\gamma^2 + V_0{}^2\alpha m \tag{23}$$

$$2q = \left(2V_0{}^2\alpha m + 4V_0V_1\alpha m - 4E_0V_0\alpha m + \frac{V_0}{2}\right) \tag{24}$$

$$E'''\Psi = E'' - V_0{}^2\alpha m \tag{25}$$

The equation (22),rewritten as

$$E'''\Psi = -\frac{\hbar^2}{2m}\frac{\partial^2}{\partial r^2}\Psi + \frac{\hbar^2}{2m}\frac{\nu(\nu-1)}{\sinh^2\gamma r}\Psi - 2q\coth\gamma r\,\Psi \tag{26}$$

2.3. Hypergeometric Method

Hypergeometric differential equation [12] can be written in term

$$z(1-z)f''(z) + [c-(a+b+1)z]f'(z) - ab\Psi = 0 \tag{27}$$

for the result

$$_2F_1(a,b,c,z) = \sum_{n\sim 0}^{\infty}\frac{(a)_n(b)_n}{(c)_n}\frac{z^n}{n!} \tag{28}$$

From equation (26) we used variable subtitution as the equation resemble to Manning Rosen potential-like. We write

$$\coth\gamma r = 1 - 2z \tag{29}$$

By multiplying equation (26) with $\left(-\frac{2m}{\hbar^2}\right)$, and we make assumption

$$-\frac{2m}{\hbar^2}E''' = k^2 \tag{30}$$

$$\frac{q}{\gamma^2} = q', \tag{31}$$

$$\frac{k}{\gamma^2} = k' \tag{32}$$

As a result, equation (26) can be rewritten as

$$\left(z(1-z)\frac{\partial^2\Psi}{\partial z^2} + (1-2z)\frac{\partial\Psi}{\partial z}\right) + \begin{pmatrix} \frac{\nu(\nu-1)}{\gamma^2} \\ -\frac{k'^2-2q'}{4z} \\ -\frac{k'^2+2q'}{4(1-z)} \end{pmatrix}\Psi = 0 \tag{33}$$

Where, to simplified the equation (33), we can named

$$k'^2 - 2q' = \alpha \tag{34}$$

$$k'^2 + 2q' = \beta \tag{35}$$

$$\frac{\nu(\nu-1)}{\gamma^2} = \nu'(\nu'-1) \tag{36}$$

Function of general wavefunction can be write in term

$$\Psi(z) = z^\alpha(1-z)^\beta f(z) \tag{37}$$

Therefore, by substituting equation (37) which consist of equation (34-36) into equation (33), we have got the Gauss equation

$$z(1-z)f''(z) + \big((2\alpha+1) - (2\alpha+2\beta+2)z\big)f'(z) + \big((\alpha+\beta-\nu'+1)(\alpha+\beta+\nu')\big)f(z) = 0 \tag{38}$$

Thus, hypergeometry parameters from the equation (38), are written as

$$a = \alpha + \beta - v' + 1 \tag{39}$$

$$b = \alpha + \beta + v' \tag{40}$$

$$c = 2\alpha + 1 \tag{41}$$

To fulfill the condition of the finite series,

$$a = -n \tag{42}$$

Energy of Schrodinger equation with minimal length presence, E, is obtained by substituting the assumption we made from equation (12),(15),(23-25),(30-32), into equation (34),and (35)

$$E = -\frac{\hbar^2}{2m}\left(\frac{\left(V_0{}^2\alpha m + 2V_0V_1\alpha m - 2V_0\alpha m - \frac{V_0}{4}\right)^2}{\left(\sqrt{l(l+1) + \frac{V_0{}^2\alpha m}{\gamma^2} + \frac{1}{4}} - \frac{1}{2} - n\right)^2 \gamma^2} + \left(\sqrt{l(l+1) + \frac{V_0{}^2\alpha m}{\gamma^2} + \frac{1}{4}} - \frac{1}{2} - n\right)^2 \gamma^2\right) + \left(\begin{array}{c} 4\alpha m\left((E_0)^2 + \frac{V_0{}^2}{4} + V_0V_1 + V_1{}^2 - E_0V_0 - 2E_0V_1\right) \\ + \frac{V_0}{2} + V_1 + V_0{}^2\alpha m \end{array}\right) \tag{43}$$

Energy for ground state, E_0, which minimal length refers to zero, is given as

$$E_0 = -\frac{\hbar^2}{2m}\left(\frac{\left(-\frac{V_0}{4}\right)^2 \frac{1}{\gamma^2}}{\left(\sqrt{L(L+1) + \frac{1}{4}} - \frac{1}{2} - n\right)^2} + \left(\sqrt{L(L+1) + \frac{1}{4}} - \frac{1}{2} - n\right)^2 \gamma^2\right) + \frac{V_0}{2} + V_1 \tag{44}$$

Thereafter, the wave function can be obtained by substituting equation (29), (39-41), into equation (37) and (28), with some adjustment, we have

$$\Psi(r) = \left(\frac{1 - \coth\gamma r}{2}\right)^{\sqrt{\frac{k'^2 - 2q'}{4}}} \left(\frac{1 + \coth\gamma r}{2}\right)^{\sqrt{\frac{k'^2 + 2q'}{4}}} {}_2F_1(a', b', c', z) \tag{45}$$

For wave function in each state is shown by Table 1

Table 1. The wavefunction for various quantum number, n

n	Wavefunction
0	$\Psi(r) = \left(\frac{1 - \coth\gamma r}{2}\right)^{\sqrt{\frac{k'^2 - 2q'}{4}}} \left(\frac{1 + \coth\gamma r}{2}\right)^{\sqrt{\frac{k'^2 + 2q'}{4}}}$
1	$\Psi(r) = \left(\frac{1 - \coth\gamma r}{2}\right)^{\sqrt{\frac{k'^2 - 2q'}{4}}} \left(\frac{1 + \coth\gamma r}{2}\right)^{\sqrt{\frac{k'^2 + 2q'}{4}}} \left(1 + \frac{-n(2v' - 1 - n)}{\sqrt{k'^2 - 2q' + 1}}\left(\frac{1 - \coth\gamma r}{2}\right)\right)$
2	$\Psi(r) = \left(\frac{1 - \coth\gamma r}{2}\right)^{\sqrt{\frac{k'^2 - 2q'}{4}}} \left(\frac{1 + \coth\gamma r}{2}\right)^{\sqrt{\frac{k'^2 + 2q'}{4}}} \left(\begin{array}{c} 1 + \frac{-n(2v' - 1 - n)}{k'^2 - 2q' + 1}\left(\frac{1 - \coth\gamma r}{2}\right) \\ + \frac{(n^2 - n)(2v' - 1 - n)(2v' - n)}{k'^2 - 2q' + 3\sqrt{k'^2 - 2q' + 2}} \frac{\left(\frac{1 - \coth\gamma r}{2}\right)^2}{2!} \end{array}\right)$

3. Results and discussions

The non-relativistic energy of Schrodinger equation for cotangent hyperbolic potential with the presence of minimal length (43) were obtained using the Octave software. By setting the variation for the system of angular momentum, potentials V_0 and V_1, and especially the minimal length, α, the numerical result can be seen in Table 2 and Table 3

Table 2. The non-relativistic energy for various quantum number, n. $\gamma = 0{,}01$, $l = 0$, $\hbar$=1, m=1

V_0	V_1		E (eV)			
		α=0	n=2	n=3	n=4	n=5
5	8		-1,9439E+03	-8,5881E+02	-4,7903E+02	-3,0325E+02
	12		-1,9399E+03	-8,5481E+02	-4,7503E+02	-2,9925E+02
10	8		-7,8020E+03	-3,4617E+03	-1,9426E+03	-1,2395E+03
	12		-7,7980E+03	-3,4577E+03	-1,9386E+03	-1,2355E+03
		α=0,05				
5	8		7,6294E+05	1,5071E+05	4,7688E+04	1,9536E+04
	12		7,6294E+05	1,5071E+05	4,7685E+04	1,9532E+04
10	8		1,2207E+07	2,4113E+06	7,6294E+05	3,1250E+05
	12		1,2207E+07	2,4113E+06	7,6294E+05	3,1250E+05
		α =0,1				
5	8		1,5259E+06	3,0140E+05	9,5364E+04	3,9059E+04
	12		1,5259E+06	3,0139E+05	9,5351E+04	3,9045E+04
10	8		2,4414E+07	4,8225E+06	1,5259E+06	6,2499E+05
	12		2,4414E+07	4,8225E+06	1,5259E+06	6,2498E+05

Table 3. The non-relativistic energy for various quantum number, n. $\gamma = 0{,}01$, $l = 1$, $\hbar$=1, m=1

V_0	V_1		E (eV)			
		α=0	n=2	n=3	n=4	n=5
5	8		-7,8033E+03	-1,9439E+03	-8,5881E+02	-4,7903E+02
	12		-7,7993E+03	-1,9399E+03	-8,5481E+02	-4,7503E+02
10	8		-3,1240E+04	-7,8020E+03	-3,4617E+03	-1,9426E+03
	12		-3,1236E+04	-7,7980E+03	-3,4577E+03	-1,9386E+03
		α=0,05				
5	8		1,2207E+07	7,6294E+05	1,5701E+05	4,7688E+04
	12		1,2207E+07	7,6294E+05	1,5701E+05	4,7684E+04
10	8		1,9531E+08	1,2207E+07	2,4113E+06	7,6294E+05
	12		1,9531E+08	1,2207E+07	2,4113E+06	7,6294E+05
		α =0,1				
5	8		2,4414E+07	1,5929E+06	3,0140E+05	9,5364E+04
	12		2,4414E+07	1,5929E+06	3,0139E+05	9,5350E+04
10	8		3,9062E+08	2,4414E+07	4,8225E+06	1,5929E+06
	12		3,9062E+08	2,4414E+07	4,8225E+06	1,5929E+06

From Table 2 and 3. The larger energy of non-relativistic is affected by the larger of minimal length and the larger of angular momentum. The change of energy also depended on the larger of V_1 is included. Where, the larger of potential V_1 makes the energy get larger as well. However, the energy will get smaller if V_1 is increasing, even the changes are not significant.

The visualization of wavefunction was obtained by using Matlab software.

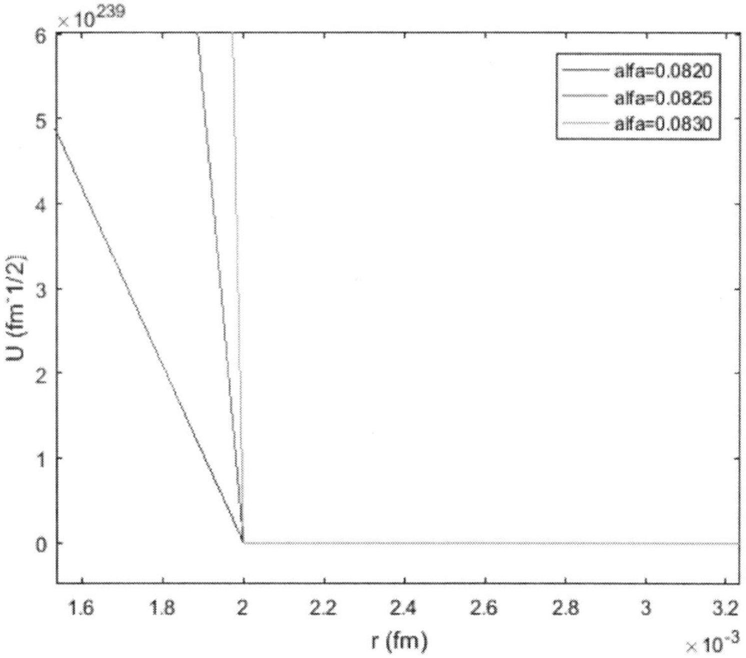

Figure 1. Plot of wavefunction for n=0 state, γ = 0,1, V_0 = 5, V_1 = 8, and l = 1

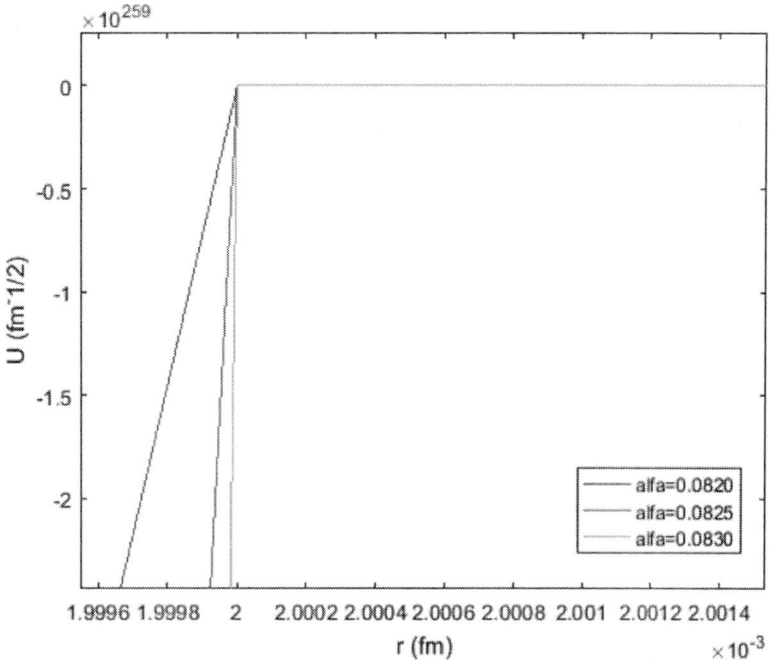

Figure 2. Plot of wavefunction for n=1 state, γ = 0,1, V_0 = 5, V_1 = 8, and l = 1

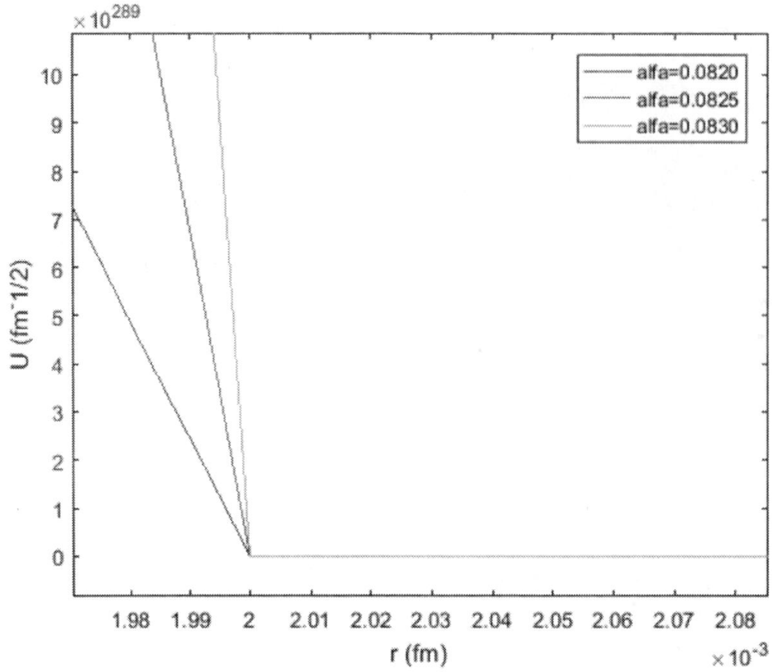

Figure 3. Plot of wavefunction for n=2 state, γ = 0,1, V_0 = 5, V_1 = 8, and l = 1

From Figure 1, Figure 2, and Figure 3. Particle on these system will be ionized after passing the radius of system core at (2,0) coordinates. The presence of minimal length gives effect for the larger wavelength, where the larger of minimal length, the larger wavelength would be. The larger of minimal length also depends by the larger quantum number. Which means, the further of the particle in this system, the larger wavelength obtained.

4. Conclusions

Schrodinger equation with minimal length for cotangent hyperbolic was solved using hypergeometric method. We can conclude, an increase from energy depends on parameters like angular momentum, width of potential, V_1, well as minimal length included. Nevertheless, the larger width of potential, V_0, make the energy decreased. Therefore, the wavelength of particles in this system will gets larger if the particle getting further from the core, and the larger of minimal length presented.

Acknowledgment
This study was partly supported by Hibah Pascasarjana UNS with contract no. 543/UN27.21/PP/2018.

REFERENCES

[1] Beiser A 2003 *Concepts of Modern Physics* vol 6, ed (New York: McGraw Hills) p 168
[2] Konishi K, Paffuti, G, and Provero, P 1990 *Phys. Letters B* **234** 276–284
[3] Garay L J 1995 *Int. Journal of Modern Physics A* **10** 145–165
[4] Kempf A and Mangano G 1997 *Physical Review D* **55** 7909–7920
[5] Hassanabadi H, Maghsoodi E, Ikot A N, and Zarrinkamar S 2013. *Advances in High Energy Physics* **2013** 1–6
[6] Azizah H, Suparmi A, and Cari C 2018 *J. Physics: Conference Series* **997** 012017
[7] Nurhidayati I, Suparmi, A, and Cari C 2018 *J. Physics: Conference Series* **997** 012016
[8] Ishkhanyan A M 2016 *Theoretical and Mathematical Physics* **188** 980-993
[9] Chang L N, Minic D, Okamura N, and Takeuchi T 2002 *Physical Review D* **65**

[10] Chabab M, El Batoul A, Lahbas A, and Oulne M 2016 *Phys. Letters B* **758** 212–218
[11] I Nurhidayati et al 2018 *J. Phys.: Conf. Ser.* **997** 012016
[12] Morales J, García-Martínez, J, García-Ravelo J, and Peña J J 2015 *Journal of Applied Mathematics and Physics* **03** 1454–1471

IOP Conf. Series: Materials Science and Engineering **578** (2019) 012096 doi:10.1088/1757-899X/578/1/012096

Application of alphazzle writing for basic school children base on android

Hartatik[1,*], P Pratiwi[1] , A Purnomo[1] and R Hartono[1]
[1]Departement of Informatics Engineering, Faculty of Mathematics and Natural Sciences, Sebelas Maret University, Surakarta, Indonesia

[*]E-mail: hartatik@mipa.uns.ac.id

Abstract. The education process is one of the human activities in which the motivational function in the education process is to generate encouragement to do activities in education. Activity can produce changes in cognitive, psychomotor and affective students. This applies to every learning in the teaching and learning process. Student learning success is determined by several factors that support the success of the teaching-learning process, including the existence of innovative learning techniques in this case is learning to write. Android games are an alternative software that is chosen, with the reason Android tends to be used on mobile drivers such as smartphones and tablet PCs. In the current technological development, children tend to be more familiar with various gadgets, even children under five are now exposed to various gadgets. With the existence of android games and supported by attractive features, it is expected that children know computer technology, lessons for following directions and rules, problem solving and logic exercises, motoric nerve training and spatial skills, can help parents to teach their children in a fun way.

1. Introduction

One alternative to help increase interest and ability in learning to write is to use learning media in the form of software (software). Android games are an alternative software that is chosen, with the reason that Android can be used on cellphones such as smartphones and tablet PCs. In addition, Android games are chosen because making gadgets or smartphones kids prefer playing games only. Even more game applications today contain elements of violence. This can be worried that it can damage a child's mental and psychic because in childhood there is a tendency to imitate the style he sees. As well as a habit that is feared to reduce the child's interest in learning to write [1,2].

Games are often accused of negatively affecting children. In fact, the game also has a function function and positive benefits for children, including: children know computer technology, lessons for review follow directions and rules, practice problem solving and logic, train motor nerves and spatial skills, establish communication between children and parents when playing Together, and also provide entertainment. In fact, for certain patients the game can be used as a healing therapy [3].

2. Experimental

The method used in this study is observation and then build applications according to user needs in the field of education. Based on the initial survey found that 80% of respondents stated that for Paud's children, kindergartens and elementary schools need applications to learn to write and read [4].

Based on the survey, the Alphazzle Application has an educational concept that is packaged in the form of an application on an Android device. In each game it has a simple way of playing so that it can be easily understood by children in preschool age. Children only need to follow a tracing board with dots that will form letters, numbers, or words from the selected image[5]. Then there is a puzzle game from what has been learned, in addition to eliminating boredom in the learning process can also train children's memory. In this game, besides being an educational media, it can also be an entertainment medium. This game can also be used to train the child's motor and brain system. For detail the stages of this research can be seen in Figure 1.

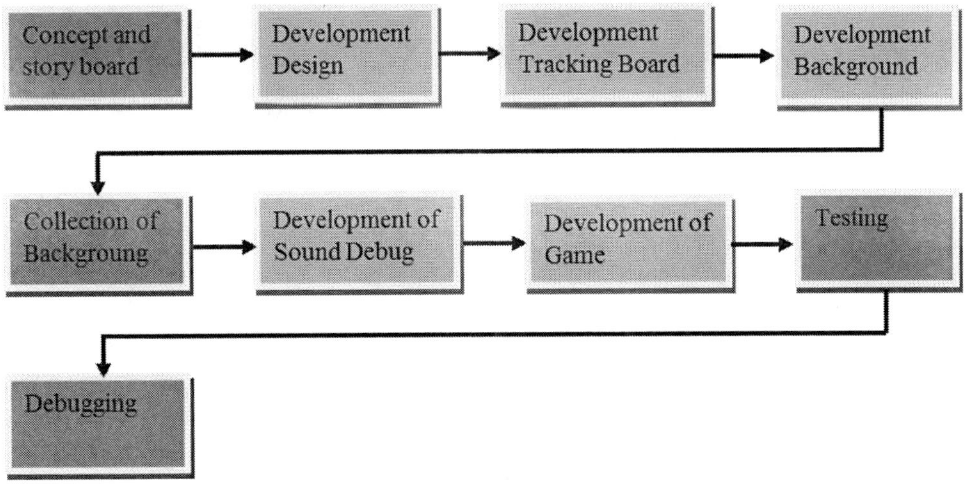

Figure 1. The flowchart of research stage.

The testing method that will be used to develop the Alphazzle application is blackbox testing. Blackbox testing is a functional test, where testing is done by observing the results of the test data execution and checking the functionalities of the game [6].

3. Results and Discussion

Gameplay of the education games from tis research are describe the concept of the Alphazzle application is an educational theme about how to write for preschoolers.In the Alphazzle application it has 2 main scenes, namely Education Scene and Learning Menu Options. With the games, players are provided with 4 (four) learning menu options, namely Alphabet, Numbers, Pictures and Puzzles. To select a player, you only need to press one of the selection buttons.

Beside that, user can be Choice of Alphabet. When choosing the alphabet learning menu, the alphabet A-Z will be displayed. Players only need to choose one of the alphabet choices by pressing it. Third option from this educations games is Choice of Numbers, in this menu user choosing a number learning menu, the numbers 0-9 will be displayed. Players only need to choose one of the number choices by pressing it.

Playability is a general measure of the quality of gameplay, the ease of play can be played, and the quantity or duration of the game can be played. The following are the playability attributes in this Alphazzle application covering Satisfaction, Learning, Efficiency, Immersion, Motivation. The following are the interface results of the educational game design in Figure 2.

Figure 2. Implementation of splash screen.

To design the level of the Alphazzle application, if you choose the alphabet learning menu, the A-Z alphabet will appear, the player only needs to select one of the alphabets and a tracing board appears to be played. In the learning menu the numbers will be provided with numbers 0-9, players also only need to choose one of these numbers. The game concept is the same as the previous alphabet learning menu. The following for the selected alphabet images can be seen as Figure 3 and for tracing board display can be seen as Figure 4.

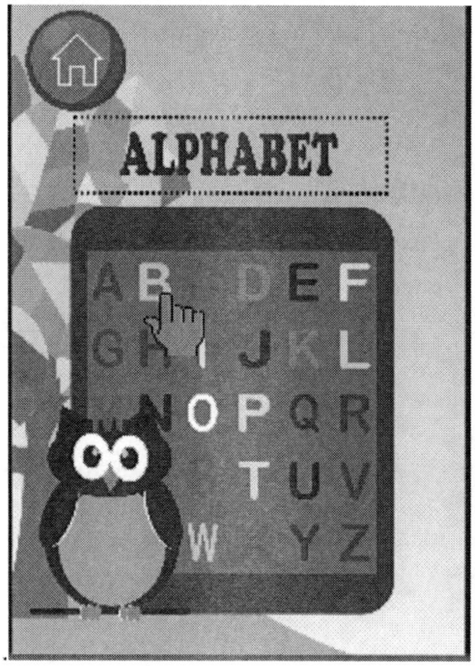

Figure 3. Choice of alphabet.

Figure 4. Interface of tracing board.

For games on the picture learning menu, an image will be provided at the bottom and at the top there is a tracing board from the name of the image. For how to play it is the same as the alphabet and number learning menu.

When pressing the Play button, it will enter Level Select. Select level is the display used to select the game learning menu. There are 4 learning menus in this Alphazzle application, namely: Alphabet, Numbers, Pictures, and Puzzles. The implementation of Level Select can be seen in Figure 5.

Figure 5. Implementation of select level.

This script is used to create Splash Screen. This splash screen will start at the beginning after opening the Alphazzle game application. This Splash Screen will display the logo used by the game. The code used to create the Splash Screen can be seen in Figure 6 [7,8,9].

```
using System.Collections;
using System.Collections.Generic;
using UnityEngine;
using UnityEngine.UI;

public class loading : MonoBehaviour {

    public Transform LoadingBar;

    [SerializeField]
    private float currentAmount;
    [SerializeField]
    private float speed;

    //Update is called once per frame
    void Update()
    {
        if (currentAmount < 100)
        {
            currentAmount += speed * Time.deltaTime;
            Debug.Log((int)currentAmount);
        }
        else
        {
            Application.LoadLevel("Awal");
        }
        LoadingBar.GetComponent<Image> ().fillAmount = currentAmount / 100;
    }

}
```

Figure 6. Script of splash screen.

The application is tested in general to 101 respondents and most of the respondents stated that it is very helpful for students to learn to write which is 60 % said that the application is very helpful in figure 7 and Figure 8 [10,11].

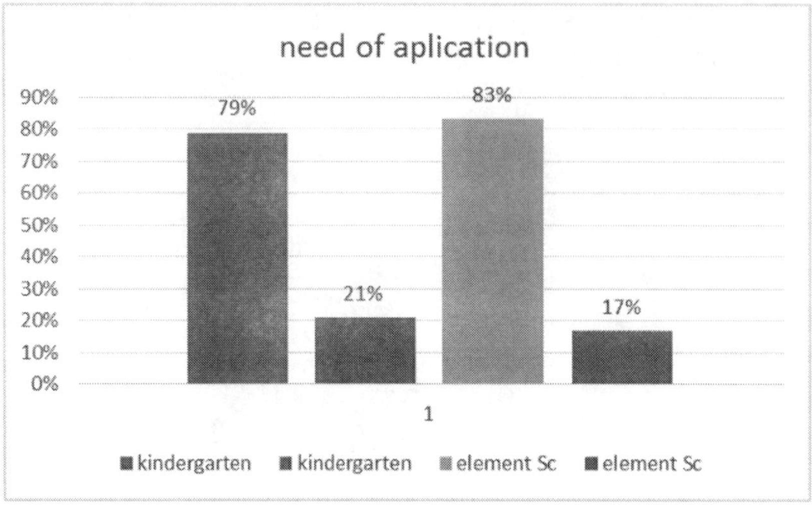

Figure 7. Public test results for need of application.

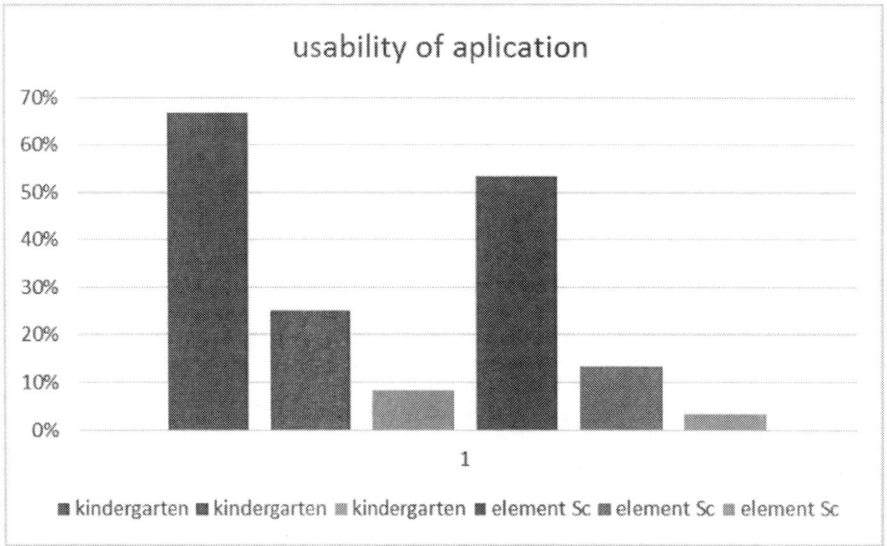

Figure 8. Public test results for usability of application

4. Conclusion

From the results of the creation and testing of the Alphazzle game, it can be concluded that the Alphazzle Game has been successfully created using the Unity 5.6.6 game engine which uses the C # programming language and the script writing uses MonoDevelop.

References

[1] Sillalahi U 2005 *Social Research Methods* (Bandung: Unpar Press)
[2] Izhar S 1998 *Pre-school Education Unit* (Jakarta: PT. Gramedia Widiasarana Indonesia)
[3] Zimmerman E 2003 *Game Design Fundamentals* (Cambridge: The MIT Press)
[4] Safaat N H 2011 *Android Programming Mobile Applications for Android-based Smartphones and Tablet PCs* (Bandung: Informatics)
[5] Henry S 2010 *Smart with Games* (Jakarta: PT. Gramedia Pustaka Utama)
[6] Agarwall, B.B, et al. 2010 *Software Engineering & Testing* (London: Jones & Bartlett Publisher,LLC)
[7] Niwan A 1996 *Programming Animations and Professional Games* (Jakarta: Elex Media Komputindo)
[8] I Made Astra, Hadi Nasbey and Aditiya Nugraha 2015 *Eurasia J. Math. Sci. Technol. Educ.* **11**(5) 1081-1088
[9] Hermawan S 2011 *Easy to Make Android Applications* (Yogyakarta: Andi Publisher)
[10] Nilwan A 1998 *Animation Programming and Professional Games 4* (Jakarta: Elex Media Komputindo)
[11] DiMarzio J F 2008 *Android Programmers Guide* (United States: McGraw Hill Professional)

IOP Conf. Series: Materials Science and Engineering **578** (2019) 012097 doi:10.1088/1757-899X/578/1/012097

Comparative Study of three non-destructive techniques for metal loss inspection in pipe walls

J L Quiroga[1], J E Quiroga[2] and P P Diaz[1]

[1]Facultad de Ingeniería Mecatrónica, Universidad Santo Tomás, Cra 27 Calle 9, Bucaramanga, Colombia
[2]Escuela de Ingeniería Mecánica, Universidad Industrial de Santander, Cra 18 Calle 9, Bucaramanga, Colombia

Abstract. Identifying and describing metal loss defects in pipe walls is an indispensable task in transporting operations, this abnormal condition mainly appears as a result of corrosion and gouging. These metal losses can lead the pipeline to unsafe conditions originating leaks, rupture of the structure and associated hazards. In this work is presented a comparative study to assess the capabilities of detecting and describing the metal loss in a pipe wall based on three non-destructive techniques: Magnetic Flux Leakage, Eddy Current Testing, and Guided Waves. These are compared by revising their physical principle and presenting a brief review of results obtained in metal loss inspection by considering some criteria as its sensitivity, depth of penetration, the range of inspection, and implementation easiness. Finally, the discussion is focused on of establishing which technique is most adequate for the inspection according to some characteristics of the metal loss defect.

1. Introduction

Magnetic Flux Leakage, Guided Waves, and Eddy Current testing are the most common on Non-destructive techniques used in the pipeline industry. These techniques have an advantage over others like liquid penetrant testing or X-ray as these allows automatic implementation and variations in designing for adapting to different conditions required.

A common characteristic of three techniques is that all of them can be generated by taking advantage of the properties of magnetic fields: One way of generating guided waves is by using magnetostriction principle, where the shape and size of ferromagnetic materials change during magnetization, eddy currents are produced due to the presence of a varying magnetic field in the test material and MFL take advantage of the leakage of the magnetic field around a metal loss defect.

This paper pretends to compare these three techniques in terms of detection and diagnosis capabilities and implementation in situ. The selected criteria to compare these techniques are based on determining high capabilities to detect and diagnose metal loss. The implementation easiness stands for the design and building of the transducer for the specific exploring condition excluding for the electronic components. The high sensitivity of the technique may produce false alarms and the relevant defects are not always the smaller ones. Therefore, a trade-off between high sensitivity and detection robustness must be considered.

2. Magnetic Flux Leakage

Magnetic Flux Leakage (MFL) is the most widely used method for inspecting cracks, corrosion and other associated defects in pipelines. MFL setup is composed by a ferromagnetic core with an energized coil or with strong permanent magnets like NdFeB to apply a magnetic field to the pipe wall close to saturation. The discontinuities presence in the pipe wall yields magnetic field leakage around the defect which is detected via an array of magnetic sensors. In this way, the screening is performed just in the pipe wall covered by the magnetic sensors. This technique provides information for characterizing the wall loss shape by estimating its width, length, and depth that are measured in the axial, circumferential and radial direction of the pipe [1]. These dimensions are estimated by analysing the signatures of MFL signals that are composed of the three-axis components of the magnetic field flowing around of the discontinuity.

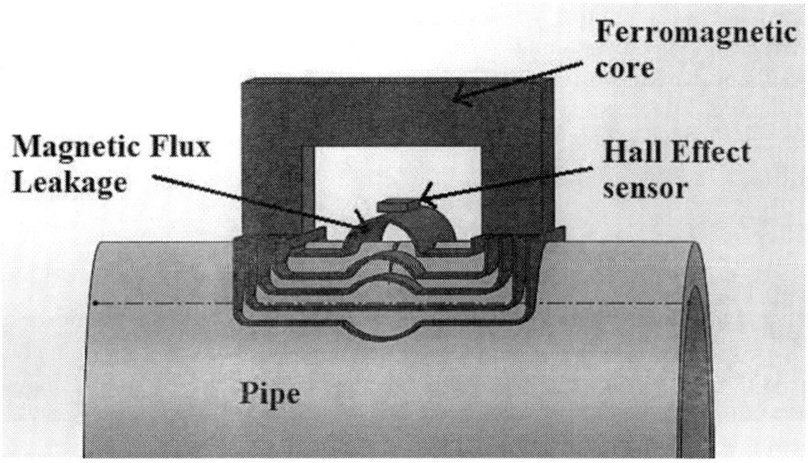

Figure 1. Principle of Magnetic Flux Leakage in a pipe with metal loss

Some features in the MFL signals are related to defect dimensions as criteria for estimating its shape. The length and width estimation of metal loss is highly affected by defect depth altering the accuracy of the estimated value of these magnitudes [2]. Some features of the selected signal as criteria are obtained experimentally [3], for example, a higher magnitude of MFL signals may be interpreted as large depths in metal loss defect [4], and also large discontinuity width. From the axial distance between intersection points of axial MFL signal is estimated the defect length and from the circumferential distance between the points where the axial signal has drop 70% of its peak value is estimated the defect width. Other criteria for estimating length and width are the peak values of radial and circumferential MFL signals [2]. Similar established criteria are proposed in other works [5]. Another challenge related to metal loss wall are the cracks with axial orientation and narrow widths that are typically almost undetectable as the magnetic field generated has the same orientation. One way of solving this difficulty is changing the direction of the magnetic field [3]. In [6] cracks of 50x5x2 mm are clearly detected. In [3] cracks of 0.1 mm with, 5% to 20% of wall thickness depth and 25 mm length were detected with high repeatability.
When the magnetic field is generated in the ferromagnetic core by induction coils; the depth of penetration depends on frequency and material properties. In this case, depth of penetration is the same as Eddy current testing [7]. MFL based on permanent magnets are unable to provide the discontinuity depth of penetration as the technique covers the whole cross-section with different intensity.
On the other hand, the implementation of the MFL system requires to verify the saturation point in the magnetization curve of the material under exploration. This point provides the necessary information of the magnetic field that must be generated. This field is generated by the magnetic circuit composed of the ferromagnetic core, the pipe wall and the corresponding source that can be permanent magnets or induction coils [8]. The behavior of the magnetic field is described by Maxwell equations that are the base of the modeling software used for this technique.
For extracting information about detection, location or sizing of metal loss from the three MFL components, it has been applied several methods like pattern-adapted wavelets, Hilbert, Hilbert Huang Transforms, artificial neural networks [4]. In addition, this technique allows designing and modeling MFL sensing systems with different software such as "Ansys" [5], "Magnet" [9], "Gmsh", "GetDP" [10] or "Comsol" [11], [] among others, applying criteria for designing the appropriate MFL sensor for the pipe.

3. Guided Waves
Guided waves are suitable for inspecting defects in pipes because of its capability for exploring long distance in the range of hundreds of meters from a single probe position [12]. This range is affected by

the operating frequency, pipe coating, insulation, depth of burial, among other factors. In contrast to ultrasonic bulk waves that usually propagates at frequencies in the order of MHz guided waves work typically between 20 and 200 KHz range [12]. For bulk waves, longitudinal and shear wave speeds in steel are 5.96 km/s and 3.26 km/s, respectively, for guided waves speed depends on frequency and mode excitation. Several type of guided waves can be generated depending on its boundary conditions, material properties, and geometry conditions: Rayleigh waves propagate along a single surface as rods and rail tracks; lamb waves propagate in solid plates and analogous to lamb waves are longitudinal, torsional and flexural waves that propagate in pipes or cylindrical structures. The behavior of guided waves in pipes can be modeled by the Navier's governing (1) wave equation with appropriate boundary conditions [13]

$$\mu\nabla^2\vec{U}+(\lambda+\mu)\nabla\nabla\square\vec{U}=\rho(\partial^2\vec{U}/\partial t^2) \tag{1}$$

Ultrasonic wave propagation without boundaries yields bulk waves instead of guided waves. By solving the equations according to a specific structure (thickness and material), it is obtained the dispersion curves. These curves show which modes can be generated in a hollow cylinder and at which velocity for a given spectrum frequency of a signal excitation. In figure 2 is presented the phase velocity dispersion curve for a one-inch steel pipe.

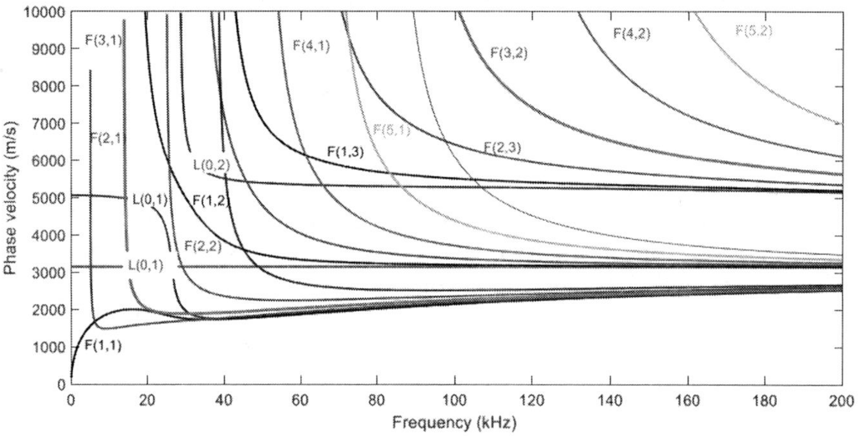

Figure 2. Phase speed dispersion curves for a 1" Sch. 40 A-106 pipe

A key element of the inspection is the selection of a single mode. Exciting several modes in the range of the bandwidth frequency of the excitation signal produce the propagation of a signal with multiple wavepackets which is much too complicated to interpret or analysis. One suitable mode is the first torsional mode T (0,1), as it is nondispersive [14], it means the velocity is independent of frequency and it is not strongly affected by the fluid inside or outside of the pipe. The second longitudinal model L(0, 2) is also nondispersive in a long range of frequency and it is faster than many unwanted modes, so this mode is received before the unwanted signals arrive [15]. Flexural modes have been also used to indicate defects near or at a weld [16]. In addition to these ones, other modes have been used for inspecting purposes [16], [17].

High sensitivity to small discontinuities has been reported in some research works using guided waves. T(0,1) were used to recognize a notch of 0.5 mm depth in a 1 mm wall thickness, while a shallower groove modeling surface corrosion of 0.3 mm depth was recognized by L(0,1) mode [12]. Artificial notches of up to 5% of thickness wall were detected by T(0,1) mode on a bent pipe [18]. The size of damage is also related to the time of flight and the instantaneous phase of the guided wave. The change in time of flight and phase of L(0,1) was used to distinguish between pristine pipe and 1.6 mm damage diameter among others [19].

Implementing the guided wave hardware requires to know the waveguide material and dimensions to select the suitable modes to be used in the specimen assessment. According to the desired excitation mode array of transducers are located with specific space between them. These sensors are typically bonded to the pipe, so it has to consider the effects of the bonding layer in the received signal [19], [20], on the other hand, EMAT does not need any bonding couplant as the guided wave is generated through an alternating magnetic field [18], [16].

4. Eddy Current Testing

This method is based on the generation of an electric current induced on the test material due to an alternating magnetic field generated by a coil pointing to the material. This behaviour is described in Faraday's electromagnetic induction law. This induced current generates a secondary magnetic field that opposes the field that produced it. This technique has several configurations for sensing the secondary magnetic field. Some of them use a receiving coil to sense the magnetic field generated by the induced current. If the induced current is interrupted by the presence of a defect in the material, the receiving coil detect a variation in the magnetic field.

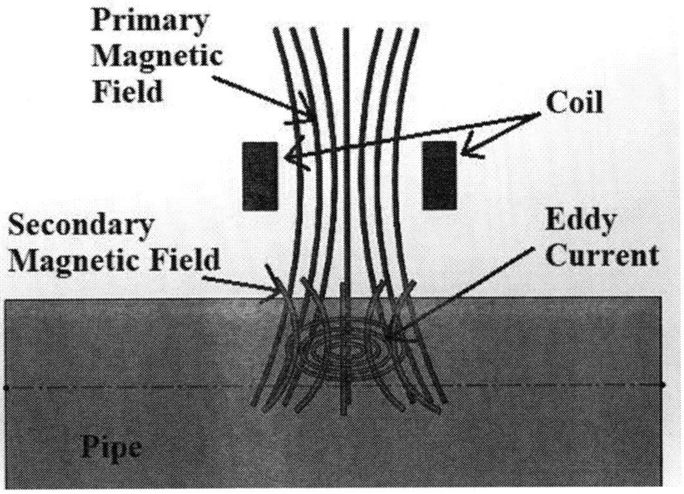

Figure 3. Primary and secondary magnetic field in pipe wall

The sensitivity of this technique is given by the size of the coils, its orientation, the electrical conductivity σ of the material to be inspected, the magnetic permeability μ and the excitation frequency. The depth of penetration affects the sensitivity of the instruments as the eddy current generated is denser near the surface of the test piece, where the probe is close. Eddy current decrease its density with the distance to the surface. In this way, a standard depth of penetration is defined where the instrument is sensitive to defects. Standard depth of penetration is described by equation (2), where the frequency, magnetic permeability and conductivity increase the depth of penetration decrease.

$$\delta = \sqrt{\frac{2}{\mu\omega\sigma}} \tag{2}$$

For a frequency of 250 KHz, using a stainless steel type 304 pipe, the standard depth penetration is 0.9 mm. [21]. The instrument sensitivity is also affected by the condition of the surface as porosity, or corrosion as these conditions have regions where no eddy current are able to flow giving false alarms. One way of improving the sensitivity of the technique is by working with pulsed eddy current instruments (PEC) [22] that generate square, triangular and other waveforms instead of single sinusoidal excitation. This improvement gives a broad spectrum of frequency that gives information about the characteristics of the hidden defects in the piece under investigation. Sensitivity is also affected by the direction of the eddy current generated. If a metal loss depicted as a crack has the same orientation than eddy currents, the instruments are not able detect them. For this reason, the exciting coils that draw the orientation of eddy current are designed forming spirals with different forms looking to detect different sizes and shapes of metal loss defects. Recent works showing the sensitivity of eddy current testing has given good results. Applying several novel configurations in the flexible exciting coils of eddy current probes is possible to detect artificial defects with 500 um depth, 200um width and 2 to 4 mm length at a frequency of 200 KHz [23]. External corrosion with dimensions between 2 and 4 mm depth, 14 to 27 mm width, and 20 to 40 mm length were shaped using Pulsed Modulated Eddy Current (PMEC) and Pulsed Eddy Current (PEC) [24]

Table 1. Comparison of techniques.

	MFL	Guided Waves Testing	Eddy Current Testing
Sensitivity	High, but is still needs improving in quantifying and shaping defects	High	High,
Depth of penetration	High	High	Depends on frequency, conductivity and magnetic permeability of the material. It is often used to detect defects near the surface
Range of Inspection	In situ	Long Range	In Situ
Implementation easiness	High,	Medium. With some transductors material designing and configuration must be specific for generating the modes required	High

Conclusions

The three techniques revised have high sensitivity for small defects but results in quantification have better results in Guided Wave testing and Eddy Current testing than in Magnetic Flux Leakage due to the complex relationship between MFL signals and dimension of the metal loss. However, in guided waves reducing frequency will reduce the capacity of detecting defects. In eddy current testing the size and shape of the picking up coil must be designed according to the kind of defect that is forwarded to detect.

The whole penetration and screening in the wall thickness of the pipe can be achieved by magnetic flux leakage and guided waves when the system is properly configured, however the Eddy Current Technique is strongly affected by the frequency and it is not always covering the whole thickness.

Magnetic flux leakage and Eddy Current testing allow to design and implement the sensors in an easier way compared with guided waves implementation when the actuation and sensing are performed with piezoceramic sensors. Ferromagnetic cores with different dimension and coils of different shapes are more intuitive for designing and building than guided waves testing sensors. Guided Waves generation typically requires the configuration of more variables such as material, frequency, transducers arrange, etc

If a long range is required in an inspecting procedure the best option is guided waves, although Eddy Current Testing and Magnetic Flux Leakage can be set up to cover long distances by displacing the system along the pipe. However, if it is required to move the instrument Guided Wave is not the best option as the transductors need to be strongly attached to the pipe.

References

[1] Shi Y, Zhang C, Li R, Cai M and Jia G, 2015, *Sens.* **15**, 31036-55.
[2] Kopp G and Willems H, 2013, *NDT & E Int.* **55**, 75-81.
[3] Liu B, He L, Zhang H, Cao Y and Fernandez H, 2017, *Meas.* **103**, 275-82.
[4] Layouni M, Salah Hamdi M and Tahar S, 2017, *Appl. Soft Comp.* **52**, 247-61.
[5] Jian F, Jun-Feng Z, Sen-Xiang L, Hong-Yang W and Rui-Ze M, 2013, *Chin. Phys. B,* **22**, 018103.
[6] Wu D, Liu Z, Wang X and Su L, 2017, *NDT & E Int.* **91**, 148-55.
[7] Wilson J, Kaba M, and Tian G, 2008, *17th World Conference on Nondestructive Testing, (*Shanghai*/NDT Internet Publishing).*
[8] Jaimes C and Roa S, 2016, *Prosp.* ***14**, 22-30.*
[9] Sorabh, Raghav V and Sengar S, 2017, *Materials Today: proc. (Elsevier/ Hyderabad, India)* **4**, 2665-72.
[10] Gloria N, Areiza M, Miranda I and Rebello J, 2009, *NDT & E Int.* ***42**, 669-77.*
[11] Person N, 2012, *Nondestruct. Aust.* ***49**, 48-51.*
[12] Tatarinov A, Barkanov E N, Davydov E and Mihovski M, 2018, *Non-destructive Testing and Repair of Pipelines, Springer, Cham, p 15-29.*
[13] Rose J, 2014, *Ultrasonic guided waves in solid media*, Cambridge university press, Cambridge, UK.
[14] Lui Z, He C, Wu B, Wang X and Yang S, 2006, *Ultrason.* ***44**, 1135-8.*
[15] Lowe M, Alleyne D, David N and Cawley P, 1998, *Ultrason.* ***36**, 147-54.*
[16] Nakamura N, Ogi H and Hirao M, 2017, *NDT & E Int.* ***87**, 78-84.*
[17] Lowe P, Sanderson R, Pedram S, Boulgourin N, and Mudge P, 2015, *Phys. Proced.* ***70**, 338-42.*
[18] Cheong Y, Kim S, Lee D, Jung H, and Kim Y, 2004, *World Conference on NDT, Montreal, Canada.*
[19] Amjad U, Yadav S and Kundu T, 2015, *Ultrason.* **62**, 223-36.
[20] Vogelaar B and Golombok M, 2016, *Mechanical Systems and Signal Proc.* **78**, 107-17.
[21] Diaz A, Doctor S, Hixon J and Mathews R, 2006, *U. S. NRC. Washington, U.S.*
[22] Ulapane N, Alempijevic A, Vidal T, and Valls J, 2017, *Sens.* **17**, 2208.
[23] Machado M, Rosado L, Pedrosa N, Vostner A, Miranda R, Piedade M and Santos T, 2017, *NDT & E Int.* **87**, 111-8.
[24] Li Y, Yan B, Li W, Jing H, Chen Z, and Li D, 2017, *NDT & E Int.* **88**, 51-8.